SYSTEMS ECOLOGY

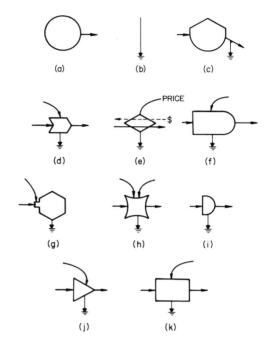

Symbols of the Energy Language. (a) Source; (b) heat sink; (c) storage; (d) interaction; (e) money exchange transaction; (f) producer; (g) consumer; (h) switching subsystem; (i) cycling receptor; (j) constant gain amplifier; (k) miscellaneous symbol for subsystems.

SYSTEMS ECOLOGY:
An Introduction

HOWARD T. ODUM

*Environmental Engineering Sciences,
University of Florida,
Gainesville*

A Wiley-Interscience Publication
JOHN WILEY & SONS

New York   Chichester   Brisbane   Toronto   Singapore

QH
541.15
.S5
O38
1983

Copyright © 1983 by John Wiley & Sons, Inc.

All rights reserved. Published simultaneously in Canada.

Reproduction or translation of any part of this work beyond that permitted by Section 107 or 108 of the 1976 United States Copyright Act without the permission of the copyright owner is unlawful. Requests for permission or further information should be addressed to the Permissions Department, John Wiley & Sons, Inc.

Library of Congress Cataloging in Publication Data:

Odum, Howard T., 1924–
   Systems ecology.

   "A Wiley-Interscience publication."
   Bibliography: p.
   Includes index.
   1. Ecology—Simulation methods. 2. Biotic communities—Simulation methods. 3. Bioenergetics—Simulation methods. 4. Biological models. I. Title.

QH541.15.S5O38 1982   574.5'0724   82–8650
ISBN 0-471-65277-6   AACR2

Printed in the United States of America

10 9 8 7 6 5 4 3 2

*To My Students
and Associates*

Series Preface
ENVIRONMENTAL SCIENCE AND TECHNOLOGY

The Environmental Science and Technology Series of Monographs, Textbooks, and Advances is devoted to the study of the quality of the environment and to the technology of its conservation. Environmental science therefore relates to the chemical, physical, and biological changes in the environment through contamination or modification, to the physical nature and biological behavior of air, water, soil, food, and waste as they are affected by man's agricultural, industrial, and social activities, and to the application of science and technology to the control and improvement of environmental quality.

The deterioration of environmental quality, which began when man first collected into villages and utilized fire, has existed as a serious problem under the ever-increasing impacts of exponentially increasing population and of industrializing society. Environmental contamination of air, water, soil, and food has become a threat to the continued existence of many plant and animal communities of the ecosystem and may ultimately threaten the very survival of the human race.

It seems clear that if we are to preserve for future generations some semblance of the biological order of the world of the past and hope to improve on the deteriorating standards of urban public health, environmental science and technology must quickly come to play a dominant role in designing our social and industrial structure for tomorrow. Scientifically rigorous criteria of environmental quality must be developed. Based in part on these criteria, realistic standards must be established and our technological progress must be tailored to meet them. It is obvious that civilization will continue to require increasing amounts of fuel, transportation, industrial chemicals, fertilizers, pesticides, and countless other products; and that it will continue to produce waste products of all descriptions. What is urgently needed is a total systems approach to modern civilization through which the pooled talents of scientists and engineers, in cooperation with social scientists and the medical profession, can be focused on the development of order and equilibrium in the presently disparate segments of the human environment. Most of the skills and tools that are needed are already in existence. We surely have a right to hope a technology that has created such manifold environmental problems is also capable of solving them. It is our hope that this Series in Environmental Sciences and Technology will not only serve to make this challenge more explicit to the established professionals, but that it also will help to stimulate the student toward the career opportunities in this vital area.

ROBERT L. METCALF
WERNER STUMM

Preface

If the bewildering complexity of human knowledge developed in the twentieth century is to be retained and well used, unifying concepts are needed to consolidate the understanding of systems of many kinds and to simplify the teaching of general principles. This book was conceived with this ideal. Initiatives from those of many backgrounds have generated a varied science of systems of marvelous diversity. In the process many new ways of thinking and representing relationships have been conceived. Using the systems languages, comparative study of contrasting systems has shown similar designs in the processing of matter, energy, and information in many realms of the biosphere.

General education can be approached through a synthetic and comparative view. This book was written as a textbook introduction to general systems based on a senior-graduate course given at the University of North Carolina during 1966–1970 and at the University of Florida during 1970–1981.

Expression of concepts and theories in the form of models and study of the implication of these models are becoming standard and possibly necessary parts of any scientific study, one of several general techniques of organizing theory and fact. Some believe that modeling and simulation must accompany successful investigations of nature and humans.

Systems of the environment, sometimes called *ecosystems*, have been much studied by students of systems ecology. Systems ecology is the study of whole ecosystems and includes measurements of overall performance as well as a study of the details of systems design by which the overall behavior is produced from separate parts and mechanisms. Computer studies of various types of systems are beginning to show us ways to recognize actions from inspection of the diagrams of system design. Phenomena studied by systems ecology are important in themselves and they constitute a basis for human partnership with nature. Ecosystems are also well suited for the teaching of general systems. Intermediate in size between the microscopic and the astronomical, ecosystems have easily recognizable parts, so that emphasis can be placed on the study of relationships. The main problem is seeing so much detail, the perennial problem of not seeing the forest because of preoccupation with the trees.

This book introduces ecological systems, while summarizing general principles of all systems, and uses ecosystem examples most frequently to illustrate generalizations about system designs and functions.

Twenty years ago, in attempting to use systems languages to compare systems, it became apparent that no one available language showed all the diverse characteristics of systems simultaneously. For example, the formulations of energetics were mainly done apart from kinetics. Diagrams using symbols developed in electrical systems were not acceptable to those in biology because many main features, such as self-maintenance, were not recognized. Most mathematical formulations ignored the constraints of energy. Even in systems sciences there was a tendency to analyze by distinction between different processes and study them one at a time. For example, one elemental cycle was modeled at a time, whereas the real world systems do all things at the same time.

In the intervening years the author and colleagues elsewhere have attempted to develop a systems language that would combine features of actual systems, drawing from other systems languages as needed. An energy circuit language of symbols and diagrams was developed combining kinetics, energetics, and economics. It does mathematics symbolically but at the same time keeps track of energy laws. In the process of its use it was realized that the diagrams are themselves a form of mathematics with emergent theorems and perceptions for the workings of the mind that extend the capacity to see wholes and parts simultaneously.

New energy principles emerged, such as corollaries to the maximum power principle and the hierarchical principle of energy quality. Proving that principles are general may not be possible, but testing with many applications shows generality. Another purpose of this book is assembly of observations that are consistent with the maximum power principle and may form corollaries. The characteristic energy chains and feedback webs of ecosystems were found in all systems and their design explained by the constraints of energy laws.

The language was found to be useful for teaching systems and comparing systems languages. It can be used intuitively prior to teaching mathematics as an aid and stimulus to learning mathematics. Two earlier books aimed at the beginner and intermediate levels used the language in a semiquantitative way to introduce systems and show the behavior of energy as a prevailing common denominator in organizing the balances of nature. In the meantime, scientific papers, graduate theses, and dissertations have applied the language as a tool for energy analysis, perceptual modeling, simulation, planning, and public policy.

Although the energy circuit language is the common thread used in this book, other systems languages and approaches are given and translated into energy circuit language to assist in interlanguage translations and the development of a common view. Further standardization of diagrams has been sought, especially in positioning symbols by the energy quality principle.

The purposes of the book, therefore, are to help in (1) teaching the nature and theory of systems, (2) introducing systems ecology, (3) introducing systems of nature of many scales of size in a comparative way, (4) using energy language to generalize and compare systems of nature and humans, (5) teaching systems languages and modeling approaches to understanding systems, (6) unifying concepts of kinetics, dynamics, energetics, environment, and economics, (7) presenting evidence for generality of energy hierarchy, energy quality, maximum power selection, and pulsing control, (8) introducing theory of energy analysis, and (9) seeking a more rigorous and holistic way to introduce and unify general ecology.

The book is arranged in four parts. Part I introduces systems and simulation with energy language. After introducing systems concepts in Chapter 1 and examining ecosystems briefly in Chapter 2, the phenomena of storage and linear systems are considered in several languages. Next, microdigital and analog computers are introduced, and in the author's courses students begin immediately to simulate the models so that they come "alive." Microcomputers with plotting programs are a very inexpensive way of providing this experience. Since courses in digital programming are now a standard part of curricula everywhere, further details are not given here. Logic systems, programmatic languages, and other languages are introduced in Chapters 4 to 6. Chapter 7 discusses principles of energy in open systems and their expression in energy language.

Part II, on design elements, includes Chapters 8–13 with various kinds of systems according to configuration as intersections, loops, series, parallel relations, and webs. Many of the basic models of population ecology, biochemistry, and electronics are included.

Part III, on organization and patterns, includes Chapters 14–18. Energy and information characteristics of systems designs are considered, including temporal, spatial, connective, and spectral aspects. Many of the generalizations of science are included.

Part IV considers systems of humanity and nature in Chapters 19–26. Systems of the earth are considered, including producers, consumers, ecosystems without humans, ecosystems with humans, economic systems, and environmental systems of larger scale, including cities, regions, states, nations, the world. The summary (Chapter 27) recapitulates the procedures of thought and study for examining any system and means for solving problems. A minimodel of the universe is suggested.

Each chapter presents study questions and some references to more complete treatments of topics and other efforts to introduce systems. There have been many excellent parallel efforts by others to generalize about systems and systems study. Part of the excitement in the area comes from the rich variety of alternative ways of organizing knowledge. Models are the simplifications of the human mind used to help understand, and modeling in its broad sense is almost synonymous with knowledge. The book organization was based on the philosophy that a teacher of the subject should introduce variety and make use of representative book summaries of the approaches. However, it may also be the teacher's obligation to emphasize one system of thought enough to provide unity, simplicity, and ease of perception.

Initiative in the study of systems has been dispersed among many languages and quantitative formulations. In varying degrees each language has had its proponents who have used them as the central common denominator in efforts to find commonality. Good texts and summaries are available for most of the popular languages. Some concepts require a special language to see them well, some languages have large volumes of published work, and some languages are part of the necessary interface with computing technology. Some have had success in practical applications, and others appeal because of inherent feelings of elegance they contribute. Some are a challenge to learn or operate. Ideally, one should not be hindered from ideas and results because symbols, intellectual formats, or ways of thinking are different. In this book other languages are introduced very briefly and efforts made to translate them into energy circuit language.

After the equivalent equations for energy diagrams have been established in the first sections, later sections use the diagrams without the redundancy of writing equivalent equations since the diagrams are rigorous as equivalent expressions.

This book was developed for a semester course but is longer in its present form. *Ecological and General Systems* was taught in a semester by making sections and most of the chapters of the last part optional, although a year's course may be appropriate if students learn programming, matrix algebra, and mathematical tools at the same time. In my course in recent years, I have required calculus and advanced standing in some field of science, engineering, or quantitative studies as prerequisites. Students in other areas do well if they have had mathematics, biology, and earth science or physics. Assignments were made from my two general books since the greater use of English explanation was helpful. After a slow coverage of Chapters 1–7, examinations were repeated until all students achieved a grade of B to bring students of varied backgrounds to a common sharing of concepts. A third of the course credit was used for a modeling-simulation project to involve a macroscopic mini-model that encompasses important issues, real evaluations, scaling, and written and oral presentations. These were kept within reason by asking that the model include no more than four storages and three interactions. Two thirds of the course credit was given for exams. Whereas a part of Chapter 5 on analog computers might be omitted if students are to use microcomputers for simulation, some analog diagrams need to be learned as a language since they have been extensively used in the literature for explanations.

For each chapter, suggested readings are given in a section at the back. These are especially important for showing alternate approaches and the riches of the field. Suggested readings include some clear explanatory texts, some classic papers, and some reference treatises.

In spite of this and other attempts to synthesize elements of systems into a coherent ecological science, much of the rapidly emerging but scattered literature is still ahead for us to study and savor.

HOWARD T. ODUM

Gainesville, Florida
September 1982

Acknowledgments

I am indebted to my former students and colleagues who have shared and contributed to the exciting process of developing ecosystem theory, energy analysis, and general systems modeling of the realms of our universe. Applications and examples were developed under projects and grants from the Rockefeller Foundation, National Science Foundation, Office of Naval Research, Atomic Energy Commission and Department of Energy, Water Resources Research Institute, and Sea Grant Administration, Department of Interior, Nuclear Regulatory Commission, and others. The author acknowledges his great teachers, Howard Washington Odum, Eugene Pleasants Odum, Robert Ervin Coker, George Evelyn Hutchinson, Warder C. Allee, and John Bugher among others.

Research and writing received encouragement and support of Department Chairman E. E. Pyatt, and others in the administration of the College of Engineering and the University of Florida. Part of the final revision was done on sabbatical leave on an Erskine Fellowship at the Joint Centre for Environmental Sciences, University of Canterbury, Christchurch, New Zealand.

T. Bullock, of the University of Florida, aided in simulation studies. D. J. Cowan, Gettysburg College, helped develop concepts of molecular order. C. Kylstra collaborated on energy analysis. T. Robertson helped with governmental public policy. Ben Fusaro, Chairman of Mathematics at Salisbury College, Maryland, read and criticized the entire manuscript. Illustrations were done by Barbara Lemont, and the manuscript was typed by Joan Breeze.

H. T. O.

Contents

PART I ENERGY, SYSTEMS, AND SIMULATION

1. Introduction to Energy Systems — 3
2. Ecosystems and Energy Hierarchy — 15
3. Storage and Flow — 25
4. Microcomputer Simulation — 46
5. Analog Computer Simulation — 53
6. Logic Systems and Other Languages — 72
7. Energy — 95

PART II DESIGN ELEMENTS

8. Intersections — 123
9. Autocatalytic Modules — 141
10. Loops — 160
11. Series — 182
12. Parallel Elements — 206
13. Webs — 223

PART III ORGANIZATION AND PATTERN

14. Energy Quality and Embodied Energy — 251
15. Spectrum of Energy Distribution and Pulsing — 269
16. Temperature — 288
17. Complexity, Information, and Order — 302
18. Spatial Distribution and Diversity — 323

PART IV SYSTEMS OF NATURE AND HUMANITY

19. Producers — 355
20. Consumers — 383
21. Ecosystems — 406
22. Succession — 443
23. Economic Systems and the Nation — 476
24. Ecosystems with Humans — 508
25. Cities and Regions — 532
26. World Patterns — 554
27. Summary: The Unity of Systems — 572

References — 584
Author Index — 615
Subject Index — 625

SYSTEMS ECOLOGY

PART ONE

ENERGY, SYSTEMS, AND SIMULATION

CHAPTER ONE

Introduction to Energy Systems

This book concerns energetics and systems considered together, especially environmental systems. General concepts of systems study are considered that will help humans to visualize complex networks of parts and processes. The environment has organisms, chemical cycles, water, air, humans, machines, soils, cities, forests, lakes, streams, estuaries, and oceans; and connecting them all are flows of energy, including that associated with matter and information. It is sometimes said that humans view the world with tunnel vision; that is, only one or two relationships can be considered at a time. In the past, tunnel vision has been well used; and advances in science and engineering have followed from learning how one or two components interact at a time. Also, up to the last two decades, our use of all the knowledge of parts and relationships has been limited by our inability to synthesize all the parts into an understanding and prediction of the behavior of larger systems. It was not critical to our survival before because we were embedded in and protected by the smooth functioning of a giant biosphere of ecological networks. Now, however, as our available energies begin to be a substantial portion of the network, we are capable of damaging our own basis for support if we make changes that we do not understand. Hence, there may be urgency that progress be made in systems science of the environment.

HUMANS AND THEIR MODELS

It has been stated as Gödel's theorem (Gödel, 1931; Nagel and Newman, 1958) that no system understands itself because more systems components are required to analyze and understand than to simply function. A doorbell rings but does not understand how this happened. However, humans, who designed and constructed the doorbell, understand it because they have sufficient mental ability to visualize the parts, functions, and external relationships so as to predict consequences of the systems structure.

Similarly, humans function but may not be able to understand themselves wholly all at once. However, we can make and understand simplified models that have enough of the essence of humans to be predictive within the limits of the model's testing. For example, we have concepts such as love, hate, sleep, and learning, which refer simply to complex phenomena.

The world of the environment is even more complex. There are many people in it along with the complexity of the rest of nature. Because humans are in the system, they see so much detail that they are often bewildered. A simplified version of a system is called a *model*. People can learn and understand simplified models even if they cannot perceive the full system at a time. Model building becomes even more necessary if the human mind is to grasp and understand the performance of the system of which it is a tiny part. Thus a book on environment systems for humans must include models along with the facts of real performance of the systems modeled.

GENERAL SYSTEMS THEORY

Many models developed to explain environmental systems are found to be the same as or similar to some invented and discovered in other fields independently. Similarly, some of the ecological models are found ready for application to systems of other scale. It has become increasingly obvious to generalists that wasteful duplication follows from separate and isolated study of similar systems in many fields, which independently discover and develop the same theorems, models, and laws that have already been discovered. Especially for the teaching of systems concepts, the principles of systems need to be stated in general terms and learned early so that repetition is unnecessary. In this book we seek to understand the principles of general systems theory along with the reality of environmental systems.

Mathematics generally has this aim, but often mathematics is taught so generally that it omits the constraints of energy, matter, and information that are a part of the real world. The new systems concepts should retain the real constraints of natural laws to form a restated new kind of mathematics. The teaching of mathematics and physical laws need not be separated if languages for their expression synthesize both these disciplines. Many efforts at generalizing systems language may be examined in the yearbooks of the Society for General Systems Theory, from suggested readings, and in the periodicals listed at the end of the book.

Definition of a System

A system is a group of parts that are interacting according to some kind of process, and systems are often visualized as component blocks with some kind of connections drawn between them as in Fig. 1-1. The connecting lines represent the interplay of the parts. The ancient concept of *holism* finds new properties "emerging" from the interactive combination of parts. As often stated, the whole becomes more than the sum of the parts when there are interactions. Such wholes that have emergent properties from the interaction of the parts we call *systems*. Sometimes emphasis is on wholeness of a system (Henderson, 1917; Smuts, 1926; Laszlo, 1972). At other times emphasis is on *analysis*, the relating of parts.

Systems within Systems

Within every realm of size there are parts that interact to form systems. For example, there are chemicals that form chemical systems, there are biological cells and organs that form biological systems, there are organisms and physical components that form ecological systems, and there are larger interactions of humanity and nature that form environmental systems. The universe is readily visualized by humans as a group of systems, each one serving as part of the next larger system. Note the three levels of systems within systems illustrated in Fig. 1-1. Principles that apply to one level of size often apply to systems at other levels, too, although there may be some aspects that change with size.

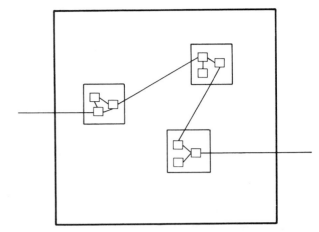

FIGURE 1-1. Connected box diagram indicates a system's connected parts. Some components are within parts. Simple box diagramming such as this does not explain processes and mechanisms and does not indicate kinetic or energetic relations.

For example, molecular forces are replaced with radiational and gravitational forces as realms shift from micro to macro scales.

Open and Closed Systems

Von Bertalanffy (1950, 1968) gives historical background of open-system concepts. An open system is one that has one or more inflows and outflows. For example, the biosphere is an open system with solar energy and cosmic matter flowing in and outgoing infrared radiation with some gaseous molecules flowing out. A sealed aquarium is open to energy with light flowing in and heat flowing out. The sealed aquarium is closed to matter but open to energy.

A closed system is conceived as one with complete isolation, with nothing flowing in or out. In the real world, closed systems with complete isolation are rare and temporary. Figure 1-2a shows an open system with water flowing in being stored temporarily and flowing out. Figure 1-2b, however, shows a closed system. Water molecules from liquid are diffusing into the gaseous (vapor) state, and simultaneously these are diffusing back into the vapor state. This happens in a sealed bottle of water.

State Variables, Steady State, and Equilibrium

The quantities stored in a system can vary. One can plot graphs of the quantities with time. Such variables are called *state variables* since they describe

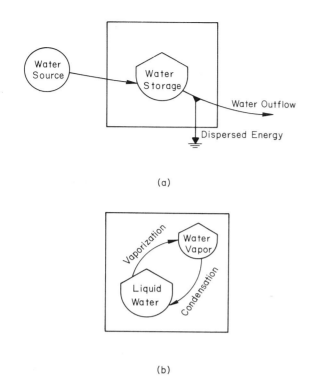

FIGURE 1-2. Comparison between open system (*a*) and closed system (*b*).

the state of the system while it is varying. The system is in a *transition state*. When the storages and patterns in an open system become constant with a balance of inflows and outflows, the system is in a *steady state*. In many common languages, the term "equilibrium" refers to any constant state, but in most branches of science and engineering, this term refers to a closed system when the storages become constant.

SYSTEMS LANGUAGES

Any written or spoken symbolism that facilitates understanding the combination of parts into systems is a systems language. There are many systems languages, some qualitative and some quantitative. Some are vague, with their meaning not clear to everyone; some are clear to those who have learned the same special definitions; and some are clear because they are defined by measurable procedures that relate the language to reality. The human mind often creatively develops ideas in soft and vague languages first, later making them more precise with measurable definitions.

Most humans retain concepts about wholes and parts in their minds in verbal languages such as English, Spanish, and Russian, for it was in those languages that they first learned to think. The clarity of verbal languages for thinking and expressing systems depends on the precision and clarity with which the words are defined and on the words that have the same meanings when used by others (Mason and Lagenheim, 1957). Other systems language invented for various aspects of systems description and expression may be more precise than verbal language, and thus we teach these, but apparently we must include verbal expressions as well as symbols to fully relate meanings to most humans.

Since language is the content by which we think as well as the means for communication between people, the more precise and general the systems languages we use, the more clarity and usefulness we find in our thinking and in communication. Some ideas are well expressed in one language and others in another so that our thinking may best be made up of thinking in many languages, especially during the present period of rapid development of systems knowledge when various groups are using different languages to express similar ideas. Thus it is an ideal of this book that we seek to learn as many of the systems languages as are useful and as there are blocks of knowledge.

Often some organization is given to preliminary thoughts and ideas by inventorying in diagram form the parts of a system and the existence of connecting relationships as yet unspecified. The blocks and lines in Fig. 1-1 are an example of a simple inventory often found in nonquantitative descriptions to assist the understanding of systems. Even for initial consideration, however, we can use languages with more rules and rigor. A list of some of the languages used in this book is given in Table 1-1. One, the *energy circuit language,* is used as the principal means for explaining concepts in this book. The performance of real world systems as quantitatively measured is the basis for evaluating the reality of the concepts.

ENERGY, WORK, AND POWER

The energy language is a way of representing systems generally because all phenomena are accompanied by energy transformations. *Potential energy* is that capable of driving a process with energy transformation from one form to another. An energy transformation driven by potential energy is *work*. In most kinds of work, one type of energy is transformed into another, with some going into a used form that no longer has potential for further work. Potential energy from outside energy sources pro-

TABLE 1-1. Some Systems Languages

Name	Chapter Where Explained
English	All
Real-world measurement	1
Energy circuit language	1 and 2
Differential (rate) equations	3
Integral (cumulative) equations	3
Integrated equations for states	3
Difference equations	3
Block diagrams in time domain	3
Block diagrams in frequency domain	3
Operational analog diagrams	5
Passive electrical equivalent circuits	3
Programmatic flow charts in English	4
Programs in BASIC	4
Forrester diagrams	6
Dynamo digital programs	6
Graphical representation of functions with time	3
Neural nets	6
Logic circuits	6
Boolean algebra	6
Probability diagrams	6
Signal flow diagrams	6
Conductivity	3
Causal word chains	6
Statistical path coefficients	6
Irreversible thermodynamics	7
Chemical reaction networks	8
Matrix algebra	6 and 13
Markov chains	6
Bond graphs	6
Graph theory	6
Art	6
Equations of motion	13
Queues	13

vides the means for keeping systems generating work. Storages within the system have potential energy that can drive work processes.

Many forms of energy are involved in different processes. There are energies in photons of sunlight, in sound waves, in water waves, in water of rivers, in chemicals which react, in magnetic fields, and in concentrations of matter. Kinetic energy is the energy of movement, as in a spinning top or traveling car.

In practice, energy is defined and measured by the heat that is formed when energy in other forms is transferred into heat. All kinds of energy can be converted into heat. *Heat* is the energy of molecular motion. The generation of heat raises the temperature. A small calorie is the heat that raises one cubic centimeter of water one degree. A kilocalorie is 1000 small calories.

The energy language keeps track of flows of potential energy from sources going into storages or into work transformations and finally into degraded form as used energy leaving the system. Pathways of the energy language are pathways of energy flow. The rate of flow of energy may be expressed in such units as calories per unit time. Figure 1-3a shows the energy flows associated with the use of potential energy in work. The rate of flow of energy into useful work is defined as *power*.

Energy Laws

The first law of energetics is the conservation law: energy is neither created nor destroyed in a system. Pathways trace energy flows through the system to storages and out again. As shown in Fig. 1-3, all energies enter from the source and ultimately leave as used energy dispersed through the heat sink.

The second law of energy requires heat dispersal from all storages and all processes. Thus tanks and process symbols have energy drains that connect with the heat sink.

Maximum Power Principle and Designs with Feedback Reward Loops

A major design principle observed in natural systems is the feedback of energy from storages to stimulate the inflow pathways as a reward from receiver storage to inflow source (see Fig. 1-3a). By this feature the flow values developed reinforce the processes that are doing useful work. Feedback allows the circuit to learn. Boltzman (1905) said, "struggle for existence is a struggle for free energy available for work." Lotka (1922) formulated the maximum power principle, suggesting that systems prevail that develop designs that maximize the flow of useful energy. The feedback designs are sometimes called *autocatalytic*. They maximize power, and theories and corollaries derived from the maximum power principle explain much about the structure and processes of systems (Odum and Pinkerton, 1955; Odum, 1971a).

In the energy language the hexagon is used as a group symbol for autocatalytic units that consume energy. If the hexagon is used, there is at least a

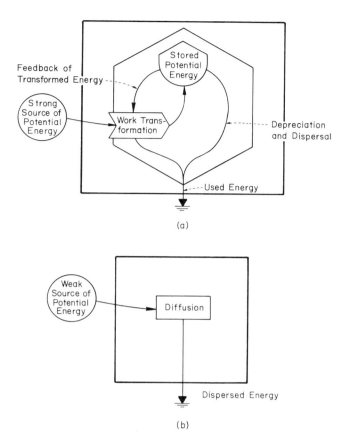

FIGURE 1-3. Comparison of energy flow with strong source of potential energy (*a*) and weak source of potential energy (*b*). The first builds storages and structure.

storage and a feedback reward loop contained along with the heat sinks by which heat is dispersed as potential energy is degraded (see Fig. 1-3). Consumers are organized in this way, and the autocatalytic symbol is sometimes called a *consumer unit*. Autocatalytic loops may be found within the producer symbol also.

When the potential energy available to drive a process is small, energy degrades without the work transformation that develops an autocatalytic loop. An example is diffusion of a concentrated substance into a more dilute state. See Fig. 1-3*b*. To develop the loop, the energy flow available must be sufficient to exceed the flow of depreciation from the storage in the loop.

For some years there was a controversy as to whether power flow in open systems tended to maximize as given in the maximum power principle or tended to minimize as suggested by Prigogine and Wiaume (1946). The controversy now seems to be resolved. When potential energy concentrations are low and insufficient to develop the autocatalytic loop, energy merely degrades without affecting power. At higher potential energy levels the power maximizing loops can be supported (Nicolis and Prigogine, 1978; Odum, 1982).

SYMBOLS OF ENERGY CIRCUIT LANGUAGE

In this book, as in earlier qualitative books on systems (Odum, 1971a; Odum and Odum, 1976), an energy circuit language is used as the common language for relating various languages, for presenting ideas, to aid in translations and calculations, and ultimately to replace some more difficult procedures. The language of symbols and pathway networks represent the flows of energy and thus the pathways of casual force actions (Odum, 1971b). The symbols and conventions of the language are introduced as needed, and summaries are included in Fig. 1-4.

For example, the flow of any quantity such as water from an outside source into a tank and from the tank through a drain is represented by the storage tank symbol in the simple system shown in Fig. 1-2*a*: The quantity indicated as stored by the energy circuit symbol for a tank may either be pure energy such as light or heat or matter carrying energy with it as with chemical substances. In Chapter 3, many properties of a storage unit are presented with translations to many other system languages to show that concepts associated first with other expressions are represented as clearly by the storage symbol.

Example of Use of Energy Language Symbols

In Fig. 1-5 the symbols of the energy language listed in Fig. 1-4 are used to portray a farm system. Solar energy is used by crops to produce food. The crop subsystem is a producer with a self-limiting chlorophyll unit that catches the solar energy and combines it with nutrients and water and human management to generate food and plant growth. Some of this storage is used by the necessary respiratory work *R* that maintains chlorophyll and other internal processes. A switching symbol indicates the system of harvest that takes place with human effort when the crop reaches a sufficient size. The harvest is sold in exchange for money according to the market price. The earned money is used to buy fertilizer. The family is shown to be reproducing according to its own inherent gain, using energy from the food

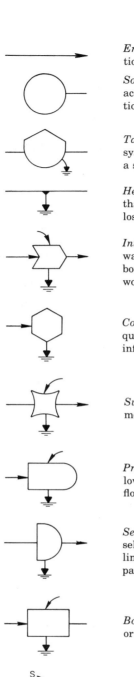

Energy circuit (Chapters 1–3). A pathway whose flow is proportional to the quantity in the storage or source upstream.

Source (Chapter 7). Outside source of energy delivering forces according to a program controlled from outside; a forcing function.

Tank (Chapter 3). A compartment of energy storage within the system storing a quantity as the balance of inflows and outflows; a state variable.

Heat sink (Chapter 7). Dispersion of potential energy into heat that accompanies all real transformation processes and storages; loss of potential energy from further use by the system.

Interaction (Chapter 8). Interactive intersection of two pathways coupled to produce an outflow in proportion to a function of both; control action of one flow on another; limiting factor action; work gate.

Consumer (Chapters 9 and 20). Unit that transforms energy quality, stores it, and feeds it back autocatalytically to improve inflow.

Switching action (Chapter 6). A symbol that indicates one or more switching actions.

Producer (Chapters 8 and 19). Unit that collects and transforms low-quality energy under control interactions of high-quality flows.

Self-limiting energy receiver (Chapter 10). A unit that has a self-limiting output when input drives are high because there is a limiting constant quantity of material reacting on a circular pathway within.

Box (Chapter 6). Miscellaneous symbol to use for whatever unit or function is labeled.

Constant-gain amplifier (Chapters 8 and 9). A unit that delivers an output in proportion to the input I but changed by a constant factor as long as the energy source S is sufficient.

Transaction (Chapter 23). A unit that indicates a sale of goods or services (solid line) in exchange for payment of money (dashed). Price is shown as an external source.

FIGURE 1-4. Some of the symbols of the energy circuit language with qualitative descriptions. For quantitative definitions, see chapter cited.

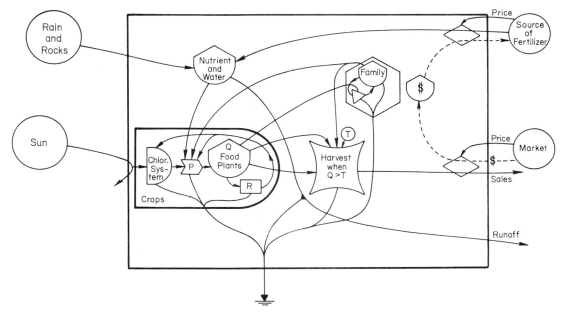

FIGURE 1-5. Diagram of a farm drawn with energy language symbols.

storages of the farm. The nutrients and water are maintained by the inflow from rain and rocks and from fertilizer. Outflow is in plant growth and runoff. Figure 1-5 shows parts, relationships, processes, energy pathways, and money relations.

Comparison of the systems diagram with a real farm suggests many ways that the diagram could have been drawn differently, perhaps showing more or less detail, including other factors, and aggregating more or less. Simpler models are retained easily in the mind for overviews. Sometimes more complex models are required for a good mental summary or computer simulation of real events. Sometimes very complex models are drawn as an inventory of everything known about a system. The purpose of the modeling may determine what is included. Sometimes modeling is considered as an art because of the individuality, creativity, and essence of experience that may go into selection of what is included.

Pathway Types and Notations

Lines represent pathways that we visualize for the flow of energy or materials and information (always with accompanying energy). Various types of pathway are recognized in the notation of the energy language as given in Fig. 1-6. Heat sinks accompany those pathways where energy is dispersed, such as in friction. Figure 1-6a shows that flow can go in either direction (bidirectional), depending on the pressure from both ends; Fig. 1-6b shows a valve that allows flow only to the right. Figures 1-6c and 1-6d have force in only one direction from the upstream storage, usually because there is an *energy barrier* as in Fig. 1-6c. Figures 1-6e and 1-6f are special pathways whose flow rates depend on *geometrical relations* or distance r; Fig. 1-6g shows an *interactive flow* requiring two different types of flow together; Fig. 1-6h shows a flow that accelerates, increasing velocity with inertial storage and backforces; Fig. 1-6i shows *additive flows* of similar type that merge together; and Fig. 1-6j shows a *diversion flow* subtracting from the initial flow. In the language, a pathway line has no storage and thus has no time delay in assuming its behavior according to forces. The flows may be expressed either in units per time or a flux in units per crosssectional area perpendicular to flow. If there is a time lag in a pathway, it is indicated as a "lag" box (Fig. 1-6k).

Linear Laws for Pathways of Steady Velocity

A plain pathway line in the energy circuit language represents the flow of some quantity that is either a form of energy or is accompanied by energy. The

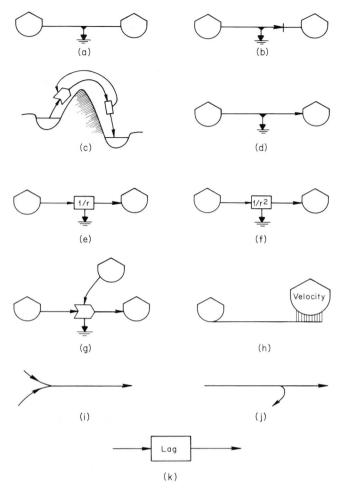

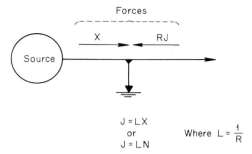

FIGURE 1-7. Steady flow where there is a balance of driving force X and frictional backforce RJ.

FIGURE 1-6. Types of pathway as drawn in energy language: (a) flow proportional to difference between forward and backforces; (b) valve; same as (a) except backflow prevented; (c) sketch of energy barrier that prevents backforce from storage from affecting flow; (d) energy diagram for flow without backforce from downstream storage; (e) pathway resistance proportional to distance r; (f) flow decreasing as inverse square of distance; (g) flow controlled by a second interacting flow of a different type; (h) pathway of acceleration increasing energy in velocity storage; (i) sum of two flows; (j) subtraction of a flow, (k) pathway with time lag.

ordinary pathway with a one-way barb is defined as following one of two linear laws relating flow to driving force. The pathway line is like the usual electrical current in a wire with the driving force balanced by a frictional back force that develops almost instantaneously in proportion to the rate of flow so that there is a balance of forces. The steady-state properties of the flow from a specified force X is expressed by Eq. (1-1), with the resistance R and is the rate of flow J (see Fig. 1-7):

$$X = RJ \qquad (1\text{-}1)$$

or

$$J = \frac{1}{R}X \qquad (1\text{-}2)$$

An example is the familiar Ohm's law that shows electrical current flow J as proportional to the electrical force X according to the resistance coefficient R describing the pathway. For another example, water flow is proportional to water pressure. A quantity L is called the *conductivity* and is the reciprocal of the resistance as in Eq. (1-3):

$$L = \frac{1}{R} \qquad (1\text{-}3)$$

Substitution of L for $1/R$ provides expression of the force-flux law in another well-known form in Eq. (1-4):

$$J = LX \qquad (1\text{-}4)$$

Similar expressions for force exerted by populations are given in Eqs. (1-5) through (1-7), where the number of parallel component processes N is the variable to which flow is proportional. Visualize, for example, 10 people pulling on a rope to pull a sled:

$$N = RJ \qquad (1\text{-}5)$$

$$J = \frac{1}{R}N \qquad (1\text{-}6)$$

$$J = LN \qquad (1\text{-}7)$$

Examples are the metabolism J of a population of animals N, the rate of decay J, or a population of radioactive molecules N. Most of the natural worlds flows are covered by one of the two linear laws [Eqs. 5. (1-4) or (1-7)] and thus the linear pathway definitions of the energy circuit language apply to most systems. In visualizing an energy network diagram such as that in Fig. 1-5, one may regard the pathways as lines and flow mainly in proportion to force action from the upstream energy sources and storages.

Consider centimeter-gram-second (cgs) units for these equations. If a flow J is given in grams per second and the force N is in grams per centimeter (quantity divided by volume per unit area of the exposed units that generate the force), the coefficient describing the pathway L is in centimeters per second [Eq. (1-7)]. As long as one is consistent, the units used do not matter since each of the quantities in Eqs. (1-1) through (1-7) is defined in terms of each other.

For other examples which fit these force-flow concepts see Table 7-1.*

System Boundaries, Forcing Functions, State Variables, Process Pathways, and Sinks

The boundaries of a system may be defined in any way. Boundaries are represented in the energy circuit diagram in Figs. 1-3 and 1-5 by the boxed lines. One separates the component units inside by drawing them within the boundary. External influences are drawn outside with symbols for sources and sinks outside affecting the system through pathway lines, and these lines are also the lines of action by forces. These are the process pathways.

An influence from outside must have an outside source of energy to create a force or a population of forces along an inflowing pathway. The outside source is sometimes called a *forcing function* and in energy circuit language is represented by a circle (see Fig. 1-7). The pattern of outside application of

* The force-flow equation [Eq. (1-2)] and its notation are those used in steady-state thermodynamics (Chapter 7), and the population force-flow equation [Eq. (1-6)] is established in many fields. These refer to a simple pathway. Earlier, Chapman (1931) in analogy with Ohm's law suggested that environmental resistance related biotic potential to production, which might be true for very simple special cases. However, these processes usually involve production intersections and storages and cannot be so easily described. See Chapters 8 and 9.

energy represented by the circle is the forcing function driving the inflow J. The forcing function might be constant or varying, and a complete description of a source requires a description of the function with time.

Within any system boundary there are components that have the property of storing. Storage compartments are sometimes called *state variables*. Figure 1-2 shows one storage compartment (tank); thus there is one state variable. The amount stored depends on the time history of inflows and outflows. At any time that one begins some calculation, the quantity present is the initial condition (initial quantitiy Q_0). For example, it may be the water in the bathtub. As shown in Fig. 1-2, the inflow tends to fill the storage Q.

Outflows from the boundaries of the system are sometimes called *sinks*. Figure 1-2 shows one sink that is an export of quantity Q. Also shown is a heat sink (the arrow into the ground symbol) that is the loss of potential energy into heat as required by the second law of thermodynamics. This law may be stated as the *energy degradation law:* energy becomes less able to do work as it becomes dispersed in the form of heat. Heat sinks must accompany all real processes and storages.

In choosing boundaries for an analysis of a system of interest, one may focus on the unit of interest by making all its influences outside forcing functions. However, this procedure is reductionistic and concentrates on details of a single unit and limits one's ability to understand the interplay of pathways to the forcing functions. We sometimes say that one must model and simulate a system that is one size larger than the one of interest. For example, to understand and predict trees, one has to model a section of forest; to understand and predict bass populations, one must model the pond; and to understand the human individual, one must establish boundaries around the larger section of the society surrounding that person.

Visualizing Changes of State with Time and Graphs with Time

The energy diagrams, because of their pictorial display of storages and locations of pumping interaction, allow the reader to conceptualize the time patterns. One visualizes the system as a network of storage basins with flows filling and draining as surges are entering or generated within.

For example, Fig. 1-2 shows a tank with one outflow pathway and one inflow. Suppose that it starts with no initial storage. Outflow is in proportion to the storage. The stored quantity increases at first until the outflow grows to equal inflow because of increasing pressure from the storage.

Part of the power of the energy language is its utility to facilitate visualization of performances with time by inspection. In later chapters, this ability develops from visualization of how performances of basic elements are verified with mathematics and by simulation.

Equations and Networks

There are many different systems that generate similar equations. For example, in Chapter 9, 11 different sets of premises are found to generate equations that can be manipulated algebraically into the form of the logistic equation. When drawn in a network language such as energy language, these systems are seen to be quite different phenomena. If translated into equation form without manipulating terms (without factoring, combining terms, multiplying and dividing by constants, etc.), there is a unique correspondence. However, traditional use of equations implies free manipulation to obtain analytic characteristics. Consequently, as generally used, an equation does not clearly define a system's characteristics. Many semantic errors occur because people express models as manipulated equations.

Oster and Desoer (1971) state these relationships of network topological languages to other kinds of mathematics: "A network diagram representation contains more information than a differential equation representation," and "a set of state, loop, node, etc. equations do not uniquely determine a network, but only an equivalence class of dynamically similar networks."

The Rectangular Box Symbol

Often there are special operations and functions for which we have no general symbol. In some circumstances the function is not known; or if it is known, we may not want to specify the function. For these, we use the rectangular box in which we can write whatever characterization is desired. We thus save the rectangular box as a miscellaneous symbol. We may use it to indicate a subsystem whose details are diagrammed on another page, and so on. Names should be given for the functions in general boxes. The black box is one in which we know the overall performance function but not enough about the details of its operation. A white box is one in which we know the details but do not want to be involved with them in the model under consideration. When the box contains a switching on–off (discontinuous) subsystem, energy language uses a concave-sided box (Figs. 1-4 and 1-5).

Money and Energy

Money is an exchange medium that flows as a countercurrent to energy (Odum, 1967a, 1971a) and can be symbolized separately in the energy circuit language. Thus money flows as a countercurrent to materials, information, and all other quantities used. Money is given out for goods purchased according to a ratio that we call *price*. The price, in turn, may have regulator mechanisms operating. For example, price regulation may be external. We represent money by a dashed flow line. Money is really a type of energy flow since it is a quantity that controls or releases other energy flows and thus has high-energy value in its interactions. The transactor symbol with an external control of price is given in Fig. 1-4 and used in Fig. 1-5. The symbol may have a heat sink that indicates energy cost of the transactions accounting or machinery for business.

SUMMARY

In this chapter we have introduced and defined the concept of a system, an energy circuit language for thinking and working with systems, and indicated some of the many other languages in which systems concepts are presented. Basic terms have been defined and examples given. Quantitative expressions were given for the relationships of flows and forces. We have begun the work of using environmental systems as examples to learn general systems concepts as well as content and principles of systems ecology. The students can check their comprehension at this point by working through the questions and problems that follow. Some of the symbols were used for environmental science examples by Turk et al. (1978).

QUESTIONS AND PROBLEMS

1. Define the following in as few words as possible:
   a. system
   b. subsystem
   c. source
   d. state variable
   e. forcing function
   f. storage compartment
   g. system boundary
   h. ecosystem
   i. model
   j. energy circuit language
   k. heat sink
   l. first energy law
   m. second energy law
   n. general system theory
   o. systems language
   p. force
   q. population force
   r. resistance
   s. conductivity
   t. flow
   u. force–flux laws
   v. energy
   w. work
   x. potential energy
   y. heat
   z. calorie
   aa. power
   bb. autocatalytic process

2. Draw energy circuit symbols for an outside forcing function, a storage compartment in the system, a heat dispersion process, a multiplicative interaction by which one pathway interacts with another, a self-maintaining component, switching actions, an exchange of money for goods or services, and a constant-gain amplifier component.

3. Use energy circuit language to diagram the following:
   a. A system of competition for food involving two species populations.
   b. A system with two sources interacting multiplicatively to produce a stored quantity that has some steady losses to export.
   c. A system with one source, one export sink, two storages, and a heat sink loss for pathways entirely within the system.
   d. A system with a feedback that acts multiplicatively on an upstream pathway intersection.
   e. A system with two water tanks and an electrical water pump between them.
   f. A system of about 10 components that is familiar to you.

4. If force X is 10 volts and flow J is 1 ampere, what is the value of resistance in ohms? What is the value of conductivity in mhos?

5. If force X is 10 dynes and water flow J is 10 grams per second, in what units is conductivity L? What is its value for that pathway?

6. What would happen to an aquarium if it were completely isolated and insulated? Would this be a closed or open system? When first isolated, would it be in transition state, steady state, or equilibrium? What would be the state after a long period of time?

SUGGESTED READINGS

Ashby, W. (1956). *An Introduction to Cybernetics,* Wiley, New York. 295 pp.

Bender, E. A. (1978). *An Introduction to Mathematical Modeling,* Wiley, New York. 256 pp.

Bennett, R. J. and R. J. Chorley (1978). *Environmental Systems,* Methuen, London. 624 pp.

Cannon, W. B. (1939). *The Wisdom of the Body,* Norton, New York.

Freedman, H. I. (1980). *Deterministic Mathematical Models in Population Ecology,* Dekker, New York. 254 pp.

Getz, W. M. (1980). *Symposium on Mathematical Modeling in Biology and Ecology, 1979, Pretoria,* Springer, Berlin.

Gordon, G. (1969). *System Simulation,* Prentice-Hall, Englewood Cliffs, NJ. 303 pp.; Chapter 1.

Grodins, F. S. (1963). *Control Theory and Biological Systems,* Columbia University Press, New York.

Hall, C. and J. Day (1977). *Ecosystem Models in Theory and Practice,* Wiley, New York.

Harbaugh, J. W. and G. Bonham-Carter (1970). *Computer Simulation in Geology,* Wiley-Interscience, New York.

Hare, Van Court, Jr. (1967). *Systems Analysis, A Diagnostic Approach,* Harcourt Brace, New York. Chapters 1–2.

Heinmetz, F. Ed. (1967). *Concepts and Models of Biomathematics,* Dekker, New York.

Kazemier, B. H. and D. Vuysje (1961). *The Concept and the Role of the Model in Mathematics and Natural and Social Sciences.* Reidel, Dordrecht, Holland.

Kobayochi, H. (1978). *Modeling and Analysis.* Addison-Wesley, Boston.

Klir, George J. (1969). *An Approach to General Systems Theory,* Van Nostrand Reinhold, New York. Chapter 1.

Koenig, H., Y. Tokad, H. K. Kesevan, and H. Hedges (1967). *Analysis of Discrete Physical Systems,* McGraw-Hill, New York. 447 pp.

Koestler, A. (1978). *Janus,* Random House, New York.

Lotka, A. J. (1945). The law of evolution as a maximal principal. *Human Biol.* **17,** 167–194.

Lynch, E. P. (1980). *Applied Symbolic Logic,* Wiley, New York. 260 pp.

Mesarovic, M. D. (1964). *Views on General Systems Theory,* Wiley, New York. 178 pp.

Mesarovic, M. D. and Y. Takahara (1975). *General Systems Theory,* Mathematical Foundations, Academic Press, New York. 268 pp.

Milsum, J. H. (1966). *Biological Control Systems Analysis,* McGraw-Hill, New York. Chapters 1 and 2.

Odum, H. T. (1971). *Environment, Power, and Society,* Wiley, New York, 330 pp.

Odum, H. T. and E. C. Odum (1976, 1981). *Energy Basis for Man and Nature,* 2nd ed., McGraw-Hill, New York. 297 pp.

Ogata, K. (1978). *System Dynamics,* Prentice-Hall, Englewood Cliffs, N.J.

Olinick, M. (1978). *An Introduction to Mathematical Models in the Social and Life Sciences,* Addison-Wesley, Menlo Park, Calif.

Patten, B. C. (1971). *Systems Analysis and Simulation in Ecology,* Vol. 1, Academic Press, New York. Chapter 1.

Poole, R. W. (1974). *An Introduction to Quantitative Ecology,* McGraw-Hill, New York. 532 pp.

Ranson, R. (1981). *Computers and Embryos, Models in Developmental Biology,* Wiley, New York. 224 pp.

Rogers, P. H. (1976). *Analog/Hybrid Computation in Biochemical Engineering, Advances in Biochemical Engineering,* Vol. 4. Springer, New York.

Rosen, R. (1970). *Dynamical System Theory in Biology,* Wiley-Interscience, New York.

Rosen, R. (1978). *Fundamentals of Measurement and Representation of Natural Systems,* Elsevier North Holland, N.Y. 221 pp.

Rubinow, S. T. (1975). *Introduction to Mathematical Biology,* Wiley, New York.

Shugart, H. H. and R. V. O'Neill (1978). *Systems Ecology: Benchmark Papers in Ecology,* Vol. 9, Dowden Hutchinson and Ross, Stroudsburg, Penn. 368 pp.

Smith, M. J. (1968). *Mathematical Ideas in Biology,* Cambridge University Press, London. 152 pp.

Vandermeer, J. (1981). *Elementary Mathematical Ecology,* Wiley-Interscience, New York. 294 pp.

Vemuri, V. (1978). *Modeling of Complex Systems,* Academic Press, New York. 449 pp.

Von Bertalanffy, L. (1950). The theory of open systems in physics and biology. *Science* **111,** 23–29.

Von Bertalanffy, L. (1968). *General System Theory,* Braziller, New York.

Waite, T. D. and N. J. Freeman (1977). *Mathematics of Environmental Processes,* Heath, Lexington, Mass. 192 pp.

Walters, C. Chapter 10. (1976). *Ecology,* 3rd ed., E. P. Odum, Ed., Saunders, Philadelphia.

Weinberg, G. M. (1975). *An Introduction to General Systems Theory,* Wiley-Interscience, New York. 277 pp.

Yearbook of General Systems Theory, Volume 1–present.

CHAPTER TWO

Ecosystems and Energy Hierarchy

In this chapter ecosystem patterns are introduced with guidelines for representing them with models. Energy flows develop characteristic webs of energy transformation, feedback interaction, and recycling. The webs form a hierarchy of converging transformations. The different kinds of energy are of different quality, and their quality is measured by the embodied energy of one type of energy required to develop another type in the work transformations. Potential for control is also dependent on embodied energy. Energy quality is recognized by the position of symbols on energy systems diagrams; thus the symbols display hierarchical patterns and control. Typical ecosystem models are given, including a simplified producer–consumer model. The chapter should guide readers in the ways of reading and drawing systems diagrams so as to portray structure and functions of ecosystems. For more details, see Chapters 15–18; 21–22.

CONCEPT OF ENERGY-QUALITY CHAIN

Typical ecological food chains illustrate the concept of successive energy quality transformation. As shown in Fig. 2-1, energy supporting phytoplankton is transformed through zooplankton and then through small fish to large fish. At each step much of the energy is used in the transformation, and only a small amount is transformed to a higher quality, one that is more concentrated and in a form capable of special actions when fed back. Real systems form webs rather than chains, but the energy changes are similar. Declining energy is accompanied by increasing quality.

Hierarchy and Spatial Convergence

Flows of low-quality energy are abundant and widely dispersed, and individual units are small in size. Higher-quality units and their flows, although less in total energy flow, are more concentrated, and each unit is larger in size with a larger territory from which it receives energy and feeds back its actions. Figure 2-1c shows the typical pattern of spatial hierarchy. In a familiar military hierarchy, soldiers report to corporals who report to sergeants who report to lieutenants, and so on, but commands exert control actions that diverge from lieutenants back to soldiers. Similarly, grass supports sheep, which in turn support shepherds, and so on, but control actions pass from shepards to sheep and from sheep to grass.

Ratios of Energy Quality

The ratio of energy of one type required to develop another type of energy in a transformation is one of the useful efficiencies often calculated to descibe energy patterns. It is a measure of energy quality since the new form of energy has abilities to amplify or control if it is fed back to interact with low-quality energies. Energy transformation ratios are derived from energy chains such as that shown in Fig. 2-1. If an inflow is a by-product from upstream or downstream, it need not be added, since this would entail counting the energy twice.

If energies of different types are to be compared in regard to energies required in their formation or their effect, they may be converted into equivalents of the same type by multiplying by the energy transformation ratio, which measures energy quality. For example, fish, zooplankton, and phytoplankton can be compared by multiplying their actual energy content by their solar energy-transformation ratios. See Fig. 2-1b. The more transformation steps there are between two kinds of energy, the greater the quality and the greater the solar energy required to produce a calorie of that type.

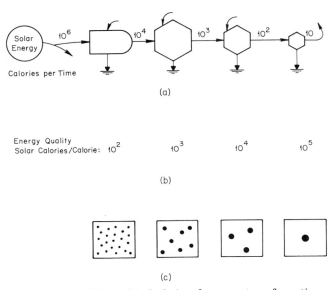

FIGURE 2-1. Hierarchical chain of energy transformations: (a) decrease of energy in successive transformations; by-product pathways are omitted; (b) energy-transformation ratios in solar equivalents; (c) spatial hierarchy characteristics.

When one calculates the energy of one type that generates a flow of another, it is sometimes referred to as the *embodied energy* of that type.

Arranging Energy Sources by Quality

The various energy flows of the biosphere such as sun, wind, rain, river flow, geologic uplift, tides, and waves, are all of different quality with different ratios of solar calories required in the world web to generate a calorie of that type of energy. Maximum utilization of energy comes from the interactions of these different quality energies so that they will amplify each other. In drawing an ecosystem so as to be consistent with the natural organization of energy, to minimize crossing lines, and to portray hierarchical relations, low quality energy is placed on the left and others arranged in order of their energy quality from left to right. A typical signature of energy inflows arranged in this manner is shown in Fig. 2-2.

Ecosystems of the world are often classified according to the different climatic or geochemical forcing functions that cause the patterns with time to differ. Where humans are involved, the ecosystems may be classified according to the main uses that are also causal forcing functions due to harvest of materials, release of wastes, and other interactions.

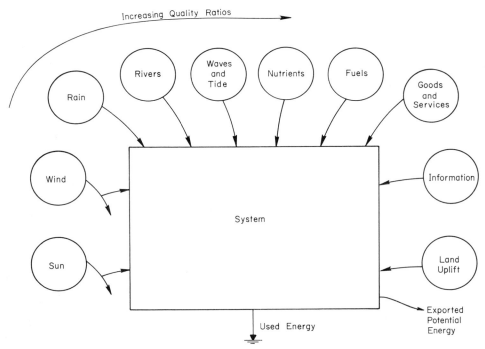

FIGURE 2-2. Typical energy sources driving an environmental system arranged in order from low quality on left to high quality on right.

Webs with Feedback Controls

Because the high-quality energies achieve their maximum effect when they are fed back as an amplifier or control action interacting with energy flows of lower quality, characteristic webs are observed. Figure 2-3 shows the characteristic pattern with converging energy and feedback contol interactions. Similar to the energy chain concept, there are successive transformations with increasing energy quality going to the right. The convergence that forms a hierarchy is recognized by the position on the paper and by using producer symbols on the left and consumer symbols on the right. The work of the consumers is feedback control actions in part.

ENVIRONMENTAL SYSTEMS AND SUBSYSTEMS: ECOSYSTEMS

An environmental system is a network of component parts and processes on the scale of the environment. Environmental systems usually include some area of the earth's land or water. Examples are forests, lakes, seas, farms, cities, regions, countries, and the biosphere as a whole. These tend to be comprised of living organisms, chemical cycles, water flows, components of the earth, and so on. The components often include humans and human-manufactured machines, units, or organization such as industry, cities, economic exchanges, social behavior, and transportation, communication, information processing, politics, and many others. Each of these components is a complex subsystem of the larger environmental systems. Biological community subsystems control the detailed performances of the many species of living organisms. What is a system, and what is a subsystem is an arbitrary characteristic of one's point of view. One person's system is another person's subsystem. There are apparently similar laws of function and mechanism operating at all levels of scale and size.

The word "ecosystem" is a short expression for "ecological system." To some people, "ecological system" has become a synonym for "environmental system." Others use the word "ecosystem" only for those systems of nature that generally do not include humans. In this book we use the terms "ecosystem" and "environmental system" interchangeably. An organized system of land, water, mineral cycles, living organisms, and their programmatic behavioral control mechanisms is called an *ecosystem*.

Examples of large ecosystems are whole forested regions, the seas, and the biosphere as a whole. Examples of small ecosystems are such associations of organisms and microenvironments, as ponds, coral heads, and aquaria in school classrooms. Small ecosystems are subsystems within the larger ones. In the broadest modern usage, systems that include humans, such as farms, industries, and cities, are also regarded as ecosystems, and there is a growing recognition that humans are incomplete without the life support of self-maintaining natural ecosystems. All these systems develop structures and processes adapted to the environmental inputs available to them.

In time through the process of trial and error, complex patterns of structure and process have evolved, the successful one surviving because they use materials and energies well in their own maintenance, competing well with other patterns that chance interposes. The parts of successful ecosystems are organized by exchanges of minerals, food chains, economic transfers, and other work services that act among the components. These pathways of work can be diagrammed to show the designs of ecosystems (Figs. 2-1–2-4). The engineering of new ecosystems designs is a new field that uses systems that are mainly self-organizing. Ecosystems will re-

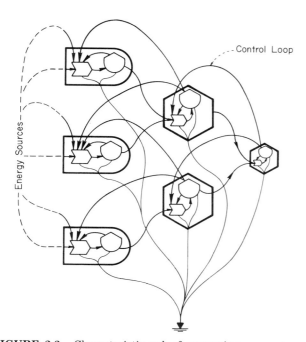

FIGURE 2-3. Characteristic web of converging energy transformations and feedback of interaction control loops found in self-organizing systems.

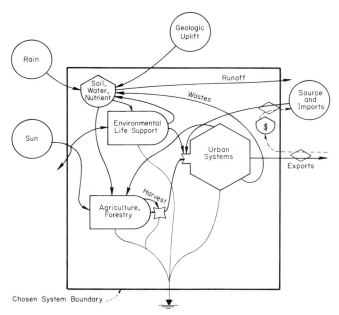

FIGURE 2-4. Diagram of an environmental system with use of symbols of the energy circuit language.

quire delicate and intelligent measures for effective use.

With our growing knowledge of ecological design, management of the ecosystems by human interactions is one of the bright hopes for humans if their large, rich sources of energy supporting their present complex urban system begins to wane.

Example of an Environmental System

Figure 2-4 is a diagram of a typical environmental system that has several energy sources, each of a different quality and a course aggregation of components to show organization in the characteristic hierarchical pattern.

In qualitative translation into English, the diagram in Fig. 2-4 expresses energy inflows in (1) sunlight, (2) accompanying rain, and (3) geologic uplift. These develop soils and interact to operate production by environmental life support systems (natural areas without humans) and production by agriculture and forestry. These supply products and services to the urban areas along with imports that are obtained by purchases with money. The money for purchasing imports comes from sale of exports. The urban systems sends wastes back to the environmental system and services of labor, fertilizer, equipment, and so on, to the agriculture and forestry sector. A switching unit controls the harvest of products for shipment to the urban systems. The life-support systems helps build soils. Outflows include the exports; the runoff; and the used, degraded energy.

Cycle of Materials

A characteristic of systems is their recycling of materials, especially those that have few inflows. Forests and plankton ecosystems recycle phosphorus, human economies recycle steel and aluminum, the biosphere recycles water, the mountain building systems recycle sediment materials, and so on. Diagrams of systems usually show this important feature; for instance, Fig. 2-4 shows a recycle of wastes (Lotka, 1925).

Material Flows as Energy Flows

Sometimes there is confusion about the energy associated with flows of materials. Since any concentration of a substance relative to its environment constitutes a small potential energy storage, all recognizable flows of materials carry an energy content. The energy is sufficient to drive the degrading diffusion of that substance away from its concentrations.

Although the actual concentrations of energy are small, the embodied energy of successive work in developing the concentrations may be very large. For example, large solar energies are used to develop the food chains to guano birds, which in turn develop island phosphate deposits. The solar energy embodied in phosphate deposits is large. The energy transformation ratio is the ratio of solar calories per calorie of phosphate concentration. Ways of calculating these are given in Chapters 7 and 14. Generally, the more critical and important a material is as a limiting factor, the more embodied energy it carries. The energy is of high quality.

It is recommended that the student become familiar with material flows and energy flows and appropriate schematic representations of the energy language pathways.

Information Flows as Energy Flows

As one passes from abundant low-quality energy through successive transformations in chains such as shown in Fig. 2-1 or through webs such as shown in Fig. 2-3, the actual calorie flows decline, but the embodied energy and energy quality increases as already described. Where the sequence of transformations is very long, only tiny fractions of the original quantity of calories is still contained, but the energy transformation ratios are very large. Customarily people don't regard these flows as energy, but usually call them *information*. Flows of genes, books, television communications, computer programs, human culture, art, political interactions, and religious communications are examples of information. However, these all have very large embodied energies and high ratios of calories of solar energy per calorie of information. It is still appropriate to consider the calories of potential energy in information, since this is what drives the depreciation and loss of information as with losses of other kinds of storage. Information cannot be stored without some concentration of substance or energy fields relative to the environment. For example, words on a page are concentrations of ink, memory on a computer disk is a concentration of magnetic field, and information in biological genes is a storage of DNA form relative to the environment.

The flows of information carry the most embodied energy and also have the greatest amplifier and control effects per calorie. They are the feedbacks of highly embodied energy that provide systems with specialized services, feeding back positive actions that "repay" their webs for the energy dispersed in the development of the information. Energy and information are discussed further in Chapter 17. Here it is necessary to recognize that information pathways are energy pathways also and are drawn in energy language as regular pathways, but usually feeding back from the far right side of the diagrams.

Money is a special kind of information flow that controls through its circulation as a countercurrent with exchange rates according to human behavior programs for reacting to prices. Energy circuit language uses dashed lines or color to help recognize this component web.

Structure and Function

Understanding structure and function is an often stated objective of system sciences. Structure is in the pattern of connections and in the storages. To simplify models, the patterns of connections of complex diagrams may be combined as a single storage in aggregating detail. Functions are flows and changes of state as energy passes through network structure. Processes are located at points of energy transformation usually at intersections.

An Aquatic Ecosystem

Figure 2-5 is an example of diagramming an ecosystem of smaller scale, a pond. The main features of most systems that we have just described are included: energy sources of different quality; hierarchies of energy transformations (food chain); feedback control loops; recycling, materials; information exchange; outflows; and the used energy sink. Three classes of plant are shown, those with underwater life competing for light and releasing oxygen from photosynthesis. The consumers utilize organic production and oxygen and regenerate nutrients with recycling facilitated by kinetic energy.

BASIC PRODUCER–CONSUMER MODEL

Although the ecosystem shown in Fig. 2-5 is simple in comparison to its real-world counterpart, there is an even simpler basic model that represents much of the essence of ecosystems, simple hierarchy, recycling, and energy source control of growth and storage. This is the producer–consumer model given in Fig. 2-6. Sometimes this is called the P–R model, with P referring to photosynthetic production and R to respiration of the consumer parts of the system. A variation of the model in Fig. 2-7 is open to sunlight,

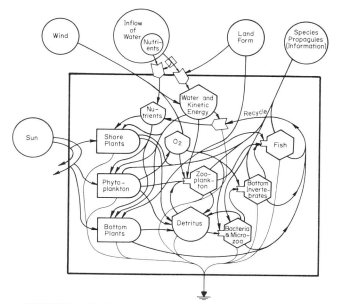

FIGURE 2-5. Energy diagram of an aquatic ecosystem.

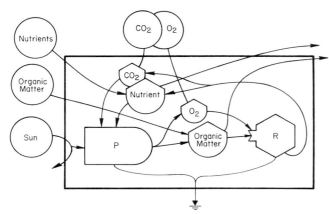

FIGURE 2-6. Main process of production, consumption, gaseous exchanges, and recycle in typical ecosystems.

is limited by its externally controlled inflow, and is closed to matter. Examples are a sealed aquarium or the whole biosphere. Compare this simple model with the more complex one in Fig. 2-5. The simple one is more aggregated but has some of the essence of ecosystem form and function. It is a good model to remember while considering new systems.

Although producer and consumer symbols as shown in Fig. 2-7 are included as a reference to the categories of hierarchy, the internal symbols are shown simpler than in most uses. Production has only the interaction process, and consumption is a linear pathway as given in Fig. 1-6d. All products from production are put into one storage and all regenerated materials, in another.

External and Internal Limiting Factors

The P–R model in Fig. 2-7 illustrates the concept of limiting factors in two ways: the external energy inflow is limited by its rate of supply from outside, so that the system inside can use only what can be pumped from the passing stream; and consequently, it can develop only as much organic storage as can be supported on the daily ration of energy flow.

The system also has a limiting factor action in the internal recycle (see dotted line in Fig. 2-7a). When energy is increased, more of the recycling materials (carbon, nitrogen, phosphorus, etc.) are bound up in the organic storage, leaving the pool of available raw material nutrients in short supply and limiting production. The graph in Fig. 2-7b shows the effect of increasing the amount of sunlight without changing the amount of internally recycling materials. At first, the sunlight increases production, but the effect diminishes and at higher light levels fails to

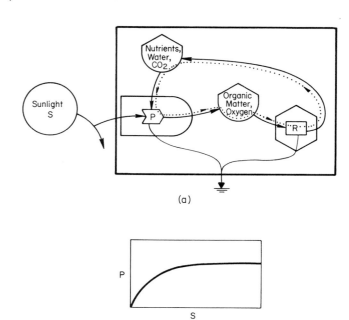

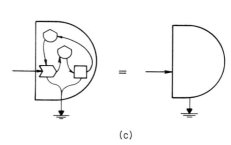

FIGURE 2-7. Basic producer–consumer model of ecosystems: (a) energy diagram with material cycle as dotted line; (b) effect of increasing energy on production; (c) symbol for systems limited by recycle of limited materials (see Fig. 1-4).

increase production because of the limitation in the rate of recycle by the consumer units.

Recall the energy symbols given in Fig. 1-4. One of these is a shorthand for systems such as the P–R model that have self-limits on energy receiving provided by limits in internal material recycle (see Fig. 2-7c).

The output of production from interaction processes may be limited by rates of supply of participating reacting flows, some from outside and some from recycling. This example should help the reader to understand limiting factor patterns as they are shown in energy diagrams.

Another version of the producer–consumer model is given in Fig. 2-8 on a different scale of size, the producers being farms and the consumers the city. The system has recycling materials in one direction

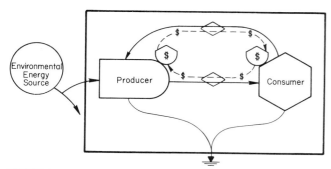

FIGURE 2-8. Producer–consumer model for the system of the landscape including humans.

and recycling money in the opposite direction. Food and fiber go to consumers; and goods, services, fertilizer, and control feedback as a reward loop; thus the division of labor of the hierarchy increases the total energy flow available to the whole system.

An interesting perception on inflation comes from this overview. Interruption of sunlight with volcanic dust in the stratosphere in 1815 caused decreased productivity. Since work was less but money circulating was similar, a money unit bought less; this is an example of energy-induced inflation.

READING DIAGRAMS

By now the student should recognize the symbols in Fig. 1-4 by their English descriptive titles and be qualitatively familiar with the functions and structures they represent. Quantitative definitions are developed in Chapters 3–13 as the most common component functions of systems are introduced.

Before we proceed to chapters on quantitative aspects of systems, with various languages, simulations, equations, and so forth, let us review with an overview of the qualitative nature of ecological systems with and without humans. The energy circuit symbols introduced in Chapter 1 showed the kinds of energy flows, principal storages, pumping intersections where energy flows interact, feedback pathways, and other configurations that are frequent components of common ecosystems of the planet (Figs. 1-5, 2-4, and 2-5).

You should read these diagrams by first visually focusing on the external energy sources from which external causal forces are derived (the circles, usually on the left) and then let your eye follow in the direction of the energy flows into places of storage, into the hexagonal units that have storage and feedbacks of work to multiplicative action on the inflow, and ultimately out of the system as dispersed heat as indicated by the heat sinks. Having scanned the pathways to see what is causing what and where the system stores the means to control its own forces, try to see if you can close your eyes and remember the system. Can you redraw it from memory? Can you make some guesses as to what would happen if (1) the energy sources increase, (2) decrease, (3) work costs of maintenance increase, (4) the system started up with zero storages, or (5) the system started up with very large initial storages?

In such intuitive scanning, you are making hypothetical visualization of the patterns that would result if you drew graphs with time. Before we finish the next three chapters, you will be simulating simple systems on computers to see what are the consequnces for the structural relationships one has energy diagrammed. We will see how the energy diagrams are another but easier way of writing the classical kinds of differential or difference equation that are often the point of departure in systems study. Can you visualize some characteristic ecosystems; production; respiration; consumption; mineral cycling; food chain stages; and the way agriculture, cities, money, and older natural systems are coupled in our biosphere and the kinds of subsystem that are important in control and regulation?

Making a Systems Diagram—Communication in a Group

Figure 2-9 shows the main steps in making an energy diagram: (1) all the extrinsic compelling functions believed to be important are listed in quality order (as in Fig. 2-9a); (2) the main state variable components are listed and placed inside the system boundaries, also in order of energy quality; and (3) the process pathways believed important between the compartments are added as in Fig. 2-9c. A heat sink is added to every real process as required by the second energy law, and outflowing, exported, potential energy pathways are also drawn.

When a group gathers around a table to talk about analyzing the main components of a new system or problem, one person can diagram for the group, enhancing the coherence of the discussion. If the symbols are understood by all, the process of discussion and drawing unites people and thinking around a task with a minimum of semantic problems about meanings. A group, collective-thinking exercise stimulates memories and draws out qualitative and quantitative knowledge from combined

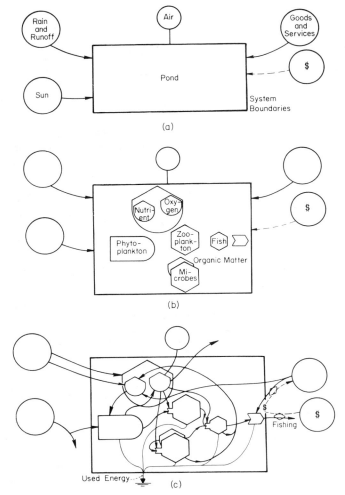

FIGURE 2-9. Steps in drawing an energy language diagram with an example, of a fish pond: (*a*) boundaries and sources; (*b*) components; (*c*) completed diagram.

experiences with the real-world system of concern. This can be an effective problem-solving and analysis activity in management, research, and classroom settings and should be attempted as a learning tool. It is a useful first step before quantitative or simulation studies.

In this book the rectangular frame representing the system's boundary is used when whole systems are being aggregated but is often omitted where mechanisms and parts are being considered.

SUMMARY

Using energy symbols, typical patterns of hierarchy, recycle, production, consumption, limiting factors, and multiple sources were considered in relation to energy quality. The P–R model was introduced as a basic minimodel for examining many ecosystems. Practice was offered in reading and drawing standard energy diagrams. For more on hierarchy see Chapters 13–18 and books by Whyte et al. (1969) and Weiss (1971).

QUESTIONS AND PROBLEMS

1. Can you reproduce diagrams from memory?

2. Can you develop an energy systems diagram for the system visited in a field trip, being careful to include the main structures, the main causal energy flows, and the main control actions, but not putting in more than readily fits on one page?

3. Can you assume that the P–R model is in steady state and write on the diagrams reasonable values of the energy flow per square meter per day? First indicate storage and rates in grams of carbon. Then indicate the energy storages also in kilocalories per square meter as if these storages were used only as a fuel.

4. Modify the P–R model in Fig. 2-6 with inflows and outflows so that it represents the following:
    a. a forestry plantation
    b. a sewage plant
    c. water hyacinths in an agricultural runoff ditch
    d. a city with relatively few parks
    e. a soil subsystem under a forest cover
    f. a spring-fed stream

    Hint: Drop out certain pathways that are relatively unimportant for some cases.

5. Draw a diagram of a prey–predator population. Can these be symbiotic with reward loops?

6. Does money accompany the main flows of human work shown in Fig. 2-4?

7. Write a definition, clearly diagram, and then mention an example for the following:
    a. food chain
    b. herbivore
    c. carnivore
    d. detritivore (pathway from organic storage pool called detritus)
    e. production

f. photosynthesis
g. consumption
h. respiration
i. balance of payments
j. purchase of goods with money
k. a process where two energy flows of different kind are both essential to the product outflow
l. a deterioration pathway that is an inherent part of structure
m. inorganic nutrient
n. organic nutrient
o. succession
p. threshold switching action
q. a self-maintaining subsystem
r. an autocatalytic subsystem

8. Can you arrange environmental energies in order of energy quality as compared to sunlight?

9. Explore your landscape and look for hierarchical relationships.

10. Two field trips are desirable during this period.
    a. *Forest system.* First the class gathers in a forest spot and digs a soil hole and then discusses the various main components of the forest system, photosynthesis, transpiration, mineral cycling, the trunk role in transport, seasonal cycles in foliage, roles of kinds of roots, litter, diversity, microbes as decomposers, microbial regulators, herbivores, switch feeders, soil horizons, leaf area index, stratification, and light fields.
    b. *Spring, pond or stream.* The class gathers at an aquatic ecosystem and does a little qualitative collecting and visual examination together, identifying the main trophic levels, the manner of inflow of light and other energy controlling forcing functions, the exports, and the organization of geomorphology and biota as a storage interacting with physical energy flows such as water waves and current.

The result of these two field trips, which take about 2 hours each, should provide a visual picture of two main kinds of ecosystem, one terrestrial and one aquatic, with an understanding of the manner by which the structures operate the processes and vice versa. Students new to ecology thus obtain some direct experience, and those with biological and ecological training may get some feel for the process of aggregating and grouping mentally so that the vast detail about life in their training does not interfere with the need in modeling to include what is important while excluding what is unimportant at the scale of consideration.

SUGGESTED READINGS

Bogner, R. E. and A. G. Constantinides (1975). *Introduction to Digital Filtering*, Wiley, New York.

Hall, C. and J. Day (1977). *Ecosystem Models in Theory and Practice*, Wiley, New York.

Henderson, L. J. (1913). *The Fitness of the Environment*, Macmillan, New York. 575 pp.

Horn, D. J., R. D. Mitchell, and G. R. Stairs (1979). *Analysis of Ecological Systems*, Ohio State University Press, Columbus. 312 pp.

Jeffers, J. N. R. (1978a). General principles for ecosystem definition and modeling. In *Breakdown and Restoration of Ecosystems*, Plenum Press, New York. pp. 85–101.

Jeffers, J. N. R. (1978b). *An Introduction to Systems Analysis with Ecological Applications*, University Park Press, Baltimore.

Jorgensen, S. E. (Ed.) (1979). *State of the Art in Ecological Modelling*, Environmental Sciences and Applications, Vol. 7, Pergammon Press, N.Y.

Kormondy, E. J. (1969). *Concepts of Ecology*, Prentice-Hall, Englewood Cliffs, NJ.

Kormondy, E. J. and J. F. McCormick (1981). *Handbook of Contemporary Developments in World Ecology*, Greenwood Press, Westport, Conn. 766 pp.

Lasker, G. E. (1981). *The Quality of Life: Systems Approaches.* Applied Systems and Cybernetics, vol. 1, Pergammon Press, New York.

Lugo, A. E. (1974). The ecological view of the steady state society. In *Man and Environment,* G. T. Van Ray and A. E. Lugo, Eds., Rotterdam Press, Hague, Netherlands. pp. 293–316.

Mason, H. L. and J. H. Langenheim (1957). Language and the concept of environment. *Ecology* **38**: 325–340.

Mitsch, W. J., R. W. Bosserman, and J. M. Klopatek (1981). *Energy and Ecological Modelling*, Elsevier, Amsterdam. 839 pp.

Odum, E. P. (1963). *Ecology*, Holt, Rinehart and Winston, New York.

Odum, E. P. (1971). *Fundamentals of Ecology*, 3rd ed., Saunders, Philadelphia.

Odum, H. T. (1971). *Environment, Power and Society*, Wiley, New York. 331 pp.

Odum, H. T. (1978). Energy analysis, energy quality, and environment. In *Energy Analysis, a New Public Policy Tool,* M. Gilliland, Ed., Westview Press. pp. 55–87. (Reprinted in *AAAS Selected Symp.* **1**, 1–110.)

Odum, H. T. and E. C. Odum (1976). *Energy Basis for Man and Nature*, McGraw-Hill, New York. 337 pp. (2nd ed., 1981).

Pattee, H. H. (1973). *Hierarchy Theory*, Braziller, New York. 155 pp.

Quastler, H. (1964). *The Emergence of Biological Organization,* Yale University Press, New Haven, CT.

Richardson, J. L. (1977). *Dimensions of Ecology,* Williams and Wilkins. Baltimore. 412 pp.

Smuts, J. C. (1926). *Holism and Evolution,* Viking Press, New York.

Steele, J. H. (1977). Spatial pattern in plankton communities. *NATO Conference Series.* Section IV. *Marine Sciences.* Plenum Press, New York.

Watt, K. E. F. (Ed.) (1966). *Systems Analysis in Ecology,* Academic Press, New York. 267 pp.

Watt, K. E. F. (1966). *Systems Analysis in Ecology,* Academic Press, New York.

Weiss, P. A. (1971). *Hierarchically Organized Systems in Theory and Practice,* Hafner, New York.

Whittaker, R. (1975). *Communities and Ecosystems,* 2nd ed., Macmillan, New York.

Wilson, L. L., Q. E. Whyte, and D. Wilson (1969). *Hierarchical Structures,* American Elsevier, New York. 322 pp.

CHAPTER THREE

Storage and Flow

Storage compartments are important and obvious parts of any system. A valuable quantity that is stored is called a *state variable*. An environmental system is full of storages such as those of air, water, chemicals, biological populations, geologic quantities, urban structures, finance, information, and people. The properties of the storage function and flows are subject to many laws and have been studied, described, and modeled in many languages. In this chapter we consider storage and flow quantitatively, introducing and using many systems languages to introduce the pertinent principles.

Various different systems languages express properties of storage in different ways, each providing the mind with some special insight. Since the various formulations are translations, one represents all the properties of storage and flow when one expresses the function in one language. In the paragraphs that follow the storage function is indicated in many of these formulations. While learning about storage in several languages, the student should begin to learn the various languages, thus developing a unity of concept about storage and flow. As in Chapter 2, we use the energy circuit language as a starting and comparing basis. Its storage symbol was given in Fig. 1-2a. An equation or model involving only one storage is sometimes said to be of first order.

INHERENT ENERGY DRAINS OF STORAGES

There are losses inherent in any storage. Any concentration of a quantity constitutes an energy source and tends to react and drive opportunistic pathways. Pathways tend to develop if the arrangement provides the pathway with more energy and thus selective advantage under conditions of pathway competition. For example, a storage of food tends to develop chemical reactions, populations of insect eaters, or whatever other pathways may be opportunistically available. If one has containers or other means for preventing energy losses from a storage, one is spending energy on this work. Energies are either lost in drains or spent in preventing the losses. In either case there is an energy drain that is part of any storage, and models that do not include it are incomplete.

Role of Energy Storages

Since storages lose energy, they are not advantageous unless storage functions are required by the system for operations that cannot be arranged without them. What are some useful or adaptive roles for storage within a system that justify the storages that we observe so commonly in the earth, air, waters, and living compartments?

Systems require forces to drive their operations, and forces cannot be developed and focused on a particular pathway point unless there is an energy storage at that point to deliver the force. Unless a force is provided from an outside energy source, the system must have storages at those points where necessary functions must be controlled.

Another role of storages is to supply materials and energies during periods when outside energy inflows are interrupted. The longer the interruptions, the larger must be the storage to keep the system continuously functioning. For example, organisms store foods for nighttime use, and polar environmental systems must store foods to last throughout the long winter. A cactus must store water between rains in the desert.

Storage of Energy and Matter

Since energy is inseparable from matter, it is appropriate for models that represent both to use the storage symbol to mean both at the same time. A storage is shown with a heat sink, which means that there is depreciation and entropy increase. The heat sink alone means that potential energy is lost into used energy, but not that matter is dispersed there. If matter is also dispersed, this flow requires a separate pathway as given in Fig. 1-2a and 3-1, where an outflow from storage branches. Even dispersed matter may have high energy value relative to another system in which it becomes a limiting factor and other energy is supplied.

Materials can be stored in a dispersed, low-energy state with less depreciation since such storage has little potential energy to drive depreciation. Open systems may store more of the recycling materials in low-energy state efficiently; this phenomenon may partly explain why more water is stored in the ocean than in elevated lakes and the atmosphere.

Statistical Property of Pathways and Storages

Most flows and storages that are identified as single entities in our concepts and models are really aggregations of parts. They contain populations—populations of molecules, animals, dollars, eddies, and so on. The populations in a flow carry energy, and populations in a storage contain potential energy. Energy generates variation, some of which is on a small scale so that it may be regarded as noise that drains off energy into dissipation. When a flow is represented by a pathway or storage by a tank symbol, a population of energy carriers may be involved with a range of values. In the aggregate there is a total flow and a total storage. The distribution of energy within the population can be considered when it is of interest or important to the model. Energy sources can be characterized by the statistical distribution of the populations that comprise their energy delivery. Models for the distribution of population properties such as normal, exponential, and log-normal equations are considered in relation to energy in Part III.

Quantity Stored

The quantity stored Q may be a pure energy such as heat or light, or it may be a material, information, or

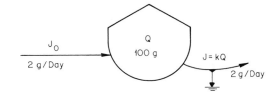

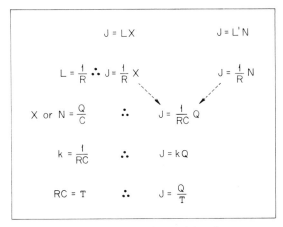

FIGURE 3-1. Summary of relationships between storage quantity Q, forces X, N, outflows J, resistance R, conductivity L, time constants T, and transfer coefficients k. All these relationships are automatically implied by the energy circuit symbol above. Data for leaf fall in a forest are given.

money with which energy is combined in the storage. When there is a flow of matter or information, there is energy flow with it, but the energy and the quantity are not always linearly proportional. For example, the energy stored in a cylindrical water tank is a function of the square of the quantity because each succeeding addition of water has to be raised higher over the water already in the tank. The outflow rate through a pipe is proportional to the force of the water pressure and thus is proportional to the quantity of water stored, but the rate of delivery of energy with the quantity is not linear since the pressure is changing. To fully represent the relationship of energy storage to forces and flows, one must consider the kind of energy storage and include an equation for the storage function relating energy to the quantity stored.

STORAGE PATHWAYS AND TURNOVER

Forces Delivered from a Storage

A storage delivers a force to a pathway in proportion to the quantity stored. Depending on the nature and geometry of the storage, there is a force X or a population of forces N that the energy source delivers to available pathways. As an example of a single force action X, a storage of electron charges in an electrical tank (capacitor) delivers an electrostatic field force measured in volts. As an example of a population of forces N, a storage of organic waste in water delivers impetus N to the parallel consumption by many microorganisms. As given for force X in Eq. (3-1) or population force N in Eq. (3-2), the driving action delivered is proportional to the quantity in the storage according to C, a coefficient that is a function of the dissected nature of the storage. If storage is in smaller units it has more surface and C is smaller. Coefficient C is proportional to the volume:area ratio:

$$X = \frac{Q}{C} \qquad (3\text{-}1)$$

$$N = \frac{Q}{C} \qquad (3\text{-}2)$$

The tendency for a quantity of storage to drive opportunistic circuits is proportional to the surface area per unit of quantity stored. The more surface, the more outflows there are or the more energies are spent preventing the flows. In electrical terminology C is the capacitance measured in farads and is the ratio of quantity of charge stored Q measured in coulombs to the force X that it delivers measured in volts.

Flow Delivered from a Storage

A flow is delivered from a storage in proportion to the driving force, and this in turn is proportional to the stored quantity. If expressions for X and N [Eq. (3-1) and (3-2)] are combined with the linear laws given in Chapter 2 [Eq. (1-4) and (1-7)], one obtains Eq. (3-3), which relates flow rate to stored quantity in terms of the capacitance coefficient C that is a function of the form of storage and resistance coefficient R that is a measure of the pathway. Where flows are proportional to storages, the pathways are said to be linear:

$$J = \frac{1}{R}\frac{Q}{C} = L\frac{Q}{C} \qquad (3\text{-}3)$$

Transfer Coefficients

A transfer coefficient is the fraction of storage that is transferred out by an outflow pathway in a given unit of time. The transfer coefficient k can be defined as the reciprocal of the quantity RC as in Eq. (3-4).

$$k = \frac{1}{RC} = \frac{L}{C} \qquad (3\text{-}4)$$

With k substituted in Eq. (3-3), Eq. (3-5) results and represents in simple form the flows delivered by a quantity stored:

$$J = kQ; \qquad k = \frac{J}{Q} \qquad (3\text{-}5)$$

As the preceding derivation shows, the transfer coefficient depends in part on the nature of the pathway and in part on the nature of the storage. Where such details are not of interest, Eq. (3-5) is the simplest equation for ordinary unidirectional pathways that are indicated in energy language with a barb. Many systems models are made entirely of these types of pathway. Because of the porportionality of flow to upstream quantity, these pathways are said to be linear. They also can be described as donor driven, since the upstream storage controls the flows rather than the recipient.

Determining Coefficients from Data

For a model of storage and flow such as that shown in Fig. 3-1, coefficients can be calculated from data with the help of the diagram. Use data for a given unit of time, or use average data. First, write numbers (data) for quantity stored in tank and for flow rates on pathways as in Fig. 3-1. All flows in or out of a tank must be in the same units, such as grams per day, people per year, and so on. The storage number is in the same unit, but without time.

Next, set the flow equal to the mathematical term for that pathway. For example, for the outflow in Fig. 3-1, $kQ = 2$ when $Q = 100$. Therefore, $k = 2/Q = 0.02$.

Metabolic Rate

In biology the flow of materials and energy associated with the overall process of maintaining life is called *metabolism*. If the outflow of products of life is

J in Eqs. (3-5), the flow is indicated as proportional to the quantity of living biomass Q. In this case the transfer coefficient k is the flow per unit quantity that is sometimes called *specific metabolism* or the *metabolism per unit weight*.

In defining the storage capacitance coefficient in Eqs. (3-1) and (3-2), we indicated the geometric fact that storage tends to drive pathways in proportion to surface exposed and that metabolic work done to protect storages is also area proportional. Where surface volume relationship is the only factor controlling the force per unit quantity, the capacity factor C is proportional to the ratio of the volume v to the surface area a and thus to the length dimension l of the storage unit as in Eq. (3-6). Here K is a function of the geometric shape. Farads, the measure of electrical capacitance, are equivalent to cgs length units:

$$C = \frac{kv}{a} = Kl \quad (3\text{-}6)$$

Combination of Eq. (3-6) with Eqs. (3-3) and (3-5) produces expression (3-7), which shows metabolism per gram k as an inverse function of size l, a property often observed in comparing metabolisms of similar proportioned animals of different size:

$$k = \left(\frac{L}{K}\right)\frac{1}{l} \quad (3\text{-}7)$$

If one is reminded that a volume V is proportional to the cube of its linear dimension, the quantity of biomass Q is also proportional to the cube of the linear dimension l with the additional factor of density ρ as given in expression (3-8):

$$Q = \rho V = K_2 \rho l^3 \quad (3\text{-}8)$$

$$l = \sqrt[3]{\frac{Q}{K_2 \rho}} \quad (3\text{-}9)$$

Combination of Eq. (3-9) with Eqs. (3-6) and (3-3) produces Eq. (3-10), which relates overall metabolism J to two-thirds the power of the biomass quantity, another law often observed approximately in comparisons among animals of different sizes and similar shapes.

$$J = K_3 Q^{2/3} \quad (3\text{-}10)$$

These manipulations show how storage ideas in one language have their equivalent in others; biological ideas on metabolism and electrical ideas on flow rates are equivalent and adequately described by the more general definitions of the energy circuit language that was invented to generalize in this way.

If growth of organisms such as fish are changing capacitance as a result of size change (volume:surface ratio), metabolism follows Eq. (3-10) instead of (3-5) (Von Bertalanffy, 1968). Gill size increases accordingly (Gray, 1954).

Storage and Time

When storage is introduced into a system, the property of time is also introduced. Without storages that can vary, there is no change and no way to recognize passage of time. Systems with small storages change more rapidly than do those with larger storages. Changing scales of storage is the same as changing the scale of time (Whitrow, 1980). See Chapter 27.

Time Constant

The time constant is turnover time. Refer to the diagram of a tank at steady state (constant level Q) in Fig. 3-1 in which the inflow J_0 is equal to the outflow rate J, which is equal to kQ. Imagine that the inflow is suddenly changed to a tagged stuff as if colored water were flowing into a water tank whose water level was constant. As the colored water moves through, displacing the original uncolored water, a time is required for the turnover displacement that is obtained by dividing the quantity in the tank by the quantity per unit time in the inflow–outflow. This turnover time is called the *time constant* τ.

If the time constant is inverted and the flow per time J is divided by the quantity Q, one obtains J/Q, the turnovers per unit time, a quantity already called k, the transfer coefficient [Eq. (3-5)]. As given in Eq. (3-1), the transfer coefficient is the reciprocal of the turnover time τ. Noting also Eq. (3-4), we see that the time constant τ is equal to RC:

$$\tau = RC = \frac{1}{k} \quad (3\text{-}11)$$

When R and C are given in their special electrically named units (ohms for resistance R and farads for capacitance C), the time constant is in seconds. For other units, other time units may result. The relationships of storage to flux, transfer coefficients, time constants, *RC,* and so on are very important to various aspects of systems study that follow in this book, and drill is desirable. The relationships are summarized for convenience in Fig. 3-1. The diagram helps us see why linear transfer coefficients have been so useful in systems analysis even though some of the processes involve single forces and others are population phenomena.

DIFFERENTIAL EQUATIONS FOR STORAGE

A rate equation for a storage lists and sums all the rates of the inflows and outflows. It has a term for each pathway affecting the storage. Equations constitute a systems language and indicate how separate terms representing component parts and processes should be combined. One type of equation, the rate equation, sums the rates of contribution to or subtractions from a storage such as that illustrated by the J symbols in Fig. 3-1. If outflow J is dependent on the quantity stored, the flow may be represented as a product of the transfer coefficient and the quantity as shown in Fig. 3-1:

$$\dot{Q} = J_0 - kQ \qquad (3\text{-}12)$$

The rate equation [Eq. (3-12)] for the storage is given in Fig. 3-2, with the energy circuit language diagram again shown. Included as heat losses from the storage and its pathways, the rate of change is the balance of inflow J_0 and outflow kQ. The rate of change of Q with time is indicated by writing a dot over the top of the letter $\dot{Q}$. Rate equations such as that shown are also called *differential equations.* Differential equations for a single tank without multiplicative pathways are said to be first-order equations, thus the tank system is a first-order system.

The energy circuit diagram shows the relationship of storages and flows pictorially and thus represents the differential equation very simply and equally rigorously with one pathway for each term in the equation.

One may evaluate the terms and totals of a differential equation with real data obtaining the rate of change that exists at the instant of evaluation. For example, suppose that in the example given in Fig. 3-2 the inflow rate J_0 is 10 liters of water per hour and that when the tank contains 50 liters, it is outflowing with a transfer coefficient of one-tenth of the

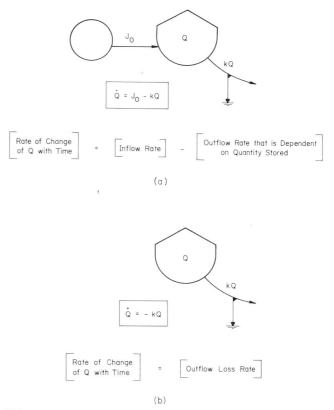

FIGURE 3-2. Rate equations associated with a storage: (*a*) system with one inflow pathway and one quantity-dependent outflow pathway; (*b*) system with a draining pathway only.

storage per hour ($k = 0.1$). The outflow rate kQ is 5 liters/hour. The overall rate of change of the storage $J_0 - kQ$ is 10 minus 5, or a positive 5 liters/hour. Since the inflow J_0 and the quantity Q may vary, the rate of change may also vary and some automatic continuous computation such as a computer calculation is required when a record of the behavior of the rate of change is desired.

Figure 3-2*b* is another example of a differential equation for the rates of change of flows affecting a storage. Here there is only one pathway and thus only one term in Eq. (3-13), one that drains whatever quantity of storage was present initially:

$$\dot{Q} = -kQ \qquad (3\text{-}13)$$

State Variable and Order of Equations

A variable that is stored is a state variable and is represented by a tank symbol. For each storage tank (state variable), there is one differential equation that can be written where two tanks are connected by relationships through which they form a system of two equations. These can be written separately, each having at least one term in common with another, or the two equations can be combined to form a second-order equation (see Chapter 11). The order of the differential equation depends on the number of storage tanks combined. Mathematically it is the highest order derivative. The *degree* is the exponent of the highest order derivative.

Integral Equations for a Storage

Another way of describing systems is with equations for accumulations of storage Q. Sometimes called *integral equations*, expressions for storage accumulation add the initial quantity Q_0 at the start of the time period ($t = 0$) to the net amounts that accumulate as a balance of inflowing and outflowing over the pathways. Whereas a rate equation describes the sum of the rates of inflow and outflow of water to a bathtub, the integral equation gives the amount of water in the tub after some time due to the net effect of flows and due to the quantity at the start. The process of summing flows over time is indicated with the integral sign $\int$ in front of the flows to be integrated (in parentheses) and dt after the items to be integated, meaning that the accumulation computation is over a given time period. Summing over time means multiplying each rate of flow times the time it flowed at that rate to obtain the volume added or subtracted. The summing process is called *integrating*.

For the single storage example shown again in Fig. 3-3, the integral equation [Eq. (3-14)] is given as the sum of that present initially Q_0 plus the net

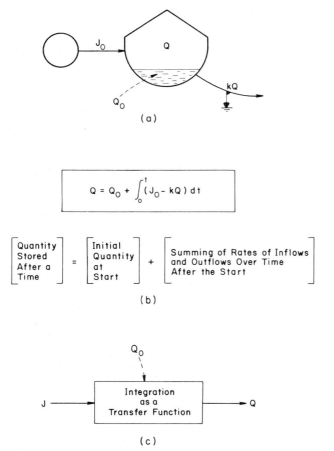

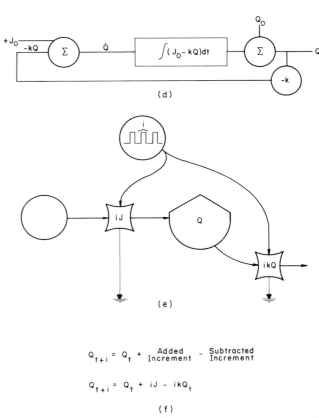

FIGURE 3-3. Expressions for the quantity of storage in a tank in several languages: (*a*) energy circuits; (*b*) cumulative (integral) equations; (*c*) box representing inputs to the system as a single function in time domain; (*d*) block diagram of equation; (*e*) energy diagram for discrete transfers in and out with time steps; (*f*) difference equation.

effect of the additions and subtractions during the time period after the start:

$$Q = Q_0 + \int (J_0 - kQ)dt \qquad (3\text{-}14)$$

Visualizing the equation in relation to the diagram (Fig. 3-3), we see that the quantity that is inflowing at rate J_0 and outflowing at rate kQ is to be summed over time and added to the amount of water Q_0 there initially to obtain the quantity present at any time Q.

The final overall effect of the accumulating process (integrating) is to determine the quantity Q when the rates are given. Since $\dot{Q}$ is the symbol for rates, we may summarize the process with Eq. (3-15):

$$Q = Q_0 + \int (\dot{Q})dt \qquad (3\text{-}15)$$

This equation says that Q is found by integrating the sum of rates $\dot{Q}$ and adding initial condition Q_0. If the rates are varying, the computation of Q from them is very difficult since it requires continuous multiplication of the passing time by the rate of each instant of time. Computer procedures are used as described in Chapters 4 and 5. Integral equations, regardless of whether they are evaluated, are alternate ways of expressing the energy circuit symbol for storage functions.

Mathematical Block Diagramming Language and the Storage Example

Another language for representing systems is the block diagram that arranges along lines on diagrams the successive mathematical operations called for by an equation. Addition, subtraction, multiplication, and so on are performed as the equation directs, one at a time and in order along a line. For example, in Fig. 3-3d the integral equation from Fig. 3-3b is block diagrammed. The inflow J_0 and outflow $-kQ$ are summed first to form the rate of change $\dot{Q}$. This result (output) is then integrated and added to Q_0 to form Q. The diagram is an alternative language for writing the integral equation. It shows the order and relationship of operations. Note that in connections to the integrator box the rate $\dot{Q}$ is on the left and the quantity Q is on the right. Block diagramming is much used in describing systems, as the diagramming of equations provides a method for seeing, remembering, comparing, and generalizing about systems.

For some, this language is the primary one in which they think. The block diagram in Fig. 3-3d expresses the same content as the other expressions for storage in Figs. 3-1, 3-2, and 3-3. Because the equation describes a function relating quantity with time, it is said to be in *time domain*. Later we use block diagramming of equations where the variable quantity is a function of the frequency; these are block diagrams in the *frequency domain*.

Notice the shape of the diagram in Fig. 3-3d with a negative feedback pathway that tends to decrease back at the start in proportion to the rise of Q to the right in the diagram. This negative feedback pathway tends to hold the model stable and is called *negative feedback stabilization*. Representation of a storage into block-diagrammed form dramatizes this property, and use of a negative, quantity-dependent term provides this stabilization. In biology these effects are called *density-dependent effects* and are similarly recognized as a stabilizing property in this field.

To summarize the use of these block diagrams, note that the sum of the rates $\dot{Q}$ is at the front of the integrator and that after addition of Q_0, the Q value emerges at the right. One diagrams the sum of rates (the rate equation) first and then runs the output through an integrator action to calculate the quantity. Sometimes the overall action of a system or an input to produce an output is called a *transfer function*; in this example it is integration.

For more examples and explanations see Jones (1973).

Integrated Equations for Storage

In Figs. 3-3 and 3-4 integral equations are introduced as another way of describing the cumulative behavior of storage compartments. The integration symbol was drawn to indicate that if the rates were integrated, the state quantity Q would result. The integral equation indicated integration but did not provide expressions for Q; it indicated integration but did not do it. Later we show means for performing the integration with computers. We could have done it by measuring the real-world example also. However, for simple systems such as this, the branch of mathematics termed *integral calculus* develops equations for Q. Procedures are learned and equations are derived so that rate equations can be integrated to find state quantities. The equations for Q that result from integration we call *integrated equations*. The converse process called *differentia-*

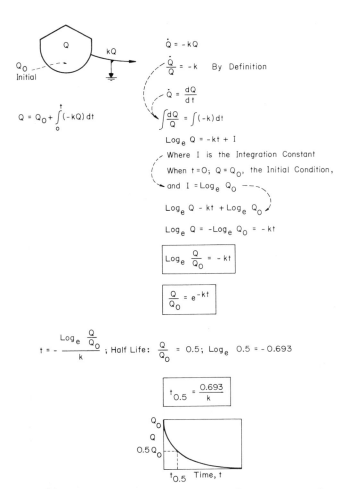

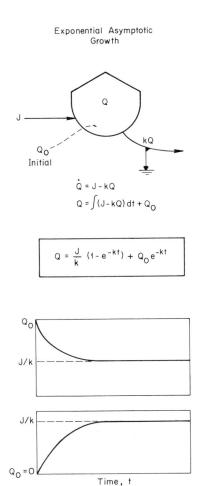

FIGURE 3-4. Integrated form of equation for storage quantity. Note exponential form, logarithmic form, expression for lifetime for half decay (half-life), and shape of curve with time.

FIGURE 3-5. Integrated form of equations for storage systems with an inflow. Note graphs of quantity with time with and without an initial quantity at time zero.

tion provides procedures for obtaining rate equations from equations for quantity. These are called *derivatives*.

Integrated equations provide another language for expressing the nature and performance of storage units. In Figs. 3-4 and 3-5 are given again two simple storage unit systems, one with an inflow and outflow and one with only an outflow. The differential and integral equations are repeated, and then below are given the integrated forms of these equations that indicate Q in terms of the properties of the system. Refer to a calculus text for more details. The integration constant is really the initial condition Q_0. Because storage units are principal parts of most systems, the student needs to learn the characteristics of these basic integrated equations and memorize them as needed language. Note where the symbols in J_0, k, and Q_0 in the energy circuit diagram are located in the integrated equations.

Graphical Representation of Functions

The graph of quantity with time that results from the real world or from the simulation of a valid model is another language that may be used to represent a system. Examples are the graphs for charging and discharging of single storage units given in Figs. 3-4 and 3-5. Since the shapes of the Q curves may vary with the magnitude of the pathway rates, a graphical representation of the many kinds of graph may require that families of curves be shown. Since there are many different kinds of system that produce graphs that look alike, the graphical representation is not definitive. One can sometimes indicate that a system is different because its graph is different, but one cannot be sure when two graphs are similar that their systems (their equations) are similar. There are many famous examples of curve fitting in which systems were judged erroneously to be involved in producing observed graphs. For example, the S-shaped growth curve may be produced by logistic curves, Gompertz curves, sine curves, cumulative probability functions, and many more (see later chapters). Learning the kinds of graph produced by systems is useful; the reverse process of trying to determine the system from the observed graph is less useful (see Smith, 1951).

Exponential Decay

Exponential decay is diagrammed in Fig. 3-4, and Eq. (3-16) is a common model for many processes in many fields, including radioactive decay, monomolecular reactions in chemistry, the decay of litter in a forest, the utilization of organic matter in a biochemical oxygen demand bottle, and any other process in which the loss is proportional to that quantity remaining; thus

$$\frac{Q}{Q_0} = e^{-kt} \tag{3-16}$$

In logarithmic form the exponential decay forms a straight line; hence a plot of quantity with time is straight on semilog paper and the slope is the outflow transfer coefficient k:

$$\log \frac{Q}{Q_0} = -kt \tag{3-17}$$

Decays are often described in terms of *half-life*, which is the time for half of the initial quantity to be used up. Substitution in Eq. (3-17) of 0.5 for Q/Q_0 produces a useful expression for the half-life $t_{0.5}$:

$$t_{0.5} = \frac{0.693}{k} \tag{3-18}$$

The time constant (turnover time) was earlier shown to be the reciprocal of the transfer coefficient k. After a time decay equal to the time constant, what fraction of the initial quantity Q_0 will remain? Substitution of $1/k$ into Eq. (3-16) for time yields a fraction Q/Q_0 of $1/e$ or 37%.

Relaxation Method

Coefficients for pathways from storages can be determined by cutting off inputs and measuring the decay time. The k values can be calculated from the half-life by using Eq. (3-18). This is sometimes called the *relaxation method*.

Exponential Asymptotic Growth and Von Bertalanffy Growth

When the storage unit is receiving an inflow and has a quantity-dependent linear drain outflow, the integrated equation is an exponential expression given in Eq. (3-19) and Fig. 3-5:

$$Q = \frac{J}{k}(1 - e^{-kt}) + Q_0 e^{-kt} \tag{3-19}$$

When represented graphically, this function either grows asymptotically up to or declines from the initial condition down to the steady-state level that is the ratio of inflow and outflow transfer coefficient J/k. This equation and form of growth is sometimes called *Von Bertalannfy growth*, after a leader in the popularization of the general systems idea who showed the widespread generality of the storage expression (Von Bertalanffy, 1962). Examples of growth patterns shown to fit the asymptotic growth model are the patterns of leaf litter on the forest floor studied in relation to latitude and analog simulated by Neel and Olson (1962). Rashevsky (1961) used the model for build up of nervous excitation.

MacArthur and Wilson (1967) used the model for the balance of species immigration and extinction on islands [see Wilson and Bossert (1971)].

In other fields the single storage and its integrated equation is called a *first-order equation* because it has only one storage compartment and no multiplier pathways. First-order systems are very stable and do not oscillate. They are self-regulating, as we showed earlier.

Substituting $\tau = 1/k$ into Eq. (3-19) for time with no initial conditions ($Q_0 = 0$), one finds that the growth level achieved in a time period equal to one time constant is 63% of the steady-state level.

Ramp

Readily visualized as a tank with a steady inflow and no outflow or backforce, a ramp has a steady rise in storage quantity making a straight-line graph of storage quantity with time. Various expressions for the ramp are given in Fig. 3-6. The ramp is sometimes used as a driving function for other processes.

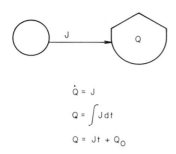

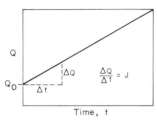

FIGURE 3-6. Ramp system in several languages, where J is inflow per unit time slope indicates the ratio of change of storage Q to change in time t.

Chargeup Against Backforce

Figure 3-7 shows the systems diagram equations and graph for a storage that is being charged by a driving force against the backforce of the storage. Flow stops when storage builds up to the point where the backforce equals the driving force. The equations are similar to the model without backforce shown in Fig. 3-6, but the system is quite different, as the diagrams show.

PASSIVE ELECTRICAL EQUIVALENT CIRCUITS FOR STORAGE

Readily related to the energy circuit language are electrical circuits of resistors, capacitors, and other units that have flow as the quantity of interest. A passive network for a unit storage is given in Fig. 3-8. These operate on a principle different from that used in standard analog computers introduced in Chapter 4. The storage unit (capacitor) stores electrical charge and expresses the resulting voltage force along pathways forward and backward. Chargeup involves backforce and outflow.

The unit of storage in passive analogs is the capacitor, which receives electrons for storage from inflowing and outflowing wires of designated resistance as shown in Fig. 3-8. The quantity of electrons (charge) measured in coulombs Q, which is stored in the capacitor, is the integrated sum of the initial quantity Q_0 and that flowing in and out on the two conducting pathways shown.

Passive analogs were the first kind of analog computer used, and they are still used for systems that have backforce and where there are many duplicate units to be included in a network as in simulating water flow in a large geographic area. Because the behavior of the resistor–capacitor networks is well understood and familiar to engineers and scientists, the wiring diagrams of these circuits are sometimes used as a systems language not only for their own networks, but also to represent other systems with similar behavior. The energy circuit language was defined originally as a generalization of the passive analogs. The elecrical network is a special case of the more general language.

Passive circuits are very cheap and have the immediate pathway by pathway correspondence to our intuitive thinking about networks as represented with energy circuit language. In passive analogs the energy circuit language loses energies into heat along the pathways in ways analogous to those of the real network being simulated. College-level exercises with the use of passive analogs for ecological teaching were prepared by Vogel and Ewel (1972).

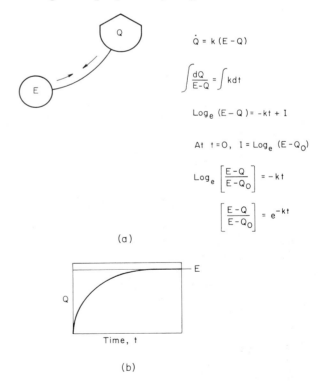

FIGURE 3-7. Charge of storage with backforce: (a) energy diagram; (b) graph of storage quantity Q with time starting at zero.

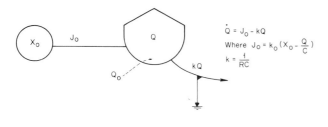

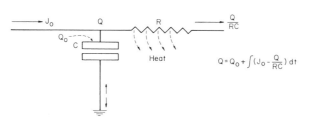

Item	Symbol	Units
Storage Quantity	Q	Coulombs
Flow	J	Amperes (Coulombs per Second)
Resistance	R	Ohms
Capacitance	C	Farads
Time Constant	RC = T	Seconds (R in Ohms and C in Farads)
Transfer Coefficient	$k = \frac{1}{RC}$	Turnovers per Second
Outflow	kQ	Amperes

FIGURE 3-8. Passive electrical equivalent circuit for a storage. Outflow pathway has no backforce in this example.

The display of current or voltage variations of a passive network must be routed through some kind of amplifier so that the recording output does not appreciably drain small energies of the passive network. The electrical current is directly analogous to the flows of the quantities of the real systems as represented by the energy circuit language. Flows of electrical current (flows of electrons) are analogous to the flows of the problem such as the flows of water, nutrients, food, money, and other materials. The electrical capacitor storages are directly analogous to the tanks of the energy circuit, and the conducting pathway wires directly correspond to the flow lines of the energy language. The resistance of the wires is directly analogous to the resistance R of the pathway. The relationships between current, conductivity, resistance, capacitance, time constant, transfer coefficient, and other variables are those summarized in Fig. 3-1.

The hardware of electrical systems has its special symbols, and Fig. 3-8 shows the same energy storage unit already much discussed in this chapter and with it the passive electrical network that is equivalent. Note the electrical storage unit called a *capacitor* and characterized by its capacitance C and conducting flow wire called a *resistor* and characterized by its resistance R.

In the electrical system the storage quantity Q is in coulombs and the quantity flow in coulombs per second (ampere). If resistance of the resistor is in ohms and the capacitance of the capacitor is in farads, the time constant of the storage in relation to the outflow pathway is in seconds. The reciprocal of the time constant is the transfer coefficient in units of the fractional turnovers per second. The outflow rate in amperes is proportional to storage kQ. These electrical properties of the passive storage are summarized and their units tabulated in Fig. 3-8. The energy relationships of the passive analog circuits also compare with that of the energy circuit language since real heat is released at every real process step.

Scaling a Passive Analog Computer Circuit

To use passive circuits as an analog computer, construct a table such as Table 3-1 using data on the system being simulated. First, decide on a scaling equivalent between the quantity being simulated and coulombs of the electrical circuit. This ratio of quantity per coulombs is also the ratio of quantity of flow per amperes. Determine the unit of time of the problem and set it equal to seconds in the electrical system. From the time constant of a pathway in or

TABLE 3-1. Example of a Scaling Table for a Passive Electrical Analog Circuit

Symbol	Item	Units in Real System	Units in Passive Electrical Analog
Q	Quantity	10^6 liters set to	1 coulomb
t	Time	Days set to	seconds
J	Flow	10^6 liters/day	ampere (1 C/s)
		10^3 liters/day	milliampere
		1 liter/day	microampere
C	Capacitance	1 m³/m² set to	1 microfarad
τ	Pathway time constant	Days	seconds
$R = \frac{\tau}{C}$	Resistance for the pathway (above)	Days per meter	ohms

out of a storage, one obtains the RC value needed in the hardware for that path and storage (in seconds). The time constant of an outflow path is the reciprocal of the transfer coefficient (Fig. 3-1).

Determine the capacity of the storage C in the real problem as the ratio of the volume to surface or by other means. Set values of capacitance C in cubic meters per square meter equivalent to a selected value of electrical capacitance in microfarads.

For each pathway, divide the electrical time constant in seconds by the capacitance in microfarads (millionths of a farad) to determine the electrical resistance for that pathway in megohms (million ohms). Repeat this process for any other flows from that storage and for other storages.

When flows of current, rises and fall of storage, and other properties are measured in the electrical circuit, the graphical recordings may be used to indicate the equivalent behavior of the analogous system being simulated. Equivalent values are given in the scaling table.

DISCONTINUOUS STATE CHANGES

Discrete Increments Added to Previous Storage

Figure 3-9 shows a model where a constant quantity is added to change state at each time step. An example is the addition of silt to a delta at the time of its annual flood. This model for a single tank is equivalent to the differential equation for a ramp (Fig. 3-6).

The convention used here in energy language for removal and substitution by switching action is shown in Fig. 3-9. The concave box symbol for discrete processes was introduced in Fig. 1-4. A switching program symbol is placed adjacent to the storage controlled by a pulse at each interval of time. To accomplish the change for the time step interval, a quantity must be transferred by the switching action that controls the pathways. The energy of the storage can be regarded as flowing almost instantaneously during the step when the switch is closed. In this example increments are transferred to storage with each unit of time, producing a stairstep graph (Fig. 3-9a). Another example is the stepwise charge-up of a single storage in Fig. 3.3e.

In using the energy language, one could translate the model diagram in Fig. 3-9b into the equivalent continuous function, which is the ramp given in Fig. 3-6. Although the overall effects may be similar, the system is operating differently, and the energy language must be used to show continuous and discrete functions when they are appropriate to the system or model of a system. Both occur in nature; often modelers use one for the other because of the type of computer they may be using.

Difference Equations for Storage

Difference equations are another kind of equation for showing the time changes in the interaction of flows and parts of a system. Quantities at any time are stated to be those at the time of the last calculation plus increments added for the short time interval i since the last calculation. An example for the stair-step increases is given in Fig. 3-9. A difference equation that describes approximately the single storage unit in Fig. 3-2 is given as Eq. (3-20) and diagrammed in Fig. 3-3e:

$$Q_{t+i} = Q_t + i(J - kQ) \qquad (3\text{-}20)$$

This states that the quantity Q after the next time interval $(t + i)$ will be the quantity at time t plus the increment that is the flow rate $(J - kQ)$ times the lapsed interval of time i. Successive calculations of the difference equation provides a running calculation of the quantities in a system.

Such calculations are done in digital programming. If the discontinuous calculation procedure is being made to represent a continuous process, a small systematic error is introduced because a single value of flow is used for each interval evalua-

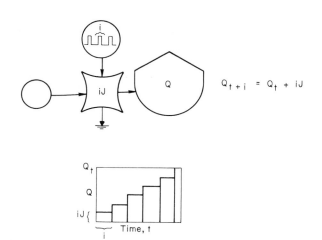

FIGURE 3-9. Energy circuit model of a ramp with discrete state changes with time steps. Compare with continuous function in Fig. 3-6.

tion when actually the rates are changing during the interval.

Future States and Time Steps

One way of perceiving and modeling systems regards patterns at one time followed at a later time interval by change to another pattern in discrete steps. Some system processes actually operate in steps, such as the bank accounts that have interest added all at the same time to produce new totals or the algae that reproduce only at dawn, advancing their number once each day. However, any process, even those that are continuous, can be regarded as a sequence of steps as a convenient approximation. Simulation of continuous time processes on digital computers involves stepwise calculations repetitively determining states followed by changes per time interval and summing again to find new states at the later time. The consideration of system processes in discrete steps has produced models that provide insights not seen with other perspectives.

States are the storages. At each time step the states become new states by being transferred, changed, multiplied, added to, or whatever is the system's process or the model's way of simulating the system. When a state is removed and replaced with a new state at each time step, this may be equivalent to a storage with pathways of negligible resistance that transfer quantities according to program instantaneously. For example, quantity Q in Fig. 3-10 is removed entirely and returned to storage. This diagram reflects one concept of systems modeling where states are transformed with each time step according to some transition formula. For emphasis the step transfer is indicated by tr.

Steps take energy for structure and process, and the heat sinks on the symbol indicate some heat dispersion and depreciation. There may be separate energy sources incoming for the steps as in the case of manipulations mediated from outside. As drawn in Fig. 3-10, there are no outside controls and sources, but the energy costs are drawn from the donor storage.

State changes can be indicated mathematically without constraints-of-energy required, one of the ways of generalizing systems mathematically (Klir, 1969). The energy language forces one to consider these in the same way real systems do when they operate step changes of state.

OTHER STORAGE PATHWAYS

Sensor of Rate of Change in Energy Circuit Language

In some situations a system may sense the rate of change of a storage, using this value to develop a force to drive some other action. For example, people read meters on their automobiles that indicate rate of change of velocity (acceleration). People sense the rate of change of the funds in their bank account. To indicate such a sensor in energy circuit language, a collar is placed at the base of the tank and all inflows and outflows to the tank are passed through the collar. This arrangement indicates that the rate is the algebraic sum of the flows (Fig. 3-11).

Decay with Second-Order Kinetics; Quadratic Decay, Increase with Time

A quadratic drain is given in Fig. 3-12. An example is the mortality caused by crowding. An interesting property of a storage with a quadratic drain is that its graph of quantity with time forms a reciprocal function. The differential equation is given in Fig. 3-12 and integrated to obtain the reciprocal of the

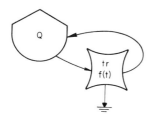

FIGURE 3-10. Diagram of state transition to the future by complete transfer of one state to another after a time interval; Q is transformed (tr) according to some function of time $f(t)$ at each time step. In this example energy is drawn from the storage.

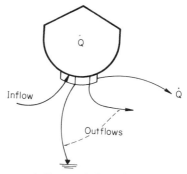

FIGURE 3-11. Collar symbol used to indicate a sensor of the rate of change $\dot{Q}$ of a storage Q.

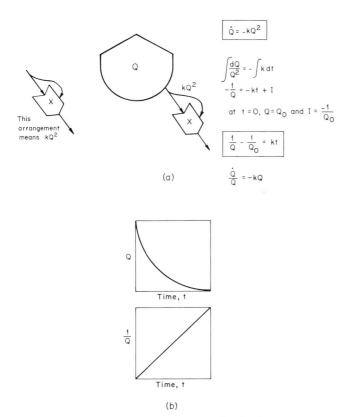

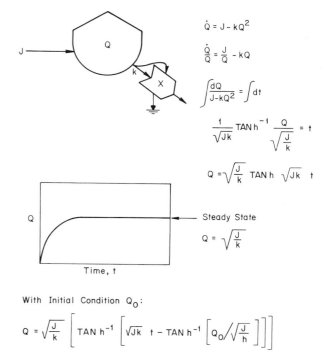

FIGURE 3-12. Quadratic decay; second-order kinetics: (a) energy diagram; (b) characteristic graphs, with Q a reciprocal function of time.

FIGURE 3-13. Design of a stable reservoir with a quadratic outflow.

quantity as a declining linear function of time. The function has a steeply declining graph, falling more rapidly than the exponential decay of a linear drain. The system is often called *second-order kinetics* in the sense of chemical terminology, which is different from that for differential equations already mentioned.

A common example of quadratic outflow is the drainage of water with canals. The cross section of the discharge pathway increases with height as does the response to pressure. The result of both effects is a quadratic effect. A ½ power drain is the Manning equation in hydrology.

Quadratic Outflow with an Inflow

Figure 3-13 shows the reservoir with inflow and the equation that results. The graph form is simple although the equation looks complex. The system is simple in energy language.

Constant Flow and Force from a Surface

Many bulky storages deliver force or forces for interactions in proportion to their surface. For example, storage of organic matter on the bottom of a lake results in buildup of the quantity without changing the surface of interaction much. The model that expresses the reaction of the storage must hold a constant force while keeping track of the quantity as it increases or decreases. In Fig. 3-14a the flow from such a storage is shown as proportional to the quantity and also divided by it again in its expression. Mitsch (1975) used this structure for the recycling of nutrients from the bottom sediments of a lake.

The constant-gain symbol gives similar kinetics and implies a control action on drain from a large reservoir (Fig. 3-14b). Another way is to provide an area storage (a tank designated with units of area) with its own inflows and outflows.

A storage with constant drain is sometimes described as zero-order kinetics and has a straight-line graph of drain. See constant rate of change in a down ramp as shown in Fig. 3-14c:

$$\dot{Q} = -k \qquad (3\text{-}21)$$

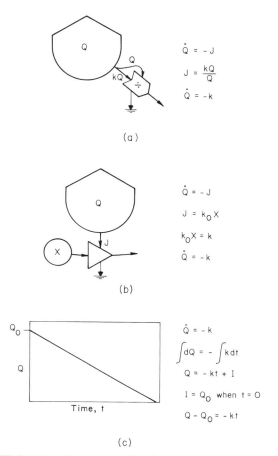

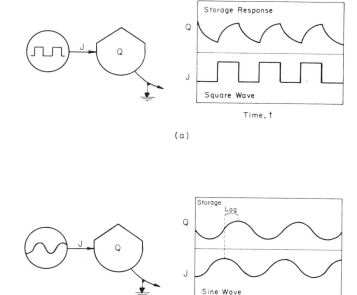

FIGURE 3-14. Constant outflow from a storage; zero-order kinetics: (a) flow from a storage is constant because of fixed surface relation; (b) flow from a storage is constant because of control by a second constant factor X; (c) graph of linear decrease of Q.

FIGURE 3-15. Response of a storage quantity Q to a repeating pulsing input source: (a) square wave; (b) sine wave.

FREQUENCY DOMAIN

Frequency Response of a Storage to Square-Wave and Sine-Wave Sources

When an input flow from energy source is turned on, a storage with outflow drain pathway charges asymptotically, as we already indicated in Fig. 3-5. The turning on of an input force or flow is called a *step function*. When the input J is turned off, the storage decays exponentially, as indicated in Fig. 3-4. There is a lag in time before the storage reaches its maximum and a time lag before it returns to its original level after the input is turned off.

If the input is rhythmically and regularly turned on and off, the pattern of input forcing function is called a *square wave* (see Fig. 3-15a). The storage is alternately filling and discharging, always lagging the input timing. The more frequently the input is turned on and off, the more the storage filling gets out of phase with the input. Also, the more frequently the input is alternated, the lower the level of storage that develops before the input is turned off again.

Another kind of repeating forcing function is the sine wave, an undulating rising and falling of input with time (Fig. 3-15b). Sine waves are common in nature, occurring with the seasonal rise and fall of the energies of the sun due to changing angles between the sun and the ground or the rise and fall of tide.

As shown in Fig. 3-16, a sine wave is generated by the projection of a point on a rotating wheel. The amplitude of the projection may be related to the spoke of the wheel according to the trigonometric sine function in expression (Figs. 3-22). The definition of the sine of any angle of any right triangle is the ratio of the opposite side A over the hypotenuse h of the triangle, and A is given by

$$A = h \sin \theta \qquad (3\text{-}22)$$

There is a triangle generated in the rotating wheel (Fig. 3-16) in which the spoke of the wheel is

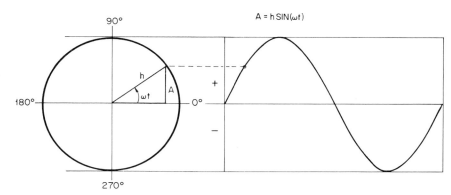

FIGURE 3-16. Sine wave generated by a point on a circle with A the amplitude, h the constant hypoteneuse, ω the angular velocity of the circle, and ωt the angle. Imagine a light shining from the left. The shadow of a nail on the rim of the rotating wheel will trace a pattern shown to the right.

the hypotenuse and the side opposite from the angle generated by the turning spoke is the amplitude A. The degrees of the angle formed by the spoke describes the position within the sine wave. Amplitude A is maximum at 90° and 270°, and A is zero at 0° and 180°.

When an undulating sine wave is the input to a storage function, the quantity stored also develops a sine wave in its storage, but with a lag in phase angle and with a diminished amplitude as illustrated in Fig. 3-15b.

Bode Plots of Frequency Response

In Figs. 3-17 there are two graphical representations called *Bode plots* that represent the response of systems as a function of the frequency ω of pulses of sine wave as an input flow (Bode, 1945). The frequency ω is the number of pulses per unit time. The upper graph uses as coordinates the logarithm of the amplitude ratio, which is defined as the ratio of the systems level of storage Q to the level of the varying input A. The graph has a line produced by measuring response to various frequencies. Notice the shape of a curve for a system of a single storage unit. The curve bends down to the right as pulse frequency is increased.

In the lower Bode graph Fig. 3-17b the phase lag is plotted as a function of the frequency of pulsing ω. Lag is measured in degrees ϕ as if both input and Q were generated by wheels such as shown in Fig. 3-16 but out of phase by ϕ degrees.

For a single storage, the maximum lag possible is 90°, when output is out of phase with input, with one maximum when the other is minimum. The frequency response graphs are a way of representing a system and thus provide a basis for another language in which a system is characterized by its amplitude lowering, phase angle lag, and frequency of applied input. The graphs are a characteristic of the system. Earlier, the graphs and equations for quantity with time were described as expressions in time domain. Sometimes the graphs or equations of the systems response expressed in frequency response form are said to be expressions in *frequency domain*. A system can be studied by varying the frequency and plotting graph of the amplitudes and phase lags. Dwyer, et al. (1978) graphically represented responses of microcosms to imposed frequencies.

Laplace Transform for a Unit Storage System

Constituting another language for systems manipulation are the equations that result from performing a *Laplace transformation* on equations for the behavior of systems with time (Jaeger, 1949). Laplace transformation is done by multiplying the regular time equation $f^{(t)}$ by e^{-st} and then integrating from zero time to infinity. This mathematical operation is summarized as

$$\mathscr{L}[f(t)] = \int_0^\infty f(t)e^{-st}\, dt \qquad (3\text{-}23)$$

For more work, the transformations may be obtained from Laplace transform tables. The results of these transformations of time equations are equations that describe the systems action when the input forcing function and system contributions are both expressed in terms of Laplace variable s. The Laplace variable S is sometimes called a *complex frequency* and is given the dual symbols $(j\omega)$ in place of (s), where j is the square root of minus one ($\sqrt{-1}$),

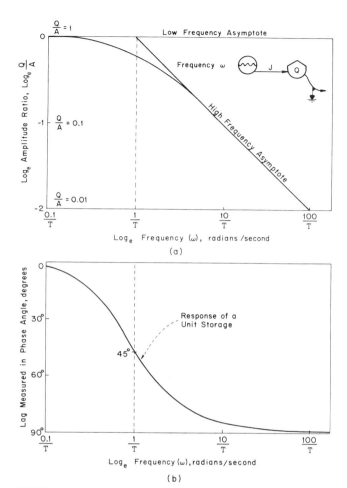

FIGURE 3-17. Bode plots of the amplitude (a) and phase lag θ of a unit storage (b) as a function of the frequency of the input sine-wave forcing function ω.

sometimes called the *imaginary quantity,* and ω is the frequency of the undulations with time. For example, on the right side of Fig. 3-18 the time domain equation for the simple storage unit considered much in this chapter is Laplace transformed, with Eq. (3-24) resulting, which gives the quantity Q as a function of the input amplitude $X(s)$ and frequency:

$$Q(s) = \left[\frac{k_1}{\tau\omega j + 1}\right] X(s) \qquad (3\text{-}24)$$

where the Laplace variable $s = \omega j$. Also derived from related procedures is an equation for the phase lag φ as a function of the frequency and time constant τ:

$$\phi = -\arctan \omega\tau \qquad (3\text{-}25)$$

In other words, after the Laplace transform, time has disappeared as a variable and frequency ω is the time-related variable. An equation for behavior of Q with time has been changed to an equation for the response as a function of the frequency (Kessler, 1950).

Transfer Functions in Frequency Domain

In Figs. 3-3c the overall effect of the time storage system on its input J was summarized by showing a large block that represented all the detail block diagrammed in Fig. 3-3d. Representing a whole system in this way is another language for representing systems. The block and its mathematical overall equation indicate the blocks effect on input in producing output. In Fig. 3-3 this is an input-output function in time domain because the equation that summarizes the overall effect is a function of time.

A system's overall effect in frequency domain is a *transfer function*. For example, the overall effect that the unit storage equation in frequency domain [Eq. (3-24); Fig. 3-18] has is given by a transfer function. The bracketed part of equation (3-24) goes in a block to indicate that it may be multiplied by input variables to achieve output variables, all in frequency domain (see Fig. 3-18).

Since any *linear* system has a Laplace transform, there may be as many transfer function blocks as there are subsystems. Each input forcing function (steps, square waves, sine waves, etc.) has a frequency domain Laplace transform. The systems also have their frequency domain equations (transfer functions). Frequency domain transfer functions have the convenient property that they may be strung together in a line of blocks such as that shown in Figs. 3-18, thus indicating that their overall effect is achieved by multiplying them together. In Figs. 3-18 we see a combination of a forcing function and unit storage. In more complex systems one may string together many blocks to calculate an overall block transfer function for the whole system in frequency domain language. Then if it is desired, the inverse Laplace transform is performed putting the equation back into time domain again (using tables in many cases). This procedure is one way of obtaining integrated equations for overall systems when ordinary calculus procedures do not succeed.

Equations in this form may be plotted on Bode plots, as done for Eq. (3-24) in Figs. 3-17. As now understood, the procedures described operate only on linear systems (ones without multipliers). Since

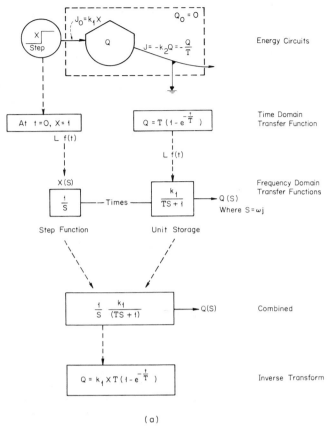
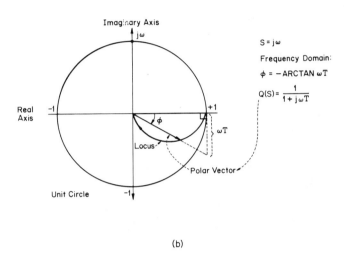

FIGURE 3-18. Storage system in frequency domain: (a) unit storage and a simple step-up input function both expressed in Laplace transformed transfer equations, then combined and inverse transformed to obtain equation for storage in time domain; (b) Nyquist plot of a single-tank system (first-order system); with graph of angle of phase lag and amplitude using polar coordinates on a complex s plane. The locus traced by the point of the vector is plotted as frequency ω is varied from zero to infinity.

most environmental systems have many multipliers associated with energy tendencies to maximize flows, we cannot often use these methods except for simplified and artificially restricted patterns. For more details and biological applications, see Milsum (1966).

Nyquist Plot

Equations for a system expressed in frequency domain may be graphed to show characteristic properties with the use of a *Nyquist plot* (Fig. 3-18b). This graph uses a complex plane (also see Fig. 11-3) that has real and imaginary coordinates where imaginary numbers are those containing $\sqrt{-1}$ referred to with the symbol j (or i in many books). The Laplace-transformed equations with real and imaginary terms may be graphed by using polar coordinates: the angle and amplitude of a vector from the origin. The locus of points generated by the vector can be plotted as frequency ω is varied from zero to infinity. Equations in frequency domain can be regarded as a vector with one imaginary and one real component. The magnitude of the vector is the square root of the sum of the squares of the sides (Pythagorean theorem). The angle of the vector is the one where tangent is the ratio of the imaginary and real components. In Fig. 3-18b the equations for a single storage used in the Bode plot in Fig. 3-17 are shown on the Nyquist plot.

In control theory the patterns of systems on Nyquist plots are used to seek stability. If the plot has an amplitude ratio larger than 1 and a phase lag of 180°, its output is being increased by resonance. It is unstable. If the plot on the Nyquist graph is outside the unit circle, it is unstable. However, first-order systems such as that shown in Fig. 3-18b are within the unit circle and are stable.

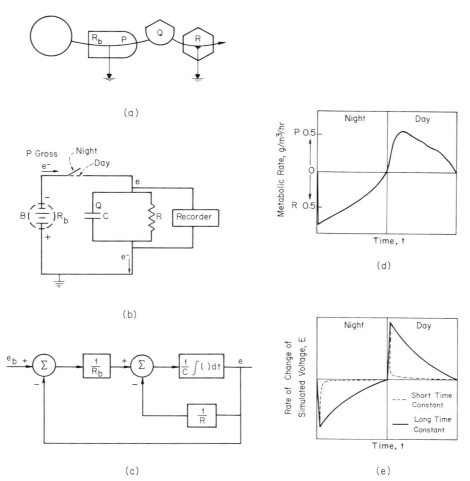

FIGURE 3-19. Charge–discharge model for diurnal variation in rates of production P and respiration R (Odum et al., 1963a), where a short time constant simulates tiny plankton cells with rapid turnover: (a) energy diagram; (b) passive analog circuit used to simulate metabolism; (c) block diagram given by Milsum (1966) in analyzing this example; (d) observed data; (e) simulated graph.

HISTORY

Understanding of the nature of a single storage and appropriate equations for rates of change and changes of state with time date back to the invention of calculus and differential equations. Lotka (1925) includes the linear model in his general treatment of kinetics of systems of mathematical ecology. Ludwig von Bertalanffy used the single tank model for many kinds of growth under open conditions starting with papers about 1934 so that the tank model is sometimes called the Von Bertalanffy growth module. His general summary (1950, 1968) reviews history of development of general systems theory around the simple linear model. Application of the linear model to the accumulation of soil storages was made by Jenny (1930, 1941) and decay of organic storages in water bottles by Phelps (1944). Kendigh (1941) related size of wintering birds to length of night without food input. Nicholson and Bailey (1935) applied difference equation models of storage to animal populations. Rashevsky (1938, 1960) uses the tank for cellular growth model. By the 1960s compartment models were general as an approach to control theory and were described in texts on tracers (Sheppard, 1962) and physiological systems (Milhorn, 1966; Grodins, 1963; Milsum 1966). Some examples using single tank models are tracing DDT in the biosphere (Harrison, 1970; O'Neill and Burke, 1971; Woodwell et al., 1971) and water in the Okeefenokee Swamp (Rykiel, 1977).

Wyckoff and Reed (1935) may have been the first to use passive electrical circuits for simulation. Monteith (1963) used resistor networks for gas flow.

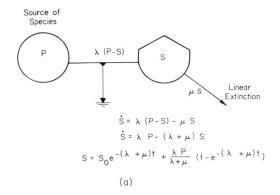

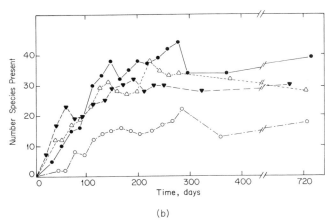

FIGURE 3-20. Model of dispersal: (*a*) balance of immigration and extinction by MacArthur and Wilson (1967); (*b*) recolonization of depauperated islands by Simberloff and Wilson (1970).

An example of early analog simulation of first-order tank models of environmental systems was given by Neel and Olson (1962) and Olson (1963a). Uptake of carbon by plants was deposited in litter and roots with losses proportional to litter storages (see Fig. 5-19).

Odum et al. (1963a) used passive analog circuits to simulate a charge–discharge model in diurnal plankton ecosystems. Figure 3-19 shows this example expressed in various ways used in this chapter. The application of a single storage model with backforce to the species balance on an island is given in Fig. 3-20 from MacArthur and Wilson (1967). Properties of this model were examined by Burton (1939).

SUMMARY

In this chapter we have considered the storage function as represented by a single tank with quantity-dependent outflows. With the unit storage as an example, we have introduced successively more systems languages such as differential equations, integral equations, integrated equations, difference equations, state transition models, passive electrical equivalent diagrams, time domain block diagrams, and frequency domain block diagrams. We have concentrated on the ramps, chargeup, and decay situations and such related concepts as half-life, exponential decay, logarithmic slope, metabolism, and size. In all examples the energy circuit language was used as the initial manner for presentation and consideration for purposes of intertranslation among languages.

QUESTIONS AND PROBLEMS

1. Define in words, write the expression, and give units for the following:
   a. capacitance
   b. transfer coefficient
   c. time constant
   d. specific metabolic rates
   e. relationship of metabolism and size; metabolism and weight
   f. differential equation for unit storage and one inflow rate
   g. differential equation for unit storage and two inflow rates
   h. differential equation for von Bertalanffy growth
   i. differential equation for exponential decay
   j. block diagram for a system with two inflows and a density dependent outflow
   k. integrated equations for items f and i
   l. half-life

2. Suppose that leaves are falling steadily in a stable forest at 2 g/(m^2 · day) and the decomposition rate is half per hundred days. What is the steady-state level of litter that will develop regardless of the initial condition? Diagram a passive electrical circuit that can simulate the model.

3. Can you show and diagram a system of two storages connected by one pathway whose flow is dependent only on the upstream quantity? Show this system in each language given in the chapter.

SUGGESTED READINGS

Bekey, G. A. and W. J. Karplus (1968). *Hybrid Computation,* Wiley, New York.

Bogner, R. E. and A. G. Constantinides (1975). *Introduction to Digital Filtering,* Wiley, New York.

Brewer, J. W. (1974). *Control Systems,* Prentice-Hall, Englewood Cliffs, NJ.

Child, G. I. and H. H. Shugart (1972). Frequency response analysis of magnesium cycling in a tropical forest ecosystem. *In Systems Analysis and Simulation in Ecology,* Vol. 2, B. Patten, ed., Academic Press, New York. 103–139 pp.

Charlesworth, A. S. and J. A. Fletcher (1967). *Systematic Analogue Computer Programming,* Pitman, New York.

Grodins, F. S. (1963). *General Theory and Biological Systems,* Columbia University Press, New York. 205 pp.

Gross, M., R. D. Gibson, M. J. O'Carroll and T. S. Wilkinson (1979). *Modelling and Simulation in Practice,* Wiley, New York. 358 pp.

Horrobin, D. F. (1970). *Principles of Biological Control,* Medical and Technical Pub. Co., Aylesbury, United Kingdom. 70 pp.

Lam, H. Y. (1979). *Analog and Digital Filters, Design and Realization,* Prentice-Hall, Englewood Cliffs, NJ.

MacFarlane, A. G. (1964). *J. Engineering Systems Analysis,* Addison-Wesley, Reading, Mass. 272 pp.

McFarland, D. J. (1971). *Feedback Mechanisms in Animal Behavior,* Academic Press, New York. 279 pp.

Mees, A. J. (1981). *Dynamics of Feedback Systems.* Wiley-Interscience, New York. 214 pp.

Milhorn, H. T. (1966). *Application of Control Theory to Physiological Systems,* Saunders, Philadelphia.

Milsum, J. H. (1966). *Biological Control System Analysis,* McGraw-Hill, New York.

Moran, P. A. P. (1959). *The Theory of Storage,* Methuen Monograph, Wiley, New York.

Nagrath, I. J. and M. Gopal (1975). *Control Systems Engineering,* Halsted Press (Wiley), New York.

Pomeroy, L. R. (1974). *Cycles of Essential Elements,* Dowden, Hutchinson and Ross, Stroudsburg, PA.

Rubinow, S. I. (1975). *Introduction to Mathematical Biology,* Wiley, New York. Chapter 1.

Shepperd, C. W. (1962). *Basic Principles of the Tracer Method,* Wiley, New York. 282 pp.

Toates, F. M. (1975). *Control Theory in Biology and Experimental Psychology.* Hutchinson Educational, London. 264 pp.

Von Bertalanffy, L. (1968). *General System Theory,* Braziller, New York. 289 pp.

Waite, T. D. and N. J. Freeman (1977). *Mathematics of Environmental Processes,* Heath, Lexington, Mass. 170 pp.

Wilson, E. O. and W. H. Bossert (1971). *A Primer of Population Biology,* Sinauer Associates, Stamford, CT. 192 pp.

CHAPTER FOUR

Microcomputer Simulation

Simulation is the process of generating the patterns with time that result from a systems operation. Aided by computers, simulation shows what a system does with time. Fantastic new microcomputers are now so cheap that they are becoming available to individuals, classrooms, and schools everywhere. Programming is so simple and feedback so immediate that they may be used as lecture demonstrations, at home, or in research to make systems designs come alive, showing graphs with time, spatial mapping, or numerical listing. In this chapter simulation of energy diagrams with microcomputers is introduced starting with the simple models of storage and flow studied in Chapter 3.

PRINCIPLE OF DIGITAL ITERATION

Simulation is accomplished by successive calculations of the quantities in the storages as they change with inflows and outflows. At regular time intervals the calculation adds the inflows and outflows during the time interval to the previously existing storage quantity and plots the point on a graph (or writes down its value). The calculation is repeated for the next time interval. The successive repetitive calculations are called *iteration*. They are digital because they involve discrete steps rather than continuously varying functions.

HAND SIMULATION

Simulation calculations can be done without computers by hand. Such a hand simulation is given in Table 4-1 to show what happens with time with a simple storage in Fig. 4-1a that has a steady inflow path and a pressure-dependent outflow. This is the same model studied at length in Chapter 3. With a starting storage of 0.1 g, the calculation for the first hour adds an inflow of 1 and subtracts an outflow of -0.05 obtained from multiplying the storage at that time ($Q = 0.1$) by its transfer coefficient ($k = 0.5$). The new value of storage, 1.05 g, is written down in the right column of the table. Then the calculation is repeated for the second hour and so on. Finally the quantities Q are plotted with time in Fig. 4-1b. See if you can reproduce the table.

Difference Equations

These calculations evalute the difference equation already introduced as Eq. (3-20). Digital simulation is the successive evaluation of difference equations that represent the system. In the iteration in Table 4-1 the time interval i in the equation was assumed to be 1.

Simulating More Complex Systems

For systems with more than one storage, there is one differential equation and program line for each. The change increments (DQ statements) should all be calculated before the increments are added to the storages. In this way calculations are all based on the states at the same time.

After each statement for a storage quantity Q one may include a limiter so that it cannot go below zero as in Eq. (4-1).

$$\text{IF } Q < 0, \text{ THEN } Q = 0. \qquad (4\text{-}1)$$

Accuracy

As shown in the stair-step graph in Fig. 4-1b, the process of simulating with incremental digital steps assumes that rates at the start of a time interval

TABLE 4-1. Successive Iteration of Incremental Inflows and Outflows of the Tank in Figure 4-1

t, hours	J	$-kQ_t$	$Q_{t+1} = Q_t + J - kQ_t$
0	—	—	0.1
1	1	−0.05	1.05
2	1	−0.05	1.53
3	1	−0.76	1.76
4	1	−0.88	1.88
5	1	−0.94	1.94
6	1	−0.97	1.97
7	1	−0.993	1.988
8	1	−0.996	1.992
9	1	−0.998	1.996

prevail during that interval, whereas the continuous process of growth being represented is changing rates during the interval. Consequently, there is a small error in calculations done this way. If the time intervals chosen for the increments are small, the error is small. There are procedures for making corrections when they are needed (see programming texts). For example, one procedure averages the rates during the interval prior to calculating the increments. For many purposes error is minimized adequately by keeping the time interval small compared to times required to fill storages. See paragraph: "Variable Interval for Iteration," p. 50. Calculating with rate at the start of the interval is the Euler method. Using average rate at start and end of the interval is the Runge-Kutta method.

FLOW CHART

Sometimes the steps in a simulation are represented with a diagram that shows the steps as connected blocks in the order in which the calculations are made. This diagram is called a flow chart. An example is given in Fig. 4-2 for the simple storage model of Fig. 4-1. As a connection of blocks, it is another

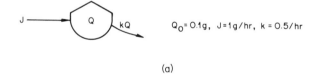

(a)

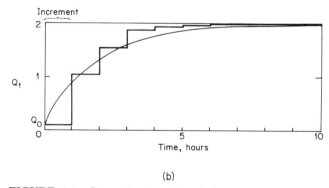

(b)

FIGURE 4-1. Successive iteration of the incremental inflows and outflows of a tank (see program in Fig. 4-2 and calculations in Table 4-1): (a) a "hand simulation" that does what the digital computer does; Q is the initial storage, J is inflow rate in grams per hour, and kQ is the outflow rate in grams per hour: (b) resulting graph when interval of iteration is 1 hour; continuous line is correct line obtained when iteration interval is 0.01 hour.

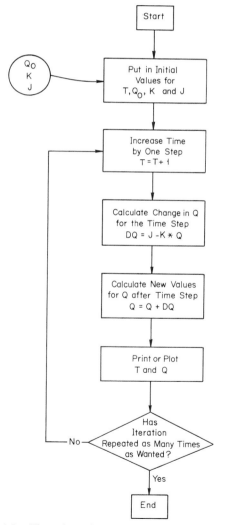

FIGURE 4-2. Flow chart for iterative calculation in Fig. 4-1. The symbol * means "multiplied by."

systems language appropriate for systems that have sequences. Figure 4-2 is a translation of the simple tank model.

PROGRAMS AND COMPUTERS

Organized instructions are a program. The flow chart (Fig. 4-2) is a program for simulating the model of a storage.

A computer stores information, makes changes in that information according to instructions received, and stores the revised information. Instructions for computers are computer programs. The new microcomputers are digital, useful for making the types of iterative computation given in Figs. 4-1 and 4-2. Instructions for analog computers are arranged by patching wires on a board and are introduced in Chapter 5.

Values for J and k and initial values for T and Q are put into memory locations. (In this example, $T = 0, Q = 0, J = 1, K = 0.5$.) Then, when instructed, a value for Q at time step 1 is calculated according to the difference equation (column 4 in Table 4-1) and the new value for Q replaces the old value. Then time is advanced by one hour, and the calculation is repeated (see Fig. 4-2). With each loop of iteration, the value of T and Q at that time is printed or plotted; k and J keep their same values since no instruction was given to change them in this example.

Computer Languages

The digital computer is made up of memory locations that have flip-flop units with two states, on and off. Whether a memory location is on (indicated by 1) or off (indicated by 0) is a "yes"- or "no"-type decision and constitutes one bit of information by definition. These are logic systems considered for their own interesting properties in Chapter 6. Built-in programs combine the flip-flops in various ways so that they can register numbers, letters, and symbols. Knowledgeable computer scientists and programmers can give the computer instructions at the level of the machine's operation dealing with bits of information. This is machine language. However, most computer users work through a "higher"-level language in which instruction is given with words to which the computer has already been programmed to respond. It takes the words it recognizes and does the various bit-by-bit operations necessary to accomplish the instruction of the programmer. The built-in programs are called *software,* although they may be fixed into the computer's memory in a way that cannot be erased by the user (firmware). Different computers may have different computer languages, just as different people have learned English, Spanish, Russian, and other languages. Commonly used computer program languages are FORTRAN, ALGOL, PL-1, DYNAMO, CSMP, SIMSCRIPT, and BASIC.

Microcomputers

Figure 4-3 shows one of the new inexpensive computers made possible by extreme miniaturization and low cost of microprocessor chips that are almost whole computers. The screen is like a television screen; there is an operating keyboard and a chamber into which memory disks can be inserted. The keyboard is like a regular typewriter keyboard but has some extra keys for special purposes. Whatever is typed on the board is listed on the screen and held in the working memory when the "enter" key is struck. The software for operating the computer through BASIC language is a permanent part of the firmware. If the line is given an initial number, it is treated as a program statement and goes into program memory. Some, like the one shown in Fig. 4-3, have devices and software for plotting graphs or other spatial displays.

The amount of memory within the computer use-

FIGURE 4-3. Photograph of a microcomputer useful for simulating systems. Radio Shack TRS-80 Model II Microcomputer. Courtesy Radio Shack, a Division of the Tandy Corporation.

ful for storing programs and data is usually between 8000 and 64,000 bytes. One byte is 8 bits.

Connected to the computer is a floppy disk unit into which small memory disks are inserted. Programs, data, and displays can be saved on the disk and later transferred back into the computer, with statements to SAVE and LOAD. The details differ somewhat depending on the make of computer. The disks are cheap and can store dozens of programs on each. Each student can have one.

As part of this chapter's work, a microcomputer should be made available, with each student simulating systems, starting with the program for one storage below.

BASIC PROGRAMS

The simplest and easiest language to learn is BASIC, and this is the language most used in the microcomputers. Its instructions are simple English words (see Table 4-2). Many programs use only a half-dozen words. Table 4-3 lists the program in BASIC for simulating the simple storage (Fig. 4-1). Compare the instructions in BASIC with the instructions partly in English in the flow chart. The BASIC program has the same instructions but in abbreviated form. Each line is numbered. The computer is programmed to keep the statements in order and not start the calculations until the user types in RUN. Then it generates a table such as that in Table 4-1 or a graph such as that in Fig. 4-1b, depending on whether a print or plot instruction is included.

The user can then change some property such as the energy source J or the initial starting condition Q and run the system again. One changes a statement by retyping it, thus putting something different in the memory locations.

Plotting Graphs on the Screen

To plot the simulation on the screen, substitute a plot statement in place of the print statement. These differ in different microcomputers. For Apple the statement is PLOT T,160-Q; for Radioshack TRS-80 the statement is SET T,20-Q; for Compucolor the statement is PLOT 2,T,Q,255. These instructions cause a point to be plotted on each iteration with T as the horizontal axis and the variable of interest, Q, on the vertical axis. One plot statement is needed for each variable to be plotted. To keep the graphs

TABLE 4-2. Some Instructions in BASIC for Simulation [a]

Instruction Word	Explanation
PRINT	Prints out in columns on the screen the values of the variables given afterward and separated by a comma
RUN	Runs the program previously stored
LIST	Lists the statements of the programs stored on the screen
GO TO ———	Sends the sequence of calculations to the line number that is indicated
IF ——— THEN ———	If the condition given after IF is true, the instruction given after THEN is activated; if it is not true, it goes to the next line
Mathematical operations:	
A + B	Add
A − B	B subtracted from A
A * B	Multiplied
A/B	A divided by B
(A + B)/C	Parentheses operation is done first and its result then divided by C
A ↑ B	A raised to power B
A = B	A becomes what B is
SQR(A)	Square root of A
END	Stops program run

[a] $\emptyset$ is zero and has a slant through it to make sure it is distinguished from the alphabetic O.

within the screen, factors may be introduced to expand or contract the graph. This is called *scaling*.

For each plot statement an amplitude scaling factor F may be included as a multiplier or divisor of the state variable Q. A statement giving the value of F is also put at the start of the program. The value selected should be one that will cause the highest value expected to be plotted near the top of the screen.

To control the time lapse represented by the width of the screen a time scaling factor G can be used as a divisor to T in the plot statement. The value of G is put in a statement at the top of the program and in the loop return statement. Increasing G includes more time on the screen by causing iterations without plotting every point.

For example, statements are given below to be added and substituted in the program in Table 4-3 to generate a graph that will fit the screen of a mi-

TABLE 4-3. Program in BASIC for Simulation of the Single-Storage Model in Figures 4-1 and 4-2

10	J = 1
20	K = .5
30	Q = .1
40	T = 0
50[a]	PRINT T, Q
60[b]	D = J − K * Q
70[b]	Q = Q + D
80	T = T + 1
90	IF T < 10 GO TO 50
200	END

[a] 50 To plot a graph, substitute a PLOT statement for the print statement. The words used for plot statements vary with the make of microcomputer and can be found from the computer manual. If the quantity is too large or too small, scale it for the screen by plotting the variable divided by a factor.

[b] 60, 70 here-means "becomes"; Q becomes what it was + J − kQ.

crocomputer which has dimensions of 100 characters by 100 characters.

$$7\ F = 10 \quad (4\text{-}2)$$
$$9\ G = 2$$
$$50\ \text{PLOT } T/G,\ Q*F$$
$$90\ \text{IF } T/G < 100 \quad \text{GO TO } 50$$

The asymptote will approach 20 on the vertical, 20% of full scale. The time represented will be 200 hours with a point plotted every other hour. The graph will resemble Fig. 4-1*b*.

A Family of Curves

The program can be fixed to run a family of curves, one after the other, by varying some property such as *J* and running it again. This is done by inserting statements (4-3) in the program as follows:

$$110\ J = J + .5 \quad (4\text{-}3)$$
$$120\ \text{IF } J < 5 \text{ GO TO } 20$$

After each run, time is thus reset and J changed by .5 and the system run again, continuing until a run has been made with J = 5.

Only a few of the BASIC commands are given in this introduction. Refer to the computer manual describing your microcomputer for others and to one of the many texts on programming in BASIC such as those listed at the end of the chapter.

Variable Interval for Iteration

In the iterations in Table 4-1 and Table 4-3 the time interval for iteration was set at one hour. As shown in Fig. 4-1, error develops because the outflow is calculated with a storage that is lagging. The error is less if the time interval is reduced. However, if the time interval is too small, the simulation may be slower than necessary. By inserting a variable for time increment in the program, simulations can be run at various intervals so as to select one that is reasonably accurate but fairly fast. For the example given in Table 4-3, add and substitute the following statements:

$$5\ I = 1$$
$$70\ Q = Q + D*I \quad (4\text{-}4)$$
$$80\ T = T + I$$

The time interval is I and can be varied by changing statement 5. Each statement for a change in storage is multiplied by I as in the example in statement 70. When the program is run with time interval I = .01, the nearly correct curve is obtained. When I = .1, the values are almost the same.

Recording Results

If more than a sketch of the shape of the graph is needed, it may be recorded if the microcomputer has a graphics recorder, or a polaroid camera may be used with a time exposure with the computer in a dark room. A portrait attachment may have to be added if the camera is an inexpensive model.

SUMMARY OF STEPS FOR SIMULATING A SYSTEM

In Fig. 2-9 procedures were summarized for making a model of a system in the form of an energy diagram, including forcing functions, components, pathways, and processes. The first draft of a model including both minor and major flows may have many components and pathways. Next, we try to group and combine units, making a complex model into a simpler one that still has main driving functions, causative pathways, and phenomena of interest. Simplification is done by aggregation, keeping the main sources and components, not by selecting small pieces to model.

For each storage unit, write a differential equation as the sum of the input and outflow pathway terms. Each term has a numbered coefficient. A system linked with connections has equations with common terms, so that each equation is linked to storage variables of other equations.

Using data for a particular time or averages write values on the diagram for storage quantities in the tanks and flow rates on the pathways. As shown on page 27, set the mathematical term for the pathway equal to the flow and substitute values for storages so as to calculate the coefficients for each pathway.

Write the computer program with a form similar to that in Table 4-2, first with statements giving values of coefficients, driving functions, and initial conditions. Then put in statements for change and increment for each storage in the model. Finally, write plot statements and a loop statement for successive iteration.

Run the model, find errors (debug), compare graphs with expectations, try different values, vary details, change the model, and work to understand the system. Seek consistency between the facts and concepts about mechanisms, the observed graphs, and the simulation performances.

COMPUTER PROGRAMS AS SYSTEMS LANGUAGE

In general, it is much easier to write computer programs than to read them. Many systems studies fail to report their models with enough detail or in a form easy to read. Journals rarely print programs or full flow charts because of the difficulty for readers in seeing the system in so much detail. Usually some diagrammatic systems language (e.g., energy diagrams) and/or differential equations are given.

Many working in systems may find themselves thinking and lecturing in programming languages. Programs put the steps of a process in time sequence even though the real system they describe may be doing many of the steps simultaneously. Thus time-based, programmatic flow charts may not be the natural language for describing many systems of nature.

There are, however, in the behavior of animals and in the seasonal programming of an ecosystem, temporally sequenced events that do seem to be operating with ordered sequences like a digital computer or like the "one-at-a-time focus" of the conscious mind. The human mind and its derived digital computer is well evolved for programmatic systems thinking.

SUMMARY

In this chapter we showed how to translate energy diagrams or English instructions to computer programs and simulate with a microcomputer. Students who have not already taken a regular course in computer programming should do so, perhaps learning Dynamo and Fortran or other simulation languages. However, the procedure given here is sufficient to simulate various models given in later chapters. Students should now be ready to do a special project with a simple model on a system in which they are interested and have data. Although costs are higher than microdigital computers, analog computers and hybrid analog-digital computers give an even faster turnaround for fully studying small- and medium-sized models. Their use is described in Chapter 5.

QUESTIONS AND PROBLEMS

1. Practice ordinary numerical operations on a microcomputer by first typing PRINT followed by expressions for adding, multiplying, finding natural log, trigonometric sine, and square root.

2. Simulate a family of curves for each model in Chapter 3 for a single tank such as ramp, exponential decay, and quadratic decay.

3. Write energy diagrams and simulation programs for the following systems, assuming values for coefficients:
   a. $\dot{Q} = kEQ - k_2Q$ (see Fig. 1-3a)
   b. $\dot{Q} = kQ - k_2Q^2$
   c. $\dot{F} = -k_0FQ$
      $\dot{Q} = k_1FQ - k_2Q$
   d. $\dot{E} = 10 \sin(T) + 10$
      $\dot{Q} = kE - k_2Q$

4. Simulate a family of curves of a tank model, using repeat and rerun statements that change the outflow coefficient by 0.1 on each run.

SUGGESTED READINGS

Albrecht, R. L., L. Finkel, and J. R. Brown (1978). *Basic for Home Computers*, 2nd Ed., Wiley, New York.

Anaud, D. K. (1974). *Introduction to Control Systems*, Pergamon, New York. 384 pp.

Bennett, W. R. (1976). *Introduction to Computer Applications for Non-Science Students (Basic)*, Prentice-Hall, Englewood Cliffs, NJ.

Bennett, W. R. (1976). *Scientific and Engineering Problem Solving with the Computer,* Prentice-Hall, Englewood Cliffs, NJ.

Brennan, R. D., C. T. deWit, W. A. Williams, and E. V. Quatrin (1970). The utility of a digital simulation language for ecological modelling. *Oecologia* **4,** 113–132.

Charles, J. J. (1980). *Basic Training for Compucolor Computers,* J. J. Charles, Hilton, New York. 192 pp.

Compucolor Corporation (1980). *Programming Manual for Compucolor,* Norcross, GA.

De Wit, C. T. and J. Gaudrain (1976). *Simulation of Ecological Processes,* Halstead (Wiley), New York. 175 pp.

Ferrari, Th. J. (1978). *Elements of System-Dynamics Simulation* PUDOC, Wageningen, Netherlands.

Finkelstein, L. and E. R. Carson (1979). *Mathematical Modeling of Dynamic Biological Systems,* Wiley, New York. 329 pp.

Forrester, J. W. (1961). *Industrial Dynamics,* MIT Press, Cambridge, MA.

Gordon, G. (1969). *System Simulation,* Prentice-Hall, Englewood Cliffs, NJ. 303 pp.

Hubin, W. (1978). *Basic Programming for Scientists and Engineers,* Prentice-Hall, Englewood Cliffs, NJ.

Kemeny, J. C. and T. E. Kurtz (1981). *Basic Programming,* 3rd ed., Wiley, New York.

Miller, A. R. (1981). *Basic Programs for Scientists and Engineers,* Sybex, Berkeley, CA. 318 pp.

Morton, J. B. (1977). *Introduction to BASIC,* Matrix, Champaign, IL.

Moulton, P. G. (1979). *Foundations of Programming through BASIC,* Wiley, New York. 271 pp.

Mullish, H. A. (1976). *A Basic Approach to Basic,* Wiley, New York.

Pegeis, E. E. and R. C. Verkher (1978). *BASIC,* 3rd ed., Holden-Day, San Francisco.

Poule, L. and M. Borchers (1979). *Some Common Basic Programs,* 3rd ed., McGraw-Hill, New York.

Sanderson, P. C. (1973). *Interactive Computing in BASIC,* Butterworths, London. 101 pp.

Sharpe, W. F. and N. L. Jacob (1971). *An Introduction to Computer Programming Using BASIC Language,* rev. ed., Free Press, New York.

Simon, D. E. (1978). *Basic from the Ground Up,* Hayden, Rochelle Park, NJ. 222 pp.

Smithe, R. E. (1971). *Discovering Basic,* Hayden, Rochelle Park, NJ. 164 + 39 pp.

Spain, J. D. (1982). *Basic Microcomputer Models in Biology,* Addison-Wesley, Reading, MA. 354 pp.

CHAPTER FIVE

Analog Computer Simulation

The standard analog computer is a device for continuous simulation of systems models. Operational analog computers manipulate voltage specified by equations. If there is to be adding, subtracting, and division, the voltages are added, subtracted, or divided by hardware devices invented for this purpose. An electronic unit, the operational amplifier, is used in various ways as building blocks to manipulate voltages as indicated by equations written for the model. Voltages are recorded to show trends with time. In this chapter we introduce some of the main components of the "operational analog computer" and indicate the steps for drawing the analog diagrams and patching the wires.

This chapter is relatively brief, and the student may need other references for reinforcement such as those by Blum (1968), Bennett (1974), or the introductory chapter in Patten (1971–1976). See Electronic Associates computer reference manual (1971). Figure 5-1 illustrates the procedure from energy diagram to simulation graph. In Fig. 3-3d the differential equation for a simple storage was block diagrammed, manipulating the steps of calculation in sequence. Figure 5-1 has the same procedure but uses analog computer symbols. The pathways of the diagram are the wires that manipulate voltage in steps according to the equation. A voltage J representing inflow is added to a voltage $-kQ$, and the sum is integrated to obtain Q, a voltage representing the quantity in the tank. The voltage $-kQ$ was obtained by multiplying voltage $-Q$ by k. A recorder or an oscilloscope shows the graph of Q with time as the voltages continuously calculate the quantities after the connections are made with a start button.

Analog computer symbols are often used to represent systems, another systems language. However, in our approach we use the equations and analog diagrams for initial wiring and computer settings. Once this is done we return to the energy circuit diagrams for study and debugging, having indicated analog computer hardware addresses in colored pencil directly on the energy diagram. The energy diagrams are also simpler with one line for a path between two tanks, and two lines on an analog diagram.

MAIN COMPONENTS

Next we introduce the main hardware units of the analog computer and their patching and operation. Figure 5-2 shows an analog computer.

The Operational Amplifier

The operational amplifier is a main building block of the analog computer. It is diagrammed as a round-back triangle symbol as in Fig. 5-3a. Whatever voltage is received on the left is amplified by millions of times (at least to the extent of available power supplies), and its sign is reversed (plus to minus or minus to plus). Normally the operational amplifier is not left free to amplify in this way, and when it is, overload lights go on. Normally, a resistor, capacitor, or other pathway is patched from the output back to the input to make the unit perform some mathematical operation (summing, integrations, multiplications, etc.). If amplifiers are not patched with feedback loops and overload lights are on, patch all amplifiers with a feedback resistor as an initial step in starting a problem. Newer instruments such as the Miniac have automatic feedback features, and no patching of unused amplifiers is required.

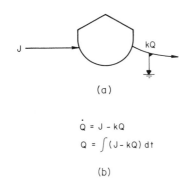

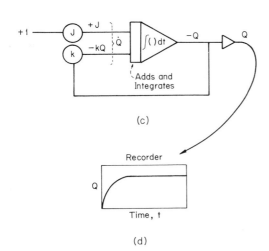

FIGURE 5-1. Example of an analog computer simulation: (a) energy diagram; (b) differential and integral equations; (c) analog wiring diagram; (d) graph generated on a recorder if initial condition is zero.

The Summer–Inverter

The triangle symbol is the analog computer summer or inverter. It indicates addition of voltages on wires patched to its front with inversion of sign (plus to minus or minus to plus) at the output. Figures 5-3b and 5-3c are more detailed wiring diagrams indicating that the summer is made up of an operational amplifier (round-back symbol) with one or more input resistors R_i and one feedback resistor R_f. The operational amplifier starts to amplify and invert any incoming voltage by millions of times, but the feedback returns a canceling voltage of opposite sign to the intersection *SJ* called a *summing junction*. By this the summing junction is maintained near electrical zero (ground). A balance is maintained so that the fall in voltage from input voltage to summing junction equals the return drop of opposite sign from output back to summing junction. If the input resis-

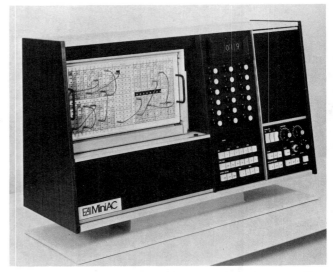

FIGURE 5-2. View of Miniac analog computer courtesy Electronics Associates, Inc., West Long Branch, NJ.

tor and the feedback resistors are the same (e.g., 1 megohm each), the output voltage is the input voltage inverted (minus to plus or plus to minus). When the summer unit is used in this way, it is called an *inverter* (Fig. 5-3b).

However, if the feedback resistor is higher than the input resistor (e.g., 1 megohm and 0.1 megohm), the output voltage builds to 10 times the input before its feedback cancels the input at the summing junction. The equation given in Fig. 5-3c shows the relationship of resistors to voltage transfer function of the summer.

Many analog computers have provision for substituting different input or feedback resistors to give factors of 10 to speed up or slow down programs or otherwise scale the output (see section on scaling).

In some computers the input hole for wires with this 10 factor is marked "10." This is given in diagram form as in Fig. 5-3f. Figure 5-3b shows the resistor arrangement for a single input multiplied by the resistance ratio. In Fig. 5-3c two input resistors are shown, one for each input voltage. In Figs. 5-3c and 5-3e the feedback and input resistors are matched and the effect of the unit is simple summing and inverting. Notice that the input wires must each have a separate input resistor to maintain their relationship to the feedback but that the outputs may be drawn from the one output, stacked on top of each other if necessary. The operational amplifier supplies the power necessary to keep the output that of the equation in spite of the loads.

Notice that analog diagramming may be done at

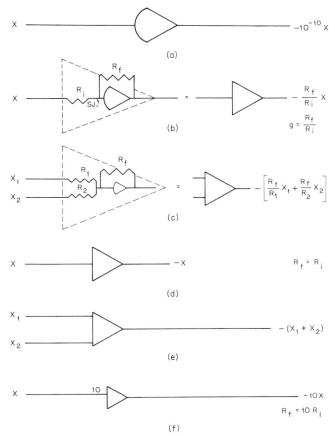

FIGURE 5-3. Operational amplifier and its use for summing and inverting: (a) operational amplifier; (b) wiring for an inverter with a gain determined by resistors used as input and feedback to summing junction *SJ*; (c) wiring for summing of two inputs; (d) abbreviated notation for inverter; (e) abbreviated notation for summer; (f) inverter with gain of 10.

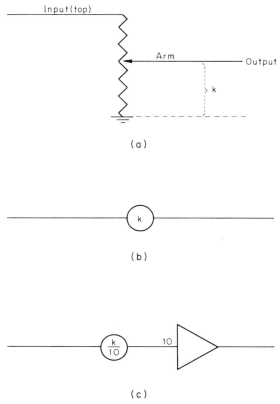

FIGURE 5-4. Analog computer pot. (a) Electrical circuit with arm on resistor; (b) symbol for pot; (c) arrangement where coefficient is between 1 and 10.

several levels of detail. Ordinarily the simple triangle is enough, but sometimes more detail is needed if special arrangements of resistors are given. Whereas the manipulation of resistors may be used to multiply a voltage by a constant, a pot is generally used for this.

The Pot

Any value that is set numerically is set with a pot. Pots are precision variable resistors that have a circuit such as that shown in Fig. 5-4a. Normally, they are indicated on analog computer diagrams as a simple circle (Fig. 5-4b), which indicates that an input voltage on the input wire is being multiplied by a fractional number. The output voltage is that fraction indicated by the pot setting. A pot connects the input voltage to ground spreading out the voltage drop linearly from the top of the pot (input) to ground. With a frictional contact, the arm receives from the resistor any voltage at that point along the resistor. If the arm is at the top, it receives the input voltage unchanged; if the arm is half-way down the resistor, it delivers half of the input voltage. If the arm is at the bottom touching ground, its output is zero. In most uses the turning of the pot moves the arm (arrow) of the output along the resistor so that the voltage is that fraction of the input voltage that the resistor is between arm and ground.

The pots are used for setting the coefficients in the equations, for setting initial conditions and forcing functions when these are constant. The value to be multiplied is set on the pot if it is less than 1. For values higher than 1, the voltage is reduced by the pot set to one-tenth of the value of k, and then the wire is put through an inverter with 10 gain (Fig. 5-4c). In Fig. 5-1 pots were used with an integrator unit.

Initial conditions and forcing functions are set by patching a wire from outside plus or minus unit voltage (one machine unit) through a pot for setting to some fraction and from the pot to the "IC" hole (initial conditions). If the output is plus the "IC" should be minus and vice versa.

A pot is used to multiply by a constant less than 1, and a pot along with an inverter with 10 or more gain is used to multiply by a constant greater than 1.

The Multiplier

Where the equations indicate a multiplication of two variables, a multiplier unit is required. In some more modern computers the operation involves two input wires, an output wire, and the simple overall multiplier diagram given in Fig. 5-5a. This kind of multiplier reverses sign as does the inverter. To determine the output sign, one first determines from the two inputs what the sign would be in normal algebra and then reverses the sign. For example, plus times plus is plus becoming minus when inverted. Or for another example, minus times plus is minus becomes plus when inverted.

Multiplication is done by a resistor network that actually processes voltages according to the quarter square equation [Eq. (5-1)]:

$$XY = \tfrac{1}{4}[(X + Y)^2 - (X - Y)^2] \qquad (5\text{-}1)$$

Each input voltage going into a multiplier must come from an operational amplifier (summer, integrator, or other multiplier units with a pointed symbol), and not from a resistor (pot) because resistors become part of (and upset the effect of) the multiplier's resistor network unless they are separated from the multiplier by an operational amplifier that uses power as necessary to hold the voltage calculated. Two voltages multiplied together give the correct value only if the operational amplifiers are next to the multiplier input.

In older analog computers more of the multiplier's network must be patched; thus more must be diagrammed to aid in checking, debugging, and so on. Figure 5-5b shows the detail required on those instruments that require patching a plus and minus input for both the input voltage wires (making four inputs altogether). The output requires a wire from the resistor network to the summing junction SJ of an amplifier (not the input resistor) and a feedback resistor across from output. In some computers the feedback resistor is 0.1 megohm for normal multiplication and 1 megohm when an additional gain factor of 10 is desired. If the X and Y input wires are

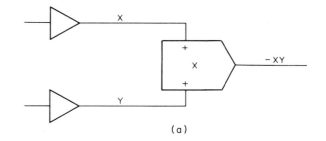

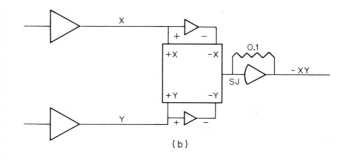

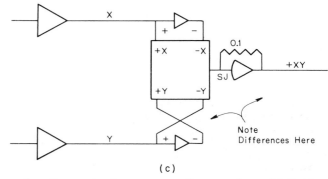

FIGURE 5-5. Analog computer diagrams for multiplication: (a) arrangement for Miniac; (b) arrangement for older quarter square multipliers and negative output; (c) same with positive output.

matched by sign with plus into plus holes and minus into minus holes, the output sign will be minus. If a plus sign is desired, one of the pairs of input wires is reversed so that the plus goes to minus and the minus to plus (e.g., see Fig. 5-5c). Note that these multipliers use three operational amplifiers, two patched as inverters and one connected at the summing junction.

The Integrator

The symbol for an integrator is given in Fig. 5-6, including an initial condition setting, and input con-

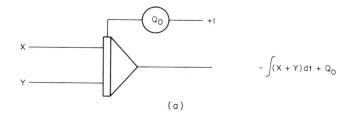

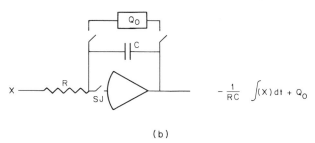

FIGURE 5-6. Analog computer diagram for an integrator: (a) general integrator symbol showing inputs, output, and initial condition with pot; (b) details of integrator showing its input resistor R, feedback capacitor C, and operations for switching from initial condition to operate.

nectors for the wires carrying voltages to be accumulated with time. More details of the wiring of an integrator are given in Fig. 5-6b showing the operational amplifier with an input resistor but with a capacitor as the feedback connection of output to summing junction. If the analog computer provides for patching of various-sized capacitors and resistors, one may determine the equation for integrator output from the value of the input resistor in megohms (R) and the feedback capacitor in microfarads (C). If both have a value of 1, the time constant is 1 sec and the integrator performs the integration with 1 sec equal to the particular time unit used in the problem.

As suggested by the switches shown in Fig. 5-6b, the integrator has provision for connecting an initial voltage across the capacitor storage when the computer is in the "initial condition" state. Then, when the "operate" button is pushed, the switches connect in the input resistors to the summing junction and disconnect the apparatus that was used to set the initial conditions. The start of the operation leaves the charge on the capacitor to be added to or subtracted from by the other newly connected pathways.

ANALOG SIMULATION OF A STORAGE UNIT

Figure 5-1 illustrates use of the integrator to indicate continuously with time the value of Q, $\dot{Q}$, or any other quantity for which there is a wire in the diagram kQ, J, etc.

The voltages desired may be selected for recording. To study a system such as this, one may add or subtract pathways by adding or removing wires or change rates by adjusting the coefficient values. One may introduce varying inputs as the forcing functions to see temporal responses. At this point, with the help of instructors, the student might make a qualitative study of the single storage example on an analog computer (Fig. 5-1), as follows: (1) wire system as diagrammed, connecting pathway of interest to display device (recorder or oscilloscope) and writing computer addresses on energy diagram (e.g., P1 for pot No. 1 or A1 for amplifier No. 1); (2) turn on machine, putting it in set-pot mode and setting pots representing initial conditions, transfer coefficient, and inflow rate; (3) put in IC (initial condition) mode; and check signs of outputs; and (4) press operate start button and watch the progress of the curve with time on the display instrument. Vary pot settings, study curve responses, and compare with the curve in Fig. 5-1d.

Amplitude Scaling of Analog Computers

Any analog computer has a maximum voltage range for its circuits and display equipment, which is called a *machine unit*. Full scale is one machine unit. There are, for example, machines with full scale of 10 volts and others with 100 volts. Any quantity to be simulated must be divided by some factor that will keep the magnitude within the scale of the machine. The routine for doing this is to divide each variable by the maximum value to be expected in the simulation so that the units of the ratio of the quantity to maximum quantity has a fraction of the full-scale maximum expected. Brackets are used to keep the fraction identified as it is used in equations and on analog diagrams. As a result of the division, graphical outputs on recorders or oscilloscope are as percent of full scale, where full scale is the divisor chosen for scaling. The procedure of converting all amplitudes to percent of the maximum is sometimes called *normalization*.

To substitute the scaled fraction into the equa-

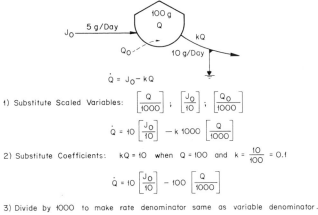

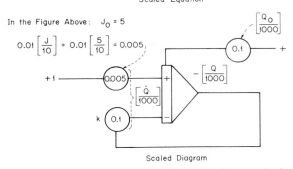

FIGURE 5-7. Procedure for amplitude scaling applied to a single-storage system.

tions without changing them, one has to multiply and divide by the same number.

For example, in Fig. 5-7 we scale the differential equation for the storage compartment in which the anticipated maximum value for Q expected is 1000 g/m^2, the transfer coefficient k is 0.1 per day, the maximum inflow rate is expected to be 10 g/m^2 · day, and the initial condition is 100 g/m^2. We first substitute the scaled variables fractions $[J_0/10]$ and $[Q/1000]$, also multiplying by the divisor in each case so that substitution will not have changed the total effect on the equation.

Next, the values of the coefficients are substituted. In Fig. 5-7 observed data are given (written on a sketch diagram of the model). The outflow is shown to be 10 when the storage driving it is 100. Solving the equation for that pathway for k, one obtains $k = 0.1$. This is substituted into the equation.

Next, the denominator for the rate of change (Q dot) must be the same as for that Q wherever it is found in the same equation, if the rate and state part of the equation are to be on the same time scale. To accomplish this, after substitution of scaled variables and coefficients, the denominator (scaling factor) of Q is divided through both sides of the equation so that it now appears in the denominator of Q dot as well as Q. This usually reduces the numerical values for each term to numbers between 0.001 and 1.0 so that they can be set on the pots of the machine. The numbers for the pots are written on the analog diagram from the equations and the scaled variable (bracketed designation) is written on the input or output of the operational amplifier by which it is generated.

If the scaled number for a variable is greater than 1, it can be represented as the product of 10 gains and a fraction. For example, if the coefficient were 6, it could be represented as 0.6 (10). Then the pot could be set as 0.6 and the wiring provided with a 10 gain. Once the wiring is patched, the operator sets the pots to the values shown. Thus pots are used to indicate the initial conditions, the scaled value for the transfer coefficient, and the setting for the scaled input forcing function J_0.

Time Scaling

The electrical integrators of an analog computer operate according to Eq. (5-2), in which R is the resistance of the input resistor that is part of the integrator hardware and C is the electrical capacitance of the capacitor used in the electrical integrator, RC is the time constant, $\dot{Q}$ is the input voltage, Q is the output voltage, and Q_0 is the initial voltage at the start:

$$Q = \frac{1}{RC} \int (\dot{Q}) dt + Q_0 \qquad (5\text{-}2)$$

If R is 1 megohm (10^6) and C a microfarad (10^{-6}), the time constant RC has the value of one second and the expression becomes that already used in Fig. 5-7. The time units used in the data become equal to seconds. In the example given in Fig. 5-7, days become seconds.

Some analog computers allow the operator to change the time constant by substituting smaller capacitors or resistors or pushing buttons that do this. In other words, one may shorten the time constant, change the time scaling, and increase the speed of the simulation by multiplying all the integrator inputs by the same factor. Faster simulation means that more real time is represented by a second of simulation.

In some problems it is convenient to substitute a number for $1/RC$ in all equations, thus changing the time represented on the machine by that factor. If $1/RC$ is 10, for example, 1 sec represents 10 times what it did before. If the problem was in years, 1 sec then represents 10 years; this is time scaling.

In other words, to time scale the whole system of equations at the start, divide all coefficients of all terms that enter integrators by a constant factor that either slows the system down or speeds it up by this factor.

Time Settings on the Miniac and Time Scale on Display

The Miniac has three items that affect the time that is represented on the display equipment—the slow–fast button, the 10-turn timer pot, and the switch for selecting a fraction of a second that is fine tuned by the timer pot (see Fig. 5-8). When the fast button is off, seconds on the machine represent the unit of time of the data (e.g., years). This is because the time constant of capacitors is 1 sec. Actual time on the oscilloscope sweep in seconds is the product of timer pot and switch. When the fast button is in, the integrators and the graph generated goes 500 times faster, and thus the seconds on the scope represent a factor of 500 more time.

If the scaling equations were time scaled by dividing through by a factor such as 10, the electrical tanks would fill one-tenth as fast. The analog goes more slowly even with the fast button in fast position (in). A second on the scope now represents one-tenth of a year.

An example of settings of the timer is given in Fig. 5-8 for the simple case of a drain from an initial storage of unity. Without time scaling, a year is a second, and the setting shown provides a horizontal sweep of 10 seconds representing 10 years.

In Fig. 5-9 the same example is time scaled by dividing equations by 100. Now the quotient for $\dot{Q}$ and Q are different by a factor 100; less hardware is used, and 1 s equals 0.01 yr. The setting shown provides an oscilloscope sweep of 10 s, or 0.1 year. In this setting the fast button is used in slow position (out).

An Example of Analog Simulation

Figure 5-10 shows a simple model of a consumer in energy circuit language. There is an outside *food source concentration* which has a food pathway that

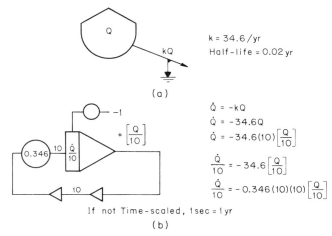

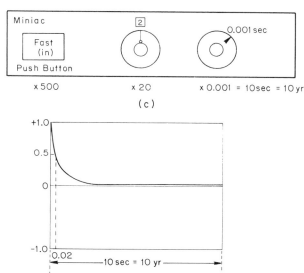

FIGURE 5-8. Example of calculation of time represented by the sweep of the oscilloscope on the Miniac; example without time scaling, from Mitsch (1975): (a) energy diagram; (b) analog diagram; (c) time controls on Miniac; (d) oscilloscope display.

is determined by an outside forcing function, the food concentration F. Food then flows into storage with the coefficient k_1 characterizing the pathway. Mass storage Q is depleted by two pathways: one is dependent on its own storage mass k_2Q representing metabolism and maintenance work; and the other is the loss due to drains by parasites. This drain is proportional to the number of parasites and their multiplicative interaction with the biomass.

Our procedure for analog simulation begins with the energy circuit diagram with its storages and flows. One first writes a differential equation summing the rate pathways contributing to or draining each tank. It may be helpful initially to write the

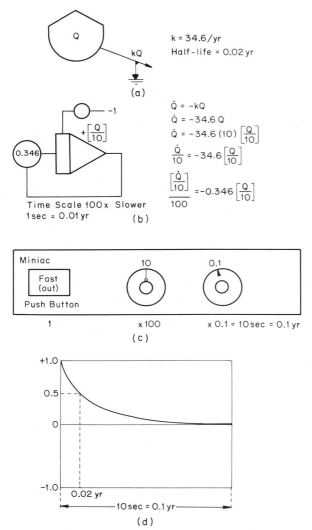

FIGURE 5-9. Same as Fig. 4-11 but with time scaling by factor of 100 times slower.

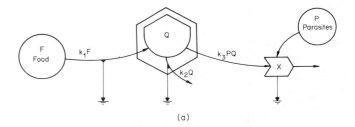

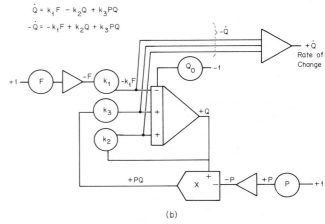

FIGURE 5-10. A system with steady energy inflow regulated from outside a single-storage unit and drain and an outflow under multiplicative pumping control of an external consumer forcing function: (a) energy diagram and differential equations; (b) operational analog computer diagram (unscaled). Circuit for reading rate of change is also included.

algebraic expressions for each pathway on the pathway diagram.

The energy diagram in Fig. 5-10a is translated next into a differential equation with three terms for the three pathways. Next we divide through by -1 to make the differential equation negative so that the integrated quantity Q can be positive for convenience in recording on chart or oscilloscope.

Next we diagram the differential equation in operational analog computer symbol language (Figs. 5-1 and 5-10). First we draw an integrator for each tank (in this case there is only one). Label its front $-\dot{Q}$ and its output Q. Then we draw into its summing front (left side) a wire and pot for each pathway term in the equation (in this case 3). Initially, ignore the problem of minus and plus signs until all the symbols and wire lines are drawn.

For convenience and checking, it is useful to write the algebraic expression on the wire line as shown. Then connect to that pot wire lines from those units that will produce that term. The term k_2Q is obtained by running a wire line from the Q that is the output to the pot k_2, thus making k_2Q. Since F is an outside forcing function, it starts with an outside voltage (reference voltage) that, when multiplied by pot F and pot k_1, produces the pathway term k_1F. In general, one does not put two pots in a row since this becomes inaccurate over some ranges. Instead, one puts a summer–inverter in between. Such arrangement also allows the recording of the F value as an output on recording equipment. In this particular example we are setting F constant so that we may not be interested in recording its variation.

The third pathway k_3PQ into the integrator has a product of two variables and thus must come from a multiplier that is drawn next. Its output is to be PQ, and so we connect Q to one of its inputs and P to another. The parasite concentration P was shown as an outside forcing function to this system, and thus we start it with a reference voltage through a pot marked P (parasites). Recalling the requirement that inputs to multiplier units be from amplifiers, we insert a summer–inverter between the pot and the multiplier. The wiring diagram is now complete since we have a pathway patched for each of the terms in the equation. Next we consider the requirements for algebraic sign and the special peculiarity of the analog units that switches sign through each summer, integrator, or multiplier.

For convenience, mark the desired sign on the integrator at the point of entry of each pathway as the differential equation indicated. Also, mark the integrator output as plus (opposite from Q dot). In this case we are using the negative of the differential equation (the second equation in Fig. 5-10). Next, backtrack along the pathway incoming to the integrator checking and adjusting signs. For the k_1F pathway, we need a minus input from the summer–inverter, which we mark minus, and we set the input reference voltage the reverse, a plus. For the k_2Q pathway, we need a plus, and we see that the pathway is already properly derived from a plus source. For the multiplicative pathway PQ, we need a plus, which means that the algebraic product of the inputs must be minus (since multipliers invert sign). (If you are programming an older style of multiplier, turn to Fig. 5-5 to arrange signs.) Since the input from Q is already set as a plus, the P should be minus. Marking the summer–inverter output minus, we then adjust the input reference voltage as plus.

If one wants to read off the rate of change itself ($\dot{Q}$ rather than the value for Q), one may sum the inputs through a summer–inverter first as a parallel unit to the integrator as shown in Fig. 5-10b. Some analog computers have automatic arrangements for reading this without patching. The one shown in Fig. 5-10b is reading the summing junction using an amplifier so as not to affect the voltages and currents while reading its value.

Calculating k Values

The first step in putting real numbers into the model (scaling) is to put these numbers and their units on the energy diagram as done in Fig. 5-7 and 5-11. These numbers can be the stocks and flow rates at any time of observation, or one may approximate by using averages.

A procedure for amplitude scaling was given in a

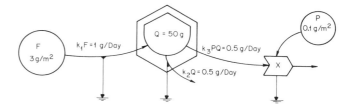

FIGURE 5-11. Starting with values for stock and flow on the energy diagram, coefficients (k values) are calculated from flows and stocks.

previous paragraph, and this procedure is applied to the model shown in Fig. 5-10 (see Fig. 5-12).

Next, substitute the values in the energy diagram (Fig. 5-11) into each pathway flow expression, and solve for the coefficient k. This is done in Fig. 5-11, producing a table of coefficients. The units for the k values vary, and normally one does not need to think about them. Substitute these coefficients into the general differential equation (Fig. 5-12).

Scaling State Variables and Forcing Functions

The maximum values to be expected are assigned for the state variables (Q values) and for the forcing functions. These are placed in denominator of a bracketed expression, and a list is prepared as done in Fig. 5-12. When these variables reach the maximum, the bracketed expression becomes unity (full scale, or 100% of chart amplitude). Both inputs to the same comparator must have the same scaling.

Next, put the bracketed variables into the equation in place of the unbracketed variables. To do this without changing the equation, one must multiply each term substituted by the same denominator as shown in the bracket. The resulting equation has numbers that can be combined, and when this is done, one usually has coefficients for the variables that are greater than 1 and thus cannot be set on pots. One next divides through by the scaling factor (denominator) to make rates and states of the tank on the same time scale. In Fig. 5-12 this factor was selected as 100. As we did with 100 in this example, routinely divide through by the same value used for the denominator scaling factor for Q. To accomplish

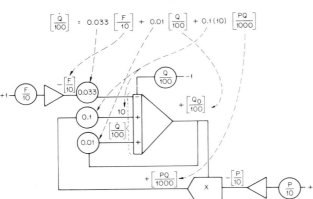

FIGURE 5-12. Forcing functions and variables in the system are scaled, substituted in differential equations along with coefficients, and a final equation is found by dividing through by a number (100) to reduce constants so that they may be set on pots. Designations are written on computer diagram.

TABLE 5-1. Coefficient and Pot-Set Table for Figures 5-11 and 5-12

Item	Value	Pot Setting
Coefficients:		
k_1	0.33 g/m² · day	0.033
k_2	0.01/day	0.01
k_3	0.1 per m²/day · g	0.1 (10)
Forcing functions set to values in Fig. 5-11:		
F_0	3 per g/m²	0.3
P_0	0.1 g/m²	0.01
Initial conditions:		
Q_0	50 g/m²	0.5

this in some examples, one must leave an expression ×10 or ×100 as part of the coefficient. In those instances one may run the input pathway wire through a gain of 10 of an amplifier for each such factor of 10 remaining. When the equations are modified by these three manipulations (entering k values and substituted bracketed variables, multiplying accordingly, and dividing the whole equation through by some factor to make rate denominator same as for Q), one is ready to run the program.

Transfer the bracketed variables and pot settings to the analog diagram as done in Fig. 5-12. The pot settings are the new coefficients of the equations (one for each pathway). Patch the wiring as indicated by the scaled diagram (which may differ from the first analog diagram by adding some 10 gains or extra summer–inverters to permit addition of 10 gains). Set the pots and test the program. Making summary table of pot settings is very useful (e.g., see Table 5-1). For more details on patching, see the appendix to Chapter 5.

The first step after patching the program is a static check with the computer in initial conditions.

Recall that in the initial condition mode the wire pathways are disconnected from the front of the integrators (tanks) but the initial conditions are connected. By turning the output selector switch to the various amplifier numbered or lettered addresses, one may read off the voltages to determine whether they are what they should be. Checking to see that the signs are in accordance with the specifications on the diagram may suffice. To do this, one needs some initial condition set on each integrator regardless of whether there is to be one in the run.

If no errors are found, push the operate buttons, preferably on "repetitive operate" (rep-op) fast setting. The program may still go off scale. Scaling does not necessarily guarantee that the model will stay on scale because there may be incompatibilities in having certain coefficients with certain values in the equation. To get a model on scale if it is possible, turn down the pots that set the energy sources and increase the negative drains that should be on each tank (since any tank has such drains in real problems).

In the example in Fig. 5-11, which the student should try on the analog computer, the mass Q grows up to a steady state. If more parasitism is added, a lower state results.

Doing Scaling Calculations Directly on the Analog Diagram

One convenient and fast way to do the calculations that we just did in scaling the equations is to do them on the diagram where the eye can readily see relationships. As shown in Fig. 5-13, write in the blank quotient expression [Eq. (5-3)] for each pathway entering an integrator. It is useful to place all

Analog Computer Simulation

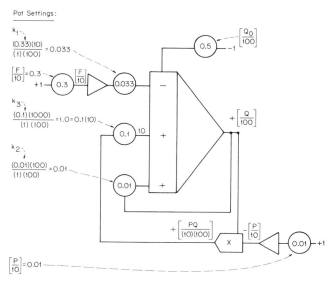

FIGURE 5-13. Example of analog scaling done directly on diagram with quotient method: two pots are used for forcing functions.

coefficients. The calculation resulting from the quotient expression is the pot setting. It is the same number found with the equation method.

Quotient for Forcing Function

The pot setting for an external driving function may be calculated with the quotient method similarly. As given in the example in Fig. 5-14, one pot may be used to represent a scaled forcing function. The downstream scaling factor and the time scaling factor are put in the appropriate parentheses in the same way as for coefficients. In the upper left parenthesis is given the numerical value of the forcing function for the initial run, which is a number less than 1, since it is the value of the scaled variable. In the upper right-hand parenthesis is written the scaling factor from the denominator of the scaled forcing function. The one pot in Fig. 5-14 represents the same thing as two pots in Fig. 5-13.

Other Forcing Functions

Instead of the constant forcing function in Fig. 5-14, variable forces and flows can be applied. Square waves, sine waves, ramps, and so on, can be generated with the use of other integrators. Use of the integrators in a line along the left margin, arranging them in a vertical column to facilitate the scaling.

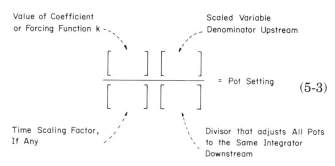

Then fill in the k values in the upper left parenthesis, and in the upper right parenthesis write in the denominator of the scaled variable bracket of that variable that is the upstream source of that pathway. If time scaling is done to get all pot settings in range that can be set without so many (10) gains, put the factor in the lower left.

If the time scaling is done, it must be done for all

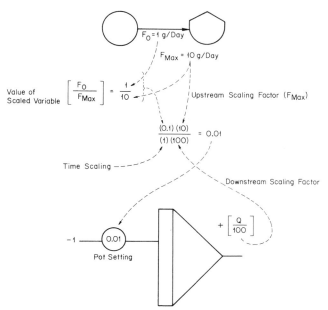

FIGURE 5-14. Use of quotient expression for scaling a forcing function using one pot, where maximum value F_{max} expected is 10 (scaling factor for F_0).

sine waves to represent diurnal and seasonal variations is given in the appendix to this chapter.

A random noise generator that supplies voltages with Gaussian distribution around zero can be applied to represent forcing functions such as seed dispersal or turbulence. With the use of a multiplier, the variation can be applied to vary the conductivity of pathways, simulating some of the population aspect that the models aggregate.

Large Range of Pot Settings

If the range among the several pathways entering the integrator is large, use 10 gains on the pathways. If there is a very tiny quotient more than 10,000 times smaller than the largest one, it is probably negligible in this problem and can be omitted. One could have decided this earlier by looking at its value in the energy diagram setting relative to others affecting the same tank. What is important at one time and space scale may be unimportant at a much larger or much smaller scale.

OTHER COMPONENTS

Limiters and Diodes

In many situations the problem requires that there be no change of algebraic sign. For example, biomass of an animal in the real world does not become negative. If the parasites are eating the animal at a rapid rate exceeding the animal's intake rate, the biomass will decrease and stop at zero. The analog program, however, does not stop at zero unless one introduces a device to stop it. Left without limiting devices, the multiplier pulls the Q value to zero and then into minus quantities. At that point the program would be unreal and meaningless. One of the most common overloads in environmental circuits when they are first put on analog or digital computers is due to the multipliers pulling variables negative from their normal range.

The problem is readily avoided by patching a limiter circuit as indicated in the computer manual setting the lower limit at zero. One may allow a variable to go off scale without overloading the electronics or affecting the rest of the system by putting a limit of 1 (machine unit) on the integrator output pathway. If there is no provision for limiters, one may connect a diode in the multiplier or other circuit so that electrical current can go only in the di-

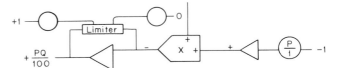

FIGURE 5-15. Inverter added to system in Fig. 5-10b with limiter circuit and adjustment of sign.

rection associated with the range from 0 to 1. A diode is a one-way electrical valve. In the example given in Fig. 5-15 a limiter was added, but some sign adjustment was required.

Reversal of sign can also be prevented by inserting a diode in the wire circuit (see Fig. 5-16), thus permitting current to flow in only one direction.

Track Store; Analog Memory

The track store unit with symbol given in Fig. 5-17a is connected to the output of another unit that it tracks, holding the same value until it receives a logic instruction (logic 0) to store. The module on store is disconnected from the input that it was tracking and holds the value it had at the time it was put on store. It will resume tracking its input when it receives a logic input of "on" (logic 1). The track store unit is sometimes described as an analog memory unit since it retains a bit of information.

An example of its use is given in Fig. 5-17b, where the switching symbol is used in the energy language as an instantaneous transfer of the value of a storage Q to a second storage V, which may be information in proportion to the original storage. Its use on the analog diagram is given in Fig. 5-17c. With suitable switching, the track store unit can be used as a time lag. In one use the value of the track store unit is used to start the initial condition at a

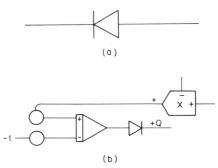

FIGURE 5-16. Diode added in wire connections so as to prevent sign reversal: (a) diode symbol; (b) diode preventing output of integrator from going below zero.

Analog Computer Simulation

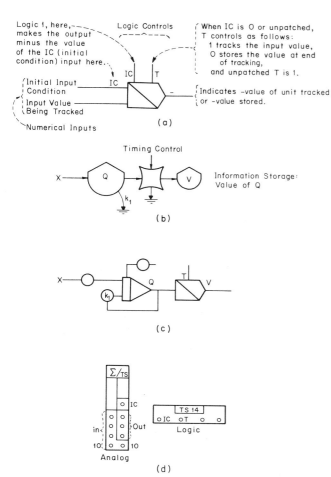

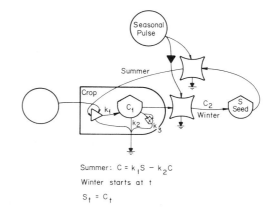

Summer: $C = k_1 S - k_2 C$
Winter starts at t
$S_t = C_t$

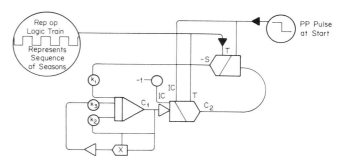

FIGURE 5-17. Track–store unit: (*a*) symbol; (*b*) in energy language; (*c*) analog diagram of part *b*; (*d*) patch holes on panel of Miniac.

FIGURE 5-18. Example of "bucket brigade" use of track–store units. Seed harvested and stored at end of summer growth period determines size of crop in the next growth period. Varying time of sweep varies level of seed developed, simulating effect of growing season. The value of S holds while C_1 and C_2 grow. Then C_1 is reset to zero while C_2 holds and its value is transferred to S. Then the cycle repeats.

later time. In Fig. 5-18 two track store units are connected to form a *bucket brigade* configuration. Crop growth generates seed that are stored and used on the next growth cycle to grow a new crop.

Some Errors to Avoid

Although many machines are protected with limit devices and fuses to avoid permanent damage to computer hardware, here are some no-no's to avoid. Avoid connecting any output to another output as this tends to overload and burn out power supplies. Do not ground power supplies or let positive and negative voltage connect.

If the patchboard is not on the machine, one can patch wires in the direction of the mathematical operations from source to output, which may be the natural way one thinks. However, if the board is on the machine, one should patch in the opposite direction so that the wire is first put into the downstream hole and then into the upstream end from which voltage is derived. If one does this, there is no active voltage on the end of the wire one is trying to insert. If one patches downstream and touches other metal surfaces or other wires, one may cause spurious surges and blow out pots, fuses, and so on.

Do not put patchboards in upside down. Do not include loops with even numbers of amplifiers unless they are intended and there are limits of operation time or voltage limiter devices attached. Even-number loops are positive feedbacks that produce rapid exponential growth of the quantity with immediate overload and danger to power supplies on some older machines.

Some kinds of electronic switch must not be used to switch voltages because they may be damaged by short circuits in the instant of their switching. If the computer does not already have obligatory arrangement of resistances in series with switches, they should be included. Switching is given in Chapter 6.

SUMMARY OF SIMULATION STEPS

In Chapter 4 a summary of steps for simulation was given starting with energy diagramming, followed by writing differential or difference equations, converting data on the diagrams into coefficients.

For analog computer simulation, translate differential equations into an analog computer diagram (wiring diagram). Scale the analog diagram and patch wiring according to the wiring diagram. The limitation of a facility to available amplifiers and integrators per model may be a useful restraint in compelling one to condense models to include important pathways only. In energy circuit language, these diagrams are readily remembered and can be written as rapidly as the hand can write. The number of coefficient combinations is very large, even with moderate size systems; thus full testing may take many hours.

Testing for steady states and transient patterns, simulate the model for sensitivity of response to varied forcing functions and coefficients and for similarities to observed time graphs (validation). Relate to issues in question.

Revise models to incorporate improvements or fit observations, and do it in a way that is explainable in terms of known mechanisms of the systems parts. Record results on recorder charts or with polaroid photographs of oscilloscope displays operated on the repeat operation mode (rep-op).

Draw conclusions about the consequences of the models and actions on them. Try to find test examples that will validate in real-world tests the experiments done on the models. Often this requires analysis of historical trends.

SIMULATION EXAMPLE WITH SEVERAL STORAGES

One of the early simulations of ecological phenomena was of organic matter and litter on the forest floor by Neel and Olson (1962) and Olson (1963a, 1963b), as shown in Fig. 5-19. Heinmetz (1966) simulated a complex general model of cell growth and genetic control with actions of injury and pathology.

SUMMARY

In Chapter 5 our objective was to take a systems model that is given in the form of an energy circuit

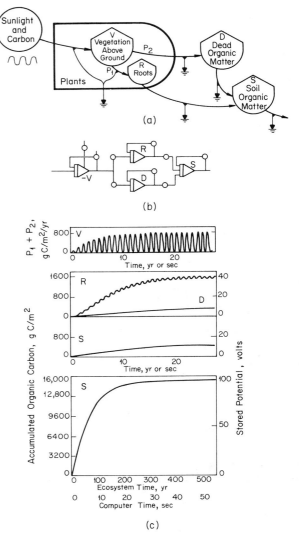

FIGURE 5-19. Simulation of organic matter accumulation in ecosystems from Neel and Olson (1962) and Olsen (1963a): (*a*) energy diagram; (*b*) analog diagram; (*c*) simulation graphs.

diagram, write the differential equations, determine coefficients from data and diagram, and patch and simulate the model on standard analog computer. At this point the student should have operated each system in the chapter and invented and executed a program with the use of some of these functions.

QUESTIONS AND PROBLEMS

1. Can you draw all the symbols (at least eight) of the standard analog computer program that you have learned and indicate the mathematical expression of the transfer function that

each accomplished between receiving the input voltage and delivering an output voltage?

2. Can you explain each of the following:
   a. Inverter of analog signal.
   b. Modes of analog computer operation including set pot, hold, IC, operate, and rep-op.
   c. Summing junction.
   d. Input resistor.
   e. Feedback capacitor in an integrator.
   f. Initial condition mode of analog computer operation.
   g. Initial conditions setting of an integrator.
   h. Location in an integrator patched circuit for deriving a voltage that is the rate of change (derivative).
   i. Feedback pathway (subtracting).
   j. Patching arrangements for producing a positive multiplier output.
   k. Arrangement for feedback and input resistors for a 10 gain on a summer.
   l. Arrangement for input resistors for a 10 gain to an integrator.
   m. Limiter.
   n. Threshold action on analog computer.
   o. Scaling a quantity variable and determining the scaling factor for its rate of change. Are they the same?
   p. Top and arm of a pot.
   q. Operational amplifier.

3. Explain the following on analog computer: use of reference voltage and pot for initial conditions and forcing functions; rate of change of a storage; and amplitude scaling, and time scaling.

4. Can you write the unscaled analog computer program for the energy circuit configurations shown in Figs. 3-12 to 3-15.

5. Starting with an energy diagram that has data written for flows and storages, write equations, draw analog computer diagram and scale coefficients by two methods.

6. Write an energy diagram, differential equations, and an analog program for the following model described with the following narrative: "Water flows from a waterfall into a mountain pool overflowing from it to the next waterfall. Small fish in this pool depend on insects being washed into the pool; this washing-in is proportional to the swiftness of the inflowing waterfall. The fish gain weight from eating the 'drift insects' that if not eaten die and deteriorate by microbial decomposition. The fish lose weight as a result of their own metabolism in proportion to their own weight. If food levels of drift insects fall below a low threshold value, the fish begin to migrate out through the outflow stream in proportion to the number remaining."

7. Can you write energy diagram, differential equations, and an analog program for the following narrative: "A zone of coastal sea produces sea shells per acre of offshore zone in proportion to the fertility of plankton inflowing from off shore. In proportion to the shells grown, some are drifted onto the beach. The stock of shells on the beach is depleted by two processes. First, shells are picked up in proportion to the number of tourists present; the number of shells is proportional to the population stock of people in the nearby region. Shells are also crushed by the passage of cars driven on the beach. The number of cars on the beach depends on the population of tourists at the beach, but there is a threshold such that when the interactions between people and cars exceed a threshold, a regulatory action stops all flow of cars to the beach."

8. If a scaled variable is $Q/10,000$ people and the oscilloscope display is half scale, how many people are indicated?

9. Most introductions to analog computers first show examples from some familiar equations for behavior of physical machinery or some standard types of differential equation. Although we are emphasizing the energy circuit concepts and environment, the student should now be able to diagram and simulate the following equations, the meaning of some of which is considered later:

$$y = kx + b \quad \dot{Q} = kQ^2$$
$$\dot{Q} = -kQ \quad \dot{Q} = -kQ$$
$$\dot{Q} = kQ \quad \dot{Q} = J - kQ$$

10. Given the equation $\dot{Q} = (J - kQ - k_2Q^2)$, diagram this equation; translate to energese. Put numbers for stock and flow on the diagrams,

and state the equations and analog diagram; Q is 100 people when J is two persons immigrating per year, when density-dependent loss is one per year and the energy loss due to interactive stress is one per year.

11. Can you back-translate analog diagrams into equations and energy language?

SUGGESTED READINGS

Bennett, A. W. (1974). *Introduction to Computer Simulation*, West Publishing Company, St. Paul, MN.

Blesser, W. B. (1969). *A Systems Approach to Biomedicine*, McGraw-Hill, New York.

Blum, J. J. (1968). *Introduction to Analog Computation*, Harcourt Brace, New York. 175 pp.

Carlson, A., G. Hannauer, T. Corey, and P. J. Holsberg. (1971). *Handbook of Analog Computation*. Publication No. 00-800.0001-4. Electronic Associates, West Long Branch, NJ. 389 pp.

Gordon, G. (1969). *Systems Simulation*, Prentice-Hall, Englewood Cliffs, NJ.

James, M. L., G. M. Smith, and J. C. Wolford (1971). *Analog Computer Simulation of Engineering Systems*, Intext Educational Publishers, Scranton, PA.

Jenness, R. R. (1965). *Analog Computation and Simulation*, Allyn and Bacon, Boston.

Johnson, C. L. (1956). *Analog Computer Techniques*, McGraw-Hill, New York.

Karplus, J. *Analog Simulation*, McGraw-Hill, New York.

Korn, G. A. and J. M. Korn (1972). *Electronic Analog and Hybrid Computers*, McGraw-Hill, New York.

Ledley, R. S. (1965). *Use of Computers in Biology and Medicine*, McGraw-Hill, New York.

McLeod, J. (1968). *Simulation*, McGraw-Hill, New York.

Milsum, J. H. (1966). *Biological Control Systems Analysis*, McGraw-Hill, New York.

Patten, B. C. (1971). *Systems Analysis and Simulation in Ecology*, Vol. I, Academic Press, New York. Chapter 1.

Ricci, F. J. (1972). *Analog/Logic Computer Programming and Simulation*, Spartan Books, New York.

Tomovic, R. and M. Vukobratovic (1972). *General Sensitivity Theory*, American Elsevier, New York.

Truitt, T. D. and A. E. Rogers (1960). *Basics of Analog Computers*. John F. Rider, New York.

Valisalo, P. E. (1976). *Analog Computation*, Department of Industrial and Systems Engineering, University of Florida, Gainesville (laboratory manual for use of applied dynamics equipment). 108 pp.

APPENDIX: USEFUL ANALOG CIRCUITS

The following are useful circuits for generating forcing functions, reading results of simulation, and dia-

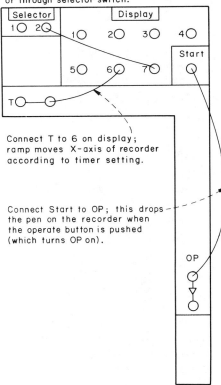

FIGURE 5A-1. Patching of Miniac control panel for slow operation with chart recorder. Fast button should be out. Turn operation on with OP and off with IC.

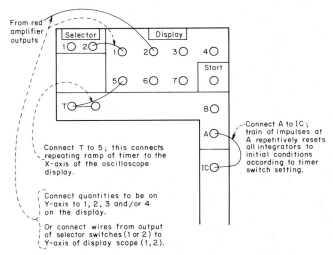

FIGURE 5A-2. Patching with Miniac control panel for fast repeat operation (rep-op) with oscilloscope. Push fast button in; turn rep-op on with PP control button and off with IC button.

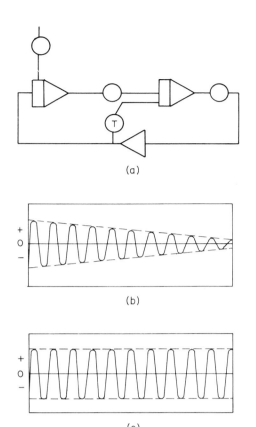

FIGURE 5A-3. Extra pot to trim sine wave: (*a*) circuit and trim pot T; (*b*) sine-wave decay due to leaking capacitors or other causes; (*c*) after trimming.

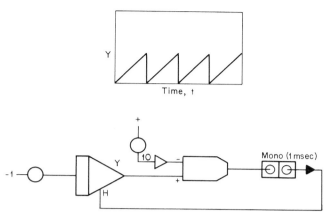

FIGURE 5A-5. Circuit for sawtooth generator. The unit labeled MONO is a monostable pulse unit. When it receives a logic pulse, it pauses with a time delay before sending the pulse forward.

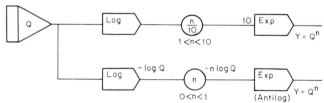

FIGURE 5A-6. Circuit for raising a quantity to a power.

FIGURE 5A-4. Use of sine-wave oscillator as forcing functions; (*a*) sine wave; (*b*) seasonal rhythms; (*c*) daily rhythms; (*d*) square wave.

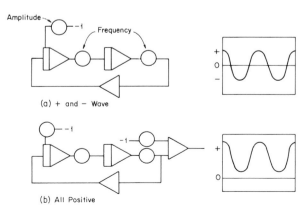

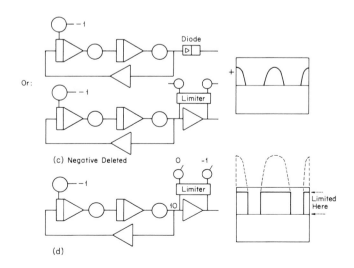

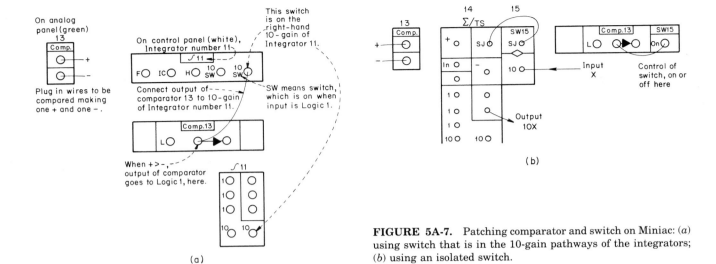

FIGURE 5A-7. Patching comparator and switch on Miniac: (a) using switch that is in the 10-gain pathways of the integrators; (b) using an isolated switch.

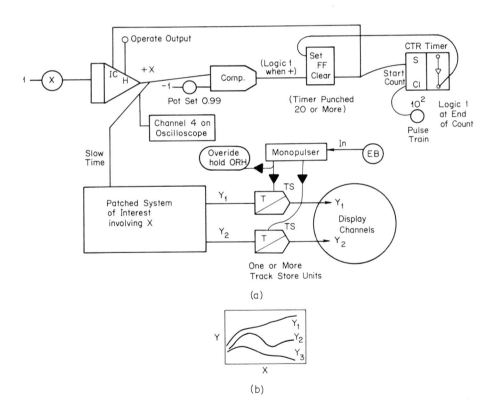

FIGURE 5A-8. Circuit for generating the steady-state results of a model run fast and rerun generating an X–Y plot in slow time. (a) Analog diagram; one integrator in slow time increases X slowly with a ramp circuit. Variables of interest (Y values) run in fast time with value at end of each rep-op stored in track store and shown on display, whereas the fast circuit resets and runs again quickly at a new position of X. The fast Y runs with time are hardly visible. (b) Final Y values as curves as a function of X.

grams of patching (see also *Miniac Manual*, Electronic Associates, 1971). Suggestions were provided by T. Bulloch, J. Browder, S. Brown, and M. Sell. Patchboard diagrams are for EAI Miniac analog computer (see Fig. 5-2).

Analog circuits are given in part because of their utility in analog simulation, but also because they show systems relationship in a graphic way.

Patching of Miniac Control Panel for Display

Figure 5A-1 shows a Miniac circuit for display with a chart recorder. This requires simulation in slow speed. Figure 5A-2 shows a circuit designed for fast simulation with the use of oscilloscope display and repeat operation (rep-op).

Sine Waves

Figure 5A-3 shows the circuit for generating a sine wave with a cross-linked pot to trim its performance if it shows changing amplitude caused by electronic leaks.

Figure 5A-4 shows ways of using sine waves to generate diurnal or seasonal driving functions or square waves.

Sawtooth Generator

Figure 5A-5 shows a convenient circuit for generating a sawtooth, which is a series of repeating ramps. A ramp raises the voltage on an integrator until it exceeds a threshold to a comparator, which sends a logic pulse to a monostable pulser that delays momentarily and sends a pulse back to reset the integrator to its initial condition again (zero), after which it can ramp again. Placement of a 10 gain on the comparator threshold reference adds sensitivity for setting the threshold. One use for the sawtooth is operation of the variable-function generator.

Logs, Exponentials, and Raising to a Power

Figure 5A-6 shows a circuit for raising a flow to any power. Logarithmic units take logs of the input voltages, which is then multiplied by a pot setting that represents the exponent. Then the result goes to an exponential unit that reverses the log. In other words, it takes the antilog producing a voltage that is in proportion to the original voltage raised to a power.

Patching Comparator

Figure 5A-7 shows details for patching comparator and switch in the Miniac for systems such as that shown in Fig. 6-4.

Scanning the Effect of One Variable on the Outcome of a Model's Simulation to Steady State

Shown in Fig. 5A-8 is a wiring diagram for the Miniac so that it will run a model on fast time, reset, run it again with one variable now increased, reset, repeat, and so on. The result is a slowly generating graph of the steady-state outcome of the model as a function of one of the input variables (forcing function or initial condition) that is increased gradually.

CHAPTER SIX

Logic Systems and Other Languages

As a result of the fertile work of the world's leading intellects, supported by this century's high-energy times, many languages have developed for the invention and examination of systems. These symbols and mathematics are widely used in scientific literature. Comprehension of these formulations provides access to the insights and a view of systems science. Many concepts developed in one language or type of system are transferable. In this chapter, starting with logic, various systems languages are introduced and related by intertranslation with energy systems diagrams (see the suggested reading list for this chapter for other efforts to overview systems languages).

LOGIC SYSTEMS

Logic systems are networks of on and off processes. Pathways are either on or off. They are chains of switches. The hardware of digital computers are logic systems. Logic operations occur in varying degree in all kinds of system. They tend to develop and be in use when energies are high and when high-quality complex actions are adaptive. In general, they may require convergence of more energy for construction and maintenance than do continuous-varying mechanisms, and they may conceivably be useful only where they can contribute to effective amplifier control of high-energy flows. Examples of digital (logic) actions include the switching of neuron functions, the structure of living populations in discrete units, the stimulus and responses of behavior of higher organisms, and threshold actions in geologic cycles. Patterns of logic actions are sequential in time, although often very fast. The steps of sequential action are called *programs*. Whereas logic systems are important accepted tools in the study of systems as in digital computer simulation, recognition of digital properties within the environmental and other systems under study as part of analysis and understanding is little developed. *Cybernetics,* the science of control, includes the study of logic and switching systems (Ashby, 1963; Klir and Valach, 1965). Many opportunities may exist for low-energy control of environment through understanding cybernetics of environmental systems.

Logic systems are expressed in various ways in different fields according to different traditions, and many sets of symbols are readily substituted. In the energy language our convention is to include logic networks in traditional symbols or in words within the concave box, which designates switching actions. As with any structural storage and process, heat sinks are required from the box to indicate the depreciation, and/or maintenance, and process energy (see example in Fig. 6-1). Within the box the details of the logic program, if known and if detail is desired, are diagrammed using the same symbols that are customary in diagramming the logic functions of the analog computer. These are given in Fig. 6-2. The concave box, by the way, is similar to a regular Japanese language character for gate. More complete listing of logic symbols is given by Greenstein (1978) and Lynch (1980).

Continuous Functions to Switching Functions; Comparator and Switch

Water in a stream may be in various levels, with the storage function varying continuously. However, if the water exceeds the threshold of the stream's levee, water overflows. The action is an on–off action, a digital process. The symbols used for this and the devices on analog computers for this function are provided by the comparator (Fig. 6-1). The unit compares two input sensors, and when one designated plus is larger than the one designated minus, the output of the comparator switches "on" and is called

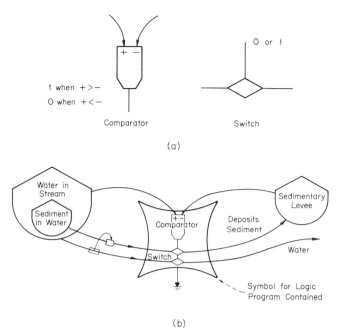

FIGURE 6-1. Comparator and switch transforms a continuously varying phenomenon to a digital action: (a) symbols; (b) example of a logic system flooding over a levee.

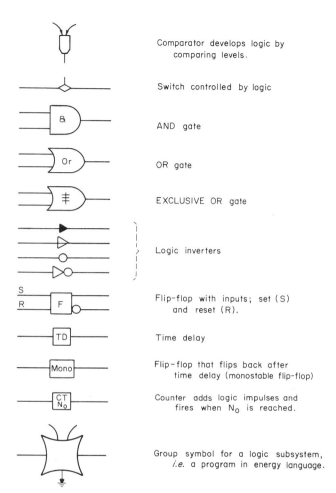

FIGURE 6-2. Common logic symbols.

logic 1. When the level of the plus is less than the minus, the output is off and is called *logic 0*; it is off. The comparator connects the continuous functions to the logic functions.

The logic output can be connected to a switch and turn on a flow within the network of continuously varying functions. The switch on a regular pathway connects logic (on–off) action to the continuous functions. These symbols can be diagrammed either as the wiring diagram for the analog computer or within the logic box to indicate the same actions in any system as part of the generalized energy language.

Figure 6-1 gives the example of the overflowing stream that exceeds a threshold that releases an overflow and also causes sediments to be deposited on the edge, making the levee higher and changing the logic threshold. Other examples are photoperiodic threshold for starting breeding in ducks, thresholds of energy storage for starting war, and thresholds of votes for change of political parties. Another example, given in Fig. 6-3, is a threshold model for behavioral response to an external impulse from Lorenz (1950). Diagramming helps make verbal concepts precise and is a step to help simulation. Hinde (1960) reviews motivation models involving buildup of "energy."

Example of Use of a Comparator and Switch to Patch Hybrid System on an Analog Computer

In Figs. 5-11–5-13 an example of analog patching was given of a storage with simple inflow and pumped outflow. To show operation of the com-

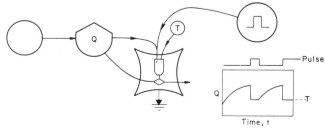

FIGURE 6-3. Energy diagram of Lorenz' (1950) model of motivation discharge that depends on both internal storage and level of external stimulus.

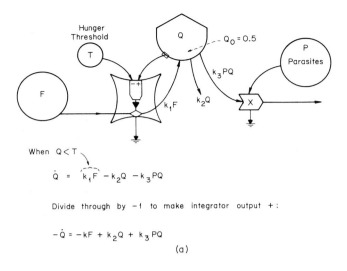

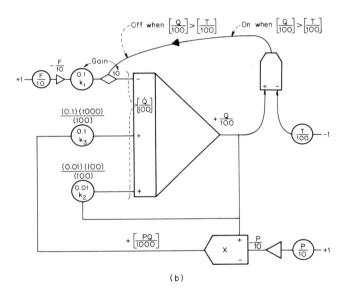

FIGURE 6-4. Energy diagram of a system with a hunger-threshold switch on its energy source, a single storage unit, and an outflow under multiplicative pumping control of an external consumer forcing function: (*a*) energy diagram and differential equations; (*b*) analog computer diagram with quotient expression drawn on the diagram for ease in scaling; (*c*) graph of storage with time generated by setup in *b*.

parator and switch with the analog computer, this same example is modified in Fig. 6-4 to include a control switch on the energy inflow dependent on the level of storage operating through a comparator controlling the switch. An example of this type of action might be hunger of a simple animal. As the food storage is reduced, feeding action is renewed. A system with mixture of continuous-varying functions and switching functions is sometimes called a *hybrid system*.

Figure 6-4 shows the energy diagram, the translation to differential equations, and the patching of the analog computer diagram. Output voltages of the integrator (storage) and a pot representing an externally set threshold are connected to the comparator so that one is plus voltage and the other, a negative voltage. When the plus is larger, the comparator puts out logic 1 and lights the panel. The logic 1 is patched to the input control of the switch.

In the Miniac some switches are separate, and some are built into the 10-gain inputs to integrators. For those patching the Miniac, see appendix to Chapter 5.

Logic Gates and Truth Tables

In Fig. 6-5 are given symbols for manipulating voltages according to some of the main logic operations that are now taught in general schools as part of the new math. There are AND gates, OR gates, EXCLUSIVE OR, logic inversion (NOT), and many others made by combination. On the left is given the symbol for the wiring arrangement. On the right is a table, called a *truth table*, that shows the combinations of outputs possible for various combinations of inputs. There are various combinations of logic gates that may have identical truth tables. Any kind of on–off process may involve combinations and chains of logic circuits to perform such operations as counting, switching, and programming according to the interplay of various thresholds and conditions and so on.

The logic gates are adopted for analog computer programming and also for the energy circuit diagramming; they are placed into the logic program box with heat sink for the energy costs of its deterioration and maintenance.

The main logic operations shown in Fig. 6-5 may be described with the following descriptions. An AND gate requires all its inputs to be on (logic 1) to deliver an output (logic 1). If any one of its active inputs is off (logic 0), its output is logic 0. Two com-

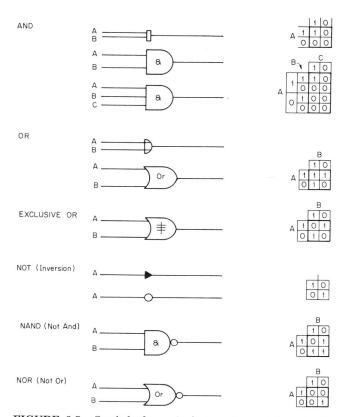

FIGURE 6-5. Symbols for main logic gates and their truth tables. Equations use notation of logic algebra.

gates. The EXCLUSIVE OR gate produces a logic 1 if one and only one of its inputs is on; otherwise, output is logic zero. Indicated are two symbols for logical inversion (the small circle in the line and the small solid triangle). Logic inversion turns logic on to logic off or logic off to logic on. This unit is sometimes called a NOT gate. When the output of an AND gate is negated by the logic inversion, it is called a NAND gate, shortened from the phrase NOT AND. Similarly, the output of an OR gate may be negated to form a NOR gate, which is an acronym for NOT OR.

From these basic gates many kinds of digital logic mechanisms are readily diagrammed and patched. Many systems of nature have various patterns of performance that are digital in nature, meaning that they have on and off actions. Much of the behavior of the higher animals and of human responses involve thresholds, if statements, and trains of successive switching actions.

De Morgan's Principle for Interconverting AND and OR Gates

Some analog computers have AND gates only, and others have OR gates only. Others have NAND gates only, and so on. One may connect gates together so as to develop other gates and vice versa. The fact that two units are identical is verified by composing truth tables. A common conversion is from AND gates to OR gates and vice versa by using logic inversion of inputs and outputs. The principle that there is an equivalence between these combinations, called *de Morgan's principle,* states that OR gates are identical in overall response to AND gates of which the inputs and outputs are negated by logic inversion and vice versa. These relationships are given in Fig. 6-6 with appropriate truth tables. Students should verify that the table is the same for both sets of units shown.

mon symbols for AND gates are given. Because all the inputs must be on, the AND gate is sometimes called *logic multiplication* since a multiplier output also requires that its inputs not be zero. On the machines there are AND gates with two inputs and others with three or more. Some hardware units (e.g., Miniac) have outputs on until the wires are patched, after which there must be logic on voltages to cause the unit to have an on output. The OR gates produce a logic 1 output (on) if any one or more of its inputs is on. Shown are two ways of writing OR

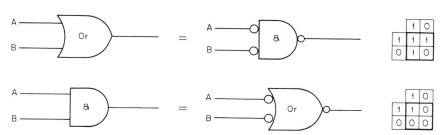

FIGURE 6-6. DeMorgan's identities useful for making AND and OR gates where analog equipment has only one kind of gate; $A \cup B = (A' \cap B')'$ and $A \cap B = (A' \cup B')'$.

Flip-Flop

Occurring in many systems is a unit that is switched by a logic input and holds its position until it is reset by a separate logic input. A flip-flop is a unit of logic memory since it holds itself in one of its two positions until it is changed. The unit holds after its first set input even when the original setting input pathway changes. The reset impulse on another pathway is required for resetting. Figure 6-7a gives the symbol for the flip-flop as it is found on an analog computer or as it is used to refer to that function even when it is done in the ecosystem as a result of switching actions or organisms.

The flip-flop can be made up of actions of other logic units as shown in Fig. 6-7. A pair of NEGATED OR gates is used, and in the second example a pair of NAND gates is used.

Figure 6-8c shows an example of the use of the flip-flop in controlling a switch. The flip-flop is operated by two comparators, sensing the levels of a storage. When the tank level reaches a high

When S is Logic 1, the flip-flop is set, output A is on (1), and B is off (0).

When R is on, the flip-flop is reset (cleared), output A is off (0), and output B is on (1).

(a)

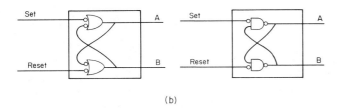

(b)

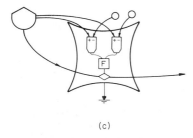

(c)

FIGURE 6-7. Flip-flop: (a) symbol and operation; (b) subsystems that are flip-flops; (c) flip-flop operated by two comparators, one with high threshold and one with low threshold.

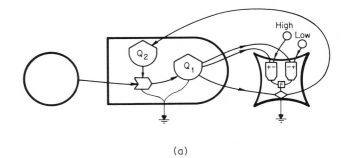

(a)

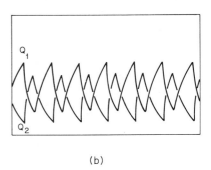

(b)

FIGURE 6-8. Model of pulsing of high-quality energy feedback actions; an example is grassland and fire. (a) model; (b) simulation (F is a flip-flop unit).

threshold, the flip-flop sets and opens the switch, draining the tank. It stays in the open switch position until the tank level reaches the low threshold. Then the low sensing comparator switches and resets the flip-flop, cutting off the outflow of the tank.

There are more complex kinds of flip-flop and combinations that can be learned from the suggested readings (Hughes, 1968). Flip-flops on an analog computer are operated by a train of pulses turned on by "run." They will not operate unless the run button is on. The flip-flops on the analog computer also have an "enable" control feature that, if patched, allows the flip-flop to operate only if receiving logic 1.

Pulsing Regenerative System

Figure 6-8 shows a type of system very commonly observed in many situations such as those with recurring forest fires, storms, earthquakes, insect outbreaks, and so on. It is a hybrid system with continuous production and storage followed by discontinuous recycle and restart controlled by the level of storage. In later discussions of energy quality and survival of systems, this class of system is suggested to be one of the most general and one that maximizes power under some circumstances. It illustrates the role of flip-flop mechanism in developing a digital action.

Sets and Venn Diagrams

Another systems language is that of sets. A set is a collection of items. One set may include all or part of another set. A set can be made up of those parts of two other sets that have a common property. Relationships of sets are sometimes given in Venn diagrams as in Fig. 6-9 with some of the main types shown. The shaded area of the diagram indicates the logic relationship of A, B, or surrounding the area (not A or B).

A set made up of shared area is conjunction, a set made up of either of two sets is disjunction, a set made up of anything that is not in another set is a complement, and exclusive or is a set made up of either, but not both. There is a subtraction set made up of one set minus the overlap of a second.

These relationships have logical equivalents in the logic operations and symbols, which we have given in pathway diagrams in Figs. 6-2 and 6-5. In Fig. 6-9 the Venn diagrams are given together with their logical equivalents.

The gate can be regarded as one spot in the Venn diagram, and it has a logic 1 output when it is in the shaded area of the Venn diagram. Thus there is AND, OR, NOT (inversion), EXCLUSIVE OR, NAND, and NOR.

Truth tables were already given for the logic gates (Fig. 6-5). These same truth tables apply to the equivalent Venn diagram. Can you derive these directly?

Logic Notation

The logic ideas are represented with a logic notation, which is a form of mathematics describing the relationships given in the sets and illustrated by the Venn diagrams. These are given opposite the diagrams in Fig. 6-9.

Boolean Algebra

Boolean algebra is another language for describing logic systems. The same actions represented by the logic gates, Venn diagrams, and truth tables are given in a special algebraic notation. These are

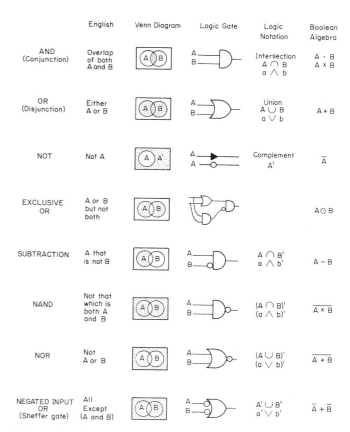

FIGURE 6-9. Venn diagrams, Boolean algebra, and logical equivalent gates. Shaded zones in the diagram correspond to output of gate and the notations.

given along with Venn diagrams and gates in Figs. 6-9. Logical multiplication is the same as AND gate action. Logical addition is the same as OR gate action. These notations can be very confusing for those used to other languages where the same symbols are used for different meanings. For example, $A + B$ in chemistry means that both A and B are required; in Boolean algebra this is written $A \cdot B$.

Notations for Verbs

A number of symbols are commonly used in equations, which have the role of verbs in regular written and spoken language. These are given in Fig. 6-10.

Logic Timers

Some units control the timing of logic pulses. The "Mono" unit refers to a monostable unit that sends out a square wave pulse with a short time delay

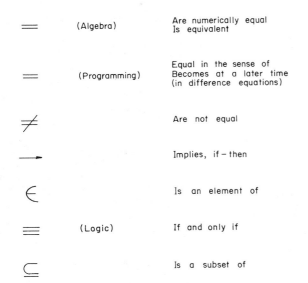

FIGURE 6-10. Symbols commonly used in equations for logic systems.

after it is stimulated by a pulse. Flip-flops can be arranged in chains so that each turns on the next in sequence so as to constitute counting circuits. Some units are called *down counters* because they provide whatever time delay is set on the counter. These digital units are available on analog computers.

Walter's Turtle; Learning Reflex

Illustrating the fairly complex patterns that can result from relatively few logic units is the mechanical learning turtle developed by Walter (1951). The machine runs along the floor scanning, going away from strong light. If weak light and sound occur repeatedly (20 times), behavior is changed for 5 minutes so that the machine approaches weak light. Some models recharged their batteries. One model is given in Fig. 6-11 [see Zemanek et al. (1961) for other more complex models]. The model involves AND gates, an OR gate, three monostable flip-flops, and a counter.

Ecosystem Nerves

An interesting but little studied question in ecosystems is: How much communication structure is there per unit area and how does it vary with succession, latitude, and environmental conditions? Studies and tabulations of brain-body weight ratios are available to make some interesting calculations. How are logic functions related to such measures?

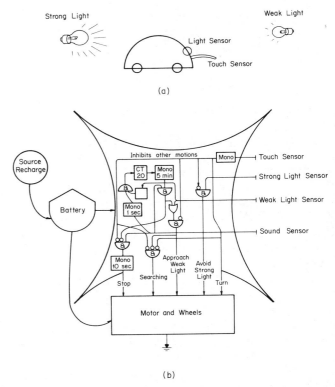

FIGURE 6-11. Walter's logic "turtle" has conditioned reflex (Walter, 1951) (a); (b) circuit modified from Zemanek et al. (1961).

Neural Nets and Symbols

Systems of logic actions and controls have been expressed in a language of neural nets using symbols used by Pitts and McCulloch (1943). Although the language and symbols were stimulated by efforts to understand nervous systems (Mueller et al., 1962), the system is more general, and some properties learned in these may apply to other systems. If these are used with the energy language, our convention would place them within the concave-box group symbol with heat sinks. As a system of on and off actions, Boolean algebra is the associated mathematics. Here the energy source is implied as more than necessary to operate the pulsing system, and the pulses constitute high-quality energy flows.

If the units shown in Fig. 6-12a have a threshold of 2, both A and B must fire together to cause the unit to fire to the right. In Fig. 6-12b the units cannot fire while A is firing. In Fig. 6-12c the second unit transmits a pulse only after two pulses in the frequency corresponding to the time delay of the first unit. In Fig. 6-12d there is a closed loop that can fire repeatedly, producing an alternation and an output that is twice the delay period of each unit.

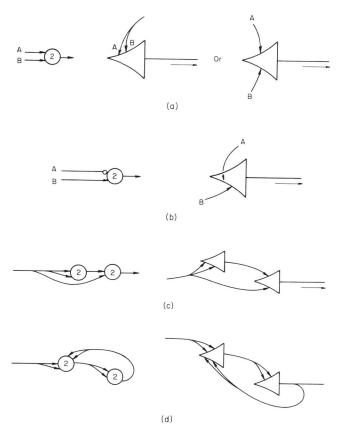

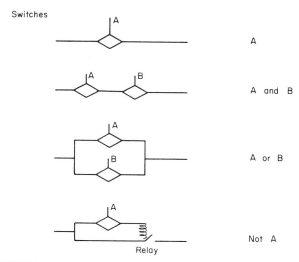

FIGURE 6-13. Switch sequences and logic equivalents of output.

FIGURE 6-12. Symbols and examples of neural nets using system of McCullouch and Pitts (1943) after Rashevsky (1938, 1960). These neurons require two input pulses to fire. An alternative notation uses small circles as logic units. The number is the number of synchronous logic pulses required to fire its output. (a) Neuron firing with a unit lag time after A and B fire together; (b) neuron inhibited with A fires; (c) system transmitting from left to right only if there are repeating pulses at the same interval as the neuron delay interval; (d) a closed loop of alternative pulses.

Switching Circuits

On–off switches may be arranged in series, in parallel, or in other combinations that have logical equivalents to AND, OR, NOT, and other operations. When an energy diagram has been drawn of a complex network such as an ecological system, it may be examined as if the units were switches and their logic equivalents estimated. The presence or absence of the action of a species, for example, constitutes a switching action. Some switching circuits are shown in Fig. 6-13.

Many aspects of ecosystems that have sharply accelerating and decelerating activities can be viewed as digital networks; for instance, consumers who eat alternative foods are OR functions, consumers who require more than one type of input are AND functions, units that eat to a certain threshold before reproducing are counters, populations that start each year with seasonal cues to migrate and turn off are like flip-flops, reproduction and inheritance of numbers is a digital process, and so on.

Turing Machine

A concept for visualizing self-reproduction machines called *automata* is the Turing machine (Turing, 1936; Singh, 1966; Berlinski, 1976). A tape has symbols arranged in a single row along its length. The machine has a reading head and responds to each symbol as it passes under its view according to a set response. The responses may be to move the tape or duplicate the symbol on a printing device on another tape or change a symbol on the same tape. The manipulation of the machine has been important in digital thinking in visualizing the possibility and mechanisms of increasing information. Digital programs were written so that one system initiates another. The Turing models affected thinking about gene replication and control actions.

PATHWAY LANGUAGES

Causal Word Chains

Causal word chains are a qualitative language. Causal relationships are denoted by strings of words and arrows indicated with either a plus or a minus to indicate respectively an increase or decrease. As a

80 ENERGY, SYSTEMS, AND SIMULATION

Quantitative Representation of Pathways

The properties of pathways may be represented in different ways with different concepts involved. Data regarding flows of matter, energy, or money may be written on diagrams as in Fig. 6-16a. As given in Fig. 6-16b, these can be represented also in matrices and in one of several available network diagramming procedures such as bond graphs and signal flow charts. Often a system is quantitatively characterized by the properties of the pathways, such as conductivities, transfer coefficients, and time constants, to mention some already considered in Chapter 3. Others include probability of state transition and statistical path coefficients.

Path Coefficients

If relations of variables can be described by linear equations, these relations can be diagrammed with a network of arrows with coefficients written on pathways. In Fig. 6-17a, for example, the network of arrows shows three quantities, each related to another according to a coefficient. The network of arrows and path coefficients constitutes another systems language.

The plus or minus pathway diagrams can be regarded as pathways with path coefficients unity (+1

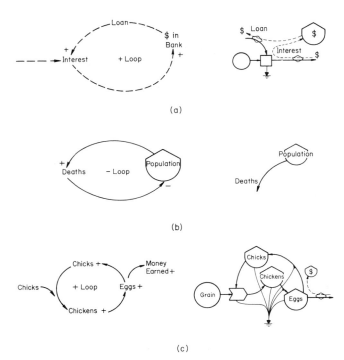

FIGURE 6-14. Examples of causal chains of words and their equivalent in energy language: (a) savings by compound interest; (b) deaths reduce population; (c) grain stimulates growth of chicken and eggs, bringing in money.

natural extension of verbal thinking, it is recommended by some as the first step in organizing thoughts in a model [see example diagrams in Fig. 6-14; see Forrester (1961) and Meadows et al. (1972) for other examples]. Compare these diagrams with energy language equivalents. Combining concepts, Fig. 6-15 uses plus and minus signs on an energy diagram to show the effects of increasing phosphorus.

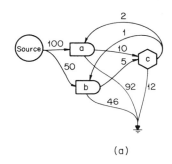

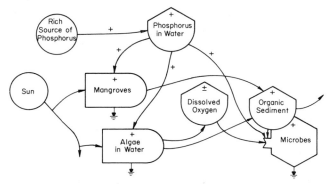

FIGURE 6-15. Use of plus and minus signs with an energy diagram to show increases or decreases to be expected with increase of an impact of phosphorus.

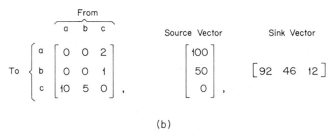

FIGURE 6-16. Example of expression of data in a network as a matrix and vectors: (a) energy diagram with pathway data; (b) matrix and vectors; column vector for source and row vector for sink.

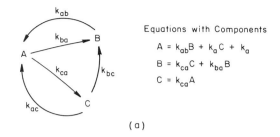

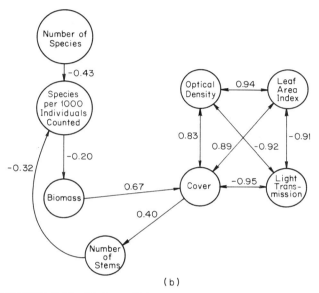

FIGURE 6-17. Path coefficients: (*a*) example of deterministic coefficients and the equivalent meaning in equation form; (*b*) example of use of correlation coefficients to relate properties in an ecosystem (Ewel, 1971).

or -1). Pathways with numerical values are sometimes called *weighted digraphs*. Roberts (1976a), for example, describes structure of energy systems with weighted digraphs.

Statistical Path Coefficients

When relations between variables are determined from observed data with statistical inference, multiple regression methods, regression coefficients, partial regression coefficients, or correlation coefficients are determined, relating each variable to others. The statistical procedure identifies components of total variation, which may be attributed to each of other correlated variables so that diagrams such as that in Fig. 6-17*b* result.

Correlation does not necessarily mean that there are direct mechanisms relating one variable to its correlated member, since correlation can result from both being causally related to a third variable. However, by showing the field of correlated connections, pathway diagrams with pathway coefficients suggest where in the network large effects may be causal (Bowdon, 1965; Blalock, 1971).

HYDROLOGY AND ENERGY LANGUAGE

Hydrology uses models of many kinds extensively, usually expressing models in some kind of compartmental diagrams to show the flows and storages of water with main details given as differential equations or simulated with difference equations. The water channel is a pathway of material and energy flow in which the driving force is pressure X, which is proportional to the weight of water. Energy language is useful for the compartmental diagrams. The linear force flux law of a linear resistive pathway is called *Darcy's law* in hydrology. As given in Fig. 6-18, the energy storage coefficient C represents the shape of the tank or the porous fraction where water is stored in soils. Flow is proportional to the hydrostatic head Z, which is related to the quantity of water storage Q according the storage capacitance C and the area A. Because storages are against gravity with height as the measure of potential, the relative height position of zero level of the tanks

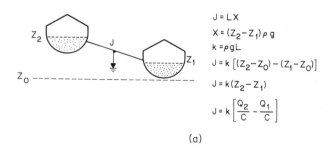

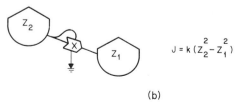

FIGURE 6-18. Hydrological systems: (*a*) Darcy's law with flow proportional to difference in water levels; (*b*) channel flow where conductivity increases with difference in water levels. Symbols: Z, height: ρ, density; g, acceleration due to gravity; Q, volume of water; A, area; X, pressure; J, flux; k, hydraulic conductivity.

determines whether they flow one way or another. Pressure X is the product of height, density, and gravity ($Z\rho g$).

Where water builds its own pathway of its own mass as with water flowing in channels, the pathway may be quadratic or have other nonlinear relationships. At steady state, the friction that develops is dependent on the eddies that generate, since these may be visioned as ball bearings on which water moves on itself at greater speeds per unit pressure. The higher the water level in the channel, the more conductivity it has for two reasons: (1) the cross section of the channel with water is higher, which is equivalent to adding more pathway; and (2) the deeper channel develops a fluid structure that increases its conductivity using some of the energy of the source in temporary angular-momentum storages.

Water carries other substances that may be represented as tanks within those of water and moved in proportion to the water flow.

Water models often form a main integrating subsystem in landscape models. A typical example is given in Chapter 24.

MATRIX ALGEBRA

Properties of networks can be written in tabular form called a *matrix* [for examples, see O'Neill (1971)]. The various pathways can be represented in an organized way as shown in Figs. 6-16 and 6-19, where the matrix is a table of properties of the pathways between inflow sources across the top and outflow sources across the side. The property of the pathway may be presence or absence of pathway, flow data on materials, energy or money over the pathways, conductivities, resistances, transfer coefficients, probabilities, and so on. There are many kinds of matrix, and they are used in many ways in systems ecology research (Halfon, 1979).

In different usages the donor and receiver are sometimes the row and column and sometimes vice versa. The expression of network properties in matrix form often allows useful operations with matrix algebra, which do duplicate operations simultaneously, and, because of standard computer programs, can be called forth to do much computation with only a few instructions (Bellman, 1960; Hohn, 1965; Searle, 1966; Chaston, 1971).

A vector is a row or a column of numbers with the same pathways involved as in the matrix. The word "vector" is appropriate since a vector can be drawn

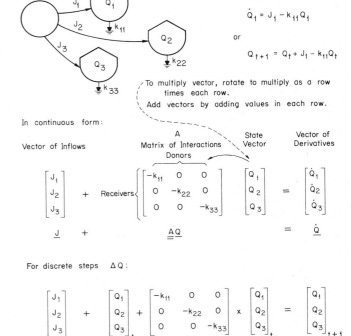

FIGURE 6-19. Expressing a system of linear state changes in vector–matrix form.

for any set of numbers that are regarded as components along reference coordinates (see Fig. 13-6). Vectors that are pertinent to the same matrix can be added when it is appropriate that each item be added. Where it is appropriate that numbers in a matrix are multiplied by other numbers, a vector can be multiplied by the whole matrix through matrix multiplication. This is done by rotating the column vector to a row vector position and multiplying by each row of the matrix. Then the sum of the products of each row is added across to obtain a new column vector that is the product.

State Transition in Matrix Form

In Figure 6-19 is given the state change for a network of storage tanks. The tanks are a column vector. The flows to and from the tanks are given as transfer coefficients in a matrix. The new state consists of the external inflows (another column vector) plus the initial conditions (states) times the transfer coefficients. In Fig. 6-19 this is done first as differential equations in matrix form, and then below the

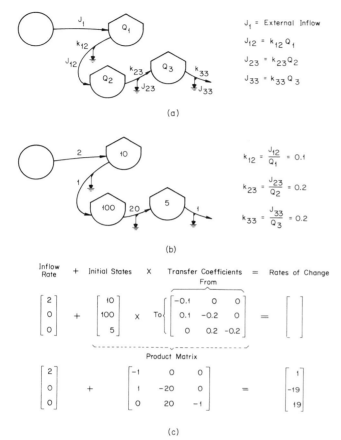

FIGURE 6-20. Numerical example of a matrix representation of a network: (a) energy design of a flow of materials; (b) data on flows and storages at one time; (c) matrix representation of flows that are also state transitions.

same operation is done again in discrete form as difference equations in matrix form. The energy language symbols for pathways with discrete step transfers were given in Fig. 3-10. In Fig. 6-20 a numerical example is given.

Matrix Notations and Systems

A matrix can represent one property of the parts of a system in an organized overall summary. However, in a systems network there are many important numbers such as storages, forces, flows, backflows, return flows, outflows, inflows, various ratios of flows, coefficients, and so on. Each of these may be represented with a separate matrix for the network. No one describes the system alone.

The requirements of various operations of matrix mathematics put severe limitations on using matrices for describing systems. Typical subscript notation defined to help the mathematical operations designates the first subscript for the row and the second subscript for the column of the number in the matrix. For example, in a 3 × 3 matrix with nine numbers, k_{23} is a number in the second row and third column.

The notations that help visualize a network may be different from the notations that accompany matrix algebra. In the simple example in Fig. 6-20 transfer coefficients are coded according to the storage nodes to which the path is connected, but such coding does not fit the coding by row and column in the matrix, one negative (outflow from Q_2) and one positive (inflow to Q_3). If coded for row and column according to matrix notation, these become k_{22} and k_{23}. Concepts are very dependent on the language in which they are visualized. It is sometimes difficult to connect thinking in matrix concepts to thinking in network concepts. The energy diagramming may be helpful in clarifying and relating these languages.

More kinds of matrix are given in Chapter 13 on webs and in Chapter 23 on economics. A matrix with 1 or 0 for presence or absence of a pathway is an incidence matrix. Casti (1979) includes food chain examples in discussing incidence matrix. Probability matrix is given next.

Pathway Probabilities

When there are several pathways from a module, flow may be represented as probabilities for a transition step from one module to the next or one state to the next state in time. The ratio of average flow on each pathway to the total of all flows from that unit is the probability of flow along that pathway. It is a fraction of the total. Probabilities can be used in place of other kinds of pathway coefficients, where flow is entirely controlled from the source. Figure 6-21a shows an example of outflows from a unit with probabilities calculated for the outflows.

If there are branches, the outflows from the branching point may be expressed as probabilities also. The probability of flow down one arm of the branch becomes the product of the probabilities. The overall combined probability is the fraction of the original outflow to reach that branch.

Often the network flows are described by a matrix of transition-state probabilities as illustrated in Fig. 6-21c. See Kemeny et al. (1966) and Malkevitch and Meyer (1974) for introduction to probability networks.

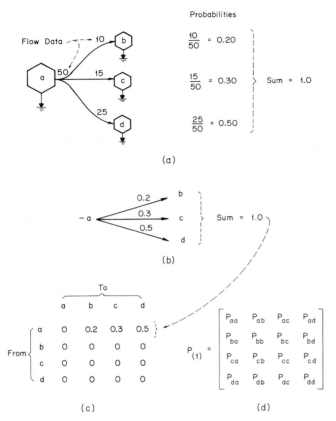

FIGURE 6-21. Pathway probabilities: (a) data on flows are divided by total flows to obtain probabilities; (b) typical form used for a transition diagram with probabilities written; (c) same system as in b given in form of a transition probability matrix; (d) general statement of transition probability matrix.

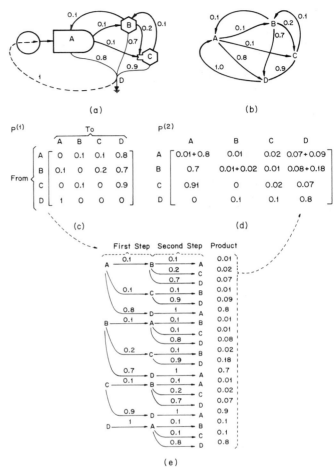

FIGURE 6-22. Markov chain: (a) energy diagram with transition probabilities for one step written on pathways; (b) traditional diagram of transition probabilities; (d) transition probability matrix = one step; (d) two-step transition probability matrix (two-step Markov chain); (e) Markov chains. Sum of rows in both matrices is unity.

Markov Chains

The probabilities shown in Fig. 6-21 and Fig. 6-22 indicate the fractional flow in one transition state. After two transition steps, flow will have moved through two segments of the system, and the probability of arrival at the end of the second transition will be the product of the two probabilities in the sequence. The total flow arriving at each point is the sum of the pathways leading to that point that have two-step connections upstream. In Fig. 6-22 transition probabilities are first written on an energy diagram on a traditional transition state arrow diagram and on a transition probability matrix [one step: $p^{(1)}$]. Then the probabilities of all the two-step segments are calculated (as products of two probabilities). Then those leading to the same point are summed. Finally, the total probabilities of the two-step transition is written in matrix form $[p^{(2)}]$ in Fig. 6-22d. A chain of steps, each according to transition probability, is called a *Markov chain*.

Provided that all pathways are part of a closed loop, the sum of all the elements in each row in the one-step transition matrix and in the two-step transition matrix is equal to unity. For an ecological system where there may be inflows and outflows, an artificial closing of the outflow with inflow by means of a single pathway can be used, representing the source sink of the environment (see dashed line in Fig. 6-22a).

Harbaugh and Bonham-Carter (1970) model spatial distribution of dominant bottom organisms in successive time intervals by the sum of influence coefficients of surrounding members affecting probabilities of survival as a Markov chain.

Diffusion spatially can be simulated by the random selection of movements which is a *Monte Carlo* process that results in a spread of diffusion away from a concentration (Hagerstrand, 1967).

Game Theory

Strategies of action and response may be expressed in matrices that describe the actions for alternate situations. The matrix describes a program of response. For example, Patten (1961) showed the successive tendencies of bird species to increase i after a previous decrease d or vice versa. Observed frequencies were recorded in a two × two matrix:

	d	i
d	.17	.67
i	.83	.33

The most frequent pattern was increase following decrease, not competitive exclusion. A systems relationship was expressed in the language of game theory.

SYMBOLIC NETWORK LANGUAGES

Signal Flow Diagrams

Another of the symbolic languages for representing systems is signal flow language (Mason, 1953), which diagrams equations along lines somewhat in the manner of the block diagramming already described (Fig. 3-3), except that only simple arrows are used. Signal flow arrows converge on small circles where the terms carried by the arrows add. These points of arrow convergence are called *nodes*. Each arrow does some operation on the quantity at its base, delivering the modified quantity to the output node. The name "signal flow" originates from the analogy of sending messages along wires from originating stations (nodes) to downstream stations that modify or repeat the message transmitting it to the next station (e.g., see Fig. 6-23a). Introductions are supplied by Robichaud et al. (1962).

The signal flow language has theorems and procedures for converting complex equations and networks into others that are equivalent but simplified.

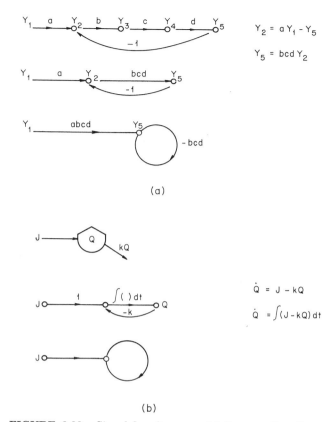

FIGURE 6-23. Signal flow diagrams: (a) diagrams for a linear equation that involves a loop (simplification can be drawn in two stages); (b) signal flow diagram for the integral equation for a storage. Compare with Fig. 3-3; note simplification.

A storage unit is given in signal flow language in Fig. 6-23b. As shown, loops may be simplified; the property of negative feedback is indicated by the loop form that results.

In Fig. 6-24 an elementary economics model often used to discuss macroeconomic relations is given and translated into signal flow form. Like any language, one learns to recognize functional characteristics from the form.

Forrester Symbols

Starting with his book on industrial dynamics in 1961, Jay Forrester used a set of symbols to show diagrammatically computational relationships of his digital programs. These symbols have been widely used in many fields, including systems ecology to diagram programs, whether expressed in differential equations, difference equations, or other

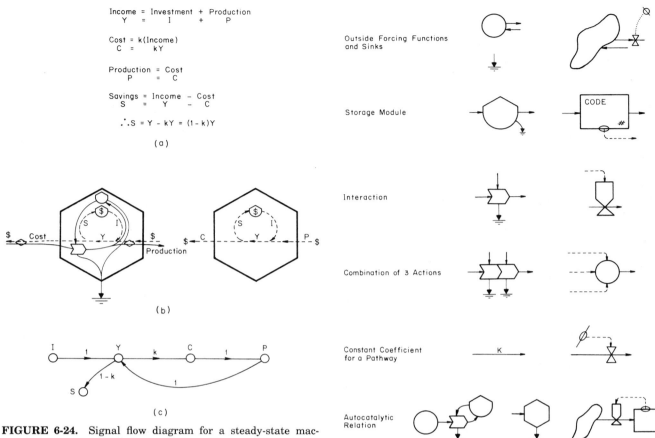

FIGURE 6-24. Signal flow diagram for a steady-state macroeconomics model: (*a*) typical equations; (*b*) money flows shown as countercurrent to an energy diagram; (*c*) same model as a signal flow diagram (Hare, 1967).

FIGURE 6-25. Symbols in energy language with computational meanings similar to those in Forrester's dynamics symbols.

form. These symbols are given in Fig. 6-25*a* with the kinetic equivalent from the energy language.

Forrester's diagrams have material quantities that are moved in and out of storages according to various control functions that are controlled by various informational coefficients. Lines of different colors are used for each material flow such as population, electricity, copper, and fish. Informational coefficients are dotted lines that control the flows with control values. Many of Forrester's models are fairly complex; thus not all the symbols can be expressed on a page. Often the functions involve various multipliers, quotients, thresholds, tabular data, and so on, the details of which are not shown in the diagrams. An aggregated circle receives a number of converging informational coefficient pathways and indicates that there are subroutines. Ordinarily, one must resort to the separate statements of the program to see exactly what is being done.

The language has a rectangle as its tank, whereas rectangular boxes have traditionally been held for black-and-white boxes. No recognition is given to pathways of materials and information as being embodied energy. Fuels and lower quality energies are represented as separate flows such as any commodity without any effort to account for energy in the sense of the first law. *There are no heat sinks*. The external sources are drawn as infinite pools whose limitations are indicated by pathway coefficient multipliers rather than as contributing inherent characteristics. Even with the many differences, one can translate an energy diagram to Forrester's symbols and vice versa just on the basis of the computations. For example, see the translation of a model in Fig. 6-26.

The use of a symbol language is a separate choice from selection of a programming procedure. Many draw in energy circuit language while simulating

Logic Systems and Other Languages

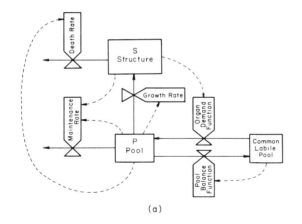

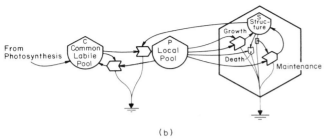

FIGURE 6-26. Example of system represented in Forrester's symbols: (a) unit growth of a plant organ with the use of sugars from photosynthesis (Cunningham and Reynolds, 1978); (b) equivalent energy diagram.

with Forrester's economical Dynamo programs. Some diagram with Forrester's symbols while simulating on analogs, and so on.

Linear Graph Theory

Network diagrams are the subject of a branch of mathematics called *graph theory,* where the word "graph" refers to network diagrams and thus uses the word in a different way from the more common meaning of a graph as a plot of one variable as a function of another. Roberts (1976b) considers many applications.

Points of junction of lines are called *vertices* (or *nodes*), and connecting lines are called *arcs* (or *edges*). A subgraph that contains all the vertices without closing a loop is called a *tree,* and its arcs are *branches.* Various matrices and matrix operations are part of the utility of the graph theory. Some of the abstract math has been applied to network diagrams of various types related to real systems such as chemical molecule graphs and electrical network graphs. See Biggs et al. (1976) for a history of the field and Harary (1969) and Wilson (1972) for concepts.

Pathway connections with signs (+ or 1) are called *signed digraphs,* and digraphs with barbs for directions are called *directed digraphs.* See ecological examples given by Costi (1979). The barb is given from predator to prey, which is reverse from the use by others in the direction of food flow.

Gallopin (1971) uses nomenclature of graph theory to describe various possible kinds of food webs to formulate expressions for combinations (complexity) for trophic distance and status.

Bond Graphs

Bond graphs are another network language, which represents forces, fluxes, configurations and energy characteristics (Paynter, 1960). The language requires more mathematical designations than energy language or electrical network diagrams. Examples are given in Fig. 6-27, where arrows indicate flows. The type of intersection or interaction is coded by letter or number at the junctions: 0 denotes additive flows of the same type; and 1 denotes multiplicative junction of different types of qualities necessarily interacting or divergence of two different by-product flows. In energy circuit language, diverging by-products from a transformation can be adopted with a general-purpose box as shown in Fig. 6-27 for divergence. Storage, resistance, and sources are shown to one side. For introduction, see Koenig et al. (1967) and Karnopp and Rosenberg (1975) and Oster and Auslander (1971).

A transformer has a product of force and flow of input equal to that of output; thus change in force is reciprocal to change in flow. Both input and output are the same type of flow. Named after the gyroscope, a *gyrator* has a flow in output process proportional to force in input process and vice versa with identical coefficients. A transformer has a different kind of energy-carrying process on the input as compared with output. Both gyrator and transformers conserve power and thus do not need heat sinks.

Oster and Auslander (1971) used bond graphs to express irreversible thermodynamics in network topological terms, which is what is attempted with energy diagrams. The uses of energy language and bond graphs in representing an enzyme–substrate reaction system are compared in Fig. 6-28.

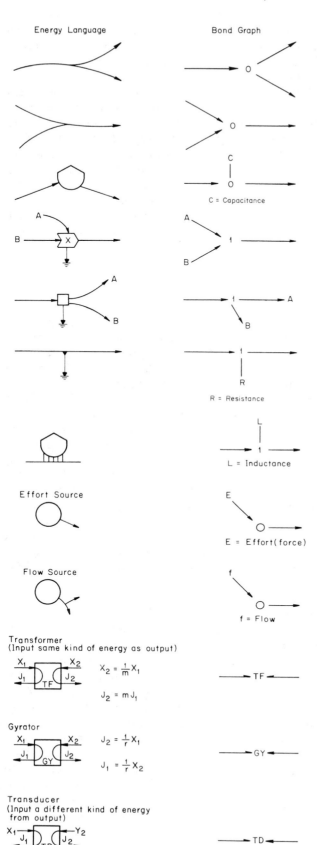

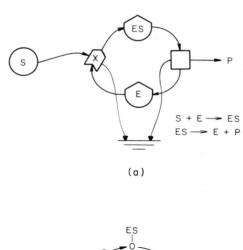

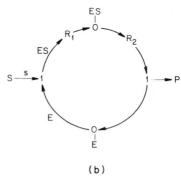

FIGURE 6-28. An enzyme substrate reaction system: (*a*) energy language (see Chapter 10); (*b*) bond graphs Bungay (1966).

Koenig Diagrams

Koenig diagrams are another initiative in combining energetics and kinetics to represent complex general systems. See Caswell et al. (1973) and Koenig et al. (1972). Energy transforming units are circles into which inflows converge and products diverge. Dispersal of heat is shown with a black dot within the circle. An example is given in Fig. 6-29.

Dominance Relationships

The relationship of units in dominating each other may be expressed in network arrow diagrams or in the matrix equivalent. Dominance relationships exist in animals and people. For example, there is the pecking order of chickens shown by W. C. Allee. A

FIGURE 6-27. Some symbols of bond graphs and comparison with energy language. Bond graphs compared with energy language, where E (effort) is the same as X (generalized force) and f (flow) is the same as J.

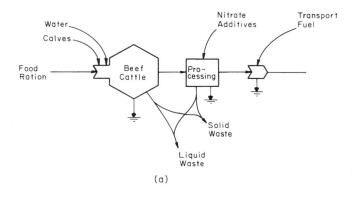

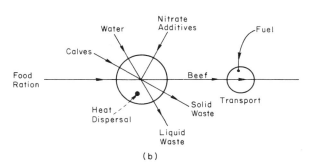

FIGURE 6-29. Cattle feedlot: (a) energy language; (b) Koenig symbols (Caswell et al., 1973).

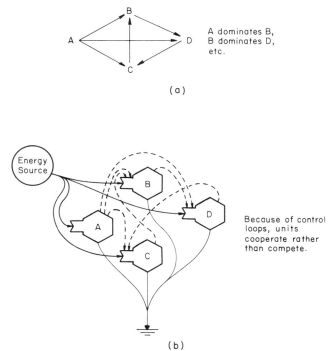

FIGURE 6-30. Dominance network: (a) arrow diagram indicating dominance relations; (b) energy circuit diagram of same relations.

newly grouped association of animals makes many tests of relationships, pecking at each other, until a dominance hierarchy is established. Thereafter, the existence of the patterns of dominance simplifies overall groups functions with only an occasional reinforcing peck to verify the relationships. Other dominance phenomena are the relationship of units in athletic competitions. An example is given in Fig. 6-30. In human affairs the arrow diagrams indicate "social power influence." See Kemeny et al. (1966).

In energy terms, dominance is a communication pathway of high-quality energy that has a control action. It is a pattern of systems interaction dependent on the existence of a basic power supply to the population so as to generate the equipment and operation of the organizational process. In Fig. 6-30b the basic energy flows and the control actions of the dominance pathways are both shown together. Without the control pathways, the units would be in competition, which could cause destructive exclusion of units.

Saaty (1968) considering arms race strategies for countries provides influence diagrams with arrows of influence.

Digraphs

Using matrix theorems, Levins (1975a,b) analyzes designs for ecosystems for stability and other characteristics. A type of flow graph diagramming language was used that forms loop shapes as given in Fig. 6-31. The diagrams indicate the presence or absence of a pathway and whether the effect is stimulatory (positive) or inhibitory (negative) to the receiving module. Each module is represented as a small circle (vertex) that is named or numbered. Positive effect is indicated by an arrow, and negative effect is indicated by line to a small sphere on the unit affected. Examples are given in Fig. 6-3; see also Roberts (1976a).

Comparison with energy language is given in Fig. 6-31b. As shown, an ordinary pathway in energy language without a loop drains one unit, contributing to the next. In the digraphs two lines are required, forming a loop. For a loop in energy language, four pathways are required in digraphs. Both types can show by inspection when some systems are unstable, going to zero or blowing up with accelerated growth as in Fig. 6-31d. Systems are unstable when units have gains without drains or vice versa.

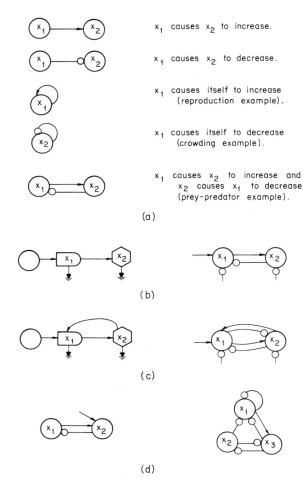
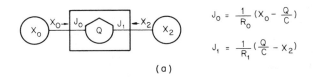
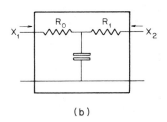
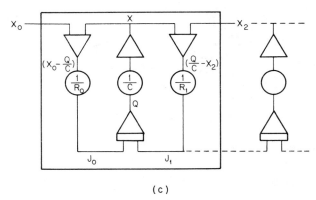

FIGURE 6-31. Use of Levins digraphs (Levins, 1975a): (a) definitions of main pathway arrangements; (b) translations of some energy diagrams without a loop in Levins language; (c) loop in energy language; (d) unstable system.

FIGURE 6-32. Analog subsystem as a module for a storage: (a) energy language with force–flux equation; (b) passive electrical circuit; (c) active analog subsystem wiring configuration. Modified from Krigman (1967) and McLeod (1968).

Flux Diagrams

A diagrammatic language for fluxes and cycles of biochemistry is used by Hill (1977) to analyze kinetics and structure. The diagram uses straight directed (barbed) arrows formed into triangles, rectangles, and so on, in representing the systems. See Fig. 10-28 (in this book), where procedures of analysis are given.

A Language of Analog Subsystem Blocks

Another approach to simulation of storage uses analog units in a standard wiring configuration (see Fig. 6-32). A unit storage and passive analog circuit are compared. The passive circuit in Fig. 6-32b is already an electrical simulation. The active circuit shown in Fig. 6-32c is sometimes called a *simulation of a simulation*. These units can be strung together to alternate the integrator line with summer line as shown in Fig. 6-32c.

Biochemical Network Language

A very simple network language is in use to represent systems of chemical reactions. Simple arrows are used with interactions represented by an arrow point connecting to the side of a pathway as shown in Fig. 6-33 (Savageau, 1976).

Verbal Languages

Verbal languages reach the most humans as this is the language in which their primary childhood

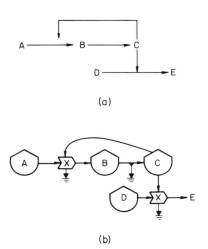

FIGURE 6-33. Biochemical network language (Savageau, 1976): (a) example; (b) energy language translation.

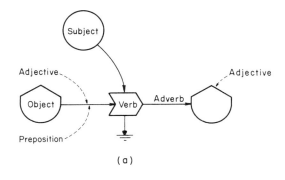

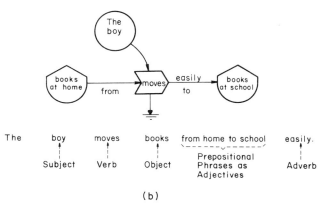

FIGURE 6-34. English sentence structure translated to energy language: (a) syntax equivalents; (b) example.

learning is communicated. As shown in Fig. 6-33, the natural role of energy in structuring that which is natural seems to include the verbal language. In English grammar, sentence structure is analyzed with subjects, verbs, adjectives, adverbs, prepositions, and so on. Possible equivalents are given in Fig. 6-34, where action flows from subjects (sources or storages) through verbs (interaction and production processes) to objects and various modifiers that indicate the nature of the pathways. Linguistic comparisons often have language composition related to that of the world it represents. Information indices for letters and words are similar to that of the real world about which they are written. Ratios of common and rare words are similar (see Chapter 17). Von Foerster (1970) compares node intersections in networks to verbs.

Art as a Systems Language

The essence of systems is sometimes communicated by means of artistic expression. For example, Dansereau uses the picture in Fig. 6-35 to describe the progression of energy hierarchy from dilute inorganic processes at the bottom through the ecological food chains to informational and controlling units at the top, expressing the energy quality sequence, which is the subject of Chapters 14 and 15 in this book. Winarsky (1980) models the role of art as a high-quality general language that can reach more people than concise languages. Judging by the very large number of people who can be reached through verbal and artistic languages, their power may be inverse to their precision (Winarsky, 1980).

SUMMARY

Twenty-five additional systems languages were introduced with examples and related to energy diagramming. The comparison of symbols may help transfer special ideas that go with these languages into the usage of the energy diagramming and vice versa, improving its role as a systems translator, making more mathematical operations available and unifying knowledge. The student should learn to understand the systems presentations of others even though the language used may vary. Supplementary reading is recommended to elaborate on the thumbnail introductions given here.

QUESTIONS AND PROBLEMS

1. Can you identify all the logic units on your analog computer, draw their symbols, indicate what they do, write appropriate equations, and indicate how to patch?

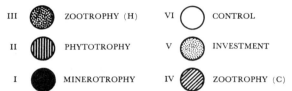

FIGURE 6-35. Example of systems relations with the use of art (Dansereau, 1975); a scheme showing a projection on the six trophic levels of the mainstream flow of energy (central part), the supply of resources (left part), and the reinvestments (right part), as well as the import (left margin) and export (right margin).

2. Name 10 logic devices in the macroscopic world. Examples are stop lights, ticket gate, sexual action, and so on.

3. Can you draw a circuit that can perform learning?

4. Draw an energy diagram for an agricultural crop with harvest followed by replanting. Use logic units.

5. Patch the models in this chapter given only the energy diagrams.

6. Draw an energy diagram for a system that grows exponentially, then crashes and grows again, forming a repeating pattern. Patch on analog computer. Program on microcomputer.

7. Is this a flip-flop?

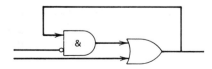

8. What does this system do with input of logic 1? Of logic 0?

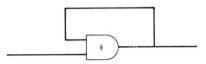

9. Make a truth table and verify that the output is on if $A \equiv B$.

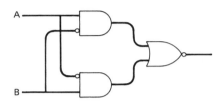

10. If a logic training device and self instruction booklet is available, work through exercises (e.g., Hughes, 1968).

11. Identify the systems language that goes with the symbols that are given in Fig. 6-36.

12. With the help of your previous mathematics texts or one of the sources given in the readings, conduct the following matrix operations: add two matrices; multiply a vector times a matrix; and explain what inversion of a matrix does.

13. Compare a matrix to its network diagram. What can be included in the diagonal of the matrix? Self-interactions? Outside inputs? Outside outflows? Can you draw an energy diagram for a matrix of inflows and outflows? Can

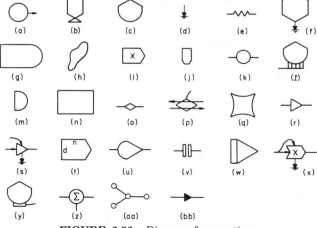

FIGURE 6-36. Diagram for questions.

you write the matrix for an energy diagram of an ecosystem minimodel?

14. Take an energy diagram of an ecosystem such as those given in Chapter 2. Suppose that there is an increase in one of the storages or external sources; put pluses and minuses on pathways and storages to indicate the response of each of these due to the impact of the change imposed.

15. Find a paper that utilizes correlations or regression to relate a unit to simultaneously changing factors. Try diagramming the relationship, and place the statistical parameters on the pathways.

16. Consider some water flow situation. Carefully diagram the water storages and flow pathways to the dispersal heat sinks. Suppose that the properties of the water storage situation were changed by addition of rocks to the storage space. What change would there be in the parameters of Fig. 6-3? What law in hydrology states that water flow is linearly proportional to water pressure?

17. What is a systems diagram with probability path coefficients on the pathways? What is the sum of all pathways from a unit? If one considers the successive probabilities of four steps along a network, how does one calculate the probability of reaching the end of one of the possible pathways that start from the same initial state? What are such sequences called?

18. In signal flow diagrams, how does one relate the arrows that converge at the same node? In going from one node to the next, what does one do with the value of the pathway? In translating an energy diagram to signal flow diagrams, does one relate pathways directly or go through another language first? Give an example of condensing a more complex signal flow diagram to a simpler one.

19. Translate a Forrester dynamics diagram into an energy diagram, substituting appropriate energy sources and heat sinks that may be needed. Translate an energy language minimodel into a dynamics diagram.

20. What are two meanings for the word graph? Take a chemical reaction ($A + B = C + D$) and convert to an energy diagram; then convert to a simple network of arrows. Write in style of biochemical networks. Then write equations for the rates of reaction. What is a tree?

21. In Koenig diagrams, how is heat dispersal indicated? In this language, do items that are necessary interacting commodities converge on a point of that interaction? Does this mean that their rates are proportional to the sum of the interacting constituents? Does this convergence mean that the energies are additive at that point?

22. In energy language, what are dominance arrows? What is the relationship of energy flow to a dominance control act?

23. Take a simple minimodel and translate it into digraph language, expressing the effect of one unit on the other as increasing or decreasing its structure function. Then take the same diagram and translate to a matrix of 1's and 0's.

24. Take a flow chart for a program and translate into an energy diagram of the system.

25. Try your hand at art with systems content, perhaps with some animated network and entities.

26. Take systems periodicals and identify the kinds and number of systems languages found of 30 articles examined.

27. Review other languages as given in earlier chapters.

28. Use letters from the group shown in Fig. 6-36 to answer the following questions:

    —— 1. operational amplifier without feedback
    —— 2. second law of thermodynamics
    —— 3. equivalent circuit for a pathway with frictional loss
    —— 4. switch in analog computer language
    —— 5. exchange
    —— 6. electrical storage
    —— 7. sums and accumulates on analog
    —— 8. state variable in energy language
    —— 9. digital program group symbol
    —— 10. signal flow language
    —— 11. Forrester's multiplier
    —— 12. black box
    —— 13. cannot receive input from a pot
    —— 14. analog to digital converting unit
    —— 15. coefficient setting
    —— 16. logic inverter
    —— 17. constant gain amplifier in energy language
    —— 18. inertial back force
    —— 19. divisor

—— 20. summing in a block diagram language
—— 21. external pool as used by Forrester
—— 22. storage and reward loop feedback
—— 23. production unit
—— 24. forcing function
—— 25. quadratic pathway
—— 26. derivative
—— 27. count the number of energy symbols in Fig. 6-36

SUGGESTED READINGS

Ashby, W. (1956). *An Introduction to Cybernetics,* Wiley, New York. 295 pp.

Ashby, W. R. (1960). *Design for a Brain,* Wiley, New York. 286 pp.

Bennett, A. W. (1974). *Introduction to Computer Simulation,* West Publishing Company, St. Paul, MN.

Blalock, H. M. (ed.) (1971). *Causal Models in the Social Sciences,* Aldine Atherton, Chicago.

Brewer, Joseph. (1958). *Introduction to the Theory of Sets.* Prentice-Hall, Englewood Cliffs, NJ 108 pp.

Brown, G. S. (1972). *Laws of Form.* Julian Press, New York.

Casti, J. L. (1979). *Connectivity, Complexity and Catastrophe in Large Scale Systems,* Wiley, New York.

Chaston, I. (1971). *Mathematics for Ecologists,* Butterworths, London.

Cooper, R. B. (1972). *Introduction to Queueing Theory,* Macmillan, New York. 277 pp.

Crowe, A. F. and A. Crowe (1969). *Mathematics for Biologists,* Academic Press, New York.

Domenico, P. A. (1972). *Concepts and Models in Groundwater Hydrology,* McGraw-Hill, New York. 405 pp.

Electronic Associates (1971). *Miniac Reference Handbook.* Publication No. 00-800.2071-0.

Forrester, J. W. (1961). *Industrial Dynamics,* MIT Press, Cambridge, MA.

Glushkev, V. M. (1966). *Introduction to Cybernetics,* Academic Press, New York. 321 pp.

Greenstein, C. H. (1978). *Dictionary of Logical Terms and Symbols,* Van Nostrand Reinhold, New York.

Harary, F., R. Norman, and D. Cartwright (1965). *Structural Models: An Introduction to the Theory of Directed Graphs,* Wiley, New York.

Hastings, N. A. J. and J. M. C. Mello (1978). *Decision Networks,* Wiley, New York. 196 pp.

Henley, E. J. and R. A. Williams (1973). *Graph Theory in Modern Engineering,* Academic Press, New York. 303 pp.

Howe, C. W. (1979). *Natural Resource Economics,* Wiley, New York. 350 pp.

Hughes, J. L. (1968). *Computer Lab Workbook,* Digital Equipment Corporation. Boston, MA. 169 pp.

Karnopp, D. and R. Rosenberg (1975). *Systems Dynamics, a Unified Approach,* Wiley, New York.

Kemeny, J. G., J. L. Snell, and G. L. Thompson (1966). *Introduction to Finite Mathematics,* Prentice-Hall, Englewood Cliffs, NJ.

Kent, E. W. (1981). *The Brains of Men and Machines,* McGraw-Hill, New York. 250 pp.

Klir, J. and M. Valach (1965). *Cybernetic Modeling,* ILIFFE Books, Princeton, NJ. 437 pp.

Lasker, G. E., Ed. (1981). *Applied Systems and Cybernetics,* Vol 1–6, Pergamon, New York.

Latil, Pierre de (1956). *Thinking by Machine,* Houghton Mifflin, Boston. 253 pp.

Lee, R. (1980). *Forest Hydrology,* Columbia University Press, New York.

Lynch, E. P. (1980). *Applied symbolic Logic,* Wiley, New York. 260 pp.

Mayeda, W. (1972). *Graph Theory,* Wiley, 588 pp.

McCulloch, W. S. (1965). *Embodiments of Mind,* MIT Press, Cambridge, MA.

Mesarovic, M. D. ed. (1968). *Systems Theory and Biology.* Proc. Systems Symposium, Springer-Verlag, New York.

Minsky, M. L. (1967). *Computation—Finite and Infinite Machines,* Prentice-Hall, Englewood Cliffs, NJ.

Moder, J. J. and S. E. Elmaghraby (1978). *Handbook of Operations Research,* Van Nostrand Reinhold, New York.

Muraga, S. (1979). *Logic Design and Switching Theory,* Wiley, New York.

Nagle, H. T., B. D. Carroll, and J. D. Incarn (1975). *An Introduction to Computer Logic,* Prentice-Hall, Englewood Cliffs, NJ.

Nahikian, H. N. (1964). *A Modern Algebra for Biologists,* University of Chicago Press.

Patten, B. C. (1970–1976). *Systems Analysis and Simulation in Ecology,* Vols. 1–4, Academic Press, New York.

Rashevsky, N. (1961). *Mathematical Principles in Biology,* Thomas, Springfield, IL.

Roberts, F. S. (1976). *Discrete Mathematical Numbers with Application to Social Biological and Environmental Problems,* Prentice-Hall, Englewood Cliffs, NJ.

Schaefer, G. (1972). *Kybernetik und Biologie,* Metzlerscher, Stuttgart.

Shearer, J. L., A. T. Murphy, and H. H. Richardson (1971). *Introduction to System Dynamics,* Addison-Wesley, Reading, MA. 420 pp.

Shugart, H. H. and R. V. O'Neill (1979). *Systems Ecology, Benchmark Papers in Ecology,* Vol. 9. Dowden, Hutchinson and Ross, Stroudsburg, PA. 368 pp.

Simon, W. (1972). *Mathematical Techniques for Biology and Medicine,* Mass. Inst. Technology Press, Cambridge, MA.

Singh, J. (1966). *Great Ideas in Information Theory, Language, and Cybernetics,* Dover Books, New York. 338 pp.

Toates, F. M. (1975). *Control Theory in Biology and Experimental Psychology,* Hutchinson Educational, London. 264 pp.

Todd, K. K. (1980). *Groundwater Hydrology,* Wiley, New York. 448 pp.

CHAPTER SEVEN

Energy

Energy storages and flows accompany all features of systems, and the scientific concepts of energy provide a commonality for understanding the resource basis of structure and process. In this chapter classical concepts of thermodynamics and newer concepts of open-system energy flow are given, with the use of energy circuit language to relate them to systems principles. Since real systems operate their energetics and their kinetics simultaneously, our energy circuit models of real systems also must include both together.

Energy sources control environmental systems according to the signature of energy inflow. Definitions are given of energy, heat, work, free energy, entropy, and temperature, and principles that control energy sources and storages. Rates of energy transformation depend on maximum power selection. In Chapters 14 and 15 we see that the energetic basis for system design is related to concepts of energy quality.

HEAT AND ENERGY

Energy is a quantity common to all processes; it flows, is stored, and is transformed in form. Energy is used as a common measure of all kinds of activity. Storages and flows with more energy represent more activity. One form of energy is heat, which is a measure of the motion of molecules. When there is more molecular motion, matter expands. For instance, the expansion of a tube of fluid indicates a greater content of heat. The degree of expansion is used as a measurement scale called *temperature,* which is a measure of the concentration of heat (molecular motion). When a human touches a hot object, some of the intense motion of molecules is transferred to the finger, causing human nerves to send a message indicating the degree of heat concentration (temperature). If the molecular motion is sufficiently active, tissue is destroyed and the person has a burn.

There are many kinds of energy, but they can all be quantitatively related by converting them into heat because all kinds of energy can be transformed into heat with 100% efficiency. In other words, energy is quantitatively defined by the heat generated when converted into the heat form of energy. For example, heat releases are measured in a bomb calorimeter.

Common units used to measure heat energy are the kilocalorie and the British thermal unit (BTU). Other kinds of energy are measured with other units. Some conversions in transforming other units into heat units are given in Fig. 7-1.

SOURCES OF POTENTIAL ENERGY

Open systems require energy from outside sources. Energy of a type that can support system processes is *potential energy*. Outside influences that cross the boundary into a system driven by forces from outside originate from sources of potential energy and are called *forcing functions*. Energy flow is always accompanied by forces and vice versa. Concepts of causal action are expressed in many ways in many fields, but in any language there is the implication that an added quantity from outside has a component that affects the energy flows. In the language of energy circuits, all outside influences to a system are energy sources, although they may be delivered according to a variety of functions, some interacting with inside structures and storages that control energy admissions.

Figure 7-2 gives several kinds of external energy programming sources commonly found in environmental and other systems. These may deliver constant or programmed variations of force, population

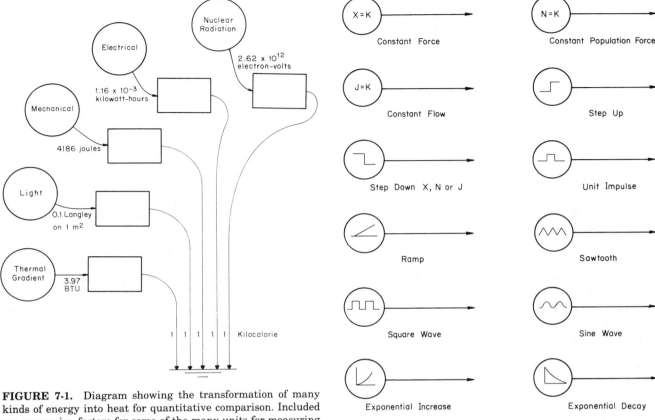

FIGURE 7-1. Diagram showing the transformation of many kinds of energy into heat for quantitative comparison. Included are conversion factors for some of the many units for measuring energy.

FIGURE 7-2. Some common forcing functions from outside energy sources.

force, or flow. The circular source symbol in energy circuit language implies that the outside source delivers the indicated energy and force contribution to the system but leaves behind on the outside the various heat releases or other losses associated with such delivery, since they are not in the system of interest. We include only the effect that is delivered by the input pathway.

Signature of Energy Sources

Any system has a combination of energy sources that characterize its external conditions. Sometimes these external connections are called *boundary conditions*. The kind of system that can develop and survive within the system boundaries depends on the combination of external inflows and outflows of energy. The combination of external energy sources can be called the system's *energy source signature*.

The kind of system developing depends on the magnitude of the energy sources, the kinds of energy forces for delivery, and the combinations of different kinds of energy.

In older ecological literature the word "biotope" was used for the combination of environmental factors impinging on a particular ecological community of organisms. There has been a long history of relating observed ecosystems, also called *biocoenosis* to causal *biotope* (energy signature). Examples are the climate and soil parameters related to terrestrial ecosystems as given by Bakuzis (1961). As shown in Fig. 2-2, energy diagramming arranges sources in quality order. In Chapter 14 the energy signature is given energy transformation ratios so that their embodied energies may be compared.

Sources with a Distribution of Energies

Some sources contribute a distribution of energies of different magnitudes. A spectral graph can be included with the source symbol indicating the frequencies of energies of different intensity of type. For example, a Gaussian normal distribution is indicated in the jitter source in Fig. 7-2. Other examples are the inflows of energies of different frequency in solar radiation as shown in Fig. 16-1. The characteristic skewed distribution shown in Chapter 15 is found to be common because of the hierarchy of energy flows in most kinds of system.

Constant-Force Source

Some outside energy sources are described by the phrase "constant force." These sources deliver a constant force (X or N) regardless of whether flow is being drawn. Very large energy sources are examples in which the flow drawn is not sufficient to change the force delivered by that storage. Examples are large dams with small water outflows, large batteries with small electrical drains, or large populations with small population migratory movements. The constant-force source can be designated as shown in Figs. 7-3a and 7-3b by writing the equation $N = K$ or $X = K$. A constant-force source is an unlimited source supplying energy as required to keep the force constant. Self-designing systems develop pathways by using such sources and tend to grow and further tap energies until the flows are sufficient to affect the force delivered. After that the source is no longer a constant-force source.

Control Action Forcing Functions

Many of the outside influences on an environmental system provide information or cues that cause processes only by interaction with second flows of energy (Fig. 7-3e). They provide relatively small parts of the total energy (heat equivalents) of a process but have higher embodied energy (see Chapter 14). The forcing action depicted in Fig. 7-3e is multiplicative, controlling an energy flow without constituting the entire source of that reaction's energies. As the energy circuit language recognizes, the energy contribution is from both reactants. Either is capable of controlling the process and thus being a limiting factor. A forcing function may be pumping a flow outward, thus constituting an active energy sink (Fig. 7-3e).

Examples of input control actions are the cues that plants and animals take from the photoperiod, behavior signals, critical materials, and information that humans receive and use to accomplish some complex work. Money and price are shown as examples of forcing functions in Fig. 7-3f.

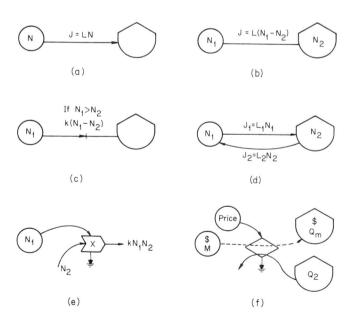

FIGURE 7-3. Some forcing function arrangements in energy circuit language: (a) uniforce; (b) with backforce; (c) backforce but backflow valve; (d) dual-path return flow; (e) multiplicative input; (f) money drive.

Constant-Flow Source

Many systems have inflowing energies that deliver a constant amount per time regardless of the use made by the system receiving the energies. If not used, the inflow may pass through the system without further interaction or may develop such accumulations of storage quantities until a utilization pathway develops. A limited constant flow delivers a set rate of energy delivery. This type has a constant flow without a backforce in its inflow pathway. Another kind of constant-flow source is one that overcomes a backforce and augments its energy to deliver a constant force even though the backforce is changing. Such energy delivery may vary with flow. Examples of possible simple constant flow sources are the rainfall to a region and the leaf fall to a forest floor.

An Energy Source That Is Flow Limited

Some external energy sources are said to be flow limited because it is their flow that is externally controlled rather than their driving force. A system using a flow-controlled energy source can draw from the source only the flow that is incoming and no more. For example, a water wheel running on a small stream can get no more energy from that stream than is inflowing from upstream. Nothing the user can do will increase the flow above the flow limit that is applied. Another example is the flow of litter to the forest floor. The soil organisms can use only the fall of litter that reaches them controlled by processes beyond their system. Figure 7-4 illustrates a flow-limited source. An equation is shown for the force that this flow generates that is a function of the amount used. What can be used is the remainder R calculated as the difference between the inflow J_0 minus that used J. In the example shown in Fig. 7-4, the following equation results:

$$R = J_0 - k_0 R \qquad (7\text{-}1)$$

Solving for R produces

$$R = \frac{J_0}{1 + k_0} \qquad (7\text{-}2)$$

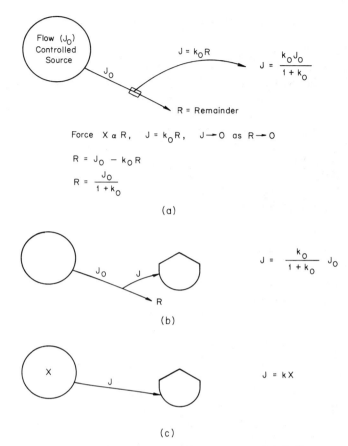

FIGURE 7-4. Forces and flows from a source that is flow controlled: (*a*) equations and energy diagram; (*b*) flow into a tank from a flow controlled source; (*c*) flow into a tank from a force controlled source for comparison.

The force is delivered in proportion to the remainder R, and the flow is in proportion to that in the arrangements shown in Fig. 7-4.

When the flow from a source is directly into a tank without interactions, the final equation is a constant times the flow, and not too different from the equation for flow into a tank from a force controlled source as shown in Fig. 7-4*c*. Both are constants times the parameter of the source. However, the pattern of control is very different when there is pumping from the source as described in Chapter 8. See Fig. 8-8.

Fluid Transported Sources

The inflow of a river may carry with it in proportion to water flow nutrients, organisms, salt, sediments, or other quantities. In this example, most of the energy of transport is from the water flow, which carries the others in proportion to their concentration. In energy diagramming these sources operate as a coupled cluster as suggested in Fig. 7-5. The small

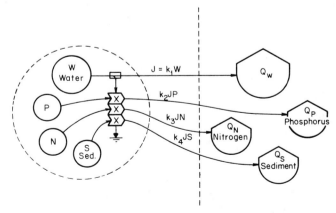

FIGURE 7-5. Energy circuit diagram for group of coupled energy sources, water carrying sediments and nutrients.

pathway block indicates that the flow rate J is the source of a force that interacts multiplicatively with concentrations in the fluid transported.

Energy Transmission as Waves

When pure energy is transmitted, it is often as waves. Thus light, sound, water waves, surface tension waves on water surfaces, earthquake deformation waves, and alternating current (AC) power lines are examples of the generalization. High-quality energy transmission used for control purposes is often called *information*, which is often sent as modulations of amplitude or frequency on other energy transmissions. Thus we modulate sound in speaking and radio waves in broadcasting.

FIRST LAW OF THERMODYNAMICS

A long history of study has shown that energy measured according to its heat equivalents is neither created nor destroyed in any process. Thus any network diagram of flows and interactions will have the sum of the flows of energy inflow equal to the change in energy storage plus the energy outflowing. When a diagram of a system is drawn with the energy circuit language, numbers for the flows can be drawn on the pathways, and the sum of inflows will equal that being stored plus that outflowing. If there is a steady state so that storages are not changing, the energy inflows equal the energy outflows. Sometimes we call the energy diagram that has inflows and outflow energy written on pathways a *first-law diagram*. The numbers on the pathways are in calories per unit of time and the numbers in the storages are stored calories. These flows are not all in heat form, but the numbers indicate that if that flow were isolated and converted into heat, it would yield that amount of heat calories. A first-law diagram is given in Fig. 7-6 for Silver Springs, Florida, a system of underwater bottom plants and a glass-bottom boat tourist economy. This diagram should replace the one in textbooks based on incomplete diagrams (Odum, 1955).

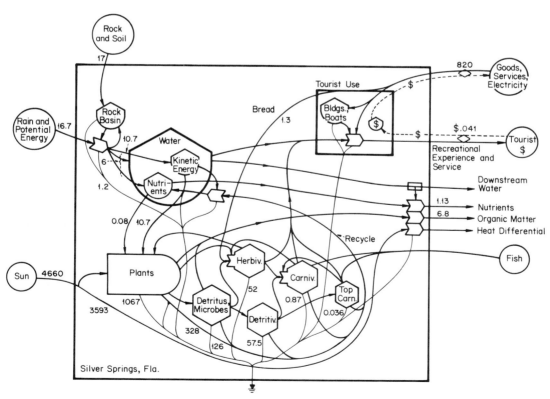

FIGURE 7-6. Energy diagram of Silver Springs, Florida. Modified by Robert Knight from Odum (1978).

WORK

Mechanical Work

When a force is operated against an opposing force for a distance, the energy that is transformed is the product of the force times the distance (Fig. 7-7). For example, work is done against the force of gravity in lifting an object, against the backforce of friction in sliding an object, or against the backforce of inertia in accelerating an object from a still condition to a moving one.

One of the great achievements of the nineteenth century in developing the general concept of energy was to show that there was always the same mechanical equivalent of heat. The mechanical work done could be converted into heat. Work has a heat equivalent and thus is a form of energy.

Sometimes in comparing processes of the same type, we say that processing more work means greater energy flow. Many physics courses turn this around and initially to define energy as the ability to do work. This is true as long as one is comparing processes and energies of the same type.

Work—a Useful Transformation

Maxwell (1877) defined work as a state transformation. In layman's usage, "work" means useful energy transformation where use can be defined as contributing to the survival of the system of which the process is a part. In some sciences work is defined as mechanical work, but a more general definition is: *Work is a useful energy transformation*. Systems and transformations that do not contribute to maximum power tend to be eliminated in competition for energy. Useful works are transformations that feed back materials and services.

Power

Power is the energy flow per time. In one sense "power" refers to any energy flow rate. Engineers restrict the term to the rate of flow of energy in useful work transformations.

SECOND LAW OF THERMODYNAMICS

All real processes are accompanied by energy flow and transformation, and some of the energy's ability to sustain work is lost. We sometimes speak loosely of energy being "used up," whereas what is really meant is that the potential for driving work is consumed, whereas the calories of energy inflowing and outflowing are the same.

The observed property of energy that it is degraded in any real process so that it can do less work is usually described as the second law of thermodynamics. Energy concentrations tend to be degraded by being dispersed in space and/or by being converted into dispersed heat, the random motions of molecules. The principle is called times arrow (Blum, 1951).

The energy language recognizes the property that most of the energies capable of driving work are transformed into a form not so capable of driving further work (while upgrading some in a work transformation). A heat sink symbol is placed on every process (see Fig. 7-8).

The second law of thermodynamics requires not

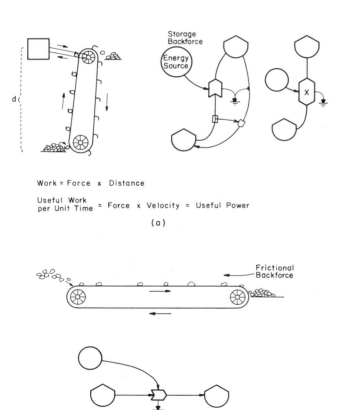

FIGURE 7-7. Energy flow in systems with mechanical work: (*a*) two ways of diagramming potential generating work storing against force of gravity; (*b*) work against friction; transporting ore on conveyor belt.

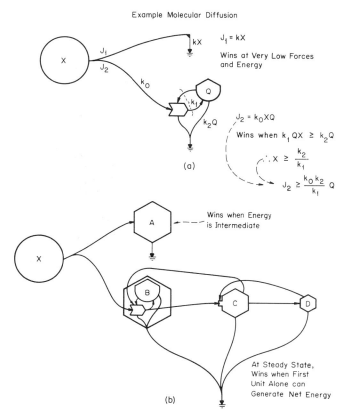

FIGURE 7-8. Designs resulting from competition for maximum power: (a) threshold for autocatalytic pumping; (b) chain draws more power because of added amplifier effect of C and D.

only that we have a heat sink on all processes and pathways indicating energy dispersal, but that all storages have such a heat sink drain to indicate that energy is being degraded by dispersal of the concentration. We sometimes call this dispersal *depreciation*. It occurs on all kinds of storage—those of pure energy, matter, information, money, and so on. Sometimes in the process of simplifying models some storage heat sinks are omitted if the rate is unimportant for the phenomena being simulated.

MAXIMUM POWER PRINCIPLE

The maximum power principle was given in Chapter 1. Maximization of useful power may be the most general design principle of self-organizing systems. Whereas other design principles have been offered, such as maximization of biomass (Margalef, 1963, 1968), reproduction rates (Wilson, 1968), minimum energy flow (minimum rate of entropy generation) (Prigogine, 1947), minimum heat (Maxwell, 1892), maximum structural entropy, maximum profit, maximum efficiency, maximum stability, etc., these can be shown to be special cases, criteria that maximize power. More consensus is developing now on the maximum energy flow principle. Prigogine (1978a,b) and Nicolis and Prigogine (1977) now recognize autocatalytic maximum power tendencies for situations further away from equilibrium (meaning with more potential energy). Oster and Wilson (1978) now use energy as a fitness criterion for social insect systems.

See Chapters 9, 14, and 17 for more corollaries and explanations to help clarify the principle, such as the relationship between hierarchy and energy flow (Odum, 1967a, 1968, 1971, 1975, 1976). It may be time to recognize the maximum power principle as a fourth thermodynamic law as suggested by Lotka (1922a,b) and use it more centrally in basic teaching of science and engineering.

Fontaine (1981) used a self organizing algorithm to select model parameters for various criteria that have been proposed as objective functions for ecosystems. Maximum power designed models matched observed data.

Useful Energy Transformation

The maximum power principle can be defined as a maximization of useful power and its relation to the second law of thermodynamics simply by use of Fig. 7-8. Since all processes disperse energy, an energy transformation would be a loss and not useful if the transformation did not also produce a product that has special value to the system, thus putting energy in a form that did not exist before the transformation. In Fig. 7-8 energy transformation (work) is shown to generate special qualities that have the property feedback as an amplifier. Although the first law diagram shows less energy remaining, the quality of that energy is now higher than before in the sense that it can amplify and accelerate other energy flows. Such transformations are among those that are useful and are selected to survive in competition (see Chapter 14).

However, when the energy source is too weak (Fig. 7-8a) to supply the force to maintain the autocatalytic storage in balance with its depreciation, the process becomes linear and sets up a steady state that minimizes energy flow and entropy generation (Maxwell, 1892; Prigogine, 1947).

Acceleration of Disorder by Ordering

In the comparison in Fig. 7-8, energy is being degraded in both of the two pathways shown: one direct and one through a process that first builds a storage of transformed energy that has the property of feeding back as an amplifier on its own inflow. Both of these seem to follow the second law of spontaneously degrading energy from a source into the heat sinks (degraded energy dispersed usually as heat). However, the second pathway, by building the extra structure and maintaining it against the storages depreciation, can pump in more energy because of the high-quality amplifier action of the feedback.

The interesting paradox often stated is whether order tends to run down as implied by the second law or to increase as maintained by students of biological and cultural evolution. Figure 7-8 shows the solution to the paradox. Energy tends to run down into dispersed form, but the system that does this the faster and thus is selected in maximum power competition is the one that builds order along the way so as to be able to pump energy faster. It disorders faster by building order and thus wins out in competition for energy flow.

Chains of Energy Quality

The action of the maximum power competition thus results in energy systems with work transformations that involve upgrading of energy quality at the cost of degrading other energy; thus one finds the systems of nature made up of chains of transformations with feedbacks of high-quality amplifier action. The natural selection process leads to differences in the quality in which energy is found, and each has a requirement in energy of another type to generate it. The calories of one type required to generate a calorie of another is defined as an energy transformation ratio. The energy circuit language, by its chains and webs of symbols, shows their relations and energy quality increases, usually diagrammed on the paper from left to right (see Figs. 2-1 and 17-7*b* and Chapter 14).

Efforts to standardize energy units along with adoption of the metric system by using joules instead of traditional heat units may be unfortunate. Most kinds of energy cannot actually be transferred 100% into mechanical energy; thus the use of joules traditionally used for high-quality energy ignores differences in energy quality (see Chapter 14).

HEAT AND ENTROPY

Converting Heat to Mechanical Work

Degraded heat, meaning heat all at the same temperature, cannot do work at the macroscopic level, but if there is a difference in temperature (a difference in heat concentration), this difference can drive other transformations. What percentage of the heat can be converted into mechanical work? What is the energy-quality transformation factor in going from a heat difference to mechanical work?

The conversion of heat to work, if done at the slowest, most efficient rate possible, is given by the Carnot cycle calculation. As shown in Fig. 7-9, a gas is moved in a circle between four states as shown with the four tanks. First, the gas is allowed to expand, thus lowering the temperature, and then heat is allowed to flow in from the environment as it expands at constant temperature. Third, the gas is compressed again, which makes it hot, whereupon as a fourth step, the heat is allowed to diffuse out as

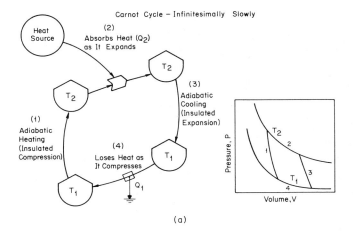

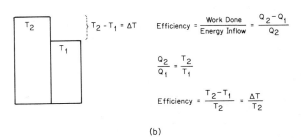

FIGURE 7-9. Carnot cycle for estimating maximum efficiency for converting heat to mechanical work: (*a*) energy diagram; (*b*) graph of usable portion of heat difference.

it compresses at constant temperature [see steps (1–4)]. This brings the fluid back to its starting point. In the process, heat Q_2 was allowed to enter and Q_1 was dispersed. The difference $Q_2 - Q_1$ went into pressure volume work.

The work done is the force times the distance; in this case it is the force per unit area (pressure) times the volume. Since heat contents Q are in proportion to absolute temperatures T, the efficiency can be expressed as the ratio of temperature changes. The ratio of the work done to the heat flow is the efficiency of conversion that comes out to be the ratio of temperature change to absolute temperature. As shown in Fig. 7-9c, the percentage of heat that can be used is the fraction that is at different temperature. The higher the temperature gradient, the greater the fraction of heat energy that is potential, meaning the greater the fraction that can be transformed in useful work. Power plants with very high temperatures have high degrees of efficiency. This calculation (Carnot cycle) is visualized as a theoretical upper limit since it is regarded at infinitesimally small rates without any real driving energy. In real processes even more energy is added to be degraded, so that the process is spontaneous and real.

Absolute-Temperature Scale

If all the thermal motion of molecules could be removed, a state called *absolute zero* would result. The temperature is zero in the ideal sense of having no heat. As heat is added, the molecular motions increase and the temperature rises. When expressed on an absolute-temperature scale (Kelvin scale), the temperatures are in proportion to the heat energy. Absolute zero is $-273°C$; freezing (0°C) is $+273$ K.

Entropy Change

Much used in energetics is a quantity S called *entropy*. In an energy transformation, the change in heat divided by the absolute temperature at which the heat change occurs is defined as the entropy change. Flows of heat can be divided by their temperatures (absolute scale) to obtain flows of entropy. When heat is added to a storage, entropy increases, and when heat is subtracted, entropy decreases.

Entropy change of state of a system is only that part of the entropy change that involves heat change within the system, thus involving changes in its state:

$$\frac{\Delta Q}{T} = \Delta S \qquad (7\text{-}3)$$

Third Law of Thermodynamics

Sometimes the premise that absolute zero exists and that its entropy is zero is said to be the third law of thermodynamics. There is no heat visualized at absolute zero. Temperatures within a fraction of a degree of absolute zero ($-273°C$) have actually been obtained.

Entropy of State

The entropy content of a storage is the accumulated heat divided by the temperature at which the heat was added. It is the stepwise integral of the reversible heat changes divided by the temperatures (see Fig. 7-10). The integration starts at absolute zero. The heat is visualized as being added infinitesimally slowly so that there is no heat processed that does not go into the storage. State changes made in this way are said to be *reversible*.

Entropy of State as a Measure of Molecular Complexity

As heat is added, more and more kinds of molecular complexities develop so that more heat is required per unit increase in temperature (Fig. 7-10). Heat capacity is the energy required to raise the temperature of a mass by 1°K. The molar characteristics are called *specific heats* (C_v, constant volume; C_p, constant pressure). The specific heat increases with temperature. At higher temperatures matter passes through a phase change such as that from solid to liquid or liquid to vapor, and even more energy is required. More molecular energy is involved in all the motion of molecules in a gas than for the same mass at the same temperature as liquid. Similarly, more energy is involved in molecular motion in liquid than in solid. The harder and more compact the crystal, the lower the entropy. In the sense that a crystal is simpler than the liquid and the liquid simpler than the gas, entropy content of state is a measure of the complexity.

Often it is said that entropy is a measure of disor-

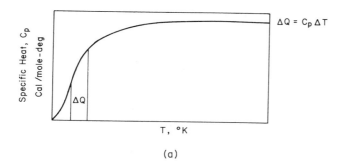

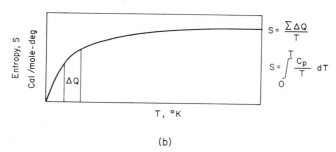

FIGURE 7-10. Increase of complexity with addition of heat starting at absolute zero: (*a*) specific heat; (*b*) entropy (Nash, 1974; Kittel, 1969).

der, where the crystal is the ordered state and the higher-temperature states such as the liquid and gas, respectively, are more disordered. However, *there may be reasons for not using these words for the increasing entropy content of more complex states as discussed in Chapter 17.*

Entropy Change of a Heat Sink

In the energy language the heat sink indicates a dispersal of degraded dispersed energy into the environment usually as heat at the same temperature as that of the surroundings. The heat sink is more fully described if its sink temperature (that of the surrounding) is indicated. With this temperature, the entropy change associated with this flow is obtainable by dividing the heat flow by the absolute temperature of the environmental sink.

Environmental Entropy Change in Real Processes

In a real spontaneous process (as contrasted with theoretical visualized processes such as the Carnot cycle or the addition of heat to a state starting at absolute zero), energy in a form more useful than degraded heat is dispersed into degraded heat. Thus we show the heat sinks on processes and tanks in the energy diagramming. All real processes are accompanied, therefore, by some heat flow that can be divided by its absolute temperature to obtain the entropy change of that flow. The heat added to the environment is a measure of the irreversible process. The associated entropy increase of the environment is another way of measuring the degree of irreversibility of the process.

Although most energy transformations in environmental systems are estimated indirectly using chemical calculations, the heat releases may be measured directly with continuous calorimeters. For example, Pamatmat (1982) measures sediment reactions directly.

ENERGY STORAGE

Generation of New Potential Energy Concentrations

The symbols in Fig. 7-11*g* illustrate the concentrating development of new potential energy when work is done against backforces that build up potential energies capable of driving these or other pathways later. Examples are water pumped into a tank against gravity, electrical charges forced on a capacitor against the backforce of the negative charges of the electrons stored there, or molecules pressured into a compressed-air tank against the back pressure of the compacted population of molecules. The relations of quantity stored and energy stored vary with the kind of system. Some potential energy storage functions are given in Table 7-1.

In many biological systems in which all details are not to be dissected and diagrammed, it may be desirable to show some potential energy storages as passively transferred in. For example, an animal eats and stores fat. However, some of the energy involves concentrating work, and heat sink tax is required for any concentrating process. The combination of both types of energy storage is shown in Fig. 7-11*h*.

Potential

The English word "potential" generally implies a capability. One particular scientific use of potential is the measurement of properties of storages and source. The potential is the potential energy per unit

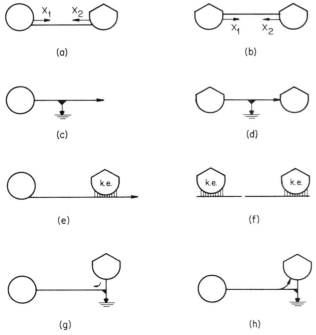

FIGURE 7-11. Energy relationships in various arrangements for flows of energy from outside source to storage or between storages: (a) reversible static balance of source and storage; (b) static balance between storages; (c) frictional steady-state flow from source to system; (d) steady-state flow from storage to storage without backforce; (e) accelerating flow from source to kinetic energy storage with inertial backforce; (f) inertial flow from one storage to another as in pendulum; (g) source concentrating energy against static backforce; (h) source supplying some energy to storage by passive transport of energy package and some by potential generating work.

of that matter stored that carries the energy. Thus there are water potentials, potentials in electrical charge storage, and chemical potentials. An alternative definition is as the ratio of power to flux of the carrying quantity.

Availability

In this and previous books (Odum, 1971; Odum and Odum, 1976, 1981) the term *potential energy* has been used to refer to the aspect of energy which is used up in the process of doing useful work during which the potential energy is degraded. A term for this used in thermodynamics is *availability*, which refers to the potential energy capable of doing mechanical work.

Net Energy of Concentration

When the food concentration is just enough to maintain a consuming unit without growth or loss, then the metabolism of that consumer is the energy required to concentrate and process the food. Hatanaka and Takahashi (1960a,b) conducted feeding experiments that located the break-even point. Available energy is that in excess of the break-even concentration. Much organic matter in the sea is below the usable concentration without auxiliary energy such as currents.

TABLE 7-1. Forces and Energy Storage Functions

Kind of Energy Storage	Quantity Stored (Q)	Force to or from Storage (X or N)	Potential Energy Stored with quantity Q
Energy inherent in storages:			
Water tank	Elevated water	Pressure $\frac{Q}{C} = X$	$\frac{Q^2}{C}$
Capacitor	Electrons	Voltage $\frac{Q}{C} = X$	$\frac{Q^2}{C}$
Heat	Heat	Percent temperature $\frac{\Delta T}{T} = X$	$\frac{Q \Delta T}{T}$
Compressed gas	Gas	pressure $\frac{nRT}{v} = X$	$RT \log_e P_2/P_1$
Solution*	Solute	Activity $\frac{aQ}{v} = N$	$RT \log_e A_2/A_1$
Macroscopic storing of food, fuel	Increments packages	Population force $\frac{Q}{C} = N$	(unit P.E.) $\times Q$
Light	Photons	Light pressure $k\left(\frac{I}{v}\right)$	$\left(\begin{array}{c}\text{Energy per}\\\text{photon}\end{array}\right) \times (Q)$

Potential Energy Functions for Storage

The last column in Table 7-1 lists the potential energy contents of storages as a function of the quantities stored there. Potential energy was generated in that storage (or moved in as packages after potential generating work done elsewhere). Potential energy inside is generated by doing work overcoming backforce with potential energy drives from outside the storage. If the inforce matches the static backforce, nothing happens and the situation is reversible. No potential energy is generated in the storage. If the driving inforce is greater than the backforce, potential energy is developed as the work done against the backforce in putting a quantity into storage. One may pump water into a tank against the back pressure of water already present or pump molecules into a tank against the pressure of molecules already there. In such situations the more quantity pumped, the greater the increase in backforce and the more potential energy there is developed per unit of quantity added. The potential energy is thus a sum of the work overcoming backforce that existed at the time each quantity was added. This is another case of cumulative summing that was described earlier as integration:

$$\text{P.E.} = \int \left(\frac{\text{energy}}{\text{unit quantity}} \right) dQ \quad (7\text{-}4)$$

where P.E. is potential energy.

For example, the sum of the potential energy in water storage is the water pressure times the volume of water stored at that pressure. The water pressure Q/C is proportional to the water quantity Q already stored, where C is a characteristic of its tank geometry:

$$\text{P.E.} = \int \frac{Q}{C} dQ = \frac{Q^2}{2C} \quad (7\text{-}5)$$

An almost identical expression results from compressing electrical charges where the backforce is proportional to the charge Q/C already stored and the potential energy stored is the sum of products of the energy per quantity times the quantity stored against that force. The potential energy stored is that due to the change of pressure at constant volume and that due to the change of volume at constant pressure. The latter is Gibbs free energy. In the biosphere many changes occur under the near-constant pressure of the atmosphere.

For populations of molecules undergoing compression in a tank, the pressure is related to volume V, number of moles n, and absolute temperature T by the familiar gas law $PV = nRT$, where R is the gas constant. The potential energy in the tank is the integrated sum of the volume of gas stored times the pressure as that unit of volume is added [Eq. (7-6)]. The product of pressure and volume is energy:

$$\text{P.E.} = \int_{P_0}^{P} \frac{nRT}{P} dP = nRT \ln \frac{P}{P_0} \quad (7\text{-}6)$$

For introduction of the dilute dissolved substances into fluids, the behavior of the molecules has been found to be mainly like that of gas molecules bounding randomly in a tank, and the potential energy content of such storage is similar to that given in Eq. (7-6).

An example of a calculation of the potential energy due to a chemical concentration difference was given for salinity differences in layers of the ocean (Levenspiel and deNevers, 1974) and for humidity differences in wind (Odum, 1967a) and fresh water relative to the sea (Odum, 1970).

In all these cases we determine and integrate the expression for potential energy per unit quantity added according to the backforce to be overcome at that level of storage. Energy stored is the product of force and displacement caused against the equal and opposite force of storage.

Backforce from Storage and Potential Energy

Following our convention, the barb on the input pathway in Fig. 7-11d indicates influence without complication from backforce from inside storages and forces. Often, however, the input forcing action interacts with the system within. Examples of this are given in Figs. 7-3 and 7-11. Figure 7-11a shows the action of backforce and possible backflow on a pathway, which has no barb or valve; Fig. 1-6b shows the backforce on a pathway with a backflow-preventing valve; and Fig. 7-3d shows a contrasting arrangement for return flow by a separate one-way path from storage to inflow.

Examples of backforces are the water pressure backforce that develop in filling a water tank and the back voltage that develops in a circuit that is charging a capacitor. Because so many complex living creatures are involved in biological pathways, it is not always clear how much backforce develops in these cases, although certain negative effects do slow down input energies after a steady state is reached.

When a flow is doing work against a backforce, it is developing and storing new potential energy, whereas a steady flow that has no backforce is doing work only against frictional forces and transporting energy downstream as restoring of preexisting, passively transported packages of energy.

Balanced Forces

Following from Newton's laws is the concept that forces are always balanced by equal and opposite forces. In energy circuit language, there is in every pathway line an equal downstream and upstream force opposing. There are three classes of force that are involved in these force balances: (1) static; (2) frictional; and (3) inertial. Illustrations of arrangements between these forces are given in Fig. 7-11a, 7-11c, and 7-11e.

Static forces are those delivered by a storage of energy according to relationships given in Eq. (3-1); frictional forces are those that oppose a velocity as illustrated in Eq. (1-1); inertial forces are those that oppose an acceleration as given in Eq. (7-7) with backforce X in proportion to acceleration $\dot{J}$, where $\mathscr{L}$ is the constant of proportionality:

$$X = \mathscr{L}\dot{J} \qquad (7\text{-}7)$$

Two forces that are pushing against each other from energy storages (Figs. 7-11a and 7-11b) may be in balance, forming a static-to-static force balance that is called *physical equilibrium*. Nothing is happening; there are no flows, there is no heat sink loss, and the system is essentially dead. Classical thermodynamics is a study of the conditions under which various static forces are balanced. Writing equations that describe the relationships of chemical and physical properties that produce balanced forces has been useful in computing various details of physical and chemical relationships.

A system that has no outside energy sources contributing flow is said to be a *closed system*. Without energy flow from outside, systems soon reach a state of balanced static forces such as that in Fig. 7-11b. This book, however, concerns open system defined as those with potential energies inflowing and outflowing.

As described in Chapter 1, steady flows are accompanied by frictional backforces that increase with velocity until a steady balance is achieved. There is a balance between driving force from storage and frictional backforce. This arrangement is the usual pattern indicated by most of the energy circuit language pathways with an energy source inflowing energy and with some of it leaving through heat sinks as a necessary part of pathways of this class (Figs. 7-11c and 7-11d).

Another kind of force balance is the inertial backforce that pushes back against any acceleration of matter or electrical current as in Figs. 7-11e and 7-11f. Following Eq. (7-7) there is an acceleration backward with an inertial force equal to the forward acceleration. Moving bodies of mass m have energy in proportion to the square of the velocity as in Eq. (7-8), which indicates kinetic energy for translation velocities v and for rotational angular velocities ω, where I is the moment of inertia the rotational equivalent of mass:

$$\text{K.E.} = \tfrac{1}{2}mv^2; \qquad \text{K.E.} = \tfrac{1}{2}I\omega^2 \qquad (7\text{-}8)$$

When potential energies are transferred into kinetic energy in situations without complex involvement with other pathways, there is no loss of potential energy to heat. As in the example of the pendulum, the kinetic energy may be converted back to potential energy with deceleration. In other words, kinetic energy is a form of potential energy relative to a frame of reference moving at different velocity. Energy transfers driven with static force against inertial force have no heat sink. Availability of energy is not lost since energy is transferable from potential energy to kinetic energy and vice versa. Another example is the toy that stores energy in a mechanical flywheel.

The transfer of one kind of potential energy into energy storage as kinetic energy is indicated by the inertial pathway and storage symbol given in Figs. 1-6h, 7-11e and 7-11f. These transfers do not increase concentrations of energy.

Einstein (1905) indicated that the conversion of potential energy to kinetic energy would seem to be the reverse—kinetic to potential—to the moving observer. Changes that are dependent on relative position of the observer are interchangeable. Schlegel (1961) reviews discussions by physical theorists who conclude that the entropy is independent of the velocity from which it is viewed.

Reversible and Irreversible Processes—Equilibria

Whereas a balance of static forces without action defines a simple true physical equilibrium, a system with exchanges of kinetic energy and potential energy such as the pendulum is a more complex form of force and energy balance that is also without appreciable loss. A population of molecules is another example of many interacting components that exchange potential and kinetic energies without appreciable loss. These are also forms of physical equilibrium not receiving or losing potential energies, although there is continual motion. All these actions are *reversible* since no energies are made unavailable. The amount of kinetic energy in microscopic molecular movements and motion is observed as heat by larger systems such as the expansion of mercury in a thermometer or the increased activity registered by human nerve sensors.

In Fig. 7-12a energy storage in the form of heat is shown. On a very microscopic scale at the dimension of separate molecules, there are collisions that occasionally produce some movements and vibrations that have higher energy of motion than the general average. The activated components constitute a potential energy storage that is quickly degraded, again dispersing its energies into the heat pool. Such a state is often regarded as equilibrium with all energies on the average similar; but locally when it generates more active concentrations of potential energy, it is not static.

Most important processes in real environmental systems differ from the equilibria just discussed in losing more concentrated, organized, potential energies into heat. Most processes in the biosphere are said to be *irreversible*, requiring new potential energies to maintain them. Real irreversible processes are accompanied by heat sinks and are so indicated in energy diagramming of their pathways. Heat sinks must be on all pathways except the static balance, purely inertial transformations situations and reversible flows of heat and radiation.

The word "equilibrium" is troublesome since it has been used in different ways in many fields. We have already recognized the static force equilibrium (Fig. 7-11b), the inertial balance equilibrium with motions (Fig. 7-11f), and the molecular heat equilibrium in Fig. 7-12a. Another is the chemical reaction balance in which the forward pathway from chemical substance A balances the return reaction from chemical substance B by another pathway (Fig. 1-2b and Fig. 7-12c) both facilitated by the equilibrium

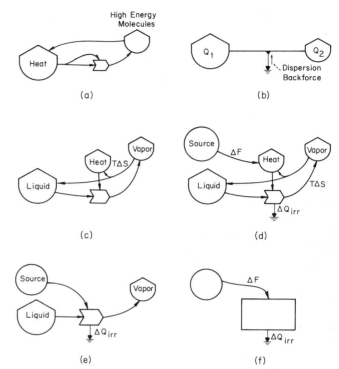

FIGURE 7-12. Energy relationships in molecular transformations: (a) molecular equilibrium; (b) diffusion; (c) water and vapor in equilibrium with reversible heat changes of state; (d) evaporation driven by use of outside source of potential energy; (e) abbreviated way of representing d without showing heat storage explicitly; (f) reversible changes aggregated in a box so as to emphasize the fate of free energy in irreversible heat of the environment (Denbigh, 1952).

motions and inertial exchanges of the heat pool interacting with molecular populations. These are all possible without outside energy sources. They are closed systems.

In many fields of study the word "equilibrium" or the phrase "dynamic equilibrium" is used for systems that have potential energy inflowing and heat outflowing and that hold internal patterns constant. Such open systems (Fig. 1-2a) may develop constant patterns or patterns that fluctuate in a regular manner. These systems work only as long as their energy inflows are intact. Thus there are steady states and oscillating or pulsed steady states. Ecological systems that reach steady state are called *climaxes*, and the transition states on the ways to climax are called *succession*. Environmental systems are open, but they may have some of the static equilibrium, inertial exchanges, and heat equilibria among their subsystems. Where they become of interest, they may be diagrammed within the whole network to show their relationships to the overall open system. It is probably preferable usage to use the word "steady state" for open system and the word "equilibrium" only for closed-system balances. The word "climax" is suitable for steady states in complex self-reproducing systems that store and transfer much information for self-programming.

Potential Energy in Chemical Sources

The chemical potential energy content associated with material storages depends on the point of reference or sink with which it is compared. A chemical substance may have one energy value in one reaction and different value in reactions with another substance, pathway, or situation. One cannot state the energy content of a storage without knowing the states that come next.

In most chemical reactions there are changes of physical state as when water evaporates or solids melt and so on. Inherent in the reaction is some heat energy absorption or release in changing the state. A molecule in a gas, for example, has more kinetic energy than a molecule in a liquid. If a reaction pathway has only those heat changes necessary due to the changes of state and no potential energy supplied beyond these energies of state change, the pathway is reversible. The process goes neither forward nor backward; as drawn in Fig. 7-12c, it is another of the reversible situations. A change of state may take up heat, cooling the temperature, or release it, raising the temperature. Evaporation of water, for example, takes up kinetic energy of molecules in making a liquid into a gas. About 0.54 kcal of heat is absorbed per gram. These energy changes are part of the reversible process and are thus not potential energy and not available to drive the process spontaneously. This energy change does not do work on a macroscopic scale, or generate dispersed heat as a part of spontaneous real flows. If this energy is the only energy available, nothing happens.

To be spontaneous, there must be potential energy in the reaction over and beyond that necessary for the reversible heat changes involved in change of state. Per unit mass, the total heat change may be summarized as in Eq. (7-9) as the sum of two heat changes: *Reversible heat change of state* is ΔH_{rev}. The heat released from the potential energy is called change in *free energy* over and beyond the reversible heat change. Total heat change is called *enthalpy* (ΔH):

$$\underset{\text{total}}{\Delta H} = \Delta F + \underset{\text{reversible}}{\Delta H} \quad (7\text{-}9)$$

Data and tables on ΔH and ΔF are usually given per mole (gram molecular weight). Gibbs free energies are those where pressure is held constant during the change; Helmholtz free energies are those where volume is held constant. It is sometimes called *arbeit* (ΔA) (meaning "work" in German).

The convention on signs in chemical thermodynamics is as follows: If heat is released and temperature rises as heat flows out to the environment, ΔH_{tot} is minus. If there is potential energy to drive the reaction, ΔF is minus. (The quantity *affinity* is defined as the Gibbs free energy with opposite sign.) Heat is also released from this to the environment. If there is heat released as part of the reversible heat changes, ΔH_{rev} is minus. If heat is absorbed, as with evaporation of water, ΔH_{rev} is plus. If there is no free energy, the reaction is reversible; the criterion that reactions proceed is that ΔF is minus. The free energy ends up in heat release as part of any spontaneous process. To proceed spontaneously, the free energy must be minus, meaning that there is potential energy to drive the process. If free energy change in the process is positive, there is potential energy available only against the reaction so that it may proceed backward if pathways permit.

An *endothermic* reaction is one such as water evaporation in which reversible heat is large and positive (absorbing heat) so that, even though there is negative free energy, the overall sum is positive. The spontaneous process is less cooled than the reversible one. Most processes that do not undergo phase changes as from solid to liquid or from liquid to gas have small heat changes of state, and for these, the overall heat change is negative (exothermic), meaning that heat is released and temperature rises (see Fig. 7-12).

In Eq. (7-3) entropy change was defined as the ratio of heat change to the absolute temperature at which the change occurs. If Eq. (7-9) is divided through by the absolute temperature, an alternate way of expressing the reversible and irreversible parts of a reaction is obtained in terms of three entropy changes:

$$\underset{\text{total}}{\Delta S} = \underset{\text{irreversible}}{\Delta S} + \underset{\text{reversible}}{\Delta S} \quad (7\text{-}10)$$

Equation (7-10) indicates that the total entropy change is made up of two contributions: (1) the reversible entropy change of state that is inherent in the structural changes and thus is a measure of them plus (2) an irreversible component derived from the spontaneous heat release of the potential energy (free energy) as it is dispersed as environmental heat while driving the process. If negative, heat is being released to the environment (exothermic process). The reversible entropy changes of state are used as a measure of structural order; more positive values are usually regarded as more random (however, see Chapter 17).

A commonly used technique in classical thermodynamics is to allow two substances or states to obtain equilibrium, whereupon the free energy change is zero. This often allows equations and values to be calibrated. For example, the free energy of rock surfaces can be determined from the free energies of the solutions with which they have been equilibrated.

ENERGY FLOW PATHWAYS

Molecular Diffusion

Any concentration of molecules above absolute zero carries its vibrations and motions of thermal energy and by this random pattern of its momentums has a characteristic temperature. The higher the temperature, the higher the concentration of local kinetic energies. If there are surroundings where the molecules are less concentrated or the molecular momentum energies less concentrated, then there is a random wandering of matter or motions from the concentrated center outward into the less concentrated zone. This wandering, called *diffusion*, is driven by its own potential energy of concentration relative to the surroundings. The expansion of the concentration of heat and molecules is a transfer of potential energy into heat with entropy increase. A diffusion pathway is represented with the single line and heat sink without a barb (Fig. 7-12b) since the net flow can be in either direction. The driving forces on diffusion are molecular concentration gradients, temperature gradients, or both. One may regard the concentrations at the downstream end of the pathway as backforce since there is a return wandering tendency that leads to equilibrium when concentration gradient of heat or molecular concentration is zero. The potential energy that existed in the gradient before diffusion disappears as the diffusion process becomes part of the general random motions of the molecular vibrations that we call *heat*. This entropy-generating process (diffusion) may be regarded as being a balance between the driving force (the concentrations) and the drag inherent in generation of new entropy. In this concept there is an energy degradation backforce visualized in Fig. 7-12b as pushing up from the heat sink. Diffusion is a balance between driving gradient and entropy generation backforce. Eddy diffusion involves an additional energy storage and source and is thus different (see Chapter 8).

Circular and Reversible Pathways

In molecular systems where forward and backward processes are visualized to be the same, a pair of arrows is sometimes used as in the A to B connection in Fig. 7-13a. In the usual way of aggregating in energy language, this type of reversible path is shown simply with a single pathway line without barb. If there is equilibrium in the sense of constant states and no differences in potential energy, there is no flow of degraded energy and thus no heat sink.

A more complex case in Fig. 7-13b has two diffusion processes coupled, such as diffusion of heat and chemical concentration, shown in the figure at equilibrium.

Sometimes offered as a chemical principle is *microscopic reversibility* (Onsager, 1931). The idea is usually expressed as Fig. 7-13c and contrasted with Fig. 7-13e, which shows the concept of circular pathways in equilibrium. The energy translations of these are given in Figs. 7-13d and 7-13f, respectively. Microscopic reversibility may not be a general principle (Hamilton and Peng, 1944). Closed chains of transitions may exist in cyclic balance (Watanabe, 1951). Consider the equilibrium between water and vapor diagrammed in Fig. 7-12c. The process in one direction is different from that in the other and one is required to diagram it that way. Morowtiz (1966), starting with microscopic reversibility at equilibrium, showed that adding an energy source made material cycling necessary in open systems since they are away from equilibrium.

Representation of pathways at the level of molecular collisions is given in Fig. 7-13g, where two molecules transfer energy to a third, which emerges at a high speed (and thus with high energy). Conservation of energy for this impact is illustrated in Fig. 7-13h, which refers to populations of such collisions. In the distribution of molecules that results, there are a few molecules with high energy and many with low energy (see Chapters 14 and 15).

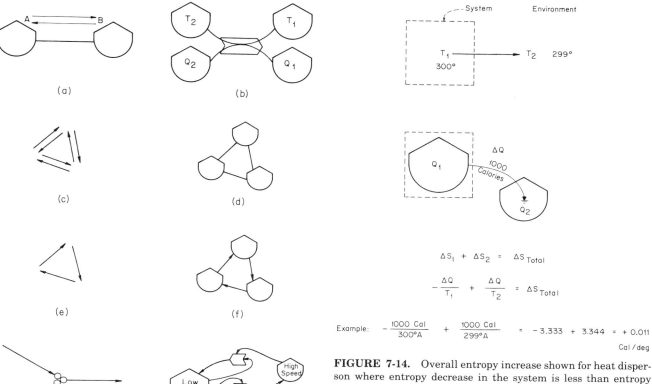

FIGURE 7-13. Concepts of circular and microscopic reversible pathways of states in reversible equilibrium as far as energy exchange: (a) diffusion pathway with A and B at equilibrium; (b) heat and chemical diffusion coupled but at equilibrium; (c) concept of microscopic reversibility; (d) energy diagram of part c; (e) circular concept; (f) energy diagram of part e; (g) sketch of molecular collisions and conservation of momentum; (h) energy diagram of part g.

FIGURE 7-14. Overall entropy increase shown for heat dispersion where entropy decrease in the system is less than entropy increase in environment.

Entropy Always Increasing

A famous expression of the second law of thermodynamics is that "entropy in any real process always increases," at least on earth (the universe is an open question). The diagram in Fig. 7-13 may be a help in visualizing this relationship. The phrase refers to entropy changes in both the system of interest and its environment. Entropy may decrease in a system as long as it increases even more in the environment as part of the process. Recalling that entropy is the ratio of heat change to absolute temperature of that change, consider that there are entropy changes in a process and in the environment around it. In Fig. 7-14 the dotted lines outline a system in which a chemical reaction process is occurring and that has the reversible and irreversible components of heat change and thus of entropy change. When heat is released and the temperature starts to rise, heat gradients develop and the heat energy is dispersed into or in from the environment. It is a property of ratios and fractions that the increased fraction is greater than the decreased fraction. The heat flowing out of a place of higher temperature is accompanied by a lower entropy decrease than the same heat produces flowing into a place of lower temperature. Thus any releases of heat such as those in Fig. 7-14 ultimately raise the overall entropy where the reaction and the environment are concerned. A numerical example in Fig. 7-14 shows the increase obtained by heat releases to the environment.

Energy Effect Dependent on Downstream Pathway and Interactions

Formulations for energy storage and transformation were given in previous paragraphs. The energy content of a storage or flow is not an indication of the energy flow that it may cause downstream (in the short run). Figure 7-15 shows various uses and in-

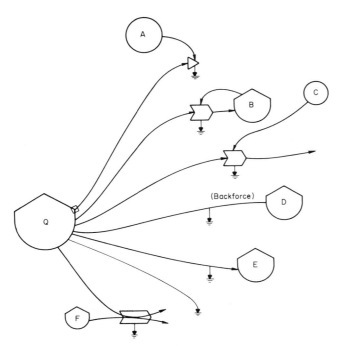

FIGURE 7-15. An energy storage with several kinds of pathways, each involving quantity Q in a different process and with the effect of the interacting quantities (A-F).

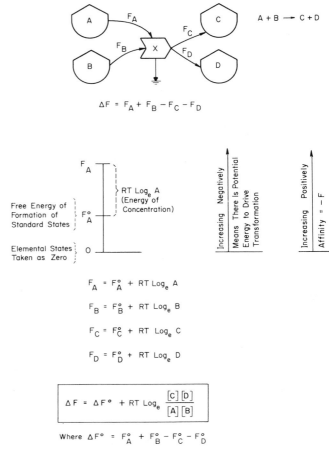

FIGURE 7-16. Free-energy change in a chemical reaction. All free energies (F values) are really energy differences from a reference state chosen arbitrarily as zero for elements in elemental state. Thus Δ values are sometimes used for all F values. In any process the energy equal to ΔF is ultimately dispersed into the heat sink.

teractions of a source with the systems components that affect the rate of energy use.

In the long run after changes are made in the systems designs to allow for maximum utilization of all available energies, the power use may become proportional to the potential of the source. Selection for maximum power may affect organization of the systems components for effective utilization of sources.

Sensor

A storage or a flow may be connected through a sensor so that it controls some other flow but does not supply the main energy for it. The tiny square indicates a sensing action if it is accompanied by an arrangement for supplying the energy from another source (see A in Fig. 7-15).

Dual Energy Source to a Reaction Process

Where two flows interact as in a chemical reaction (see Fig. 7-16), the source of energy is in neither of the two reactants, but in their relationship. Often we fall into the habit of thinking of one of the interactive sources as the energy source. For example, since oxygen in the air is very large and constant, we think of wood as the energy source in the fire since it is the item that runs out, stopping the reaction. If the oxygen were scarce and the organic fuel large and constant, we would regard the oxygen as the energy source.

In chemistry the question as to which reactant is the energy source has been avoided by measuring energy relative to some arbitrary points of reference that are given zero values. Elemental substances such as oxygen or iodine are given zero values, and energy of reactions is measured relative to the elements as zero.

Other references may be more pertinent to environmental studies such as the biosphere average state. New work is needed in this direction.

Calculating Potential Energy of Chemical Reactions

The energy in a compound is that energy required to generate the compounds from the elements. If the heat change (ΔH) observed in the reaction (formation or decomposition) is known, and if the entropy contents of state are known for the component molecules, the free energy (chemical potential energy) can be estimated by using Eq. (7-9).

The chemical potential energy in a reaction is the difference between the free energy of formation of the reactants and that of the products. As shown in Fig. 7-16, the free energy of formation is measured from elemental states as a reference. The free energy of the reaction is the difference between the initial energy and the final energy.

From Eq. (7-6) we see that the chemical potential energy (free energy) in a chemical substance is related to the activities of the molecules as RT times the natural logarithm. The equation for adding free energies is one of adding logarithms of activities. Addition of free energies is done in Fig. 7-16, where standard free energies are added and logarithmic terms are first added and then the terms combined.

Overall Equation for Metabolism

An important example of energy transformation in chemical reactions is the breathing metabolism of humans, plants, animals, microorganisms, and equivalent reactions of many machines. Carbohydrate fuel reacts with oxygen to ultimately release carbon dioxide and water. Sometimes this is called *respiration;* it is the reverse of photosynthetic production. The relationships of the reaction to energies are shown in Fig. 7-17.

The overall equation for photosynthesis of normal green plants is well known to most grammar school students as the light-driven reaction of water and carbon dioxide to produce carbohydrate organic matter plus oxygen. Actually, there is not one reaction, but hundreds that make up the detailed biochemical systems for which many pages are required. Thermodynamics, however, deals with overall energy computations so that it is possible to describe free energy F, enthalpy H, and reversible entropy changes of state S independent of the details of the pathways. Given in Fig. 7-17 is the chemical equation for overall carbohydrate production, the thermochemical relationships, and an energy circuit diagram that summarizes the process. The energy

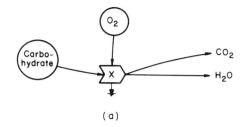

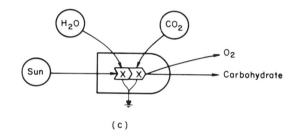

FIGURE 7-17. Overall reaction for spontaneously driven respiration and the reverse sunlight driven photosynthesis: (*a*) respiration; (*b*) overall equation; (*c*) photosynthesis; (*d*) thermochemical relationship where products and reactants are in standard states (bases at one atmosphere pressure; solutions at one molar).

relations in Fig. 7-17 are shown with products and reactants in standard state ΔF_0. For the real environment, concentrations are different and energy changes are somewhat different.

As Brody (1945) indicates, the free energy for metabolism of glucose is 708 cal/mole with carbon dioxide gas at 0.0003 atmospheres, water in liquid state, oxygen at 0.2 atmospheres, and the sugar as solid (see Fig. 7-18).

As written in Fig. 7-17, the free energy is nega-

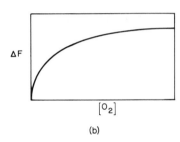

$$\text{Glucose}_{(s)} + 6\,O_{2(g)} \rightarrow 6\,CO_{2(g)} + 6\,H_2O_{(g)}$$

(s) Solid
(g) Gas

$$\Delta F = \Delta F_0 + RT\,\text{Log}_e \frac{[\text{Products}]}{[\text{Reactants}]}$$

$$\Delta F = \Delta F_0 + RT\,\text{Log}_e \frac{[CO_2]^6 [H_2O]^6}{[\text{Glucose}][O_2]^6}$$

$$\Delta F = \Delta F_0 + RT\,\text{Log}_e \frac{[0.0003]^6 [1]^6}{[1][0.2]^6}$$

300 ppm CO_2, Oxygen at 0.2 Atmosphere, Liquids and Solids with Activity of 1

Given: $\Delta F_0 = -685$ Calories/Mole

$RT\,\text{Log}_e = 1.364\,\text{Log}_{10}$

So: $\Delta F = -685 + 1.364\,\text{Log}_{10}\left[\dfrac{7.2 \times 10^{-22}}{6.4 \times 10^{-5}}\right]$

$\Delta F = -685 + (1.364)(-16.95)$

$\Delta F = -685 - 23$

$\Delta F = -708$ Cal/Mole

(a)

(b)

FIGURE 7-18. Calculation of free energy in metabolic reaction when concentrations are those of the terrestrial environment: (a) equations; (b) graph of free energy of metabolism in aquatic environment where oxygen may range from zero to an atmosphere of pressure or more.

tive, which by convention means that the reaction goes spontaneously to the right and can go to the left only if there is other potential energy added. In photosynthesis the reaction is driven to the left by coupling light energy to the process. This is an example of multiplicative intersections, an example of thermochemical relations, and important to memorize because of its generality in summarizing activities of whole ecosystems, including the biosphere as a whole. An overall reaction is really a kind of model simplifying into one or two multiplicative interactions a chain of processes that has hundreds. It is not a bad model for many purposes. For example, the relationship of metabolism to oxygen concentration is given in Fig. 7-18b as predicted by the model. Others are given in later chapters.

IRREVERSIBLE THERMODYNAMICS

An earlier effort to develop a system of equations for open-system energy flows produced a set of principles that has come to be called *irreversible thermodynamics* (Prigogine, 1947; Denbigh, 1951; De Groot and Mazur, 1962). *Irreversible thermodynamics* is another systems language concerning the coupling of energy processes. Some of its principles are as follows.

First, any process is regarded as driven by a thermodynamic force X, and the resulting flux J is in proportion to the conductivity L as we have already given in Eq. (1-4). However, in this special irreversible thermodynamics usage, forces are selected so that the product of force and flux is power [P in Eq. (7-11)]*:

$$P = JX \qquad (7\text{-}11)$$

For coupled processes, two force-flux equations are written, each with two components of force action:

$$J_1 = L_{11}X_1 + L_{21}X_2 \qquad (7\text{-}12)$$

$$J_2 = L_{12}X_1 + L_{22}X_2 \qquad (7\text{-}13)$$

See energy diagram, which is F in Fig. 7-15; Fig. 7-13b shows coupling in an equilibrium state. For three coupled processes, there are three equations of three terms each.

Equations (7-11) and (7-12) may be verbalized as follows. The first flux J_1 is being driven by force 1 (X_1), which acts in proportion to a conductivity L_{11} that relates force 1 to flow 1 plus additional drive from force X_2 acting in proportion to conductivity L_{21} that relates force 2 on flow 1. For the second equation, flow 2 (J_2) is driven by force 1 (X_1) through conductivity L_{12} that relates force 1 to flow 2 and an additional drive from force 2 acting through a conductivity L_{22} that relates force 2 to flow 2.

For coupled processes, conservation of energy re-

---

* The requirement that JX equal power caused the developers of this language to select potential energy as the driving force for chemical reactions, but by so doing, J no longer became proportional to the selected force. As we have already seen in Chapter 1, we made a different choice in our energy circuit system by selecting all forces so that the fluxes are proportional regardless of whether JX is power. Thus we define a population force N for chemical and other population processes rather than call chemical potential a *force* (Odum, 1972). See Table 7-1. In this latter convention JN is not power.

quires that the sum of potential energy flows and the heat dispersed into the environment with raised entropy be zero:

$$J_1 X_1 + J_2 X_2 - T\frac{dS}{dt} = 0 \quad (7\text{-}14)$$

By combining Eqs. (7-11), (7-12), and (7-13), it is possible to derive various interesting equations of physics and chemistry where these component equations apply correctly. (They are not correct for chemical reactions because J is not proportional to the force if force is taken as chemical potential.)*

Onsager's Principle

In using the equations for force and flux [expressions (7-12) and (7-13)], Onsager discovered that the conductivity by which force 1 was affecting flow 2 was equal to the conductivity of force 2 affecting flow 1:

$$L_{12} = L_{21} \quad (7\text{-}15)$$

This identity simplifies many equations. Probably this is a version of Newton's principle of equal and opposite forces. The effect of train 1 dragging train 2 on an adjacent track is equal to the reverse force of train 2 on train 1. In energy language, two coupled linear processes are diagrammed as in Fig. 7-13b and F in Fig. 7-15. In biological systems diffusion processes are often coupled (Katchalsky and Kedem, 1962).

Carnot Ratio Indicating Potential Energy in Heat Gradients

In the macroscopic world there is potential energy available in heat energy only when there is a difference in temperature so that there is a gradient in the random motions of the molecular populations. It is the difference in molecular kinetic energies that can be harnessed. The potential energies available to drive other work are in ratio as the absolute temperature. The potential energy available is found by multiplying the heat flow J_h by the Carnot ratio $\Delta T/T$, which is the fraction that the temperature gradient is of the absolute temperature of the heat dispersion:

$$J_{\text{P.E.}} = J_h \frac{\Delta T}{T} \quad (7\text{-}16)$$

The Carnot ratio $\Delta T/T$ is called a *thermodynamic force* since it indicates the intensity of delivery of a potential energy flow from a storage of heat relative to one at lower temperature. See Fig. 7-9.

Heat flow J_h is proportional to the temperature gradient ΔT as in Eq. (7-17):

$$J_h = K \Delta T \quad (7\text{-}17)$$

Substitution of Eq. (7-17) in (7-16) gives Eq. (7-18). Thus the power delivery of available potential energy is a square function of the temperature gradient:

$$J_{\text{P.E.}} = K \frac{(\Delta T)^2}{T} \quad (7\text{-}18)$$

These equations for the special behavior of heat in generating potential energy are implied in any heat storage placed in energy circuit language. The energy available is the Carnot fraction of the heat quantity stored Q as indicated in Eq. (7-19):

$$Q \frac{\Delta T}{T} = \text{stored potential energy} \quad (7\text{-}19)$$

For example, wherever there are heat gradients in the environment, there is potential energy available according to Eq. (5-7). If the efficiency of a process such as a thermal power plant is measured, it should be referred not to total heat energy, but to the Carnot fraction of it, since this is all that is available even if the process were 100% efficient in using potential energy. Real processes sacrifice efficiency for speed so as to deliver more power. See below.

Efficiency and Power When Output Load Is Varied

The methods of irreversible thermodynamics can be used to relate efficiency to power and consider efficiency at maximum power (Odum and Pinkerton, 1955; Odum, 1972; Tribus, 1961; Caplan, 1966). Where an energy transformation is developing a new storage doing work against backforce, the rate of energy transformation and the efficiency of transformation depends on the loading. See Fig. 7-19; consider the way transformation varies as the backforce loading is varied. At one extreme there is a balance between backforce loading X_2 and the driving force X_1. There is no flow and thus no power transformed.

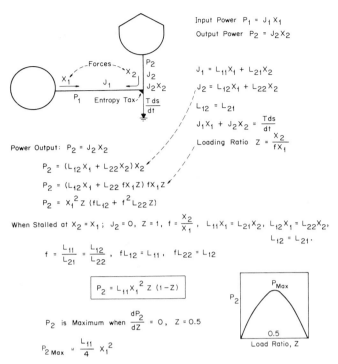

FIGURE 7-19. Power transformation and efficiency as a function of load ratio when input force X_1 is held constant. Maximum power is found at a loading ratio of 0.5; X_2 is taken as positive from left to right.

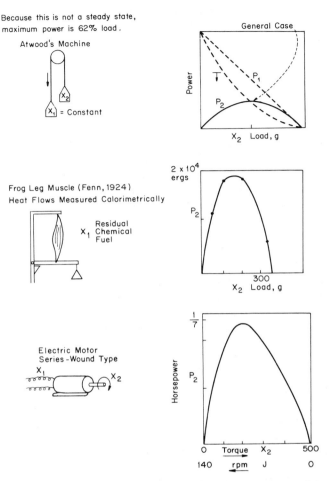

FIGURE 7-20. Examples of power loading in potential generating processes at constant input force.

The system is stalled. When the loading is almost as great as the driving force, the flow is small and the work done is almost as much as the energy delivered.

At another extreme there is no backforce, and no arrangement is made to store energy as output. In this situation all the potential energy delivered with the driving force goes into dispersed heat. Without backforce, the speed is maximal, power utilized is maximal, but no useful power is transformed. Thus no power is delivered at either extreme: stalling or flow without loading.

Given in Fig. 7-19 are equations for the transformation in terms of loading, and graphs are drawn of the transformation and efficiency as a function of the loading. At steady state, maximum power transformation into storage is found at 50% loading and 50% efficiency of energy output.

According to Lotka's maximum power principle, systems tend to develop designs that maximize power and thus may be expected to develop loadings less than the most efficient. At maximum power half of the input energy must be dispersed with a corresponding entropy increase. Sometimes this heat use is referred to as *entropy tax*. Sugiyama and Shimazu (1972) extend the derivation to graph efficiency as a function of pathway characteristics. See also Fig. 23-37 (examples are given in Fig. 7-20). Milsum (1966, p. 405) provides a graph with maximum power for human elbow muscle at an intermediate force.

Efficiency and Power When Input Load is Varied

The loading relationships of many units in nature are set as a part of their structural nature, whereas their input energy sources vary, delivering different driving forces at different places and times. Figure 7-21 shows the relationship of efficiency and power when only the input load is varied. At one extreme on the left the input force balances the backforce loading, efficiency approaches 100%, and the process is stalled. At another extreme on the right the input force is much larger than the set backforce X_2. The

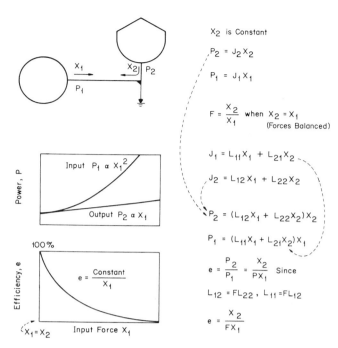

FIGURE 7-21. Power transformation $P_1 \to P_2$ and efficiency e as a function of driving force X_1 where backforce loading X_2 is held constant.

input power required is very large and wasteful, and the efficiency is very small. Efficiency is inverse to driving force. Curves of efficiency of energy transfer with this hyperbolic shape are often obtained in experimental studies of energy transformation with varying light or other types of energy transformation. The loading for maximum power to which the input driving force can be adjusted so that the process is neither so fast as to be too wasteful nor so slow as to generate too little transformation was given in Fig. 7-19.

Power in Oscillating Steady States

A major unsolved question is whether oscillating steady states develop more useful power than do constant steady states. Pulsed and oscillating states are possibly the more general patterns in nature. What frequency of oscillations maximizes power?

To maximize outputs in engineering, computer optimization methods have been used to determine frequencies for maximum output [see reviews by Guardabassi (1976) and Guardabassi et al. (1974)]. Whether a more general thermodynamic principle can be discovered remains to be seen (see Chapters 15 and 22). Richardson and Odum (1982) found an optimal frequency for maximum power in a general model of production, consumption and recycle.

Energy Storage in Light Flux

Some energy storages are tied to matter storages Q in different ways and by different laws as with chemical potential energies or water storages against gravity (Table 7-1). Other storages are almost pure energy.

Light energy passing through a system is very dilute, and the per volume rate is only about 10^{-15} cal/cm^3 volume (light intensity divided by speed of light). This tiny storage still delivers a force, as do all potential energy storages. *Light pressure* is a real property observed in astronomical systems. Because light per unit volume is so dilute, much energy is required to go into heat to concentrate it. Visions of great energies to humans from sunlight beyond those achieved by plants are not realistic because they ignore the requirements that increased concentrations of energy be accompanied by dispersals of much of the energy into unavailable heat. Efficiencies of photovoltaic silicon cells and those of plant chloroplast semiconductors in the first step of photosynthesis are similarly 10–20% and depend on load as explained in Fig. 7-21. Widespread belief that the hardware has a higher degree of efficiency than the chloroplast cells comes from comparing gross conversion of hardware with net photosynthesis of the plant after it has used the initial conversion energy to maintain the many transformations that are part of converging and upgrading the initial chemical conversion to concentrated and higher-quality products (see Chapter 19).

CORE CONDUCTORS

A special kind of energy pathway is the wave that moves along a cylindrical core conductor. The nerves of living organisms fit the model of modulations on a core conductor following the Hodgkin–Huxley concepts. As given in Fig. 7-22 in energy language, a potential is maintained between outside and inside of the tubular organized "cable" with an energy source supplying a source of potential. Because of the storage capacity and the source of energy, a deformation in the surface storage can be transmitted as wave moving down the tube with a restoration process rebuilding the storages ready for another impulse. Simulation of the action potential by Dubois and Schoffeniels (1974) included equations

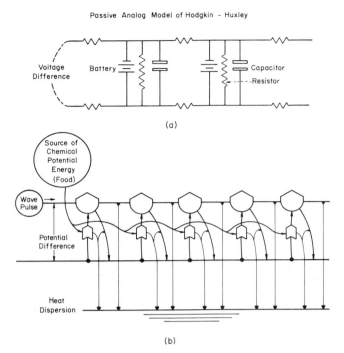

FIGURE 7-22. Energy transmission as a wave on a core conductor: (*a*) equivalent circuit language; (*b*) energy language.

for the conductances in terms of concentrations of sodium, potassium, calcium, and acetyl choline.

PROPOSITIONS OFFERED TO EXPLAIN RATES OF FLOW OF REAL PROCESSES

Starting in Chapter 1, the principle of maximum power competition is suggested as responsible for adjusting of rates in the real world to be at that optimum speed and efficiency that is neither as fast nor as efficient as they might be, but intermediate in speed and efficiency where power is maximum. The maximum power principle is an old concept. Lotka (1922a) quotes Boltzman (1886) "that the fundamental object of competition in the life struggle—is available energy" and Ostwald (1892) "of all possible energy transformations, that one takes place which brings about the maximum transformation in a given time." Lotka stated "natural selection tends to make energy flux through the system a maximum."

Sugita (1951), using examples such as stream capture in land drainage, suggested that systems tend to maximize power in their organizational work. This appears to be an independent statement of the maximum power principle. Prigogine and Wiaume (1946), in a statement of minimum entropy generation, described the return of a displaced steady state to the steady state. The entropy generation rate during the transition is the regular one plus that of the return process. However, this principle said nothing about the loading and rate of the steady state chosen among alternative steady states. Surviving systems don't tend to minimize entropy except when they are displaced toward the side of faster entropy generation rate than the one that maximizes useful power. If one really minimized entropy generation rate, one would stall and stop the process. Experimental simulation showed linear dissipative pathways compete with autocatalytic pathways only when the energy available is small (Odum, 1982).

In chemistry, LeChatelier's principle was used to predict which way an equilibrium will be shifted when a reactant is increased. This is equivalent to adding an input from an energy source so that potential energy exists to drive the process in the opposite direction (see Chapter 8).

Leopold and Langbein (1962) related steepness and meander of stream beds to the tendencies of entropy to increase at some rates other than the maximum possible. In other words, the cutting of the stream develops land structure that helps it to do work at a rate that is more adaptive to survival than if the water were dumped with all its energy suddenly released at the end. Streams, as high-quality energy by interacting over broader areas with land and sunlight, help to maximize power of a region's total energy flows.

Sometimes given as a principle for energy flow is the Hamiltonian principle, which refers only to potential and kinetic energy and their intertransformation under conditions in which there is a conservation of energy. The motion goes so that the difference between kinetic and potential energy is minimized (see p. 186).

USING ENERGY AS A MODELING VARIABLE

In the next chapters basic systems designs are considered with models in energy circuit language. How is energy related? After data are tabulated and referred to a model in the various units that may be convenient, an option remains. Can the data on flows and storages be expressed in various units of matter or converted entirely to units of energy? For certain substances such as biomass, energy is proportional to the measures of matter, and thus conversion only changes coefficients. Flow is proportional to storage in either unit.

For some others such as water and electricity, energy is the square of the quantity of matter; and for others of molecular nature, the energy is a logarithmic function of the quantity (see previous sections of the chapter).

Some of the flows, however, also change when converted from matter units to energy units. For water and electricity, for example, the flow of energy is a square of the quantity flowing, so that the flux of water-borne energy is proportional to the storage of water potential energy so that a linear law still remains. Similarly, for chemical storages, energy is a logarithm of the quantity (concentration) and the flow is proportional to the concentrations, but the energy flowing is a logarithm of matter flowing and energy flow is proportional to energy storage. Thus *in these cases changing to energy units for storages makes the flows change by the same function so that flux is still proportional to the quantity chosen to measure storage,* regardless of whether it is in original mass terms or in energy terms. Apparently, units used to model and simulate can be either energy or other state variables that accompany energy. If some case is found where the kinetics are different for energy than for the energy-carrying state variable, the model would have to be changed to accommodate the shift in units.

SUMMARY

In this chapter concepts of energy storage and flow were introduced, including energy laws, entropy, classical concepts, open-system thermodynamics, maximum power selection, and expressions in energy circuit language. Supplementary reading may be needed in texts on chemical energetics as listed.

QUESTIONS AND PROBLEMS

1. Give a quantitative definition or explanation of each of the following:

    Three energy laws
    Gibbs free energy
    Reversible entropy change
    Enthalpy
    Entropy of state
    Irreversible energy flows
    Irreversible entropy
    Endothermic process
    Exothermic process
    Absolute zero
    Meaning of bomb calorimetry value
    Approximate range of bomb calorimetry values for different substances
    Definition of Calorie; kilocalorie (kcal)
    Helmholtz free energy
    Affinity
    Chemical potential energy
    Chemical potentials by chemical component
    Entropy as a measure of molecular uncertainty
    Onsager relation
    General language of coupled reaction in steady-state thermodynamics
    Potential energy for a population of molecules under compression
    Relationship of free energy and molecular concentrations
    ΔH, ΔF, and $T \Delta S$ values for the overall metabolic reaction
    Absolute temperature in terms of entropy of state
    Difficulty of use of chemical potential as thermodynamic force
    Prigogine's minimum postulate
    Sugita's maximum principle
    Maximum power postulate
    Relationship of electric potential and chemical potential energy
    Energy coupling
    Feedback of free energy to maximize rates
    Shape of efficiency and power transfer curves: case 1 input force constant, varying output load; case 2 output load constant, input force varying
    Meanings of load; backforces: frictional, potential derived, inertial
    Carnot ratio for reversible conversions of heat gradient energy to other potential energy
    Standard free energy and total free energy of the same reaction
    Error in using heat changes in bomb calorimetry to estimate potential energies
    Microscopic reversibility

2. Diagram the following relationships so that energy sources and sinks are given correctly according to the first and second laws and the criterion for increased quality with work:

    Evaporation of water
    A reaction that has a positive Gibbs free energy
    A chemical reaction with a decrease in entropy of states
    A linear coupling according to Onsager relationships

A true equilibrium with no energy flow
A potential to kinetic energy transformation without entropy increase
Flow across an energy barrier that utilizes free energy to overcome a high energy of activation
A chemical reaction with no irreversible heat
Ultimate fate of the free energy in a real process
Diffusion
Turbulence generated by flow from a potential
Pumping out heat to cause a liquid to crystallize
Pendulum
Electric transformer
Light pressure

3. What is the relation of power and efficiency for different loadings in potential generating energy transformation? What is the curve of efficiency and power where the output load is constant and the input force varied? How is maximum power selection related?

4. How does energy for mechanically reorganizing objects on earth vary with speed of that organization? What speed is selective?

5. Write an equation for the rate of generation of entropy in terms of forces and fluxes.

6. Give three different types of relation of energy storage to storage of state variables that are associated with and carry that potential energy. Explain why these differ.

7. What is the condition of matter at absolute zero and as energy is added to raise the temperature? What happens to complexity, heat capacity, and entropy in this sequence of changes?

8. Review energy diagramming of various types of pathway from one storage to another, the situations with and without backforces, and the types of backforces.

9. What are the main types of external source? How do they respond to demand?

SUGGESTED READINGS

Brody, S. (1945). *Bioenergetics and Growth*, Rheinhold, New York. 1023 pp.

Calow, P. (1977). Ecology, evolution and energetics. A study in metabolic adaption. *Adv. Ecol. Res.* **10**, 1–63.

Denbigh, K. L. (1951). *Thermodynamics of the Steady State*, Methuen, London.

Dickerson, R. E. (1969). *Molecular Thermodynamics*, Benjamin, New York.

Gates, D. M. (1962). *Energy Exchange in the Biosphere*, Harper and Row, New York.

Grodzinski, W., R. Z. Klekuwski, and A. Duncan (1975). *Methods in Bioenergetics*, IBP Handbook No. 24. Blackwell Scientific Publishers, Oxford.

Heinz, E. (1978). *Mechanisms and Energetics of Biological Transport*, Springer-Verlag, New York. 159 pp.

Hill, T. C. (1977). *Free Energy Transductions in Biology*, Academic Press, New York.

Hoffman, E. J. (1977). *The Concept of Energy*, Ann Arbor Science, Ann Arbor, MI.

Katchalsky, A. and P. F. Curran (1965). *Nonequilibrium Thermodynamics in Biophysics*, Harvard University Press, Cambridge. 248 pp.

Kittel, C. (1969). *Thermal Physics*, Wiley, New York. 418 pp.

Klotz, I. M. (1967). *Energy Changes in Biochemical Reactions*, Academic Press, New York.

Lewis, C. N. and M. Randall (1961). *Thermodynamics* (revised by K. S. Pitzer and L. Brewer), McGraw-Hill, New York.

Lehninger, A. L. (1965). *Bioenergetics*, Benjamin, New York. 452 pp.

Linford, S. H. (1966). *An Introduction to Energetics*, Butterworths, London. 223 pp.

Lotka, A. J. (1925). *Elements of Mathematical Biology*, Dover Books, New York.

Morowitz, H. J. (1968). *Energy Flow in Biology*, Academic Press, New York.

Morowitz, H. J. (1970). *Entropy for Biologists*, Academic Press, New York. 195 pp.

Morowitz, H. J. (1978). *Foundations of Bioenergetics*, Academic Press, New York. 344 pp.

Nash, L. K. (1972). *Elements of Statistical Thermodynamics*, Addison-Wesley, Reading, MA.

Nicholls, D. G. (1982). *Bioenergetics*, Academic Press, New York. 200 pp.

Nicolis, E. and I. Prigogine (1977). *Self Organization in Non-Equilibrium Systems*, Wiley, New York.

Patten, A. R. (1965). *Biochemical Energetics and Kinetics*, Saunders, Philadelphia.

Peusner, L. (1974). *Concepts in Bioenergetics*, Prentice-Hall, Englewood Cliffs, NJ. 298 pp.

Phillipson, J. (1966). *Ecological Energetics*, Edward Arnold, London, 57 pp.

Prigogine, I. (1955). *Introduction to Thermodynamics of Irreversible Processes*, Wiley-Interscience, New York.

Rashevsky, W. (1940). *Advances and Applications of Mathematical Biology*, University of Chicago Press, Chicago. 214 pp.

Scott, D. (1965). The determination and use of thermodynamic data in ecology. *Ecology* **46**, 673–680.

Spanner, D. C. (1964). *Introduction to Thermodynamics*, Academic Press, New York.

Scientific American. (1971). Energy and power. W. H. Freeman, San Francisco. 224 pp.

Thorndike, E. H. (1976). *Energy and Environment*, Addison-Wesley, Reading, MA.

Tribus, M. (1961). *Thermostatistics and Thermodynamics*, Van Nostrand, New York.

Wesley, J. P. (1974). *Ecophysics*, Thomas, Springfield, IL.

PART TWO

DESIGN ELEMENTS

CHAPTER EIGHT

Intersections

Systems pathways interact in intersections with many different relationships, including additions, multiplications, divisions, and switching actions. Most interactions involve energies of different type in work transformations that generate productive output. In any intersection the sum of the energies inflowing must equal those outflowing, and any intersection that is not an equilibrium situation has a heat sink through which degraded energy is dispersed as part of spontaneity. Since intersections are the means by which one energy flow controls another, they are often control points for system regulation either from within the system or by outside control actions. In this chapter we consider ways of representing the various functions concerned with intersections, outputs of interactions affected by limiting factors, and models for intersections with more than two flows.

ADDITIVE ENERGY FLOW PATHWAYS

Illustrated in Fig. 8-1a is an additive intersection by which two energy flows add and their material flows add; the outflow is the sum of the two inflows. To be added, the two flows must be of similar nature in their contribution to the next process downstream. The two flows must be able to substitute for each other so that either one or both may serve. Examples are converging water pipes, converging streams of people, electrical junctions, and duplicate food sources.

LINEAR COUPLING OF TWO FLOWS

Figure 8-1b shows a pair of linear flows that are different in nature but coupled energetically, with one flow affecting the other but without the interactions of multiplicative and more complex nature. For example, a thermocouple has flows of heat and electrical current; evapotranspiration involves a diffusion of heat and a molecular diffusion. The flows are proportional to their driving forces *plus* some positive or negative contribution from the other flow's force. The expressions of irreversible thermodynamics were invented for this class of energy coupling (Luikov and Mikhailov, 1965) (see Chapter 7).

CONSTANT-GAIN AMPLIFIER INTERSECTION

An important, frequently occurring intersection of energy flows is where the energy from one source is being controlled to deliver output in proportion to the other source but not drawing much energy from it. This is the property of a constant-gain amplifier (see Fig. 8-1c). The operational amplifier symbol is adapted from the analog computer language to the energy circuit usage with source and heat sink added and without any sense of positive or negative sign switching. The output is proportional to the input control B according to the gain factor g, and energy is supplied as needed and thus also in proportion to B from the power source A. If the demands on the power source A begin to exceed its ability to make the required inflow, the module is no longer appropriate, and the system begins to be a multiplier action or other configuration. An example is a microphone amplifier; another is the reproduction of some higher organisms that have set rates of increase as long as the food supplies are not limited.

INTERACTION SYMBOL

A double-pointed block symbol (introduced in Chapter 1) is used for energy transformation with inter-

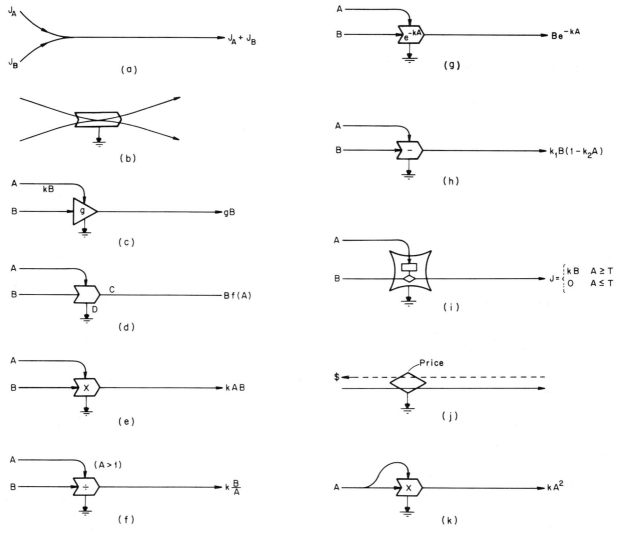

FIGURE 8-1. Types of intersection between two energy flows: (a) additive flows; (b) linear coupling; (c) constant-gain amplifier; (d) undefined work gate with backforce (no barb); (e) multiplier without backforce; (f) divisor; (g) exponential; (h) negative multiplier, drag; (i) switching control; (j) price-controlled transaction; (k) quadratic self-interaction. Equations for (b): $J_A = L_{AA} + L_{BA}$; $J_B = L_{AB} + L_{BB}$.

sections of two or more flows, both of which are required for the process (Fig. 8-1d). The interaction symbol may involve different functions for the kinetic and energy relationships. The specific process of interaction determines the energy relationships and the kinetics by which the driving forces on the interacting pathways generate the outflow. For example, Fig. 8-1e shows a common interaction intersection with a multiplier action, the output proportional to the product of the driving forces on the interacting pathways. If one of two intersecting flows becomes zero, the process stops.

WORK OF INTERSECTIONS

Most interactive intersections between pathways of energy flow involve work done by one flow on the other in various arrangements and functions. Such intersections are said to be *coupled*. Work is defined as a transformation, and many kinds of work are measured as the product of a force times the displacement over which that force has acted. Pathway intersections where one flow is doing work on another involve one pathway that exerts forces on the other and receives backforces from it. For coupled work intersections, the nature of the flows is normally different. The process requires something from both A and B. In energy circuit language the double-pointed block is a work intersection or work gate.

The general class symbol in Fig. 8-1d does not indicate the nature of mathematical function produced by the interaction. It does show that the sum of the inflow potential energies A and B must be equal to outflows C and D. The symbol has no storage in it and thus is instantaneously responding without time lags.

If there is flow, there is usually a loss of some potential energy into dispersed environmental heat as the second energy law requires, and this loss is indicated by the heat sink as part of the symbol. There are many common types of intersection, some of which are indicated next.

Proportional Conductor, Multiplicative Intersection

If one energy flow is controlling the conductivity of another flow according to a linear effect of its driving force, Eq. (1-4) may be combined with Eq. (8-1) to obtain the overall multiplicative expression [Eq. (8-2)]:

$$J = LX_1 \qquad J = LN_1 \qquad (1\text{-}4)$$

$$L = kX_2 \qquad L = kN_2 \qquad (8\text{-}1)$$

$$J = kX_1X_2 \qquad J = kN_1N_2 \qquad (8\text{-}2)$$

In other words, the pathway with a conductance proportional to another energy drive delivers a potential energy flow that is proportional to the product of the two input forces. The work intersection is multiplicative. As drawn in Fig. 8-2a with an "X" symbol, the intersection is shown to deliver the product function and, in proportion, a flow of dispersed heat.

The simplest version for a multiplier as drawn in Fig. 8-2a utilizes energy from the two sources in proportion to the output of the work process, although the coefficients are different. Shown in Fig. 8.2a are the four algebraic expressions that are implied.

One pathway may be of the single force–flux type, and the other may be of the population force–flux type. In this situation the algebraic expressions are

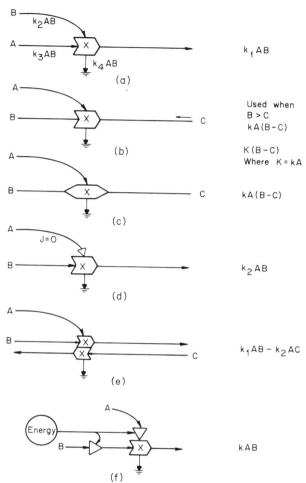

FIGURE 8-2. Multiplier intersections: (a) one-way multiplier without backforce; (b) multiplier with backforce; (c) reversing gradient multiplier; (d) multiplier with enegy drawn mainly from one source B; (e) pumping forward and backward; (f) energy source different from interaction controls.

of mixed X and N multiplier type as given in Eq. (8-3):

$$J = kXN \qquad (8\text{-}3)$$

EXPONENTIAL ATTENUATION

Where one pathway is causing an exponential effect in changing a second flow, it may be represented with an interaction symbol in which the exponential relationship is written within the symbol (see Fig. 8-1g). Attenuation of light and action of temperature are examples sometimes represented in this way.

Haggett (1966), summarizing various functions used to relate migrants and other influences moving out spatially from center, found exponential attenuation appropriate (Johnson, 1952).

Exponential processes are often the mathematical summary of systems actions involving several units, and in those examples the process can be shown in more detail to express the mechanisms. Examples follow in later chapters.

DRAG—RETARDING INTERACTION

We described the interaction in which one energy flow increased the conductivity of another as a multiplier. If the intersection decreases the conductivity, it retards the flow, subtracting a multiplicative term. We indicate this arrangement in Fig. 8-1h with a negative sign in the work gate and call it a *negative multiplier*. An example of this action is the effect of aquatic plants in blocking a water flow channel or the effect of picket lines in blocking delivery of goods.

QUADRATIC PATHWAY

When energy is lost from a storage because of some process that involves interaction within the quantity stored, the flow may be quadratic. It is proportional to the square of the quantity stored as shown in Fig. 8-1k. For example, a very crowded population of animals may be interfering with each other through wastes, encounters, or other interaction where the effect is proportional to the frequency of interaction, which depends on the square of the quantity. Other examples are effectiveness of military fire power, energy flow for information retrieval, and group cooperative behavior.

Multipliers That Do Not Drain Their Sources

If the intersection of energy flows draws all its energy from one source but operates in proportion to the product of them both, we show this with a combination of the multiplier work gate and a connecting amplifier symbol as given in Fig. 8-2d. Although there must be some energy drain for an effect to be transmitted, it can be negligibly small, with the energy for its amplification coming from the other inflows. As shown in Fig. 8-2f, arms of the interaction receive energy from a source different from that controlling the multiplicative process.

Work Gates with Backforce

Equation (8-1) presented the multiplicative work gate as a conductivity-varying action of one force on the flow from another. Consider the case where the flow is subject to forward and static backforce from a downstream storage as well as to the conductivity varying action of the multiplier. The expression for the differential backforce is given in Fig. 8-2b as coefficient k times the difference in forces B and C times the force A. If the flow is always forward, the pointed block is used. The barb along the multiplier input or output indicates absence of backforce.

Reversing Multiplier, Dual Pathways

If the flow through the multiplier with backforce is subject to a reversal of direction when the downstream force is greater than the one upstream, a double-pointed block is used (Fig. 8-2c). Examples are actions of force A in controlling the flows in pipes without valves. The reversing multiplier is similar to the system shown in Fig. 8-2e with two counter directed pathways, each a multiplier whose conductivity is controlled by the same multiplier variable. However, the dual-pathway system has two coefficients.

Eddy Diffusion and Transportation

A special example of reversing a multiplier is the eddy diffusion pumping by which fluid eddies mix concentrations from both ends with a net effect of moving a flow in the direction of the gradient. A similar case is the circulation of trucks and trains between cities that transport a net exchange of groceries or other goods in the direction of the gradient. In these examples the more energy from A going into circulation, the more conductivity there is. Eddy diffusion coefficients K thus are made up of the outside pumped energy source A and its flux relationship k. They are not pathway constants since they depend on an energy source.

CHEMICAL EQUATION LANGUAGE

A familiar scientific language is that for chemical reactions in which the reactants are listed with plus signs, indicating that they are required ingredients, and then connected to products with an arrow indicating direction of process. The equations for rates are written separately according to the principles that dilute reactions react in proportion to the number of reacting units (i.e., concentrations). Thus in Figs. 8-3a–8-3c are given zero-order, first-order, and second-order reactions in their usual chemical notation with reaction rate equations and energy language translations. Figure 8-3c shows a forward reaction in the direction in which there is free energy (ΔF is minus); Fig. 8-3d shows a back reaction where the pathway is different from that in Fig. 8-3c; Fig. 8-3e has both pathways; Fig. 8-3f shows a self-reaction; and Fig. 8-3g includes the back reaction. When a reaction has two by-products from one flow, it is shown as Fig. 8-3h or 8-3i.

Additive Chemical Potentials in Reaction Intersections

In Chapter 7 free energies of chemical substances (chemical potentials) were shown to be logarithmic functions of the concentrations of molecules. Rates of reaction are proportional to the product of concentrations. Since the chemical potentials (u values) are logs and thus exponents, they are added when the concentrations are multiplied. Therefore, to obtain the free energy of a multiplicative intersection involving molecular population reactions, add the component free energies and subtract that of the products as in Eq. (8-4), which refers to reaction in Fig. 8-3c:

$$\Delta F = u_A + u_B - u_C \qquad (8-4)$$

Quality and Position of Interacting Arms

If A and B are identified in order of their energy quality with the sense of amplification implied by the product symbol, the position of B indicates that high quality is controlling low quality (A), yielding production (see Fig. 8-3c). If inputs were reversed, it would mean that high quality (B) was being used for low-quality purpose with a loss of production. See Fig. 8-4a.

DILUTION AND DIVISOR ACTION

A common interaction of two variables involves the dilution of one by the other. For example, water may dissolve a stock of sugar, land area may be the receiver of seeds dispersed over an area, and the energy concentration of a total quantity of plankton is defined inversely to the volume over which it is dispersed. These examples involve the division of one variable by another and with the dispersion energy flows into heat as part of the free energy driving the process. These divisor actions are diagrammed in energy language as in Fig. 8-1f, with the divisor at the top of the work gate A acting to dilute the flow B passing from the left.

If one section of a model concerns total quantities but another involves variables related as concentrations to area or volume, the pathways from the first part of the model to the other involve a divisor action and return flows a multiplier action between concentration and the area or volume (see Fig. 8-5a).

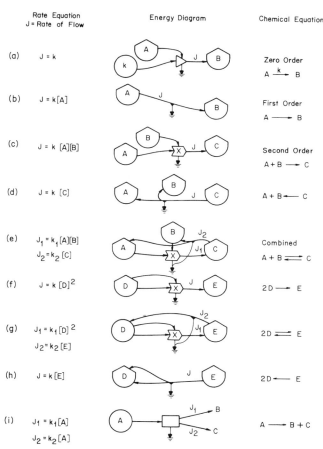

FIGURE 8-3. Chemical reactions.

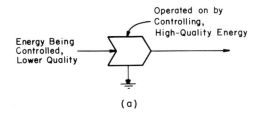

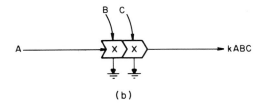

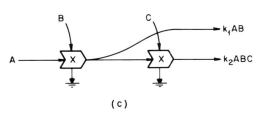

FIGURE 8-4. Conventions in energy diagrams: (a) positions of high- and low-quality flows; (b) aggregated diagram of a process that has an overall output as the product of three flows; (c) arrangement for split of flow within a series of multiplier intersections.

An example of this use is given in Fig. 8-5b in translating a single tank model of pollution.

FLOW SENSORS AND DIVERGING PATHWAYS

Whereas the intersections discussed so far in the chapter concerned converging energy flows, configurations by which energy flows diverge are also important. If an energy flow diverts from another flow, it draws some force from the source flow. Given in Fig. 8-6 is an arrangement in which the lateral flow is in proportion to the main flow. Some energy is diverted with the flow.

If this diverging flow is connected to a work gate action, the diagram indicates that there is a force derived in proportion to the flow, but the flow, if any, depends on the function of the work gate. Thus, as shown in Fig. 8-6c, the multiplier is affected by the force from the flow, but the flow that the multiplier

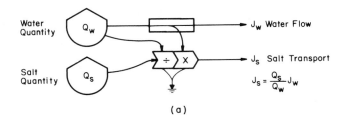

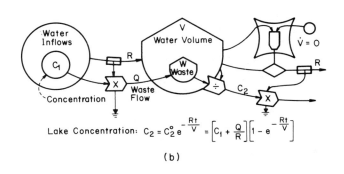

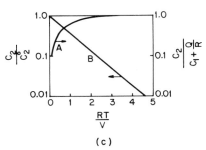

FIGURE 8-5. Intersections that represent concentration of a solute with a divisor followed by flow as the product of water flow and concentration: (a) energy diagram for example of salt concentration in water; (b) example of a model of pollution (Rainey, 1967); (c) graphs of effect of water displacement RT/V on concentration during pollution and afterward.

draws is proportional to the product of input forces B and kJ_0).

If the arrangement delivers a flow regardless of whether the multiplier is operating, then a bypass arrangement is shown as in Fig. 8-6d.

If the energy for the process taking a measure of the flow is from another source, the flow is sensed without drawing energy from it, and the amplifier symbol is appropriate as given in Fig. 8-6e.

A common example in which a flow is in proportion to another flow is the transport of solutes and sediments by a carrier fluid flow. In Fig. 8-5, for example, the transport of salt by a tidal current de-

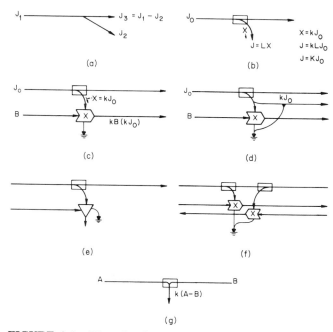

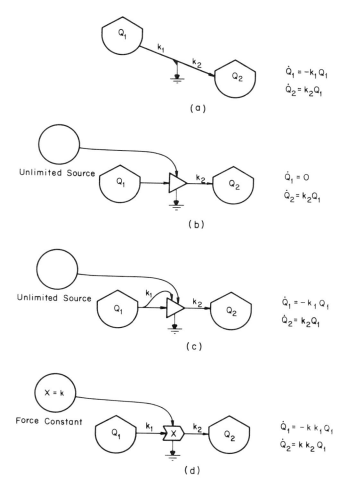

FIGURE 8-6. Diverging flows and sensors: (a) subtraction of a flow without transformation; (b) diverging flow with force proportional to main flow J_0; (c) multiplier operating with one force dependent on a flow J_0; (d) combinations of parts a and b; (e) a flow sensor with amplifier; (f) different flows for reversing pathway (see also Fig. 8-3e); (g) output of difference in force between two states.

pends on the concentration of salt in water times the flow of water. Since the concentration of salt is the total salt:total water ratio, the relationships involve the combination of a divisor and a multiplier.

The delivery of lateral force from a flow shown in Fig. 8-6b indicates force when the main flow J_0 is from the left. In Fig. 8-6f provision is made for a force from a reversed flow as well, and in the configuration shown, it drives a multiplier to the left different from the one pumping to the right.

In Fig. 8-6g a force is developed by sensing the difference in two states.

LINEAR WORK PATHWAYS WITH AND WITHOUT EXTERNAL ENERGY

Linear pathways are those with flow proportional to upstream (donor) force. Figure 8-7 shows pathways of several types, each of which is linear and has energy transformations. In the first (Fig. 8-7a) energy is supplied from the upstream donor tank source of driving force as discussed in Chapter 3. The second

FIGURE 8-7. Linear pathways with different energy arrangements: (a) energy from storage; if flow is matter and Q_1 and Q_2 are quantities of the same type, $k_1 = k_2$; (b) constant-gain amplifier, external source, no material flow from Q_1; (c) external source used as needed to drive material; (d) external drive held constant.

(Fig. 8-7b) is a constant-gain amplifier with control from the storage but energy from an outside source. The third (Fig. 8-7c) is the same, except that a pathway is included to show matter flowing from the storage as controlled by the storage. In the fourth (Fig. 8-7d) the flow is proportional to the product of the storage and the force from the outside energy source. If the driving force from outside is constant, this pathway is also linear.

PUMPING FROM A SOURCE

Storages within the system are often used to interact with energy sources entering from outside the

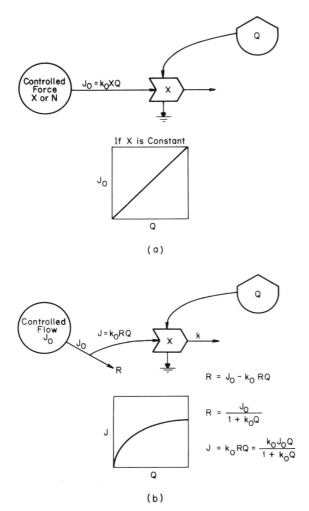

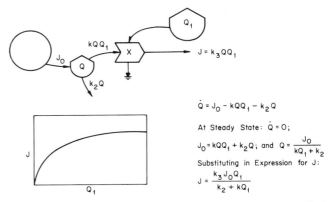

FIGURE 8-9. Pumping from an overflowing storage supplied by a source.

FIGURE 8-8. Interactions between a storage in a system and an external source from which it is pumping: (a) pumping from a source with controlled force; (b) pumping from a source with a controlled flow.

system so as to pump and accelerate the inflow as shown in Fig. 8-8. If the external source is the type that supplies a constant force (X or N constant), the flow is in proportion to the pumping force of the storage Q. The graph in Fig. 8-8a is a straight line.

Pumping from a source or another storage is sometimes called a *demand*.

Source-Limited Flow

If a source is a type that supplies a regulated flow to a system, pumping can supply up to, but no more than, the flow supplied from the source and controlled by the source J_0 (see Fig. 8-8 for a comparison of the two kinds of source). The curve of energy use has a diminishing slope with demand. The force available to the pumping interaction is proportional to the remainder R that is not yet pumped (Fig. 8-8). Equations result for the limiting factor graph.

Demand from a Flow-Limited Source with a Small Storage

The source-limited flow described in Fig. 8-8b can be visualized as the extreme case of a flow into a tiny storage out of which there is a pumping demand that generates the flow (Fig. 8-9). There is a differential equation for the small storage as the balance of a source-limited flow and a varying demand. With a tiny storage, steady state is reached almost instantly so that inflow equals outflow. The equations in Fig. 8-9 show a type of hyperbolic curve in response to increasing pumping force from Q that is found in limiting factor situations of several types in this and later chapters (see Figs. 8-8 and 8-9 and Chapter 10).

The Limiting-Factor Configuration

It was Liebig's law of the minimum (Liebig, 1840) that formulated well the idea of complex processes being limited by scarce reactants, which become less important as their relative quantities increase. Limitation situations are given in Figs. 8-8 and 8-9. Another given in Fig. 8-10 has a multiplicative interaction between a constant force source and a storage that is receiving a constant flow. As indicated by the implied algebraic expressions for the pathways, the more flow there is through the multiplier, the more drain there is on the quantity storage; thus as the

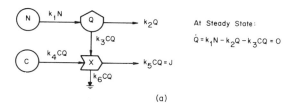
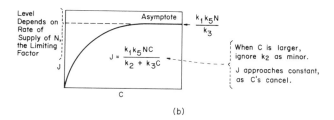
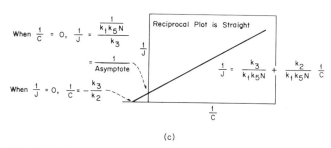

FIGURE 8-10. Limiting-factor configuration expressed in various languages: (a) energy diagram; (b) rectangular hyperbola relating steady-state output as a function of one source (c) where N is limiting; (c) Lineweaver–Burk plot (reciprocal plot) of equation in part b.

level of quantity Q drops, it becomes limited by rate of resupply from its source N. An example is the interaction of carbohydrate food with oxygen diffusing in from a source with the storage representing the oxygen at the site of interactive multiplicative process. A configuration where there is an outside limiting inflow such as that in Fig. 8-10 is sometimes called a *Monod model* (Monod, 1942).

Given in Fig. 8-8–8-10 are similar equations for similar limiting-factor situations. The equations for these systems at steady state are limiting-factor hyperbolas that have similar graphical shape with decreasing slope. Output is strongly affected at low concentration but becomes less affected as some other factor becomes relatively more limiting. The law of diminishing returns as used in economics is an example involving interaction of commodities, labor, capital, and so on.

In Chapter 10 similar equations are found in configurations that have limiting-factor action in recycling loops. These loop-limit equations were first used by Michaelis and Menten (1913), and sometimes all limiting factor equations are referred to as Michaelis–Menten kinetics since their algebraic properties are similar. However, sharp distinction should be made between external limits and those due to internal recycle. The latter are more easily eliminated than those where energy from the outside is limiting. Shinozaki and Kira (1956, 1961) fitted data on growth of plants to limiting factor models.

Lineweaver–Burk Graph

Especially for fitting data, it is convenient to plot limiting-factor hyperbolas in the form given in Fig. 8-10c following Lineweaver and Burk (1934). Both coordinate quantities are plotted as reciprocals. A straight line results. Data can be more easily fit to the line to test whether the model is appropriate, and regression methods may be used to determine coefficients from the intersections of the line with ordinate and abscissa (see Fig. 8-11). An example is the uptake of organic matter by bacteria given by Hobbie and Rublee (1977).

Standard Form of Limiting-Factor Equations

When limiting-factor equations are derived from fundamental energy diagrams, they have more coefficients than do the aggregated equations sometimes found in scientific literature for fitting to observed data. In Fig. 8-11 the limiting-factor equation for an external limiting factor from Fig. 8-10 is converted into the common form found in much environmental literature so that one may use either and interconvert.

The simple equation in Fig. 8-11 uses $V_{\max}$, the maximum rate of the output when the variable factor c is large and k is the concentration when the output is half the maximum rate.

Problem of Control of Flow into Intersections

A common error is writing equations for intersections so that the flow is controlled from outside at the same time that the intersection function is also supposed to control the flow. It is not possible to have both flow equations for the same flow, but one can add a branch so that both can operate at the same time. Figure 8-12 shows the correct and incorrect versions.

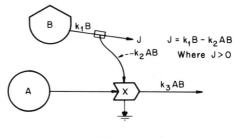

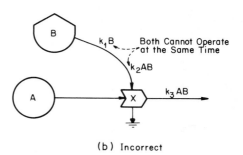

FIGURE 8-12. Configuration for an intersection involving a flow controlled at its source: (*a*) correct version; (*b*) incorrect because flow control indicated for two contradictory actions.

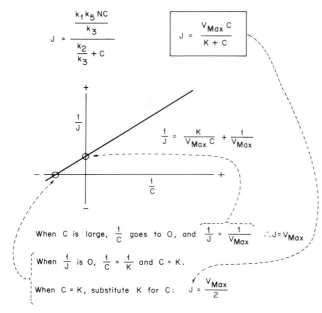

FIGURE 8-11. Traditional constants used with limiting-factor data equation in reciprocal form (Lineweaver–Burk plot), which makes a straight line and helps fitting data.

Switching Intersections

Pathways switched on and off by control intersections were introduced in Chapters 1 and 6 and Fig. 8-1*i*. The effect of a limiting-factor configuration in increasing output can be almost as abrupt as a threshold switching as shown by similarity of graphs in Fig. 8-13. Sharply limiting factors are almost digital in effect.

Conservation of Matter

If there is a conservation of matter that is being processed through an intersection, this constraint can be recognized by setting coefficients accompanying the matter into the intersection equal to that coming out with the matter. For example, in Fig. 8-12, setting k_2 and k_3 equal would mean that the flow was a quantity conserved in passing from the inflow to the outflow.

THRESHOLD LIMITED INTERSECTIONS

Source with Constant Force but with a Limit

If a source provides a constant force over a lower range and then reaches an upper limit after which it supplies a source limited maximum flow, the relationship of flow to pumping demand is as given in Fig. 8-14*a*.

Threshold Limit to Energy Use

Many units have threshold limits to their use and transformation of energy. Given in Fig. 8-14*c* is a unit with a limit to use of energy inflow such as light not used beyond a threshold. Thornley (1976) calls this a *Blackman response* (Blackman, 1905), recognizing historical developments in plant physiology. Although logically simple, other mechanisms for approaching thresholds are usually found and may be energetically less expensive to operate, such as the limiting-factor hyperbola.

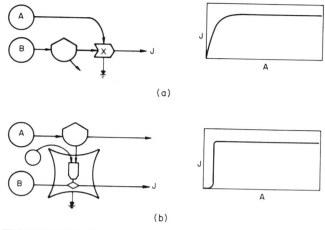

FIGURE 8-13. Comparison of limiting action of a multiplier and a threshold switch: (a) Monod response as in Fig. 8-9; (b) switch-on response as A passes a threshold.

Interaction with Threshold-Limited Unit Flow

A different model for limiting-factor action involves an exponential equation given in Fig. 8-15. One of the earliest applications of this type of equation was by Mitscherlich (1909) for the response of plant photosynthesis to light increase. The efficiency of production was reduced according to the difference from threshold. However, plants may use the automatic natural action of limiting factors (as modeled by the hyperbola) rather than setting up threshold controls.

Watt (1959) and Ivlev (1961) derived the equation for the effect of increasing concentrations of prey on predators, and that derivation is given in Fig. 8-15. The principle used is that the feeding rate j of each individual unit slows down as it approaches its individual capacity threshold T. It can also be used for action of predators on prey (see Fig. 9-16).

A translation of the model from equation to energy diagram form is given in Fig. 8-15a, whereas Fig. 8-15c diagrams it as an aggregated box.

INTERSECTIONS WITH TWO OR THREE VARIABLES

Resistance Model of Factor Interactions

The steps involving productive interaction along a process of energy transformation can be viewed as

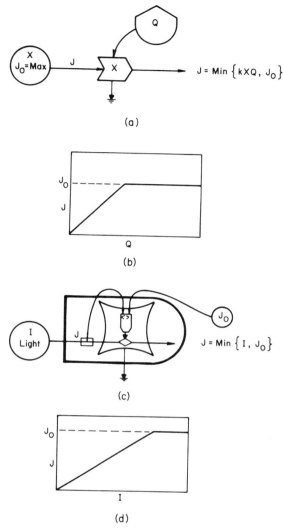

FIGURE 8-14. Models for threshold control of energy inflow: (a) source has property of cutting off when user unit draws more than threshold J_0; (b) graph of inflow as a function of demand by Q; (c) user unit has mechanism of cutting off when source would cause a flow greater than threshold J_0; (d) graph of inflow as a function of source pressure I.

the sum of two resistive pathways as diagrammed in Fig. 8-16. The onflow and action of two resources, y and z, each add to the conductivity of its segment of the pathway. The resulting equation is the reciprocal of the sum of the reciprocals, or after algebraic manipulation it becomes the product divided by the sum. Among those using this algebra for limiting factor interactions was Rabinowitch (1951) and O'Neill (1976) [see Thornley (1976)]. A graph of output as a function of increase in one factor has the characteristic limiting-factor graph.

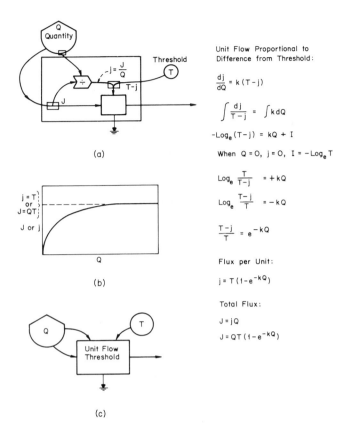

FIGURE 8-15. Diagram of intersection with threshold-limited flux (Ivlev, 1939; Watt, 1959): (a) energy diagram with flow proportional to the difference between unit flux j and a threshold T; (b) graph of flux and quantity; (c) summary diagram and shorthand.

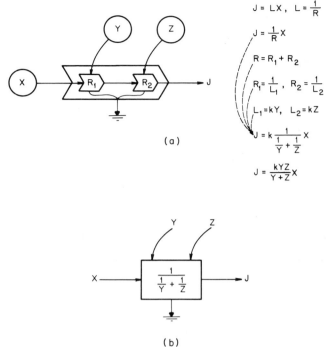

FIGURE 8-16. Model of interaction of two forces, each increasing conductivity of one part of a pathway (derivation provided by L. Burns): (a) diagram corresponding to the model of resistance control; (b) diagrammed as a black box.

Models for Two Limiting Factors

Models in Figs. 8-6–8-15 concerned output of intersections where one of two limiting factors was varying, the other held constant. Models in Fig. 8-16—Fig. 8-18 are ways of relating output of intersections to two limiting factors, both varying. The resistance model (Fig. 8-16) is one.

Rashevsky (1938, 1960) considered such models mathematically and experimentally for metabolism where limiting factors such as sugar and oxygen were diffusing to organisms where the interaction was determining rates of metabolism. See Fig. 8-17 and discussion of pertinent energy diagrams elsewhere (Odum, 1972).

Some other models, instead of being derived from interaction systems, were derived by multiplying together expressions for each factor as derived for one variable limiting action. Translations of these are shown in Fig. 8-18, where the blocks are shown multiplied like transfer functions. Baule (1917) combined nutrient effects by multiplying Mitscherlich equations (Fig. 8-18b).

One of these (Fig. 8-18c) combines two internal recycle limit symbols. In Chapter 10 similar algebraic expressions are found for recycle limitations as for the external limiting actions that were given in Figs. 8-8–8-10. Therefore, when the internal cycle actions are multiplied, the output equation resembles the equation that was intended for representation of external limiting factors.

All these models have somewhat similar, but not identical, surfaces when factors are varied. Which model seems to be the natural or simple one following from mechanisms depends on which systems language is used to visualize the system. For some simulation purposes, fitting models of one type or other to empirical data for limiting actions is often done without necessity for indicating which mechanisms are involved. As discussed in Chapter 10, many limiting-factor pathways that we model simply in large-scale aggregation are really comprised of dozens of limiting-factor intersections in chains and loops. It remains an open question as to which of

Intersections 135

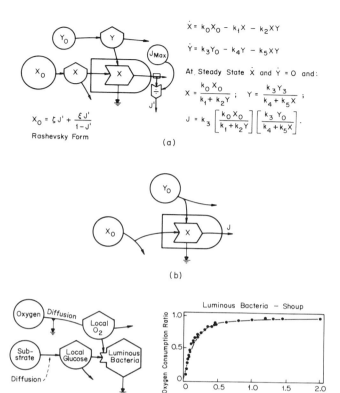

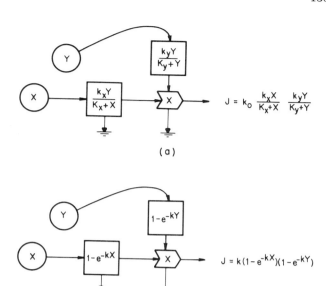

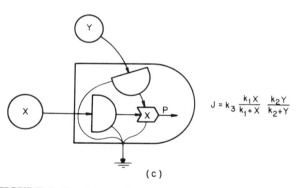

FIGURE 8-17. System of two limiting factors: (a) energy diagram and equation derived by Rashevsky (1938, 1960); (b) variation without local storages; (c) example of luminous bacteria modeled by Rashevsky with the use of data in d; (d) metabolism indicated by gas exchange as a function of oxygen concentration (Shoup, 1929).

these simple models best represents complex chains of limiting-factor action.

Three Limiting Factors

Given in Fig. 8-19 is a comparison made by W. J. Mitsch of three alternative models for connecting three variables. The graphs show the effect of increasing one variable and the family of curves that result. The patterns are different.

An interesting special case studied by Gilliland (1973) for the system given in Fig. 8-20 is the effect of a great excess of one limiting factor, which so depresses the running steady-state storage of another limiting factor that the total process output is inhibited. Examples may be the effect of very high phosphorus content in some Florida estuaries and very high calcium levels in Michigan marl lakes.

FIGURE 8-18. Energy diagram translations of models for interaction of two limiting factors where expressions for each are simply multiplied: (a) two hyperbolic equations for limitation due to external supply; (b) two Mitscherlich equations; (c) two Michaelis–Menten internal cycle limits (see Chapter 10).

DIAGRAMMING DERIVATIVE RELATING TWO STATES

When a process draws from interaction of storages or flows and has an output that is proportional, the rate of change of one with respect to the other, it may be represented as a Fig. 8-21. For example, if the behavior of humans in setting prices is in proportion to the rate of change of a production flow as a function of inflow, the interaction can be shown as a derivative. Some energy is used and some energy is degraded, even if the process is one of processing and calculating the information.

Another example is given in Fig. 8-21d. Here

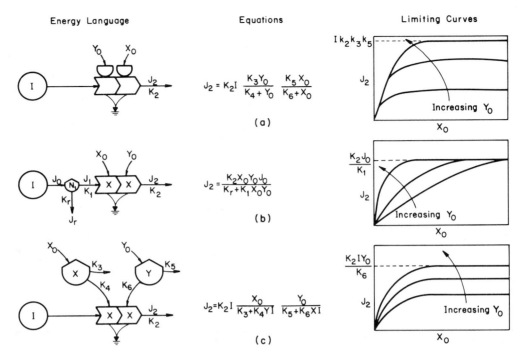

FIGURE 8-19. Comparison of limiting-factor models from Mitsch (1975).

time derivative processes are interacting as a quotient. This is equivalent in ultimate effect to the simpler design in Fig. 8-21c.

Allometric Relations

Relative growth of two parts of a system is called *allometric*. If the time rate of change of each is di-

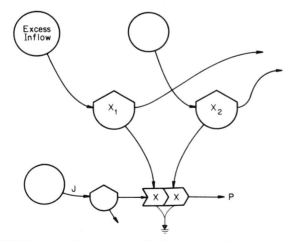

FIGURE 8-20. Excess of one limiting-factor X_1 draws storage of another factor X_2 so low that output P is less than when it is not so much in excess. Gilliland (1973, 1975) found nitrogen limiting when phosphorus was in excess.

vided as in the equations in Fig. 8-19, time drops out and the two parts to be related become so by a logarithmic function with a coefficient of allometric growth. Examples are the different parts of a body or the different concentrations of chemicals in a lake. If both processes are growing with a constant although different percent growth rate so that each considered separately is in exponential growth, the ratio of the two is straight on a log–log plot as the equation indicates. In energy language the ratio between the rate of change of the two quantities may be as shown in Fig. 8-22. Allometry was developed by Huxley (1932).]

Intersections Involving Acceleration and Fields

Since any pathway in the energy language has a balance of forces as given in Chapter 1, some systems are described by their force balance from which useful equations are derived. Illustrated in Fig. 8-23 are energy diagrams for interactions of kinetic energy, fields of gradients, and balances of forces, including backforces from acceleration. These are common relationships concerned with fluid motions. Sensors that read difference in pressures are used (see Fig. 8-6g).

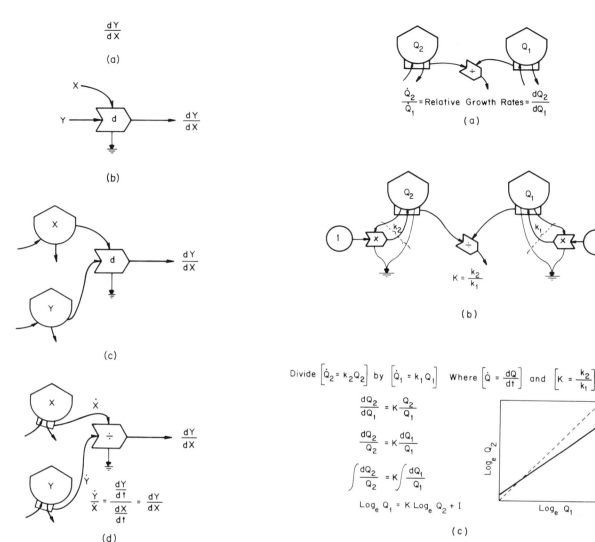

FIGURE 8-21. Symbols and configuration to represent a derivative: (a) calculus form; (b) energy symbol; (c) an example; (d) same output as in part c, where relative rate of change results from ratio of two time derivatives.

FIGURE 8-22. Energy diagram of relative growth: (a) intersection that reads observed relative growth (rate quotient); (b) example of relative growth that is straight on log–log plot. Slope K is relative growth coefficient; (c) derivation of relative growth equation and graphical representation.

Stochastic Pathways

Variation is a property of most pathways, and parameters of the distribution may be indicated on the pathway as in Fig. 8-24b for the special case of Gaussian distribution measured by variance. The energy of the variation storage may be from the pathway's own energy (Fig. 8-24b) or imposed from another flow at an intersection as in Fig. 8-24c. In general, variation frequency is higher for the lower quality inputs from the left compared to the higher quality ones that control from the right.

Statistical properties can be added to pathways and sources in simulation (Mihram, 1971; Gall and Richter-Dyn, 1974).

SENSITIVITY ANALYSIS

The rate at which two or more variables change in relation to each other is called *sensitivity* (Tomovic, 1963; Tomovic and Vukobratovic, 1972). Percent responses of the variables of a model can be tested readily with an analog computer by scanning suc-

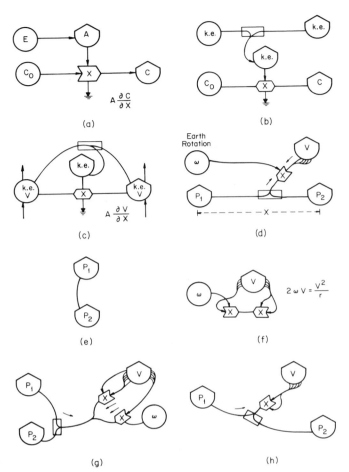

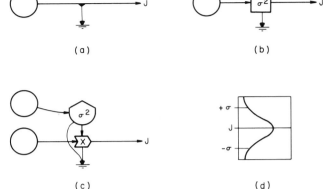

FIGURE 8-23. System of force relationship in fluids that show energy sources: (a) eddy diffusion of a chemical substance driven by an external energy source E; (b) eddy diffusion of a chemical substance driven by momentum diffusion that generates eddies; (c) eddy diffusion of momentum with self-generated eddies; (d) geostrophic balance of pressure gradient force and Coriolis force maintaining velocity V; (e) hydrostatic balance with weight balanced by upward pressure gradient force; (f) inertial motion with balance between Coriolis force and centrifugal force; (g) meander, gradient wind balance between pressure gradient force, Coriolis force, and centrifugal force; (h) cyclostrophic balance between pressure gradient force and centrifugal force.

FIGURE 8-24. Stochastic properties of pathways that carry populations of units such as wind, heat, and migration of people: (a) variations generated from pathway's own energy flow are implicit in all pathways (see part e); (b) energy of variation in a pathway flow explicitly indicated; (c) variation from a storage supplied by a second source; (d) Gaussian distribution of components of flows in parts a–c.

cessively the change in each variable as pots are varied. With digital computer scanning programs are available or families of output curves may be run. Many stable models have smaller amplitude of responses than of the test. For example, see Parker's (1975) model of the Kootenay Lake, British Columbia, test.

Patten (1972) finds that in linear systems sensitivity of one variable in response to control by a second is quantitatively inverse to the sensitivity of the second relative to perturbations of the first. Sensitivities may be determined for isolated relationships and for the relationships with the whole system in operation.

SUMMARY

In this chapter various kinds of intersection were diagrammed with care regarding kinetic and energetic meaning, and thus processes of system interactions were summarized. The actions of limiting factors in controlling productive transformations were considered with traditional models involving intersections. Included were interactions of three or more commodities, interactions of two flows, and linear pathways involving other converging and diverging pathways. The mechanisms described are building blocks for the design elements in the next five chapters.

QUESTIONS AND PROBLEMS

1. How many types of intersection can you diagram? Give the kinetic equation for the output of each. Explain the energy behavior of each.

2. Explain the concept of limiting factors by drawing the smallest network that exhibits this phenomenon. Derive the equations for steady-state output as a function of external driving forces.

3. Compare a Monod rectangular hyperbola with a Mitscherlich exponential relationship.

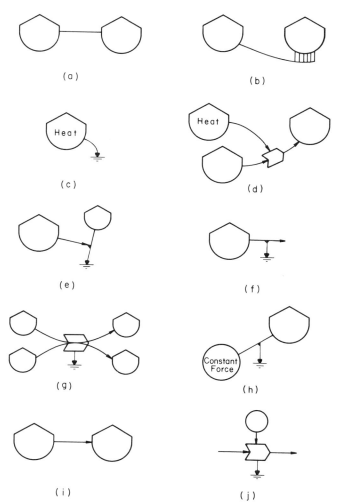

FIGURE 8-25. Diagrams for questions.

4. Explain the law of diminishing returns.

5. What graphical axes are used to fit observed data to a rectangular hyperbola? What are the intercepts used for?

6. Where is the higher quality of two intersecting flows joined to the intersection symbol? How does the output quality compare?

7. What is the "resistance" model for limiting interaction?

8. Can you give the Ivlev–Watt derivation for the exponential limiting factor relationship?

9. Diagram the flow of water out from a reservoir carrying a dissolved constituent. Include total quantities and concentrations in the diagram.

10. Which intersection functions are sometimes used for effect of temperature and the effect of turbid shading of light penetrating water?

11. Compare a sharply limiting factor interaction and a threshold switch limiting action.

12. Draw a network in which an excess of a limiting factor can reduce output because of interaction with a second limiting factor.

13. What are coordinates used to represent allometric relationships? How can this be represented in energy diagrams?

14. Diagram a derivative; give an example where it becomes a force in another part of the network.

15. How is a spatial gradient diagrammed by use of tanks as discrete compartments? Using a general-purpose rectangular symbol, indicate a force derived from the spatial gradient, such as the pressure gradient force.

16. How does a storage of kinetic energy associated with velocity and represented diagrammatically by a tank express an inertial force against an acceleration? Diagram centrifugal force. Diagram coriolis force.

17. Draw the following intersections: dilution of one flow by another; addition of flows; subtraction of flows; control of one flow by another but appreciable energy not supplied; retardation of the effect of one flow by another flow, resulting in an output that is a declining straight line function; chemical reaction with two products; quadratic self-interactive process; product of one force times a gradient so as to pump in the direction of the gradient; and source that supplies both the controlling force and the energy source but by a different pathway.

18. Use analog computer or microcomputer to simulate output as a function of limiting factor, using models given in this chapter.

19. Explain energetics, force relations, and write equations for diagrams in Fig. 8-25.

SUGGESTED READINGS

Eyring, H., S. H. Lin, and S. M. Lin (1980). *Basic Chemical Kinetics,* Wiley, New York. 512 pp.

Huxley, J. S. (1932). *Problems of Relative Growth,* Methuen, London.

International Association for Theoretical and General Limnology

(1978). *Experimental Use of Algal Cultures in Limnology,* Mitt. Internat. Verein. Limnol. No. 21, Stuttgart.

Jammer, M. (1957). *Concepts of Force,* Harper, New York. 269 pp.

Malek, I., organizer (1970). *Predication and Measurement of Photosynthetic Productivity,* Center for Agricultural Publications, Wageningen, Netherlands.

Nicolis, G. and I. Prigogine (1977). *Self Organization in Non-Equilibrium Systems,* Wiley, New York.

Poole, R. W. (1974). *Introduction to Quantitative Ecology,* McGraw-Hill, New York.

Prigogine, I. (1980). *From Being to Becoming,* Freeman, San Francisco. 272 pp.

Rashevsky, N. (1980). *Advances and Application of Mathematical Biology,* University of Chicago Press, Chicago. 214 pp.

Rose, J. (1961). *Dynamic Physical Chemistry,* Wiley, New York. 1218 pp.

Stewart, W. E., W. H. Ray, and C. C. Conley (1980). *Dynamics and Modelling of Reactive Systems,* Academic Press, New York. 413 pp.

Watt, K. E. F. (1968). *Ecology and Resource Management,* McGraw-Hill, New York.

Wilson, R. J. (1972). *Introduction to Graph Theory,* Academic Press, New York.

CHAPTER NINE

Autocatalytic Modules

Many naturally occurring units in the real world store energy and then feed it back internally to facilitate in the inflow of other energy. The feedback acts as a control, often as a multiplier, and catalyzes the inflow. Such units are sometimes termed *autocatalytic* (see Fig. 9-1). The process of storing and using the storage to pump additional energy tends to accelerate growth and maximize power. Such modules are frequent in all kinds of system. This chapter considers autocatalytic modules and their characteristics, energetics, and growth curves. Other kinds of feedback design are given in Chapter 10 on loops.

MAXIMUM-POWER PRINCIPLE AND FEEDBACK

The maximum-power principle introduced in Chapters 1 and 7 suggests that natural selection operating on a variety of system designs tends to select those that generate maximum useful power. Systems that gain more power have more energy to maintain themselves and their habitat and to overcome any other shortages or stresses and are able to predominate over competing units. To maximize power, the energies stored tend to be fed back to pumps, gates, diversity, and so on to accelerate the inflow of more energy either from the same source or from other sources. In other words, systems that survive tend to develop *autocatalytic units* because that design processes energy in a competitive way, automatically increasing the inflow from the energy source when it is possible. Many other system designs that prevail seem to be explained by the maximum-power principle, and these are pointed out in other chapters.

BASIC DESIGN FOR AUTOCATALYTIC MODULES

The basic design for *autocatalytic* modules, one of the most important parts of real systems, is given in Figs. 1-3 and 9-1. It has a storage of structure and energy reserves with a depreciation flow required by the second energy law, a feedback by which energy in the storage pumps and controls the inflow of new energy from the outside from the left. The hexagon symbol (Fig. 9-1b) is the symbol of the energy circuit language used for this module and all others of this class that have at least one storage and one interaction feedback pumping in energy. Often the maintenance module is drawn with details within the hexagon as in Fig. 9-2. The hexagon group symbol is used when the autocatalytic process is organized into a subsystem.

Examples of maintenance modules are populations of consumer organisms, cities, beaches, and fires. All these build up a storage of energy of a special type capable of interacting with the energy source to pump in more energy. As we shall see, the maintenance module is a model for the most common units of function of the real world. It is capable of maintaining itself and in part controlling some of its destiny. Like the simple tank, this unit can grow or decline in storage and flow. Often in nature the processes of storing and self-maintenance are closely linked and combined into a unit that is discrete in the way that a single organism is discrete. Often we begin the diagramming of a system by drawing a web of connected maintenance modules. As given in Fig. 9-1, the exact energy and kinetic relationships are not defined since we have not shown what kind of interactive feedback there is. Consider next some types of autocatalytic module.

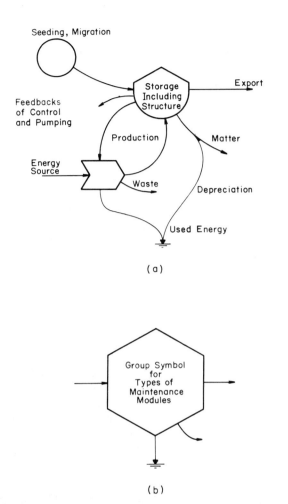

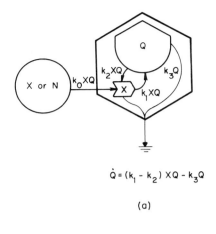

$$\dot{Q} = (k_1 - k_2)XQ - k_3Q$$

(a)

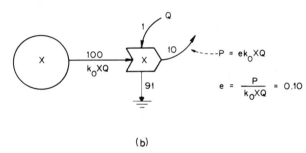

$$e = \frac{P}{k_0 XQ} = 0.10$$

(b)

FIGURE 9-2. An accelerating autocatalytic module: (a) mathematical terms and differential equation; (b) use of efficiency coefficient e.

FIGURE 9-1. Autocatalytic maintenance model in its simplest form: (a) essential components; (b) group symbol for abbreviation of this class of modules.

Exponential Module

An autocatalytic module that interacts as a multiplier with the driving force or forces of the input energy source is shown in Fig. 9-2. This is an accelerating loop that uses the storage to pump in more energy, growing with increasing rapidity until restrained by the source. The accelerating loop is indicated in several languages, where X or N is the driving force or forces and Q is the quantity stored. As we have said before, placing the mathematical terms on the pathways to express the flows is done for teaching and is not necessary for those who know the language since the symbols carry the necessary conventions to make this clear. In 1798 Malthus recognized the nature of accelerating growth and the increasing load it placed on the resources. Sometimes we call this a *Malthusian growth module*. Volterra (1926) used this equation for the predator in his prey–predator model. Growth is exponential (straight on semilog paper) when input force is constant.

Growth with an Exponential Module and Constant Force

The type of growth shown by an accelerating-loop module depends on the kind of driving function from the outside energy source. If there is a constant driving force maintained (X or N constant; see Fig. 9-3), the accelerating loop grows with an exponential growth curve. It is exponential because the integrated form of the differential equation (Fig. 9-3) contains an exponential term.

If the rate of production from the output of the interaction is less than the rate of depreciation, the module will not grow but will decay to zero. This module cannot start unless it is seeded with some of the structure Q required to operate the production

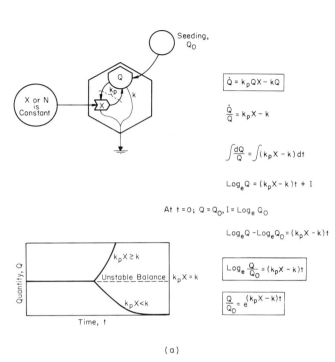
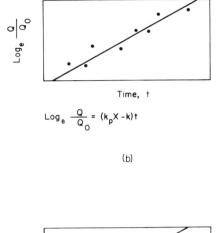
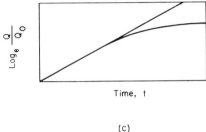

FIGURE 9-3. Growth of an accelerating loop with constant driving force (unlimited energy): (a) energy diagram; (b) semilogarithmic graph and data of exponential growth; (c) logistic growth of semilog plot turning away from straight line.

interaction. This may be as an initial storage or as a regular inflow. For example, there may be some regular migration or dispersal that seeds growth. The rate of growth depends on both the effectiveness of the feedback and the magnitude of the input force.

If the rate of production were exactly equal or could be maintained equal to the rate of depreciation, the growth curve could level, but any slight deviation above or below this balance would lead to exponential growth or decline. Because of this instability, any model with the accelerating loop tends to be unstable unless it is source limited or has other regulatory features.

The accelerating loop is one of the properties of all life that stores, maintains, and grows. For this reason, life is inherently unstable without additional features of control within or from outside the loop system.

Use of a Coefficient for Efficiency of Energy Transfer in Growth

In simulation production coefficients can be inadvertently adjusted so as to generate more energy than is received. Instead of using various k values to indicate coefficients of energy transfer in the production of an autocatalytic module, an efficiency e can be used for the output as shown in Fig. 9-2b. Coefficient e can be kept as a fraction of the inflows appropriate for that energy transformation.

In biological experiments in which living organisms are seeded into situations with large food supplies that are maintained at the same concentration, the organisms can express their inherent ability to grow or reproduce (or both). It is sometimes said that the growth rate under these conditions is the *intrinsic rate of natural increase*. It is expressed as the percent rate of change as follows:

$$\text{Percent rate of change} = \frac{dQ/dt}{Q} = \frac{\dot{Q}}{Q} = r \quad (9\text{-}1)$$

The intrinsic rate of natural increase r is sometimes called the *specific growth rate*, whereas $\dot{Q}$ is the regular growth rate. When derived and explained as in this paragraph, there is no consideration of energy levels, except that there are not supposed to be any source limits.

Compare Eq. (9-1) with the equation for accelerated growth in Fig. 9-3. What is called *intrinsic rate r* is equivalent to $k_pX - k$, where k_p is new production and k is depreciation. In living things k takes the form of death and tissue destruction that must be repaired:

$$\text{Intrinsic rate } r = k_p X - k \quad \text{or} \quad (9\text{-}2)$$
$$r = k_p N - k$$

So long as the energy source provides a constant force or population of forces as when there is a constant concentration of food, the two approaches are similar. The "intrinsic rate of natural increase" really is not intrinsic but requires an unlimited extrinsic energy source that maintains a constant impetus to growth.

Graphical Test for Exponential Growth

Exponential growth may be expressed by differential equation, exponential equation, or logarithmic equation as shown in the derivation in Fig. 9-3. Since the logarithm of the quantity in ratio to initial quantity is a linear function of time, exponential growth is straight on a semilogarithmic graph as shown in Fig. 9-3b. Plotting of data on semilogarithmic graphs allows comparison to the graphs of a model for validation. Many steeply accelerating curves appear to be exponential but are not.

A Model with Autocatalytic, Discrete Growth Steps

Exponential growth is given in discrete form in Fig. 9-4a, where the transition from one state to the next at a later time increment is by multiplication of a transition factor times the state variable at the earlier time. Where there is a constant transition factor, this is equivalent to constant percent change per unit time, which is an exponential growth or decline.

Graph of a Variable as a Function of Its Value at a Previous Time Step

A useful graph plots N_{t+1} as a function of N_t. Others use $\log N_{t+1}$ as a function of $\log N_t$. These have been called *Moran plots*.

A Growth Model with Continuous Loss and Discrete Growth

Figure 9-4b shows a population storage draining continuously with a linear pathway. At discrete

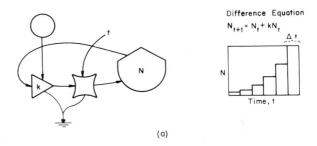

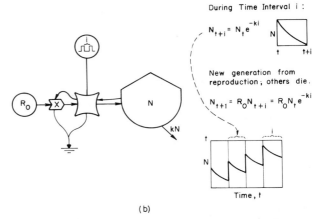

FIGURE 9-4. Diagram of a state change model with discrete steps for exponential growth: (a) growth in steps; (b) step reproduction with continuous mortality, where i represents the time interval between reproduction pulses.

time steps the state storage is removed, multiplied by a constant increment from an energy storage representing reproduction, and returned to the storage where it declines again until the next time step. The model has step reproduction and continuous mortality. This was used by Nicholson and Bailey (1935) as appropriate for an insect population.

Limits in Simulation

When accelerated growth modules are simulated without source limit or other controls, the growth quickly goes off scale and may exceed the limits of the simulation device. For analog simulation, procedures for attaching a limiter are as shown in Fig. 5-16, and in a BASIC digital program a statement can be inserted "IF Q > A THEN Q = A," where A is the limit. Also, limits at zero are needed for storages that make sense only if positive. Models that pump from a tank can draw it below zero if there is no arresting mechanism.

Growth of an Autocatalytic Module on a Controlled Flow Source

Whereas the autocatalytic module grows with increasing rapidity when it is pumping from the unlimited energy of a source with constant force, the pattern is different with the limited energy of a controlled flow source. As shown in the constant-flow case in Fig. 9-5, growth accelerates with upward curvature at first and then levels off as the system becomes limited at its source. Many of our natural energy sources such as the sun are renewable but limited at the point of origin. Growth based on these sources diminishes as more and more area is used, until some steady state is reached. Figure 9-5 includes the model in energy language and appropriate equations as well. Where the source is a controlled flow but varying, the user unit is continually tracking its source limit. For a discussion of equations used to describe source-limited models in population biology, see Gallopin (1971). Use of an external resource limit introduced stability into models otherwise unstable when resource forces were held constant (unlimited resources).

Growth on an Unrenewed Storage

Figure 9-6 shows autocatalytic growth driven by a storage that is being depleted by its use. Growth goes through a maximum and back to zero. We

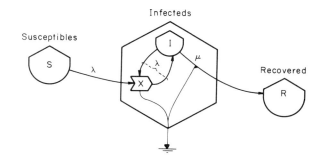

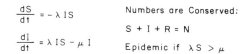

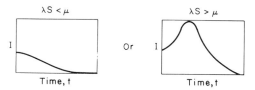

FIGURE 9-6. Autocatalytic growth on an unrenewed storage source. Equations are those for a model for infection by Kermack and McKendrick (1927), after Watt (1968).

sometimes call this a "dead-log" model as insects and fungi develop during its consumption. An early example was the model of infection given in Fig. 9-6, which has an additional constraint of conservation of the flow of human population.

Growth of an Accelerated Module on Constant Force with Backforce

In the autocatalytic modules shown thus far there is no backforce expressed from the storage against the input pumping process, although in many processes in nature there may be backforces against the pumping that accompany storing energy. For example, there is a backforce of gravity when water is stored in a tank, the backforce of pressure when air is stored in an automobile tire, and the backforce of chemical reaction gradients when one stores energy in biochemical chains (see also Fig. 8-1d and explanation of backforce in Chapter 1). In the energy language, backforce is indicated by removing the barb.

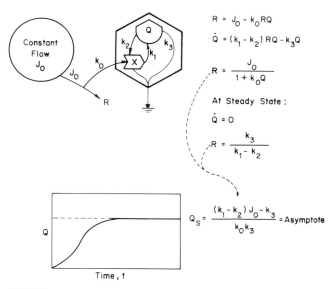

FIGURE 9-5. Accelerated growth from a constant flow source; R is unutilized remainder of flow J_0.

146 DESIGN ELEMENTS

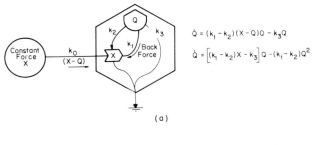

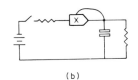

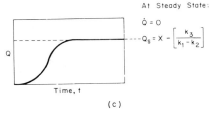

FIGURE 9-7. Autocatalytic growth on a constant-force source with backforce is logistic: (a) energy diagram; (b) passive analog prepared in collaboration with J. O. Baltzer (Odum, 1967a, 1971b); (c) growth graph.

When the accelerated module is pumping from a source into a storage with backforce, the equations are different and the growth curve levels off instead of continuing to accelerate (see Fig. 9-7a). A passive electrical analog circuit such as that shown in Fig. 9-7b was used to simulate the unit by using a voltage-controlled variable resistance as multiplier.

AUTOCATALYTIC SYSTEM WITH QUADRATIC DRAIN; LOGISTIC UNIT

Shown in Fig. 9-8 is an autocatalytic module with a quadratic drain pathway such as that due to crowding (Fig. 8-1k). Compare this system with the plain autocatalytic module operating on a constant force as given in Fig. 9-3. Whereas the autocatalytic module without the quadratic drain is unstable in the sense of either growing exponentially or dying out, the system with the quadratic drain levels off to a stable asymptote, even with an unlimited energy source (constant force source). Sometimes in modeling, a quadratic drain is added to stabilize a system.

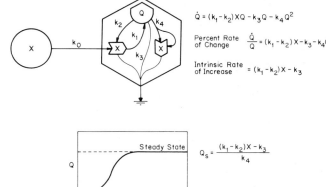

FIGURE 9-8. Logistic system formed from an autocatalytic feedback, constant-force source, and a quadratic drain.

Presumably, real systems develop quadratic drains for this and other reasons—for maintaining enough stability to survive. Notice that the level at steady state Q_s is a function of the force from the energy source X.

Examples are growth of yeast (Richards, 1937) and duckweed (Ikusima, 1962). The latter showed the effect of initial condition on growth (Fig. 9-8b) and agreement with logistic evaluations by Shinozaki and Kira (1956), who wrote equations for effect of initial condition.

Logistic Equation

The equation resulting in Fig. 9-8 is one of several that emerge in this chapter that are sometimes classed as *logistic* growth curves because they have the quadratic form

$$\dot{Q} = aQ - bQ^2 \qquad (9\text{-}3)$$

This is the form used by Volterra (1926). Verhulst (1845) developed the logistic curve for populations. A historical summary of the logistic equation and its application is given by Hutchinson (1978). The percent rate of change depends on the quantity Q stored:

$$\text{Percent rate of change } \frac{\dot{Q}}{Q} = a - bQ \qquad (9\text{-}4)$$

and the growth reaches a steady state $\dot{Q} = 0$ in which

$$\text{Steady state } Q_s = \frac{a}{b} \qquad (9\text{-}5)$$

Compare Eqs. (9-3)–(9-5) with the equations in Fig. 9-8 and confirm that the form of the equation is logistic. Identify what constants are a and b. The characteristic S-shaped curve with time shown in Fig. 9-8, sometimes called *sigmoid*, is generated by the logistic equation. Examine Fig. 9-7 and verify that its equation is also logistic.

In contrast, Fig. 9-5 has a sigmoid curve of growth but for a different reason, because it is limited at its source. Its equations are not logistic.

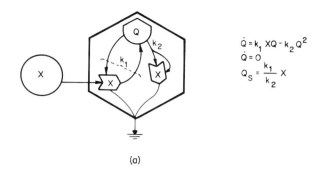

Energy Required for Diversity

A special case of quadratic drain is the energy flow associated with diversity of units stored. The state variable in this case is the number of types of unit n, which is substituted for Q in Fig. 9-8. If the energy for organizing and preventing disordering interactions is dependent on the possible combinations among the units stored n, the energy that is drained from the storage due to diversity is a square function ($\alpha\, n^2$) of the number of units. Energy is either lost as a result of confused interactions or is required to be used to prevent it. Either way, there is a drain (see Fig. 9-8).

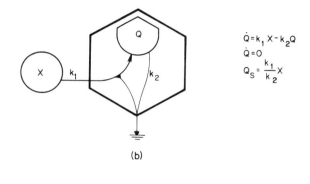

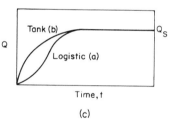

FIGURE 9-9. Use of simple tank as a substitute for autocatalytic logistic modules: (a) model being substituted for; (b) tank within hexagon; (c) comparison of curves.

Use of the Tank as an Approximation for a Logistic Module

Autocatalytic logistic modules generate S-shaped curves that level off asymptomatically as described in this chapter. Sometimes in modeling a tank model simpler than the autocatalytic one can be substituted where shape of rapid growth phases is not important. Substitution of the model in Fig. 9-9b for that in Fig. 9-9a saves hardware and makes it unnecessary to add the pumping multipliers. As shown in Fig. 9-9c, the graphs are differently shaped at the start but level off similarly. Where the simpler model is being used as a simplification for the autocatalytic module, the tank model can be indicated within the hexagonal autocatalytic symbol, as in Fig. 9-9b.

Refuge and Population Pools; Quadratic Migration

In many systems involving consumer sequences, the rise and fall of stocks due to seasonal changes in energy or to the prey–predator oscillation effects drives one or more species to extinction because there is no reservoir for species to begin an upturn when energy conditions and predator and competing populations permit. Two mechanisms are shown in Fig. 9-10: one is a reservoir within the system; the other is an external pool that stabilizes systems against extinction effects.

A linear outflow pathway does not stabilize exponential growth, but a quadratic one does, and the model in that instance is logistic. Interactive animal behavior that may be quadratic in its stimulus to migration are suggested by group behavior of lemmings, snowy owls, grasshoppers, and so on. Terborgh (1974) suggested migratory dispersal was quadratic.

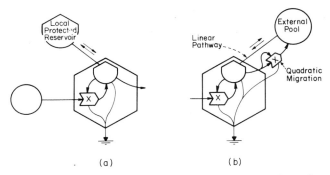

FIGURE 9-10. External and internal population pools used to start and stabilize autocatalytic models with large fluctuations: (a) pool within system; (b) outside pool with linear and quadratic migration pathways.

Alternative Meanings for Similar Equations; Logistic Modules

In Chapter 1 many different mechanisms were found to generate similar equations. The logistic equation is a good example of a type of equation that has many forms and has had many suggested physical interpretations. Jensen (1975) gives two designs for the logistic equation. Here we give many more (Figs. 9-7, 9-8, 9-10, 9-12, 10-18). Many different real-world phenomena can generate autocatalytic growth characteristics that are mathematically of the form of the logistic equation. Principles include backforce, quadratic drain, pumping in proportion to difference from threshold, decrease in unit growth rate due to density, and conservation of recycled quantity.

Special tests are required to determine which mechanism may be responsible for logistic type of equation. Growth data alone do not help distinguish which mechanism was responsible.

People using different forms of the equation to visualize real meaning are considering entirely different phenomena without a language to avoid semantic misunderstanding. Diagramming supplies clarity. Actually, if constants and terms are not manipulated, an equation has a unique translation to energy language.

Logistic Equation Formulated in Terms of Resource Used

The logistic equation is widely used in biology and ecology to describe growth, but the traditional reasoning used to derive, explain, teach, and relate the equation to real situations is usually presented in a manner different from that given so far in this chapter. Growth is stated to be at its intrinsic rate that occurs when there are no limits to resources (Q is small); but as the quantity increases, the resource is diminished by the fraction of the full capacity occupied. Following Gause (1934), for example, one may visualize a culture of protoza Q in which the full capacity at steady state is K and the fraction of full capacity that is unused at any time is $(K - Q)/K$. Then the rate of growth is given in Eq. (9-6) as the intrinsic rate times the fraction of full capacity not yet utilized:

$$\frac{\dot{Q}}{Q} = r\frac{(K - Q)}{K} \qquad (9\text{-}6)$$

where K is the asymptote (see Fig. 9-12b). Manipulation of the equation puts it in the form of Eq. (9-7) for comparison with the logistic equation given in Eq. (9-3):

$$\dot{Q} = rQ - \frac{r}{K}Q^2 \qquad (9\text{-}7)$$

If $(k_p X - k)$ is substituted for the intrinsic rate of natural increase r as in Eq. (9-2), this form of logistic becomes

$$\dot{Q} = k_p XQ - kQ - \frac{k_p X - k}{K}Q^2 \qquad (9\text{-}8)$$

where $k_p X$ is the rate of production per unit and k is the per unit rate of loss. This procedure shows how an energy source is really implied in the customary classical presentation of the logistic expression. When integrated, the logistic equation becomes

$$Q = \frac{K}{1 + C \exp^{-rt}} \qquad (9\text{-}9)$$

where K, the full capacity at steady state, depends on the initial condition.

Modules with Constant Gain

The symbol for designs where one energy source (usually small) drives a process while drawing energy as necessary from a second source was given in Chapters 1 and 8. This arrangement is a constant-gain amplifier; a voice amplifier is an example. The unit operates only within the range of power that

Autocatalytic Modules

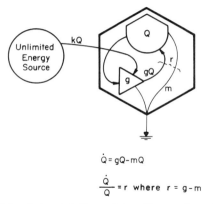

$$\dot{Q} = gQ - mQ$$

$$\frac{\dot{Q}}{Q} = r \text{ where } r = g - m$$

FIGURE 9-11. Autocatalytic module with use of constant gain amplifier, which assumes an unlimited energy source. As in the module in Fig. 9-3, growth is exponential.

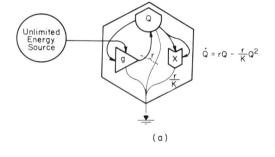

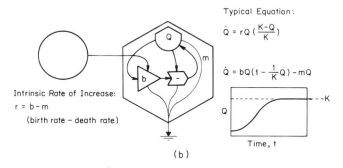

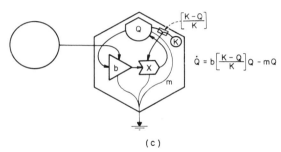

FIGURE 9-12. Three different autocatalytic models that have logistic equations, each based on a different principle: (a) constant gain and quadratic drain; (b) unit growth rate diminished by density; (c) unit growth rate proportional to difference from saturation.

the energy source can supply, and the amplifier symbol cannot be used outside this range. Some of the equations and systems of this chapter can be represented by using the constant-gain amplifier design in place of the feedback multiplier action on the source. The expression for exponential growth with the use of a constant-gain amplifier is represented in Fig. 9-11 (compare with Fig. 9-2). The gain is the percent growth rate (or reproductive rate).

Figure 9-12 gives logistic growth models, also with the use of constant-gain module. The first (Fig. 9-12a) has a quadratic drain like the external energy model in Fig. 9-9. The second (Fig. 9-12b) has a density-dependent drain on production and is a translation of the logistic in the form of Eq. (9-6). The third (Fig. 9-12c) has threshold-affected production.

Logistic Loop with Conserved Storage

Figure 10-18 (next chapter) shows an autocatalytic growth module that draws on and uses a necessary quantity that is conserved in a loop. As growth accelerates, the available quantity Q_0 becomes involved in Q_1 and decreasingly available for production. An example is the growth of a human economy with limited land. Another example is a population of birds with limited nest sites. Another is the growth of a unialgal culture on recycled nutrients without bacteria or other consumers. As the equations show, addition of the constraint for recycle changes the system into another type of logistic equation.

Shown in Fig. 10-16b, with similar kinetics, is a system with an internal recycle of limiting resources. These modules fit the usual way of teaching the logistic equation where the term $(K - Q)/K$ from Eq. (9-6) is visualized as a limit on the input production process.

SUPERACCELERATED GROWTH

If the feedback pumping is done quadratically, the rate of growth is super-accelerated (see Fig. 9-13). The pumping in of new energy as well as the drains from the storage by such work are quadratic. The work of each individual unit is increased by having more present. The work may be described as

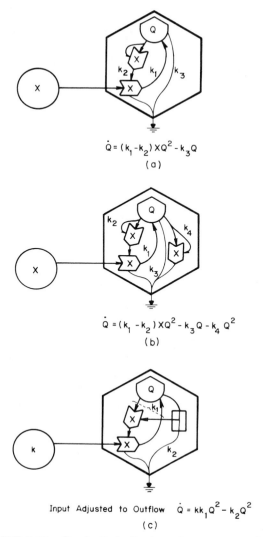

FIGURE 9-13. Quadratic feedback produces super accelerated growth that is steeper than exponential growth: (*a*) linear drain only; (*b*) quadratic production and drain; (*c*) pumping proportional to drain.

"cooperative." Nicolis and Prigogine (1977) describe a Schlögl chemical reaction with this form. An examination of the economic growth of the United States by Von Foerster et al. (1960) showed that the period of most rapid development could be fitted to a line generated by this function, although Pearl and Reed (1920) attempted to fit the logistic. During periods when there are unutilized resources, such growth is favored and takes more of the energy resources.

This module, although favored in competition during periods when growth is possible, is energy wasting when energy supplies become a limiting factor.

Many of the features of the rapidly growing urban economy were cooperative and quadratic such as advertising, close coordination between organizations, and an emphasis on transportation. Quadratic pumping favors growth but does not reach a stable steady state without additional features. Ordinary exponential growth (Fig. 9-3) levels when quadratic drains are added, but superaccelerated growth does not. This is a special case of the mathematical principle (supplied by M. Hale) that the production term must be of one power less than the drain. Figure 9-13*c* gives a design studied by W. H. B. Smith (1976) in which the inflow production is adjusted to be controlled by the outflow depreciation. The equation that results is of the quadratic growth type.

COOPERATIVE AUTOCATALYSIS WITH OPTIMUM PERCENT GROWTH

Figure 9-14*c* contains a graph of growth in relation to density Q studied by Allee with many papers and books under the heading "cooperation" (Allee, 1938). Many field studies revealed populations that had maximum net growth rate per individual at an optimum density, with a lower rate of survival at both lower and higher densities. Contrast the hump-shaped graph of percent growth rate (growth rate per unit) that has an optimum in Fig. 9-14*c* with the linear declining one for the regular autocatalytic logistic curve in Fig. 9-14*b* and constant rate for exponential growth in Fig. 9-14*a*. Hutchinson (1947) suggested an equation.

If populations have the optimum property shown in the third graph in Fig. 9-14*c*, their percent rate of change graph is quadratic. The equation for growth rate of the quantity as a whole is cubic. A module with this relationship is given in Fig. 9-15. Whereas the quadratic accelerated systems in Fig. 9-13 are rapid growing but unstable, the model in Fig. 9-15 is stable. Odum and Allee (1953) showed a stable point (Fig. 9-14*c*) where surviving populations tend to be regulated. Does this model apply to human affairs?

AUTOCATALYTIC MODULES WITH LIMITS ON UNIT GROWTH

Range of Operation of Multiplier for Food Consumption

Where the autocatalytic system is a population of living organisms, a question is raised frequently as to whether the simple feedback multiplier is appropriate or is too simple for the feeding functions. If

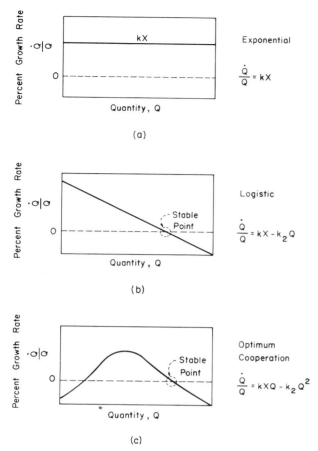

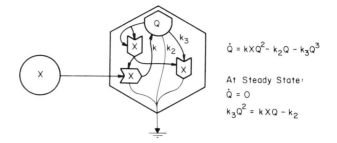

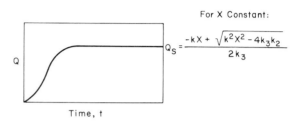

FIGURE 9-15. Optimum cooperation model; an autocatalytic module that has an optimum percent growth rate with density as defined in Fig. 9-14c.

FIGURE 9-14. Percent growth rate as a function of quantity (density) for three autocatalytic modules: (a) exponential growth; (b) logistic growth; (c) optimum cooperation (Odum and Allee, 1953).

food is processed as the product of food concentration force X and the number of organisms Q as in Fig. 9-3, the flow of food consumption is kXQ and the flow per unit of consumer is kX. By this function, increase of the food X indefinitely would increase the consumption per individual indefinitely, which seems unrealistic. Most organisms have a limit to the food they can process per individual (e.g., see Fig. 8-15). Consequently, many autocatalytic models for organisms are given thresholds and ranges over which the product function operates.

Whereas this may be necessary in modeling a system where such ranges of food relative to consumers are expected, it may not be generally necessary if it is a characteristic of real systems to substitute species and change designs so that the system is maintained in a range where it can respond readily to increased energy availability. It may be reasoned from the maximum power hypothesis that surviving designs will be those that maintain a range of operation to maximize all energy availabilities. Letting Q be biomass is one way (see Chapter 20). This remains an open and controversial question.

Controlled Feeding Rates

Five autocatalytic modules with controlled feeding rates are given in Fig. 9-16. All these modules permit reduction in feeding when the food source is large either directly or by stopping feeding following growth to a threshold. The first has a threshold and comparator that turns off feeding when the storage quantity reaches a threshold; the second pumps with decreasing effort as the storage approaches a threshold. The third pumps in lesser effort as the concentration of the source increases. In a model that may translate as Fig. 9-16b, Wangersky and Cunningham 1957a,b) had seasonally varying threshold and thus an oscillating asymptote. In fourth and fifth units the rate is controlled on a unit flux basis with flow proportional to its deviation from threshold; in the fourth the controlling flux is related to consumer quantity Q and in the fifth the controlling flux is related to food quantity force X (see explanation of the intersections with threshold limit to unit flux in Fig. 8-15).

Wiegert provides threshold switching mecha-

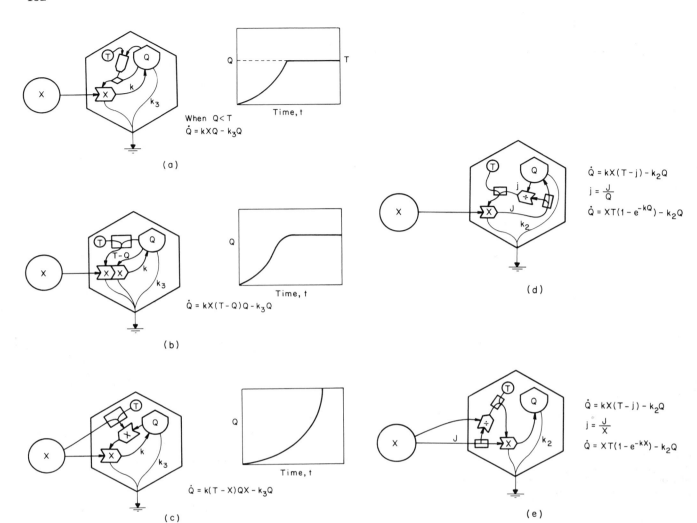

FIGURE 9-16. Autocatalytic modules with controlled feeding rates: (a) hunger threshold; (b) work in proportion to deviation from threshold (logistic); (c) feedback in proportion to difference between driving force and threshold; (d) growth based on deviation in flux per storage unit from threshold; (e) inflow based on deviation of flux per inflow unit from threshold.

nisms that provide a drag on intake, limiting growth (see Fig. 9-17).

Constant Growth

A rare case is the constant growth when the rate of change is constant. As shown in Fig. 9-18, such growth occurs when the growth rate per unit is inversely proportional to the quantity. Two ways of showing the system in energy language are suggested.

Autocatalytic Growth and Self-Dilution

An interesting model developed by Milsum (1966) feeds back a divisor action representing the dilution of its energy source. This has a steady state that is dependent on the level of energy source E (see Fig. 9-19).

Action of Source Diluting Drains

Equations given by Leslie (1948) and Shimazu et al. (1972) express the dilution effect of a divisor acting on the negative drains as shown in Fig. 9-20.

AUTOCATALYTIC MODULES WITH TIME DELAYS IN FEEDBACK

In addition to the time involved in the storing of the main structure Q, there may be additional pathway time delays in which a flow, determined by states at

Autocatalytic Modules

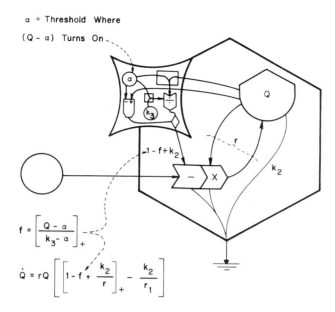

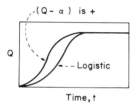

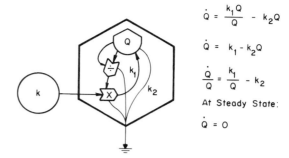

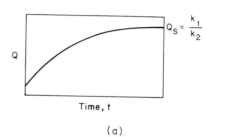

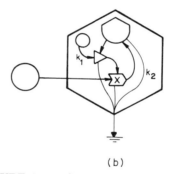

FIGURE 9-17. Switching autocatalytic system due to Wiegert (1974) where thresholds determine when exponential growth shifts to leveling growth.

FIGURE 9-18. Constant production module: (a) using division; (b) using constant gain unit.

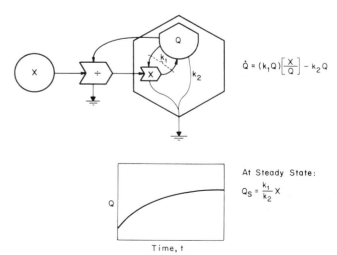

FIGURE 9-19. An autocatalytic design that dilutes its resource (modified from Milsum (1966)).

one time passing along pathways, arrives after an interval of time i. Hutchinson (1948, 1954) suggested that logistic models with time delays in their negative terms overshoot the asymptotic steady state level, producing an oscillation. Wangersky and Cunningham (1956, 1957a, 1957b) simulated logistic growth and prey–predator models, studying the effects of time lags. They found that the oscillation depended on the size of the product ri of intrinsic growth rate r and time delay i. As shown in Fig. 9-21a, small values produced of ri caused growth to approach the asymptote, medium values (0.7–1.8) caused a damped oscillation, and larger values caused a stable oscillation limited by an envelope (limit cycle). In other words, the size of the oscillation was greater with greater energy (since r is a function of energy).

In other simulations the time delays were put on different terms of the equations. In energy language this means that time delays are on different path-

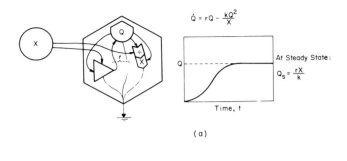

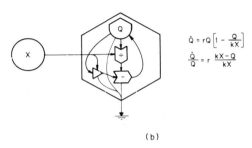

FIGURE 9-20. Designs with input force a denominator limit: (a) equation for a predator from Leslie (1948) with energy source affecting drain; (b) one way of translating the equation for a predator after Shimazu et al. (1972).

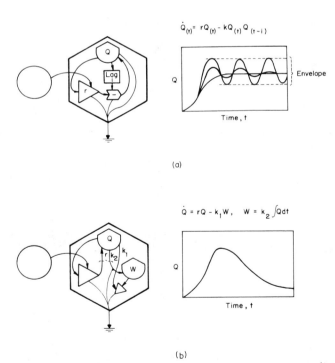

FIGURE 9-21. Autocatalytic systems with time delays: (a) logistic with pathway time delay (Hutchinson, 1954; Wangersky and Cunningham, 1956; May, 1973, 1974). Graphs are for three values of ri, where i is the time lag on the negative term; (b) lag in development of negative drain due to by-product storage accumulation W (Raston, 1962).

ways when diagrammed [see review by Wangersky (1978) and Goel et al. (1971)].

Time Lag in Discrete Systems

May (1974) shows that discrete generations may also generate time-lag, zigzag oscillations in models that would not oscillate in continuous functions (e.g., logistic). These may be real in systems that are discrete. However, zigzags often appear as artifacts in digital simulation of continuous-function systems when flows are large relative to storages. For example, growth in one generation pulls down resource storages that are not restored in time for the next time step, so that population drops and resource overshoots, and so on (see also Fig. 9-4). In Chapter 6 time delays in discrete pulses were given as elements in logic circuits such as "mono" units. With increasing energy oscillations between states occur called *bifurcations* which may generate secondary bifurcations and chaotic-looking time series (May, 1976, 1978). Mackey and Glass (1977) found oscillating physiological processes and "chaotic" shifts with time delay models.

Autocatalytic Growth with Cumulative By-product Drain

The model in Fig. 9-21b has been used for populations of microorganisms in crowded cultures in which harmful wastes accumulate that drain the main population in proportion. Energy may be unlimited and decay of wastes may be omitted as relatively small. No steady state is possible, and the population grows and declines as they often do in batch cultures. The waste accumulation acts as a time delay in expression of population losses. An example is the accumulation of alcohol in sugar fermentation by yeast. The time lag requires an extra storage, and the models may have such storage as information or in subsystems. Models with two storages are discussed in much further detail in Chapters 10–12.

Integrodifferential Equations and Lag

In the model in Fig. 9-21b one of the flows is controlled by storage. When expressed in equation form, $\dot{Q}$, the rate of change of Q becomes a function of the storage that in equation form is the integral

DIFFUSION-REGULATED GROWTH

Another autocatalytic model is given in Fig. 9-23, where autocatalytic production is balanced by spatial diffusion. As diagrammed in Fig. 9-21b, diffusion is well known as being proportional to the gradient. This may be expressed in its continuous form as the partial derivative of the quantity Q with respect to a space variable or in discrete form as proportional to the difference between the quantities divided by the distance between the centers. Also well known for diffusion is the property that the rate of change of the quantity in one place is proportional to the curvature of the quantity with space. Curvature is the change of the slope of the gradient with distance. Since diffusion causes inflows as well as outflows, the rate of change is proportional to the difference between inward and outward diffusion, and thus the net effect of diffusion depends on the difference between the gradient inflowing compared to that outflowing.

The model for balance between reproduction and diffusion was described by Kiersted and Slobodkin (1953) for application to populations of poisonous red

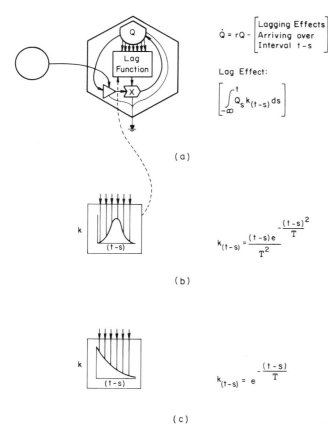

FIGURE 9-22. Lag functions used by Cushing (1977) for delayed limiting actions on logistic growth. Values of S are times before present to time t from which two actions are still arriving with their action: (a) energy diagram; (b) Gaussian lag; (c) exponential lag.

summing the waste accumulation. The differential equation contains an integral equation and is sometimes called an *integrodifferential equation*.

Starting with Volterra (1931) and summarized by May (1973) and Cushing (1977), equations of this type were used to study time lag in logistic, prey-predator, and other population models. Instead of the simple interruption lag given in Fig. 9-21a, actions are delayed by being transmitted through functions that spread the effect over a time interval according to the shape of that function with time. Figure 9-22 shows this type of model in diagram form. Various indices have been derived to indicate when these oscillate, mostly when the lag function is large and long. If the time lag is greater than time constant $1/r$, the model is unstable, with large, widening oscillations. Most of this work is done on models with constant growth coefficients and thus without the normal constraints of energy sources.

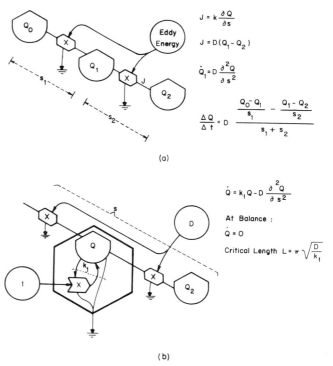

FIGURE 9-23. Diffusion regulated population models: (a) diffusion proportional to eddy stirring and rate of change proportional to curvature of population, where s is distance; (b) population a balance of reproduction and diffusion after Kiersted and Slobodkin (1953).

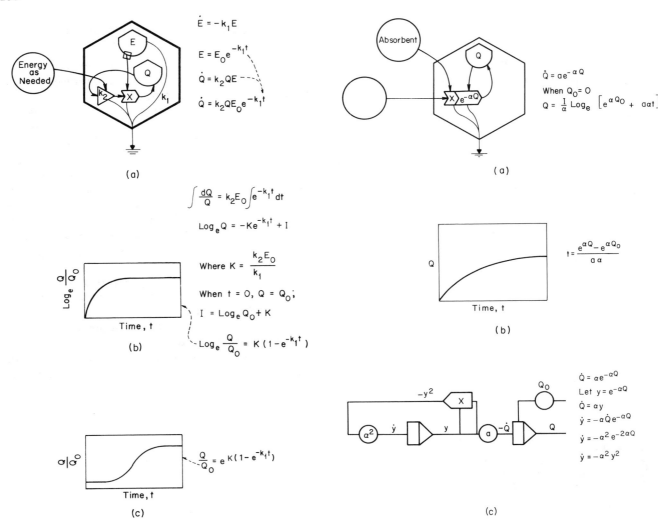

FIGURE 9-24. Gompertz growth. Autocatalytic system with linear decay k_1 of an essential element E provided at the start, generating Gompertz equation: (a) energy diagram; (b) graph on semilog plot; (c) graph on linear plot.

FIGURE 9-25. Elovich equation for rate of chemical adsorption (Krol, 1969); (a) energy diagram; (b) analog circuit (Krol, 1969), using property of exponents that rate of change is proportional to the square (see appendix to Chapter 5); (c) alternative energy diagram; (d) graph with time.

flagellate organisms in quiet seas near Florida. A criterion for development of a plankton bloom in a spot was the excess of reproduction over diffusion.

The model is unstable either growing or dispersing. When the equation for steady state was solved, a critical length was obtained. A plankton patch smaller than this is dispersed by turbulence. In a bounded system an oscillatory steady state pattern can develop (see Fig. 18-31).

The model may also apply to maintenance of human growth center in the presence of physical disturbance energies, attrition due to war, or other diffusion losses. Shimazu et al. (1972) used the model to express the spread of photosynthate from leaves in a terrestrial forest model.

GOMPERTZ GROWTH

Figure 9-24 shows an autocatalytic system in which growth depends on essential structure provided at the start only and that depreciates linearly. Depreciation of the growth mass itself is omitted. The model generates a Gompertz growth equation (Gompertz, 1825) where growth is an exponential of an exponential. On semilog graph the curve has no inflection. On linear coordinates, the graph has an S shape resembling the logistic. On a plot of log of a log as a function of time, the graph is straight.

The model has been used for weight gain of plant cells where there are initial storages of nutrients that disperse and initial storages of structure that undergo senescent loss of utility (Thornley, 1976).

ELOVICH MODEL

A model used for adsorption of chemical substances is the Elovich equation, where flow is dependent on an exponential approach to equilibrium (see Fig. 9-25). The exponential property results when the negative square of the derivative is the acceleration as shown in the analog circuit.

AUTOCATALYTIC SYSTEMS WITH ACCELERATION AND INERTIAL FORCES

There are autocatalytic systems that maximize energy flow in developing structure using acceleration and the inertial backforces.

Figure 9-26 is given a model for a star. It is consuming stellar dust drawn in by gravity and transformed there into the growing mass, releasing its energy of potential against gravity into kinetic energy and heat. As temperatures rise, thresholds for atomic reactions develop, generating atomic reactions and a redistribution of matter, including export of matter and energy into space. Simulation of this model shows a progression of stages analogous to succession in ecosystems and corresponding to the current ideas about the evolution and life of stars. If the dust sources are sufficiently large, gravity may produce a black hole in this model. When energy of atomic reactions accumulates more rapidly than gravity to contain the system, explosions may result. See also the Van der Pol oscillator in Fig. 10-30.

OTHER CONSUMER MODELS

Autocatalytic models with two sources are given in Chapter 13 on webs, and those with demand pathways for yield in Chapter 11. Versions that include money and are minimodels for world and national growth are presented in Chapter 26. More complex consumer models are given in Chapter 20 on consumers and in Chapter 22 on succession.

STATISTICAL MODELS

As described for Fig. 8-24, pathways and storages of a model may be given population variations reflecting the real world's pathways (see also Fig. 17-15). Stochastic versions of basic autocatalytic models

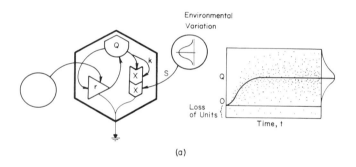

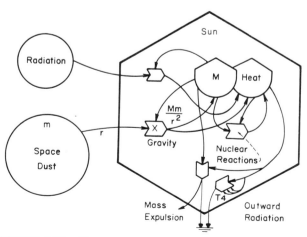

FIGURE 9-26. Energy diagram of a star.

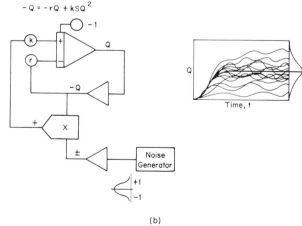

FIGURE 9-27. Logistic model with stochastic variation of equilibrium coefficient from environmental S is Gaussian distribution of values: (a) energy diagram; (b) analog circuit and oscilloscope traces.

may involve any or all of the pathways and one or more distributions such as Gaussian variation (May, 1975; Bartlett, 1960; Pielou, 1977; Bartlett and Hioms, 1973; Goel and Richter-Dyn, 1974). An example is shown in Fig. 20-23.

Large stochastic variations in logistic equilibrium coefficients (Fig. 9-27a) cause so large a range of variation that the equilibrium population is reduced by units going to zero and the population becoming extinct. Gaussian distributions may not be as typical as are more hierarchical ones related to energy patterns (see Chapter 15).

Figure 9-27b gives an analog circuit for a stochastic logistic with the use of a noise generator that varies voltage so as to affect coefficient for equilibrium.

SUMMARY

In this chapter we examined simple designs with autocatalytic feedback pumping and controlling the flow of input energy. Because they tend to maximize power toward unit persistence, maintenance, and survival, these modules have many of the properties of living and nonliving units that use energy to maintain their own structure. There is a minimum energy concentration for survival. The characteristic autocatalytic process upgrades energy in an interactive productive process and stores it as structure and reserves, finally using it in feedback and export and losing some in depreciation.

The kinetics of autocatalytic modules depend on whether there is some feature that limits growth. In all cases growth is at first accelerated. The growth of quadratic feedback is more rapid than that of simple feedback that generates exponential growth. Growth levels off to a steady state if the source is limited by inflow externally, if there is sufficient quadratic or cubic drain, if there is a limiting, necessary recycle, or if there are control mechanisms involving thresholds that limit growth.

Traditional equations for population growth assume an unlimited energy source and feedback as constant-gain amplifiers. These traditional formulations are compared with similar ones that include the external energy source through its driving force or concentration. Models of the latter type respond to energy availability, as the maximum power principle suggests real systems do to survive. Whereas a model of the autocatalytic properties of an inflexible species realistically may have intrinsic limits, the unit that ultimately results after adaptation, substitution, or other redesigning response may follow the extrinsic equation.

In general, the simple models of this chapter are useful for clarifying concepts of growth. However, they omit many features of living populations that may be important in representing ecosystems. More complex population models are given in Chapter 10.

QUESTIONS AND PROBLEMS

1. Starting with simple exponential growth, arrange modules according to their complexity, powers of feedback, and drain pathways. Can you draw the growth curves for each with the correct shape?

2. Use an analog computer or microcomputer to simulate two of the modules.

3. Write differential equations to accompany the diagrams in Chapter 9 by inspection.

4. Draw an autocatalytic model, adding calories so that they are reasonable according to the first and second energy laws. Then calculate efficiencies, various kinds of production and growth rate, and calculate the energy transformation ratio.

5. Which modules maximize power during periods of growth on unlimited energy?

6. Give two alternate ways of diagrammatically representing exponential growth.

7. Give three alternate systems that generate logistic growth equations.

8. What quadratic effect drains and levels growth but does not negatively affect life?

9. Using analog computer diagrams, identify a system that has both positive and negative feedback but no negative loop in energy language.

10. What change occurs with time that causes growth to be of the Gompertz type?

11. What are the minimum features for a module to be classified as autocatalytic?

12. What types of system tend to develop because of the maximum power principle?

13. What is similar and different about the growth curves for a Von Bertalanffy chargeup and logistic growth?

14. Diagram an autocatalytic process where the growth steps are discrete (digital) steps.
15. How does a star grow autocatalytically? What are the force relationships involved?
16. What is the energy criterion that determines when an autocatalytic unit replaces a linear process?
17. How does overall effect of diffusion depend on the spatial curvature? Indicate the situation with a three-tank diagram. What model balances an exponential growth unit whose stock is being dispersed by diffusion with a curvature, causing outward dispersal? What happens when the diffusion decreases below a threshold?
18. What model has its inflowing resources diluted by the growing stock?
19. Give models that involve the limitation of autocatalytic growth by thresholds.
20. Give models that involve the limitations of autocatalytic growth by switching actions.
21. What module is cooperative in the sense that the specific growth rate has a maximum?
22. If the autocatalytic feedback loop is fourth power and the drains are fifth power, will the growth curve level (where energy is not limiting)?
23. Explain the energetics of the intrinsic rate of natural increase and why models using this concept become inaccurate as energy resources dwindle.
24. How does backforce cause an autocatalytic loop to be logistic?
25. What property of autocatalytic modules is often implied when a system is referred to as "unstable"?
26. Draw an energy diagram for each of the following equations, and then select a letter from the key list to accompany each of the numbered phases: (a) $\dot{Q} = J$, (b) $\dot{Q} = J - kQ$, (c) $\dot{Q} = -kQ$, (d) $\dot{Q} = -k_1 E Q - k_2 Q$, (e) $\dot{Q} = +k_1 Q^2 - k_2 Q$; (1) chargeup, (2) exponential decay, (3) U.S. growth before 1973, (4) exponential growth, (5) ramp, (6) stable steady state, (7) lack of growth without initial condition, (8) starvation curve, (9) multiplier feedback in energy language, (10) linear feedback in analog diagram.
27. Using analog or digital computer, simulate a selection of the models given in this chapter. Then vary force from energy source and observe sensitivity of responses.

SUGGESTED READINGS

Bartholomay, A. F. (1972). Stochastic Models. In *Foundations of Mathematical Biology,* Vol. II, *Cellular Systems,* R. Rosen, Ed., Academic Press, New York, pp. 23–213.

Begon, M. and M. Mortimer (1981). *Population Ecology.* Sinauer Associates, Sunderland, MA, 216 pp.

Cushing, J. M. (1977). *Integro-Differential Equations and Delay Models in Population Dynamics,* Lecture Notes in Biomathematics, No. 20, Springer-Verlag, New York.

Emlen, J. M. (1973). *Ecology an Evolutionary Approach, Indiana University*, Addison-Wesley, Reading, MA.

Gause, G. F. (1934). *The Struggle for Existence,* Hafner, New York (reprinted 1964).

Goel, N. S., S. C. Maitra and E. W. Montroll (1971). *Non-linear Models of Interacting Populations,* Reviews of Modern Physics, Academic Press, New York. 145 pp.

Haken, H. (1978). *Synergetics,* Springer-Verlag, New York.

Hutchinson, G. E. (1978). *An Introduction to Population Ecology,* Yale University Press, New Haven, CT.

May, R. M. (1973). *Stability and Complexity in Model Ecosystems,* Princeton University Press, Princeton, N.J.

Mclaren, T. A. (1971). *Natural Regulation of Animal Populations,* Alberta Press, New York. 195 pp.

Murray, B. G. (1979). *Population Dynamics Alternative Models,* Academic Press, New York. 211 pp.

Poole, R. N. (1974). *Introduction to Quantitative Ecology,* McGraw-Hill, New York.

Slobodkin, L. B. (1962). *Growth and Regulation of Animal Populations,* Holt, Rinehart and Winston, New York.

Wangersky, P. L. (1978). Lotka–Volterra population models. *Ann. Rev. Ecol. Syst.* **9**, 189–218.

Watt, K. E. F. (1968). *Ecology and Resource Management,* McGraw-Hill, New York.

CHAPTER TEN

Loops

Loops are designs in which a pathway circles back from a downstream unit and connects with the pathway incoming from upstream (see Fig. 10-1). Sometimes these return pathways are called *feedback*. The autocatalytic models studied in Chapter 9 included a feedback loop from a storage to a multiplier or to a constant-gain amplifier. In this chapter other basic loop designs are considered, especially those involving two or more storage or interacting units.

A feedback is positive if it increases the flow at its point of reentry and intersection. It is negative if it decreases flow at that point. Without the clarification of diagrams or equations, the word "feedback" is misleading since it can refer to entirely different phenomena. Feedback in one system's language is a different phenomenon from that in another language. For example, the storage shown in Fig. 3-3a had no feedback in energy language, but when drawn in block diagrams (Fig. 3-3d) or analog computer symbols (Fig. 5-1c), it did have a negative feedback. Some feedbacks add, some multiply, and others have other effects. We can avoid semantic misunderstanding by stating which language is being used and then presenting the design configuration in a diagram for clarity.

CIRCLE OF STORAGES THAT DRIVE THEIR PATHWAYS

Figure 10-2 shows a simple loop of linear flows between two storages with the force or forces for each pathway derived from the energy storages. An example is a population of cows going back and forth between barn and pasture in proportion to their numbers, if one ignores other properties of cows except their walking movement. Figure 10-2a shows an energy source flowing into one storage, and in the example of the cows this may be the inflow of grass food. Energy drains out from the storages as depreciation and from the pathways as a friction and transport process. Another example is a chemical reaction in which there is back reaction.

The example in Fig. 10-2 has linear pathways driven by forces from storages. These do not always go together, as the next designs will illustrate. If the example were one in the macroscopic world (larger than molecular dimensions) and if there were no outside energy source, energies would drain away and the circulation would stop, one storage draining into another with lower potential energy. At the molecular level, where energy of degraded heat resides in the motions of the molecules, circulation may continue although chemical equilibrium may exist. Here energy is supplied from the thermal motions of molecules, and the loop may have some circulation, even when there is no outside energy source. See Fig. 7-12 for discussion of energetics of reversible flows.

Where energy drains and transformations occur in the pathways, coefficients may be different in the segment leaving one storage as compared to the segment entering the next storage. Even the dimensions of the units may be different. For example, burning wood decreases wood in one storage and increases nutrients in another storage.

Figure 10-2 shows the time lags that follow from inflows to one tank that then charges another tank. More is given on these kinetics in Chapter 11 on chains and series.

Cycles of Materials; Overlay Diagrams

In many systems there are circulating quantities that are conserved, meaning that the inflows equal the storages plus outflows. Examples are the circulations of matter and money. It is often convenient to draw diagrams of the matter flow omitting the energy drives and sinks, which has the advantage of

Loops

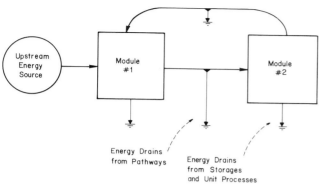

FIGURE 10-1. Generalized pattern for loops with energy sources and sinks.

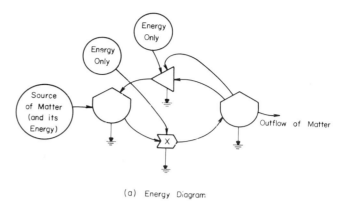

(a) Energy Diagram

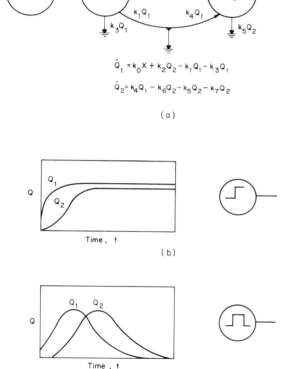

FIGURE 10-2. Loop of storages driving one-way pathways: (a) energy diagram with pathway terms; (b) typical response to step function in outside source; (c) typical response to source pulse from energy source.

(b) Overlay Materials Diagram

FIGURE 10-3. Example of a materials diagram overlay to the more complete energy diagram.

showing only the pathways of the type of matter being considered, sometimes with numbers written for the flows observed as a way of summarizing a network, and the disadvantage of separating the unit of energy, matter, and money in one's thinking. Erroneous models may result that omit the driving forces that produce characteristic flow. A recommended suggestion is to draw the complete diagram with energy and matter first and then make an overlay labeled as a material or money diagram, leaving out those flows that do not include the matter being considered. This diagram can retain the symbols of the more composite diagram kept in the same position on the paper so that one can be overlaid on the other mentally or by using semitransparent paper. Dotted lines are suggested for distinction of material diagrams. An example is given in Fig. 10-3. Where money is used, dashed lines are suggested.

Loops in a Matrix

In a matrix representation of a network such as that in Fig. 6-16, loops are formed by units located as symmetrical pairs on diagonals from lower left to upper right.

Circle of Storages with Linear Pathways

Especially where mineral cycles are of interest, models are studied involving a circle of storages connected with linear pathways. See Fig. 8-7 for several kinds of linear pathways. The example given in Fig. 10-3 illustrates three outside energy sources and conservation of matter. In Fig. 10-3b the materials diagram indicates that each storage has a content of the same kind of material that is cycling.

Equations for Modeling Mineral Cycles with Conservation of Matter

Where flows of matter are concerned, the rates may be simplified by the conservation of matter. For pathways such as those in Fig. 10-3, matter flowing out of one tank flows into the next. If the variable is matter, the outflow coefficient must be equal to the inflow coefficient.

If equations are written for a mineral cycle that passes around the loop in Fig. 10-2, the two transfer coefficients on each pathway become equal ($k_1 = k_4$; $k_2 = k_6$), and the terms for the drains to heat sinks are omitted (since no matter is drained). The equations for Fig. 10-2 then become

$$\dot{Q}_1 = k_0 X + k_2 Q_2 - k_1 Q_1$$
$$\dot{Q}_2 = k_4 Q_1 - k_6 Q_2 - k_7 Q_2 \quad (10\text{-}1)$$

Lotka Loop; Steady State with Rates Inverse Proportion to Storage

While considering the circulation of mineral cycles, Lotka (1925) considered the relationship of the storages and the conductivities of the pathways in a closed loop of linear pathways such as that in Fig. 10-4. In this example there are three storages in a circle connected by linear pathways, and an energy source is involved at one place in the figure.

At steady state the rates of change of storage are zero and the product of storage and transfer coefficient of each pathway is equal to that of the other pathways, or the storage is inversely proportional to the coefficient of transmission of the pathway just ahead (see equations in Fig. 10-4). Large storages develop when rate processes are slow. The large

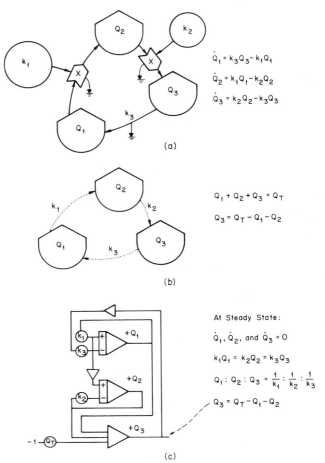

FIGURE 10-4. Linear circle of storages in which conservation of matter is appropriate: (a) energy diagram; (b) matter diagram; (c) analog diagram (note one less integrator than storages because of conservation of matter).

mineral deposits in the earth are in those parts of the earth cycles that have slow specific rates of outflow.

Michaelis–Menten Loop

Given in Figs. 2-7 and 10-5 is an important design that is prevalent in processing energy and many systems because of its special properties for stability. There is a multiplicative interaction between an inflowing energy source and a quantity of recycling matter that produces a storage of higher energy quality. This storage makes a return circle and in the process may interact to transfer its energy to other units downstream. The cycling material goes from receptive to activated state and back. The material quantity that makes the cycle is conserved,

Loops

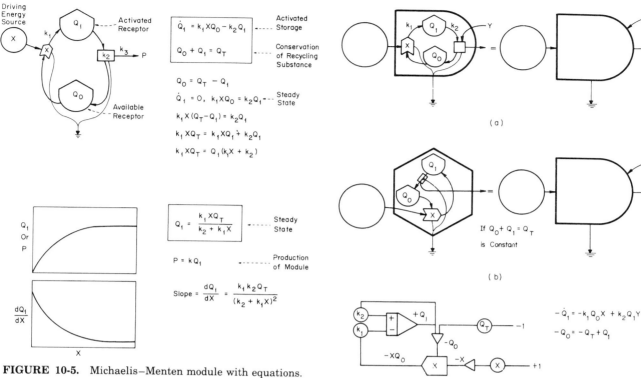

FIGURE 10-5. Michaelis–Menten module with equations.

FIGURE 10-6. Symbols for Michaelis–Menten module: (*a*) detailed and aggregated designation; (*b*) autocatalytic unit becomes Michaelis–Menten unit when matter conservation is added; (*c*) analog program.

and as shown in Fig. 10-5, the equation for conservation of cycling receptor material gives the module many special properties.

The properties were first studied by Michaelis and Menten for the reaction of a recycling enzyme with its substrate (substrate is the substance carrying energy catalyzed by the enzyme causing a reaction after which the enzyme is reused). The equations for the module are given in Fig. 10-5, where the recycle pathway is linear and in the module in Fig. 10-6, where the return pathway is an interaction with another variable Y. Mineral raw materials are also recycled as shown in Fig. 2-7.

When the energy is in steady state, a very characteristic algebraic expression results that we have already studied in Chapter 8 involving external limiting factors (Figs. 8-8–8-10). Often this type of algebraic expression is called the *Michaelis–Menten equation*. When the system is one of recycling receptor as in Fig. 10-5, the name is appropriate. When similar algebra results from pumping from a constant flow source (Fig. 8-8*b* and 8-9) or from source limited flows (Fig. 8-10), it refers to a nonlooping, different system and may not be as appropriately called Michaelis–Menten.

Because the cycling receptor system is so common, a summary symbol is suggested as shown in Figs. 2-7*c* and 10-6*a*, a short bullet-shaped symbol. Where it is used, there is a multiplier and two tanks in a loop with the cycling quantity conserved ($Q_1 + Q_2 = K$).

Many autocatalytic processes involve a feedback from storage to a multiplier as given in Chapter 9. If that feedback is in the form of a material as shown in Fig. 10-6*b*, the design can be recognized as similar to that just described as the Michaelis–Menten loop. However, it is kinetically the same only if there is a conservation of the recycling material that comes with having a closed material loop. The amount of recycling matter is constant at least over the time periods required for the module to enter steady state in response to a start of the external energy source.

Some examples of the wide variety of systems that operate as Michaelis–Menten loops are given in Fig. 10-7. Some of the most important are reception of light in photometers in first steps of photosynthesis and in photochemical reactions in industry. Also included are examples of pumped storage,

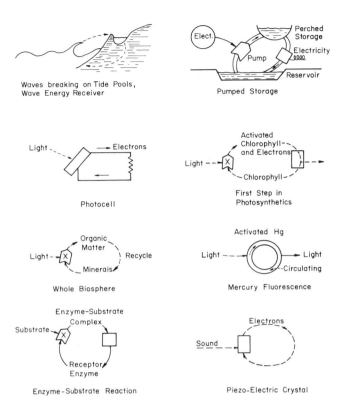

FIGURE 10-7. Examples of Michaelis–Menten loops.

mechanical devices, and biochemical reactions. The most important is the P–R model, which applies to whole ecosystems, including the whole biosphere as given in Fig. 2-7.

The time graph of the storage and flux through the Michaelis–Menten system after a step on input is given in Fig. 10-8. It has an asymptotic shape, essentially similar to that of the simple storage chargeup.

The time graph of a Michaelis–Menten system using an initial unrenewed storage is given in Fig. 10-9, an example of which was analog simulated by Rogers (1976). If the storages within the module are small enough to be close to steady state so that the steady-state formula $S/(K + S)$ may apply, the diagram is appropriate for the equation.

Two-Tank Loop in Exponential Growth

If the conservation of materials constraint is removed from the Michaelis–Menten module running on a constant-force source, a two-tank autocatalytic system remains that is more slowly growing (or dying) than the simpler one-tank exponential module

(see Fig. 10-8c). This is not a Michaelis–Menten unit.

F. E. Smith (1969) simulated population models with consumers, as in Fig. 10-8c, without a restraint of conservation of population within the unit (i.e., $Q_0 + Q_1$ were not conserved), finding unrestrained growth.

Symbiosis of Ordered and Disordered Materials

In the modules in Fig. 10-5, materials flow from an ordered high-energy state to a disordered lower energy state. Yet the lower energy state achieves as much through multiplier action when it interacts with new outside energy. In one sense, disordered material available for the ordering process is as valuable as the order when the disorder is fed back as a scarce commodity. There can be a shortage of either ordered or disordered materials. Development of more disordered materials with recycling may stimulate total power flow if the recycle process is limiting (see Chapter 17).

Loading Characteristics of the Michaelis–Menten Loop

The Michaelis–Menten loop has excellent energy transformation characteristics that tend to maximize the incorporation of energy from a varying source. These characteristics may explain the generality of this type of system in nature. First, refer back to Chapter 7 and Fig. 7-21, where a varying input energy source was found to be far from the optimum loading for maximum power most of the time. Since the backforces and loading of the receptor system were fixed, the energy transformation ranged from being too efficient and thus too slow to being too inefficient (and too rapid) for maximum power transfer.

In contrast, compare the performance a Michaelis–Menten system supplied under conditions of varying input force as shown in Fig. 10-5. Here the property of saturation caused by the transfer of the receptor reservoir into the activated storage stops drawing power once the structure is loaded in the vicinity of maximum power transfer. Energy can then pass to the next receiver, thus permitting a stacking of Michaelis–Menten modules, each loading toward optimum and then allowing energy to load the next and thereafter the next, and so forth.

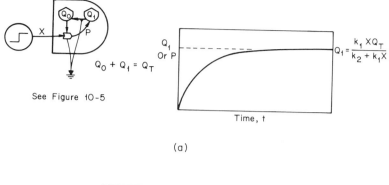

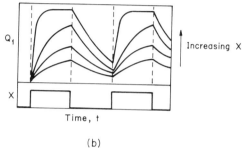

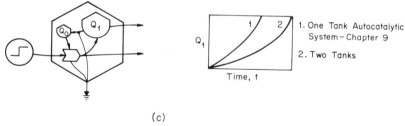

FIGURE 10-8. Temporal response of Michaelis–Menten loop to startup: (a) responses of unit to a step input; (b) family of curves produced by increasing source X in successive runs; (c) without the constraint of $(Q_0 + Q_1)$ constant, exponential growth results. However, growth with two tanks is slower than with one (Fig. 9-3).

The chlorophyll receptor units of the chloroplast of plants are stacked in this way, first in the grana, and then the chloroplasts are stacked in the cell; then the cells are stacked in the tissues of leaves or algae; and finally the leaves or algae are stacked in the ecosystem (see Fig. 19-21).

Regulation of Amount of Recycling Receptor Material

The loading of a Michaelis–Menten receptor depends in part on the quantity of recycling material (e.g., chlorophyll), or in a selenium cell, on the quantity of recycling electrons. A graph of the quantity of recycling receptor needed to maximize power for different levels of input energy drive is given in Fig. 10-10. At very low energy inflow, more receptor is required to generate maximum power for the same receptor structure. The curve was derived as shown in Fig. 10-10 from the Michaelis–Menten equation. Well-established examples are the adaptations of many plant systems to increase chlorophyll that is recycling where conditions are shady. One example is the adaptation of chlorophyll in culture plates of algae stacked on top of each other with light entering from the top. Other examples are the adaptations to shade in some lower forest trees. See Chapter 19 for more information on plant production

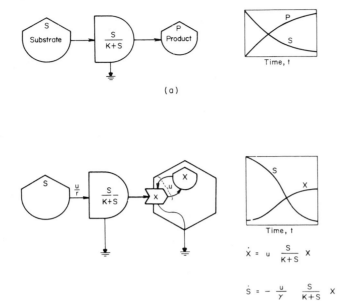

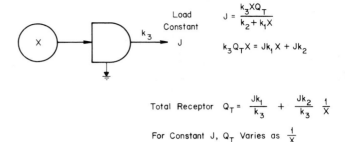

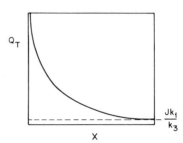

FIGURE 10-9. Simulation of Michaelis–Menten process using unrenewed storage sources (Rogers, 1976): (a) Michaelis–Menten pathway; (b) autocatalytic use of Michaelis–Menten pathway. Ordinates are on different scales.

FIGURE 10-10. Quantity of cycling receptor for maximum power where input drive varies. Examples of data with this form are given for *Chlorella* by Phillips and Myers (1953).

systems. Groden (1977) used a log function instead of the hyperbola to describe shade adaptation.

Michaelis–Menten Loop with Source Inhibition

Figure 10-11 illustrates source inhibition of energy transformation of the Michaelis–Menten loop at high energy levels. Many other equations have been offered to account for light inhibition of plant photosynthesis that are empirical or based on other designs (see Chapter 19).

Michaelis–Menten Loop with Flow-Limited Source

Figure 10-12 shows a system with a source-limited flow and a Michaelis–Menten loop. Depending on coefficients, either can limit, one with an external limit and one with a recycling limit. Two mechanisms with similar effect are a common configuration in real systems. However, simulation models can be simplified to include only one limiting

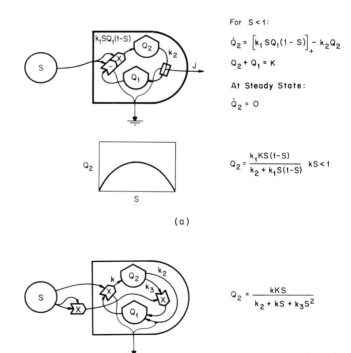

FIGURE 10-11. Michaelis–Menten module with high-energy inhibition (see Chapter 19): (a) inhibition of inflow; (b) accelerated stress drain (Haldane, 1930; Andrews, 1968).

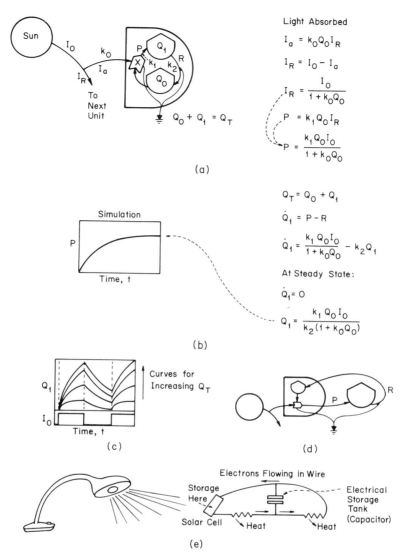

FIGURE 10-12. Michaelis–Menten module using a limiting source with outside flow control: (a) energy diagram; (b) simulation with I_o constant; (c) response to square waves for different quantities of recycling matter Q_T; (d) example of P and R from Fig. 2-7; (e) inexpensive passive analog circuit.

mechansim if the structure of the system does not need to be shown.

A Two-Storage Loop with Two Energy Sources

Figure 10-13 shows a loop interacting with two outside energy sources. Where power is to be maximized and limits may develop in recycling, energy can be applied to the recycle part of a loop. Examples are urban renewal, automobile body recycling in industry, and generation of more human mobility in tightly ordered society. Lumry and Rieske (1959) give kinetics as applied to isolated chloroplasts under loading with added substances as in Fig. 10-13. One of the simplest models of our economy that runs on sunlight and fossil fuels has this configuration. If the forces from the energy sources are held constant, each flow is in proportion to its storage in a linear manner. If the material circulating is conserved, $k_1 = k_4$ and $k_2 = k_3$. The storage is self-adjusting, with higher storages developing upstream from the smaller transfer coefficient. It is, therefore, another case of a Lotka loop proportionality given in Fig. 10-2. If one of the energy sources

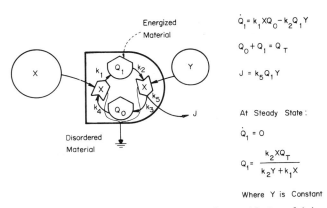

FIGURE 10-13. A Michaelis–Menten loop with two driving sources.

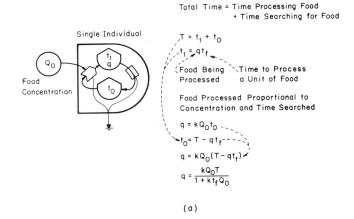

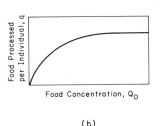

FIGURE 10-14. Time of the individual as a limiting factor in consumption generates a Michaelis–Menten response to increased energy supply (Holling, 1965): (a) energy diagram; (b) graph of food consumed per individual.

is constant and the other is varied, the module becomes a Michaelis–Menten loop. The force Y may be from downstream system loops (see Fig. 11-25).

SMALL ENERGY REQUIRED FOR DOWNHILL PUMPING

When energy is used to pump disorder into order, an efficiency can be measured. If the same kind of energy is used to pump in the direction that the process would go spontaneously with its own stored order as its energy source, the pumping is downhill and, similar to jujitsu, can accomplish more per calorie. Since relatively little energy is required to relieve a recycling block, down pumping mechanisms are easily found.

Both inputs may originate from the same source, one pumping up and one pumping back (see discussion of photosynthesis and photorespiration symbiosis in Chapter 19 and also Fig. 10-11b).

RECYCLING OF A CONSUMER'S TIME

A variation of the Michaelis–Menten internal limitation is the feeding of some animals studied by Holling (1965) [see discussion by Emlen (1973)]. As the animal feeds, it processes the food for a period of energy combination and activation and then is recycled to the ready state to interact with food stocks again. The population's inherent recycling time provides an internal limiting factor to feeding due to increases in food conservation. Holling used the time involved in food processing as a way to calibrate the system. The total time for feeding was con-

served and was the sum of time spent in search of food and time spent in processing food (see Fig. 10-14). The relationship fits the Michaelis–Menten module diagram, where the single animal is recycling. Energy is partitioned in proportion to time spent in two states. As stated colloquially, "time is energy."

REPRESENTING CYCLES OF MATERIAL IN ENERGY MODELS

Where a material such as a chemical nutrient is recycled between its available pool (M_1 in Fig. 10-15) and a combined state, more than one model may be used to represent materials (e.g., recycling nutrients) explicitly. In Fig. 10-15a one tank Q is used for stored biomass from production and its enclosed content of the nutrient. The conservation of matter is indicated by use of the Michaelis–Menten short bullet symbol or by the equation. In Fig. 10-15b the

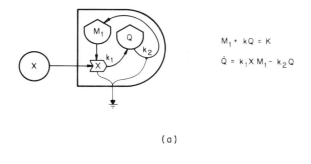

(a)

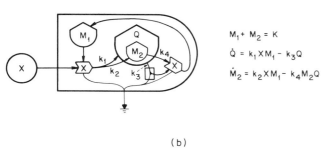

(b)

FIGURE 10-15. Alternative diagrams for loops with conservation of materials (M); (a) Q includes material M; (b) material within Q explicitly separated.

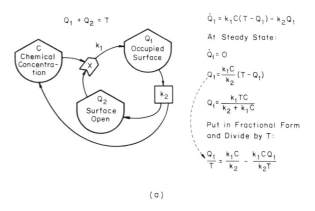

(a)

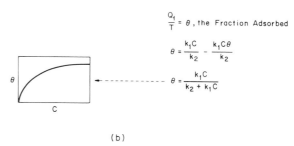

(b)

FIGURE 10-16. Langmuir adsorption "isotherms" equations, a chemical equilibrium example of a Michaelis–Menten loop: (a) a loop (see Chapter 8); (b) graph of fraction of space adsorbed as a function of concentration.

nutrient that is incorporated in biomass is given a separate tank M_2 within the biomass tank Q, thus allowing linked equations for matter and energy in the same system. Other efforts to combine energy and matter are discussed in the following paragraphs; see also Chapter 6 [bond graphs, Paynter (1960) and Oster and Desoer (1971) and coupled graphs, Koenig et al. (1967)]; and Chapter 13 [webs; Ulanowicz (1972)].

THREE-STORAGE NESTED LOOPS; LANGMUIR MODEL

The adsorption of one substance by another has been modeled with a loop system such as that given in Fig. 10-16 (Langmuir, 1921). It is usually applied as a reaction that goes from an initial state with potential energy to equilibrium with balance of forward and return reactions. Materials are conserved. As shown, the equations are similar to those of the Michaelis–Menten loop. These reactions are often visualized as reversible pathways rather than as circular, differences that do not affect kinetics (see microscopic reversibility in Chapter 7).

Verhoff and Smith (1971) used the flow of matter in a three-storage loop in Fig. 10-17 as the basic unit

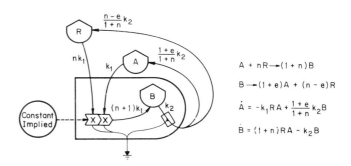

FIGURE 10-17. Three-storage loop models for a trophic level with conservation of mass only (Verhoff and Smith, 1971); R is available mass conserved in recycling, A is feeding population, and B is nonfeeding population. As part of the translation, an energy source was added with value unity.

model for trophic levels in ecosystems. Although operating as an open system, the implied energy source was constant (and thus unlimited) so that the open and closed system had similar equations. One tank was identified as free nutrients (recycled resource), another comprised of the new reproductives (containing matter also), and the third was the combination of material and reproductives generating the biological storage. Michaelis–Menten response was found in simulation, and the model was expressed in matrix form. Conservation of matter was handled by using only two independent coefficients, with the others derived from these with fractional multipliers.

MATRIX FORM

With two or more storages and equations, it may be convenient to represent a model with the use of matrices (see Chapter 6 and 13). Since the terms of the pathways in an energy diagram are the terms of the set of differential equations for the model, the model may be represented as a vector and a matrix that represent the set of equations. The column vector containing the rate of change is equal to the matrix that contains the coefficient terms times the vector representing the quantities. Loops are identified with the diagonals of the matrix since these pathways are reciprocally connected (i.e., Q_1 to Q_2 and Q_2 to Q_1). An example of representing equations in this form is given in Fig. 13-20 with discussion of complexities.

AUTOCATALYTIC CONSERVATION LOOPS (LOGISTIC)

Slightly more complex than the Michaelis–Menten loop is the common module in Fig. 10-18, where there is an autocatalytic feedback pumping receptor into ordered state and a linear return. Like the Michaelis–Menten module, there is a conservation of cycling reactant and the system is asymptotic to an increased driving force and very stable because of the self-limiting role of the recycling receptor. Without the latter provision, the system would be an exponential growth module like that given in Chapter 9. With time, the system first accelerates with upward curvature before leveling off, thus generating a sigmoid curve. The combination of the rate equa-

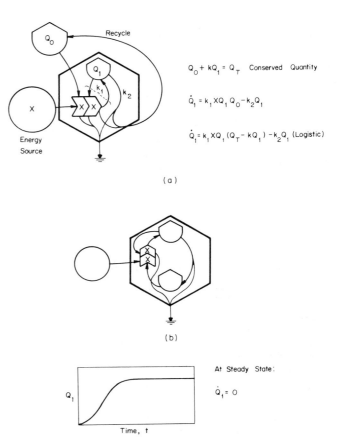

FIGURE 10-18. Logistic loops between autocatalytic module and a necessary quantity recycled and conserved: (*a*) external loop; (*b*) internal conservation; (*c*) graph with time.

tion and the conservation constraint generates another logistic expression where the percent growth rate declines linearly with quantity. The hexagon is used for the whole system since it has an autocatalytic feedback.

An early model of infection with these characteristics (Fig. 10-19) but with a third storage was given by Martini (1921) and Lotka (1925).

Chemical Loops

A catalase enzyme system simulated with analog computer by Chance et al. (1952) is a similar one with a recycling loop that provides stability (Fig. 10-20*a*). Prigogine et al. (1977) give a chemical reaction system with quadratic autocatalysis (Fig. 10-20*b*), which can oscillate with a limit cycle on a phase plane.

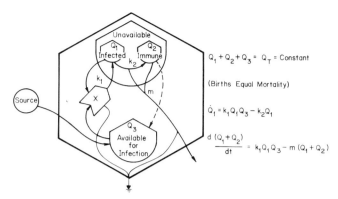

FIGURE 10-19. Autocatalytic model of infection after Martini (1921) that has a cycling, conservation type logistic equation (after Lotka, 1925).

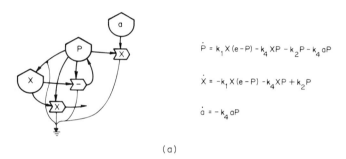

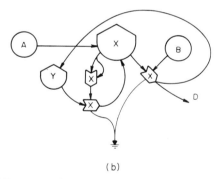

FIGURE 10-20. Complex biochemical loops: (a) simulation of catalase enzyme system (Chance et al., 1952); (b) quadratic autocatalytic model with recycle loop from Prigogine et al. (1977) quoting Lefever (1968)—a reaction called *cross catalysis*.

Ross Malaria Equation

An early population equation given by Ross (1911) (Fig. 10-21) contains two Michaelis–Menten loops coupled together. Lotka's analysis of the system in Fig. 10-21b showed a surface of population stock that has stable and unstable zones depending on parameters.

STEADY STATE AND PHASE PLANE GRAPH

Systems discussed here and in following chapters that contain two or more tanks are mostly nonlinear because they have interactive functions such as multipliers. Many of these systems have steady states that are constant or oscillate. On a phase plane graph (x–y plot) they approach a point or a circle or ellipse pattern, sometimes called a *limit cycle* (e.g., see Fig. 10-22). These are two kinds of "stability." A stable point is sometimes called a singular point (see also p. 182).

The conditions for which a system approaches a point are found by setting rates of change equal to zero, as done in Chapter 9 for a one-storage system. Then one has to determine if the point is stable in the sense that it will return to that point if slightly displaced (see Chapter 11).

Production–Consumption Loop Containing Michaelis–Menten Pathways and Conservation of Mass

Figure 10-22, from Quinlan and Paynter (1976), shows a production consumption loop with conservation of mass, autocatalytic feedbacks, plus Michaelis–Menten modules acting on the connecting pathways. Energy sources are implied as constant, but the model is stable over much of its range because of the loop constraints of the whole system and those within each Michaelis–Menten loop (see also discussion of limiting-factor pathways in Chapter 8).

If A is large and B is small, the Michaelis–Menten pathways disappear and the Lotka–Volterra prey–predator model results (see Chapter 11). Different characteristics of the Michaelis–Menten pathway constants lead to steady states with stable points, limit cycles, or damped oscillations as shown in Fig. 10-22b.

Oscillation Control

In this and many models of later chapters, changes in rate adjustments can cause oscillation or steady, steady states. Whether systems oscillate or not is not a compulsive property but readily adjustable in self-organization. Energy-related reasons for retaining oscillations are given in Chapters 14 and 26.

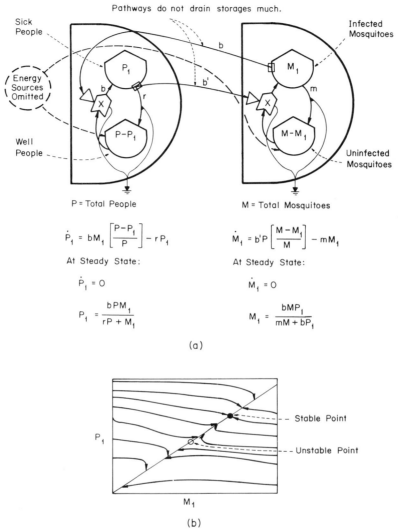

FIGURE 10-21. Parasite–host model with symbiosis of sick people and malaria, from Ross (1911): (a) equations and energy diagram; (b) state diagram after Lotka (1925).

LOOPS WITHOUT STORAGE

Multiplier Loops without Storage

Some multiplier loops without storage and thus without time are given in Fig. 10-23 and a feedback to its own interaction is given in Fig. 10-23a. An example is a power plant that generates electricity that is fed back without much storage. It is possible (stable) only when the driving force and efficiency are inversely related.

Figure 10-23b illustrates a symbiosis of intersections without the storage that was in an otherwise similar diagram in Fig. 10-13. An example is an electric pump for water Y that generates the cooling water for the pump motor, where X denotes the source of electricity. Without a regulation by coefficients these configurations are instantaneously explosive.

Additive Loop without Storage

An additive loop without time, that is, without storage, is given in Fig. 10-24. Therefore, the output is a constant function of the input. Examples are common in electronics where voltage is the force X and voltages are fed back, with algebraic addition being

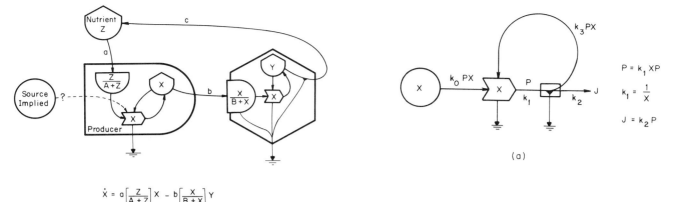

$$\dot{X} = a\left[\frac{Z}{A+Z}\right]X - b\left[\frac{X}{B+X}\right]Y$$

$$\dot{Y} = b\left[\frac{X}{B+X}\right]Y - cY$$

$$\dot{Z} = cY - a\left[\frac{Z}{A+Z}\right]X$$

$X + Y + Z = m = $ Constant

Singular Points:

$$\hat{X} = \frac{cB}{b-c}$$

$$\hat{Y}^2 + \left[\hat{X}\left(1 - \frac{a}{c}\right) - (m+A)\right]\hat{Y} + \frac{a}{c}\hat{X}(m - \hat{X}) = 0$$

$$\hat{Z} = m - (\hat{X} + \hat{Y})$$

(a)

(b)

FIGURE 10-22. Production, consumption, recycle loop with Michaelis–Menten pathways with energy implied and matter conserved (Quinlan and Paynter, 1976): (a) energy diagram; (b) simulations for various values of Michaelis–Menten pathways.

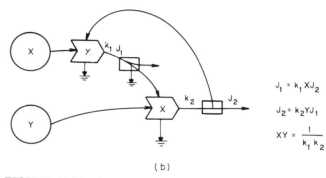

FIGURE 10-23. Multiplier loops without storage: (a) feedback to self-interaction; (b) symbiosis between interactions.

sometimes positive and sometimes negative. The boxes at B and g may or may not have outside special energy sources. If these boxes do contain storages, the algebra applies to the steady state condition when the rate of change of storages is zero. When B is positive, the system is self-reinforcing and overloads. An analog computer version is given in Fig. 10-24b. When B is negative, the feedback is negative and tends to dampen out growth tendencies.

CHAIN REACTIONS

Where a system has closed-loop feedbacks with large amplifier factors, unlimited or large energy pools from which to draw energy, and rapid rates of reaction, processes may grow very rapidly, accelerating and sometimes drawing from more than one energy source, each reinforcing the other. Such loops are called *chain reactions* in chemistry. Figure 10-25 shows the types as given in a review by Dainton (1956, 1966); two examples are given in Fig. 10-26.

Chain reactions can exist in rapid growth states that are explosive. Whether they explode or not may

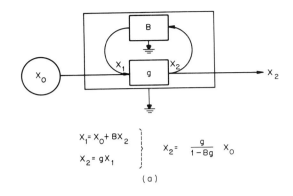

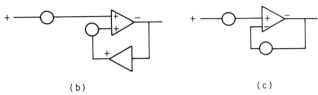

FIGURE 10-24. Additive loop without storage; the transfer equation is given for the unit: (a) energy language; (b) unstable positive summing loop in analog circuit; (c) stable negative feedback.

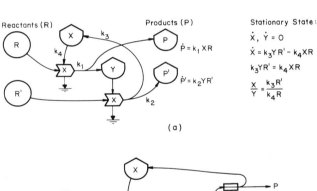

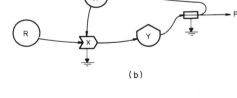

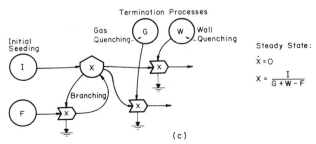

FIGURE 10-25. Types of chain reaction after Dainton (1956, 1966): (a) two sources and two products; (b) one source; (c) initiating and quenching pathways included.

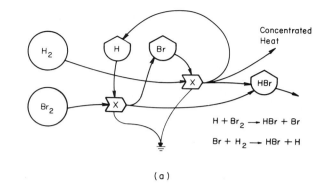

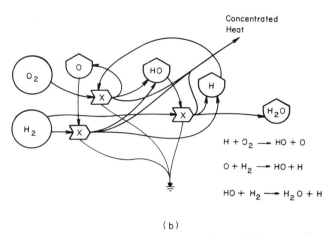

FIGURE 10-26. Energy diagram of chemical chain reactions: (a) hydrogen and bromine; (b) hydrogen and oxygen.

depend on the negative terms that drain energy and concentrations away from reaction centers (storages). These quenching reactions, sometimes called *terminating processes*, include reactions with walls, outward diffusion, and dilution.

The total energy involved in explosive reactions may not be as great as in slower processes, but the special qualities of the reactants allow fast reactions. In other words, the energy thresholds for reaction are low. Also, the design of the interaction is double autocatalytic.

The form of the system design does not suffice to account for the special explosive actions of chain reactions since there are many systems that operate with these designs in biological systems without the explosive growth and release of energy. It may be the high quality and energy previously invested that makes such systems different. Delivery of fast action is a high quality, capable of high amplifier action, for which much energy may be required to generate the ingredients. Refer to Chapter 14 on energy quality and the role of high-quality pulsing.

Some chain reactions move into stationary states, as they are limited by the rate of supply of source energy. Some reactions terminate by self-interactive reactions that are quadratic. These occur with and without feedback. The result is the logistic growth curve. Some involve growth by branching from the stored center of activity so that the process is regular exponential growth. These are described in the chemical literature as linearly branched chains with quadratic termination. Figure 10-25 shows various chain reactions given by Dainton. Most of these are actually the autocatalytic and loop designs already considered with examples from other fields.

LOOPS IN BIOCHEMICAL SYSTEMS

More complex loops are familiar in biochemistry. Roberts (1977), following Cleland (1963), classifies enzyme loops according to the number of input sources (called *substrates*) and output products (see Fig. 10-27). For simplicity, the reverse reactions accompanying each pathway are omitted.

For these types of system, if all but one source were held to a constant force, the energy diagrams would reduce to the Michaelis–Menten loop insofar as permitting estimation of the coefficients for the production–multiplier output function (called *velocity* in biochemistry) or for steady state.

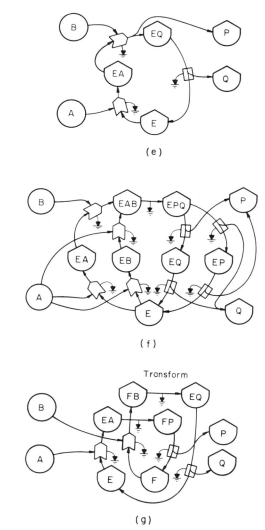

FIGURE 10-27. Types of recycle loop classified by Cleland (1963) after Roberts (1976) with enzymes E as examples; here reversible pathways exist but are omitted: (*a*) unit–uni; (*b*) bi–uni; (*c*) uni–bi; (*d*) bi–bi; (*e*) example from Thorell and Chance; (*f*) choice order ("random"); (*g*) "ping-pong."

Inspection Method of Writing Steady State Equations for Storages

Roberts (1977) summarizes a procedure for obtaining steady-state equations from King and Altman (1956) in which arrow diagrams of reactions pathways are dissected so that all the pathways producing each quantity (storage) are isolated for counting. The rate expression (omitting driving storages) is written on each pathway. The terms are assembled into the general equation.

$$\frac{\Sigma Q_i}{\Sigma Q} = \frac{\text{sum of the terms in the arrow diagrams for the quantity being considered}}{\text{sum of all the terms in the arrow diagram for all quantities}} \quad (10\text{-}2)$$

Figure 10-28 gives an example for a simple loop.

The energy language has the same pathways as the arrows of the King–Altman procedure, and thus the procedure is easily applied to energy diagrams. The expressions placed on the pathways and included in the equation are the transfer coefficients; or if there is an interaction (multiplier) with an external quantity, it is included as a variable.

Enzyme Kinetics and Inhibition

Much studied in biochemistry are the systems of enzyme and substrates of which the Michaelis–Menten loop is the simplest. Energy translations of types of observed reaction systems summarized by Roberts (1977) are given in Fig. 10-29. The first four diagrams are loops that recycle one or two materials (enzymes or modifiers) processing substrate S to generate output of products.

The last four diagrams in Fig. 10-29 show systems of competitive inhibition of increasing complexity. They differ in their place of competitive reaction and the degree of recycle. Roberts gives appropriate equations derived from these models in convenient tables. Inhibitors change the slope of the Lineweaver–Burk plots of the original Michaelis–Menten loop.

In analog simulation of a methane production system with a chemostat such as that shown in Chapter 11, Andrews (1968) used an inhibition model from Haldane (1930). This formula can be derived from a modification of the Michaelis–Menten unit (Fig. 10-11).

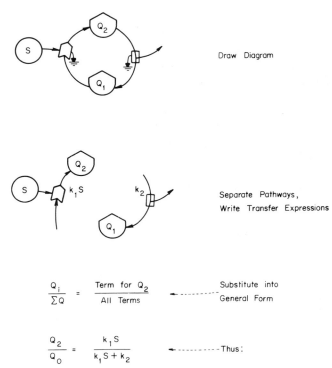

FIGURE 10-28. Procedure for writing a steady-state equation from King and Altman (1956). Compare with long procedure in Fig. 10-5.

VAN DER POL OSCILLATOR

Originally made for radio oscillator systems is the model shown in Fig. 10-30, where a system has opposing inertial force, a driving force, and a frictional force. The system is a loop with oscillating characteristics because of the inertial energy storage (see the inertial oscillator in Fig. 11-7 also).

Other systems generate oscillations for which Van der Pol equation is a model (Pavlidis, 1973).

TIMING OSCILLATORS

Many loops have the property of oscillating, and these may be adaptive properties if pulsing increases function performance (see Chapter 15). Chance et al. (1967) used analog simulation of biochemical reaction loops to study observed pulsing in glucose metabolism. Such pulsing may be used as biological clocks by the systems of which they are a part. Weaver (1965) found the Van der Pol equation

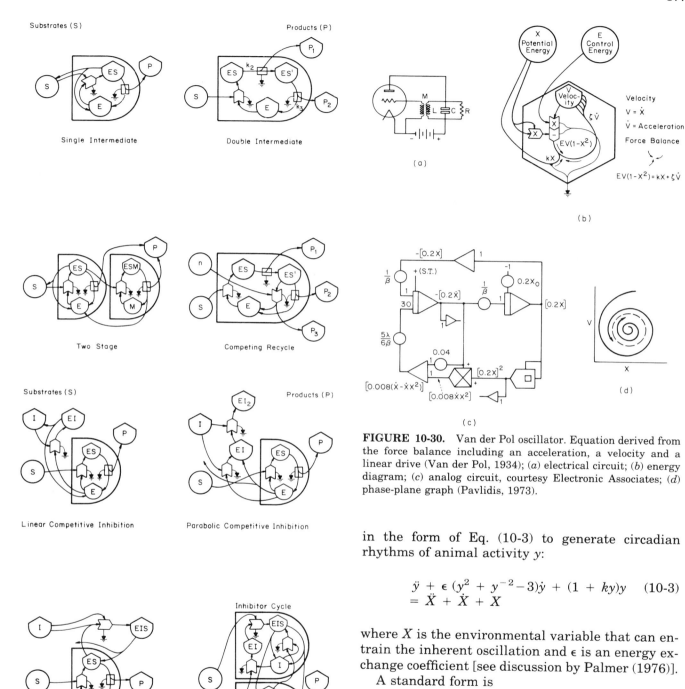

FIGURE 10-29. Translations of models given by Roberts (1977).

FIGURE 10-30. Van der Pol oscillator. Equation derived from the force balance including an acceleration, a velocity and a linear drive (Van der Pol, 1934); (*a*) electrical circuit; (*b*) energy diagram; (*c*) analog circuit, courtesy Electronic Associates; (*d*) phase-plane graph (Pavlidis, 1973).

in the form of Eq. (10-3) to generate circadian rhythms of animal activity y:

$$\ddot{y} + \epsilon (y^2 + y^{-2} - 3)\dot{y} + (1 + ky)y = \ddot{X} + \dot{X} + X \qquad (10\text{-}3)$$

where X is the environmental variable that can entrain the inherent oscillation and ϵ is an energy exchange coefficient [see discussion by Palmer (1976)].

A standard form is

$$\ddot{X} + k(X^2 - 1)\dot{X} + X = 0 \qquad (10\text{-}4)$$

which can be compared with the following general differential equation:

$$\ddot{X} + \dot{X} + X = 0 \qquad (10\text{-}5)$$

(Rosen, 1970).

REPRODUCTION CIRCLES

Most living populations undergo recognizably different stages, one or more of which feed back to the start with some form of reproduction. Such patterns of life history are sometimes called *life cycles*. Figure 10-31a gives a typical pattern for a life history with three stages followed by reproduction. In the fairly typical example shown, energy is used from the environment for the three stages, but energy for the reproductive products (e.g., seeds or eggs) is stored. Mortality and other depreciations are shown. More complex population models are given in Chapter 20. Kranz (1974) provides simulations of life cycle models involved in parasitism. Figure 19-34 shows a moss cycle.

Since energy is used to transform each stage to the next, each surviving unit is gaining in energy quality and often is serving to become more concentrated in valuable energy and information storages serving a larger area. Then the concentrated products are dispersed with the reproductive products and are thus diluted.

Each stage may occupy a different location in the environment and thus draw on a different energy source. The transfers from one stage to another may be controlled by external or internal control cues and may involve migration. The source of control for transfer from one stage to the next is omitted from Fig. 10-31a but is shown in Figs. 3-3 and 3-9.

Figure 5-18 shows switching control of reseeding of a two-tank model of agriculture. Labine and Wilson (1973) provide a model of *Daphnia* populations that has a loop between juveniles and adults, both drawing food energy from a logistic production model of algae. Food pathways contain Michaelis-Menten modules, and four mortality pathways are included.

Origins of Self-Reproduction as a Loop

Self-reproduction is an autocatalytic loop process. Jacobsen (1958) developed hardware devices that performed a circular process of self-reproduction. Figure 10-31b gives a model suggested by Morowitz (1967) and Odum (1967a) for primitive self-reproduction and origin of life. Miller (1953) showed early stages of the process of amino acid formation in microcosms exposed to sparks and ultraviolet light.

Pattee (1966) and Pattee et al. (1966) describe stepwise polymerization and division that constitutes essence of self-reproduction on which stepwise natural selection can develop increasing order if there are continual energy sources [see author's diagrams of origin of life given elsewhere (Odum, 1971a)].

Using the name "hypercycle," Eigen and Schuster (1979) simulated and generalized the properties of closed loops and their ability to mediate evolution. Oscillating and level steady states were observed.

Quadratic Loops

Figure 10-32 shows loops that include a quadratic pathway, one a high quality transformation and one a recycling drain. Farned (1974) used the pattern to model structured order. There are stable configurations. They are an open-system equivalent of the closed-system equilibrium loops given in Fig. 7-12.

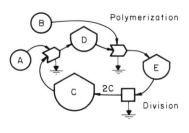

FIGURE 10-31. Life cycles with reproductive products and stages of growth and life history: (a) energy diagram of typical cycle; (b) biochemical self-reproduction model for origin of life.

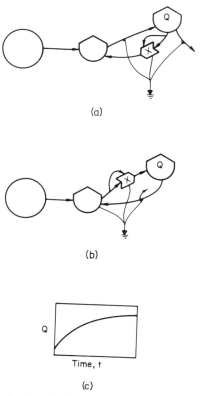

FIGURE 10-32. Quadratic loops: (*a*) with quadratic outflows from high-quality storage; (*b*) with quadratic flow to high-quality storage; (*c*) chargeup with time.

Loops in Other Languages

There are loops in other languages that are not loops in energy circuit language. Some of these are given in Fig. 10-33 with their translations.

Stability of Loops

Loop design elements provide stabilizing mechanisms when recycling quantities are conserved, whereas many other loops are chain reactions and explosive. Loops are an efficient mechanism of receiving energy because they receive energy only in an efficient transformation range. Maximum power through loops may be facilitated by applying pumping energies to any energy-limiting link.

SUMMARY

In this chapter units were considered that contain closed loops, connecting two or more components in

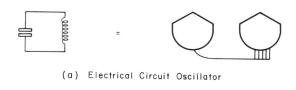

(a) Electrical Circuit Oscillator

(b) Negative Feedback - Block Diagram

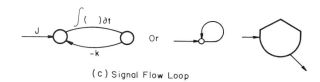

(c) Signal Flow Loop

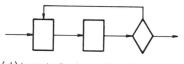

(d) Loop in Program Flow Chart

(e) Logic Loops

FIGURE 10-33. Loops in other languages that are not loops in energy languages (see Chapter 6).

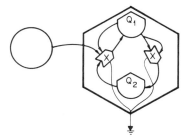

FIGURE 10-34. Diagram for question 1.

a circle. When there is also a constraint of a limited recycled quantity, the loop becomes an internal limiting factor, causing the system to level, even when external sources are not limiting. Michaelis–Menten modules and logistic modules of the type

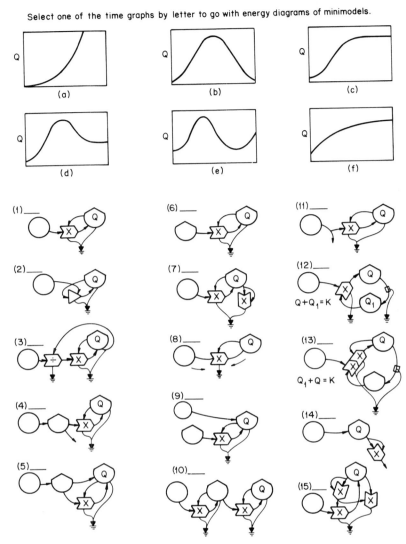

FIGURE 10-35. Diagrams for question 17.

restrained by a recycling resource are common examples. Many other loop designs were studied originally with biochemical enzyme systems in mind. Other typical loops are in the life cycles of organisms, the linear additive feedback systems of electrical engineering control, multiple-staged loops of biogeochemistry, and the alternating multiplier–storage designs in explosive chain reactions. Loops predominate in most systems because of the feedback–reward feature of designs that remain adapted and maximize power. Most systems have material cycle loops driven by external energy sources and controls. Both open and closed systems, such as the Langmuir adsorption system, recycle.

QUESTIONS AND PROBLEMS

1. Is the following loop system shown in Fig. 10-34 logistic if the sum of $Q_1 + Q_2$ is constant (conserved)? Does its growth curve level?

2. Arrange storages in loops using different kinds of connective pathway and at least one external energy source. Consider the kinetic behavior of the system for different types of pathway.

3. Draw and derive equations for the Michaelis–Menten module. How does the algebra at steady state compare with that of an external limiting factor as given in Chapter 8? Where is

the limiting factor in a Michaelis–Menten loop? If you are given only the equation, can you unambiguously determine whether external or internal limit is involved?

4. What happens to growth of a Michaelis–Menten system with time (where there is a constraint on recycled materials)? Is the growth curve with time S-shaped with a reversal of curvature? If additional recycling material is supplied from outside and an outflow is also supplied, what does the module's growth curve do?

5. How are flows of "pure" energy, such as light, sound, and water waves, received and incorporated into systems involving matter?

6. What is a Lotka geochemical loop system? How are storages related to transfer coefficients?

7. In a Michaelis–Menten module, how does the amount of cycling receptor material necessary to maintain constant output as the input driving energy forces declines vary? This question concerns the changing adaptation of the module by changing the total material recycling.

8. How is enzyme inhibition diagrammed? Take some biochemical system from a biological textbook and diagram its patterns in energy language form with special attention to kinetics, energy sources, sinks, and energy quality position on the paper.

9. What kind of growth curve results when a feedback loop is added to a Michaelis–Menten module?

10. What kind of feedback intersection is usually involved in the type of feedback control loop found in electrical engineering systems?

11. Diagram a chain-reaction network.

12. Compare a closed-system Michaelis–Menten loop with the typical open-system module.

13. What is an order–disorder loop? What are its kinetics? To which side of the loop is more energy required per unit material flowing? On which side of the loop is high-quality energy effective? What is the criterion for application of energy for system effectiveness in contributing to survival?

14. What are the positions of two arms of a loop in a matrix of the system?

15. What modules are involved in the Ross malaria equation?

16. Explain whether loops in one systems language are loops in others.

17. See Fig. 10-35.

18. Simulate systems in the chapter.

SUGGESTED READINGS

Banks, H. T., Ed. (1975). *Modeling and Control in the Biomedical Sciences, Lecture Notes in Biomathematics*, No. 6. Springer-Verlag, New York. 114 pp.

Dainton, F. S. (1956, 1966). *Chain Reactions, Methuen's Monographs*, Wiley, New York.

Gold, H. J. (1977). *Mathematical Modelling*, Wiley, New York. 357 pp.

Goodwin, B. C. (1963). *Temporal Organization in Cells*, Academic Press, London. 163 pp.

Hill, T. L. (1977). Free energy transduction in biology, Academic Press, New York.

Kranz, J. (1974). *Epidemics of Plant Diseases, Ecological Studies* No. 13, Springer-Verlag, New York.

Lam, C. F. (1981). *Techniques for the Analysis and Modelling of Enzyme Kinetic Mechanisms*, Wiley, New York. 416 pp.

Pavlidis, T. (1973). *Biological Oscillators, Their Mathematical Analysis*, Academic Press, New York.

Pielou, E. C. (1977). *Mathematical Ecology*, 2nd ed., Wiley, New York.

Rashevsky, N. (1938, 1960). *Mathematical Biophysics*, 2nd vol., Dover, New York.

Roberts, D. V. (1977). *Enzyme Kinetics*, Cambridge University Press, Cambridge.

Rubinow, S. I. (1975). *Introduction to Mathematical Biology*, Wiley, New York.

Savageau, M. A. (1976). *Biochemical Systems Analysis*, Addison-Wesley, Reading, MA. 379 pp.

Thornley, J. H. M. (1976). *Mathematical Models in Plant Physiology, Monographs in Experimental Botany*, Vol. 8. Academic Press, New York.

Walter, C. (1965). *Steady State Applications in Enzyme Kinetics*, Ronald Press, New York.

Watt, K. E. F. (1968). *Ecology and Resource Management*, McGraw-Hill, New York.

CHAPTER ELEVEN

Series

Units in series are those in which one energy-storing module is connected to another so that the energy flow from one supports the next. There may be a series of modules connected to form a chain. Various types of coupling provide varying yields of one unit to the next. Systems with two or more storage units may oscillate. Units in sequence perform work transformations so that the second member has higher embodied energy and usually greater potential for amplifier action. Some characteristics of two units in series have already been given in Chapter 10 on loops. In this chapter we consider the designs, kinetics, and energetics of other series-connected pairs and chains. Models generate graphs with time for two or more states.

PHASE-PLANE GRAPH

A system with two state variables (two storages) may also be represented graphically with a plot of one state as a function of the other. This plot is called a *phase-plane graph*. Examples were given in Figs. 10-22b and 10-30. Graphs are generated by computer or analytically by dividing one time equation by the other and integrating. When there is a regular stable oscillation, closed elliptical rings are generated. When there is a constant steady state, a point results. A third type of performance has curves that stay within a donut-shaped envelope and are called limits cycles. See also p. 171.

Lyapunov Function

When there is a positive function such as available energy that decreases to zero as states go to equilibrium it is a Lyapunov function. For example, following displacement in Fig. 10-30, the oscillation moves toward the closed-cycle loop, after which it is bounded. A system is stable if there is some function toward which the system's variable trends, reducing amplitude. One approach to determining stability of nonlinear systems is to search for a limiting trend function (Harte and Levy, 1975).

LINEAR CHAIN OF STORAGES

A sequence of storages connected by linear pathways is given in Fig. 11-1a. Sometimes called a *cascade*, this simple series arrangement has no feedback pathways nor transformations of energy type. Flux is proportional to storage, and the energy flow is proportional to the energy storage in each unit. Energy is used in frictional processes along the way. An example is a series of water storages with flow through a porous conduit such as sand. The capacitance C describes the shape of the tank in the water example.

Another example is a series of electrical capacitors joined by resistors. Often the electrical network can serve as an analog of the water system. In these examples the pathways are not over energy barriers, and backforce is expressed, with the flux proportional to difference in forward and backforces. The system can flow backward. Since energy is neither transformed nor concentrated, the substances and energy quality of each storage are similar. As in the Lotka loop of Chapter 10, storage at steady state is inversely proportional to the conductivity of the outflow pathways (p. 162).

Second-Order System on Bode Graph

The Bode plot was used in Chapter 3 to represent the amplitude and lag of pulses passing through a stor-

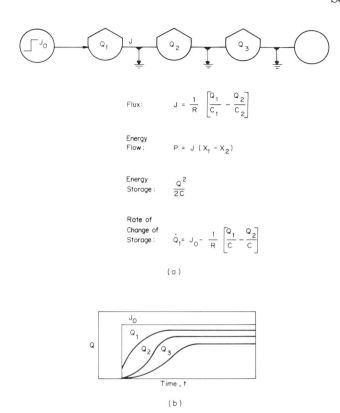

FIGURE 11-1. Sequence of linear storages: (a) energy diagram; (b) chargeup curves in response to step up function from the left.

age as a function of the frequency. For the simple case of a single storage, the equations for the graphs were obtained by Laplace transformations of the differential equation and the pulse train.

Networks of linear storages can be expressed in frequency domain (see Chapter 3) after Laplace transformations, thus allowing calculation of the time lag and amplitude of the response to input. If the separate differential equations for each tank are combined and the combined equation integrated, the equation for the storages ultimately contain one exponential term for each contributing storage. For example, the equation for one tank Q_1 driving another Q_2 with a path without backforce becomes

$$Q_2 = \frac{k_0 k_1}{k_3 - k_4} (e^{-k_3 t} - e^{-k_4 t}) \quad (11\text{-}1)$$

where k_0 and k_1 are inflow coefficients and k_3 and k_4 are outflow coefficients (Milsum, 1966). Networks of such linear storages can be described as a matrix as given in Chapter 13.

Second-Order Equation

A system with two storages such as that shown in Fig. 11-2 is a second-order system. Its two differential equations have at least one common term since the storages are connected. Consequently, the two equations can be combined to form one equation that has a first and a second derivative. Either the combined equation or the two separate equations may be diagrammed by using analog symbols that contain two connected integrators. Figure 11-2 does not have backforce.

The frequency response of two-storage systems can be represented on a Bode plot also. For the simple example of two connected tanks, the responses can also be derived by Laplace transformation. The resulting family of curves, amplitude, and phase lag are as given in Fig. 11-2c. Recall that phase lag is expressed in degrees of rotating circle that is visualized as generating the fluctuating input. Whereas the maximum lag for the tank was 90°, the cumulative lag for two storages may be 180°. The amplitude may be larger if energies within the system are sufficient.

If the output of a two-storage system is connected so as to reinforce an incoming pattern, the system may oscillate, especially if energy is supplied within the loop. The possibility of oscillations in two-unit systems is then considered. In some high-energy systems the amplitude may grow in the oscillation. Such an expanding oscillation is regarded as unstable.

STABILITY OF LINEAR SYSTEMS

To obtain the equations for the quantity in storage, the differential equation was made into a quadratic algebraic equation by substituting r for $\dot{Q}$ and r^2 for $\ddot{Q}$:

$$ar^2 + br + c = 0 \quad (11\text{-}2)$$

This algebraic equation is called a *characteristic* equation and may be solved for its values of r when t is zero (its roots). The roots of a quadratic equation are found by substituting in the general formula

$$r = \frac{-b \pm (b^2 - 4ac)^{1/2}}{2a} \quad (11\text{-}3)$$

The roots r formed by this procedure are the r values

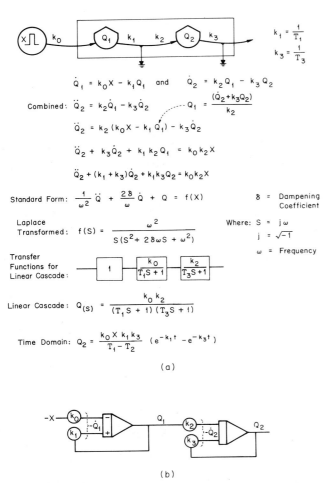

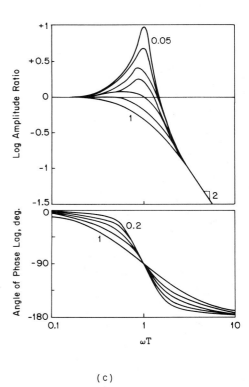

FIGURE 11-2. Linear cascade of two storages (second-order system): (a) energy diagram and equations in various forms; (b) analog diagram; (c) Bode plot of frequency response of standard second-order system to pulse train.

in the exponential terms of the time equation. They are the *poles* of the system:

$$Q_2 = C_1 e^{r_1 t} + c_2 e^{r_2 t} \quad (11\text{-}4)$$

The nature of the r values indicates the kind of graph Q_2 takes with time as given in Table 11-1. The part of Eq. (11-3) that is within square root parenthesis is called the *discriminant* and determines whether r is positive, negative, or imaginary (containing $j = \sqrt{-1}$). When an exponential term is negative, it decays with time; when it is positive, it grows with time; and when it is imaginary, it oscillates (see Table 11-1).

Complex Polar Graph, S Plane

A complex number (i.e., one with real and imaginary parts) may be represented by a graph that plots

TABLE 11-1. Response of a Two-Storage System to a Step Impulse

Result of Calculating[a] $(b^2 - 4ac)^{1/2}$	Term in Time Equation	Stability Consequence
Negative, real number, $-r$	ce^{-rt}	Damped, stable
Positive, real number, r	ce^{rt}	Grows, unstable
Zero	c	Constant
Pure imaginary, $j\omega$	$c_1 e^{j\omega t}; c_2 e^{-j\omega t}$	Harmonic oscillation[b]
Complex, $\alpha \pm j\omega$	$c_1 e^{(r+j\omega_1)t};$ $c_2 e^{(r-j\omega_1)t}$	Growing oscillation Damped oscillation

[a] In $[-b \pm (b^2 - 4ac)^{1/2}]/2a$.
[b] Euler's identity, $e^{j\omega t} = \cos \omega t - j \sin \omega t$, shows that a sinusoidal graph follows with time.

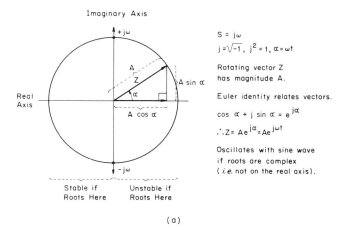

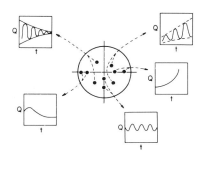

FIGURE 11-3. Use of a complex *s* plane and roots of characteristic equation for considering stability of second-order systems: (*a*) rotating vector *Z* generates sine waves when roots of second-order systems are complex (having real and imaginary parts) (roots are stable if located in the left quadrants, because exponential terms are negative as a result of dampening drains; if roots are on the right, the system is unstable (i.e., either grows exponentially or decays to zero); (*b*) time graphs corresponding to roots.

real numbers on the *X* axis and imaginary numbers (containing $j = \sqrt{-1}$) on the *Y* axis. This graph is called a *complex plane* (see Figs. 3-18*b* and 11-3*a*). The roots of the characteristic equation may be plotted on the complex plane showing graphically the stability status in Table 11-1. The left side of the diagram is stable.

The rotating vector *Z* may be visualized as the sum of two component vectors, each a projection of *Z* on an axis. Magnitudes of these vectors are related by Euler's identity. When the roots of the characteristic equation are complex, having real and imaginary parts, the rotating vector exists generating sine waves in time domain.

To develop control stability, a system may be studied, its values changed, or its design modified so as to obtain roots within the stable zones of the polar plane diagram. If no oscillation is desired, the roots should be negative and on the real axis.

As done in Fig. 3-18 with the first-order system (single storage system), the system equations, after being Laplace transformed to frequency domain may be graphed on the complex plane to form a Nyquist plot (Milsum, 1966); Milhorn, 1966). Second-order systems trace a locus of amplitude ratios outside the unit circle for some frequencies, indicating instability of resonant growth.

CASCADED INTEGRATION

Forrester (1961) shows the effect of a chain of cascaded storages, each controlling the input to the next through sensors, but not drawing from their own storages. The model in energy language is given in Fig. 11-4. The first step is a simple ramp (Chapter 3). Each of the next stages curves upward

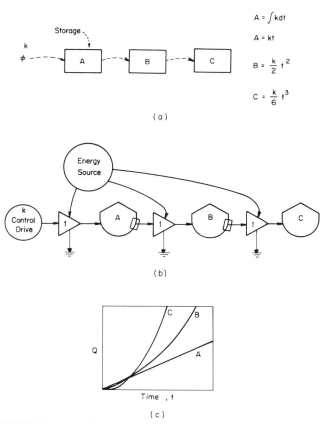

FIGURE 11-4. Cascaded integrators: (*a*) as given by Forrester (1961); (*b*) energy language; (*c*) time graphs.

to a greater degree, successively a square and cubic function of time.

MAXIMUM POWER TRANSFER WITH TWO RESISTANCES IN SERIES

Well known in electrical systems but more general in principle is the matching of resistances for transmission of energy without unnecessary loss. Illustrated in Fig. 11-5 are linear pathways between units connected in series. If resistance of one is lowered, current increases, and the increased friction in the other resistance dumps energy there unnecessarily. Power is maximized when the two resistances are matched.

If the second pathway resistance is a useful resistive process such as grinding corn or arranging books, then power is maximized by adjusting resistance so that the frictional loading of the backforce is half of the input force.

Often the first pathway resistance is that within the energy source unit (for example, internal resistance of batteries) and the second resistance is in the utilization unit. Examples are given in Fig. 11-6.

In electrical systems the word "impedance" is used for several kinds of pathways that have backforce. In addition to resistive pathways, there are those with inertial backforce when flow is accelerated. Impedance matching is a design principle for utilizing energy sources well and maximizing power.

HAMILTONIAN PRINCIPLE

The Hamiltonian principle concerns systems that shift energy between potential and kinetic energy with conservation. For example, the shift illustrated in Fig. 11-7 may be an oscillating spring, a pendulum, or a charged particle moving in a magnetic field. No dissipative processes with heat sinks are included. The quantity L is called the *Lagrangian* and is the difference between kinetic energy T and potential energy V:

$$\text{Lagrangian } L = T - V \quad (11\text{-}5)$$

The Hamiltonian principle states that the sum of the Lagrangian between two points in time for energy conservative processes is unvarying. In other words, the following integral is constant:

$$I = \int_{t_1}^{t_2} (T - V)dt \quad (11\text{-}6)$$

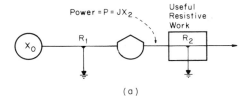
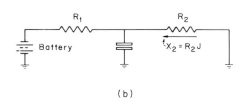
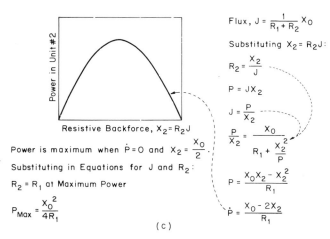

FIGURE 11-5. Maximum power transfer in resistive pathways: (*a*) energy diagram; (*b*) electrical example; (*c*) graph of power transfer to second unit as a function of frictional load.

The least-action principle states further that action of these processes tends to be minimized where action is the sum of the products of momentum and distance. Action in this sense is the flow of power transformation between kinetic and potential energy and vice versa. In other words, the difference between kinetic and potential energy (above integral) is minimized. Conversion of potential energy moves to eliminate differences. This may be consistent with the maximum power principle, which stipulates that systems that use potential energy are maximized.

A system that minimizes the difference tends to be neither all in potential form nor all in kinetic

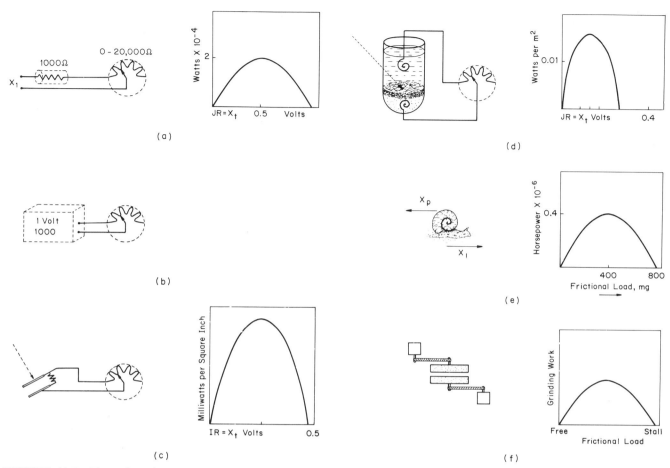

FIGURE 11-6. Examples of maximum power transfer as a function of resistive load; maximum power for resistive work is with matching of resistances: (a) two resistors in series; (b) battery with internal resistance loaded with available resistor; (c) photovoltaic cell with a variable resistor as a load; (d) living blue–green algal mat and resistor (Armstrong and Odum, 1964); (e) muscular power in useful dissipative work of snail resisting water current (Jobin and Ippen, 1964); (f) grindstones.

form, but in between. For many systems, this means oscillation. Minimizing potential to kinetic transformation for a given period insures that both kinetic and potential are maintained in similar magnitude and thus the energy transformers will continue.

INERTIAL OSCILLATOR

Important in physical sciences is the two-storage oscillator in which energy flows from a resting potential energy storage accelerates motion into a kinetic energy storage thus generating a back inertial force against the potential driving force. An oscillation results. The kinetic energy storage is shown in Fig. 11-7a with its special symbol (see Fig. 1-6h). The equation for the system is obtained by setting the driving force equal to the inertial force. Notice that there is no heat sink as part of the acceleration process, although there may be one as part of the regular potential energy storage (not shown here). Einstein (1905) wrote what appears to be potential and what appears to be moving (kinetic energy) depends on the moving frame of reference of the observer (Schlegel, 1961). Thus, if this is true, neither state is more degraded than the other, since it is a matter of point of view. Thus kinetic energy and still potential energy are of equal potential for work.

Frequently cited examples are the oscillating weights on a spring or a pendulum and the surge of electrical current from a capacitor through an inductance coil where it generates a magnetic field and inertial backforce during its acceleration (Fig. 10-33a). These systems oscillate back and forth with little dispersion of energy. There is dispersion

188   DESIGN ELEMENTS

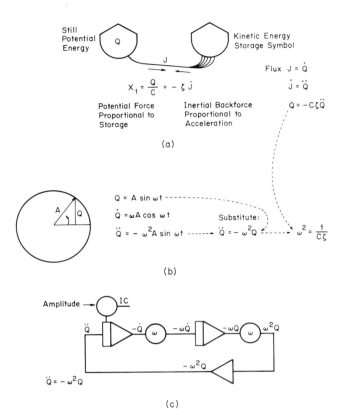

FIGURE 11-7. Inertial oscillator: (a) energy diagram and oscillatory equation resulting from balance of potential and inertial forces; (b) equation generated from projection of a rotating wheel; (c) analog computer diagram for oscillator equation showing pots that represent frequency (angular velocity w).

caused only by friction in air, and so on. The nature of traveling photons of light involves a movement of an electromagnetic wave oscillating between potential energy and magnetic field storage.

The equation for oscillators is often expressed as a system with acceleration negative to displacement. The projection Q of the radius A of a rotating wheel generates a similar equation as shown in Fig. 11-7b, in which the acceleration is a negative function of the quantity stored and ω is the angular velocity in radians per time. Thus the frequency is a function of the storage capacitance and the inertia.

The projection of a rotating wheel generates a sine wave (Fig. 3-16). The two-tank oscillator with similar equation also does so. When the equation is translated to analog computer language, two integrators again result (two tanks). The sine-wave generator, as this design is called, may be memorized because of its usefulness in providing ready driving functions for diurnal, seasonal, and tidal phenomena. See appendix to Chapter 5.

Notice that where there is a second derivative, it is formed in the analog computer by connecting two integrators (Fig. 11-7c). The integral of acceleration $\ddot{Q}$ becomes velocity $\dot{Q}$, and the integral of velocity is the quantity Q. By adding pots that each represent the angular velocity and setting them to be the same value, the frequency of the oscillation ω can be set according to that needed to simulate a problem. To oscillate, there must be an initial condition. The amplitude of the oscillation is controlled by the amplitude of the initial condition. The analog unit generates the three equations given in Fig. 11-7b (see also appendix to Chapter 5, Fig. 5A-4).

Although the examples usually given are from simple physical systems, the oscillator may be more general. For example, if the response of human groups to attempts to persuade them in a given situation is a stubborn negative action in proportion to the acceleration attempted, the system could oscillate.

Oscillators are very useful subsystems in larger networks. Thus we use clocks, timers, regular mechanical devices for some kinds of control work. This property in a component can be used in energy language also by using a labeled rectangle.

To express the period of the oscillation T in terms of the angular velocity of the wheel rotating with that period, recall that there are 2π radians in a circle. If angular velocity ω is in radians per second, 2π radians (full circle) will result from the product of that velocity and the time T of one revolution. See Eq. (11-7) or (11-8). These equations are used to set period of oscillation T by setting pots for ω in the circuit in Fig. 11-7c.

$$\omega T = 2\pi \quad (11\text{-}7)$$

$$T = \frac{2\pi}{\omega} \quad (11\text{-}8)$$

EXPONENTIAL PUMPING FROM A STORAGE

Figure 11-8 shows the system of an autocatalytic growth module in series and pumping from a tank that has an energy source and separate outflow. Although growth of the second unit starts out exponentially, it is limited by the first storage that drops to that level at which the pumping balances the inflow minus tank outflow. The configuration is similar to that of the flow-limited source shown in Fig. 9-5, and

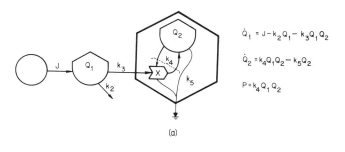

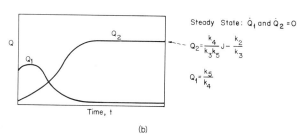

FIGURE 11-8. Exponential growth pumping from a storage: (a) energy diagram; (b) growth graphs.

the steady states are similar following a transient adjustment period.

STORAGES DOWNSTREAM FROM GROWTH UNITS

In a common series pattern storages receive outflow from autocatalytic growth units. The pattern generated in the first unit is passed to the second with some time lags. The first example in Fig. 11-9 shows Ph.D.s generated by universities being distributed outside. The second example illustrates penicillin generated by microbes with a logistic model.

PUMPING YIELD FROM EXPONENTIAL MODULE

In Fig. 11-10 an exponential module Q_1 draws power from a constant-pressure source and yield is being drawn from it by pumping by Q_2 in proportion to the product interaction. As shown in Fig. 11-10b, there is no stable steady state. Growth is either accelerating or declining to zero depending on the initial conditions. If the system were regulated to operate at steady state (by means of additional controls), it

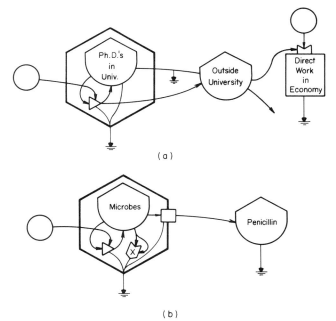

FIGURE 11-9. Examples of storages in series with growth units: (a) model of Ph.D. production (Bolt et al., 1965); (b) logistic module generating a storage [Rogers (1976), after Constantides et al. (1970)]. The second tank accelerates and levels, but at a later time.

would not be at maximum power. The system maximizes power and yield by growing and thus is adapted for a transitional role. The total harvest is the product of Q_1 and D. If the harvesting is being done by another population Q_2, then D is proportional to the second population ($D = kQ_2$).

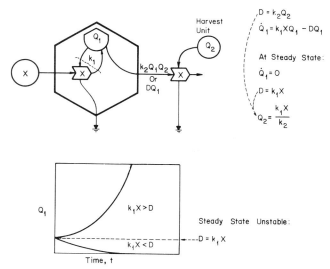

FIGURE 11-10. A system that harvests a module in exponential growth. Depreciation is omitted as small: (a) energy diagram; (b) growth or decay if energy is small.

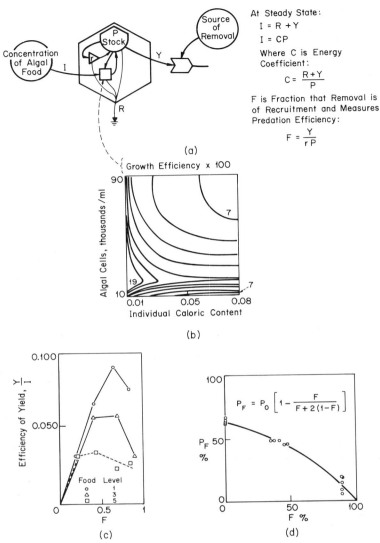

FIGURE 11-11. Yield studies by Slobodkin (1962a, 1962b, 1962c): (a) yield notation related to energy diagram; (b) growth efficiency of *Daphnia* as a function of *Chlamydomonas* concentration and size, using data from Armstrong (1960); (c) efficiency of yield as a function of food levels; (d) effect of predation F on standing stocks P_F.

Figure 11-11 illustrates studies on yield by Slobodkin (1959, 1960), who withdrew water fleas to simulate fishing in microcosm. Several coefficients were defined and related to maximizing yield. Maintaining fishing pressure also kept stocks younger with less biological senescence and thus more efficient in terms of reproduction and net gain. Richman (1958) measured energy transformations by these cladocera.

The harvest of exponential growth may yield the maximum biotic potential from the first population for a given energy source X. Agriculture and continuous cultures tend to be adjusted toward this goal. Note that if the stock is maintained at such low densities and ages that there are negligible drains the energy costs of harvest are raised. Also note that the energy diagram correctly shows two energy sources: the food energy source and the energy source to the harvester. This system requires some controls to keep it stable. For example, in the model in Fig. 11-12 harvest flip-flop is the control.

When harvest is less than the growth coefficient $k_1 X$, it grows until the model is no longer appropriate as depreciation and crowding become important

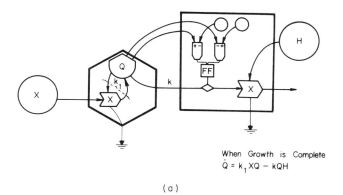

When Growth is Complete
$\dot{Q} = k_1 XQ - kQH$

(a)

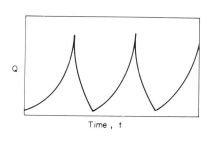

(b)

FIGURE 11-12. Periodic harvest of exponential growth module: (a) energy diagram with comparator–flip-flop control system; (b) J-shaped growth curves.

and maximum potential is not achieved. If harvest is more than the coefficient for growth, the stock goes to extinction.

A system of yield from logistic growth models is given in Fig. 11-13 (see also a model with economic price with similar kinetics in Fig. 23-36). As fishing pressure increases, yields reach a maximum and decline again, forming a parabola. There is also an optimum standing stock for maximum yield. Equations for maximum yield at steady states were obtained by setting rates of change equal to zero to obtain a steady-state equation for yield and then setting these equal to zero to find maxima. Maximum yield per unit stock is half the basic net production rate [see the algae yield example by Benoit (1964)]. Seip (1980) showed advantages of lower yields for long-range harvesting of brown algae growing with a logistic model.

Whereas these yield models have been important for the concepts of fishery management in past years, it is all too apparent that maximizing yield without at the same time feeding back supporting energy to the productive ecosystem is fallacious—driving out the species of interest to be replaced by others. See Fig. 23-27.

CHEMOSTAT

One system with harvest properties such as that in Fig. 11-13 is the chemostat. A chemostat is a liquid microbiological culture operated in steady states with a balance between one or more inflowing energy sources and controlled outflow pumping. Inflow and outflow pumping volumes are equal since the level of the culture fluid is maintained constant. The inflow and outflow are expressed as fraction of the volume D turned over per time. Generally in simpler arrangements one species is present. The pumping tends to wash out contaminating species. The system is operated to prevent attachments to the wall eliminating other storages so that the system will operate as diagrammed.

The equations for the food level and the population level at steady state are obtained by setting the rates of change equal to zero. The yield is a product of the harvest fraction rate and the population stock. The equations omit depreciation pathways as small relative to exchange.

The graphs in Fig. 11-14 show the variation in storages and yield as the turnover of rates of harvest, inflows, and stocks are increased. The arrangements for removal of storages at the rate of inflow supply provides a mechanism for maintaining a steady state within a wide range. Maximum yield is at an intermediate rate of harvest when the concentrations are half that of the source. Because inflow resource is supplied in proportion to the yield drawn, there is a feedback so that gross production is not diminished by the harvest.

When harvest reaches the unit reproductive rate, the population is swept out since it cannot replace itself. Some chemostats may follow these kinetics, as data of Jannasch (1969) suggest. However, chemostat kinetics usually have limiting factor actions that modify relationships as described next.

Chemostat with Limiting Factor Action

Since the medium rarely, if ever, has the exact ratio of nutrient requirements needed by the population, increased flushing, although adding main requirements, causes some limitations to develop, either in internal cycles or within the medium. The model for

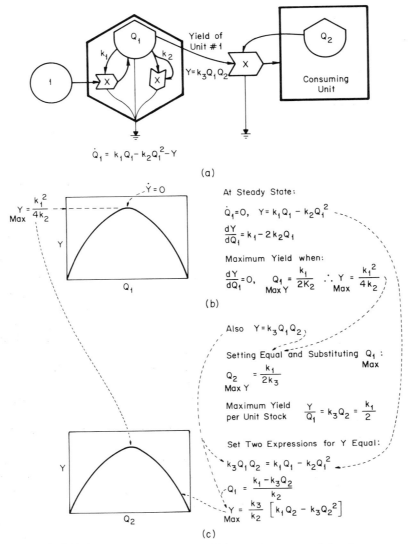

FIGURE 11-13. Maximum yield from a logistic module after Schaefer (1954, 1957), Watt (1968), and Emlen (1973): (a) energy diagram; (b) yield as a function of quantity; (c) yield as a function of consumers.

the chemostat system shown in Fig. 11-15 has a limiting-factor module as part of its kinetics (Herbert, 1958, 1962; Herbert et al., 1956; Malek, 1958). There is an internal limiting module (Michaelis–Menten loop), although the algebra for limiting element in an external medium is the same. External limiting factors are involved in Fig. 11-16. In Novick and Szilard's initial experiment in 1950 there was a shortage of an amino acid, even though the main energy supply was in excess. The system was used to study mutation of species between one that needed the requirement and one that did not.

Differential equations for the energy source Q_1 and the stock Q_2 are given in Fig. 11-14. At steady state the rates of change become zero, and two equations result that describe the levels of the population and the level of source quantity. Yield equation is obtained by multiplying flushing rate times population equation.

Consider next the effect of increasing the flushing rate D starting with zero. The resulting graphs are given in Fig. 11-15b. Without flushing, there is no energy source and no stocks. At higher flushing rates, stocks are higher as medium is renewed as used. The level of the medium is kept low and limits population levels. At higher levels of flushing, the

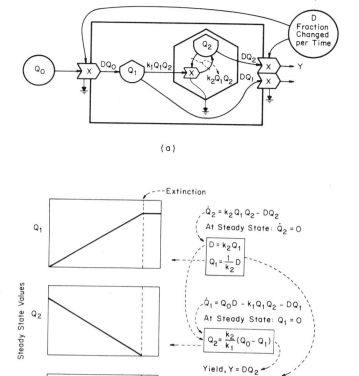

FIGURE 11-14. System of coupled supply and harvest without other limiting factors: (a) energy diagram; (b) graphs of steady state as a function of flushing rate D.

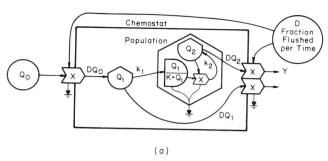

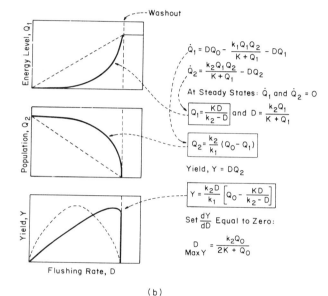

FIGURE 11-15. Chemostat system with limiting factor: (a) system; (b) graphs of steady state as a function of flushing.

medium concentration rises as population becomes more limited by its flushing out. Finally when flushing rate reaches a high level, washout exceeds growth and all cells are flushed out.

The chemostat system may be compared with the simpler one without limiting factors in Fig. 11-14. Dotted lines given for the latter are included in graphs of Fig. 11-15.

Jannasch (1962) compares biostats to microbiological realms in nature, reasoning that species may be absent because the rate of loss of their species (as in flushing) may be greater than the critical value of Q_0 necessary to establish a steady state. Growth of a marine spirillum cultured in a biostat under generally low nutrient concentrations that prevail in the organism's natural environment was related to the concentration of asparagine nitrogen in the input.

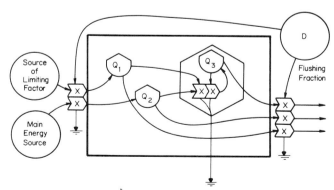

FIGURE 11-16. Diagram of chemostat with a limiting factor different from the main energy source, both supplied at the flushing rate (Novick and Szilard, 1950).

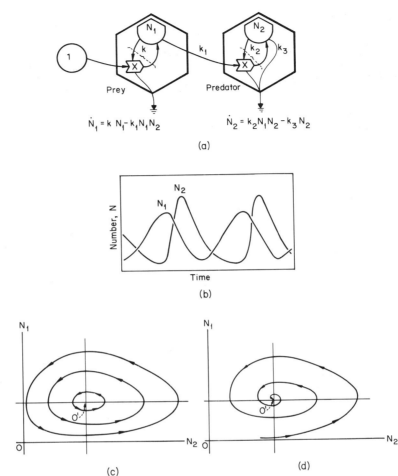

FIGURE 11-17. Lotka–Volterra oscillating model of prey–predator interaction: (*a*) equations and their energy language translation; (*b*) typical time graph showing oscillation; (*c*) state diagram of typical oscillation; (*d*) damped oscillation when there are higher-order negative pathways as in Fig. 11-20*a* (Lotka, 1925).

LOTKA–VOLTERRA PREY–PREDATOR OSCILLATOR

Classical equations in Fig. 11-17 were given independently by Lotka (1925) and Volterra (1926) for a predator consuming prey. Translation of these equations to energy language can be made with either constant-gain amplifier or constant-force energy source multiplier as discussed in Chapter 9. Whereas the model has too little of the real aspects of prey–predator relationships to serve as a full description or prediction model of usual ecological relationships, its simplicity has helped to establish the principle that prey–predator and parasite–host relationships can oscillate with an internally regulated frequency. Note that the prey has a constant percent growth but that the only drain is the autocatalytic pumping by the predator. The predator has only a linear drain. There are no self-stabilizing terms other than the depletion of its food.

Time graphs are given in Fig. 11-17*b*. Properties of such oscillators are often represented by tracing quantities of prey and predator on a *phase-plane graph* of one as a function of the other as in Fig. 11-17*c* and 11-17*d*.

The model is a favorite one for demonstrations (Electronic Associates, 1965). Apparently, prey–predator oscillations do occur in simple laboratory ecosystems and something like this in the simple natural ecosystems of the Arctic (Elton, 1926).

Prey–predator oscillator systems are another way of stabilizing the exponential harvest system shown in Fig. 11-10. Near exponential growth is allowed, and then harvest is taken before the system is given another cycle. Gallo and Rinaldi (1977) applied the Liapunov method to prey-predator models using an energy argument from Volterra.

Prey–Predator Oscillation in Chemostat

Tsuchiya et al. (1972) studied a food chain of amoebae eating bacteria that were eating sugar in continuous culture. They found a prey–predator oscillation. The system was fitted to the model in Fig. 11-18, which has Monod limiting-factor action on both bacteria and on amoeba. The oscillation was greater at high energy than at low energy. Van den Ende (1973) found prey–predator oscillation between bacteria and protozoa in chemostat for a while, after which some physiological changes caused adaptation that eliminated the cycle.

Simulation of host–parasite cycles in chemostat models reproduces patterns such as those shown in Fig. 11-17 [Rogers (1976) after Canale (1969a,b)]. Increase in the dilution rate damps the oscillation and reduces the average population level. In effect, the stress of diverting energy lowers the energy level and with it the oscillation. For other host parasite models see Usher and Williamson (1974).

Parker (1976) simulated the stabilizing effect of diffusion on Lotka–Volterra oscillation in plankton systems. Here diffusion and advection are comparable to chemostat flushing.

LOGISTIC UNITS IN SERIES

If each of the autocatalytic units in sequence includes a quadratic drain, so as to be logistic, a model such as that shown in Fig. 11-20 results. This system also may oscillate. Garfinkel (1962a–c) simulated this system.

SERIES PAIRS WITH POSITIVE FEEDBACKS

Theoretical reasons were given for the development of positive "reward-loop" feedbacks of high-quality energy acting in the manner of an amplifier. Many traditional prey–predator models such as the Lotka–Volterra oscillator do not have this feedback pathway. Examples of models with feedback loop are given in Figs. 11-19 and 11-20. The feedback multiplier pathway changes the relationship from one where the downstream unit is draining the first to a symbiotic one in which each is a positive action to the other.

In the first example (Fig. 11-19) both units are exponential growers if their energy sources are constant in force. The pair will grow exponentially. In Fig. 11-20b both units are logistic and the system of the two will level, just as the pair will do if operating separately.

PULSING RECYCLE SYSTEMS

Consumers who can pulse their use and recycle actions after periods of storage in their energy source may have competitive position and may contribute to a more effective energy chain in the long range (see Chapter 15).

Mechanisms of pulse triggering are given in Fig. 11-21 (see also the switching action shown in Fig. 11-12).

Ludwig et al. (1978) developed a pulsing model for spruce–budworm oscillations (Fig. 22-16). See also Fig. 13-31a.

SURFACES AND CATASTROPHE THEORY

A set of equations may define a surface in space of coordinates. For example, a three-dimensional plot of three components may form the same kinds of surface found in land with ravines, ridges, saddle points, cliffs, and overhangs. Going from one position to another on such a surface may be a catastrophe in the same way that a person falls off a cliff. Thom (1970, 1975) gives many equations for such surfaces of different shape. It will be interesting in the future to find which of these equations will be found to be identifiable with and generated by known mechanisms of interactions of real systems. Scanning and comparison of such equations provides a new initiative to creative exploration of models. An example of a surface was given in Fig. 10-21b.

CONSUMER–HOST SERIES WITH DISCRETE GENERATIONS

Nicholson and Bailey (1935) give a parasite–host model for a fly consumed by parasitic wasp (one that lays eggs in the dead body of a host). Diagrammed in Fig. 11-22, the general design is that of Lotka–

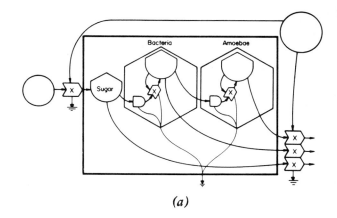

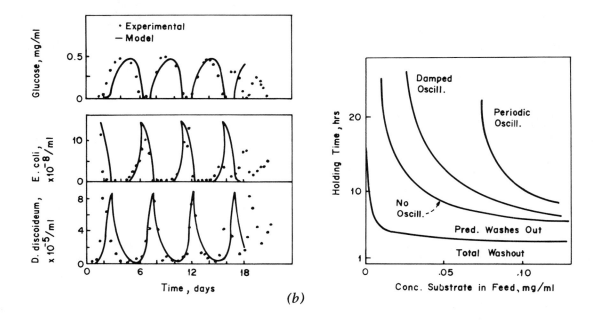

FIGURE 11-18. Consumer oscillation in continuous culture with limiting factors: (a) energy model; (b) observed relations (Drake et al., 1968; Tsuchiya et al., 1972).

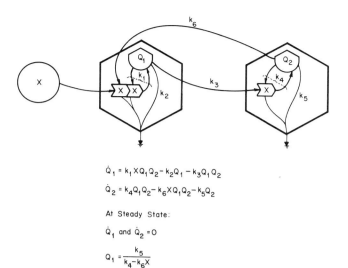

FIGURE 11-19. Exponential pair coupled with controlling feedback.

Volterra prey–predator cycle, except that the reproduction is in synchronized generations and the equations used were difference equations for a process within a generation followed by step transition to the next generation. See Fig. 9-4. For insects, there is little carryover of stocks from the season of one generation, and the reproduction may be seasonally

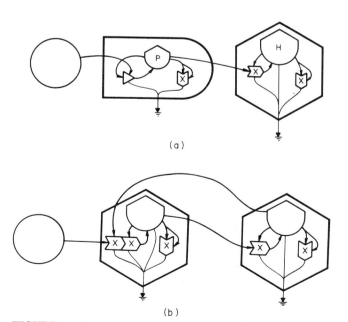

FIGURE 11-20. Logistic pairs; (a) without feedback multiplier action; (b) pairs connected with positive feedback multipliers.

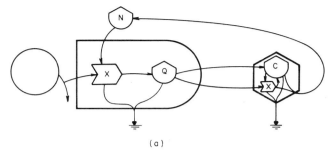

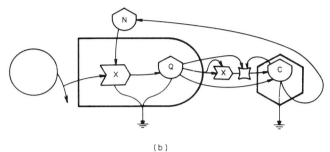

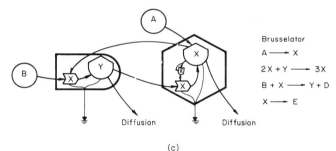

FIGURE 11-21. Pulsing consumer that is self-triggering in rapid autocatalytic use and recycle when level of storage Q is higher than threshold: (a) triggering consumer model from Alexander (1978); (b) epidemic model (Odum, 1976a) with threshold switch action on quadratic path; (c) diagram of the Brusselator chemical reactions given by Nicolis and Prigogine (1977) illustrating an oscillator with bifurcations and with oscillations dependant on energy source B.

synchronized. This example illustrates the relationship between energy language and models in difference equation form. Bailey (1967) and Hassell and May (1973) review models of this class with a consumer searching function combined with discrete population increase steps.

SIMULATION OF A CHAIN OF CONSUMERS

Patten (1966), using analog computers, simulated a string of consumer units with linear connections and

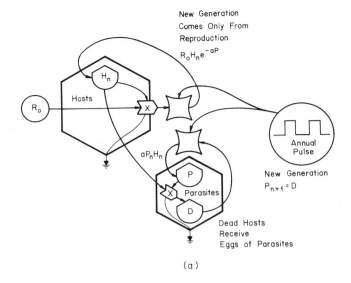

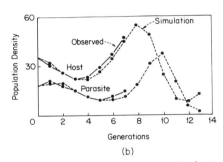

FIGURE 11-22. Model of discrete parasite host generations modified from Nicholson and Bailey (1935) (Emlen, 1973): (a) energy diagram; (b) housefly parasite oscillation by Debach and Smith (1941) and simulation after Poole (1974).

without energy constraints, finding damped oscillations (see Fig. 11-23b for one energy language translation).

Several attempts have been made to simulate consumer chains with models of different variations. Smith (1969) and Verhoff and Smith (1971) connected a chain of autocatalytic loops similar to that in Fig. 10-17, but without any energy constraint and only the constraint of recycling materials. Each level had two storages, one for active feeders and one for inactive feeders (see Fig. 11-23). They reported very unstable oscillations. Rescigno (1972) found a prey predator chain more stable.

Brown (1980) simulated chains of consumer units with the constraint of flow-limited energy and with and without means of feedbacks between units (to storage or to production interaction process, etc.). One example is shown in Fig. 11-24. The time constants of the chains were adjusted so that the lower trophic levels, although larger in quantity of processed energy, had smaller time constants (because of the small individual units as discussed in Chapter 15). The system was fairly stable. The hierarchical distribution of stocks at steady state are graphed as an energy spectrum in Fig. 15-10. Another example includes a pulsing of the second source that transmits some of the period of its variation back to the left through its feedback control actions.

MICHAELIS–MENTEN SERIES

Ross Malaria Equations; A Michaelis–Menten Pair; Parasite–Host with Positive Feedback

The Ross malaria equations and their appropriate energy diagrams are given in Fig. 10-21. This model was one of the first differential equation models for populations (Ross, 1911) and has the feature of symbiotic feedback between the upstream host (sick people) and downstream parasite (infected mosquitoes). The variables are numbers of people and numbers of mosquitoes. Diagramming of the equations shows that the modules are a pair of connected Michaelis–Menten modules. Setting the rate of change at zero gives the typical Michaelis–Menten algebra. In the model the people are held constant in total number with neither inflows nor outflows of people or energy. The state graph for the model is given in Fig. 10-21b from Lotka (1925).

Chain of Michaelis–Menten Loops

Many systems such as those of biochemistry of living organisms consist of chains of loops in which chemical substances, enzymes, and structural actions feed back to interactions. Each loop alone constitutes a Michaelis–Menten module with an inherent internal limit to flow when the input drives are increased (e.g., see Fig. 11-25).

The overall transfer action of the chain is similar to that of one Michaelis–Menten chain, namely, a diminishing returns curve as shown in Fig. 11-25d. Often in modeling one Michaelis–Menten unit is used to represent the whole chain since the output functions are similar and the mechanism is the correct one even though simplified to one loop from multiple loops.

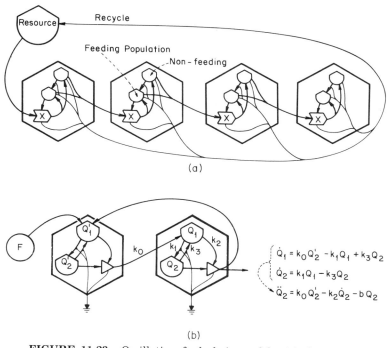

FIGURE 11-23. Oscillating food chain models: (a) chain of growth units with a recycling constraint simulated by F. Smith (1969) (see unit model in Fig. 10-17), where instabilities occur without an energy limitation or sink; (b) energy diagram for chain of second-order equations simulated by Patten (1966) with damping drains.

Sometimes models aggregated into one Michaelis–Menten equation are evaluated from physical experiments on the real complex chain. The coefficients found in this way can generate the correct shaped curve but do not represent the coefficients of any process necessarily.

Complicating the modeling are the adaptive mechanisms that tend to adjust the quantities of circulating materials and rate processes to generate the maximum flow. Ideal models would either include the adaptive mechanisms or use an aggregated type of model that does not have internal limits. In other words, the Michaelis–Menten equation is not a good model for interactive systems that adapt so as to maximize the energy yields of the external drives. Figure 13-10 illustrates a web of loops alternating with external sources.

Michaelis–Menten Pair

Thornley (1976) considered pairs of Michaelis–Menten processes in series at steady state and found the combined process with a graph of production as a function of driving energy to be similar to that of a single Michaelis–Menten unit.

Thornley (1976) also represented transport across a membrane with a model that alternates Michaelis–Menten units with storages that also have direct diffusion pathways (see Fig. 11-26a).

Michaelis–Menten Unit with Autocatalytic Consumers

Dubois (1979) simulated Michaelis–Menten units with autocatalytic consumers and found the systems stable with oscillations damped.

Chemostats in Series

Chemostats in series have been studied experimentally and theoretically (Malek et al., 1966). Curves of stocks and yields as a function of dilution rate are very similar to those for one chemostat unit (Herbert, 1964).

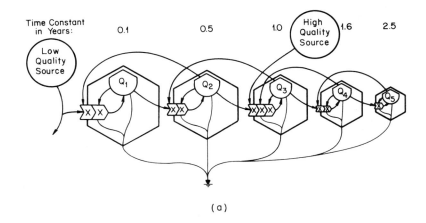

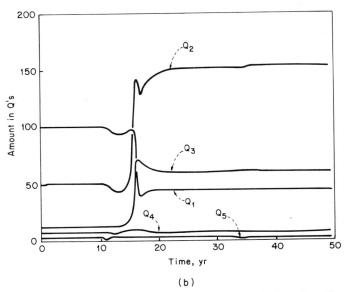

FIGURE 11-24. Hierarchical chain of autocatalytic units with feedback action on production processes, with time constants increasing to the right (numbers above) and a second high-quality energy source used in midchain (M. Brown, 1980): (*a*) energy diagram; (*b*) simulation starting without the second source that is added at time 10. The energy spectrum at steady state is given in Figs. 15-10*c* and 15-10*d*.

SINUSOIDAL OSCILLATORS IN SERIES

Sine waves are generated by the analog network, in Fig. 11-7*c*. Exploring concepts of linear food webs in Fig. 11-7*e*, Patten (1966) connected five such oscillators together by adding a linear decay path to each. This is tantamount to connecting five generalized linear second-order equations as a system. The simulation was self-damping with amplitude of oscillations dependent on initial conditions. Whereas the inertial example in Fig. 11-7*a* is not relevant, energy diagramming of the equations with inclusion of energy sources provides an ecosystem model to accompany these equations in Fig. 11-23*b*.

FEEDBACK INHIBITION INTERACTION

Goodwin (1963) used loop networks for enzyme networks involving gene controls. The unit model has two storages in series, with the second feeding back

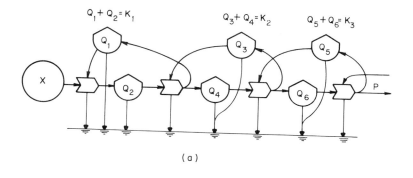
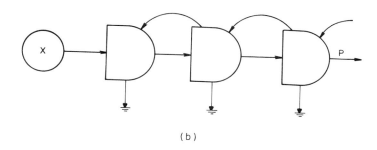

FIGURE 11-25. Chain of loop reactions with circulating materials conserved: (a) details; (b) summary representation using Michaelis–Menten symbols; (c) use of one unit to represent a chain; (d) response of all systems to varying input drive.

as an inhibitor interacting as a divisor. Rosen (1970) studied the stability of its oscillations by use of linearization methods (see Fig. 11-26b).

REINFORCEMENT EFFECTS ON EVERY OTHER UNIT IN A CHAIN

Several authors have shown the property of chains that causes an energy drain to have alternative effects of stimulating and depressing populations along a chain. A chain in which the end member on the right is depressed by stress S is shown in Fig. 11-27. This reduces the energy pumping by D on C so that C can develop a higher steady state. In turn, the effect of more C is to depress the steady-state level of B. The effect of depressed level of B is to permit a higher level of A. If the disturbance to D is positive, the effect on the chain is reversed with higher levels of B and D and lower levels of A and C.

Whereas the effects on a simplified model such as that in Fig. 11-27 is clear, the effect may not be observed much in real systems because of other features not shown in Fig. 11-27. The reward-loop feedbacks usually found in energy chains and webs are not included in Fig. 11-27.

If the theory of adaptation for maximum power is correct, surviving patterns adjust their rate coefficients to build those stocks that maximize power. Thus the reinforcement effect on alternate units may be only temporary.

A COMPLEX SERIES

A simulation model due to Garfinkel (1962b, 1965) is shown in Figs. 11-28 and 11-29. Plants have lo-

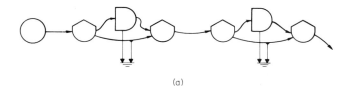

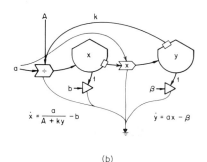

FIGURE 11-26. Other loops: (a) active and passive transport across membrane (Thornley, 1976); (b) unit model of Goodwin networks (1963) studied by Rosen (1970).

gistic characteristics, and consumers have linear and exponential consumption and growth patterns. Increasing role of rabbits cause a consumer sequence oscillation. Addition of a third stage (fox) adds a second oscillation. When fox numbers were large, there was a surge in plants since rabbits were kept low. See also Garfinkel and Sack (1964).

ENERGY DIAGRAMS OF OTHER SERIES MODELS

Energy diagrams of some other models proposed by several authors in equation form are given in Fig. 11-30. In models these quotients, square roots, quadratics, and exponentials were added to give growth kinetic characteristics. When translated in diagram form, complex patterns result that may not have the correct mechanisms and energetics. See Fig. 6-26 for a series relation with a product consumption for maintenance plus a quadratic effect of the pool on growth.

Royama (1971) reviewed prey-predator models with alternate food transfer mechanisms. Shimazu et al. (1972) simulated those with a limiting factor consumer function and found limit cycle behavior.

ENERGY QUALITY IN SERIES

Intuitively from experience, we are accustomed to the increase in quality of products as one passes

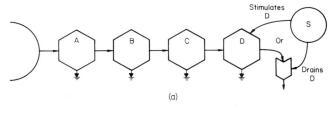

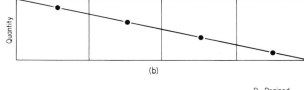

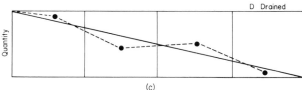

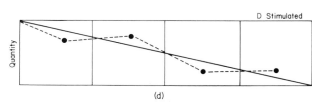

FIGURE 11-27. Response of a chain to external effects on one storage: (a) energy diagram; (b) steady state before addition of S; (c) S draining D; (d) S increasing D.

along an energy chain by using an increasing amount of the potential energy but producing items at the end that are recognized as high in information or energy concentration or are otherwise valuable. These high-quality items are valuable because they require much energy of lower quality to develop and maintain and also because they have the ability to

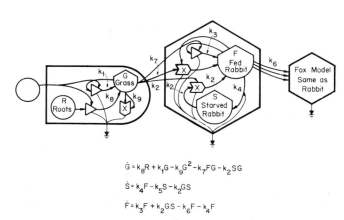

FIGURE 11-28. Consumer sequence model given by Garfinkel (1962b).

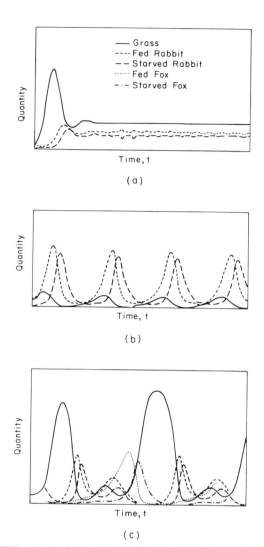

FIGURE 11-29. Simulations of model of rabbits and grass in Fig. 11-22 (Garfinkel, 1962b): (a) small rate of consumption by rabbits; (b) larger coefficient of consumption by rabbits; (c) same as in part b, with fox added.

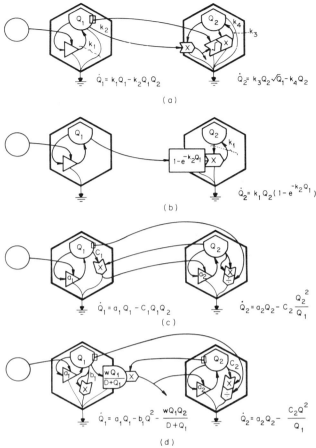

FIGURE 11-30. Energy circuit diagrams for some other paired unit designs in published literature: (a) model for prey–predator protozoa, *Paramecium* eaten by *Didinium* (Gause, 1934); (b) insect *Phormia* used by parasitic wasp *Mormoniella* (Gause, 1934) (see also Fig. 8-15); (c) host–parasite model from Leslie and Gower (1960) after Pielou (1977); (d) host–parasite model from Tanner (1975) after Pielou (1976).

amplify and control lesser energy flows. The real world is usually in the form of energy webs rather than simple chains, but the simplified chains form subsystems and are useful in helping to understand the more complex networks (see Chapter 14).

SUMMARY

In this chapter patterns of flow were considered for units in series. Phenomena resulting include time lags, cascade amplification, and oscillations—some stable and some unstable. Yield from the first unit to the second was studied for various designs and demand arrangements. Maximum sustained yield occurs at intermediate demand, but washout develops at higher rates of removal. Some characteristic series are parasite–host pairs, prey–predator pairs, Michaelis–Menten pairs, potential–inertial oscillators, and chemostats. Most mechanisms were those considered with single units in earlier chapters, but special behavior patterns emerge from the combinations. Oscillations involving only potential and inertial forces generate sine waves and have energy distribution given by the Hamiltonian principle. Some pairs have regular patterns of growth and pulsing recycles that may be important in survival, stability, and useful power transformation. Sequences have zigzag adjustments alternating relative storages, up in one when the next in sequence is down, and so on. Energy transformation in each unit causes the output to be of higher quality. Sequences may be important as control elements in more complex units, providing oscillators and timers.

QUESTIONS AND PROBLEMS

1. Draw ten types of series, indicating for each a real situation where the model may be relevant.
2. Draw the trajectory of an oscillating pair on an x–y graph in which axes x and y are magnitudes of storages in the two units.
3. In a chemostat, how does increasing flushing rate affect stock, production, and stores of available energy resources?
4. Which logic switching mechanisms provide pulsing recycle?
5. Which consumer designs have such rapid growth that they cause pulsing recycle as sharp as those produced by logic switching?
6. Draw an analog diagram that generates sine waves. Show that the first and second derivative is proportional to the negative of the quantity. Draw an energy diagram of a two-unit system with this property.
7. How does the graph of output vary as a function of input in a Michaelis–Menten unit? How does this compare with a series of several Michaelis–Menten loops?
8. How many exponential terms are in an equation for a tank at the end of a three-tank series?
9. Which two basic units were coupled in the Ross malaria equation?
10. Can steady pumping cause an exponentially growing autocatalytic unit to level its growth?
11. Will an exponentially growing unit level if it is drawing from a simple storage that is fed by a steady inflow?
12. How does the discrete Nicholson–Bailey parasite–host model compare with the Lotka–Volterra continuous-function model?
13. How do you describe a second-order system in energy diagramming, differential equations, and in analog diagrams?
14. What is the largest time lag expressed as phase angle of the projection of a spoke of a rotating wheel observable between input and output of a single tank? How much lag is there with two tanks in sequence?
15. Give examples of resistive, dissipative processes that are useful in the sense that their energy utilization is a necessary part of a larger system's survival functions, where resistive processes are in series and at least one is useful. Describe the relationship of the two resistances that maximizes useful power processing.
16. Describe the behavior of a chain of autocatalytic units.
17. Draw the shape of the output–input curve for a chain of Michaelis–Menten loops that are coupled.
18. Can you draw mechanisms for four kinds of oscillator system, each involving a sequence of two modules (from this or previous chapters)?
19. Simulate basic models in this chapter using a microcomputer or analog computer.

SUGGESTED READINGS

Bartlett, M. S. and R. W. Hiorns, Eds. (1973). *The Mathematical Theory of the Dynamics of Biological Populations*, Academic Press, New York. 247 pp.

Capellos, C. and B. H. J. Brelski (1980). *Kinetic Systems*, Krieger, Huntington, New York.

Emlen, J. M. (1973). *Ecology and Evolutionary Approach*, Addison-Wesley, Reading, MA.

Gold, H. J. (1977). *Mathematical Modelling*, Wiley, New York. 357 pp.

Goh, B. (1980). *Management and Analysis of Biological Populations*. Vol. 8, Developments in Agricultural and Managed Forest Ecology, Elsevier, Amsterdam, The Netherlands.

Harris, L. D. (1961). *Introduction to Feedback Systems*, Wiley, New York.

Hutchinson, G. E. (1978). *An Introduction to Population Ecology*, Yale University Press, New Haven, CT. 256 pp.

Lotka, A. J. (1925). *Elements of Mathematical Biophysics*, Dover, New York.

Margalef, R. (1978). General concepts of population dynamics and food links. In *Marine Ecology*, Vol. 4, *Dynamics*, D. Kinne, Ed., Wiley, New York. pp. 617–704.

May, R. M. (1923). *Stability and Complexity in Model Ecosystems*, Princeton University Press, Princeton, NJ.

May, R. M. (1976). *Theoretical Ecology*, Blackwell Scientific Publishers, Oxford.

Minorsky, N. (1962). *Nonlinear Oscillations,* D. van Nostrand, Princeton, NJ. 714 pp.

Milsum, J. H. (1966). *Biological Control System Analysis,* McGraw-Hill, New York.

Murray, B. G. (1978). *Population Dynamics, Alternative Models,* Academic Press, New York.

Nayfeh, A. H. and D. T. Moore (1979). *Nonlinear Oscillators,* Wiley, New York.

Nisbet, R. M. and W. S. C. Gurney (1982). *Modelling Fluctuating Populations,* Wiley, New York. 416 pp.

Pielou, E. L. (1977). *Mathematical Ecology,* 2nd ed., Wiley, New York.

Poole, R. W. (1974). *An Introduction to Quantitative Ecology,* McGraw-Hill, New York.

Poston, T. and I. Stewart (1978). *Catastrophe Theory and Its Applications,* Pittman, Boston. 491 pp.

Rubinow, S. I. (1975). *Introduction to Mathematical Biology,* Wiley, New York.

Sudo, F. M. and J. R. Ziegler, Eds. (1978). *Lecture Notes in Biomathematics*, No. 22, *The Golden Age of Theoretical Ecology 1923–1940,* Springer-Verlag, New York.

Waltman, P. (1974). *Lecture Notes in Biomathematics*, No. 1, *Deterministic Threshold Models in the Theory of Epidemics,* Springer-Verlag, New York.

Watt, K. E. F. (1968). *Ecology and Resource Management,* McGraw-Hill, New York.

CHAPTER TWELVE

Parallel Elements

This chapter considers designs with parallel elements. Some parallel units draw on the same energy sources; others feed into common outlets. Some are parallel processors of energy of similar quality without having common sources or sinks. They may interact with special connecting pathways. Parallel arrangements of several types are shown in Fig. 12-1.

COMPETITION

Sometimes the word "competition" is used to describe the relationships of parallel units, and many different phenomena are included under this category. One meaning of this word is the striving for resources in short supply. Often the word "competition" suggests negative actions by each competitor on the other as when they divert resources otherwise available to the other. More complex negative actions may include organized pathways to depress competitors (Fig. 12-1c). However, some parallel units are symbiotic. Relationships are termed *symbiotic* when there are beneficial effects of one on another (Fig. 12-1c, plus effects). Thus competition applies only to some parallel units.

Cooperative Competition

Competition may be beneficial to the competitors. For example, Allee (1938) described the competition of students for grades as cooperative competition because the process improved their performances and thus was beneficial even to those receiving the lower grades. He also described the competition for negative effects such as poisons, where the competition caused the negative influences to be shared and thus less detrimental.

Intraunit Competition

In most models, different kinds of unit are drawn separately, whereas units of the same type are aggregated as a population and represented as one module. Since variation is inherent in any energy flow, differences exist between members so that natural selection can help that population to maintain functions at maximum. Competition exists among members of the population. When we draw a pathway on an energy circuit diagram, we imply that these intrapathway competition processes are involved and that they are part of the means for adapting designs to maximum power utilization.

PARALLEL STRUCTURE IN UNIFIED FUNCTIONS

Connections That Are Simultaneously Parallel and in Series

As stated in other chapters, diagrams provide clarity about relationships, whereas the commonly used words such as "competition" or "symbiosis" cover too many different classes of relationships to be precise. Although we classified arrangements into parallel units (this chapter) or series (Chapter 11) for convenience, many units are simultaneously both. Notice the two diagrams in Fig. 12-2, one of which resembles a series whereas the other is drawn on the paper to appear parallel. As far as the information given is concerned, the two are the same. If units A and B are similar in energy quality, Fig. 12-2a is appropri-

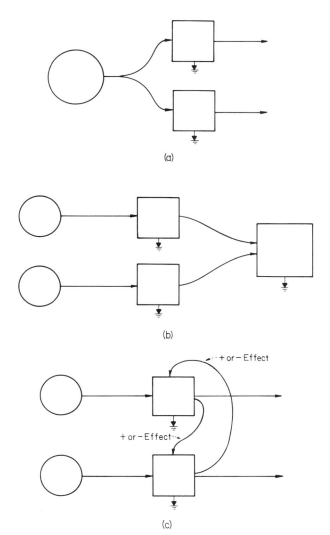

FIGURE 12-1. Arrangements between two units in parallel: (a) common source; (b) common sink; (c) affecting each other's functions by special pathways.

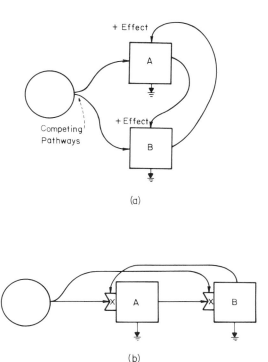

FIGURE 12-2. Designs in which units A and B have both competitive and symbiotic pathways at the same time: (a) in parallel because units A and B are similar in energy-quality processing; (b) arranged in series because of increase in energy quality.

ate; if B is of higher quality than A, Fig. 12-2b is appropriate.

Power in Parallel Pathways

Pairs of pathways in which maximum power is generated from the constant driving force when both pathways are flowing are shown in Fig. 12-3. The arrangements show some of the ways the conductivity may be controlled by intersecting controls.

Linear Programming Maximizes Parallel Use

Where there are several kinds of necessary material being used and several kinds of user population in parallel, each with a different ratio in their use of the materials, there is a relative ratio in populations that will process all the available materials without any leftovers. Presumably, this pattern, by using all the resources, is maximizing energy flow and competitive ability. A procedure for determining the optimum ratios is called *linear programming*. Where choices are available, the natural system is self-optimizing since any resource that is not being as fully utilized accumulates temporarily, stimulating the population that uses that resource more than others.

Linear programming finds the optimum point where usages have no remainders. It concerns steady states and can be diagrammed with amplifier elements in Fig. 12-4 to indicate which ratios of use are optimal.

Examples of linear programming are economic production in which maximizing dollar flow (probably also maximizes energy flows) is the common denominator. Another example is maximization of reproductive production by adjusting ratios of population groups as done by E. O. Wilson (1968). Another example is maximization of productivity by adjusting ratios of plant species (Odum, 1971a;

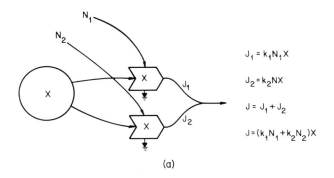

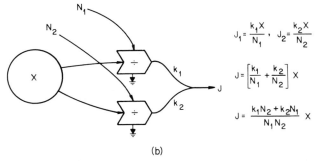

FIGURE 12-3. Parallel interactions: (a) multiplicative intersection; (b) intersection that is related by division.

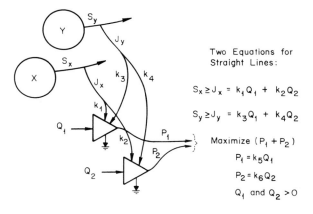

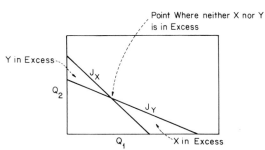

FIGURE 12-4. Example of linear programming. Two parallel units, Q_1 and Q_2, utilize two resource flows (S_X and S_Y). Procedure locates quantities of units that maximize combined production.

Kilham, 1971). Titman (1976) conducted experiments with mixed algal populations and found evidence for nutrient coexistence mechanisms.

The diagram in Fig. 12-4 shows an example of the analysis of linear programming to find the optimum point that the system may also tend to find on its own. The energy diagram shows the way linear programming can involve the regular web of production or consumption of most ecosystems. Figure 12-5 shows a usual ecosystem with parallel species processing more than one required material. It may be reasoned that recycle serves to cause self-linear programming as part of adaptation at the system level (see Chapter 22 on succession).

Tracers as Parallel Systems

Tracers such as radioactive elements are substances that may substitute for main flows but, because of their ease of recognition, can be used to trace processes and estimate rates. A good tracer is processed by a system with the same processes and interactions as the flow that it is tracing. In any tracer experiment a model is required to identify the processes of tracer flow and the rates that can produce observed uptake curves. One method for diagramming a tracer along with the main system being traced is illustrated in Fig. 12-6. Each compartment of the real system is duplicated with storage, flows, and interactions for the tracer alongside. Equations for the tracer thus are handled independently and analog simulations set up in parallel up until the point where the ratio of the tracer to the

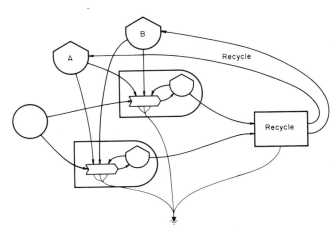

FIGURE 12-5. Parallel use of two commodities, A and B, by two units linked by common recycling.

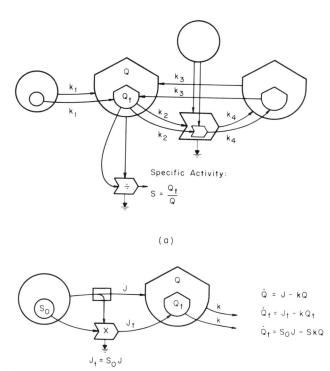

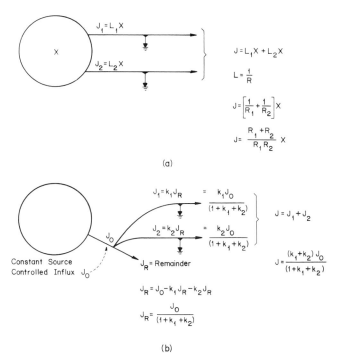

FIGURE 12-6. Diagrams of tracer and carrier both with same pathways and coefficients: (a) example that can be simulated; (b) specific activity equation.

FIGURE 12-7. Linear pathways in parallel; relationships are fixed: (a) constant force with energy as required; (b) source-controlled inflow in which competing pathways affect each other.

main flow (sometimes called a *carrier*) is to be estimated.

The ratio of tracer Q_t to carrier Q is called the *specific activity* S (see Fig. 12-6a). One may derive combined equations for the behavior of the specific activity, and an example for one compartment is included in Fig. 12-6b. However, the procedure of taking the ratio from the simulation may be easier. In a circular system at steady state, a tracer distributes itself among the compartments in the same ratio as its total quantity is in proportion to total quantity of carrier.

When a tracer is first added, its graphs with time are the chargeup curves like those described for storages in Chapter 3. Even though the main system is operating with various kinds of interaction and nonlinear flow, the tracer, as it catches up with the carrier, follows a chargeup curve in its graph of specific activity.

PARALLEL USERS OF ONE SOURCE

Linear Competition Versus Autocatalytic Competition

The competing pathways shown in Fig. 12-7 are linear, and their relationships are set by the characteristics of the pathway. Neither is changing its own conductivity or the percentage of energy that it is draining. Its competitive role is fixed. Figure 12-8 includes storages with similar results. In these pairs neither drives the others into extinction.

In Fig. 12-9b, however, the parallel units are autocatalytic and feed more energy back into additional energy gathering, accelerating growth, and taking an increasing percentage of the energy source.

Effect of Type of Source on Parallel Units

Where energy sources maintain a constant driving function, each parallel unit can flow according to its own characteristics without the drain of the other unit affecting its energy source. In other words, sources that maintain constant force are unlimited. Unless there are other complications by which the two interact, each behaves as though the other source were absent (Figs. 12-3, 12-8a, and 12-9a).

However, most sources are limited, and competition reduces the energy available to the other (Fig.

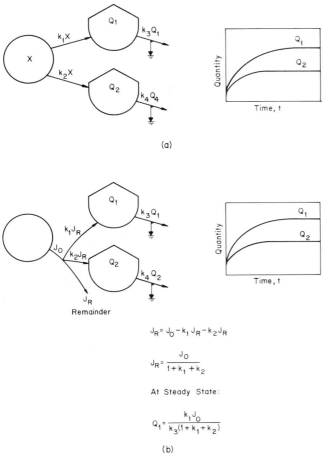

FIGURE 12-8. Competition between storage units with linear pathways (there is no competitive exclusion): (a) constant-force source; (b) constant-flow source.

12-9b). The source has a limited flow from which the parallel units draw.

Competitive Exclusion

The process by which one unit accelerates growth, taking an increasing amount of energy from the other, is sometimes called *competitive exclusion* because it leads to complete elimination of one unit by the other as when one species of weed drives another out of a field by overgrowth (Gause, 1934; Hardin, 1960, 1963). The model in Fig. 12-9b may be appropriate for two pioneer species competing for common resources. Competitive exclusion is useful to a system in early succession but is overridden by more complex designs when diversification replaces growth as a means for applying resources to maximize power.

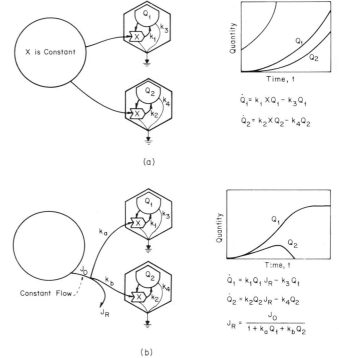

FIGURE 12-9. Pairs of autocatalytic units drawing on a common energy source: (a) energy source delivering constant force or forces; (b) energy source delivering a stream controlled at its source from which units Q_1 and Q_2 compete.

Effect of Greater Initial Condition

An autocatalytic process requires an initial stock to feed back and initiate the acceleration of energy pumping. If two competing systems such as those shown in Fig. 12-9b are otherwise similar, the one with the larger initial condition wins out in exponential growth. The difference between the two accelerates, merely because one started ahead of the other. This is a nice demonstration for analog or microdigital computer. Ikusima (1962) derived the mathematics for this and demonstrated the effect in competition among populations of duckweed.

Paired Logistic Modules

A number of different autocatalytic designs were given in Chapter 9 and found to have logistic equations that level regardless of type of energy source. When paired on unlimited energy, they behave as they would separately. However, when they draw on a source with restricted flow, one may reach steady state at a level less than that reached with unlim-

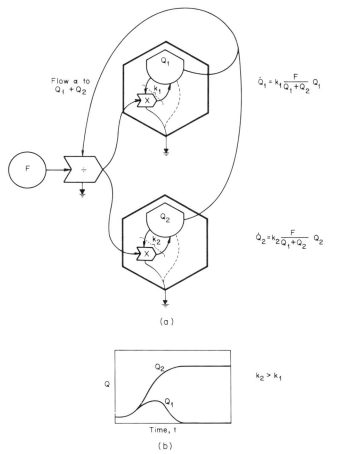

FIGURE 12-10. Competition model based on resource dilution by two populations from Milsum (1966) (depreciation was added with dashed line): (a) energy diagram; (b) growth with competitive exclusion.

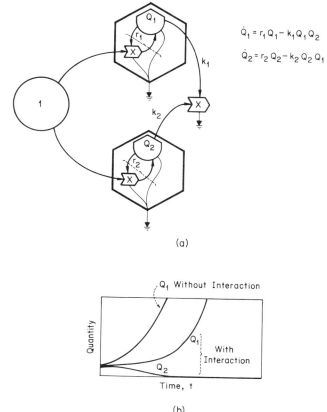

FIGURE 12-11. Parallel pair with unlimited sources and negative interaction: (a) energy diagram; (b) pattern of competitive exclusion in which growth is delayed until one is excluded.

ited energy. Competitive exclusion is possible where energy flow is limited.

Competition for Pressure

In Fig. 12-9b user pairs draw on a common resource stream, *subtracting* according to the demand of both growing units. In Fig. 12-10, however, the units *divide* up the resource pressure according to their quantity.

Parallel Units with Interacting Drain

The two parallel units shown in Fig. 12-11 are not competing for inflowing resources because the source is unlimited, maintaining constant pressure on both. There is a draining interaction in which each loses storage in proportion to the product of their stocks. This may be a simple model for two countries at war or two species that fight when they encounter each other. The model has no self-limiting terms. One excludes the other but is delayed in its exponential growth until the second one is zero.

Lotka–Volterra Competition Equations

A classic pair of equations describing the relationships of two species is the Lotka–Volterra competition equations in Fig. 12-12. Figure 12-12 is a phase-plane graph of the Lotka–Volterra model. Other ways of translating equations and adding necessary energy sources were given in Chapter 9. Van der Vaart (1978) found that these competitive interactions could not generate stable limit cycles. Discussion of stable points on these graphs was given by MacArthur and Connell (1966).

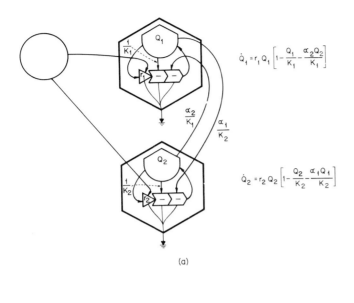

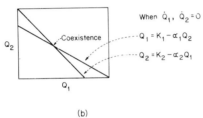

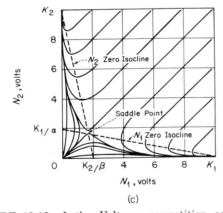

FIGURE 12-12. Lotka–Volterra competition equations and one energy diagram that generates them: (a) energy diagram; (b) state diagram with crossing lines of zero growth rate that applies for the special case of coexistence; (c) simulation results (Patten, 1964).

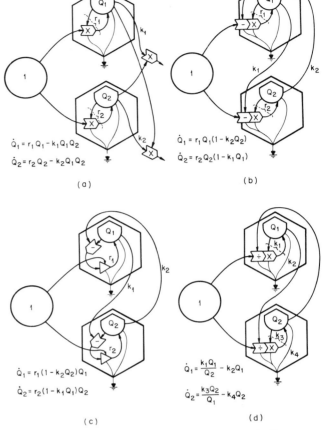

FIGURE 12-13. Parallel designs that generate similar competition equations: (a) two external drains; (b) paired drags on input flows; (c) paired drags on feedbacks to constant gain production pathways; (d) diluting actions.

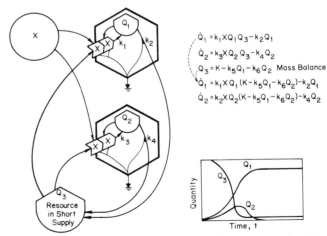

FIGURE 12-14. Design of two autocatalytic units competing for a recycling quantity in short supply such as space or raw materials; K is total resource ($Q_3 + k_5 Q_1 + k_6 Q_2 = K$).

212

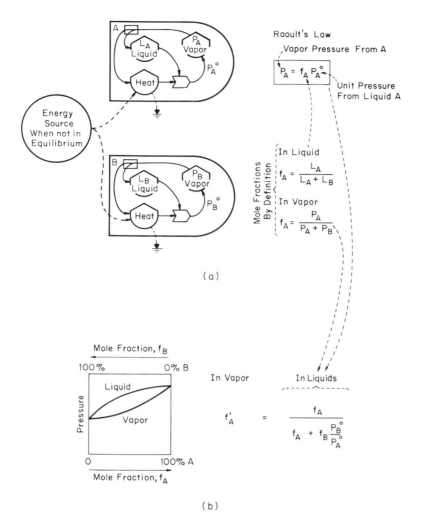

FIGURE 12-15. System of two liquid–vapor pairs that also applies to two stock–reproductive pairs: (a) energy diagram with energy shown in dashed lines since outside flow is not essential to some arguments; (b) equilibrium distribution of A and B in liquid and vapor phases.

Figure 12-12b is a state diagram plotting the trajectory of proportions of the two species. Theoretical analyses and simulations show the manner in which the growth rate of the two species depends on the coefficients, sometimes leading to competitive exclusion and sometimes to coexistence. Many studies of populations in the laboratory have shown outcomes covering the range of theoretical conditions.

The design in Fig. 12-12 has separate pathways by which each unit decreases flow of the other, each with a coefficient of interference. When the rates of change are set to zero, equations for two lines are obtained for the state diagram (Q_2 as a function of Q_1). If they cross, there is an equilibrium point at which there is coexistence. If they do not cross, there is competitive exclusion. This occurs when the negative effect of stock on another species is greater than the negative effect of stock on its own growth.

The diagram of negative interactions in Fig. 12-12 is only one of several ways in which one population may interact. Others are given in Fig. 12-13, which includes one drawing away from production and one increasing attrition.

Competition for a Recycling Quantity

Figure 12-14 shows a pair of autocatalytic units competing for a quantity in short supply that is incorporated and recycled. This quantity may be a ma-

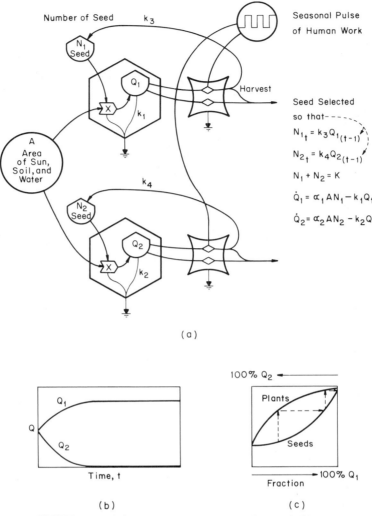

FIGURE 12-16. Competition in a system of crops and harvest studied by De Wit (1960); system also applies to fractional distillation of two liquids: (a) energy diagram; (b) time curve; (c) distribution of plants and seeds between species.

terial or a space. The resulting equations are similar to the Lotka–Volterra competition equations, but with an additional linear drain term. Depending on coefficients, there is a coexistence or competitive exclusion.

Competition in Vaporization; Raoult's Law

De Wit (1960) showed similarity between vapor–liquid relationships and population reproduction in which vapor from liquid is analogous to the seeds from a crop. Planting of seeds is analogous to condensing vapor back to liquid. See Fig. 12-15a for energy diagrams; Fig. 12-15b has curves for fractions of the two kinds of liquid or species in two phases. Raoult's law, well known in chemistry, relates the production of each type of vapor (or seeds) to the fraction of each type in the liquid (or vegetation) phases. The chemical example develops an equilibrium distribution without an external energy source, whereas the population example requires a steady inflow (see also Fig. 10-16).

A Model of Competition between Crops

Illustrated in Fig. 12-16 is the competition between wheat and oats studied by De Wit (1960) and De Wit

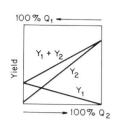

When in pure stand, yields are M_1 and M_2.

Yield Independent, no Crowding; Analog to Raoult's Law:

$$Y_1 = N_1 M_1$$
$$Y_2 = N_2 M_2$$

Yield in Mixture:

$$Y_1 = \frac{N_1}{N_1 + N_2} M_1$$
$$Y_2 = \frac{N_2}{N_1 + N_2} M_2$$

(a)

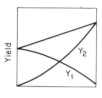

With Crowding, Differential Effect Measured by K:

$$\frac{Y_1}{N_1 M_1} = K \frac{Y_2}{N_2 M_2}$$

$$Y_1 = \frac{KN_1}{KN_1 + N_2} M_1$$

$$Y_2 = \frac{N_2}{KN_1 + N_2} M_2$$

(b)

FIGURE 12-17. Competition with and without crowding (De Wit, 1960): (a) independent yields; (b) crowding effects.

and Goudrian (1974). Each generation was planted with the same number of seeds, but in proportion to the harvest from the previous generation's growth. Resources were available in constant proportion per seed. This is a form of autocatalytic growth. De Wit obtained competitive exclusion experimentally and with the model. This was without any competitive interactions other than competition for choice of seeds in planting. In Fig. 12-17 the yield of seeds is shown with each species acting without any interaction.

Competitive Exclusion Analogous to Fractional Distillation

Ingeniously De Wit showed that the separation of one species from the other was analogous to fractional distillation in which two vapors are separated by successive vaporization followed by condensation. In the plants the seeds are analogous to vapor, and liquid is the growing crop (see Fig. 12-16c).

Competition with Crowding Interaction

Yields of seeds for various proportions of two species where there is crowding are shown in Fig. 12-17b. Compare this with the situation without the crowding (Fig. 12-17a), where De Wit found the crowding effects in planting of grain crops. He suggested the logistic curve for growth function. The energy diagram in this situation can be translated to that in Fig. 12-14. The crowding has more effect on one species than the other. This kind of competition has indirect negative effects of one on the other rather than a direct effect.

Competition Density (CD)

The higher the density of competing plants the less weight there is for each. When data are graphed, a reciprocal relationship is found between plant weight (w) and density (d).

$$\frac{1}{w} = Ad + B \qquad (12\text{-}1)$$

The product of individual weight and density is yield (Y). Dividing through by d Shinozaki and Kira (1956) obtained a model in equation (12-2) which fitted experimental data.

$$\frac{1}{wd} = \frac{1}{Y} = A + \frac{B}{d} \qquad (12\text{-}2)$$

This is a limiting factor hyperbola. There is a diminishing return for increased density of planting. After growth causes intense crowding, weights are found proportional to the inverse of density to the 3/2 power.

$$Y = Kd^{-3/2} \qquad (12\text{-}3)$$

Theories relate this to the surface volume ratio. See summary of these models by Hutchings and Budd (1981).

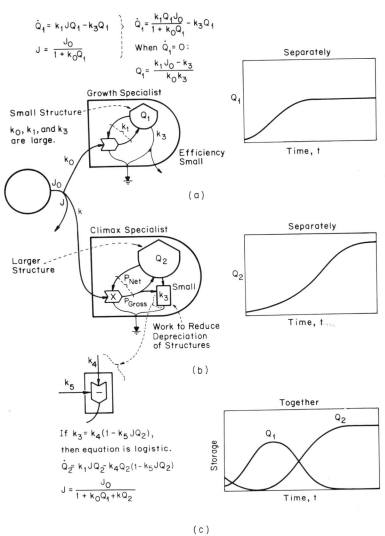

FIGURE 12-18. Competition between growth specialist and steady-state specialist: (*a*) growth specialist and growth alone; (*b*) steady-state specialist and growth alone; (*c*) growth together.

SPECIALISTS FOR GROWTH AND STEADY STATE

Maximization of power requires different specializations in different situations. Where there is an underutilized potential energy source, power is maximized by the unit that grows at the maximum possible rate by feeding back all its production possible into net growth. This is achieved by using a structural product with short life and by not using energy resources for maintenance.

In contrast, the unit adapted to maximize power in the steady state puts more energy into permanent structure into maintenance and ultimately supports more structure. Each of its units is less effective at capturing energy than the growth specialist, but collectively the massive structure of the steady state captures and uses more energy flow.

Examples studied in the rain forest are the successional trees, *Cecropia, Balsa,* and Kadam (*Anthocephalus*), which rapidly cover the ground with a leaf and litter cover, but their structures are very soft and air filled, and even the wood lasts only about 30 years.

In contrast are the steady-state specialist trees that grow slowly but last 200–1000 years with di-

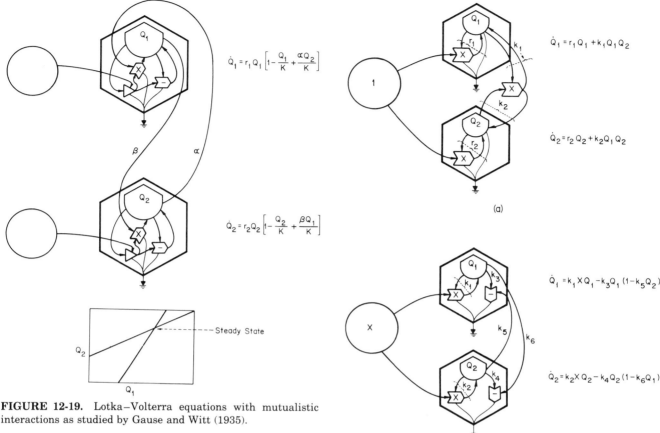

FIGURE 12-19. Lotka–Volterra equations with mutualistic interactions as studied by Gause and Witt (1935).

FIGURE 12-20. Parallel pairs with symbiosis that is optional for the existence of one alone: (a) mutualism in the generation of additional structure; (b) mutualism by inhibiting drains.

verse, dense, and complex forest structures that grow only to replace fallen units. More energy goes into respiration that is the work of maintenance.

Examples of the two kinds of specialist are given in Fig. 12-18. Both are shown with autocatalytic designs drawing on a source-restricted energy flow (e.g., sunlight and rain). In the first the coefficients of production, feedback, and energy incorporation are large, but so is the depreciation since the energy is diverted into the production process rather than in the maintenance processes. The growth rate is steep, but the steady state level of structure is smaller (when grown alone without competition from other units). A steady state is reached because of the flow limits to the energy source.

In the second unit energy is diverted from net growth to work of reducing depreciation of structure. Consequently, its growth coefficients (k_0, k_1, k_2) are lower. When growing alone, this unit grows slower but reaches higher steady state since depreciation is less.

When the two are in competition, the growth specialist wins out during the growth period but then is displaced by the steady-state specialists whose higher structural level allows it to use more of the incident energy. After the initial period, the high growth coefficients have no value since the only growth needed is replacement.

r AND K SPECIALISTS

In models of the logistic type (Fig. 9-12) r, the coefficient of unit growth (intrinsic rate of increase in some terminology), was found to be higher in those species that were pioneer, whereas others found in steady-state situations were observed with lower r values but higher steady-state levels (see K

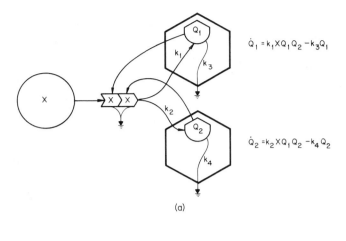

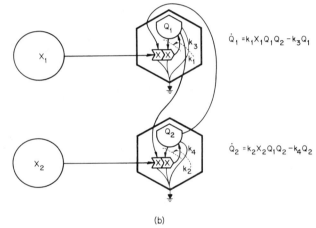

FIGURE 12-21. Design of a parallel pair in which mutualism is necessary to survival of each: (*a*) one interactive process; (*b*) two separate mutualistic interactions.

in Fig. 9-12*b*). Those with high *K* values were referred to as "*K* specialists."

PARALLEL UNITS IN COOPERATION

Avoiding Competition

Competition was indicated as having positive beneficial effects when it was intraspecific (i.e., among duplicate units) since it improved the system's efficiency. It was useful in early growth to help maximize growth. However, competition among units where one is better adapted and drives the other out in competitive exclusion can constitute a waste of the resources. Many systems have control networks that anticipate the effects of poor adaptation by preventing the competition situation and thus saving the energy of the eliminated population, saving them for places and times when they are adapted.

For example, steady-state trees in the rain forest don't germinate in the situations where light is too bright for their competitive survival. One can visualize situations among *cheap* organisms, like microbes, where the competition may be the cheapest way to achieve the adapted result, whereas among larger organisms with more energy invested in seeds and structure that energy is saved by operating a control system.

Symbiosis

Figure 12-19 gives Lotka–Volterra equations with mutualistic relationships where each member of a parallel pair contributes positive effects on the other. Gause and Witt (1935) studied the outcome of this system, finding coexistence at steady state. The case of commensalism with one interaction positive and the other negative was also considered.

The kinds of design that can generate the symbiotic relationships and equations of similar type can be entirely different as the examples in Fig. 12-20.

The mutualism of the pair illustrated in Fig. 12-21 is necessary to the survival of both. Neither exists without the other. This is obligatory symbiosis.

Figure 12-22 shows the competition of a pair as the sources for downstream consumer processes. This has the effect of symbiosis with the presence of one aiding the stocks of the other. Parallel species that have common consumers are symbiotic.

The growth outcome of mutualistic pairs on infinite energy is sustained, thus accelerating growth. If the energy source is flow limited, leveling occurs with a steady state resulting.

MODEL OF DEFENSE STRUCTURE

Competition may involve the building of defense structure as in the example of competing countries. Richardson (1960*a, b*) gave the linear model equations in Fig. 12-23 that construct armaments based on a sensing of the armaments of a competitor. When the model is diagrammed, no energy restraints are found since the model has unlimited constant action according to its coefficients. The model has mutually positive feedbacks so that the system as a whole can grow in accelerating manner.

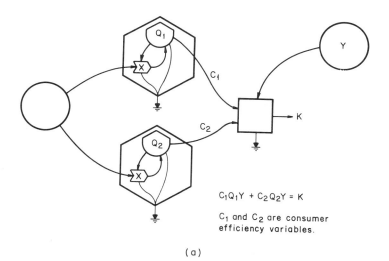

FIGURE 12-22. Mutualism by competing for consumption by downstream processes y: (a) pairs complementary in supplying constant outflow; (b) each pair inhibiting outflow consumption.

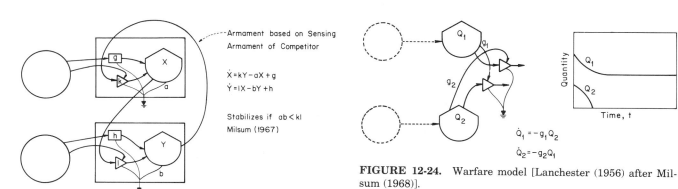

FIGURE 12-23. Model of defense armament from Richardson (1960); equations and energy diagram with energy sources added.

FIGURE 12-24. Warfare model [Lanchester (1956) after Milsum (1968)].

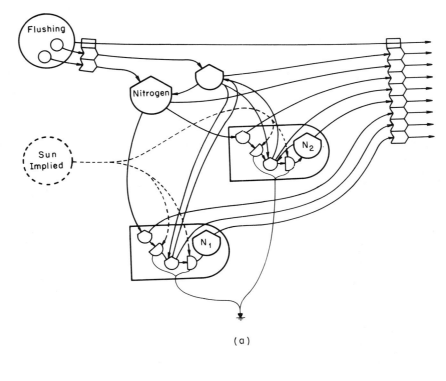

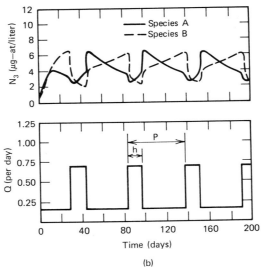

FIGURE 12-25. Simulation of two parallel species of phytoplankton with use of nitrogen units in all storages including biomass (N_1 and N_2). Flushing causes the model to resemble a chemostat (Grenney et al., 1974): (*a*) energy diagram; (*b*) simulation with alternative conditions maintaining coexistence.

A more realistic model has an energy ceiling or coefficients proportional to available energy supply.

WARFARE MODELS

Some of the competition models with negative cross-linking may serve as models of warfare. A linear model of warfare was given by Lanchester (1956) (see Fig. 12-24). The war action is based on storages that are rapidly discharged in proportion to the other. More complex models of war are given in Fig. 24-8, Fig. 24-14 and Fig. 25-27.

MICHAELIS–MENTEN UNITS IN PARALLEL

Thornley (1976) considered Michaelis–Menten units in parallel at steady state, finding their combined action with a curve of process similar to that of a single unit. Two parallel species of phytoplankton are simulated with the use of nitrogen units in Fig. 12-25.

SUMMARY

In this chapter different designs were examined for elements in which common resources are used in parallel. Depending on the form of the interaction and type of energy source, parallel units can displace each other in competitive exclusion or operate in symbiotic relationship. Since the use of more than one resource maximizes power of one unit, two elements that share the same resources may maximize resource use developing more power flow with self-regulating, linear programming. Since parallel structures with small changes in design can be organized either for survival of both or for exclusion, the question of what kind of pattern results may depend on selective reward to the collective design. During growth, where the rate of growth is a premium for survival, exclusion tends to prevail. During steady-state periods, where efficiency is a premium for maximizing energy flow, designs that utilize the diversity of parallel units may generate more power and reward loops for survival.

QUESTIONS AND PROBLEMS

1. Draw various kinds of relationship for parallel elements. Include common sources, common downstream consumers, cross-linked positive and negative amplifiers of various kinds, units with dual use of two sources, shared recycle loops, and an element contained within another element.

2. List several definitions of competition as related to diagrams in question 1. Which parallel elements exhibit competitive exclusion? Give examples of some that don't. How is energy quality related?

3. Which arrangement automatically self-organizes a parallel pair of users so that two resources are utilized without leftover maximizing power? Which mathematical procedure is used to find the ratio of the two elements that optimizes the system for maximum power?

4. What is the effect of a higher initial condition on the competition of two exponential growing units on the same flow-limited source? On an unlimited source that supplies constant pressure?

5. What is the classical Lotka–Volterra model? Give its energy diagram and relate this to the classical equations that do not explicitly include energy sources or flows. Substitute alternative autocatalytic modules that do have energy sources as variables.

6. What kinds of graph do Lotka–Volterra models generate on the phase-plane graph relating the two stocks?

7. Give an example of competition in a closed chemical system. How is fractional distillation comparable with competitive exclusion?

8. Discuss competition between growth and steady-state specialists. Draw minimodel pairs that most simply show these properties. When these concepts are used with logistic equations that have unlimited energy, the coefficients r and K are appropriate for numerically representing the growth and steady-state specialists. Explain. How does this pair in parallel simulate some aspects of the succession of events in time when life is colonizing a new area?

9. Draw a diagram representing war between two units. Draw a diagram representing armament of two countries affected by perception of each other (mutual recognition).

10. Diagram a tracer, including absolute quantities and specific activity.

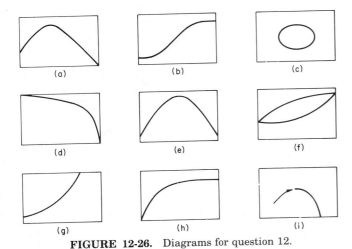

FIGURE 12-26. Diagrams for question 12.

11. Give criteria for stability in the sense of coexistence in systems of parallel units.

12. Match the letters in Fig. 12-26 with the following numbers:
    ___ 1. Yield versus effort
    ___ 2. $\dfrac{kx}{k+x}$
    ___ 3. Michaelis–Menten
    ___ 4. Ivlev–Watt
    ___ 5. Holling—time for feeding
    ___ 6. Chemostat–washout
    ___ 7. Stress effect
    ___ 8. Phase-plane prey–predator
    ___ 9. Raoult's law
    ___ 10. Adsorption
    ___ 11. Competitive exclusion x–y plot
    ___ 12. Cooperative density effects
    ___ 13. Lanchester warfare
    ___ 14. Linear programming
    ___ 15. Gompertz growth (see Chapter 20)
    ___ 16. Time integral of the output of a tank that has ramp growth
    ___ 17. Power and efficiency
    ___ 18. Allometric growth

SUGGESTED READINGS

Berryman, A. A. (1981). *Population Systems,* Plenum Press, New York. 222 pp.

Chapman, D. G. and V. Gallucci (1979). *Quantitative Population Dynamics,* Statistical Ecology Series **13,** Cooperative Publishing House, Burtonsville, MD.

Christiansen, F. B. and T. M. Fenchel (1977). *Theories of Populations in Biological Communities,* Springer-Verlag, New York.

Emlen, J. M. (1973). *Ecology, An Evolutionary Approach,* Addison-Wesley, Reading, MA.

Fredrickson, A. G. and G. Stephanopoulos (1981). Microbial Competition, *Science,* **213,** 972–979 pp.

Gause, G. F. (1934). *The Struggle for Existence,* Hafner, New York.

Gold, H. J. (1977). *Mathematical Modelling,* Wiley, New York.

Hazen, W. E. (1964). *Readings in Population and Community Ecology,* Saunders, Philadelphia.

Hutchinson, G. E. (1978). *An Introduction to Population Ecology,* Yale University Press, New Haven, CT.

Keyfitz, N. (1968). *Introduction to the Mathematics of Population,* Addison-Wesley, Reading, MA.

Kostitzin, W. (1939). *Mathematical Biology,* George Gittarap, London.

MacArthur, R. A. (1972). *Geographical Ecology,* Harper, New York. 268 pp.

MacArthur, R. and J. Connell (1966). *Biology of Populations,* Wiley, New York.

Pielou, E. C. (1977). *Mathematical Ecology,* 2nd ed., Wiley, New York.

Poole, R. W. (1974). *An Introduction to Quantitative Ecology,* McGraw-Hill, New York.

Richardson, L. F. (1960). *Arms and Insecurity,* Stevens, London.

Rubinow, S. I. (1975). *Introduction to Mathematical Biology,* Wiley, New York.

CHAPTER THIRTEEN

Webs

Webs are systems with energy flows connecting units in series and in parallel, usually with more than one external energy source. Apparently, most real systems that develop under conditions of competition for energy are webs with self-organizing relationships for maximizing power. Typical webs were introduced in Chapter 2. Whereas Chapter 11 considered modules connected in series and Chapter 12 in parallel, this chapter examines *types of web* and formulations for representing them. Webs are given in Fig. 13-1. The first one (Fig. 13-1a) has three external sources and a 3 × 3 matrix of energy flow with lateral pathways. For some purposes, systems are visualized as such regular patterns. However, real systems are more like the one shown in Fig. 3-1b, with hierarchical relations, recycle and feedback loops as given in Chapter 2. See, for example, the typical web in Fig. 2-3. Some numerical examples are given in Fig. 13-2.

For purposes of simplification to facilitate understanding, we often aggregate a web into a trophic chain (*trophic* meaning energy or food) (Fig. 2-1), and sometimes we aggregate a web to form parallel sequences to consider competition concepts. However, real-world systems seem to have the complex webs because this design permits multiple energy sources to be used in a flexible way to maximum power. Sometimes understanding is difficult because the pathways are invisible and are being switched on and off.

Systems ecology was much stimulated by interest in chemical element movements in ecosystem webs during 1950–1970 when funds were available for study of fallout and radioactive wastes. Tracers were used to evaluate pathway flows, representing data with linear models. Webs of phosphorus in microcosms were given by Whittaker (1961) and Sebetich (1975).

CONVERGING AND DIVERGING CHAINS WITHOUT LOOPS

Figures 13-3 and 13-4 give comparison of designs with converging and diverging flows where the units are autocatalytic and can pump from their sources, but no feedback loops are included. In the converging intersection in Fig. 13-3a two contributors are used interchangeably. The converging intersection in Fig. 13-3b provides the terminal consumer alternative pathways, and with this design, the energy flow is shifted to the feed pathway that is heavier, but it also permits the terminal consumer to drive other units to lower levels. Rescigno (1977) found coexistence in simulating these models.

The diverging pattern (Fig. 13-3b) has competing units, but one species cannot drive out the second without lowering its own source. Both contributors are required in Fig. 13-3d. More complex webs involve controls. Figure 13-3e shows learning loops that increase conductivity with use, Fig. 13-3f shows stabilizing control consumers on each contributor, and Fig. 13-3g shows consumption of two contributors retarded by stock of one unit. One pathway has a poisonous effect and one is retarded by mimic of the poison type.

Converging and diverging systems with more than three units are illustrated in Fig. 13-4. The converging example is like most ecosystems in which smaller sizes feed larger sizes so that more types of units are possible on the left where they are small.

The diverging example (Fig. 13-4b) is like some systems where the habitat is severe, requiring special adaptations for the units on the left with less energy for variety. Examples are salt marshes and mangroves. The dependent consumers on the right, however, are beyond the actions of stress and develop a diversity of types.

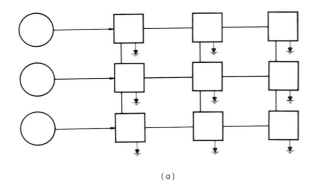

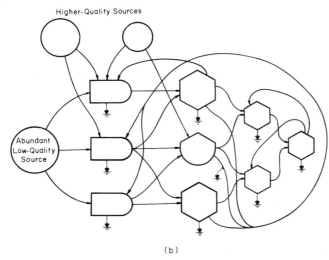

FIGURE 13-1. Webs: (a) regular matrix in the way humans sometimes visualize systems in their quest for simplicity and order; (b) more characteristic system with recognition of role of several sources, hierarchy of varying energy quality, reward loops, convergence, recycle, and heterogeneity.

ENERGY SIGNATURE AND BIOTOPE

In Chapter 2 the available energy sources to a web were called an *energy signature* and represented in order of energy quality as given in Fig. 2-2.

An older term for the combination of environmental factors is *biotope*. For example, the biotope of a marine ecosystem includes its solar radiation, current, inflowing nutrients, and so on, usually including only the nonliving sources. The energy signature refers to all the forcing functions to which the web becomes organized. An older term for the living ecosystem web that prevails is *biocoenosis*.

NICHE

Most well-energized webs contain autocatalytic units that have action roles. One definition of *niche* is the role a component plays in its web. Its role depends on both the energy signature available to it and its own capacity for utilizing inputs and outputs. The niche includes the connections and the unit as shown in Fig. 13-5.

This concept of niche as an "occupation" is that due to Elton (1926, 1953, 1958). The subsystem unit in its niche thus occupies some smaller part of the total realm occupied by the larger ecosystem that is of concern. The role of the subunit in the larger system is shown in energy diagrams since all the connections that the subunit shares with other parts of the larger system are diagrammed and have the values and mathematical relationships indicated.

A few units have similar functions with different connections. For example, compare Figs. 9-11 and 23-13c.

Niche as a Hypervolume in Multidimensional Space

Another way of describing the niche occupied by a subunit following Grinnell [see Hutchinson (1978)] is to define the range of properties of the larger realm of the larger system into which the subunit fits and interacts. Following Hutchinson (1944, 1978), the various properties that develop in a system can be graphed as coordinates as suggested in Fig. 13-6a. On paper only three properties can be easily shown, but the principle of visualizing a multidimensional space with all the properties is the same. The range of each property in which the subunit is operating defines the multidimensional "volume." Hutchinson's concept of niche is the "hypervolume" of range of characteristics of the species.

The properties that constitute the coordinates of the niche space are those developed within the system by the external energy inflows through various interactions and storages. Examples are temperature, concentrations of chemicals, concentrations of subunits, geometric properties, and other variables.

The range of operation of a subunit gives the range of properties in which it is found operating and describes the pattern but does not indicate the structure and function of the species in its role as a subunit.

Niche Vectors

The point that represents the corner of the hypervolume that represents lowest limit of each variable can be represented by a vector from the origin as

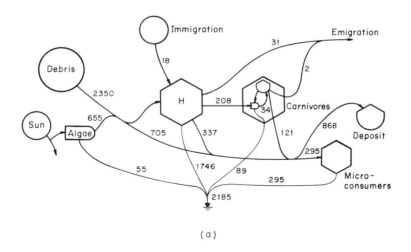

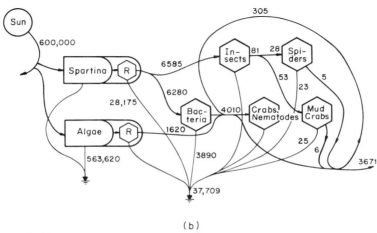

FIGURE 13-2. Energy translation of aggregated and evaluated ecosystem webs: (a) ecosystem of root spring (Teal, 1957); (b) ecosystem of *Spartina* salt marsh (Teal, 1962).

shown at P_1 in Fig. 13-6. Similarly, the corner of the volume representing the uppermost range of each factor in the hypervolume P_2 can also be represented as a vector. The niche is measured by the difference between the two vectors, which is a bound vector V, representing the niche (P_1 to P_2). The vector V is shown with its coordinates in Fig. 13-6b.

There are three component factors, each with its own coefficient; these coefficients collectively describe the vector. Description of the niche requires specification of magnitude, direction, and origin of the vector.

A vector is the resultant of the component vectors expressed along the coordinates. Stringing the component vectors together, keeping their angles unchanged, produces the resultant as in Fig. 13-6c. The resultant V is the vector sum of the component coordinates.

In Fig. 13-6d, consider the projection of a component vector projected on the resultant vector V. The magnitude of the component projected in this way is given with cosine. The sum of the three projections by the coordinates is the magnitude of the resultant vector. The angle α may be regarded as an index of influence of a factor.

When the factors are visualized as a vector, the implication may be that the factors are interacting in a linear way, each contributing in proportion to its magnitude as projected according to the angle. This concept of niche as a vector sum does not recognize the way component factors can interact with different functions. More than three dimensions cannot be graphed or easily visualized geometrically, but the relationship of a component to the vector is the same regardless of the number of dimensions.

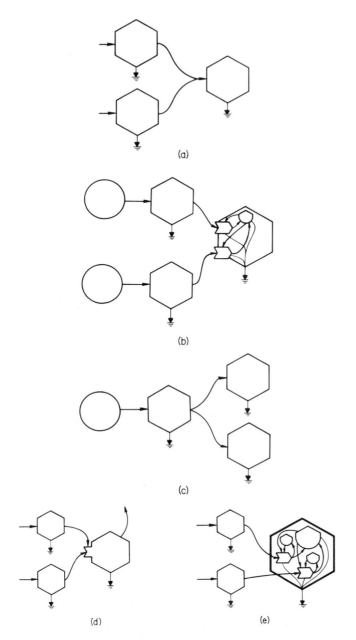

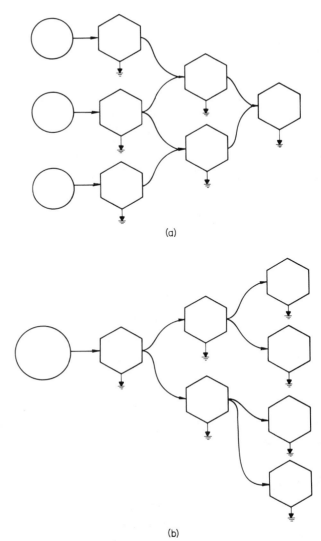

FIGURE 13-4. Webs of autocatalytic units without feedback loops: (*a*) converging; (*b*) diverging.

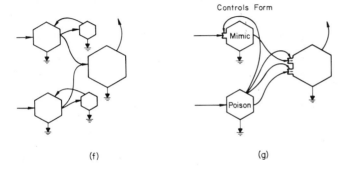

FIGURE 13-3. Converging and diverging web designs connecting autocatalytic units without feedbacks: (*a*) converging flows that sum; (*b*) converging flows from separate sources with independent production processes; (*c*) diverging flow to parallel consumers; (*d*) top consumer requiring both inputs; (*e*) learning loops in top consumer accelerating use of the type which is more abundant; (*f*) each unit stabilized by a rapidly tracking consumer (such as a parasite) that can accelerate if populations rise; (*g*) poison in one unit retarding consumption of itself and of the mimic whose pattern becomes coupled to that of the unit with poison.

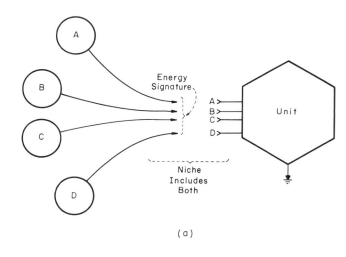

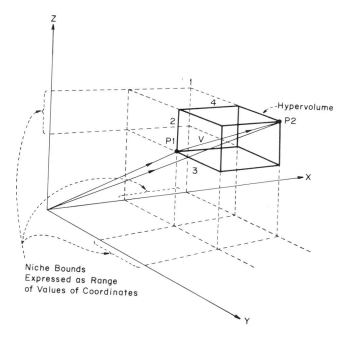

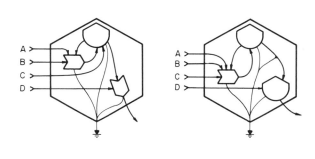

FIGURE 13-5. Niche visualized as a unit and its energy signature: (a) definitions; (b) examples of two units with the same energy connections but different internal functions.

Units in Niches

When a niche is visualized as some part of the realm of environmental properties, the question is raised as to how many types of unit can occupy the space. A generalist occupies a wide range of these properties, whereas a specialist occupies a small range. The web of specialists is more competitive collectively, although where conditions restrict the diversity, the generalist may prevail.

One idea is that the space of the environment is limited and that species occupying niche space pre-

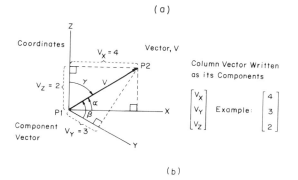

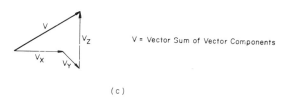

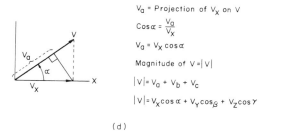

FIGURE 13-6. Niche as the hypervolume described by the range of properties of the system and represented by a vector (Hutchinson, 1944, 1978): (a) hypervolume of three properties represented on axes X, Y, and Z; (b) niche vector representing hypervolume with its point of origin p; (c) resultant vector as the vector sum of three component vectors; (d) projection of one component V_X on the resultant vector.

228   DESIGN ELEMENTS

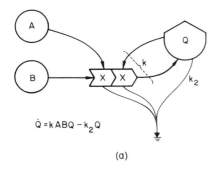

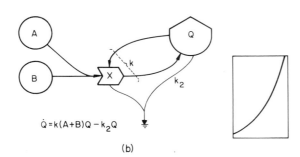

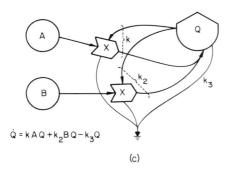

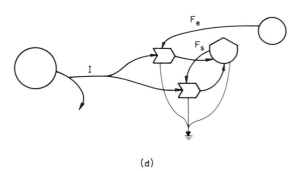

FIGURE 13-7. Alternatives for use of two kinds of energy, all of which accelerate growth: (a) interactive; (b) summing; (c) parallel autocatalytic loops; (d) external flow of high-quality energy F_e competes for matching low-quality energy with feedback from storage F_s.

clude this use by others so that occupancy is additive. A different concept that is related to the pathway niche concept represented in Fig. 13-5 is that each species provides new structures and combinations that generate even more niches so that niche space increases with species and is not additive (see concept of increased order with information content in Chapter 17).

When a niche is visualized as a configuration connecting a subsystem as in Fig. 13-5, two different types of unit cannot be in the same niche by definition, since if they are different, their niches are rendered different. They can pull on the same energy inputs, and one may drive the other out, depending on their type and relation as discussed in Chapter 12. They may also coexist.

COMPARISON OF MODELS WITH TWO SOURCES

If the interaction of high-quality energy with lower-quality energy maximizes power, webs will tend to develop that interact with more than one energy source. Consider alternative designs for interaction of two sources with a user web.

Where there is more than one source, a system has alternative ways of using that energy, such as feeding back to pump the second source, giving priority to one source over another, switching sources, and redesigning amplifier relations. Some models and their simulation results are given in Figs. 13-7–13-9. Some have better energy competition characteristics than others. For example, in Fig. 13-8a the system levels, leaving a constant-force source without full potential utilization. For other configurations with two source inputs, see Chapter 10.

DIRECT PERCEPTION OF SYSTEM BEHAVIOR

At this point in the book we hope that mastery of knowledge of the energy language has progressed to that point where time graphs of many simple configurations can be written directly from the diagrams without the necessity of equations or simulations. See if you agree that the graphs in Figs. 13-7–13-9 are of representative shapes. In other words, visualize the progression of filling of tank Q and its actions as it grows and in Fig. 13-9 as nonrenewable storages are drained.

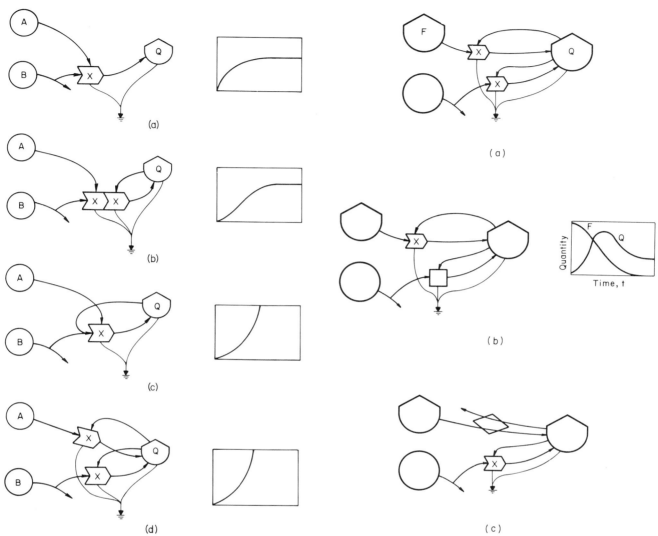

FIGURE 13-8. Designs for interactions of high-quality source A with source-controlled flow: (a) with one flow limited; (b) same but with feedback multiplier; (c) feedback adds to B so that it is not limiting; (d) dual feedback uses high-quality feedback to maximize use of both sources.

FIGURE 13-9. Designs in which the high-quality energy is from a nonrenewable storage; (a) feedback accelerates energy use; (b) renewable source is linear and source controlled; (c) nonrenewable source is obtained by exchange of high-quality energy.

SOURCE CONNECTIONS TO CHAINS

Models for the connection of energy sources with an energy chain are given in Fig. 13-10. In Fig. 13-10a sources interact as multipliers, which is appropriate in allowing limiting-factor action. Observed systems tend to have storages along the chain so that each interaction involves only one interaction at a time as in Fig. 13-10b. Note the loops that are required to recycle carrier materials and to feed back high-quality actions.

In the increase of quality that is developed in a chain, a source of valuable material may be required and enter as an interaction in more than one stage of transformation. For example, phosphorus is required in plant production, but additional phosphorus may be required to build more concentrated high-quality structures of consumers down the food chain (see Fig. 13-10c).

TRIANGULAR WEBS

Among simple web designs with parallel and convergent properties are triangular patterns as given

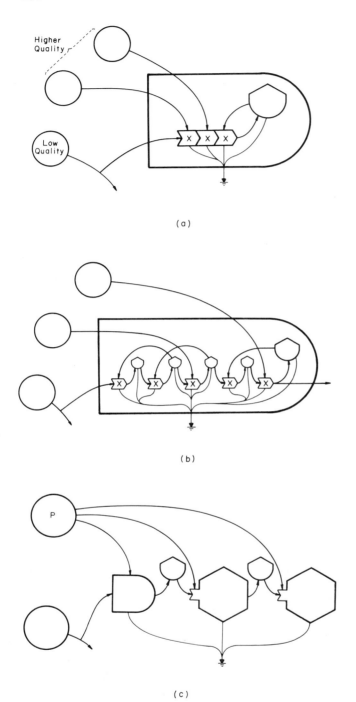

FIGURE 13-10. Webs of interaction of energy sources with chain: (*a*) direct multiplication; (*b*) with a chain of storages and feedback loops; (*c*) high-quality source *P* participates in chain of upgrading quality; example of phosphorus in plants and bacterial detritus and detritivore animals.

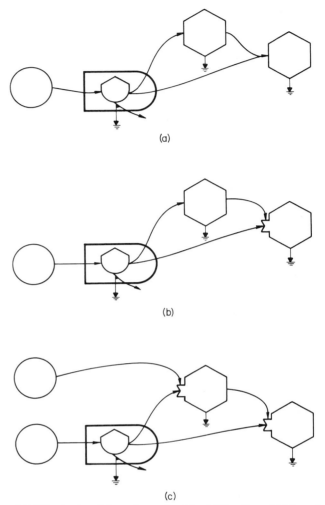

FIGURE 13-11. Triangular webs: (*a*) additive flow of high and low quality; (*b*) interactive intersection of high and low quality; (*c*) same as part *b* with a second source.

in Fig. 13-11. By providing alternative pathways and control actions at two levels in energy chains, designs with stable and flexible features result. For example Harris (1974) finds Michaelis-Menten properties in prey-predator webs.

A triangular web with a control system added is shown in Fig. 13-12. The top consumer (fishermen) take the member of the sequence that yields highest assets economically. The higher quality salmon has a higher price even though the weights are less. The system operates as a flip-flop. Once fishing is based on herring, the salmon food supply is diverted. A large surge of production, however, generates enough salmon to flip production to salmon after which the unharvested herring flows to support salmon and hold fishing in this mode. Limburg (1981)

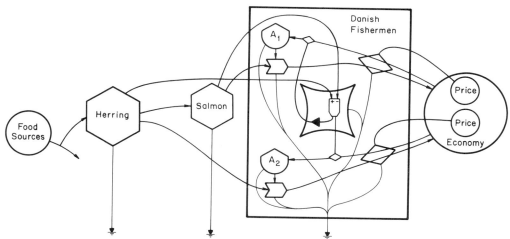

FIGURE 13-12. A switching consumer catches those members of a sequence that yield the higher quantity of assets (A_1 or A_2). Concept of relations in the Baltic Sea suggested by Bengt Owe Jansson and Anna Mari Jansson (Jansson and Zuchetto, 1978).

simulated models of Baltic Sea relationships including role of seals in fishery yields.

IDENTITY OF AGGREGATED AND SEPARATED PRODUCTION FUNCTIONS

Figure 13-13 gives two models that have the same overall differential equations but quite different *energy diagrams*. The first has separated production functions; the second has one production function. Both have the same interacting forces and energy flows. The convention used in diagramming has low-quality flows on the left and higher-quality flows to the right. Each storage is necessary to the production functions whether separated or aggregated. For example, A may be transportation, B industrial operations, and C population and housing. Each is required in the operation of the other in a city system and can be regarded as separate autocatalytic operations or as one aggregated system since they have the same differential equations.

MECHANISMS OF MAINTAINING MAXIMUM POWER PRODUCTION BY CHOICE OF FEEDBACK

A number of mechanisms are observed in systems for maximizing limited production where a number of factors interact. One mechanism involves the operation of selection on a number of units in which there is variation and choice in the coefficients of

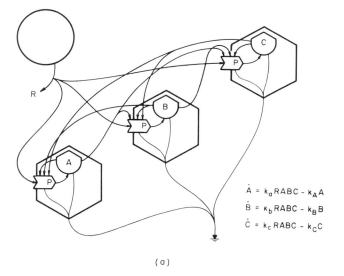

$$\dot{A} = k_a RABC - k_A A$$
$$\dot{B} = k_b RABC - k_B B$$
$$\dot{C} = k_c RABC - k_C C$$

(a)

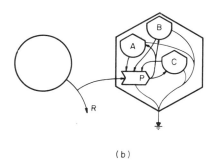

(b)

FIGURE 13-13. Model of structural interactions as in a city: (a) model in disaggregated form with three subsystems; (b) same model in aggregated form (equations are identical).

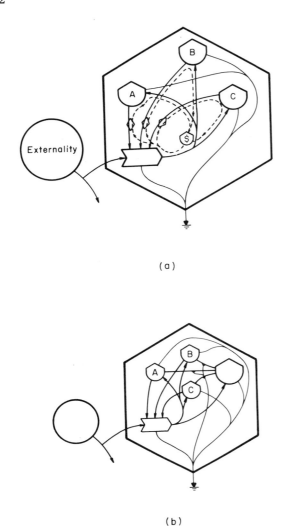

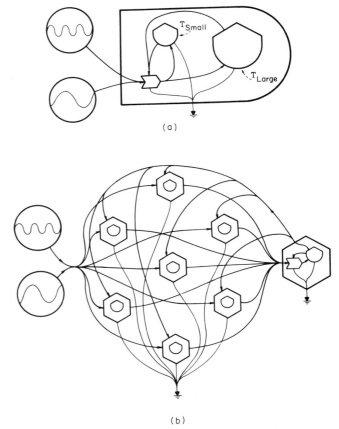

FIGURE 13-14. Mechanisms of assigning production to limiting factors so as to maximize power: (a) money flows to pathways which are limiting (money, dashed lines); (b) energies from a pool storage flows most to A, B, or C as may be limiting.

FIGURE 13-15. Systems with high- and low-pass filters capable of utilizing energies of different frequencies: (a) in parallel; (b) converging.

productive allocation among A, B, and C in Fig. 13-14. The one with more production develops more total power and grows or otherwise dominates.

Some regulation of the limiting action of A, B, and C occurs within one unit. If one is very limiting, it will grow relative to the others because its depreciation is less than its input from production.

A fast-tracking control action is supplied by the monetary system as given in Fig. 13-14a. If money is sent to A, B, or C according to the one that is most scarce, its price must be regarded as highest, and human behavior has been programmed to be thus responsive. The scarce one receives more money, which it feeds back to the production allocation, drawing more resources until it is no longer limiting.

Another mechanism involves a common pool of energy storages that are fed back to any unit that is short. In some systems simple diffusion will supply such a mechanism since the gradient is more (see Fig. 13-14b). Such a function is analogous to one governmental function. In other systems pools of resources of intermediate quality may serve this function, allowing any demand pull representing shortage to have a priority on common resources.

PLANKTON PARADOX AND LINEAR PROGRAMMING

In the discussion of parallel modules in Chapter 12 the procedure for optimizing use of resources called *linear programming* was given, showing that parallel units that process more than one necessary resource in interaction could self-design for the op-

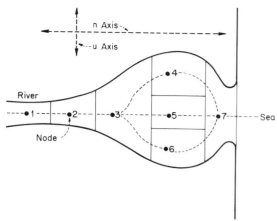

FIGURE 13-16. Modeling water quantity and motion; compartmental divisions and nodes for repetitive simulation of unit models.

timum ratio of populations for maximum power. For example, Fig. 12-5 gives a minimodel for an ecosystem with parallel producer units coupled to an indiscriminate consumer and recycle process. This model is similar to the plankton system of lakes and seas where nutrients are recycled. The design has linear programming inherent in the parallel units.

Sometimes a question is raised (Hutchinson, 1961) as to whether plankton are in the same niche without displacing each other in competitive exclusion. The model shows that the linear programming is self-correcting, tending to stimulate any type that can use an excess resource best. This mechanism tends to retain each species and overrides tendencies for competitive exclusion that would occur if the web were simpler. A numerical example was calculated using nitrogen and phosphorus (Odum, 1971a). Supporting this concept are data and discussions given by Kilham (1971), Titman (1976), and Tillman (1977).

UTILIZATION OF MULTIPLE-FREQUENCY ENERGY SOURCES AND FILTERS

In discussion of series in Chapter 11, storages were shown as means of receiving and utilizing energy arriving in pulses and waves. To absorb a pulse, the time constant of the storage receiving the pulse must be of the same magnitude as that pulse. If the storage time of the receiver is too small, a large wave will fill it quickly and pass through it, transmitting most of the energy that is in the wave. If the storage time is too large, the depreciation losses are unnecessarily great and response time is too slow. Storages that absorb waves are called *filters* in electronics.

Illustrated in Fig. 13-15 is a web with a larger and a smaller filter so that short-term and long-term stabilities can be obtained, with the smaller storage able to contribute to necessary short-term recycling and the larger one able to receive, utilize, and stabilize annual and multiple-year variations.

Just as the properties of many pathways can be represented in either matrix or network form, so the pathways acting as a filter can be tabulated in matrix or network form. The transformations can be written to mathematically represent performance of some real pathway entity, or the properties of transformation can be imposed by a computer that manipulates the web. Advanced systems of technology and of nature may develop digital and/or continuous function operations that filter because of the adaptive role of the result in survival.

SPATIAL–TEMPORAL MODELING

A body of water such as a stream or estuary can be divided into a web of connected compartments as shown in Fig. 13-16. Each compartment is given a model and evaluated as though it were operating at a point connecting with surrounding compartments by imports and exports to their points. These points are the numbered nodes in the figures. Most modeling of physical motions, chemical transport, and so on, is done in this way. The diffusion network in Fig. 13-17 is a simple example. Simulations generate spatial changes as well as graphs in time for each node. Any model, however complex, could be simulated in this way, although costs and practical difficulties increase rapidly as the number of entities in the model increase. The smaller the compartment, the more accurate the model and the more costly. For examples see Hamilton and Macdonald (1980).

WEB OF TWO-WAY PATHWAYS CONNECTING STORAGES

Some webs are simpler than the general type. A web of storages of similar quality material can be represented with a network of tanks as in Fig. 13-17 connected by linear pathways in which backforce is expressed. An example is the web of water storages in lakes, groundwaters, and streams on the landscape.

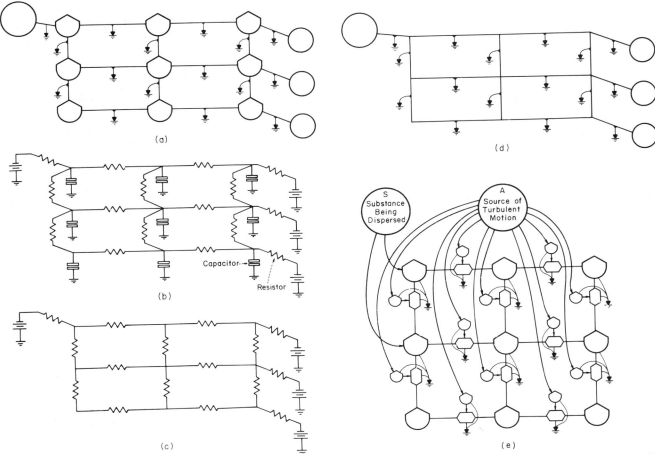

FIGURE 13-17. Web of storages and linear pathways: (*a*) energy language, with depreciation of storages included as part of interconnecting heat sinks; (*b*) passive electrical analog; (*c*) resistor network with same steady state as in part *b*; (*d*) energy diagram for part *c*; (*e*) eddy diffusion with active energy source to eddies (see also Fig. 8-23).

There is relatively little energy-quality change in flow between pathways. Kinetics are modeled by a network of electrical resistors and capacitors, which can be scaled to represent water storage, and resistance (Leopold and Davis, 1966). See Chapter 3.

Storage and the Steady State

When a network is at steady state, the storages (state variables) are unchanging and the flows become independent of the value, as they are a balance of inflow and outflow. Steady-state flows are identical to the flows without storage. For example, a network of water storages at steady state has flows independent of the water storages. A steady-state network's steady-state alternatives can be simulated with a resistor network without storages (i.e., without capacitors) (see Fig. 13-17*c*). Similarly, a network of energy language pathways may be used without tanks to describe such a network (Fig. 13-17*d*). Thron (1972) reviews means for reading stability and performance of linear webs by use of matrix criteria.

Web of Energy-Driven Conductivities

The connecting pathways may have pathways driven by another energy source *A* as shown in Fig. 13-17*e*. For example, turbulent motions in atmosphere or oceans from a source of kinetic energy may increase the effective conductivity of exchanges between tanks. As given in Fig. 7-7*a* and 8-2*c*, this may be represented by the double-pointed intersection symbol. If the source *A* is constant so that the

Kirchoff Law and Steady State

For flows of materials that are conserved in a system at steady state with $\dot{Q} = 0$, the sum of the inflows to a component equals the sum of outflows. The algebraic sum of flows at one unit is zero. In a matrix describing a system such as that in Fig. 13-17, a row includes all of the pathways to and from that unit. The sum of the row is zero at steady state.

A related principle is Tellengen's theorem, which states that the sum of the power flows into a unit is equal to that flowing out if there is no storage change (Tellengen, 1952; Oster and Desoer, 1971).

Linear System of Energy Flow in Matrix Form

As shown in Fig. 13-18, a system connected by linear pathways is readily represented in matrix form. Just as one pathway flow is the product of the transfer coefficient and the upstream storage Q, so the whole system of such relationships is the product of the vector $\underline{Q}$ and the matrix of coefficients for the connections between the storage units. This matrix is the transfer matrix. It is a table of transfer coefficients. By matrix procedures, the product of a column vector and the matrix is found by rotating the vector 90° and then successively multiplying each item in the vector by the equivalent item in the first row. Then the row vector is multiplied by the second row, the third row, and so on, until the resultant table of products, the product matrix of transfers, is found (see examples in Fig. 6-20 and Fig. 13-18).

The time rate of change $\dot{Q}$ of each storage is the sum of the inflows and outflows. For each unit, these include inflows from outside sources, exchanges transferred between units, and outflows from the system. For each unit, it is the sum of the appropriate term in the inflow source vector plus all the terms in the row of the product matrix. This set of sums is a resultant column vector representing the rates of change of storages due to transfers. In Fig. 13-18, to obtain the rates of change, the external sources E must be added to the product matrix obtained from multiplication of transfer matrix and state vector also. See also Walters (1976) and Jeffers (1978).

The matrix notation is often used as a condensed way of indicating a whole system where the interactions, storages, and rates of change are vectors and the interactions are matrices. Because computer programs can do repetitive calculations easily, ex-

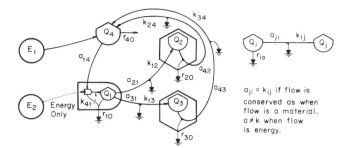

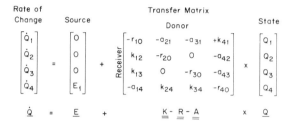

FIGURE 13-18. Equations for linear web of material or energy flows represented in matric form with transfer coefficients. Multipying the 4 × 4 transfer matrix by the 4 × 1 column state vector yields a 4 × 1 column vector. The r terms on the diagonals are sinks or exports.

effect on the pathways can be constant (at steady state), the web is still linear and can be represented with the electrical resistor–capacitor network shown in Fig. 13-17b as given by Karplus and Adler (1956).

MATRICES FOR WEBS OF UNITS OF THE SAME TYPE

Properties of webs and their units and pathways may be expressed collectively in tables called *matrices*. Matrices and some of their operations were introduced as another systems language in Fig. 6-16 and 6-19. Discussion of additional kinds of matrix and operation are appropriate here in the consideration of webs. Each row or column of a matrix is a vector.

FIGURE 13-19. Inversion of matrix $\underline{\underline{K}}$ generates state variables (Q values) when multiplied by rates of change ($\dot{Q}$ values). This result is verified by multiplying the interaction matrix $\underline{\underline{K}}$ by its inverse $\underline{\underline{K}}^{-1}$ to obtain the identity matrix $\underline{\underline{I}}$.

pression of systems in vector and matrix notation is a way of giving the computer instructions to perform the same calculation many times for the different parts of the system and express the result in a convenient tabular form. Matrix notations result in a convenient tabular form. Matrix notations now represent one of the most common ways of representing systems that are homogeneous and where one kind of process or relationship is studied for the whole system at a time. By reconstructing the equations from the energy diagram by inspection, the student can verify that the sum of each row in the product matrix is the differential equation for the component. If the purpose is mere description, the diagrams are easier.

Most ecosystems are nonlinear but can be regarded as linear when the system is little changing in its structural storages. A system of radioactive tracers or a flow of trace elements that do not affect the main structure and processes are examples of linear systems where matrix algebra is used easily.

The notation for the transfer coefficient k_{ij} and a_{ij} means that the coefficient is the effect of unit i on the flow into or out of unit j. The direction of flow is not indicated by the subscripts necessarily but is clarified by diagrams. However, usage varies on this aspect.

The a coefficients in Fig. 13-18 represent the effect of the downstream unit on the flow from the upstream element. In this use the term a_{ij} is the effect of the downstream unit j in providing a pathway outlet for flow from i. Since the flow is from i to j, the coefficient is negative. The donor is the unit that contributes a path, which is not necessarily the one from which flow is derived. This is confusing because donor refers to the source of materials and energy in other usages.

Inverting Rate Matrix in Simulation

In simulation the state variables are continuously graphed with time after the rates of change are continuously determined by multiplying transfer coefficients by state variables (Chapters 3–5). For linear systems, these calculations can be done by inverting the transition matrix and adding to the state as given in Fig. 13-19. Although tedious by hand, the matrix operation done by computer allows multiple unit systems to be simulated with relatively few computer instructions if the computer is equipped with matrix inversion. Figure 13-19 verifies that an inverted matrix is correct by multiplying the matrix by its inversion to obtain an identity matrix.

COMPARISON OF CONCEPTS INVOLVED IN INTERACTION MATRICES

Several different matrix representations are given in Fig. 13-20, for the same system. The subscripts are given in Fig. 13-20a with a vector for inflow, one for export, and an interaction matrix for linear effects of each state unit on each other or on itself. The lowermost diagram (Fig. 13-20e) is the mathematical one without regard to matter or energy constraints. Each transition coefficient is a black box. One coefficient exists for each pathway, whereas

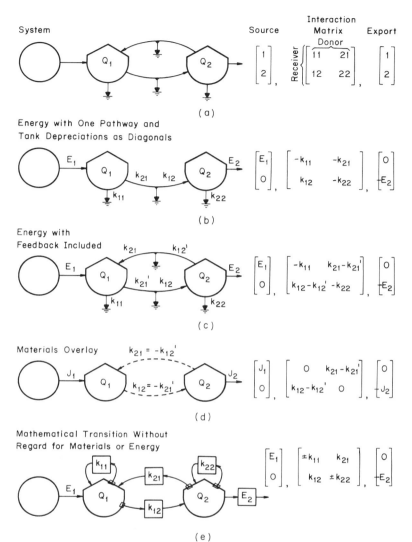

FIGURE 13-20. Concepts and matrix representations for the same linear system (a); (b) single-path energy transfer; (c) feedback energy transfer; (d) material transfer; (e) mathematical black box state transition. Prime (i.e., k'_{21}) identifies outflow pathway as different from inflow in exchange between two units. Notation k_{21} indicates effect of Q_2 on Q_1, which is to draw a flow to Q_2 from Q_1. Donor means source of influence, not source of flow, which in this case is from Q_1 to Q_2. In other uses the reverse may apply, depending on author's definitions.

real matter and energy considerations often require two. Each transition coefficient is a black box.

In Fig. 13-20d materials are modeled with conservation of matter in passing from one unit to another. The coefficient for a pathway leaving k_{21} is the same for entering the next tank k_{12}. Interior processes are not shown, and the self-effects on material storage are zero. There are two coefficients for each interunit place in the matrix.

Energy is represented in Figs. 13-20b and 13-20c. Heat sinks and changes of coefficients are required in going from one tank to another. There are twice as many coefficients on connecting pathways (two for each). If the system has no feedback as in Fig. 13-20b, there is only one coefficient to each matrix position, but in Fig. 13-20c with feedback, there are two for each connecting pathway. Since energy is lost in depreciation from all storages, there is a negative coefficient for each self-interaction, which form the matrix diagonal. Not included and even more complex are those with backforce, linear coupling, and nonlinear pathways.

Usage also varies in regard to algebraic sign. Plus and minus were given explicitly in Fig. 13-20, but in

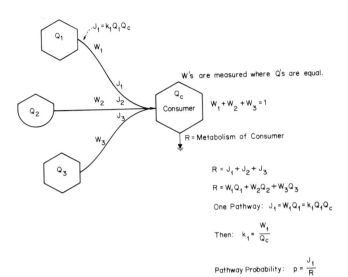

FIGURE 13-21. Consumer preference coefficients after O'Neill (1969) and their use to estimate transfer coefficients, k.

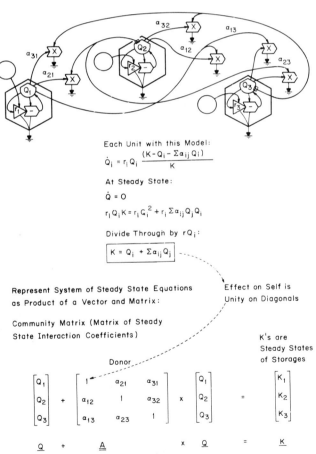

FIGURE 13-22. Linear matrix for a nonlinear ecosystem at steady state; community matrix after Levin (1968).

other usages letter designations of coefficients are given without signs, and plus or minus is given only when numerically evaluated.

From these comparisons one can see considerable ambiguity in using mathematical translations without specifying their relationship to matter and energy diagrammatically.

Consumer Preference Coefficients

Data on food preferences may be used to evaluate coefficients for a web. If the consumption of prey by predator is dependent on the concentration of prey Q_1 and on the tendency for consumption by the predator Q_c, then kQ_1 is the consumption per unit predator (see Fig. 13-21). For several different prey, the size of the coefficients (k values) indicates the relative tendency of the same predator to consume the prey, where prey is available in equal concentrations.

For cases where these k values are not available, O'Neill (1969) introduced consumer weighting factors w as the relative preference of a consumer for prey in laboratory tests where the prey are offered in equal quantity. The w values are proportional to the k values.

To obtain the flows without the k values, the product of each weighting factor w and its prey concentration Q is set equal to the expression for the inflows to the consumer that contains k and solved for k.

The probability (fraction of a total of 1) is that pathway's flux compared to the total influx to consumer R.

Nonlinear Systems That Are Linear in Steady State

With linear webs in Fig. 13-17 at steady states, the flows were found to be independent of storages once their levels had automatically adjusted so that inflows were equal to outflows. Some nonlinear systems reaching steady state become independent of storage and thus become linear.

Figure 13-22 shows a nonlinear system in which each unit operates with a model such as that in Fig. 9-12b with exponential growth on constant energy source with quadratic self-leveling terms and product interactions from other units similar to Fig. 12-12. Levins (1968) showed that at steady state the state variable partly dropped out so that the community model at steady state could be represented by a linear matrix. Quadratic terms and products of state variables also dropped out (Fig. 13-22).

Linearizing Models

The logistic model in Fig. 13-22 was linearized by constraining it to the steady state. Sometimes this is done for the purpose of making applicable such linear theories as Laplace transforms and stability theory (e.g., see Fig. 11-3).

Linear models have been used by Patten for examining nonlinear systems (Patten, 1975). Simulations are done by successive stepwise recalculations of steady states. This approach is common in economic modeling (see Chapter 23). Results are justified as valid for perturbations near the steady state. In Chapter 1 (Fig. 1-3) linearity was correlated with energy levels being small. The procedures for linearizing dynamic models may apply to low-energy perturbations only.

Substitute Matrices (Eigenvalues and Eigenvectors)

For many interaction matrices, it is possible to substitute a simpler matrix that has the same final effect when multiplied by the state vector. One substitute matrix that is useful is one in which there are only terms on the diagonals. (The diagonal has the terms for self-interaction.) In other words, by multiplying only one constant by each storage quantity, the substitute matrix produces the same answer as does the more complex matrix by multiplying by many interactive terms.

The simplified matrix has the property of producing the next state by a multiple of its previous state. The vector that is multiplied by this simplified matrix reproduces itself and therefore is an unchanging one. It is the steady-state vector. Finding the simplified matrix is one way to find the steady state.

A value of a substitute matrix is called an *eigenvalue*. The vector that produces itself when multiplied by the substitute matrix is called the *eigenvector*. Methods for writing these items are given in Fig. 13-23.

Relation of Eigenvalues to Time Equations

In a linear web the state equations with time contain a series of exponential terms [e.g., see Eqs. (11-

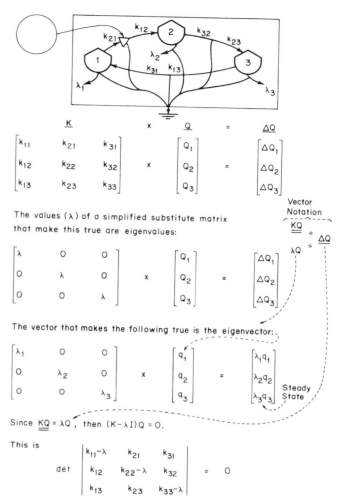

FIGURE 13-23. Substitute matrix that has eigenvalues and defines eigenvector. Example has an outside energy source.

1)]. The rate of change of an exponential function is a modified exponential as in the following example:

$$\text{If } Q = Q_0 e^{rt}, \text{ then } \dot{Q} = rQ_0 e^{rt}; \quad (13\text{-}1)$$
$$\therefore \dot{Q} = rQ$$

The matrix of exponential terms representing the state equations of a linear web is related to the matrix of exponentials that represent rates of change of that web by constant multipliers.

The eigenvalues are decay coefficients of the exponential terms in equations for the storages. The eigenvalues are the coefficients by which a set of states at one time are transformed to a new state at a later time.

Since the diagonal matrix of depreciation coefficients (r values) are values that will generate a multiple of themselves if multiplied by the state vector Q, they are eigenvalues.

WEBS OF MOTION

The self-designing flows that maximize power generate webs that include kinetic energy mixed with potential energy. Combinations of potential and kinetic energy storage may oscillate as already described in Fig. 11-7. The traditional way of representing and understanding systems of kinetic and potential energy is with equations for acceleration that have terms for the various forces that cause acceleration. These equations are called the *equations of motion* and are the starting place in examining flows of water or air in oceanography and meteorology. Translation of an equation of motion into energy circuit language produces a web such as that in Fig. 13-24. See pathway symbols in Fig. 8-23. Potential energy of elevated water against gravity is expressed as pressure in compartments P_1 and P_2. Kinetic energy storages are marked V for velocity, and inflows and outflows are the contributions to acceleration of each term in the equation of motion. Energy losses accompany frictional processes but not inertial transformations. Forces that are opposed to each other are connected in opposite direction on the same pathway. Thus Coriolis force and centrifugal force are shown opposing pressure-gradient force.

Physical Webs Comprised of Similar Unit Models

When physical processes are modeled spatially, as in the example of Fig. 13-16, the unit model is the equation of motion as given in Fig. 13-24. It is repeated in each compartment and connected to the next compartment through cross-compartment forces, exchanges of momentum, and other means of energy transfer.

Modeling the physical equations of motion, continuity, hydrostatic relationships, and so on, integrating and iterating with replication of many nodes, has generated good representations of many features of ocean and atmospheric circulation on several scales of size, although very large computer memories and costs have been required. For example, simulation of a web of connected physical and heat-balance equations for a hurricane has generated vertical distributions of wind, temperature, and humidity like that observed [see review volume (Goldberg et al., 1977)].

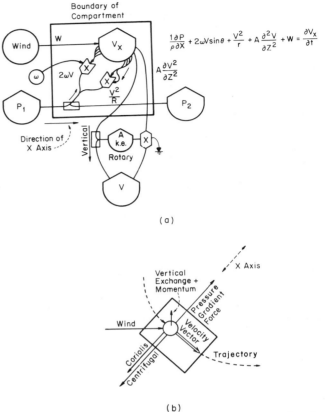

FIGURE 13-24. A web in energy language representing the equation of motion in side view for one compartment of surface water in the sea. The terms in order of their appearance in the equation are: pressure gradient acceleration; Coriolis acceleration; centrifugal acceleration; and acceleration due to eddy diffusion of momentum and wind stress. (*a*) Energy diagram; (*b*) vector diagram of force balance viewed from above the sea surface; X is left to right and Z is vertical (out of paper); acceleration vectors are single lined.

Aggregated Physical Models

Whereas the models are formed by connecting a grid of unit models to show micromechanisms and their consequences, they do not really explain in terms of the next level of organization why the system of energy flow organizes the larger-scale structures. It may be like modeling the human stomach, using a grid of equations for the forces of weight, food flow, and cellular behavior without asking whether there is an overall function (digestion) toward which the details are self-organized. An open question that needs more consideration in physical modeling relates to whether there are overriding adaptive criteria that are required by energy laws and organization of the biosphere. Consideration of fluid systems

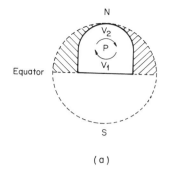

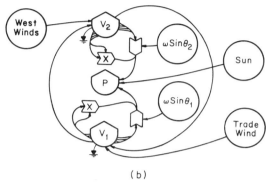

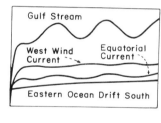

FIGURE 13-25. Aggregated minimodel of Gulf Steam with two boxes: (a) divided ocean in two parts; (b) energy diagram; (c) simulation of model with sinusoidal inflow program for seasonal winds and heatings; V_2 is westwind current and V_1 is equatorial current [McKellar and Odum (1972) from Nihoul (1975)].

at a more aggregated scale may provide simpler principles for modeling of physical systems. Modeling with large degree of aggregation is sometimes called *box modeling*. Everything is included within a few boxes and represented by overall equations.

For example, consider circulation of an ocean such as the Atlantic. Some modeling has been done with a grid of nodes, each of which represents a unit model similar to that in Fig. 13-24. However, the more aggregated method is given in Fig. 13-25, where the oceanic gyral is represented in two compartments, one the northern half of the ocean and the other the tropical half to the equator.

In the energy diagram the two compartments are connected in three ways: (1) through pressure-gradient force of elevated sea surface at P pressing northward and southward along with centrifugal force balanced against Coriolis forces of the gyral flows V_2 and V_1; (2) flow southward on the right (east); and (3) northward flow of the Gulf Stream on the left (west). All values are similar except Coriolis parameter's latitudinal term that causes energy to flow south in the center. The result of simulation in Fig. 13-29c shows the strong Gulf Stream required for steady state where there are two pathways of energy transfer southward.

Where webs consist of units of the same type similarly connected, the web repeats basic unit designs. For these, structure and behavior are often represented with matrix formulations, especially those with linear relationships between units. Some non-linear webs are linear at steady state. Matrix inversion is used to determine states from rates. Some webs can be represented by simplified substitute webs of the same performance as eigenvalues and eigenvectors.

Spatial distribution is represented by repeating unit models, each representing a section of the landscape or of a body of water. Equations of motion constitute such systems.

INPUT–OUTPUT MATRIX

Flows of material or of money through a web of connected units can be represented with network diagrams or with the equivalent matrix of flow data as in the example in Fig. 13-26. If the system is linear, flows may maintain fixed ratios to each other that can be represented by coefficients. Steady-state, average, or uniformly growing systems may maintain fixed ratios even though the ratios vary when the system is changing structural proportions.

Input–Output Coefficients

The ratio of inflow to the outflow of a compartment is an input–output coefficient as given by Leontiev (1966). The matrix of these coefficients was introduced for circulation of money as in the example in Fig. 23-32. Similar ratios have been calculated for the ratios of flows of minerals in ecosystems by Hannon (1973a,b; 1976) using data from Silver Springs, FL (Fig. 21-42a), Finn (1976), and Patten and Finn (1979). Richey et al. (1978) compared matrices of

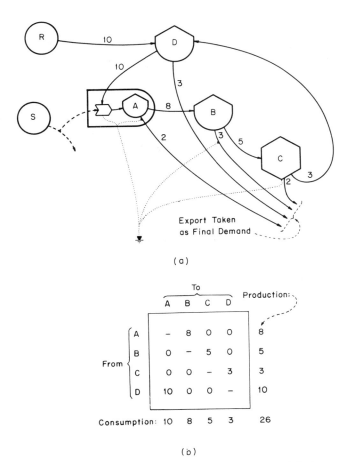

FIGURE 13-26. (a) Input–output model for materials in an ecosystem given on a network diagram; (b) input–output data matrix (assuming linearity); (c) cycling index from input-output matrix with s_{ij} relating input to output (Richey et al., 1978).

input–output coefficients for carbon flows in lakes as a means of comparative analysis of structure. The fraction of flows that were recycled was calculated to be in the range 0.03–0.66 [see Fig. 13-26c, which is notation used by Hannon (1973a,b)].

Boolean Matrix of Web Structure

Presence or absence of a pathway may be represented by matrix of ones and zeros to indicate "on" and "off" (e.g., see Fig. 13-27b). Cohen (1978) summarizes Boolean webs for thirty examples.

Plus–Minus Matrix

Directedness of pathways may be represented in matrix form with the signs + and − respectively indicating to or away from donor and 0 indicating no pathway (see loop analysis given in Fig. 6-31).

OTHER PROPERTIES OF WEBS

Many properties of systems can be shown collectively with either matrix tabulation or numbers on pathway diagrams. Menshutkin (1973) gives coupling factors that are defined as the percent change of one unit of the system for a percent change of another:

$$\left[\frac{\Delta B_i / B_i}{\Delta B_j / B_j} \right] \quad (13\text{-}2)$$

The matrix of these coupling factors shows the pattern of change holistically. See sensitivity, p. 137.

Patten (1978) represents causality by tracing external influence by way of flows through units of an ecosystem, using a matrix representing one stage, another representing two steps, another for three steps, and so on (see Fig. 13-27). The causality is also represented by the percent of external inflow identifiable in successive stages, decreasing stepwise as causality is diluted by confluence with other causes.

Measures of Interaction

The community matrix $\underline{\underline{A}}$ in Fig. 13-22 contains coefficients of interaction that are linear at steady state, but coefficients are product interactions when the system is away from steady state. Following Levins (1958), Lane and McNaught (1970) evaluated a community matrix of species interactions in Great Lakes zooplankton, where interaction coefficient was defined as the sum of the products of the frequencies of the two species in that environment divided by the square of the frequency of one of them. This was a measure of probabilities of encountering a species other than itself.

A quantity termed *niche breadth* was defined as the reciprocal of the sum of the squares of the frequencies of that species in all environments, and *niche dimensions* denoted the sum of information content of the frequencies with which that species was found in several environments. There was a wide range in interaction values interpreted as indication of complex, organized relationships rather than random individualistic growth responses to separate actions of the environment.

The ratio of population to its steady-state level without community interaction (Q/K in Fig. 12-12) was given by Lane (1978) as an index of competition success.

STABILITY

Methods for examining stability of linear webs include those given for the simple case of two units in series in Chapter 11, such as the root locus (Fig. 11-3) and Nyquist plots (Fig. 3-18) [see review by Van Voris (1976)]. Stability is determined by estimating the degree of return to a steady state after a displacement or under perturbation of varying sources. This is called *resiliency* (Holling, 1973).

Shugart et al. (1976) found the steady-state matrix of calcium flows in a popular forest stable on the Nyquist diagram apparently because of the large storages in soils and trees. This analysis does not apply if perturbations are sufficient to change the storages, thus disturbing the linearity that exists only near steady state.

For the usual world of hierarchical nonlinear systems, there appear to be no general methods for determining stability. However, see Atherton (1981).

For simple linear webs of storages, density dependent flows such as depreciation, friction, and metabolism help to damp out fluctuations. In other words, the flows caused by the second law of thermodynamics induce stability. For the simple unconnected web such as that in Fig. 6-19 the linear outflow drains affect only the donor compartment. The matrix version has coefficients only on the diagonal. The more drains there are on the diagonal, the more stable the system.

For more complex webs, the matrix is more complex, but with the substitute matrix, eigenvalues constitute an equivalent diagonal. Jeffries (1979) uses the negativeness of the mean eigenvalue as a measure of system stability.

Steele (1976) criticizes May (1973) for using negative terms on the diagonal of a food connection matrix as evidence for stability. Because of the second law there are always negative terms on the diagonal (respiration in the case of organisms), but not all models are stable. Attempts to use linear stability criteria either assume linearity or presume close proximity to steady states in order to be linear. However, it is stability under substantial perturbation that is of interest.

Stability Tests of Linear Webs

As explained in Chapter 11, a sequence of two storages or more may oscillate if there is a negative effect of the second back on the first. The oscillator in Fig. 11-7d is an example. Patten (1965) simulated a web of related oscillators (Fig. 11-23b).

A related system with a damping pathway (depreciation) more nearly resembles the connected compartments of ecosystems, and efforts have been made to test stability of webs by measuring the time required for a displacement to return to steady state and the deviation sensitivity to displacement. Harwell et al. (1977) simulated linear webs and also tested a shorter procedure, using a two storage model as an index. One was not an indicator of the other. Since real webs are not generally linear, these tests of stability may not apply to many.

Control of Consumption

O'Neill (1976) studied the potential control actions by consumers, adjusting their consumption in a

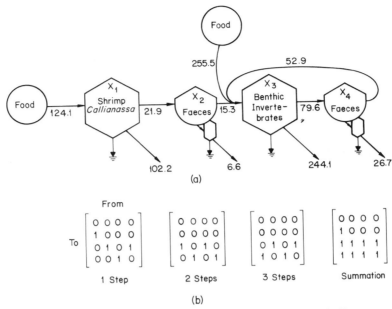

FIGURE 13-27. Causal analysis of a burrowing shrimp by Patten (1978) using unpublished data by W. G. Cale and P. R. Ramsey: (a) carbon flow values on an energy diagram in g/m² · day; (b) matrices indicating causal pathways.

triangular web model in Fig. 13-28. Consumers were capable of stabilizing perturbations. With the use of data on storages and time constants of various ecosystems, the aquatic plankton systems with greater turnover responded more rapidly to perturbation of productivity, thus maintaining less variance as compared with terrestrial ecosystems with lower turnover times. More consumers were found in the more rapid turnover systems.

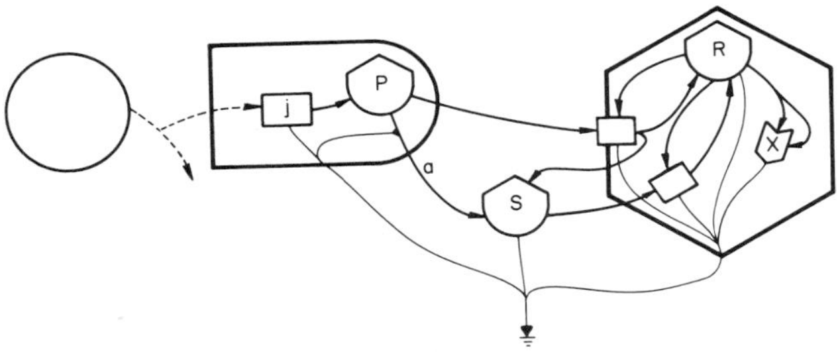

QUEUE DIAGRAMS

Another systems diagramming language for webs facilitates the analysis of production where production depends on the timing sequence and allocation with which raw materials and components are combined. In Fig. 13-29 the time required for each unit task in a sequence is written on each pathway segment (in a small square). Where pathways converge, the one with the longest time requirement for performance is indicated as the critical pathway with a double bar. The earliest time required to begin the task is the cumulative time from the start to a node through the longest (critical) pathway and is given in the lower right part of the circle at the node. The latest time required to start the task is the cumulative time counting backward from the end through the critical pathway (longest work time) and is indicated in the lower left part of the node circle. An energy translation of the process is shown as Fig. 13-29c.

Statistical queueing theory applies probability distributions to processes in sequence such as the sequence of births and deaths. Probabilities for lengths of time are determined (Cooper, 1972).

$$\dot{P} = j - aP - \frac{bcPR}{cP+R}$$

$$\dot{R} = \frac{(1-f)bcPR}{cP+R} + \frac{gqSR}{qS+R} - dR(1+R)$$

$$\dot{S} = \frac{fbcPR}{cP+R} + aP - \frac{gqSR}{qS+R} - hs$$

FIGURE 13-28. Model used to study role of quadratic-limited consumers controlling production only by rate of consumption (O'Neill, 1976). The interactive functions in black boxes can be derived by setting flow proportional to conductivities due to consumer and food (see Fig. 8-16; j is net production input available to storage P).

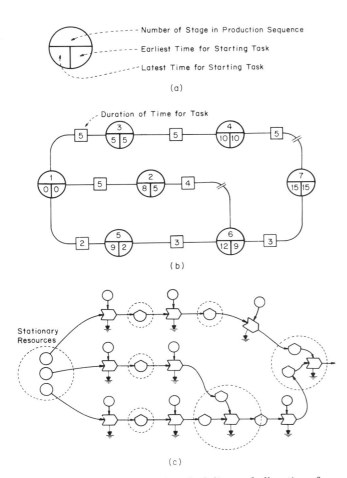

FIGURE 13-29. Diagrams for scheduling and allocating of resources (Martin, 1965): (a) explanation of numbers; (b) queuing diagram; (c) equivalent energy diagram.

SIMILARITIES BETWEEN WEBS OF SMALL, RAPID-RESPONSE POPULATIONS TO LOGIC NETWORKS

Relative to other events in ecosystems, enzyme or microbial components have very rapid turnover times with rapid growth, rapid control actions, and rapid disappearances when conditions for their growth are on. A network of such organisms has some properties of logic networks as shown by Sugita (1953, 1961, 1963; Sugita and Fukada, 1963). The sudden turnon and turnoffs of the rapidly responding populations are like switches in the more sluggish time realm of the ecosystems (Fig. 8-13). Very little has been done with the use of logic networks for microbial ecosystem simulation. One example of use of digital networks as analogs for ecosystems is given in Fig. 22-12.

Consumers who eat alternative foods are OR functions. Consumers who require more than one type of input are AND functions. Units that eat to a certain threshold before reproducing are counters. Population starts each year with seasonal cues for migration and turnoff are like flip-flops. Reproduction and inheritance of numbers is a digital process, and so on.

Positive and Negative Reinforcement Control Web

Webs of logic elements can provide information-processing mechanisms for self-organization by using feedback to reinforce properties that are adaptive. In organisms these processes are called *learning* and in ecosystems, *succession* (see Chapter 22). The reinforcement of successful performance can be a positive reward of success or a negative inhibition of failures. See mechanical turtle in Fig. 6-11.

Figure 13-30a shows a learning control system for humans simulated by Loose (1974) with positive and negative reinforcement actions in parallel. With information storages scaled in percentage of full action, learning curves were obtained like some observed. See simulation example in Fig. 13-30c, where the positive reinforcement was more effective than the system of negative inhibition, developing motive to avoid failure.

Informational control webs can operate if there is a stable energy base supplying power as needed. So far as energy supply is concerned, the system of learning in a nervous system is analogous to electronic information systems based on stable electrical power supplies.

Webs of digital logic systems are abbreviated with the concave-box symbol. Generalities about digital logic networks in environmental systems remain for the future.

Other Properties of Webs

Other webs are given in Fig. 13-31. These illustrate observed patterns of inhibition and feedback. Figure 13-31c represents one of the early simulations of living control networks. For other properties of webs, see discussion in Chapter 14 on energy quality, in Chapter 17 on complexity, and Chapter 18 on diversity.

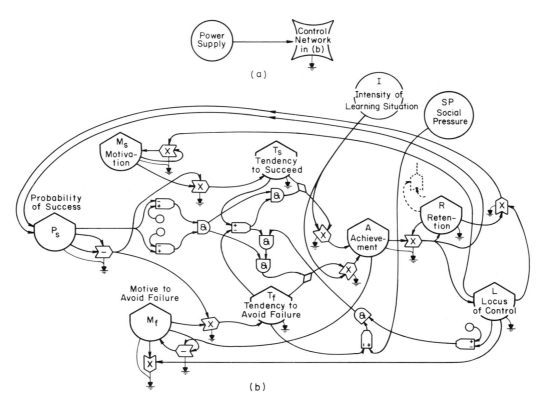

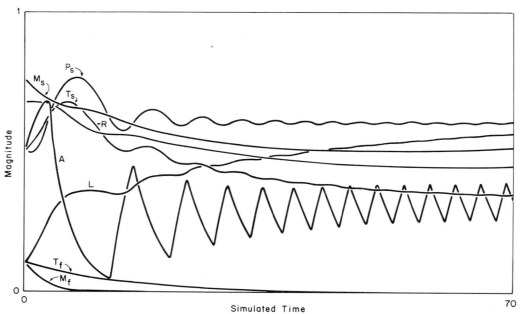

FIGURE 13-30. Simulation model of human motivation and learning (Loose, 1974): (a) energy relation; (b) energy diagram of the control network; (c) simulation results.

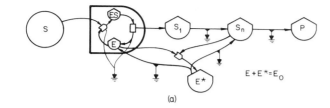

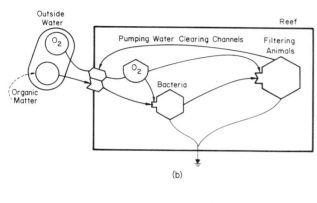

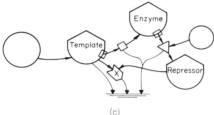

FIGURE 13-31. Other webs: (*a*) model of inhibition that oscillates [Schnakenberg (1977) after Hunding (1974)]; (*b*) animal–bacterial symbiosis within a coral reef (DiSalvo, 1969); (*c*) template, inducer, and enzyme actions simulated by Heinmetz and Herschman (1962).

SUMMARY

In this chapter webs of producer and consumer units were considered with series and parallel structures together. Triangular series–parallel arrangements provide stability of self-organizing ability to shift in use of sources. The relation of each unit to the web is the subject of niche theories. Typically, webs have feedback loops and recycle of materials between units, which can be represented as input–output models. Most real webs are heterogeneous, with the unit models varying within the web. The webs for organisms, cities, and other self-maintaining organizations with multiple energy sources have characteristic autocatalytic loops nested and woven so that by-products feed back as contributors.

Where webs are made of units of the same type similarly connected, the web repeats basic unit designs. For these, structure and behavior are often represented with matrix formulations, especially those with linear relationships between units. Some nonlinear webs are linear at steady state. Matrix inversion is used to determine states from rates. Some webs can be represented by simplified substitute webs of the same performance as eigenvalues and eigenvectors.

Spatial distribution is represented by repeating unit models with equations of motion, each representing a section of the landscape or of a body of water. Other webs include queues, logic networks, and passive electrical networks each related to others by translating into energy diagrams.

QUESTIONS AND PROBLEMS

1. Draw webs of three modules that have the property of parallel elements and series elements at the same time.

2. Draw configurations that draw from two energy sources such that energy from one can feed back into processing of another.

3. In what way do energy quality considerations limit the pathways among units based on a single energy source?

4. Compare a web of homogeneous units and pathways with one that has many kinds of unit and types of pathway. Which is readily represented in matrix form? Give examples.

5. How do production functions connect webs of units? How are recycling materials included?

6. Give an example of a single production function for two interacting component storages, which has the same kinetics and web connections as two separate units?

7. Draw a web with external energy sources and with dollar flows.

8. Draw a web of three units connected by linear pathways. Write the matrix that describes the transfer coefficients for the web. Describe the matrix operation that determines the states (stocks of the storages) from the transfer coefficient matrix.

9. Considering energetics, what must be drawn on any flow from one tank to another? How many transfer coefficients are required for that

pathway? Where the flows are given in materials units (not energy) and matter flows from one storage to another without loss, how many transfer coefficients are required? Use diagrams to distinguish between a forward and a return pathway.

10. Give equations for a web of logistic units with unlimited energy sources and at steady state. Is the web under these constraints linear? How can coefficients that relate the pairs be used in a matrix describing interunit structural relations?

11. What are eigenvalues and eigenvectors? Explain what use these calculations have for predicting the state of a web system.

12. Write the typical equation of motion used in aquatic or atmospheric sciences to define the factors that affect each compartment of fluid. Diagram these in energy language, using the storage of velocity (and its kinetic energy).

13. What does a filter do for passage of a pulse through a single pathway? For a web of pathways?

14. Describe a web of interconnected spatially replicated unit models that are customary in spatial studies.

15. Because of the various properties of energy, matter, kinetics, and so on required for adaptive survival of system, real webs—when considered in their entirety—seem to show some similar characteristics. Draw a typical web with some of these properties.

SUGGESTED READINGS

Anderson, R. M., B. W. Turner, and L. R. Taylor (1980). *Population Dynamics,* Halstead Press, Somerset, NJ.

Bellman, R. (1970). *Introduction to Matrix Analysis,* 2nd ed., McGraw-Hill, New York. 403 pp.

Casti, J. (1979). *Connectivity, Complexity, and Catastrophe in Large-Scale Systems,* Wiley, New York.

Chadwick, M. J. (1975). The cycling of materials in disturbed environments. In *The Ecology of Resource Degradation and Renewal,* M. J. Chadwick and G. T. Goodman, eds. Blackwell Scientific Pub., Oxford, England. 3–16 pp.

Christiansen, E. B. and T. W. Fenchel (1977). *Theories of Populations in Biological Communities,* Springer-Verlag, New York.

Cohen, J. E. (1978). *Food Webs and Niche Space, Monographs in Pop Biology,* No. 11, Princeton University Press, Princeton, NJ. 189 pp.

Cooper, R. B. (1972). *Introduction to Queueing Theory,* Macmillan, New York, 276 pp.

Funderlic, R. E. and M. T. Heath (1971). Linear compartmental analysis of ecosystems. Oak Ridge National Laboratory. ORNL-IBP-71-4.

Halfon, E. (1979). *Theoretical Systems Ecology,* Academic Press, New York. 515 pp.

Hutchinson, G. E. (1965). *The Ecological Theater and the Evolutionary Ploy,* Yale University Press, New Haven, CT.

Hutchinson, G. E. (1978). *An Introduction to Population Ecology,* Yale University Press, New Haven, CT.

Jacquez, J. A. (1972). *Compartmental Analysis in Biology and Medicine,* Elsevier, Amsterdam. 237 pp.

James, A. (1978). *Mathematical Models in Water Pollution Control,* Wiley-Interscience, New York, 420 pp.

Jennings, A. (1977). *Matrix Computation for Engineers and Scientists,* Wiley, New York.

Kleinrock, L. (1975, 1976). *Queueing Systems.* 2 vol. Wiley-Interscience, New York, 417, 549 pp.

Matis, H., B. C. Patten, and G. C. White, eds. (1979). *Compartmental Analysis of Ecosystem Models.* Vol. 1. Statistical Ecology Series, G. P. Patil, ed. Cooperative Publ. House, Burtonsville, MD.

Neustead, G. (1959). *General Circuit Theory,* Wiley, New York.

Nicolis, G. and I. Prigogine (1977). *Self-Organization in Non-Equilibrium Systems,* Wiley, New York.

Phillips, Don T. and A. Garcia-diaz (1981). Fundamentals of Network Analysis, Prentice Hall, Englewood Cliffs, NJ. 474 pp.

Poole, R. W. (1974). *An Introduction to Quantitative Ecology,* McGraw-Hill, New York.

Shugart, H. H. and R. V. O'Neill, Eds. (1979). *Systems Ecology, Benchmark Papers in Ecology,* No. 9. Wiley, New York.

Swamy, M. N. E. and K. Thulasiraman (1981). *Graphs, Networks, and Algorithms,* Wiley, New York, 592 pp.

Usher, M. B. and M. H. Williamson (1974). *Ecological Stability,* Chapman and Hall, London.

Van Voris, P. (1976). *Ecological Stability: An Ecosystem Perspective,* ORNL TM 55D, Environmental Sciences Division, Publication No. 900.

Zadeh, L. A. and C. A. Desoer (1963). *Linear Systems Theory,* McGraw-Hill, New York.

Zaret, T. M. (1980). *Predation and Freshwater Communities,* Yale University Press, New Haven CT. 208 pp.

PART THREE

ORGANIZATION AND PATTERN

CHAPTER FOURTEEN

Energy Quality and Embodied Energy

Flows of energy develop hierarchical webs in which inflowing energies interact and are transformed by work processes into energy forms of higher quality that feed back amplifier actions, helping to maximize power of the system (see Figs. 2-3 and 13-1b). The small amounts of energy resulting from the conversion to the new forms carry the embodiment of larger amounts of lower-quality energy used in the transformation process. Tracing of embodied energy through webs enables flows and products to be quantitatively related to energy sources. Higher-quality flows require more embodied energy and have greater amplifier effects when they feed back. Consequently, embodied energy is a measure of value, in one of the meanings of the word "value." There are several concepts of embodied energy, but it is possible to compare and clarify the various measures by using diagrams. Study of the way energy flows support production and consumption is sometimes called *energy analysis*. This chapter considers the energy flows, embodied energy patterns of webs, and ratios for evaluating energy quality. We begin with definitions of one usage (Odum, 1975a, 1976b, 1978, 1981; Odum and Odum, 1976, 1981; Odum et al., 1982) and then introduce other approaches.

EMBODIED ENERGY

As energy flows through webs of successive work transformations, it changes in form, concentration, and ability to feed back and produce amplifier effects. Whereas the actual flow of calories decreases as energy is used and dispersed, the quality of that energy can be said to increase. The flows become either very concentrated or very high in information content, in either case capable of controlling and causing work that would not be otherwise possible. The special transformations develop special quality. Energy used in developing energy of higher quality is embodied energy. It is the required energy contribution (see explanations in Chapter 2 also).

Whereas energy is a measure of ability to do work when energy flows of the same type and quality are being compared, heat equivalents are not measures of their value, cost, or ability to do work of different types. For example, one can compare heat releases of automobiles carrying different loads and by this measure their work, in this case the same kind of work. However, the heat content in flows of sunlight, water, coal, electricity, salmon, human services, and books do not represent their value, cost, or ability to cause work because these flows are each of different quality.

Energy Transformation Ratio

The simple idea of work using some energy to transform some other energy to higher quality is given in Fig. 14-1. The energy of one type required to generate a flow of another type is the embodied energy of type A required for type B. In Fig. 14-1 the energy embodied in outflow B is 100 calories of type A. The ratio of input (100) to output (10) is 10.0 and is a measure of the energy required for the transformation. It is the *energy transformation ratio* (quality factor) that relates one type of energy to another. It measures embodied energy of one type inherent in another. The energy transformation ratio is expressed in units of calories per calorie. The reciprocal is the efficiency $10/100 = 10\%$.

In Fig. 14-1b energy is stored, some being dis-

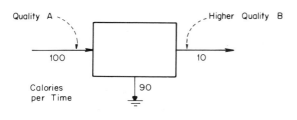

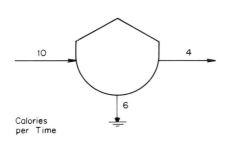

FIGURE 14-1. Energy transformation ratio, the energy used for transformation to another form of energy: (a) definition; (b) energy transformation through the process of storing energy.

TABLE 14-1. Embodied Energy Equivalents (Energy Transformation Ratios)

Type of Energy	Global Solar Calories per Calorie
Solar energy at earth's surface[a]	1
Winds[a]	315
Gross photosynthesis[a]	920
Coal[b]	6,800
Tide[b]	11,560
Water[a]	
Rain potential over land, 875 m	3,870
Chemical potential energy over land	6,900
Potential organized in rivers	10,950
Electricity[b]	27,200
Human service in the United States[c]	815,000
Work of land uplift[a]	1.5×10^{12}

[a] Based on energy web shown in Fig. 26-3. For calculations, see original source (Odum, 1978). See also Odum et al. (1982).
[b] Global sunlight contributes to land processes by providing wind, rain, waves, and so on. Using ratio of global area to land area, energy transformation ratios based on direct sunlight were converted to global equivalents by multiplying by 3.4. For example,
(2000 direct solar equivalents) (3.4)
= 6800 global solar calories/Calorie
[c] Gross National Product multiplied by global energy/dollar ratio of the United States and divided by population.

persed because of the inherent depreciation of any storage. The energy emerging on the right is of higher quality, having been transformed by storage from a more variable flow to a steadier one, capable of doing more in control actions.

Examples of energy transformation ratio are given in Table 14-1, arranged in order of embodied energy.

Types of High-Quality Energy

High-quality energy may take many forms. Some are concentrations of actual energy, such as high temperatures of a furnace or the cold of a refrigerator. Some are structures of large size such as a pyramid or a skyscaper. Others are tiny information-containing objects such as genes, computer programs, and political symbols. What these various items have in common is the large amount of energy used in their generation and the large amplification effects they may have. Land has high values estimated by the erosion rates (Judson, 1968) balanced by isostatic readjustment. Water carries several available energies—the Gibbs free energy of its chemical constitution, its geopotential, and its thermal gradient potentials (Odum, 1970).

Embodied Energy as a Measure of Value

An energy theory of value is based on embodied energy. If items and flows have value because of the effects they can exert on a system, and if their abilities to act are in proportion to the energy used to develop them (after selective elimination of those that do not), the value is proportional to the embodied energy in systems emerging from selection process. The energy transformation ratio, by giving the embodied energy per unit of actual energy, provides an intensive factor for value in the way that temperature is an intensive factor for heat. Ultimately, embodied energy may measure value because it measures the potential for contributing effects to maximize power and ensure survival. Those who survive regard that as valuable.

Thermodynamic Limits for Energy Transformation Ratios

Theory presented in Chapter 7 suggested that there is a rate that maximizes the transfer of power and that this rate is one that evolves under competitive conditions of real systems. The most efficient energy transformation that is possible with maximum power is the one that is both competitive and most transmissive of energy. The energy transformation ratio of the most competitive system at maximum power is at the inherent thermodynamic limit for conversion. Under these conditions the energy transformation ratio measures the inherent requirement of one energy to generate another. The theory suggests that any other ratio will either be less efficient or too slow to compete.

It is not difficult to observe and measure energy transformation ratios, but whether the observed ones are close to the inherent thermodynamic maximum possible is not easily known. We sometimes assume that the ratios observed in ancient systems with millions of years of operation like many in the biosphere are good numbers with efficiencies (at maximum power) not likely to be exceeded. On the other hand, energy transformations of new industrial processes may well be much less efficient when their systems are first started compared to those after years of competitions with trial and error in efforts to improve efficiency of the processes. Often, technological advance is the hidden application of additional high-quality energy and is not really an improvement in efficiency.

Examples of energy transformation ratios are given in Table 14-1, including some that may be near the thermodynamic maximum and others that are derived from single observations and may be superseded when more efficient ones (at maximum power) are found. An observed energy transformation ratio is a useful descriptive index of a system; an energy transformation ratio suspected as being near the thermodynamic limit helps to describe inherent relationships of kinds of energy.

SPACE EQUIVALENCE OF TIME IN CONCENTRATING ENERGY TO HIGHER QUALITY

Energy may be concentrated spatially by using some of it in the process of concentration, or it may be concentrated by adding to some storage in the same place over time. In 10 units of time, one may develop the same concentration as converging from a ten-times larger area. If addition is to be equivalent to converging, time used in one concentrating process is equal to geometric concentration factor in the other. There is a space–time equivalence.

ACTUAL ENERGY AND EMBODIED ENERGY EQUIVALENTS IN WEBS

To work with energy quality requires keeping two kinds of energy unit clearly straight and separate: (1) actual energy in degraded heat equivalents (calories), which is the usual measure of energy; and (2) embodied energy equivalents that are the energies required of a reference type (e.g., sunlight) to generate the type of concern. To keep these straight, three diagrams are suggested as in Fig. 14-2. First, place numbers for flux of the heat calorie equivalents on the pathways. According to the first law, these should sum at each junction and at each storage where the inflows equal storages plus outflows.

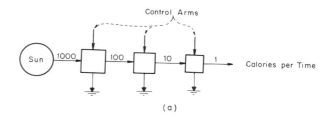

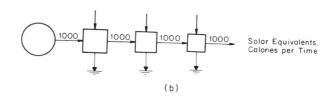

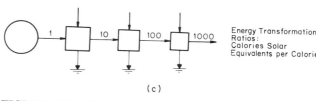

FIGURE 14-2. Concept of embodied energy and quality factors in a chain. Energy of control arm is omitted, which is correct procedure if it is a by-product of the flow.

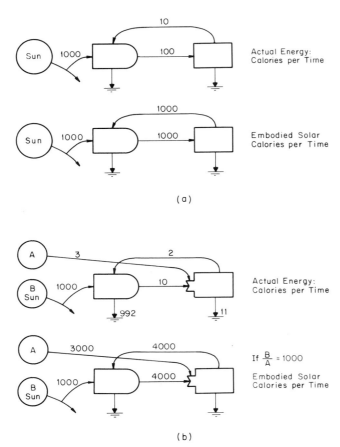

FIGURE 14-3. Evaluation of embodied energy equivalents in closed loops: (a) one source; (b) two sources. Source A has an energy transformation ratio of 1000 calories of B equivalent to one of A.

At steady state, inflows equal outflows. Often steady-state diagrams are used for energy estimations. These are "first law" diagrams if all values on the pathways are given in actual calories.

Figure 14-2a gives an example of an energy chain with actual energy values shown. They decrease as energy is used so that the last pathway has the least actual energy (heat equivalents) that would be regarded as negligible energy flow in some usages.

The second diagram (Fig. 14-2b) has the solar equivalents required to generate the flow of each pathway. If the system is operating at maximum power and minimum possible waste, each pathway is essential to the system and cost directly or indirectly 1000 calories per unit time to generate.

The ratio between the solar equivalents required (the inflowing amount at the source) and the heat equivalents of the remaining flow is given in Fig. 14-2c. These numbers are energy transformation ratios and are a measure of the quality of the flow. Notice that they increase as energy is converged and transformed. If the control arms were generated as a nondiverting by-product, they need not be evaluated as to their extra energy contributions.

One Source Supporting an Equal-Value Loop

An important reference case is the closed-loop design (Fig. 14-3) in which producers support a consumer whose sole output is a feedback, aiding the production process. The energies embodied in all pathways are the same, all traceable to the same source. Since the embodied energies are the same within the flows of the loop, it is called an *equal-value loop*.

Servizi et al. (1963) correlate population of decomposers and rate of decomposition of Gibbs free energy of the organic matter in wastes.

Evaluating Embodied Energy and Transformation Ratios from a More Complex Energy Web

Figure 14-4 gives an energy diagram of a web running on one source. In Fig. 14-4a the energy flows through transformation processes in which energy is used to generate flows of higher quality downstream. These feed back to the left, interacting to make the processes go. From left to right the flow rates decrease, but the quality increases. This is a first-law diagram, and all inflows balance outflows at any point.

Next, we ask how many calories of the type of the source such as sunlight are required for each pathway. Since each pathway is required for every other pathway and each pathway is a by-product of every other pathway, the energy required in units of calories of the type entering from the source is the same for each pathway. Figure 14-4b is an embodied energy diagram where all numbers are embodied Calorie equivalents of the same quality.

In Fig. 14-4c are plotted the quotients obtained by dividing embodied energy equivalents from Fig. 14-4b by the heat calories flowing from Fig. 14-4a. These ratios are, therefore, the energy transformation ratios. They indicate the energy at the start required to develop the energy flow of the quality flowing. The procedure is simple and clear when there is only one primary source, as in this case, or in the general flows of energy through the renew-

Energy Quality and Embodied Energy

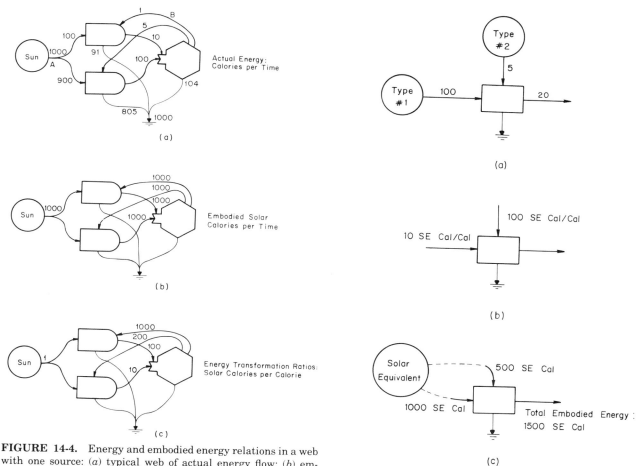

FIGURE 14-4. Energy and embodied energy relations in a web with one source: (*a*) typical web of actual energy flow; (*b*) embodied energy in solar equivalents; (*c*) energy transformation ratios calculated by dividing values in part *b* by those in part *a*.

FIGURE 14-5. Evaluation of embodied energy of a process with two sources each of a different quality: (*a*) actual energy in heat equivalent Calories per unit time; (*b*) energy transformation ratios for these two kinds of energy derived independently; (*c*) embodied energy in solar equivalent Calories obtained by multiplying number in part *a* times those in part *b*. Output is sum of inputs, both expressed in Calories of the same quality.

able resources of the biosphere as shown in Fig. 26-3, where most of the energy is from the single solar source.

Embodied Energy in a Process with Two Independent Energy Sources

When two input arms of a process originate from two different energy sources instead of being by-products ultimately from the same source, the embodied energy contribution to the transformation is the combination of the two. If both are of different quality, they cannot be added since their interaction involves the amplification of one by the other.

However, if both are expressed in embodied energy equivalents of the same type of energy, such as in calories of solar equivalents, they can be added to ascertain the ultimate solar energy requirement for the process (e.g., see Fig. 14-5). The mechanics of making the calculation consists of (1) writing the actual energies on the diagram and (2) using energy transformation ratios from previous tabulations to multiply both to convert to embodied energy equivalents of one type (Fig. 14-5*c*). Then the inputs may be added.

If one has a very local view of a process, one does not know if an input arm is a feedback (or feedforward) by-product of the same source or an input involving a different source entirely. Energy analysis calculations are difficult without an overview diagram of the whole system.

Evaluating Energy Transformation Ratios of a Local Process

Whereas energy transformation ratios may best be estimated from the world web as already described, ratios may also be observed for local processes without knowledge of the whole web. Consider the process shown in Fig. 14-6, where C is an input of high-quality control energy. If arm C is believed to be wholly a by-product, it may be ignored in the calculation of energy transformation ratio as in Fig. 14-6a.

However, if arm C is from a separate source, it is evaluated first in actual energy units (Fig. 14-6a). Then energy transformation ratios determined elsewhere are used to convert actual energy flow to embodied energy equivalents of either type A or type B. Then flow of input C is added to A (Fig. 14-6b) to get the total embodied energy of type A for generating type B.

An alternate way is in Fig. 14-6c, where energy equivalent B' for input arm C is subtracted from the embodied energy output of quality B. The energy transformation ratios in Figs. 14-6b and 14-6c are larger than in Fig. 14-6a because more energy (that of two sources) is required than in Fig. 14-6a, where part of the work is done with by-products.

Closed Loop Supported by Two Sources

When there is a closed loop supported by two sources of differing quality, the flows around the loop have the embodied energy of both input arms since both are necessary (e.g., see Fig. 14-3b). The first law diagram is given first, followed by a diagram of embodied energy of solar quality that was calculated by multiplying the energy type A by its energy transformation ratio from solar energy B.

USEFUL RATIOS AND TERMINOLOGY

Energy Amplifier Ratio

The effect that a flow of energy has is not an inherent property, but depends on the energy flows with which it interacts. However, it may be reasoned that surviving systems develop designs that receive as much energy amplifier action as possible. The energy amplifier ratio is defined in Fig. 14-7a as the ratio of output B to control flow C, both expressed in actual energy (heat-equivalent calories). The ratio is 10 (Fig. 14-7b).

Figure 14-5a illustrates an interaction between energy flows of different quality in which the high quality one (type No. 2) is amplifying the lower-quality one (type No. 1). A measure of this energy effect is the ratio of output of actual energy (20 cal) in equivalent heat calories to actual input energy flow also in equivalent heat calories. In Fig. 14-5a this is 4.

A property of the simple closed loop with one energy source is the equality of energy transformation ratio through the consumer and the energy amplifier ratio of the feedback's effect (see Fig. 14-7c). The energy used to develop feedback is reinforced by the feedback.

Theory suggests that in surviving systems, the *amplifier effects are proportional to embodied energy*. Full empirical test of this theory remains for the future. Such evidence will help confirm whether em-

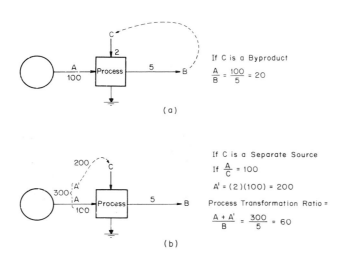

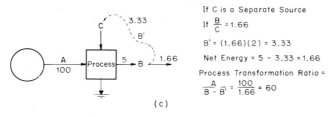

FIGURE 14-6. Estimation of energy transformation ratios for locally observed process: (a) transformation process in which C is a by-product; (b) C evaluated in Calories of type A and added to input to obtain A; (c) C evaluated in Calories of type B and subtracted from output to obtain B.

Energy Quality and Embodied Energy

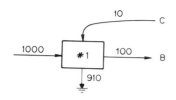

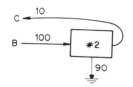

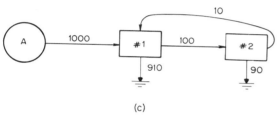

FIGURE 14-7. Energy amplifier ratio and energy transformation ratio: (a) definition of energy amplifier ratio for flow C; (b) energy transformation ratio for generating C from B; (c) special case of closed loop where energy amplifier effect of C on B equals energy transformation ratio of C from B. Energy flows are actual Calories (heat equivalents).

bodied energy is a measure of value. These ratios are not independent of low-quality inflows that are also controlled by the feedback.

The amplifier is a special case of sensitivity response.

Ratios of Embodied Energy Flow of the Same Quality in Processing an Energy Source

A very common element of energy webs given in Fig. 14-8 is a three-arm transformation process involving processing of an external energy source under control of a feedback flow. The three energy flows are the resource inflow, feedback, and output flow. There are three useful ratios often calculated

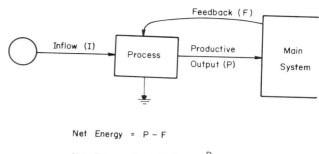

FIGURE 14-8. Definitions of energy ratios concerned with processing sources. The above quantities are calculated after all flows are expressed as energy flows of the same quality by multiplying actual energy flows by energy transformation ratios.

from the three-armed diagram as defined in Fig. 14-8. After the actual energy flows are converted to embodied energies of the same type, these flows may be compared by using ratios.

The *net energy yield ratio* is the ratio of the process production rate to the feedback energy flowing from the main system. It is a measure of the strength with which a source may contribute energy to an economy. For example, the yield ratio is 1500/500 in Fig. 14-5 and 5/3.33 in Fig. 14-6c. Whereas the energy amplifier ratio uses actual calories, the net energy yield ratio uses embodied energy of the same type. The *energy investment ratio* is the ratio of the feedback energy flow to the resource inflow that it helps process (e.g., 3.33/100 in Fig. 14-6c). The *energy added factor* is the ratio of production generated per unit of low-quality resource processed (e.g., 1500/1000 in Fig. 14-5c). It is a measure of the effectiveness of the resource in attracting high-quality matching energy.

Traditional Ecological Terminology for Gross Production, Net Production, and Net Energy

An energy transformation unit is shown in Fig. 14-9 and used to identify customary ecological terminology. The output of an interaction process is gross production G and is measured by the flow of the carrying state variable (e.g., carbon), by the heat calories in the energy upgraded, or by embodied energy of the production flow.

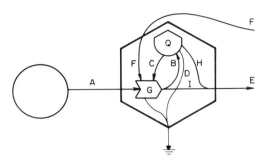

FIGURE 14-9. Ecological definitions of production and net energy usually calculated in actual energy units: gross production G; gross production stored B; overall net production in export (yield) E; net production in gain $dQ/dt = \dot{Q}$, net production in gain and export $\dot{Q} + E$; overall net production including feedbacks $E - F$ (net energy if in units of same quality); net production before storage $B - C - F$; assimilation A, F; export from storage H; directly exported production I.

Some of the gross production goes into storage Q, and of this some depreciates D, and some is fed back as part of the interaction C. Some of the production is fed further downstream (pathway E) to other units with their storages and feedbacks. At any stage downstream from the gross production, one may measure the remaining flow and call it *net production* for that point. Net production is the production remaining after some processes necessary to the production are subtracted. Because the definition depends on the place in the network and because it is also very dependent on the time interval chosen where the production is varying, net production is very ambiguous. Clarity requires that an accurate energy diagram with the kind of net production that is meant be defined by designating the pathway (see Fig. 9-2 and Fig. 14-9). Traditionally in ecology, the between-unit feedbacks F of high quality have not been included in net production measurements, even though they are essential and require very high energy. Another kind of net production is net rate of gain in storage (net growth $\dot{Q}$). Figure 14-9 gives some alternate kinds of net production. See Olson (1964).

The concept of net production in ecology has usually been calculated by use of actual energy units, whereas the net energy concept defined in Fig. 14-9 is calculated by using embodied energy units of the same quality. This is the one used elsewhere in this chapter. In deciding whether the energy yielded at a point is greater than the feedback, since the two are of different quality, they must be converted into energy units of the same type, comparing embodied energies (e.g., see $E-F$ in Fig. 14-9).

Many efficiencies may be calculated by using different ratios of pairs of flows as in Fig. 14-9. For example, efficiency of gross production is usually calculated as G/A. The efficiency of net conversion of the yield in ecology and industry is often calculated as E/A. These are ratios of actual energy flow.

SELF ORGANIZATION OF ENERGY WEBS

Uses of Net Energy

When a source yields net energy beyond its inflow process, this energy is available to further maximize power in the system in several ways shown in Fig. 14-10:

By causing downstream growth so that feedbacks are increased and more energy is processed from the same source, if that source will support increased pumping (Fig. 14-10a). The effectiveness of this alternative may be judged from the net energy of the additional increment of pumping, that is, the marginal effect (see Figs. 8-8–8-11 and Figs. 23-10–23-11).

By helping process a second source that may have a lesser net energy when considered alone (Fig. 14-10b). The effectiveness of this alternative is evaluated with the energy investment ratio that indicates the additional embodied energy drawn in for the amount fed back to the processing.

By exchanging net energy in trade for commodities of higher- or lower-quality energy so that the net effect is an increase in embodied energy, increased amplifier effect, or increased efficiency (Fig. 14-10c). The effectiveness of the exchange alternatives is evaluated with the embodied energy ratio of the exchange where both are expressed in units of the same quality.

By supporting new structures with special characteristics that increase efficiency and conservation of energy use. Examples are storages, diversity, and control systems (Fig. 14-10d).

Net energy may be regarded as temporary because its energy that is not fed back is wasted until it generates more feedback. Surviving systems develop structure and uses for their net energies after which there is no net energy in the system considered overall. In a complex web, net energy in one part of the web subsidizes other processes elsewhere in the web.

Energy Quality and Embodied Energy

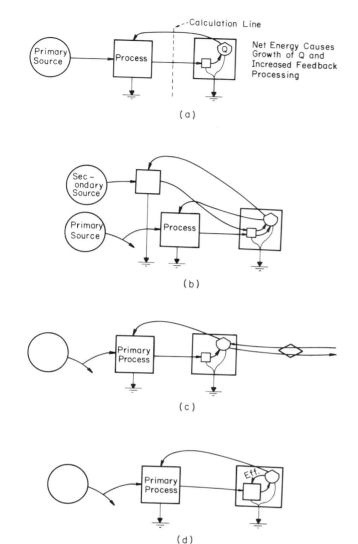

FIGURE 14-10. Alternative uses of net energy: (a) for growth that facilitates more processing from the source; (b) for subsidy to processing from a secondary source with less net energy; (c) for exchange for more energy; (d) for mechanisms of conservation and efficiency when other sources are not available.

general with services, recycles, and chemical compositions usable throughout. In webs of the economy, electricity, human services, computers, information, and so on are highly flexible and may be fed back for amplifier effects and controls throughout the web.

Because of the varying flexibility of energy with position of the energy chain, the divergence of production products in the lower part of an energy web (on the left in the diagrams) tends to resemble the left column in Fig. 14-11. Each diverging by-product carries the same embodied energy. In systems that maximize power, all by-products are fed back to have similar amplifying effects. There is no waste.

In the terminal flexible part of the web, energy flows produced are general and when diverged, represent a dividing of the embodied energy. Notice the different way of indicating pathways. This important principle allows use of money to evaluate embodied energy of high-quality feedbacks such as human labor in Chapter 23.

The manner in which energy is transformed successively in chains and webs to higher qualities with special amplifier abilities is discussed in more detail in Chapter 15.

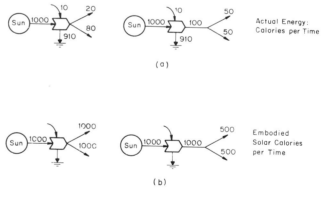

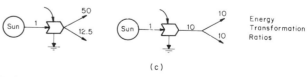

FIGURE 14-11. Comparison of diverging flows of unequal quality (on left) with a diverging flow of equal quality (on right): (a) actual energy flow; (b) embodied energy flow in solar equivalents; (c) energy transformation ratio; values in part b divided by those in a. Diverging flows that carry the same embodied energy should never have their values added since their sources of embodiment are the same.

Flexibility of High-Quality Energy

One of the properties of higher-quality energies (to the right in our diagrams) is its flexibility. Whereas lower-quality products tend to be special, requiring special uses, the higher-quality part of a web toward which the web often converges is of a form that can be fed back as an amplifier to many different units throughout the web. For example, the biochemistry at the bottom of food chains in algae and microbes is diverse and specialized, whereas the biochemistry of top animal consumer units tends to be similar and

Energy Matching and Investment Ratio

High-quality energy is most effective if used as an amplifier on lower-quality energy. If high-quality energy is used alone, it has nothing to amplify (see Fig. 14-12b); it is used for low-quality purposes. Thus *in surviving designs a matching of high-quality energy with larger amounts of low-quality energy is likely to occur*. Low-quality energy flows thus have the potential for attracting high-quality energies, maximizing useful power flows. Low quality can generate its own feedback as in Fig. 14-3 or 14-4, but if another source of high-quality energy is attracted, it may augment the self-generated feedback. However, with larger amounts of feedback, lower-quality energy becomes in shorter supply. When either interacting flow is limiting, resources generate less then their potential. A competitive system operates with its interacting inflows equally limiting. See derivation from Costanza (1979).

The energy investment ratio has been defined as the ratio of high-quality feedback energy to low-quality energy flow with both expressed in embodied energy equivalents of the same quality (see Fig. 14-8). The investment ratio that develops may be determined by that available to competitors. A system that has a low ratio can attract energy away from a system with higher ratio since power is maximized in this way.

Unbalanced Arms Input to Production

Where there are two sources of different quality such as fossil fuel and sunlight, the high-quality one may be in excess of what is required for best matching by low-quality energy flow. Since sunlight can generate its own matching high-quality energy, the addition of more high-quality energy preempts the control and increases the production some, but with less efficiency because of diminishing returns. See Fig. 14-13 for the relationship of increasing high-quality energy relative to matching lower-quality energy. The upper graph shows increasing production P with increasing investment feedback, but at a declining rate. In the left section the feedback is developed from its own net energy. Beyond an in-

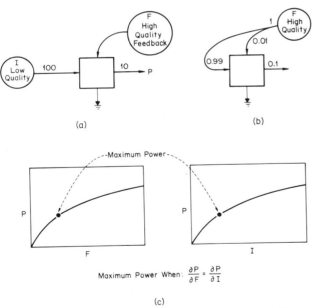

FIGURE 14-12. Feedback and matching of high-quality energy: (*a*) good matching of high-quality amplifying for lower quality; (*b*) poor use of high quality energy for low-quality purpose; (*c*) maximum power of the larger systems facilitated by good energy matching with neither high- nor low-quality energy in relative excess.

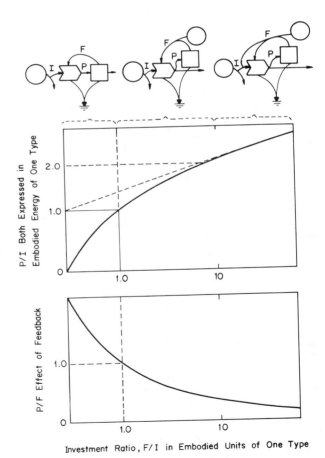

FIGURE 14-13. Effect of increasing feedback energy on production and efficiency.

vestment ratio of 1, an outside energy source is required to increase F. When F is very large, it has no low quality to match it and uses itself for low-quality purposes.

The lower graph shows the amplifier effect. Below F/I equal 1, the production P has a high amplifier action and is net energy yielding (left model). With more high-quality energy, production increases, but more slowly and with less potential than if the high quality were better matched. At very high excesses, the high energy can add further by being used for low-quality energy, but this gives a very small efficiency.

Multiple Sources; Energy Signature in Embodied Energy Units of One Type

In Fig. 2-2 the set of incoming energy sources was described as an energy signature with sources arranged in diagrams in order of energy quality. In some fields these are called *boundary conditions*. The actions caused by source pathways crossing the boundaries are also called *forcing functions*. It was postulated that the more embodied energy, the greater the ability to control the system. By expressing the flows of the sources first in actual calories per time and then multiplying by energy transformation ratios relative to one type of energy, the energy sources are expressed in embodied energy flowing of the same type. For example, Kemp (1977) calculated the energy signature for a region on the west coast of Florida peninsula (Fig. 14-14a). In Fig. 14-14b actual energy was multiplied by energy transformation ratios expressing the energy signature in embodied energy in coal equivalents. The graph shows more similarity among sources when expressed in embodied energy, possibly suggesting self-organization of the system toward energy matching. Where high-quality and low-quality energies are flowing with similar embodied energy, the high-quality energies may be well matched to low-quality energies. How typical this is remains to be seen.

Increase in Value with Scarcity

Our culture is accustomed to increased scarcity being identified with increased value. It is one tenet of economics that human behavior is the root of the increased value of scarce items. However, a basic energy concentration reason may also have been re-

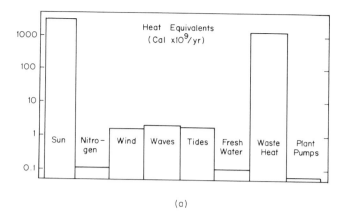

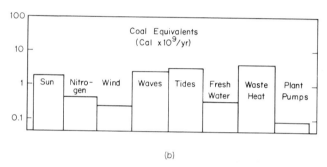

FIGURE 14-14. Signature of energy source flows for a region defined by a power plant: (*a*) expressed in actual energy, heat calories per time; (*b*) expressed in embodied energy, coal equivalents per time (Kemp, 1977). Area is that served by Florida Power Corporation.

sponsible. The programming of the flexible human attitudes may have followed as part of the necessary characteristics developed by surviving societies.

As an item becomes scarce, its position on the limiting-factor curve moves to the left with increasing marginal effect. It has greater value. A commodity must be scarce because there are greater energy requirements for its existence. More of the general energy resources of the system converge to a rare item simply because it is rare. The system retains a rare item, using it for greater-amplifier actions.

The energy quality of categories of material may, therefore, be considerably altered by the relative scarcity with which the item is used in the surviving system. This may be an energetic restatement of the concept of decreasing cost with mass production.

Ice in the tropics is a high-quality item, with high embodied energy flows leading to refrigerators required for its existence. Conversely, warm water at the poles is a high-quality item, requiring convergence of food chain energies for its existence.

In Florida cypress ponds are high-quality wetlands to which the water converges. However, in the Okeefenokee Swamp, cypress mounds are high-quality elevations to which organic processes converge in the common realm of water. Elevation is scarce in one and depression in the other.

Systems build hierarchies but may use one material for the convergence and high-quality control in one system that is the common low-quality item in another.

Energy Self-Organization and Maximum Power

Recognition of the different qualities of energy and the matching amplifier actions that maximize power helps to account for the webs observed in self-organizing systems of nature and the human economy. By feeding back net energies to amplify inflow to an interconnected web, a signature of energy sources can flexibly and automatically generate a distribution of growth and efficiency, augmenting energy uses and thus drawing more total power with all net energies used further to maximize power. See Fig. 7-8 for an example of the way feedback amplification helps to draw more energy. For different energy signatures, there are different resulting webs, but it may be speculated that all have the characteristics of a hierarchy of converging energy quality transformation feedback amplifiers and energy matching.

If the self-organization process utilizes all by-products in useful feedbacks, the emerging system obtains more energy transformation by making many products and using them than by simple energy processing of one product at a time. The reductionist tendency in human plans to solve energy problems item by item runs counter to the maximum power principle, requiring whole system plans so that by-products are fed back into the system. No flow is wasted in a well-organized system. The embodied energy in a so-called waste product is as great as the products regarded as the purpose for a process.

Self-Organization Derived from Second Law or Vice Versa

The idea of energy-quality control helps to explain why systems build assets and storage to survive. With higher-quality energy to feed back, the inflow can be pumped better than if the energy were left to flow by simple diffusion driven by its own gradient. Thus, to use energy more rapidly and degrade it more rapidly, the system must build order, provided that the energy levels are higher than the minimum necessary to keep up with the depreciation of the storages. This reasoning ties the second law and the maintenance of structure together as the same principle and explains why there are characteristic autocatalytic loops and other feedback arcs where energies are above some minimum levels. See Chapter 7.

ESTIMATION OF MAXIMUM POWER FROM ENERGY SIGNATURE

If one knows the energy sources (signature) to a system, how can one calculate the maximum power that a well-adapted, self-organizing system can generate from the sources? Estimation of the power flow is tantamount to evaluating the economic potential for the special case of human systems (Chapter 23).

Some approaches to the problem of estimating power of a web from its sources are suggested in the diagrams in Fig. 14-15. These abbreviations of the actual webs are *index models*, which retain some essence of the whole web, generating an index that may be convenient for practical purposes of estimating energy–economic potentials. Which of these or other indices will be most useful remains to be determined by empirical testing.

The following are ways of estimating potential production given the energy signature as illustrated in Fig. 14-15.

Measurement of the Dominant Productive Process (Fig. 14-15a)

Because webs involve loopbacks and interactions of each unit in other units, a central dominant productive process feeds back and receives flows interacting from most other units. Thus, measurement of the energy flow of the dominant production process and multiplication by its transformation ratio provides an integrated index of production potential. Adding in alternative productions that are secondary, whether upstream or downstream, would involve double counting. This system of counting would not work where the webs are not well organized.

Sum of Heat Dispersed (Fig. 14-15b)

When the web develops more interactions, amplifications, and energy capture, the total flow of en-

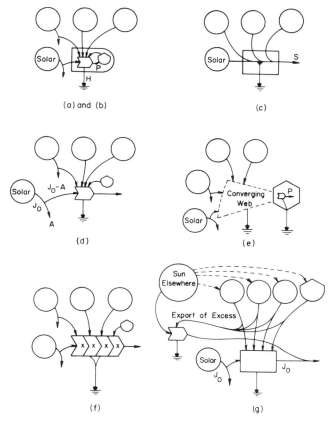

FIGURE 14-15. Methods of calculating the power generating potential of a combination of input energies of different quality and quantity: (a) measure observed production P; (b) sum energy dispersal H; (c) calculate sum S in units of embodied energy and divide by number of inflows; (d) measure unused low-quality flow A and subtract from J_0 as a measure of effectiveness of P; (e) measure production of top consumer and multiply by energy transformation ratio; (f) multiply driving forces; (g) evaluate flows in embodied solar equivalents and add that excess greater than J_0.

procedure gives greater weight to flows locally greater than the average in the biosphere. Geological flows are also expressed in solar equivalents (Odum, 1978).

Measurement of Total Low-Quality Energy Used (Fig. 14-15d)

The more higher-quality energy that is well used, the more low-quality energy that becomes incorporated as matching energy, the less unused low-quality energy (e.g., sunlight) that passes unused. Measurement of the albedo is a way to measure the effectiveness of the production process.

Measurement of Production of Consumers High in Energy Chain (Fig. 14-15e)

If webs are organized so that the top consumers represent a converging of embodied energy of all sources in useful production, evaluation of flow through top consumers can represent the total energy system. For example, conservationists often use large dominant species as indicators of the ecosystem in which they live. This index measures energy flow through terminal units of the web. The Gross National Product (GNP) is a measure of flows through the final demand sectors of the economy (see Fig. 23-19).

Multiplication of Driving Forces (Fig. 14-15f)

Where two flows of different quality intersect, output is often proportional to product of the driving or population force of each (see Chapter 8). Webs of interactions of many flows involve a series of interactions of this type, although usually interdispersed with storages and other units in the web. The Cobb–Douglas production function developed empirically from correlations in data is a product function, with exponents (see Fig. 19-6 and Fig. 23-16).

Addition of Excess Solar Embodied Energy (Fig. 14-15g)

Evaluate low-quality energy and add the embodied energy of the higher-quality flows that is in excess of the low-quality flow. As shown in Fig. 14-15g, these additions represent the energies that the high-quality excess could generate if exported in trading so as to have full matching of solar energy elsewhere, the productive products being returned in trade.

ergy increases in high- and low-quality flows alike, so that the total heat flow from the web in the combined heat sink also increases. This includes the energies attracted in by exchange and investments.

Sum of Energy Flows, Each Expressed As Embodied Energy of One Type (Fig. 14-15c)

Each flow is multiplied by its transformation ratio to convert flows into solar equivalents (or other common type of energy). These are added. Since the inflows are all by-products of the same solar processes of the biosphere, addition of converging components that diverged earlier involves multiple counting of the potential of the same original solar energy. Hence we divide the sum by the number of pathways added to obtain a mean solar embodied energy. This

Evaluation of Largest Inflow Only

Where several sources originated as by-products of one process, double counting can be avoided by evaluating the embodied energy of the largest one only. For example, rains, winds, and waves reaching land are by-products of the same global oceanic process. The embodied solar energy in the largest one includes the energies embodied in the others.

Use of Investment Ratio Between High-Quality Energy and Lowest-Quality Energy to Evaluate Production

After evaluating all flows in embodied energy of one type, calculate the ratio of the sum of high-quality flows to the low-quality flow and estimate production from Fig. 14-13.

Evaluation of a Change in Production Due to Change in External Sources

Changes can be evaluated by sensitivity analysis of a simulation model. Energy analysis calculations can also evaluate change in embodied energy and the effect it has on the total power as visualized through drains A–E in Fig. 14-16. Lavine, Butler, and Meyburg (1979), considering impact of development on the interdependent web of the environment, used the largest impact as an index of the total energy change since the impact site affects most other parts of the system through feedbacks. The idea is that lesser impacts will be less of a limiting factor than the largest impact and to add impacts would be to double-count the impact on the productive energy flow.

COMPARISON OF CONCEPTS OF EMBODIED ENERGY

In Figs. 14-3 and 14-4 each pathway was given the embodied energy equal to the total flow because each was necessary to the others. With this concept of embodied energy, the energy is not apportioned among pathways and feedbacks are not additive at their amplifier intersections. This procedure partitions energy only on diverging, flexible, high-quality energies that feed back (see Fig. 14-11). Compare this concept with others in Fig. 14-17.

Where energy transformation ratios are available from independent studies, they may be multiplied

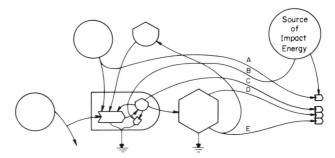

FIGURE 14-16. Production change estimated as the change in embodied energy of the maximum impact factor. Largest embodied energy diversion among A, B, C, D, E, and so on is taken as estimate of total impact on production.

by actual energy flows of a web to obtain embodied energies. These may differ from those in the web if it is more or less organized for maximum power than those of the source of data.

Embodied Energy Assigned by Matrix of Dollar Flows

Another definition of embodied energy uses the input–output matrix of dollar flows of the economy to assign the inflow energies to pathways (Herendeen, 1973, 1981; Bullard et al., 1976; Krenz, 1976). The inflowing energies considered are usually only the fuels (not the renewable energies of environment), which are assigned in proportion to the dollar flow, using a set of embodied energy per dollar coefficients (called energy intensity coefficients). The dollars that produce a contribution to final demand are multiplied by energy dollar ratios to determine the energy contributions to final demand. Input output coefficients are used. See Figs. 23-34. The calculation often leaves off the feedback from the final demand. See Fig. 14-17b, which shows how the web is abbreviated as compared with Fig. 14-17a. Thus energy that flows from fuel to B to C and then to A in Fig. 14-17a is omitted from calculations of energy reaching A according to the procedure in Fig. 14-17b. IFIAS (1974) omitted embodied quality and energy of labor in their attempt to write guidelines.

Additive Embodied Energy Partition Using Whole Web

Another definition of embodied energy illustrated in Fig. 14-17c was used by Costanza (1979, 1980). The whole web is used and environmental energies are included. To make a symmetrical matrix, degraded energies are connected to sources to make a closed, conserved circulation. Numbers are obtained by the matrix inversion method (see Fig. 23-34).

In this use of embodied energy the totals are partitioned and are additive, whereas in Fig. 14-4b embodied energy is not partitioned. The second and third concepts in Fig. 14-17 may underestimate the energy actually required for the flows.

Previous Use of Energy for General Evaluations

The concept of energy as a measurement of process related to heat was adapted to scientific usage from ancient vernacular usage about 1842 by Robert Julius Mayer, Herman Helmholtz, and Prescott Joule (Thirring, 1968; Cook, 1976). Marx (1867, see translation 1977) used human labor as a metric for measurement of useful works.

Expressions of mechanical work in terms of heat energy were demonstrated as the mechanical equivalent of heat established by Joule (Wood, 1925). The concept of work as an energy transformation was given by Maxwell (1877). The application of energy to measure total process spread into different scientific fields; for example, quantification of potential energy as ability to drive processes in chemistry (Gibbs, 1901). Energy transformations at infinitely slow rates were calculated to relate energy transformations necessary to changes of states (Carnot, 1824; Gibbs, 1901). The theory that energy could be a common denominator to measure all useful works was proposed widely with statements by (to mention only a few) Boltzmann (1905), Oswald (1907, 1909), Soddy (1912, 1922, 1933), and Cottrell (1955).

Boltzmann (1905) visualized a concept of embodied energy when he wrote: "If mental energy can be measured in such units that the amount developed was always exactly equal to the physical energy lost should we be entitled to speak of mental energetics."

Lotka (1922a,b) provided the maximum power principle as an extension of natural selection. Onsager (1931), de Groot (1952), and Prigogine (1955) formulated descriptions of linear open-system energy transformations. Energy analysis describing observed embodied energy in transformations was attempted in various ecological systems by Juday (1940) and Lindeman (1942).

In the depression of the 1930s a national organization, Technocracy, advocated various economic policies based on beliefs in an energy theory of value

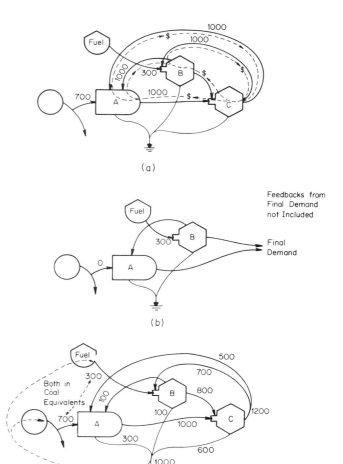

FIGURE 14-17. Comparison of alternative concepts of embodied energy: (a) embodied energy assigned to all pathways is the sum of all necessary energy required; (b) embodied energy is assigned in proportion to dollars but omitting feedbacks from final demand sector; (c) embodied energy is assigned in proportion to dollars including feedbacks and environmental energies (Costanza, 1979).

but one without energy quality or a useful role of money in stimulating energy flow (Scott, 1933; Parrish, 1933). Coon (1954) wrote: "Human beings convert energy drawn from outside their own bodies into social structure, and the greater the amount of energy consumed, all else being equal, the more complex the social structure."

Drawing diagrams of energy flow and evaluating the actual energies flowing has long been practiced in many fields. In industry such first law analysis accompanied by the study of efficiencies is called process analysis (Schmidt and List, 1962; Boustead and Hancock, 1979). Attempts to determine how much of an energy input was available for work was sometimes called second law analysis. Odum (1971) and Hannon (1973) joined those proposing energy as a standard of value, whereas many regarded value as a function of effective action or a property of human free choice. Different kinds of meaning were involved.

Other Energy Quality Concepts

The Carnot ratio (Fig. 7-9) is often regarded as a measure of the energy quality of heat gradients, measuring its ability to be converted into useful work of mechanical quality. The Carnot ratio estimates the efficiency at reversible condition (stalled). The efficiency of conversion of heat-gradient source to mechanical energy at maximum power is half the Carnot ratio (Chapter 7). This is a special case of energy transformation ratio.

Information is examined in Chapter 17; high- and low-quality information content is not distinguished by calculating bits of information, but is by embodied energy. Embodied energy per actual calorie increases with the bits per actual calorie, with the latter as a measure of quality of information (Tribus and McIrvine, 1971).

Essergy and Exergy

Another concept offered as a measure of energy quality is *essergy* (Gibbs, 1873; Evans, 1969). Essergy is a concept defined to evaluate the ability of energy sources and combinations to do work. It is the "available energy" to do work. It is the sum of the energies, each multiplied by the fraction of each energy that can be converted into *mechanical work*. For those energy types of quality lower than that of mechanical energy, it is a measure of theoretical efficiency. It does not consider efficiencies at maximum power. Energy flows of higher quality than mechanical work are not given greater value per calorie. *Exergy* is used for some of the components of the potential energy included in essergy. It is actual energy equivalents in units of mechanical work (Ahern, 1980; Sussmann, 1980).

Jorgensen and Mejer (1977) use an evaluated lake model to evaluate the buffering capacity of an ecosystem to phosphorus loading. This is related to the sensitivity concept: change in loading absorbed per unit change in state variable. This is useful as one measure of stability. The buffering capacity was found to be proportional to the stored exergy.

Jorgensen and Mejer (1979) use exergy, which is actual energy of one type, to evaluate structure. Embodied exergy may be more appropriate.

SUMMARY

In this chapter useful work was defined as those transformations of energy that contribute to maximum power and survival of the system because of the system designs. Energy transformations, by means previously selected under competition for the best possible efficiency commensurate with maximum power, define the inherent thermodynamic energy of one type necessary to generate another type. Ratios of energy of one type to generate another under these conditions are usable for predicting maxima. To compare the relative contribution of energies of different types to potential value, energies are converted to embodied energy equivalents of the same type by use of these transformation ratios.

Embodied energy was defined as a way to measure cumulative action of energies in chains and webs. Embodied energy provides an alternative theory of value, is useful for tracing sources, estimating net energy, determining relative importance of components, and comparing free items that are not covered by money. The study of relative importance of sources, feedbacks, and alternative designs is facilitated by energy diagramming followed by energy evaluations using transformation ratios and embodied energy. Such energy analysis has been applied to human problems and the energy crisis but is more generally applicable to all systems.

Embodied energy, energy quality, and the various ratios used to evaluate system configurations provide techniques of energy analysis and independent approaches to understanding value. Whereas real, self-designing systems develop complex webs to utilize the signature of available energy sources, each of different quality, apparently maximizing combined power, it is not yet clear which of a number of simplified models best predicts the combined production potential. Recognition of a scale of energy quality provides new principles of energy use such as matching of high and low quality, the requirement that control of hierarchies be cascaded, and the requirement that net energy be reinserted as a feedback. At present there is no consensus on embodied energy measures and their appropriate use. See bibliography by Frankena (1978). A manual for evaluating environmental embodied energy was provided by Odum et al. (1981b).

QUESTIONS AND PROBLEMS

1. Define *embodied energy* and *energy transformation ratio (quality factor)*.
2. How do embodied energy and energy transformation ratios vary along a typical web?
3. In counting the energy sources to a network, is it correct to add the external sources and the feedbacks or feedforward pathways within the web?
4. Discuss the hypothesis that an adapted system's by-products increase the power performance as compared to a network that leaves the by-products without a feedback or recycle.
5. In a diagram, show the meaning of embodied energy in a pathway and the energy effect of that pathway in its next intersection. Under what conditions are these equal? Discuss possible reasons why surviving systems may be selected so that these two properties are related.
6. What is net energy? Can net energy be evaluated with energies expressed in their regular heat equivalents? In what units is net energy evaluated? Is there net energy in a surviving loop in which all consumer work is fed back to pump in the energy source of the producer?
7. If two different by-products come from a process, do they have the same embodied energy? Do they have the same quality?
8. In an energy diagram, distinguish in concept and in pathways between the divergence of an output energy flow all of the same quality and two output flows of heterogeneous nature that have the same embodied energy but different energy quality.
9. Draw a web and place steady-state energy flows on the pathways to form a first law diagram; then put embodied energies on the same diagram; and finally, write energy transformation ratios on the diagram.
10. If two sources contribute to the same loop, what is the embodied energy within the loop?
11. What theories relate energy quality to ease of transport? To ability to amplify?
12. Does the energy quality of a flow change by passing through a storage tank? How do you calculate the change? In what way does the output of the storage have superior control properties usable toward adaptation and survival.
13. What is disadvantageous about using high-quality energy in places of low-quality energy? How is very high quality energy best coupled to the base of a long chain of energy transformations? Explain why energies develop more power in interaction than in separate use.
14. What is the ultimate use of net energy that is competitive?
15. Define gross production, growth rate, yield, net yield over control feedbacks, net production of the interaction production process, net production of the whole autocatalytic unit including its storage and its feedbacks from outside, net production including yield, and growth minus feedbacks from outside.
16. What is the principle of matching energy between feedbacks and lower-quality energy sources? How do lower-quality energy flows enable prediction of investment of higher-quality energy feedbacks? How is this a carrying capacity principle? If flow of high-quality energies into a limited area is increased, what is the shape of the curve of production that results (where the low quality renewable energy in that area is source limited and constant)?
17. What is an energy signature? How can this be expressed in embodied energy of one type?
18. Give several ways for estimating the maximum power flow to be obtained, given the energy signature expressed in calories of the various types of incoming energy.
18. Distinguish between different concepts of value, effect, embodiment, marginal effect, and human choice. Explain how these may be interrelated.
19. What is the Carnot ratio? How is it calculated? How is it used to estimate energy flow of mechanical quality from heat gradients? What factor is used to convert reversible efficiency to efficiency at maximum power (see also Chapter 7)?
20. What is a definition of useful work in terms of energy quality? What is exergy?

SUGGESTED READINGS

Ahern, J. E. (1980). *The Exergy Method of Energy Systems Analysis,* Wiley, New York. 295 pp.

Bayley, S., J. Zucchetto, L. Shapiro, D. Mau, and J. Nessel (1977). *Energetics and Systems Modelling: A Framework Study for Energy Evaluation of Alternative Transportation Modes,* Institute of Water Resources. Constract Report 77-10. 173 pp.

Berndt, E. R. and B. C. Field, Eds. (1981). *Modeling and Measuring Resource Substitution,* M.I.T. Press, Cambridge, MA. 314 pp.

Boustead, I. and G. E. Hancock (1979). *Handbook of Industrial Energy Analysis,* Wiley, New York.

Bullard, C. W., P. S. Penner, and D. A. Pilati (1976). *Energy Analysis: Handbook for Combining Process and Input–Output Analysis,* Energy Research and Development Administration. ERDA 77-61 UC-13, 93.

Costanza, R. (1980). Embodied energy and economic evaluation. *Science* **210,** 1219–1224.

Curran, S. C. and J. C. Curran (1979). *Energy and Human Needs,* Wiley, New York. 330 pp.

Frankena, F. (1978). *Energy Analysis/Energy Accounting,* Vance Bibliographies, Box 229, Monticello, IL.

Fluck, R. C. and D. C. Baird (1980). *Agricultural Energetics.* Avi, Westport, CT.

Gilliland, M. (1978). Energy analysis and public policy. *Science* **189,** 1051–1056.

Herendeen, R. A. (1973). *The Energy Cost of Goods and Services,* Oak Ridge National Laboratories. ORNL-NSF-EP-58. 115 pp.

Herendeen, R. A. (1976). Use of input-output analysis to determine the energy cost of goods and services. In *Energy Accounting as a Policy Analysis Tool.* Prepared for the Subcommittee on Energy Research, Development and Demonstration of the Committee on Science and Technology, U.S. House of Representatives, 94th Congress, 2nd Session, 68-391. U.S. Government Printing Office, Washington, D.C. pp. 101–110.

Hoffman, E. J. (1977). *The Concept of Energy,* Ann Arbor Science, Ann Arbor, MI.

Krenz, J. H. (1976). *Energy Conversion and Utilization,* Allyn & Bacon, Boston.

Lavine, M. J., T. J. Butler, and A. H. Meyburg (1979). *Energy Analysis Manual for Environmental Benefit/Cost Analysis of Transportation Actions,* Vols. 1 and 2. Project 20-11B. National Cooperative Highway Research Program, Transportation Research Board, National Research Council, 122. 229 pp.

Myers, C. (1977). *Energetics: Systems Analysis with Application to Water Resources Planning and Decision-Making,* Institute for Water Resources, IWR Contract Report 77-6. 1–42 pp.

Odum, H. T. (1978a). Energy analysis, energy quality, and environment. In *Energy Analysis, a New Public Policy Tool,* M. Gilliland, Ed., Westview Press, Boulder, CO, *AAAS Selected Symp.* **9,** 55–87.

Odum, H. T. (1978b). Energy quality and carrying capacity of the earth. *Trop. Ecol.* **16** (1), 1–9.

Schnakenberg, J. (1977). *Thermodynamic Analysis of Biological Systems,* Springer-Verlag, New York.

Slesser, M. (1978). *Energy in the Economy,* Macmillan, New York.

Steinhart, J. S. and C. E. Steinhart (1974). *Energy: Sources, Use and Role in Human Affairs,* Duxbury, North Scituate, MA.

Wang, F. C., H. T. Odum, and P. C. Kangas (1980). Energy analysis for environmental impact assessment. *J. Water Resources Plan. Manage. Div.* **106,** 451–465.

CHAPTER FIFTEEN

Spectrum of Energy Distribution and Pulsing

Because systems of energy flow develop hierarchical webs, large flows of low-quality energy are transformed into and support small flows of high-quality energy as in Fig. 2-1. These energy distributions can be represented on graphs called *energy spectra* in which quantity of energy flow is plotted as a function of the energy quality. See Fig. 15-1, where an energy transformation chain generates a typical spectrum. Because of the similar behavior of their energies, spectra are similar for a wide range of systems from astronomic to atomic. In this chapter energy spectra are related to energy sources, chains in closed and open systems, maximum power selection, and pulsing.

HISTORICAL NOTE

Berry et al. (1976) quote Al-Muqaddasi from year 985, giving a theoretical scheme for hierarchy of human settlements and control. Aggregation of the food webs of ecology into a pyramid was done by Petersen (1918), Juday (1940) and Lindeman (1941), calling the stages *trophic levels*. This simple model has been the basis for general teaching in ecology of energetics, although the system being represented with the model is actually always a web.

COORDINATES OF SPECTRAL GRAPHS

Energy spectral graphs relating quantity and concentration have been long used in many sciences to describe the distributions of activity in systems in whole or in part. If the number of flows of each intensity is multiplied by the power per unit flow at that level, the product is power, and sometimes the graphs are called *power spectra*. Many power spectral graphs have energy quality implied, but rarely recognized explicitly. The horizontal axis may be in energy concentration or related variable and the vertical axis may indicate the quantity of energy of that type, sometimes called "probability density."

For the most general comparisons of systems, quantity of energy flow may be graphed as a function of the energy transformation ratio in units of one common form of energy on the X axis. The Y axis also indicates the total energy per interval of energy quality, which is the area under the curve for that interval. If the coordinates of Fig. 15-1 (energy per interval and energy transformation ratio) are both plotted on logarithmic scales, the line of bars is straight.

In another kind of "power spectrum" generated in time series analyses, variance is a function of frequency. Where energy quality can be related to frequency, these spectra may also reflect hierarchical energy patterns.

EXPONENTIAL THEORY OF ENERGY HIERARCHY

Energy Transformation Chain Maximizing Power from One Source

A chain of energy transformations such as that in Fig. 15-1 is an aggregated simplification of the usual web system of flows usually observed. Chains and webs are a form of system specialization that may be explained by maximum power selection. Although energy is dispersed in each energy transformation step, the system is benefited if the downstream consumer unit develops flows of special ability that feed

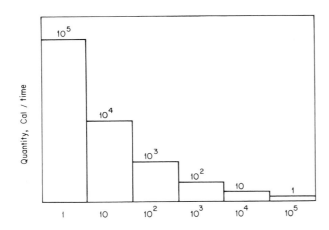

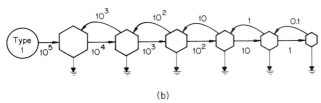

FIGURE 15-1. Typical spectrum of energy distribution in open systems with one source: *(a)* graph of energy quantity as a function of energy quality; *(b)* energy transformation chain, an aggregated simplification of webs (Odum, 1975).

back actions with as much or more energy amplifier action as energy was used in developing the feedback (see Fig. 7-8). The theory suggests that selection for maximum power generates energy transformation hierarchies. Since each energy transformation step requires utilization and dispersion of potential energy, the total energy flows decrease as one passes work from step to step. Conversely, the embodied energy per unit of actual energy increases down the chain. This is the value paradox of energy. *The less actual energy remains, the more embodied energy has been involved.* Adams (1975) presents theory and evidence that social power is generated by hierarchical convergences. Duckham et al. (1976) evaluated a graph of declining energy and a graph of accumulating embodied energy in the agricultural food chain from sunlight to human use of food (see Chapter 14).

Spatial Convergence

As shown in Fig. 15-2*a*, energy transformations down the chain involve an energy convergence in space from a more dilute form of energy in abundance to a scarce form of energy in concentrated form. The feedback of service from the high-quality unit back toward the lower-quality unit is spatially directed outward. For example, solar energy is converged into the bass or hawk, but the work of the bass or hawk is fed outward and distributed over the area from which its energy was derived. This makes sense with the requirement for maximizing power to feed back special amplifier actions to the source units.

Derivation of Exponential Spectrum from Energy Chain Model

Many spectra such as that in Fig. 15-3 fit an exponential function. When the percent decrease of Y is a constant times X, the curve is a declining exponential. The energy chain model provides for a step down in energy with each step to the right to higher energy quality and higher energy per unit. Thus, where energy is inflowing from the low-energy end on the left, the chain model generates an exponential curve where E is a measure of energy quality. If E is chosen as the logarithm of the energy transformation ratio, a generalized graph results for study of hierarchy. As shown in Fig. 15-3, exponential spectra are straight on semilog plot and also on double-log plots. Some observed spectra are given in Fig. 15-4.

Sometimes spectral graphs are drawn with discrete discontinuous steps as in Fig. 15-2. In others the energy spectra can be drawn as a continuous curve as in Fig. 15-3.

In many hierarchies, the concentration of actual energy can be identified as E in the exponential spectrum model (Fig. 15-3). This relation suggests a proportionality between actual energy concentration and logarithm of energy transformation ratio, both increasing with quality.

Examples of Energy Spectra and Chain Models

If the theory is correct, models of energy-quality chains can be recognized in all kinds of system that have long-surviving patterns of energy flow. High flows of low-quality energy on the left generate small flows of high-quality energy on the right.

The broadly spread source of raindrops converges into rivulets, small brooks, small streams, and so on, ultimately converging as major rivers. Position in

Spectrum of Energy Distribution and Pulsing

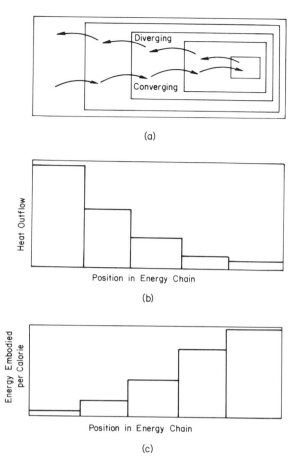

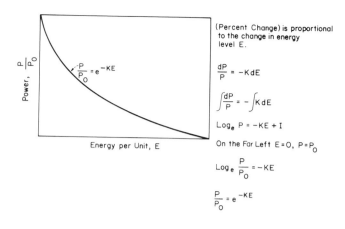

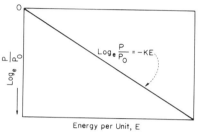

FIGURE 15-3. An exponential power spectrum and its derivation from premise of constant percent decline with energy step.

FIGURE 15-2. Energy hierarchical distributions for an open-system transformation chain based on one energy source (Fig. 15-1): (a) spatial convergence and diverging feedbacks actions; (b) decrease in actual energy; (c) increase in embodied energy per calorie (Odum 1975).

this hierarchy is the stream order in the Horton (1945) concept of stream classification. Figure 15-4a, given by Leopold and Langbein (1962), Langbein and Leopold (1964), and Leopold et al. (1964), shows that the energy in streams is an exponential power spectrum.

In Fig. 15-4b the patterns of earthquakes form a power spectrum with many small ones and only a few larger ones where more energy has been converged and concentrated.

Figure 15-5 shows the aggregated model for the flow of energy through plant cells converting low-quality sunlight to intermediate-quality adenosine triphosphate (ATP) and then into even higher-quality structure and informational storages with the genetic material deoxyribonucleic acid (DNA) as the end member. Adenosine triphosphate is sometimes called a high-energy compound in biochemistry.

Figure 15-6 on a larger scale models the convergence of energy in transformations of food chains from aquatic plants and detritus through small animals and fishes to wood storks (Browder, 1978). An unusual degree of energy convergence is brought about by the drying up of temporary lakes, causing fishes to be concentrated in pools available to storks.

HIERARCHIES IN OPEN AND CLOSED SYSTEMS

Hierarchical energy distributions are found in both open and closed systems, and similarities between the two may require some explanation. An open system and a closed system for water and its atmospheric vapor are compared in Fig. 15-7. In the closed system rates of flux into vapor balance those returning. No energy enters from outside, and there is no degrading of potential energy as there is no entropy change; the system is at equilibrium (see also Figs. 1-2 and 7-12).

The open system, however, although somewhat similar, has potential energy inflowing, driving the evaporation of water, some of which returns to complete a closed material cycle with energy being de-

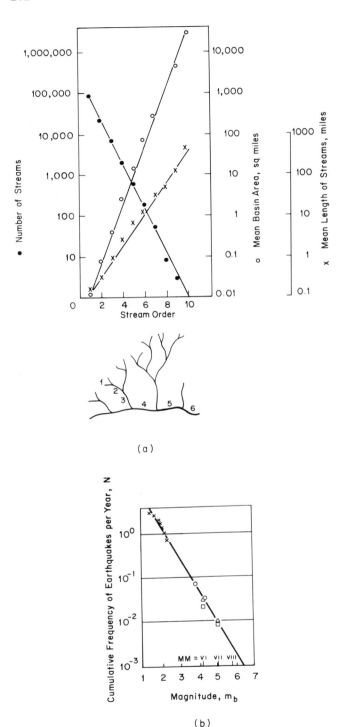

FIGURE 15-4. Energy hierarchy spectra: (a) energy spectrum of streams arranged by order of branching (Leopold and Miller, 1956); (b) energy spectrum of earthquakes redrawn from Aggarwall and Sykes (1978); m_b, Richter scale; mm, Marcali scale.

graded and entropy increased as the potential energy is used.

What is interesting is that both systems have similar states, with most of the water in liquid form and relatively little in the higher-quality vapor state. There is an energy-quality hierarchy in both open and closed systems, and they resemble each other in their states. Similarly, Sillén (1961, 1967) found chemical composition of the sea, an open system, to be near that predicted if the sea was a result of equilibrium. An explanation may be suggested. By holding the main mass of water in the low-energy state, the receptor for collecting incoming solar energy is maximized, thus maximizing power to the system. Also, less energy is dispersed in depreciation when most of the mass of the cycle is stored at zero relative potential energy.

Skewed Spectrum in Open Systems Where Low-Energy Zone Is Included

Populations of molecules exchanging momentum and energy are compared in Fig. 15-8. The population in Fig. 15-8a is receiving an outside energy source. Populations in Fig. 15-8b are in equilibrium. In both cases hierarchies of energy transformation develop, with the low-energy molecules interacting to generate fewer high-energy molecules. The closed-system hierarchy has an exponential energy distribution. The open system receives potential energy of slightly higher quality than the lowest-quality energy present, giving a maximum to the energy spectrum.

If the graph of energy distribution includes the range of energy quality below that of the incoming energy source, an asymetrical spectrum is found (see Fig. 15-9). The left-hand tail is the zone where energy is being degraded, dispersed, and removed from the system (Fig. 15-10a). It is degraded in percentage steps and is exponential (MacCready, 1953).

Many energy distributions resemble the skewed distribution typical of an open system with a single source of low-quality energy. For example, consider the distribution of rain, which represents the occurrence of storms of various sizes (Fig. 15-9).

Maximum Entropy Distributions

When potential energy capable of driving processes in a system is used up and there are no incoming energy sources, equilibrium results, as in the ex-

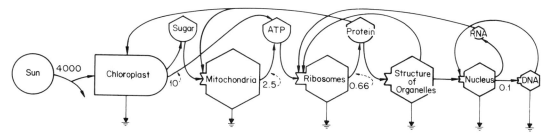

FIGURE 15-5. Examples of energy-quality chains. Energy in plant cells, kcal/m² · day.

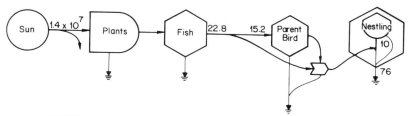

FIGURE 15-6. Energy chain to wood storks (Browder, 1978). Energy quality of Wood Stork:

$$\frac{14 \times 10^7 \text{ kcal solar radiation}}{10 \text{ kcal of bird}} = 14 \times 10^6 \frac{\text{kcal solar equivalents}}{\text{kcal bird}}$$

ample of water and vapor (Fig. 15-7a). Overall entropy increases while there are open-system conversions of potential energy into more degraded form, but when equilibrium (not to be confused with steady state) is reached, a state of maximum entropy is also reached. To calculate the maximum entropy distribution, one must locate the static state devoid of potential energy for further change.

In many distributions studies the maximum entropy state is calculated as a way of determining what states are stable, even though the situation involves open-system states. However, for reasons of power maximization just given, if open steady states resemble closed equilibrium states, there may be a reasonable basis for calculating maximum entropy distributions as a reference for examining open systems (see formulations of entropy and complexity in Chapter 17).

Energy Spectrum with Medium-Quality Source

When the energy source driving an open system is of medium quality, two exponential arms may develop. The arm on the left represents the stepwise decay

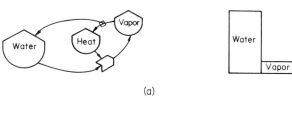

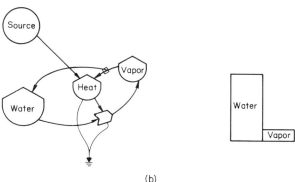

FIGURE 15-7. Similarities in energy distribution in open and closed systems for water and vapor: (a) equilibrium with most of mass in water phase; (b) open system of the earth's oceans and atmosphere with most matter in water phase.

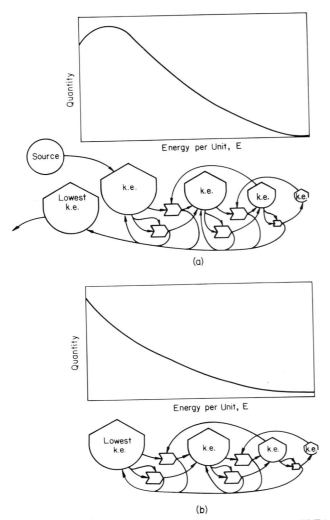

FIGURE 15-8. Comparison of models of kinetic energy (K.E.) chains and spectra for open and closed systems: (a) open system; (b) closed system.

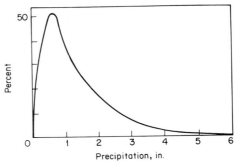

FIGURE 15-9. Example of energy spectra: summer rainfall in Kansas (Eagleman, 1976). Rainfall is a function of energy concentration in weather and thus an index of energy quality.

and dispersal of the energy to lower-quality states; the other arm on the right represents development of the hierarchical chain with transfers to higher quality. In the example of turbulence in Fig. 15-10a, the left arm of the spectrum is called the *Kolmogorov range*, where eddies decay into smaller and smaller ones and finally into viscous dimensions, with momentum dispersing into heat (Kolmogorov, 1941). The right arm represents upgrading transformations with decreasing energy and increasing quality that is also exponential as already discussed. This zone represents feedback of higher-quality energies interacting as amplifier actions on lower-quality energies. For example, eddies may help to process heat differences into mechanical work.

Hierarchy Poems

The convergence involved in chains is often described as a hierarchy. The similarity of concepts between energy flow in ecology and in fluid dynamics is evident in two poems, the first of which is a modified version of one written by Jonathan Swift[*]:

> The little fleas that do us tease,
> have other fleas to bite them.
> And these in turn have other fleas,
> and so *ad infinitum*.

The second excerpt is from L. F. Richardson (Hess, 1959):

> Great whirls have little whirls,
> that feed on their velocity;
> and little whirls have lesser whirls,
> and so on to viscosity.

The first refers to the upgrading right arm of the spectrum, whereas the second poem refers to the degrading left arm of the hierarchy spectrum, both with tendency to be exponential.

---

[*] Jonathan Swift (1733) on poetry:

> So, naturalists observe, a flea
> Hath smaller fleas that on him prey;
> And these have smaller fleas to bite'em;
> and so proceed *ad infinitum*.
>
> Thus every poet, in his kind
> Is bit by him that comes behind.

Spectrum of Energy Distribution and Pulsing

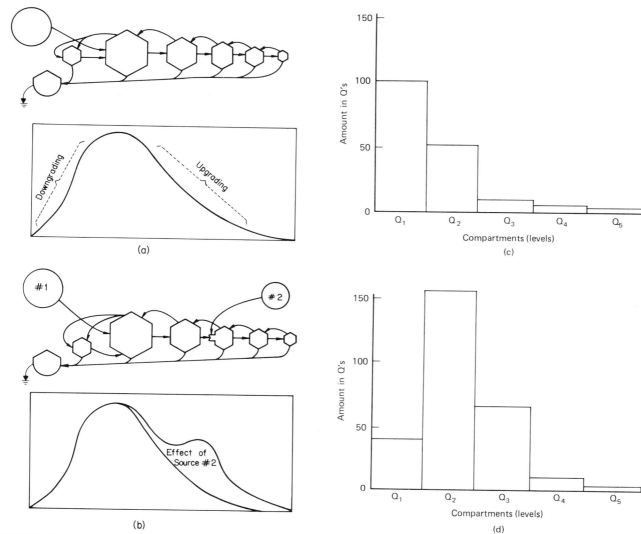

FIGURE 15-10. Energy models and energy spectra for open systems: (a) one energy source of quality lower than graph's origin; (b) two energy sources of different quality; (c) results of simulating a model of the type in part a (Brown, 1980) (see Fig. 15-1); (d) results of simulating a model of the type in b (Brown, 1980) (see Fig. 11-24).

Energy Spectra with Two Energy Sources

When there are two external sources, more complex double peaks result (Fig. 15-10b). For example, waves on the sea may have power spectra with dual peaks. Our world economy receives low-quality energy from sunlight and high-quality energy from fossil fuels, which shifted the peak. Brown (1980) simulated a spectrum of this type with a model in Fig. 11-24 similar to that in Fig. 15-10b with the result given in Fig. 15-10d.

If the patterns of energy inflow can be predicted, the spectra of energy may be predicted without actually knowing the details of the mechanisms.

NORMAL DISTRIBUTION OF PROPERTIES WHEN ENERGIES ARE SQUARE FUNCTIONS

In this chapter exponential energy distributions have been explained by the action of maximum power selection developing energy chains. As mentioned in Chapter 7, some carriers of energy have energy contained as their square. Stored water, electrostatic charge, and items with moving velocity are examples in which energy is stored as the square of their ordinary descriptive state variables. Let X represent the state variable whose square is proportional to energy. If power selection determines the spectra, it also determines the distribution of their state variables X as the square root. In Fig. 15-11, substitute kX^2 for energy E to obtain an equation for the distribution of the population of units according to the state variables X. The result is the equation for a normal distribution. By this reasoning, the so-called random distribution is found as a consequence of energetic processes rather than the cause.

The usual approach to the study of energetics of the molecular world starts with an assumption of random (normal) distribution of molecules and then derives the exponential distribution of energy. Our approach suggests that energy selection generates exponential energy distribution first and predicts normal distributions for any state variables where the energy is a square function.

GENERATION OF EXPONENTIAL HIERARCHIES BY SUCCESSIVE DIVISION

Mandelbrot (1977) generates hierarchical geometry by successive fractionation of surfaces and spaces. Exponential spectra result. MacArthur's broken-stick model for rank (see Fig. 18-18) also generates an exponential spectrum (Pielou, 1977). There is successive division of remaining segments, even though the spatial divisions are subject to random chance rather than by even division. This is comparable to the degrading range of turbulence where initial energy is successively divided into smaller eddies, generating an exponential spectrum. Thus mechanical algorithms and random divisions can simulate the subdivision of energy in energy chains that have dynamic adaptive roles in maximizing power and survival.

ENERGY ABOVE A QUALITY THRESHOLD

Many processes depend on the population of energy flows and units above an energy threshold. Figure 15-12 gives an energy spectrum with an exponential

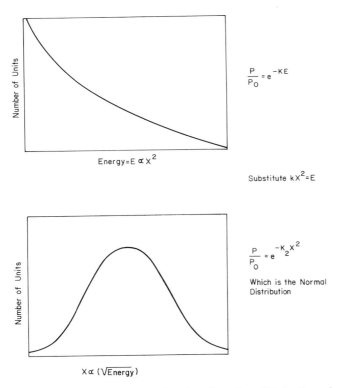

FIGURE 15-11. Diagrams showing Gaussian distribution of variables X that result when the energy E has an exponential distribution; (a) exponential energy spectrum, possibly explained by maximum power selection; (b) Gaussian distribution of X when related to velocity or other variable that is proportional to the square root of the energy.

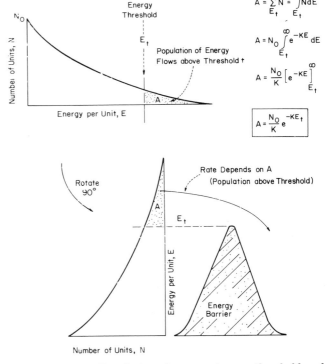

FIGURE 15-12. Population of energy above a threshold and the manner in which rates are determined.

distribution. This may be either a closed- or open-system distribution with complexities on the left shielded, so that the exponential model may apply to the zone of qualities of interest. The energy quality of the threshold E_t is of concern. For example, energies above the threshold may participate in processes with rates dependent on the energy flows at this level. In Fig. 15-12 the spectrum has been rotated so that the population of flows above the threshold can be visualized as passing over an energy barrier.

For example, in the atmosphere increased heating of the air raises the height to which air is pushed and increases the volume that can cross a mountain range. The high-quality energy is in the form of elevated air capable of doing special actions because of their height.

For another example, Strehler and Mildvan (1960) develop mortality theory for humans according to the population of energy stresses on people above the threshold of the human resistance. Medicine pushes the threshold to the right, and human senescence reduces it gradually with age (see Fig. 20-10).

Following a procedure used in statistical mechanics an expression for the population exceeding a threshold by integrating the area under the curve beyond the threshold is derived in Fig. 15-12. Since the integral of an exponential is an exponential, the expression relates threshold processes to the threshold E_t and also to the total incoming energy. Figure 15-13 shows the rapid increase in over-threshold population with increases in the total incoming energy.

Energy Minima for High-Energy Units

Many high-quality units have a minimum flow of high quality to operate. If there is less than this flow of high-quality energy, there is not enough base for any of these at all. Units whose role involves large size are examples. Tigers, grizzly bears, and other large animals or high-energy specialists such as some medical services require a minimum base for their support, one that generates enough high-quality energy. If energies decline, some high-quality functions drop out entirely. In Fig. 15-13, a horizontal line indicates the minimun energy for operating a unit of that quality. The point where that line intersects the exponential curve is the point beyond which high-energy chain functions requiring minima are cut off.

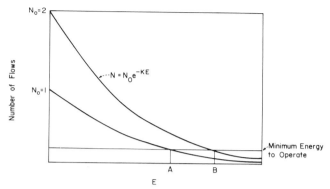

FIGURE 15-13. Effect of eliminating high-quality types with reduction of total energy. Reduction of energy by half eliminates high-quality units between B and A. Abscissa E is log of energy transformation ratio.

If the total energy incoming is reduced by half from $N_0 = 2$ to $N_0 = 1$, the chain of energy quality is shortened, with some units dropping out because the total quantity of energy exceeding the necessary quality threshold is insufficient to maintain a minimum for operation.

CHEMICAL HIERARCHIES

Maxwell–Boltzmann Distribution

The distribution of molecular energy in populations of molecules containing kinetic energy is an example of the exponential distribution explained in Fig. 15-3 on the basis of a kinetic energy chain at equilibrium and properties discussed in reference to Figs. 15-8b. The equation for this distribution is a special case of the general exponential in Fig. 15-3.

Eyring and Urry (1965) provide a simple derivation of the Maxwell–Boltzmann distribution by deriving the pressure distribution of the atmosphere as in Fig. 15-14. Here in rotated position (also see Fig. 15-12) the high-energy molecules are at the top and dependent on the number of lower-energy molecules feeding momentum into them. The potential energy in each level of the free atmosphere develops a balance between gravity pull and molecular outward pressure.

The number of molecules is proportional to the pressure. By substituting the gas law and integrating, one obtains the energy level achieved by compressing the molecules. (A similar integration was performed in Eq. (7-6) to find the logarithmic relationships of molecular populations to energy storage.) Conversion from logarithmic form to exponen-

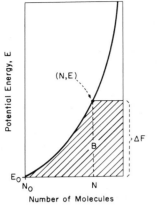

FIGURE 15-14. Maxwell–Boltzmann distribution derived using atmosphere case modified from Eyring and Urry (1965). Curves are representative of one temperature.

tial form gives the final expression for the distribution of population for the energy chain of molecules as a function of their potential energy.

Chemical Reactions

Most chemical reactions involve molecular collisions that must overcome energy barriers (zones of repulsive forces) to get to close proximity where stronger but shorter-range forces can bind the reactants. As already shown in Fig. 15-12, chemical reactions depend on the population of higher-energy molecules above the energy-barrier threshold. In chemistry the energy barrier is called *free energy of activation* ΔF^*. If the integration in Fig. 15-14 is carried out on the chemical form of the exponential distribution, the flow rate for some chemical reactions is found. From Fig. 15-14 for the energy barrier ΔF^*, we obtain

$$N = N_0 \exp\left(-\frac{\Delta F^*}{RT}\right) \qquad (15\text{-}1)$$

Integrating the area above the barrier as in Fig. 15-12, we have

$$A = \int_{\Delta F^*}^{\infty} N_0 \exp\left(-\frac{\Delta F^*}{RT}\right) d\Delta F \qquad (15\text{-}2)$$

$$A = -N_0 R \exp\left(-\frac{\Delta F^*}{RT}\right)\Big|_{\Delta F^*}^{\infty} \qquad (15\text{-}3)$$

$$A = N_0 R \exp\left(-\frac{\Delta F^*}{RT}\right) \qquad (15\text{-}4)$$

Where rates are proportional to these active molecules, the expression results for rate of reaction that matches one determined earlier empirically and often named after the originator, the Arrhenius equation:

$$J = kA \qquad (15\text{-}5)$$

$$J = kN_0 R \exp\left(-\frac{\Delta F^*}{RT}\right) \qquad (15\text{-}6)$$

Quantum Shells

The equilibrium distribution of quantum shells in matter is similar to the Maxwell–Boltzmann distribution, but with discrete stages. The shells are hierarchical and energy based, and the low-energy shells are abundant and converge to relatively few high-energy shells at higher quantum energy states (see Fig. 15-15). Quantum states are comparable to steps in the food chain.

BIOLOGICAL HIERARCHIES

Age Distribution as an Energy Spectrum

Among populations of organisms, age distribution can be represented on a graph that starts on the left

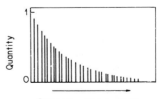

FIGURE 15-15. Energy spectrum for quantum levels in chemical molecular states (Nash, 1974).

Spectrum of Energy Distribution and Pulsing

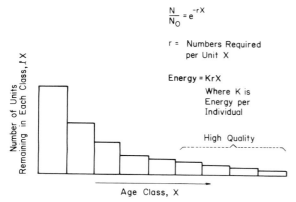

FIGURE 15-16. Age classes as spectrum of energy quality. Equation is for a constant percent change per age step.

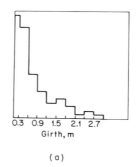
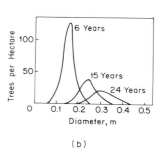

FIGURE 15-17. Examples of hierarchical age distribution in tropical forest from Whitmore (1975): (a) hierarchical pattern with many small trees and few large ones; (b) even-aged pulse of plantation trees passing up the hierarchy with remaining individuals accumulating embodied energy (*Anthocephalus chinensis* in Indonesia after Socharlan, 1967).

with many and declines to a few old individuals on the right (Fig. 15-16). A large number on the left is necessary for generation of a few on the right.

The population graph of increasing age and decreasing number is also a graph of convergence of accumulating embodied energy and is another example of energy hierarchy spectrum. If there is a steady state, there is a standing condition, with each age producing the number of the next age. This is called a *stable age distribution*.

Where the horizontal axis is weight, estimations of area between two time steps is the productivity of the age state transformation. Use of the graph for calculation of production is sometimes called the *Allen method* (Allen, 1953) in fisheries: the ratio of energy production in one interval to that in another is the energy transformation ratio of the age increase (see Fig. 20-12).

In forestry, the stable age distribution has been described by an exponential function (Meyer, 1933) given by Assmann (1970) (see rain forest graphs in Fig. 15-17a). In Fig. 15-7b graphs are given for pulses where a dominant-year class accumulates embodied energy with age. It has similar pattern to the stable one but with each class at a different time. Exponential forms like Fig. 15-3 have been used to calculate timber stand tables (Czarnowski, 1961).

Size Related to Energy Realm and Territory

Figure 15-16 shows that as animals become older, larger, fewer by age class progression, they contain more embodied energy and have higher-quality roles. Trees also begin as numerous seeds with small energy realms and sizes that gradually increase as survivors become fewer. The distribution of smaller and larger trees has the characteristic shape of the energy spectrum (see Fig. 15-17).

In general, area of the trunk cross section (basal area) measures the concentration of the unit, whereas area of the crown represents the territory of energy processing of the sun and the associated root-consumer system. The two sizes are often correlated (Desmaeris and Vasquez, 1970).

Energy Progression

In Chapter 11 simulations were reported of energy progression through food chain hierarchies with time constants based on size. Silvert and Platt (1978) generate an equation for the movement of an energy pulse along the web.

Increased Sizes of Units with Quality in Energy Chain Hierarchies

The property of increasing size of units in the convergence of low quality to high quality in the energy hierarchy was introduced with Fig. 2-1. The theory predicts increases in size as a way to accomplish feedback of high-quality amplifier actions and controls. At the higher-quality end of the spectrum, units receive from a larger territory and must feed back to a large area. One of the ways for doing this is to be larger.

For a century or more, quantitative comparisons of size and functions have been documented in living and nonliving worlds in the tradition of Darcy Thompson's *Growth and Form* (Thompson, 1942). Most of these comparative studies show increasing size of units at the high-quality end of scales of order and hierarchy. Often larger sizes are correlated in these studies with greater capacities and more informational processing, importance, value, evolutionary culminations, and so on. Recent reaffirmation of the size trends in hierarchical food chains in the sea is given by Steele (1977) and McNaught (1979) (see Fig. 18-3b).

The energy-quality concept allows units to be placed in relative position on energy-quality scale according to the energy transformation ratio from sunlight for their steady-state existence. The larger units have the larger embodied energies.

Sometimes parasites are regarded as an exception with units of diminished sized along the food chain. However, parasites are units larger than the populations of biochemical substances and the populations of food cells from which they draw. The food chain is from the small units of a host to the larger parasite, not from the whole host to the parasite.

Anderson and May (1978) give a spectrum with many hosts with few parasites and few hosts with many parasites. The dense parasite hosts may represent high-quality convergence and pulsing. Paperna (1964) provides a spectrum.

Time Constant Increase with Size and Hierarchical Position

Turnover time was already shown to increase with size in the case of a simple storage [Eqs. (3-10)] because of the inherent geometry of packaging.

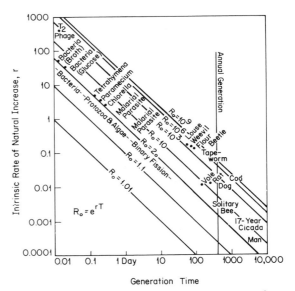

FIGURE 15-19. The relation between intrinsic rate of natural increase r, generation time (the abscissa), and increase per generation R_0 for a variety of animals. Adapted and redrawn from Smith (1954).

Graphs that show the larger time constants of larger structures are as those given in Figs. 15-18 and 15-19. Included are some examples within organisms and some comparisons between organisms. These graphs seem to be manifestations of the energy-quality hierarchy. The longer turnover times involved with larger units gives them lower depreciation rates and helps keep energy maintenance requirements reasonable (Fenchel, 1974).

Some occupants of high positions of the hierarchy are not large in size, although the *size of their realms of action are large*. Their time constants are large.

Energy Quality and Transport

It may be that higher quality units are more readily transported. Ability to transmit energy and embodied energy increases with position in the energy hierarchy. Parsons and Harrison (1981) graph distance traveled as a function of energy utilized.

Adaptations to Longer Existence

Larger units have the problem of maintaining internal structure for longer times. More redundancy is required since errors cannot be easily corrected until there is another generation of numerous offspring with natural selection. More internal work and re-

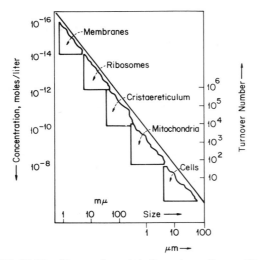

FIGURE 15-18. Size and metabolism in cell constituents (Weisz, 1973).

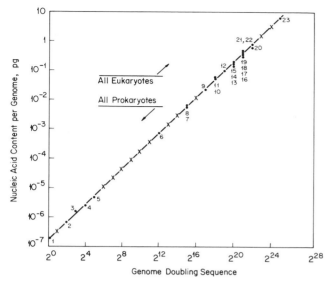

FIGURE 15-20. Increasing DNA with increasing size and length of life (Sparrow and Nauman, 1976).

pair is required with larger sizes. As shown in Fig. 15-20, the component of DNA content of organisms increases with size. Smaller units require less DNA because more rapid reproduction gives more rapid selection against errors.

Kaplan and Moses (1964) correlated sensitivity of microbes to ionizing radiation with size and DNA. Where stress energy is sufficient to eliminate higher-quality members, remaining components form a shortened hierarchy spectrum.

DNA and Information

As shown in the cellular energy chain in Fig. 15-5, DNA as the material associated with information coding of living cells is a very high-quality material (with its information) in the same way that computer memory tapes can contain high-quality control information. Can the DNA that is being maintained for biological operations be used as an index of information storage and processing? Figure 15-20 shows higher DNA content where organisms are higher on the evolutionary and complexity scale. Where more time is required to develop an organism, more DNA is observed in use to transmit the necessary information to the next generation. There are cases where DNA in cells is high without being related to information processing, such as in the lampbrush cells of amphibians or the salivary chromosomes of fruit flies, but here DNA apparently has other functions.

For ecosystems as a whole, Canoy (1970) compared DNA per area, finding similar levels for tropics and temperate communities. Only weedy stages of successional plant communities contained less DNA. Apparently communities with temporal programming requiring a diversity of programs to fit seasons had as much DNA as did systems with little temporal change and much species diversity.

THEORY OF HIGH-QUALITY CONTROL AND PULSING

If surviving systems have feedback actions commensurate with embodied energies required to generate these actions, the ability to control can be theorized to be a function of the embodied energy. System control rests ultimately with the highest-quality energies.

If delivery of high-quality actions controls a system, more control is exerted when these controlling energies are delivered in a short time. Controlling units that accumulate their high-quality embodied energies and then express their feedback control actions in short bursts may prevail over alternative controllers, at least during the periods of the bursts of action. Thus there are energetic reasons why systems may pulse with periods set by the controlling consumers. For the examples given in Figs. 15-4 through 15-6, pulsing is observed in the actions of the high-quality controlling units of the systems. Gene actions are pulsed, breeding and growth cycles in storks are pulsed, and earthquakes and geological work of large rivers are pulsed. Carnivores are pulsed (P. Kremer, 1976).

Consistent with energetics is the distribution of time constants and size of units. At the end of the spectrum, low-quality energy is abundant. There are many small units with short time constants whose pulses are readily absorbed in converging their energy outputs to the next level of the hierarchy. At the upper end of the hierarchy, individual sizes are larger as required to adequately accept and redistribute actions to the larger territory represented per individual. Consequently, time constants are large, and control actions are sufficient to impose the period of the controlling consumer on the rest of the system. The theory of pulsing consumers is another corollary inferred from the maximum power principle that accounts for many observed situations such as Brownian motion, fires, epidemics, earthquakes, storms, and human conquests. Examples of pulsing models that may be reasonable simplifications are Figs. 11-12 and 11-21.

If pulsing is a high-quality control phenomenon as suggested by the maximum power reasoning, it is a form of high-energy organization manifested in temporal patterns. Prigogine (1978a) and Nicolis and Prigogine (1977), considering the emergence of order from molecular chemical processes, suggest oscillating reactions as manifestations of order. Naroli and von Bertalanffy (1956) find decreasing pulse rate with size in hearts of different animals.

Pulsing Timers

If pulsing systems generate more power than do steady ones, the biological rhythms so characteristic of living organisms begin to make sense. These rhythms are generally kept in oscillation by some kind of internal oscillator that is entrained by environmental events to be in some phase relation to external periodicities of sun, tide, and so on (Palmer, 1976). The models proposed for the internal regulatory oscillations and clocks involve loop systems such as those observed in biochemical networks that have been shown to oscillate (Chapter 10). If a pulsing mechanism is generally adaptive in systems, we can look for some of the pulsing components in ecosystems and human societies to serve similar roles in ensuring that the whole system continues to maximize power by means of pulsing. See Chapter 27 and Pavlidis (1973).

Graphs of Variance as a Function of Frequency

Graphs of variance as a function of frequency are much used to represent repeating influences in time series. Data in the form of time series (graphs with time) are converted to the frequency domain graphs with computer routines that take Fourier transforms (Box and Jenkins, 1976). For example, if there are daily and seasonal regularities in phenomena, a graph such as that in Fig. 15-21a results when the data are expressed in frequency domain. Notice the two frequencies at which there are large deviations from random noise. White noise is noise that is independent of frequency and is shown as a band across the graph. In this example peaks above this noise occur at time periods of a day and a year. Since period is the reciprocal of frequency, a period scale can be shown on the graph.

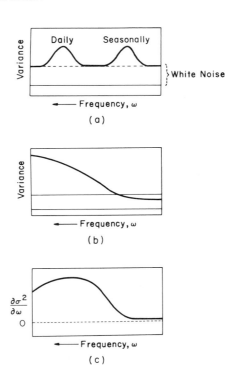

FIGURE 15-21. Graphs of variance as a function of frequency used to represent repeating phenomena in frequency domain: (a) graph with daily and seasonal pulses; (b) hierarchical system; (c) rate of change graph; (d) autocorrelation decreases with frequency.

The measure used on the vertical axis is statistical variance and is the mean square of the deviations from the mean. Where motions in a fluid are calculated as such, the process of squaring makes the variance proportional to energy since energy is proportional to a square of velocity. Thus these graphs are called *power spectra*. Sometimes the name is used for the mathematics, even when applied to time series where the square is not proportional to energy.

When frequency analysis transformations of time series involving hierarchies in nature are made, there are more phenomena with small time constants (high frequencies) than with low frequencies because of the converging nature of most energy hierarchical patterns. Consequently, some graphs of variance may decrease as frequencies decrease.

Sometimes this is called *red noise* (see Fig. 15-21b). These graphs, like the other spectral graphs in this chapter, portray energy hierarchies. High-quality energy is usually the lower-frequency phenomena of larger size. Examples are water waves, turbulent eddies, storms, and animal rhythms. High-quality energies are believed to have pulsing nature as already described and thus are readily observed in frequency domain.

When Fourier transformations are made, the dissected component oscillations are those which generate the observed data by addition. In most interactions, varying inputs interact multiplicatively or in other non-linear ways. New procedures are needed to separate out components factorially.

Because variance is a squared function, patterns of larger dimension have larger variance than do smaller ones. "Power spectra" used in time series analysis of hierarchical systems have increasing variance with lower frequencies as in Fig. 15-21b.

Because of the inherent long time constant of the larger units that cover more area in energy hierarchies, time series analysis will normally show a series of peaks, one for each level in the hierarchy. Because the astronomical systems impose diurnal, seasonal, 11-year sunspot, and longer energy pulses on the earth system, these energies are to be expected. Adaptation theory suggests that a scale of user systems with natural periods that fit these outside energy pulses are to be found, and these entrain the frequency distributions of the ecosystem hierarchies.

Electromagnetic Spectra

For spectra of electromagnetic waves such as light and radio, high-quality energy accompanies high frequency, the reverse of the relationships for most phenomena. Short-wave radiations characteristically have high concentrations of energy that represent a higher quality, requiring more energy concentrations in generation.

The Spectrum of Energy-Driving Sources

The spectrum of energy distribution not only describes the systems structure, but may represent the spectrum of energy sources that are entering a system. The energy signature plotted in Fig. 14-14 constitutes a driving energy spectrum to which the adapting system organizes its own energy hierarchy. The energy of sunlight is really a multiple source described by its spectrum. The spectrum of physical energies of varying time periods impacting life in the sea has been expressed with frequency by Ozmidov (1965), Platt (1971), Platt and Denman (1975), Wroblewski, O'Brien, and Platt (1975), Monin et al. (1976).

Ecosystem Adaptation to Frequency Signature

Analysis of times series of various physical, chemical, and biological phenomena in the environment shows a hierarchy in the energy sources that are available to the production and consumption systems of ecosystems. To take advantage of energy sources that have regular oscillatory rhythms, structures must develop with similar periods to be able to utilize maximum power. Without the tuning of the ecosystem to the frequencies, energies act as stresses or are passed without use. To some extent an ecosystem is like an FM radio, able to analyze and adjust loadings to receive all frequencies, developing a characteristic pattern of function ultimately traceable to frequencies of the driving functions and the frequencies of the high-quality controlling members. For example, a wetland has species capable of utilizing the varying frequencies of rains of the climate. Platt (1971) and Platt and Denman (1975) correlate phytoplankton periodicity with the periodicity of the physical oscillations in the sea from which the phytoplankton must be drawing auxiliary energies and avoiding stress (see Fig. 15-22).

If ecosystems are adaptive filter systems, there are many opportunities for application of filter concepts already developed in electrical engineering, computer science and other fields (Bogner and Constantinides, 1975; Terrell, 1980; Blinchikoff and Zverev, 1976).

Time Series Analysis and Autocorrelation

Other graphs used in the study of power spectra in frequency domain are included in Fig. 15-21. Figure 15-21c gives the graph of rate of change of variance with frequency. Another is the autocorrelation as a function of frequency. Autocorrelation is defined as the products of data at one time with data at a later time divided by the square of the data later. The larger the units and time constants, the more correlation there is between one time and the next.

284 ORGANIZATION AND PATTERN

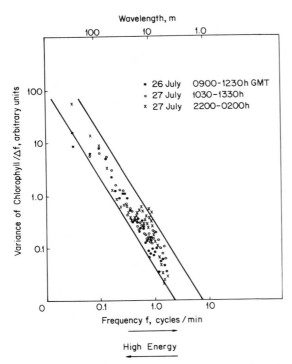

FIGURE 15-22. Frequency spectrum of phytoplankton in the sea that has a similar distribution to that of turbulent energy (Platt, 1971).

Where hierarchical data are represented, there is more autocorrelation with the lower frequencies (high quality). For these approaches see Blackman and Tukey (1958).

Maximum Entropy Principle in Finding Functions for Time Series Data

One way to obtain functions from noisy time series data is to use the maximum entropy principle. With the use of computer algorithms, functions are sought so that the deviations between the function and the data have maximum uncertainty. The sum of the logarithms of the deviations is maximized. Maximum entropy occurs when deviations are given equal probability. The function thus selected has no potential energy between it and the data. In other words, the function is closest to the data without generating any structure beyond the data. Introductions by Burg (1975), and Jaynes (1957) and many explanatory examples and explanations are collected in a volume edited by Childers (1978) (see Chapter 17).

MECHANICAL ENERGY CHAIN AND TIME

An interesting example of an energy chain is the mechanical clock diagrammed in Fig. 15-23. It increases quality of energy and terminates in energy pulsing with a high-quality process that feeds back information. Energy is passed from the coiled-spring energy source to large slow cogwheels that connect with smaller and smaller sheels and finally to a switching device that ticks off the seconds.

A timing device requires a pulsing reference. If most systems develop chains and develop high quality with pulsing delivery of the feedbacks, a clock is automatically provided so that the system can utilize timing functions.

Many hierarchical processes have been compared to cogwheels; see Fig. 15-24 for food-chain comparison (Clarke, 1946; Margalef, 1962).

OTHER MATHEMATICAL EQUATIONS FOR HIERARCHICAL SPECTRA

Characteristic hierarchical distributions are often found with skewed shape from many low-quality en-

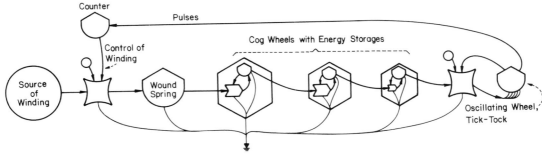

FIGURE 15-23. Energy system in a mechanical clock. Energy from wound spring generates low-energy, but high-quality, pulsing timer.

Spectrum of Energy Distribution and Pulsing

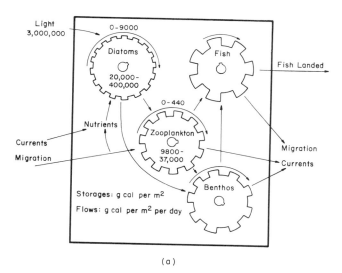

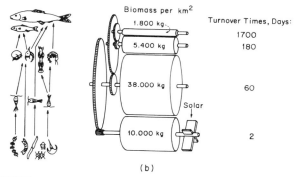

FIGURE 15-24. Cogwheels used to represent food chains: (*a*) Clarke, 1946; (*b*) Margalef, 1962.

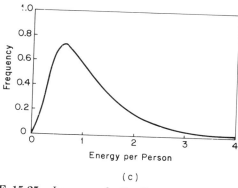

FIGURE 15-25. Log-normal distribution used by Brookes (1974) to consider energy allocation to people.

tities of flows on the left to a few high-quality items on the right. Various mathematical forms have been used to describe these, including the negative exponential already introduced in Fig. 15-3.

There is a peak in the energy distribution of open systems (Fig. 15-10), and the log-normal distribution has been used to fit observed patterns (see Fig. 18-17). Prediction of energy-quality distribution may be possible by construction of a log-normal distribution with its peak corresponding to the dominant energy input. Brookes (1974) used the log-normal distribution to consider the number of people moving up in standard of energy from left to right in Fig. 15-25. This is comparable to organisms moving up the quality spectrum with age (Fig. 15-16). Income distribution curves are similar and may be explained by the hierarchy of embodied energy similarly. The Pareto distribution, applied to describe income distributions, is given in Chapter 18 along with other rank order curves that may also be a manifestation of hierarchy.

The Weibull (1951) distribution given in Eq. (15-7) is another distribution with an exponential shape or with the skewed shape of hierarchical pattern such as that in Fig. 15-9, depending on c:

$$\text{Probability density } Y = \frac{cX^{c-1}}{b^c} e^{-(X/b)^c} \quad (15\text{-}7)$$

The cumulative probability form is

$$Y = 1 - e^{-(X/b)^c} \quad (15\text{-}8)$$

Equation (15-8) gives the cumulative percent as a function of the intensity of the phenomenon as measured from a threshold. An application to oil spills by Mikolaj (1972) showed straight lines with coordinates of log–log versus log. Catastrophes of nature and humans have larger amplitude and lower frequency corresponding to the tail area of the distribution.

Another distribution is the gamma distribution used to describe frequency distributions of floods. For values of $c = 2$, a skewed shape results with maximum number of instances at low energies and a few instances at extreme high-energy tail:

$$Y = \frac{(X/b)^{c-1} e^{-(X/b)}}{b\Gamma(c)} \quad (15\text{-}9)$$

where

$$\Gamma(c) = \int_0^\infty e^{-u} u^{c-1}\, du$$

Another distribution for spectra of this shape was developed for income by Petersen and von Foerster (1971) with an energy argument (see Fig. 23-18).

RELIABILITY

The distribution of failure is described by the same kinds of functions as those used for mortality. Curves of skewed shape are used, and the exponential, log-normal, Weibull, and gamma distributions for remaining failure free elements decrease with time (Sinha and Kale, 1980). Frequency of failure is ultimately related to hierarchical energy distribution.

REPRODUCTIVE SPECTRA

The embodied energy varies in reproductive propagules; embodied energy is small in spores and small seeds that are produced by the millions and is much greater in large seeds and offspring of organisms cultivated with a high degree of care. Collectively there is a spectrum of reproductive information emitted by each system as part of its necessary contributions to reproduction and dispersal. Wells (1976) provides a spectrum of seed variety and weight (Fig. 15-26). There are many low-energy-species seeds and few heavy seeds that presumably have a high degree of embodied energy. The graph may be another manifestation of the energy hierarchy. The actual number of successional species with small seeds is probably much smaller than the graph would indicate since the ability to be transported is included in the history of seed formation and distribution.

SUMMARY

A characteristic spectrum of energy distribution is associated with hierarchical organization of webs and chains and can be represented in several kinds of power spectral graph, including graphs of energy quantity and quality and graphs of energy manifestation in frequency domain.

Examples include stream networks, food chains, quantum shells, age–class distributions, gear chains, thermal distributions, turbulence, biochemical organization, and information systems.

Similarities between closed and open systems were explained as self-organization by the maximum power principle. Open systems have peaks in the spectra corresponding to external energy sources. Control by terminal members of an energy quality chain was related to their embodied energy. Delivery as pulses may maintain controls that maximize power.

Gaussian distributions of some variables are derived from exponential energy spectra instead of vice versa, thus relating the normal distribution to the adaptive functions of energy hierarchy.

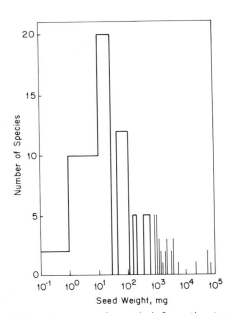

FIGURE 15-26. Spectrum of genetic information transformation in seed availability from Wells (1976).

QUESTIONS AND PROBLEMS

1. Compare an energy web to a simplified energy chain in regard to its configuration, energy-quality transformation, and total energy flow.

2. For a single source of low-quality energy, what shape graph results with number of units and total calories plotted as a function of actual energy per unit and energy transformation ratio?)

3. Compare energy spectrum of a closed system that is circulating energy, as in a molecular population, and an open system with source and sink added.

4. Enumerate various kinds of energy spectrum that can be identified with the energy hierarchy spectrum.

5. What is the shape of the energy hierarchy spectrum where there are low- and high-quality energy sources?

6. Where the energy transfers from one level to the next in the hierarchy chain are in constant proportion, which equation describes the spectrum? What proportion of the total energy exceeds some thresholds of energy quality? Give the equation for this.

7. Where energy is distributed exponentially and is a square function of some other property, how is that property distributed? If energy is distributed according to hierarchies because of the adaptive utility of such organization, what is the distribution of the energy carrying property? According to this theory, is the Gaussian distribution a consequence of system organization or inherent disorganization? Discuss the more classical alternate view that chaos generates order and hierarchy.

8. What are coordinates used in spectral analysis of time series? How is energy quality related to frequency in spectral graphs of this type? How are electromagnetic spectra different from those of other kinds of energy? How is variance related to energy quality?

9. How do age-distribution graphs qualify as embodied energy spectra?

10. How does the earth's atmosphere serve as a special case of the Maxwell–Boltzmann distribution?

11. How are reaction rates in closed and open systems related to the energy power spectra and threshold for transformation?

12. Turbulence spectra represent medium quality energy inflows and have a peak. Explain the different processes that occur on the high- and low-energy tails of the distribution.

13. Give the locations of the following relative to each other on the energy-quality hierarchy: sunlight; wind; electricity; human service; and high political influence.

SUGGESTED READINGS

Adams, R. N. (1975). *Energy and Structure. A Theory of Social Power,* University of Texas Press, Austin.

Bloomfield, P. (1976). *Fourier Analysis of Time Series,* Wiley, New York. 258 pp.

Bonner, J. T. (1965). *Size and cycle, an essay on the structure of biology,* Princeton University Press, Princeton. 219 pp.

Box, G. E. P. and C. M. Jenkins (1970). *Time Series Analysis and Forecasting,* Holden-Day, San Francisco.

Gerlsbakh, I. B. and Kh. B. Kordonskiy (1969). *Models of Failure* (English translation), Springer-Verlag, New York.

Glansdorf, P. and I. Prigogine (1971). *Structural Stability and Fluctuations,* Wiley-Interscience, Chichester.

Haken, H. (1977). *Synergetics. A Workshop,* Springer-Verlag, Berlin. 276 pp.

Hammen, C. S. (1972). *Elementary Quantitative Biology,* Wiley, New York. 144 pp.

Hastings, N. A. J. and J. B. Peacock (1974). *Statistical Distributions, a Handbook For Students and Practitioners,* Wiley, New York.

Iberall, A. S. (1972). *Toward a General Science of Viable Systems,* McGraw-Hill, New York. 414 pp.

Kendall, M. G. (1973). *Time Series,* Griffin, London.

Lam, H. Y-F. (1979). *Analog and Digital Filters,* Prentice-Hall, Englewood Cliffs, NJ.

Meixner, J. (1966). Network theory in its relation to thermodynamics. *Proc. of Symposium on Generalized Networks,* Polytechnic Press, Polytechnic Institute of Brooklyn, New York. pp. 13–25.

Miller, G. T. (1971). *Energetics, Kinetics, and Life.* Wadsworth, Belmont, CA. 357 pp.

Milsum, H. J. (197?) The hierarchical basis for general living systems. In *Trends in General Systems Theory,* G. J. Klir, ed., Wiley, New York. pp. 145–187.

Monin, A. S., V. M. Kamenkovich, and V. G. Kort (1977). *Variability of the Oceans* (English ed. by J. J. Lumley), Wiley, New York.

Otnes, R. K. and L. Enochson (1978). *Applied Time Series Analysis,* Wiley, New York. 449 pp.

Parker, R. A. (1975). Environmental periodicity and ecosystem stability. *Ecological Modelling* **1,** 91–103.

Pattee, H. H. (1973). *Hierarchy Theory, the Challenge of Complex Systems,* Braziller, New York. 156 pp.

Platt, T. and K. L. Denman (1975). Spectral analysis in ecology. *Ann. Rev. Ecol. Syst.* **6,** 189–210.

Scheidegger, A. F. (1970). *Theoretical Geomorphology,* Springer-Verlag, Berlin.

Shugart, H. H. ed. (1978). *Time Series and Ecological Processes.* SIAM Institute for Mathematics and Society, Philadelphia. 303 pp.

Sinha, S. K. and B. K. Kale (1980). *Life Testing and Reliability Estimation,* Halstead (Wiley), New York. 196 pp.

Steele, J. H. (1977). *Spatial Pattern in Plankton Communities,* NATO Conference Series Section IV, *Marine Sciences,* Plenum Press, New York.

Stewart, W. E., W. H. Ray, and C. C. Conley, eds. (1980). *Dynamics and Modelling of Reactive Systems,* Academic Press, New York. 413 pp.

Von Uexküll, J. (1926). *Theoretical Biology,* Harcourt Brace, New York. 362 pp.

Weiss, P. A. (1971). *Hierarchically Organized Systems in Theory and Practice,* Hafner, New York.

Winfree, A. T. (1980). *The Geometry of Biological Time,* Springer-Verlag, New York. 530 pp.

CHAPTER SIXTEEN

Temperature

Radiation and heat are an ever-prevalent part of all systems. Where temperature differences exist, potential energy is available for driving processes. Where temperatures are uniform, energized molecules in the thermal hierarchies cross energy barriers, facilitating chemical and biological reactions. In this chapter we consider models for temperature action in systems.

TEMPERATURE AND HEAT

Temperature is a measure of the average kinetic energy of molecules. With higher temperatures, molecular actions increase pressure, causing materials to expand. A new size equilibrium develops with a balance between expansion pressure and constraining forces. Expansion of fluids as with a thermometer is an easy way to read temperature. In the Kelvin (absolute) scale of temperature, zero is taken as the point where molecular motions are about zero. This is called *absolute zero*. As energy is added and temperature rises, water melts at 273 K and water boils at 373 K. As temperature rises, structure becomes less rigid with more alternative positions for molecules developing; more heat is required per unit rise in temperature. Whereas calories are the measure of heat quantity, temperature is the intensity factor of heat, measuring its concentration and tendencies to flow. For gases, multiplication of Boltzmann's constant R by the absolute temperature gives the energy contained. Heat flows in proportion to the difference in temperature. Heat is transmitted by many pathways, including diffusion (conduction), convection, radiational exchange, and transport with matter.

RADIATION

Everywhere in the universe heat energies are transformed into radiation and radiation is absorbed, transforming its energy into heat. The generation of radiation by hierarchically distributed molecular energies produces the Planck distribution of radiation as shown in Fig. 16-1. Transformation of a unit of molecular energy to a photon of radiation is reversible. The electromagnetic radiation as an oscillator falls in the class of transformations that individually have no losses and no heat sink [see Eq. (7-8) and discussion of reversibility of energy transformations that have a change of velocity]. For a population losing energy by radiation, however, there is energy degradation due to divergence of radiation.

Alexander (1978) suggests that the heat in a population of molecules can be used to estimate the embodied energy in thermal radiation. Since the temperature is proportional to the required heat, embodied energy of radiation of different wavelengths is a function of the temperature.

HEAT-BUDGET MODEL

The temperature of any system is the result of the balance of inflowing and outflowing exchanges of heat and radiation. A typical model of heat budget is given in Fig. 16-2. A tank is used to represent heat, and the temperature is indicated as the appropriate intensity variable. Radiation goes in and out as the fourth power of temperature. Convectional exchange of heat and water vapor is driven by kinetic energy of wind, which also renews storages in air near leaves. Vaporization ties up much heat (0.54 cal/g) into gaseous structure in converting water to vapor. Sunlight received by the photosynthetic chloroplast units after its transformations is mostly passed on to leaf heat. A small amount goes into chemical structure and returns with its decomposition. Simulation of heat-budget models has been very successful in accounting for short range temperature trends (Gates, 1980).

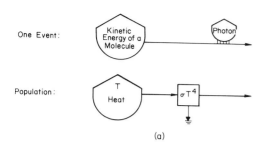

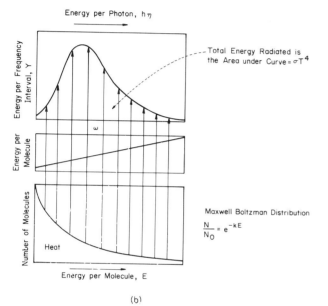

FIGURE 16-1. Energy transformation from heat distribution to radiation: (a) energy diagram; (b) distributions; product of values in the two lower graphs generates the upper graph. A peak results because the combination of numerous photons and energy per photon is largest in the intermediate range.

EFFECTS OF TEMPERATURE ON OTHER PROCESSES

Temperature affects most processes, especially those involving chemical reactions. As described in Chapter 15, chemical reactions involve the population of molecules above a threshold called the *energy of activation*. See Fig. 15-12. Increase in the temperature increases the population above the threshold according to the equation in Fig. 16-3a (see derivation in Eqs. 15-1–15-6). Such reactions can be indicated with the Arrhenius (1889) interaction as in Fig. 16-3a. The shape of the interaction with rising temperature curves up sharply as in Fig. 16-3b. Sometimes, by recognizing this shape but simplifying the arithmetic for simulation purposes, one can substitute a positive exponential term (exp T) (Fig. 16-3d).

Test of Exponential Action of Temperature

Where rates are suspected to be a function of the temperature function, a test can be made by graphing data on coordinates of log rate versus reciprocal of temperature. If the function is consistent with the data, there will be a straight line with declining slope as in Fig. 16-3c. These coordinates are called a Van't Hoff plot.

Implied Temperature Rate in Descriptive Modeling

Temperature pathways may also be considered as a temperature-controlled conductivity. The reciprocal of conductance is called *chemical resistance* (Fig. 16-3a). If no temperature interactions are drawn in a model, it is nevertheless assumed that an ordinary line pathway has a conductivity (Chapter 1) that is subject to temperature control; but if it is to be explicitly modeled and simulated, interaction symbols are shown with temperature function as in Fig. 16-3.

Pathways Controlled by Temperature

Because many of the pathways of the biosphere involve chemical processes and molecular reactions according to the models in Chapter 15, energies flow faster as temperature increases. Pathways that are affected include chemical reactions in nature, most of the biochemistry of living organisms, and part of production processes. The disordering and depreciation flows of storages and structures are partly microscopic and chemical and thus also increase with temperature.

The heat sink carries used degraded energy into the environmental sink. Regardless of whether it is written explicitly, the heat sink defines the system boundaries quantitatively by the sink temperature. Energy flows through the sink by diffusion so that no potential energy remains, and the energy becomes part of the thermal molecular motion of the environment. By definition of entropy, energy dispersion is measured by the heat flux through the sink divided by the absolute temperature of that sink.

Maximum Process at Optimum Temperature

Thompson (1942) and Johnson et al. (1942) showed many biological processes with a hump-shaped

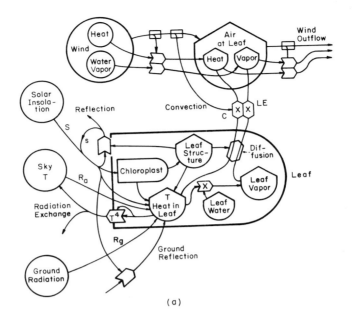

(a)

Energy Balance Equals

Income: $a_s\left[\dfrac{(1+r)(S+s)}{2}\right] + a_t\left[\dfrac{(R_g + R_a)}{2}\right]$

Minus Outgo: $-\xi_l \sigma T^4 \pm C \pm LE$

(b)

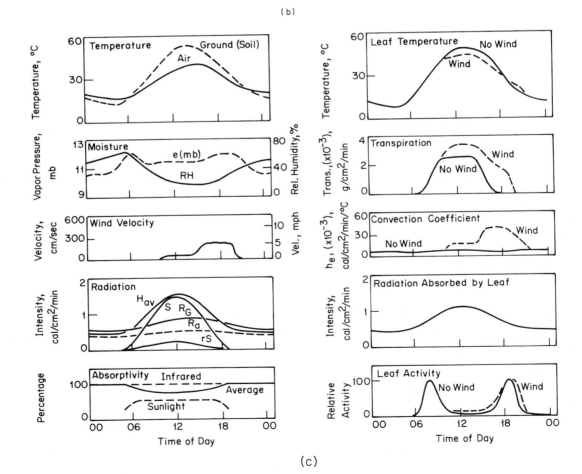

(c)

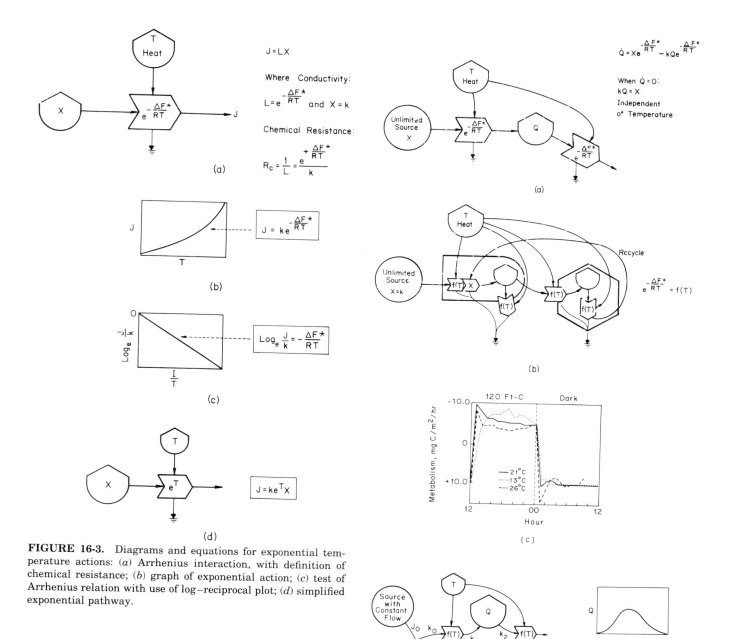

FIGURE 16-3. Diagrams and equations for exponential temperature actions: (a) Arrhenius interaction, with definition of chemical resistance; (b) graph of exponential action; (c) test of Arrhenius relation with use of log–reciprocal plot; (d) simplified exponential pathway.

FIGURE 16-2 Model of heat flows in leaf as an example of heat budget [only flows affecting heat are shown]: (a) energy diagram; (b) and (c) equations and simulation from Gates (1965) r = ground reflectance].

FIGURE 16-4. Effects of temperature dependent on type of energy: (a) unlimited source; level remains constant at steady state (Kelly, 1971); (b) closed loop; unlimited source, levels constant; (c) diurnal curve in terrestrial microcosm with little difference in metabolism with temperature (Cumming and Beyers, 1970); (d) maximum response where energy source is source limited and response has a maximum (see Fig. 8-8 for derivation of effect of source-limited flow).

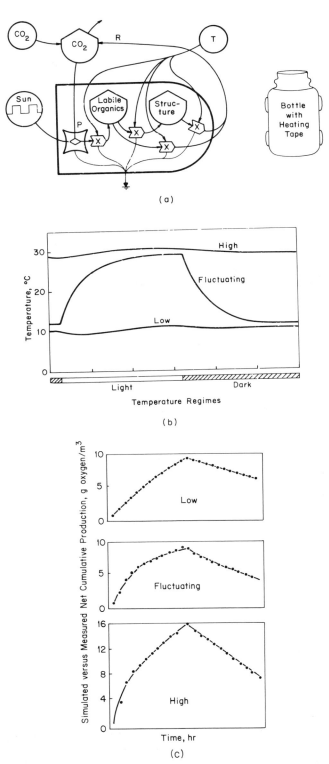

FIGURE 16-5. Energy diagram for bottle microcosm and minimodel of diurnal oxygen change with three temperature ranges (low, high, and fluctuating once a day from low to high) (Kelly, 1971): (a) energy diagram; (b) daily temperature regimes; (c) simulated daily responses and data points.

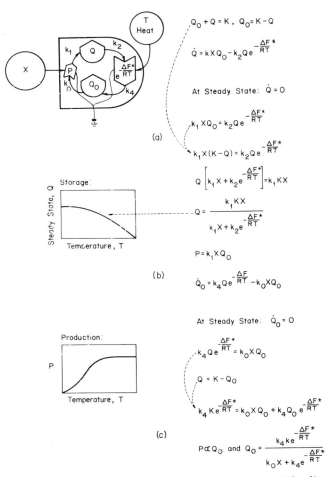

FIGURE 16-6. Effect of temperature on a P–R Michaelis–Menten model with only recycle temperature affected: (a) model; (b) storage with temperature at steady state; (c) production with temperature at steady state.

curve of response to temperature. Processes were maximum at an intermediate temperature. Models were generated that accounted for these maxima in terms of differential actions of temperature.

Effect of Type of Source on Temperature Effects

The effects of temperature on thermal–chemical pathway configurations are shown in Fig. 16-4. In Fig. 16-4a there is an unlimited source (constant force X). Temperature increase accelerates inflow and outflow to Q. The net effect on the steady state at Q is zero, but the system as a whole now processes much more energy. The production balancing deteri-

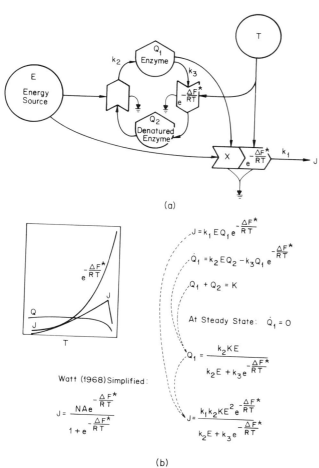

FIGURE 16-7. Energy diagram and simulation of temperature effect on a pathway and on participating enzymes (Eyring and Urry, 1965; Johnson, Eyring and Polissar, 1954; Odum, 1974): (a) energy diagram; (b) analog simulation.

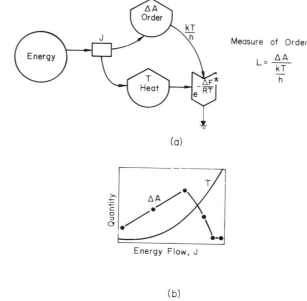

FIGURE 16-8. An order–disorder model provided by Morowitz (1968): (a) analog simulation of steady states prepared by T. Gayle, k is Boltzmann's constant per molecule, h is Planck's constant, and Δ is Helmholtz free energy.

oration evident in Fig. 16-4a is an example. Beyers (1962, 1974) found little difference in the P–R processes in closed microcosms exposed to different temperatures (Fig. 16-4c).

In Fig. 16-4b there is a loop with temperature accelerating each pathway in a circle, and the energy source is not limiting. As Schmidt (1960) pointed out, temperature effects on levels within a loop such as that in Fig. 10-24 may cancel, and such loops might be important in temperature-independent regulatory machinery.

In Fig. 16-4d the energy inflow is flow limited. Increasing temperature pulls more energy until no more can be drawn and further temperature effects reduce Q.

Kelley (1971) studied bottle microcosms in growth chambers with three temperature regimes (Fig. 16-5b). Simulation of the model in Fig. 16-5a produced the diurnal curves of oxygen observed (Fig. 16-5c). The model has a limited recycling resource and two storage levels both in sequence and in parallel.

OVERVIEW MODELS OF TEMPERATURE IN SYSTEMS

A minimodel in Fig. 16-6 illustrates the ecosystem's response to temperature where there is recycling of materials in production and consumption. The conversion of photons by light is mainly temperature independent, but the rest of the pathways of storage and recycle are chemical in nature and respond to temperature. If temperature is increased in the model shown, the recycling increases so that the system produces more than at lower temperatures since recycling is accelerated. Storage at steady state becomes less and less. Successively fewer materials are needed for recycling because rates are high.

Many simple biological and biochemical processes accelerate with temperature until a threshold is reached, after which the machinery is sufficiently

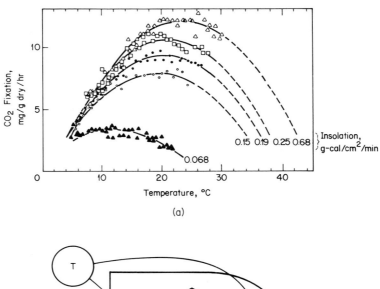

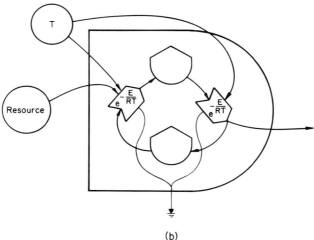

FIGURE 16-9. Family of curves for effect of temperature in which resource is also limiting: (a) photosynthesis by Alders (Webb et al., 1974); (b) model simulated by Quinlan (1978) to generate graphs of this class.

damaged that the given process declines with any further increase in temperature. Johnson et al. (1954) developed such a model (Fig. 16-7). The exponential temperature function accelerates the functional process J, but the enzymes are also denatured at accelerating rate by the same function. The system uses its energy resources to reconstitute the enzymes, but at a more linear rate. An analog computer simulation generates the typical temperature response curve of such simple components. Note similarity with Fig. 16-6 because both involve a balance between linear and exponential processes.

Morowitz (1968) gives the model that is diagrammed in Fig. 16-8a. Biological or other useful storage structures of the biosphere increase as a balance is reached between a source and the denaturation due to temperature. Arbeit (ΔA, Helmholz free energy) is used as a measure of storage potentials. As the energy source is increased, more structure is generated at first, but the increased drain eventually builds up the heat and accelerates the drain more than the source increases it. T. Gayle provided a simulation (Fig. 16-8b). The ratio of ΔA to kT/h was defined as an L *function* as a measure of order.

Quinlan (1978), examining various data for biological temperature responses including fish, plants, and microbes, considers Arrhenius' effect of temperature accompanied by limiting-factor supply of sources for the processes. She finds a general pattern that is somewhat parabolic, depending on the amount of resource available. Figure 16-9a gives one such graph of temperature effects for different availabilities of resources. Quinlan generated families of curves of the class observed by combining

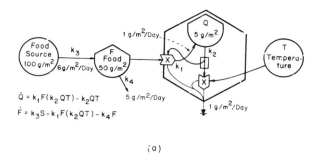

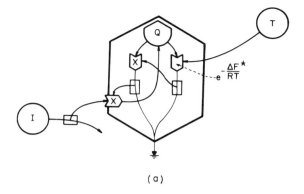

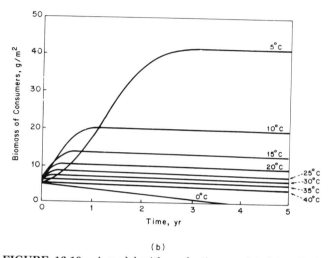

FIGURE 16-10. A model with production regulated to adjust to the temperature-controlled depreciation; energy source in flow limited at the source (W. H. B. Smith, 1976): (a) model with numbers used in scaling; (b) analog simulation.

the Michaelis–Menten limiting-factor equation with Arrhenius' relation as given in Fig. 16-9b. In another paper she gave the family of curves for the oxidation of nitrite by nitrobacter in pure culture.

One of many examples relating combined effect of temperature and energy source is given by Webb et al. (1974) in Fig. 16-9a. These authors used a parabolic equation for the temperature model and a Mitscherlich expression (Fig. 8-15) for light response. The combination gave excellent fit to observed data of metabolism of alders with varying light and temperature.

MODELS FOR REGULATION ADJUSTING TO TEMPERATURE EFFECT

In most environmental systems, temperature affects production and consumption. In addition to the in-

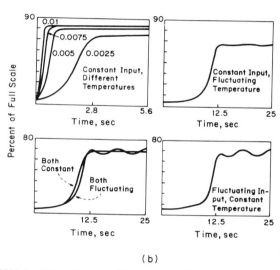

FIGURE 16-11. Model of temperature from Kelly (1971) with feedback to production adjusted by sensing depreciation: (a) model; (b) representative simulations.

herent actions of temperature on the chemical pathways, there are organized control pathways among the more complex organisms adjusting processes in response to temperature (Rose, 1967).

In Fig. 16-10a a model is provided in which production within the system is adjusted in proportion to the depreciation by a sensor. The depreciation is driven by a temperature function. If the food level is high, the production can balance the losses, but as temperature rises, the food—which is ultimately source limited—provides a limit and stocks are reduced. At higher temperatures the unit processes more of the available food; its production is high, but its net production is less. W. H. B. Smith (1976) provides a simulation in Fig. 16-10b.

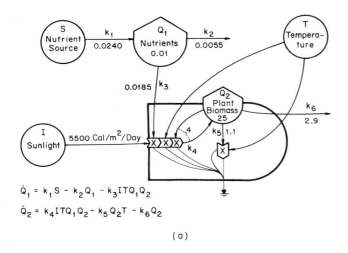

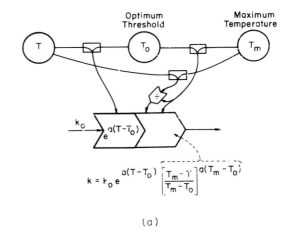

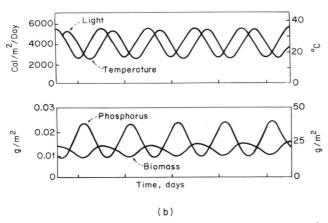

FIGURE 16-12. A model with temperature action (simple product) in push and pull scaled with numbers appropriate to an estuary at Crystal River, Florida: (a) model (W. H. B. Smith, 1976); (b) simulation.

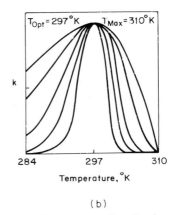

FIGURE 16-13. Temperature function in a model of biological growth (Lassiter and Kearns, 1974): (a) energy diagram of function k; (b) family of curves.

In Fig. 16-11 Kelley (1971) simulated a model in which the temperature-regulated depreciation was sensed and used to augment the feedback of storage into more energy processing. In effect, this model makes the action of storage quadratic, increasing the costs in a density-dependent way and providing a stable model. Representative simulations are shown. Fluctuations in steady state resulted from fluctuation in either temperature or energy.

Push–Pull Effect of Temperature

The "push–pull" effect, which can maintain storages if energy inflow can increase with temperature, is illustrated in Figs. 16-4a–16-4c. As shown in Fig. 16-4d, if there are fixed inflows of requirements, there is a temperature optimum for storage and a ceiling for energy consumption provided at the source.

Smith, in Fig. 16-10, provides a push–pull minimodel with scaling for values appropriate to an estuary in which there is a limiting rate of nutrient inflow. Note that steady state is reached much more rapidly at the higher temperature. The simulation in Fig. 16-12a shows a similar pattern.

Lassiter and Kearns (1974) used a model with a central peak to the response of temperature with the equation given in Fig. 16-13.

Ramsey (1976) and Kemp (1977) in Fig. 16-14 provide a push–pull model and simulation, but here

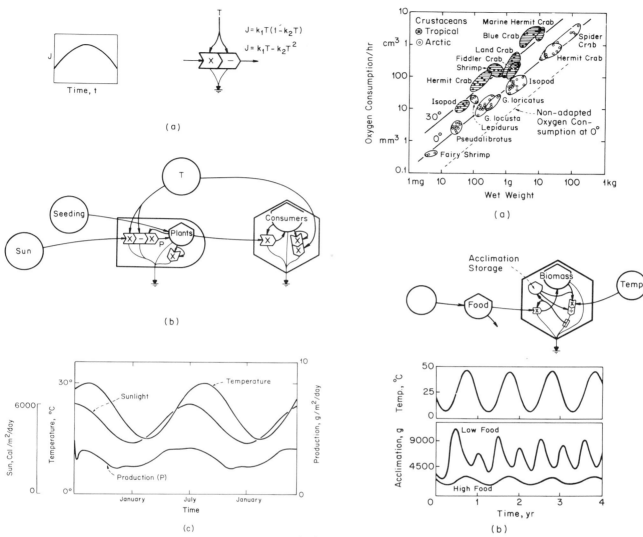

FIGURE 16-14. Effect of temperature on productivity in which temperature action is parabolic (Ramsey, 1976; Kemp, 1977): (a) temperature interaction; (b) model with temperature interaction; (c) simulation.

FIGURE 16-15. Temperature regulation by organisms so that their metabolic rates are compatible with energy supplies: (a) data due to Scholander et al. (1953); (b) model of regulation by Homer (1974).

the action of temperature was made a parabola, causing optima to occur in the temperature actions of both production and consumption, which interacted to produce the overall effect of balance seen in Fig. 16-14c.

McKellar (1975, 1977) found a small percent of increase in metabolism in simulations of push–pull effects on plankton metabolism in shallow estuaries receiving thermal effluent at Crystal River [see also simulation of area with predominant benthic metabolism by W. H. B. Smith (1976)].

Temperature Regulation by Organisms

Temperature regulation within organisms is even more organized, causing arctic animals to run faster and tropical ones slower (see Fig. 16-15a). Thus metabolic rates can fit energy supplies, overriding inherent temperature tendencies. One regulation model is given in Fig. 16-15b.

Quinlan (1981) studies seasonal shifts in the optimum temperature that adapt organisms to maximum performance.

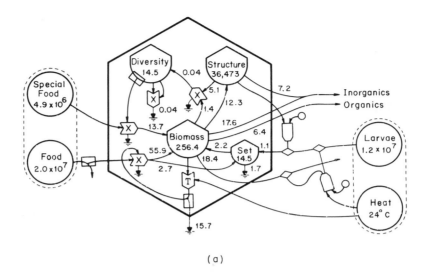

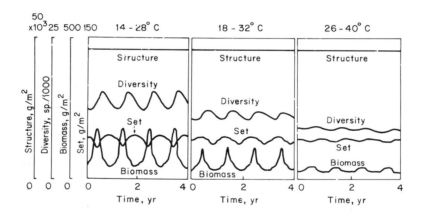

FIGURE 16-16. Model of thermally affected oyster reef system (Lehman, 1974): *(a)* energy diagram [flows are g/(m² · day)]; stocks are g/m²; diversity is in species/thousand (stock); and addition of diversity is in species/thousand/day (flow); *(b)* simulation results for three temperature ranges; temperature variation during a year was supplied as a sine wave.

Size

For the same energy supply, less tissue is required to do metabolic work; conversely, smaller sizes are necessary at higher temperatures where energy supplies are the same. Young (1975) and Hornbeck (1979) found more and smaller blades of marsh grass *Spartina* adapting to heated waters with similar productivity and respiration.

OYSTER REEF ECOSYSTEM

Even for a population model, the main factors may be external to the population in the energy sources and sinks. For an example of a population model studied in its natural setting, see Fig. 16-16, showing an oyster population model of a reef in estuaries at Crystal River, Florida. This study was in relation to power plant effluent (Lehman, 1974). Temperature was an important varying driving function along with food supplies and the actions of consumers on the oysters. The diversity of reef associates is another factor that serves to increase the total reef energy utilization but increases the energy in work processes even more. As shown in the example, varying temperature regimes produced decreasing biomass and diversity at higher temperatures but higher metabolisms and turnover. Physiological functions take precedence over those that support

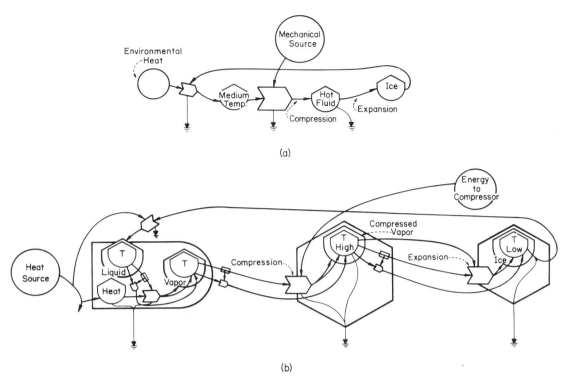

FIGURE 16-17. Energy diagram of refrigeration system: (*a*) sketch; (*b*) more complete energy diagram.

diversity. The model includes larval set and reef shells (structure). The model does not separate oyster numbers from mass. Temperature is shown accelerating food production processes and respiratory losses. Food was supplied at a source-limited rate, varying with season.

REFRIGERATOR SYSTEM

As given in Fig. 7-9, a cycle of materials driven by successive heating and cooling can transform energy of a heat gradient to mechanical work, and such a cycle done reversibly was used to derive the Carnot efficiency from conversion of heat to mechanical work.

The same cycle can be operated in reverse, converting the mechanical energy of compression and expansion into a heat gradient as either a heat pump or a refrigerator, making cold fluids and ice. Pockets of low temperature or ice constitute a high-quality potential energy, requiring considerable energy of mechanical quality to maintain it. The system of glaciers is operated by the refrigerating cycle of the earth.

As the diagram in Fig. 16-17 shows, there is a difference in quality in the heat cycle of refrigerating system, whether in technological machinery or in the large atmospheric systems. The flow to the right involves concentration, requiring energy. Higher embodied energy is developed by flow to the right, first as high temperature and then as low temperature. The expansion process that forms the ice is adiabatic (i.e., without heat exchange) and involves no change in entropy since the expansion and cooling have canceling effects. Adiabatic expansion is one that is insulated, neither importing nor exporting heat. An adiabatic change has no entropy change and thus no heat sink (e.g., see Fig. 16-17*a*). Flow back to the left is divergent, to lower quality.

DIAGRAMMING TEMPERATURE INTERACTIONS

Temperature action on pathway flows may be diagrammed with intersections of control action from heat storages or sources as shown in Figs. 16-3*a*, 16-3*d*, or 16-5. Usually interaction is a nonlinear function or the source is shown as nonlinear (i.e., exponential) interacting as a multiplier.

Where temperature actions are to be shown on

many pathways of an already complex diagram, the connecting lines may be omitted and only the starting and ending segments of the pathways shown. These line stubs may be identified with small Ts.

SUMMARY

Radiation budgets and the distribution of heat are part of the state of systems and their processes. The balance of thermal flows is an important subsystem in most systems of the biosphere. Process rates involving temperature often involve the thresholds of activation of heated populations of molecules according to exponential Maxwell–Boltzmann distribution. The action of temperature on processes that interact with sources produces different results depending on systems design.

In one sense the temperature is itself an internal energy source because it generates high-energy molecular populations that can facilitate the energy flow of pathways. Temperature differentials are energy sources that provide potential energy that drives heat engines of the environment. Many systems use energy to regulate their temperature and adapt to temperature sources so as to limit the role of temperature that might interfere with their main functions. In general, higher temperatures cause faster gross flows and greater energy demands for maintenance of structure, smaller storages, and less net yields.

Ecosystems often lose function with temperature changes to which the overall organization is not programmed.

Where resources are limiting, increasing temperature at first generates a stimulus and then negative effect. Thus temperature has a "parabolic" effect, which is the simplest general model for effects of temperature.

Loops and more complex adaptive mechanisms facilitate homeostasis and maximum power utilization.

QUESTIONS AND PROBLEMS

1. What is heat? What is temperature?

2. Diagram the main processes contributing to heat and removal of heat in the typical vegetated surface of the biosphere.

3. How is heat involved in state changes? Is the minimum heat required for a state change the same as that for the reverse change? Under what conditions is this true?

4. For a process to be real and spontaneous, what heat changes must there be in addition to the heat changes of state in the system?

5. What is the Arrhenius relationship?

6. How does the threshold concept of reaction generate the Arrhenius relationship from the Maxwell–Boltzmann distribution?

7. What is the free-energy charge of a process? What is the free energy of activation of the process? Show it on the energy hierarchy spectrum of molecules.

8. Draw models with push–pull action of temperature. What kind of response does increasing temperature produce when the push–pull units are drawing on source-limited energy flows?

9. What kind of system has a sharp crash in performance as temperature reaches a high threshold? Draw the model.

10. In the Morowitz model of utility of chemical structure, does the structure diminish with increasing energy supplies?

11. Draw models that have the following properties: parabolic temperature effect; temperature action on loops; energy input controlled to equal temperature controlled depreciation; temperature detracting from diversity; and system performance decreasing with temperature.

12. How does the spectrum of the energy distribution of molecules without a peak generate the Planck distribution of radiation with a peak? Is there a heat sink on a single energy transformation of molecular energy to radiation photon energy?

13. Discuss how size is related to temperature mechanistically and adaptively.

14. How is heat an energy source? Consider situations both with and without access to a heat gradient (hot source and sink).

15. What is temperature adaptation? Describe the mechanisms that may be operating in responses of higher organisms.

SUGGESTED READINGS

Alexandrov, V. Ya. (1977). *Cells, Molecules, and Temperatures* (translated by V. A. Bernslam), Springer-Verlag, Berlin. 329 pp.

Andrewartha, H. G. and L. C. Birch (1954). *The Distribution and Abundance of Animals*, University of Chicago Press.

Brock, T. D. (1978). *Thermophilic Microorganisms and Life at High Temperatures*, Springer-Verlag, New York.

Campbell, G. S. (1977). *An Introduction to Environmental Biophysics*, Springer-Verlag, New York.

Collier, B. D., G. W. Cox, A. W. Johnson, and P. C. Miller (1973). *Dynamic Ecology*, Prentice-Hall, Englewood Cliffs, NJ.

Esch, G. W. and R. W. McFarlane (1976). *Thermal Ecology II*, Technical Information Center, Energy Research and Development Administration.

Gates, D. M. and R. B. Schmerl (1975). Perspectives of biophysical ecology, *Ecological Studies,* **12,** Springer-Verlag, New York. 609 pp.

Gibbons, W. and B. Sharitz (1974). *Thermal Ecology*, Symposium Series, CONF-730505, Technical Information Center, U. S. Dept. of Energy.

Goguel, J. (1976). *Geothermics* (English ed. translated by S. P. Clark), McGraw-Hill, New York.

Halldin, S. (1979). *Comparison of Forest Water and Energy Exchange Models*, American Elsevier, New York.

Johnson, F. H., H. Eyring, and M. J. Pollisar (1954). *The Kinetic Basis of Molecular Biology*, Wiley, New York.

Monteith, J. L. (1973). *Principles of Environmental Physics*, Edward Arnold, London.

Morowitz, H. J. (1978). *Foundations of Bioenergetics*, Academic Press, New York.

Mount, L. E. (1979). Adaptation to Thermal Environment, University Park Press. Baltimore. 332 pp.

Palmer, J. D., F. Brown, and L. N. Edmunds (1976). *An Introduction to Biological Rhythms*, Academic Press, New York.

Quinlan, A. V. (1981). Thermochemical optimization of ecological processes. In *Energy and Ecological Modelling*, W. J. Mitsch, R. W. Bosserman, and J. M. Klopatek, Eds. Elsevier, Amsterdam, pp. 635–640.

Rose, A. H. (1967). *Thermobiology*, Academic Press, New York.

Smellie, R. M. S. and J. F. Pencock, Eds. (1976). *Biochemical Adaptation to Environmental Change*, Biochemical Society, London. 239 pp.

Thelkeld, J. L. (1962, 1970). *Thermal Environmental Engineering*, Prentice-Hall, Englewood Cliffs, NJ. 495 pp.

Witwell, J. C. and R. K. Tiner (1969). *Conservation of Mass and Energy*, McGraw-Hill, New York.

CHAPTER SEVENTEEN

Complexity, Information, and Order

Complexity is a property of systems concerned with component parts and their connections. Complexity is measured as permutations, entropy, information content, and statistical parameters and by energy flows. This chapter concerns ways of describing system complexity, the nature of depreciation, microscopic and macroscopic states, the relationship of energy and organization, the distribution of order and disorder, and the adaptive roles of structure.

COMPLEXITY OF UNITS, ARRANGEMENTS, AND CONNECTIONS

Specification of Units

A progression of situations of increasing complexity due to the addition of more units is shown in the diagrams in Table 17-1. Addition of each unit requires energy for its maintenance (see Table 17-1). Examples are populations of organisms of the same type. The complexity is in the number of units.

Morowitz et al. (1978) suggested a way to compare ecosystems using data on the taxonomic compositions. The presence or absence of species in ecosystems can be represented with a binary coding, with presence being indicated with logic 1 and absence with logic 0. Similarity of systems is measured by numbers of matching species. Since each comparison is equal to 1 bit of information, the similarities are expressed in information terms.

Permutation of Units in Sequence

The diagram in Table 17-2 shows a progression where the permutations of units is considered as a series, each one identifiably different. There is a more rapid increase in the numbers of permutations as more units are added. The permutation of the group is the number of possible arrangements. Letters in a message or sequence of seeds blowing to a plot with the wind are examples. Whether all the units (a to d) are the same, are identified, or are unknown, the possible permutations measure the complexity of the situation, the potential complexity in permutations. Possible permutations are given in factorial expressions.

Connections and Complexity

In the diagrams in Table 17-3, increasing numbers of units are connected with lines indicating functional–structural relationships. Sometimes connections are permanent as when roots are grown, but more often connections are made when the functions operate. For example, an animal eats its prey. As the number of units increase, the number of possible connections increases very rapidly. For the case where there are two connections between each unit (one forward and one feedback) and one connection of each to itself, the number of pathways equals the square of the number of units n^2. Without the pathway to itself, the formula is $n^2 - n$. With only one pathway between each, the formula is $(n^2 - n)/2 = C(n,2)$, the combination of n items taken in pairs (pathway connections). The complexity involved in connections is quadratic. Complexity of web connections is sometimes called *connectivity*. Valentinizzi and Valentinizzi (1963) calculated information for the connecting bonds of chemical molecules. Jacobson (1959a, b) calculated information in electrical circuits.

Using a matrix describing presence (1) or absence (0) of pathways (adjacency matrix), Bosserman (1980, 1981) studied a complexity index defined as the ratio between the total direct and indirect pathways and the square of the number of units between which pathways are connected. The indirect path-

TABLE 17-1. Information for Specification of *a Particular Unit* Among a Group of Unspecified Units

n	Number of Units and Hierarchy of Steps in Specifying One	Steps to Choose t	Possibilities $N = 2^t$	Probabilities $p = 1/N$	Information[a] for Identification Bits, $I = \log_2 N$	Bits per Individual $H = I/n$
1		0	1	1	0	0
2	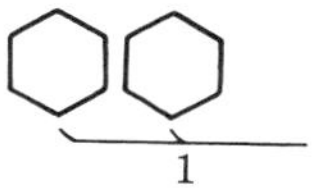	1	2	0.5	1	0.5
4	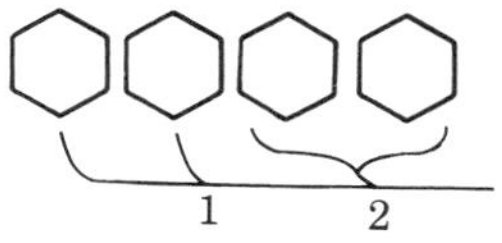	2	4	0.25	2	0.5
8	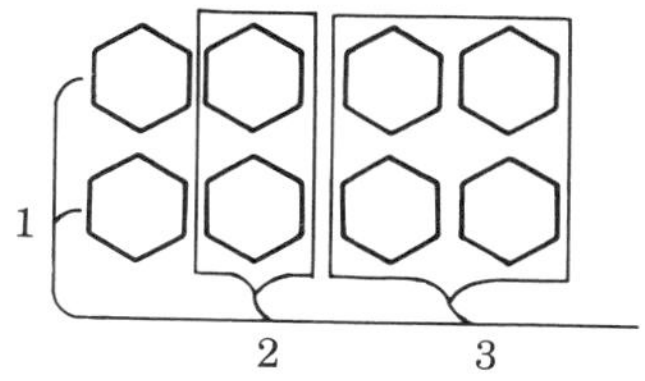	3	8	0.125	3	0.375
16	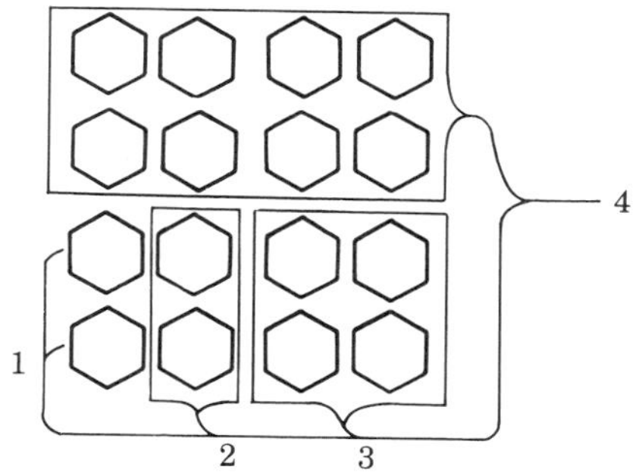	4	16	0.0625	4	0.25

[a] Number of binary decisions necessary to specify one among the units.

ways are connections reached by passing through one or more other pathways first (see Fig. 13-27). A normalized index was obtained by dividing by the number of units and ranges from 0 to 1. The effect of adding or removing a pathway is calculated as an index of pathway "importance."

INFORMATION THEORY

Whereas the word "information" refers to many related quantities and ideas, one specialized use of the word can be defined according to information theory. Here information is defined as the *logarithm of the possibilities*. It is the logarithm of the complexity. Various possibilities are indicated by number in the diagrams in Tables 17-1–17-3. The logarithm of this number is the information content. Since logarithms by definition are exponents of some base number, they increase more slowly as the number of possibilities increase. The log increases linearly as the possibilities increase exponentially. Adding exponents has the effect of multiplying the numbers. Possibilities resulting from combining independent systems are the product of the possibilities in the combined system. Addition of information content has the effect of multiplying the possibilities. Thus information is an additive function of complexity. Given the total information, one may subtract the information associated with a subset of the total and have the remainder of the information associated with the remaining group of possibilities.

TABLE 17-2. Permutations and Information[a] Inherent in a Series of Units

Number of Units n		Permutations of Units $n!$	I = Information = Bits = $\log_2 n!$	Bits per Unit $H = I/n$
1	⬡	$1 = 1\ (a)$	0	0
2	⬡⬡	$2 \times 1 = 2\ (ab, ba)$	1	0.50
3	⬡⬡⬡	$3 \times 2 \times 1 = 6\ (abc, acb, bac, bca,$ $cab, cba)$	2.585	0.86
4	⬡⬡⬡⬡ etc.	$4 \times 3 \times 2 \times 1 = 24\ abcd, acbd, adbc, abdc,$ $acdb, adcb, bacd, badc,$ $bcad, bcda, bdac, bdca,$ $cabd, cadb, cbad, cbda,$ $cdab, cdba, dabc, dacb,$ $dbac, dbca, dcab, dcba$	4.58	1.14

		Possibilities
[a] ⬡ to pick the first is one of four possibilities (see Table 17-1) =		4
⬡ to pick the second is one of three remaining possibilities		3
⬡ to pick the third is one of two remaining possibilities		2
⬡ the fourth is specified when the third is chosen		1

The possibilities involved in arranging the four are their product: $4 \cdot 3 \cdot 2 \cdot 1$, which is written 4 factorial, 4!, the permutation $P(4,4)$, four different things taken four at a time.

Several kinds of information are used, depending on the base number. Where the logarithm is to the base 2, the unit is called a *bit*. One bit represents the information in a decision between two choices as in the flipping of a coin. There are two possibilities and one choice. Log of 2 to the base 2 is 1. The presence or absence of a pathway represents 1 bit. As shown in the diagrams in Table 17-1, the number of bits is the number of decisions required to specify one possibility.

For other purposes, the base $e = 2.718$ is used and the logarithms are called *natural logs*. The unit of information measured as log to base e is called a *nit* and to the base 10 a *Hartley*.

The more matter involved, the more information there is. Therefore, information concentration is important and sometimes expressions in space are used such as bits per area or bits per volume. Table 17-4 summarizes information formulae as log of various expressions for possibilities.

Information as Uncertainty or Certainty

We have defined complexity of a system as the number N of possible patterns (connections, arrangements, etc.) of a system. Complexity may also be measured as $\log_2 N$, the syntactical, geometric, or topological information of the system. Neither indicates whether a pattern exists. The numbers measure uncertainty or certainty without distinguishing which.

The same number measures the degree of disorganization if the components are not organized, the degree of organization and order if the components are organized or ordered, the lack of information if

TABLE 17-3. Information Inherent in Potential Connections Between Units

Number[a] n^a		Combinations of Connections n^2, Including Presence and Absence	I = Information $\log_2(n^2)$, Bits	Bits/Unit $H = I/n$
1		$1^2 = 1$	0	0
2		$2^2 = 4$	2	1
3		$3^2 = 9$	3.17	1.06
4		$4^2 = 16$	4.00	1.0
5		$5^2 = 25$	4.64	.93

[a] Number of units being connected.

the situation is unknown, or the degree of knowledge if the pattern is known. There is no way to tell merely from the numbers whether a situation is known or unknown, organized or disorganized, or ordered or unordered. Information measures the complexity of the situation without indicating whether there is realization of combinations, connections, or knowledge of them. Gallagher (1968) terms the difference between total uncertainty and the information of the known part of a system the *conditional entropy*.

Probability: Reciprocal of Possibility

Probability is the fractional expectation. A coin with two sides has two possibilities. The probability of any one side is 1 of 2, or 1/2. A second coin is added also with two possibilities and a probability of 1/2. For the two coins, the possibilities are now 4, and the probability of any one combination 1/4. With a third coin, the possibilities increase to 8 and the probability to 1/8.

Probability is the reciprocal of the number of symmetric possibilities. Since information is the logarithm of the possibilities, it is also the logarithm of the reciprocal of the probabilities. Since the log of a number is the negative log of the reciprocal, information is the negative log of the probabilities. Relationships are given in Fig. 17-1. When there is only one choice, the possibilities and probabilities are one and the information is zero. The log of one is zero.

Negentropy

Sometimes information and entropy have been used with the sign reversed. Sometimes this has been

TABLE 17-4. Partition of Components of Information[a]

Information Component	Specify the First	Specify the Group (the message)[b]	Connections
Uncertainty in all units n	$\log n$	$\log n!$	$\log n^2$
Uncertainty about types	$\log N$	$\log N!$	$\log N^2$
Uncertainty in a group of one type	$\log n_1$	$\log n_1!$	$\log n_1^2$
Uncertainty remaining after specifying groups[c]	$\log \dfrac{n}{\Pi n_i}$	$\log \dfrac{n!}{\Pi n_i!}$	$\log \dfrac{n^2}{\Pi n_i^2}$
Calculations given in	Table 17-1	Table 17-2	Table 17-3

[a] There are N number of types and $n_1, n_2, n_3 \ldots n_N$ members of each type. The total number of individual members $n = n_1 + n_2 = \cdots n_N$. The symbol Π denotes product of items of the series.

[b] Stirling's formula converts factorial $\log_e n! = n \log_e n - n + \tfrac{1}{2} \log_e 2\pi n$

conversion of logarithms:

$\log_e x = 2.30 \log_{10} x$

$\log_2 x = 3.32 \log_{10} x$

$\log_e x = 0.69 \log_2 x$

[c] Dubois (1973) gives an expression for the information in the relationships of units and connectors after that in the units n and connections L is subtracted:

$$I = \log \frac{(n_i + L_i)!}{(n_i!)(L_i!)}$$

called *negentropy* (Schrodinger, 1947; Brillouin, 1962). As shown in Fig. 17-1, its graph is a mirror image of regular information and entropy. Both magnitudes increase as one increases possibilities (e.g., with addition of heat-raising temperature). Some have regarded negentropy as order and entropy as disorder—this is incorrect. Both increase in magnitude with temperature. Both are most orderly at absolute zero according to traditional thinking. The logarithm of one possibility is zero. Information and entropy of either sign are zero when choices are eliminated. Different authors use information defined with a reversed sign and use negentropy in place of entropy (Brillouin, 1962). The entropy increase and decrease associated with thermal diffusion in Fig. 7-14 may be referred to as negentropy decrease and increase, respectively. Both measure log of possibilities and thus measure order or disorder without discriminating, only with a different sign.

Information per Individual

If the total information in a situation is divided by the number of units, the information per individual is found, or the number of bits per individual (e.g., see Tables 17-1–17-3). For species in an ecosystem, this measure varies from 1 to 7 bits per individual. It is an intensity measure of information and is often given the letter H defined in Eq. (17-1):

$$H = \frac{I}{n} \qquad (17\text{-}1)$$

Formulas for Information in Populations of Several Types

In Fig. 17-2 systems are shown with two types of individual with a number of individuals of each

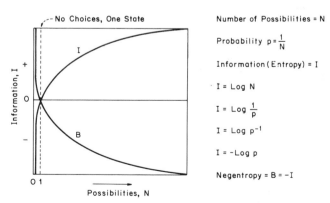

FIGURE 17-1. Some relationships of possibilities N, probabilities p, entropy or information I, and negentropy B.

type. Examples are individuals of species or numbers of letters of each type in a message.

Equation (17-2) is a form of the definition of information commonly used in enumerating information in numbers and kinds of unit (see explanation in Fig. 17-2). Here we begin with the definition that entropy is the logarithm of the possibilities and consider units of two types. The result is the commonly used expression (Shannon and Weaver, 1946) for estimating information in an assemblage of several types in different frequencies:

$$H = -\sum_{1}^{N} p_i \log p_i \qquad (17\text{-}2)$$

The log of the number in each group is the information necessary to specify each in its group. The total information is the information to specify each, times the number of each type. Division by the total number of units and substitution of probabilities for possibilities gives the expression for information in terms of probabilities. When generalized to many types, the Shannon–Wiener expression results. (It is sometimes called Shannon–Wiener expression because of independent use by Norbert Wiener.) The formula is used as an index of diversity for assemblages of species of different types (Margalef, 1956). The numbers of organisms are like the letters in a message. Margalef also arranged species as the column of a matrix and successive census counts of individuals as rows. Calculation of information content provided a measure of temporal organization.

Partitioning Information

The total information in an assemblage of units is the log of the number of units. However, many of these are duplicates. The information content in the relationships among types (species) is the logarithm of the number of types. The information in diversity and organization or lack of it in assemblages of species of different numbers is the total information minus that associated with duplicate units of the same species. Thus, as given in Table 17-1, one may partition the total information to isolate components that accompany different parts of the assemblage (Quastler, 1953).

Information and Symmetry of Numbers

As shown in Fig. 17-2, the information content is plotted as a function of the ratio of numbers of individuals of two types. Information content is highest when the numbers are in equal ratios. A similar derivation can be made to show that this is true where there are many types. More complexity and energy is required for addition of species when the species are in equal ratios than in unequal ones. Patten (1972) found ecosystems to have unequal ratios of pathway flows so that information was much less than the maximum. Ecosystems do not maximize information of their webs so far as the numbers of individuals of each species is concerned.

Evenness: Ratio of Information to That Where Numbers of Each Type Are Equal

To measure the relative symmetry of the numbers of individuals of different unit types, a ratio is calculated between the observed distribution and that of the same number of types with equal numbers (H_M). In ecology this has been called *equitability H'* [Eq. (17-3)] or *relative entropy*:

$$H' = \frac{H}{H_M} \qquad (17\text{-}3)$$

Lloyd and Ghelardi (1964) provide calculation tables.

Redundancy

Redundancy is calculated as

$$\frac{H_{\max} - H}{H_{\max}} \times 100 \qquad (17\text{-}4)$$

It is one minus the relative entropy ($H/H_{\max}$). Gatlin (1972) found redundancy to increase with size. See also DNA redundancy in Fig. 15-20.

Information in the Partition of Energy Flow

The flows of energy in a web can be related to the total energy flow to express them as probabilities (as fractions of the total). The information content may be computed from the sum of the probabilities of the pathways by using Eq. (17-2). MacArthur (1955), using the model in Fig. 17-3a, suggested that more pathways increased the total information and that this was, therefore, a measure of the stability of the web, reasoning that alternative pathways provided stability for the system as a whole in case of interruption to one pathway.

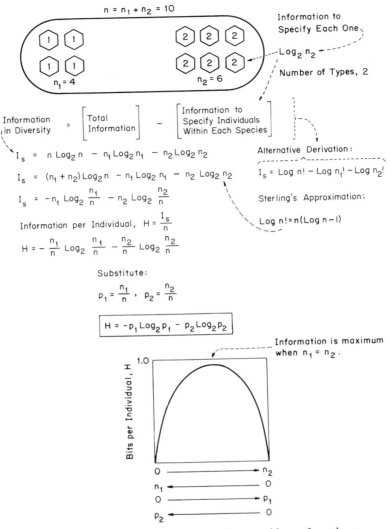

FIGURE 17-2. Information to specify assemblage of numbers of units of two types.

Rutledge et al. (1976) estimate the "metabolic throughput" information as an index of stability of having pathway choices. Figure 17-3b shows this index as a function of the sugar added to an Oregon stream by Warren and Doudouroff (1971). Increase of this supplementary energy source increased the diversity of metabolic pathways (see energy diagram in Fig. 21-21).

Van Voris et al. (1980) found correlation between complexity and stability in microcosms.

ENERGY BASIS FOR COMPLEXITY AND INFORMATION

In closed systems at equilibrium the hierarchy of energy distribution in molecules is related to the amount of stored molecular energy. The addition of energy to a closed system caused the hierarchy of molecules to increase at different energy levels, and thus to increase in complexity and information content (Fig. 7-10b).

In open systems the basis for structure was found in the flows of potential energy replacing that lost in deterioration and depreciation. The addition of more energy flow through an open system produces more structure with complexity and information.

Energy of Arrangements; Optimum Speed

Energy for organization is used in units, connections between units, arrangements, and other aspects where the energy, in doing the work of organization,

Complexity, Information, and Order

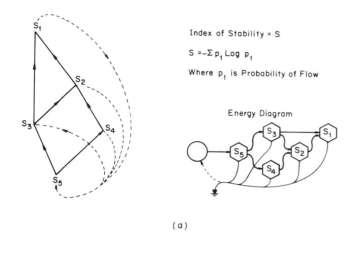

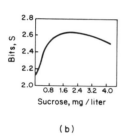

FIGURE 17-3. Information of pathway probabilities: (a) model used by MacArthur (1955) to relate stability to pathway alternatives measured by information; artificial closure of degraded energy back to energy source was made to use Markov concepts (dashed lines) (see Fig. 6-22); (b) effect of an extra energy source on metabolic information (Rutledge et al., 1976) (S = bits per path).

passes through into heat with only small amounts stored. As with other kinds of work considered in Chapter 7, the organizational work of arrangement has an optimum speed that contributes maximum useful power. For example, the trucks in a parking lot can be rearranged at different speeds. At higher speeds the energy of fuel goes into more kinetic energy before brakes are applied because energy increases as the square of the velocity. Thus more energy is required to park the trucks more rapidly. If the rearrangement leads to some energy effect on the system (and one wouldn't have organized them at all if it didn't), then if done too slowly, competitors will capture the role; if done too rapidly, the energy is needlessly wasted in braking. In such organizational work power dissipation is a cubic function of the speed (Odum, 1972). Tucker (1975) evaluates human motion.

If the process could be done with no accelerations, but only work against friction, the energy used would be less for the same speed.

Energy Flow to Units and Connections

The energy required for a system is that for the units plus that for the organization of these units. Organization may be between individuals as in the social organization of people (see Fig. 17-4a). Or it may be between types of units, but not between the individuals of one type as in organization of species where individuals are identical (see Fig. 17-4b). Here the individuals can be substituted in function for each other. Tendencies for competition may be negative to some individuals but favorable for the overall function of the population of that type.

As more units are added and more connections are developed (diagrams in Tables 17-1–17-3), the energy required for maintenance and processing increases accordingly. The energy for the units increases in proportion to number, but the energy for arrangement or connective organization increases more rapidly since the combinations possible do also. The energy required depends on the formula for connections.

The energy required may increase as the possibilities of pathways regardless of whether those organizational connections have taken place. There are either connections, or there is the energy drain that results from not connecting them due to frictions, competitions, and other interferences.

Energy of Organization as the Square of Number Organized

Representative of the formulae for connections is the square of the number of units. As shown in the diagrams in Table 17-3, the number of connections increases as the square if there is a pathway from units to themselves and one going each way between all units. The energy required for this organization is also proportional to a square of the number, if all pathways require about the same energy.

If the energy required for the pattern of organization increases as the square of the number of units, the total energy is the sum of that required to operate the units plus that necessary for the organization (see separation of these two kinds of flow in Fig. 17-3).

As shown in Fig. 17-5, the rapidity with which units can be added may be considered as the base of

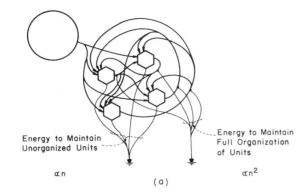

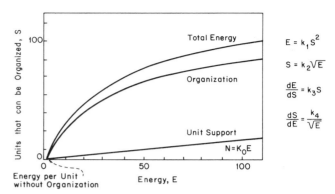

FIGURE 17-5. Units that can be organized and added as a function of the energy added. For example units may be species(S).

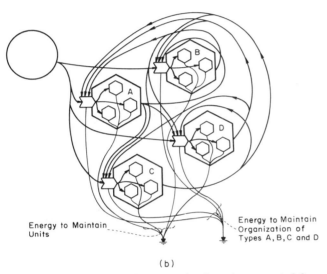

FIGURE 17-4. Energy flow to maintain units separated from that to maintain organization of units: (a) units are organized; (b) types are organized but not units.

energy for the organization is broadened to accompany successful addition. The graph has a diminishing slope as an increasing amount of energy is required for each added unit. Organizational possibilities increase more rapidly than the number of units. The addition of units where the organizational cost is quadratic is as the square root of energy. The sum is an estimate of total energy required.

Haggett (1966) gives a graph from Kansky (1963) relating connectivity of transportation networks to energy consumption. More networks were suggested with greater energy, but at a diminishing rate.

Observed graphs of species found as a function of increasing area of the ecosystem examined have declining slope, similar to the graphs in Fig. 17-5. Since the energy realms available to ecosystems in these cases are more or less proportional to the area of surface, the graphs are descriptions of species units found relative to energy regime. They have the same general form as those resulting from theory about energy of organization.

At one time some workers regarded the curve of decreasing slope as one that was approaching a level asymptote (Cain, 1930). It was assumed that one could sample until the slope was at some percentage such as 10% and infer what the total number of types was. However, the curves never level; rather, when put on semilog plot, they continue upward with accelerating curvature. This is due partly to inclusion of new subsystems as one samples in ever-widening quantity.

Information as Log of Supporting Energy Flow

As illustrated in Fig. 17-5, energy needed for organization may be postulated as proportional to the possible connections (Odum, 1967a, 1970). Information content of the possible connections is the logarithm of the possible connections which is proportional to $\log N^2$. Therefore, information is proportional to the log of the energy going into organization, and energy is an exponential function of the information content in the organization or potential organization. On semilog paper there is a straight-line relationship between energy flow and information according to this theory (Fig. 17-6b). Here the energy is proportional to the area of land counted since sunlight, rain, and other energy inflows tend to be on an area basis.

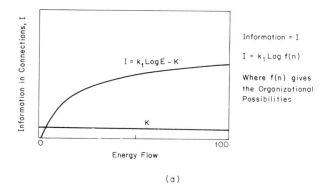

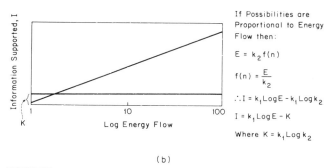

FIGURE 17-6. Information as a function of energy for organization: (a) arithmetic coordinates; (b) semilog coordinates.

Pippenger (1978) found the number of units in digital electric networks required to transmit information to increase as the log of the information.

MOLECULAR COMPLEXITY AND INFORMATION; ENTROPY

The information content of molecular states is *entropy* S. It is proportional to the log of molecular possibilities and it can be expressed as bits by using log to the base 2. The Boltzmann equation is

$$S = K \log N \qquad (17\text{-}5)$$

However, where the log is to the base e and the constant K is the Boltzmann constant (3.35×10^{-27} kcal/mole·deg) (Boltzmann, 1896), the entropy is in calories per degree Kelvin per mole (see Chapter 7). Although some people use the word "information" for macroscopic items and the word "entropy" for molecular ones, others use the terms interchangeably, since both have the same formula (the logarithm of the possibilities). What is different is an enormous difference in scale.

In Eq. 7-3 a change of entropy was defined as a change in heat divided by the temperature at which the change occurred. These two definitions can be connected.

When molecules are without energy of motion at absolute temperature, the molecules are in neat rows as a crystal as one determined state. There is no uncertainty, there is one possibility, and the logarithm of one is zero. The information content is zero, and the entropy is zero.

As energy is added as heat, the molecules develop motions and alternative states. The more states developed, the more heat that can be absorbed per degree rise in temperature. The possibilities and complexity rises and the information and entropy increase (see Fig. 7-10). The cumulative sum of the energy added is the cumulative complexity. Increments of percent change are summed. Each increment is the heat change divided by the temperature of the change, proportional to percent heat added. Summing is done by integration. The integral of a percent is a log. Summing the heat changes divided by the temperature of the change is entropy, a log function of molecular states. Information in a closed molecular system is a logarithm of probabilities.

Following Gibbs (1901), free energy is a logarithm of molecular populations, whereas at the same time energy is proportional to the square of molecular velocities. In the macroscopic relation illustrated in Fig. 17-5 species units are connected in proportion to energy so that information is a log function of energy that is a square of the number of components. Because of the nature of logarithms, information can be proportional to a quantity and to its square with appropriate change in coefficients.

Maxwell Demon

The Maxwell demon was an idea of a tiny fictitious person of molecular size who could utilize the energy of molecules one at a time by making a decision to open a small door to let molecules going in the same direction pass through while others are excluded. If this were possible, molecular energies could be upgraded in concentration contrary to the second law.

Brillouin (1962) showed that the information in bits required to concentrate this energy required as much energy as would be collected. In other words, the Maxwell demon was not possible.

However, in a system of larger scale it may be possible to concentrate energies from large populations so as to exert a local selection. Natural selection is such a choice mechanism from energy distributed in alternatives. Natural selection by one level on units of another may be a Maxwell demon (see questions regarding the universe in Chapter 27).

Spetner (1964) calculated bits added by natural selection to cellular proteins as the log of the trials N minus that of the mutations P [Eq. (17-6)]:

$$\text{Rate} = \frac{\log_{10} N}{0.352 - \log_{10} P} \quad (17\text{-}6)$$

The energy required for natural selection is that of the discarded choices (Haldane, 1957; Odum, 1971).

Microscopic and Macroscopic Entropy

The entropy of microscopic states is many orders of magnitude larger than that entropy calculated similarly for the much fewer macroscopic states of items in the biosphere of size range of ecological and anthropological entities. For instance, 6.02×10^{23} items of microscopic complexity are contained in one item of macroscopic realm, which is composed of one mole of matter (6.02×10^{23} molecules). The calculated entropy for the mole of components is 1.377 cal/degree (6.02×10^{23} bits), whereas one item of macroscopic realm is one bit (Linschitz, 1953).

The relationship of microscopic to macroscopic may be related as are the energy quality ratios. Consequently, there is high- and low-quality entropy that, like their energies, are related as the ratio of one inherent in the existence of the other.

Entropy Changes in Energy Sources

An energy source has potential energy to degrade. Stated another way, in relation to its use and sink, it has lower entropy before than after it is used. The potential energy degraded divided by the temperature is the entropy increase. It is a source because of differences in complexity and entropy. It can be either simpler or more complex than its fate after action. It can be a cold or hot source, for example. A hot source has very high entropy that decreases when used, but the entropy increase of surroundings is even greater. A cold source has a low entropy that increases more than the surroundings decrease in entropy (see Fig. 7-14).

Kinds of Entropy Change

Entropy of a state can change in several ways. Addition of heat to a population of molecules is an entropy increase. Removal of heat is an entropy decrease. Increase in the quantity of matter is an increase in complexity and entropy. Expansion of a population of units increases the entropy. However, expansion of a population of molecules also spreads the same energy over a larger volume, lowering temperature, an effect that cancels the effect due to expansion of matter. If expansion takes place without addition or removal of heat (adiabatic change), the entropy is unchanged. One can add heat and keep temperature unchanged during expansion, but this would be an entropy increase. Division of units into more, smaller units increases entropy. Entropy increases in going from solid to liquid and liquid to gas, and there is appropriate heat uptake into molecular motions that make it possible.

DISPLACEMENT FROM EQUILIBRIUM

When an equilibrium state is displaced, this is tantamount to adding a potential energy source in the

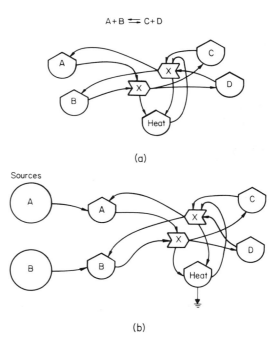

FIGURE 17-7. Energy diagram of chemical equilibrium to illustrate Le Chatelier's principle; a displacement is addition or subtraction of an energy storage: (*a*) closed system; (*b*) open system.

displacement that the system uses to return to its equilibrium state. The return of a displacement to equilibrium is a spontaneous process. Thus a lifted or depressed cork on water returns to its equilibrium level.

An open steady state can also be displaced from its steady pattern (stationary state called *equilibrium* in some fields). The extra energy added in the displacement contributes to the return to the steady state after which the energy flow through the steady state is as before. During the return from displacement, the energy flow is higher, equal to that of the steady state plus that of the extra energy to displacement.

During the return to steady state from displacement the overall entropy increase is also higher, with more potential energy going into degraded heat. This situation was described by Prigogine and Wiaume (1946) with the suggestion that steady states tend to minimize entropy generation rate, since the rate of entropy generation rate was less at steady state than in displaced state. This statement has been widely misused. Nothing is implied about the rate of generation of the steady state's regular sources. If entropy generation rate were really minimized, there would be no energy flow.

In chemistry a favorite principle for discussing equilibria of closed and open systems is the Le Chatelier's principle, which indicates displacement away from an added concentration and toward a removal. As diagrammed in energy language in Fig. 17-7, the principle is equivalent to the statement that flows go in the direction of forces from energy sources and storages.

DEPRECIATION

A structure or a storage is a displacement from equilibrium. It has potential energy and a lower entropy than if dispersed in its surroundings. It acts as a local source and spontaneously exchanges with the environment until it loses differences in its states, probabilities, and possibilities. The natural tendency is depreciation. Its loss of potential energy and increase of entropy is an expression of the second law. Depreciation is a spontaneous process.

Depreciation is more rapid where the surface:volume ratio is larger as in smaller sizes [see Eq. (3-7)]. It is accelerated by higher temperature (see Fig. 16-3).

In Fig. 17-8 depreciation is shown for structures more complex than their surroundings and for structures simpler than their surroundings. Both spontaneously lose their stored differences with an accompanying overall entropy increase.

MAXIMUM ENTROPY PRINCIPLE

The maximum entropy principle states that systems tend to develop states of maximum entropy. As reviewed by Jaynes (1978) and Tribus (1979b), the Shannon equation [Eq. (17-2)] is maximized (Levine and Tribus, 1979).

This is tantamount to recognizing that equilibrium states result from depreciation and degradation of potential energy according to the second law. The resulting equilibrium states are those of Gaussian distribution of molecular motions and skewed, Maxwell–Boltzmann hierarchical distributions of energy. These are closed-system equilibria.

The maximum entropy principle has also been applied to open systems as described in Chapter 15. Open-system patterns resemble maximum entropy patterns expected for closed systems; this may be because power is maximized (see Fig. 15-7 and discussion in Chapter 15). Entropy is maximum when units are to be specified as in Fig. 17-2, but if the units carry energy, it generates a hierarchy, and maximum entropy is the exponential distribution (see Chapter 15).

Hiramatsu and Shimazu (1970) simulate the development of a drainage basin with characteristics of Horton's hierarchy by a computer program that selects segments at random, but erases those that don't have good energy properties (e.g., forming loops). This is tantamount to a choice generator followed by maximum power selection. The information content of the network was calculated as lower than the maximum entropy that would occur for equal frequencies of segments.

Di Toro (1969) predicted salinity distribution in an estuary by maximizing the entropy calculated from diffusion probabilities.

ENERGY FLOW TO MAINTAIN INFORMATION

The increase in energy flow in recent centuries has fostered a culture with faith in the limitless potentialities of human roles. Examples are Buckminster Fuller's (1969) belief that continued accumulation of

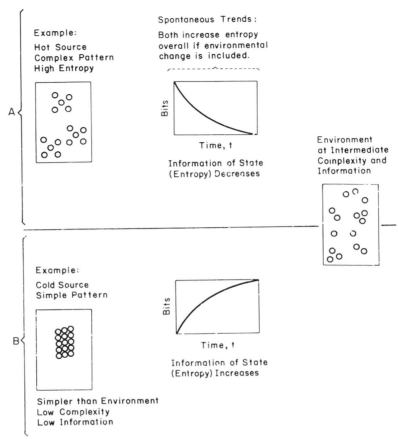

FIGURE 17-8. Spontaneous processes eliminate differences in information (entropy) with overall increase in entropy: (A) initially more complex than surroundings; (B) initially simpler than surroundings.

knowledge allows more to be done with less resources. Another example is Teilhard de Chardin's concept of increasing complexity and consciousness (O'Manique, 1969). However, information, like any other storage, has depreciation, and the embodied energy loss is very great. Large energies are required for replacement and maintenance of information. Much growth of information storage may not be possible without growth of energy flows.

MODEL OF ENERGY AND INFORMATION

In Fig. 17-9 energy flows in from low-entropy-source materials to make high-quality patterns whose information content is high. Energy flow maintains structure by keeping some of the component materials at low entropy (simple) and some of the higher-quality organization at high information (complex) (high entropy). The simple structural materials and the complex patterns both depreciate with trends toward the entropy of their surroundings as in Fig. 17-9. The first has entropy increase and the second has entropy decrease, but the overall entropy change is an increase since the first one is numerically larger.

Because of the high-energy quality and embodied (accumulated) energy of the high-quality complexity, depreciation there represents more calories of replacement value per bit than does depreciation of the lower-quality materials. Bits alone are not a measure of energy flow or energy quality. The comparison of bits may require multiplication by their energy transformation ratios.

PERSPECTIVES ON ORDER AND DISORDER

A general model of order and disorder is given in Figs. 17-10 and 17-11a. A source of potential energy interacts with disordered materials to generate

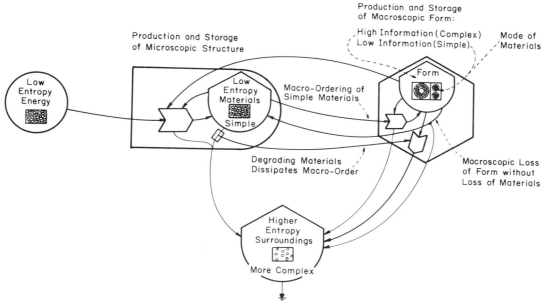

FIGURE 17-9. Model of useful molecular and macroscopic structure; one low in information, one high in information. Both require energy flow to maintain.

structure. This structure is regarded as order because it was transformed by an energy flow and is a storage capable of depreciation. The storage may be low entropy or high entropy, but its state will generate an increase in entropy when it is degraded. The degraded materials are available for upgrading again. The materials move in a closed circle. See Fig. 2-7a.

If materials are not circulating well because recycle is slow, a source of potential energy may be used to pump an orderly state to a disorderly one. Because downpumping is in the direction of diminished order and in the direction of depreciation, less energy is required than in production of the order. This model defines order and disorder as related to potential energy and material cycles. What is produced in the "ordered" storage may be either simple or complex, hot or cold, or low entropy or high entropy, provided that its structure constitutes potential energy storage relative to the surroundings and relative to the state of "disordered" materials tank on the left. The order–disorder cycle shown may also imply a spatial concentrating action in production and spatially dispersal action in the recycle of disordered materials.

Structures maintained by energy flow include art which has chemical materials apart from equilibrium capable of amplification and with information content (Arnheim, 1971).

Experimental Study of Order and Disorder

Some experiments we conducted to relate energy requirements for ordering and disordering and the action of disordering pulses in causing a subsequent stimulus to the ordering process are diagrammed in Fig. 17-11. Studies were made of terrestrial and aquatic microcosms. After metabolism of production and consumption was established, they were pulsed with destructive disordering. This was followed by a surge of production as the regenerated materials became available to stimulate the production again. The model in Fig. 17-11a is almost the same as the generalized one in Fig. 17-11. Calculations showed more energy required to develop order than to disorder. For other examples, see discussions of urban renewal, war, and political models in Chapters 24 and 25. Vela and Peterson (1969) discuss models of ultraviolet bacterial inactivation and reactivation with white light for which Fig. 17-10 is relevant.

Ambiguity of the Word "Order"

As we have seen, order sometimes has two opposite meanings in the same sentence. On the one hand, it refers to simple structure as in a low-entropy crystal. On the other hand, it refers to complex, high-entropy structures maintained away from equilib-

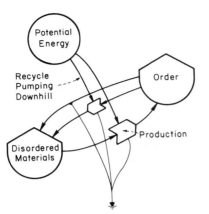

FIGURE 17-10. Concept of order and disorder and the relationship of potential energy sources. No heat sink is required on disordered materials since they are at equilibrium with surroundings.

rium because of their use in fostering processes. In Fig. 17-9 it refers to both together. Perhaps this word should be discarded for the purpose of discussion of structure.

Sometimes the word "entropy" is used to mean disorder in a vague collective sense disregarding and sometimes incompatible with its quantitative definitions (Rifkin, 1980). The possibility that the molecular states gain functional structure as their entropy is increased (opposite from classical views) is considered in Chapters 18 and 27. When entropy is increased, molecular complexity increases, which may be adaptive whether or not at equilibrium.

Our culture often displaces complex, highly organized nature for simpler lawns, gardens, and row crops which seem more structured to the simple human understanding. Which is to be regarded as the

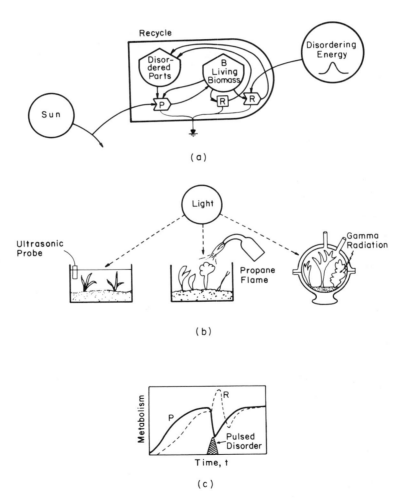

FIGURE 17-11. Microcosms used to show the high-amplifier action to high-quality disordering energy pumping toward disorder, compared to the slower cumulative transformation of low-quality energy into order: (a) energy diagram; (b) examples from Richey (1970) and Odum and Lugo (1970); (c) aquarium metabolism with ultrasonic pulse (Richey, 1970).

more ordered depends on which of the two meanings is intended (simpler or more away from equilibrium). Complex natural ecosystems are more ordered in the latter sense.

NOISE AND VARIATION

The word "noise" is applied to populations of disorderly energy flows and storages that accompany flows of energy. There are many distributions of the variations and fluctuations, which are compared to some mean or expected value. Variance is often used as a measure of noise. Graphs can be drawn as in Fig. 15-21 for the distribution of variance with other variables such as energy or size.

Separation of fluctuations due to noise from variations that have explanations in models that are part of the orderly surviving system can be done by subtracting components of variance as done in such statistical procedures as the analysis of variance. Noise is regarded as the residual variation after identified causal effects are subtracted. Noise may also be regarded as the sum of the hierarchy of pulses of many smaller levels of organization (see Chapter 27).

POWER FLOW AND VARIANCE

Another hypothesis relates power and variation. Variation was represented as a storage in Fig. 8-24 and used as a stimulation feedback factor in Fig. 17-12. The greater the flow of energy, the greater is the variation generated. Taylor (1961) found the variance of populations to be proportional to the fractional power of the arithmetic mean (Elliott, 1977). Noise of various kinds increases with power flow as a spinoff from channel energy flows. The variation may be useful for systems for such adaptive purposes as creativity, providing choices, and assistance with recycle. New structures, evolution, and progress require variability for a system to interact and choose. A model for the formation of variation and its feedback to aid the main power flow is given in Fig. 17-12. Notice the energy storage marked as variance, representing the variance being maintained by the power flow. Variance has dimensions of energy when the quantity that is varying has energy in proportion to its square as with molecular velocities or atmospheric turbulence. In the latter field turbulence is described with velocity deviations squared (Reynolds stresses).

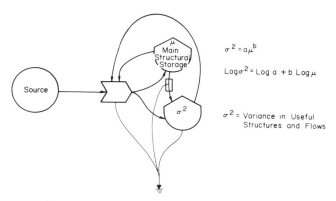

FIGURE 17-12. Variance as an energy storage with properties that feed back to aid production such as creativity, choices for selection, and useful disordering actions. In populations variance is related to stock by Taylor's (1961) equation given here.

VARIANCE AND INFORMATION (BITS)

Variance has been used as a measure of information by Fisher (1921) since it measures how effectively a mean can indicate the presence of populations of variation producing uncertainty. Greater variance indicates greater disorder and uncertainty associated with statistical measures and transmission of information.

The two measures, variance and information (log of possibilities), are related. The information content in bits can be calculated from statistical measures. For a normal distribution, the information content is the log of the normal distribution (Shannon and Weaver, 1949):

$$H = \log \sqrt{2\pi e}\, \sigma \qquad (17\text{-}7)$$

where σ is the *standard deviation* that is the square root of the variance in normal distributions. See Fig. 18-9.

The information content of a normal Gaussian distribution is maximum (Goldman, 1953). Thus a hierarchical distribution of energy for which the variables are normally distributed has maximum entropy for that number of units. See Fig. 15-11.

OVERALL RATIOS OF SUPPORTING ENERGY FLOW TO STRUCTURE

To maintain structure, the energy flow depends on the inherent rates of disordering. In Eq. (3-1) we gave the greater driving forces from a storage that has more area per unit volume, with volume per unit area expressed by the coefficient C. Smaller items depreciate more rapidly and thus their energy flows have to be greater. The ratio of energy storage to energy flow is the turnover time τ. In biological theory, turnover time has often been related to the higher qualities recognized in large organisms. Large turnover time means much structure per unit energy flow required. See Fig. 15-18.

Schrödinger (1947) helped many see the way structure was maintained by continuous inflow of even more structured fuel supply. Low-entropy structure was kept low by a continuous inflow of low-entropy fuel that was spontaneously increasing in entropy. Exportation of high entropy resulted in the maintenance of low-entropy structure inside. See also Trincher (1965).

Oparin (1962) quotes T. Matsunoya, using respiration as a measure of structure. The ratio of respiration to structural biomass may be appropriately called a *Schrodinger ratio* (Odum, 1960, 1968). If the energy flows are divided by the temperature to obtain entropy change, Eq. (17-8) results. This describes the entropy generation rate J_S necessary to maintain low entropy of the structure S relative to surroundings:

$$\frac{\text{Entropy generation rate}}{\text{Entropy of structure}} = \frac{\text{respiration}}{\text{free energy stored}}$$

$$= \frac{\text{respiration}/T}{\Delta F/T} = \frac{J_s}{S} \quad (17\text{-}8)$$

Valentinizzi and Valentinizzi (1963) estimated bits of information developed in organic chemicals per unit heat released to the environment there generating entropy (10^{13} bits/mole required 2400 kcal/mole). Leopold (1975) suggests a plot of reciprocal of entropy and turnover. Since turnover is reciprocal of specific metabolism, the slope would be the Schrodinger ratio defined previously.

Morowitz (1968), recognizing the balance between storing and depreciation, suggested a ratio to measure ordering tendency, which he called *L function*. Figure 16-8 includes this procedure. This is the ratio of free energy stored and the energy of thermal motion kT, where k is Boltzmann's constant. The numerator is potential energy and the denominator is the energy in molecular action, visualized as a measure of disorder. Thus the ratio was one of order to disorder. His L prime ratio was the storage to disordering flux that is turnover time of depreciation. The ability to maintain potential energy of order depends on a flow of potential energy to counter effects of thermal disorder. Helmholtz' free energy was used instead of Gibbs' free energy. Helmholtz' free energy does not include pressure–volume energy of expansion under constant pressure of the atmosphere that is included in Gibbs free energy [Eq. (7-6)].

Entropy Evaluation: Tropical Forest Model

Based on studies of a rain forest in Puerto Rico, metabolism was determined for near climax forest and for an exotic yield plantation, both growing on adjacent land (Odum, 1970a). The models in Fig. 17-13 summarize the results and differences. Energy of the plantation goes into yield with little structure or diversity maintained, whereas the wild forest puts its energy into structure and diversity, thus capturing more insolation, recycling better, and generating higher gross production. This study helped to understand how ecosystems achieve maximum power. Energy goes not into maximizing net production, storages, reproduction, or yields, but into the diverse means for maximizing gross productivity (power).

The energy increase due to structure and diversity may be estimated as the difference in gross production of the two ecosystems: $131 - 118$ kcal/($m^2 \cdot$ day) of organic matter energy quality. The energy required to maintain structure and diversity is the difference in respiration, which is about $[(131 - f_1 - f_2) - (28 - f)]$ kcal/$m^2 \cdot$ day, which is about 103 kcal/$m^2 \cdot$ day, since actual kilocalories of feedback (f values) are small. The energy flow of maintenance respiration divided by the Kelvin temperature gives the entropy change:

$$\frac{\text{Respiration}}{T} = \frac{103 \text{ kcal}/m^2 \cdot \text{day}}{(273 + 22)\text{deg}} \quad (17\text{-}9)$$

$$= 0.349 \text{ kcal/deg} \cdot m^2 \cdot \text{day}$$

The stored energy of structure in the forest from biomass studies is

$$\frac{\text{Biomass}}{T} = \frac{170{,}000 \text{ kcal}/m^2}{(273 + 22)\text{deg}} \quad (17\text{-}10)$$

$$= 576 \text{ kcal}/m^2 \cdot \text{deg}$$

The Schrödinger ratio for the forest according to Eq. (17-6) is estimated by using the numbers calculated:

$$\frac{\text{Entropy generation rate}}{\text{Entropy of structure}} = \frac{0.349}{576} \qquad (17\text{-}11)$$

$$= 0.00061 \text{ per day}$$

On a percent basis, therefore, the entropy increase required to maintain the low-entropy structure and diversity is 0.061% per day. Higher temperatures cause greater depreciation but also affect rates of maintenance metabolism as discussed as push–pull effect in Fig. 16-4b.

Disorganization Rate from Maintenance Rate

Each physical environmental situation constitutes boundary conditions that collectively exert a disorganization rate on the ecosystem contained. The disorganization rate is measured in terms of the conversion of organized circuits and microscopic free energy into disarray, irreversible dispersed heat, and the generation of irreversible entropy (entropy increase of the environment) (see Fig. 17-10). This disorganization rate depends in part on the amount of thermal randomness and hence is a function of temperature. In macroscopic-sized systems the disorganization rate also depends on disordering processes of larger dimension such as weather, frictions, destructive forces, and noise errors in network systems.

Since the climax steady state involves a balance of the disorganizing processes and maintenance processes measured by respiration, one may use the latter to measure the former. Respiration per unit of free energy of the structure was already identified as a measure of structure in climax systems. It is a measure of the severity of the environmental realm also.

Where the destructive tendency of the environment is small, more structure can be maintained with the same climax respiration for maintenance. Where the destructive tendency of the environment is great, less structure can be maintained with the same respiration for maintenance. Consequently, the maximum weight of biomass and organic storage per unit respiration possible is a measure of the destructive tendency of the environment of the ecosystem where energy sources are similar.

One may suppose that for each environment, there is a particular climax state that has the maximum metabolism possible. By measuring climax

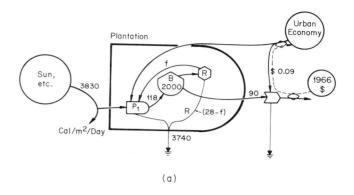

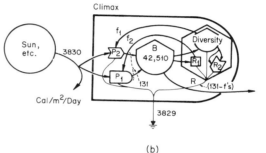

FIGURE 17-13. Comparison of ecosystem generating yield with one developing organizational diversity and structure, both operating on the same sources (Odum, 1970a), where numbers are actual calories per square meter per day and biomass B includes organic matter of wood and soil: (a) plantation of one rapidly growing exotic tree species (Cadam, *Anthocephalus chinensis*) grown for yield; (b) diverse wild vegetation, almost climax Tabanuco forest. Model includes diversity with a quadratic energy use; P_1, production without diversity based on productivity of monoculture plantation; P_2, additional gross production due to full diversity and structure; R_1, respiration of monoculture; R_2, additional respiration due to full diversity and structure.

power demand as the respiration rate of the climax, one measures the potential limits of the environment in energy units. At climax the useful energy input equals inherent dispersive energy drain of the system and its environment. Using climax respirations, we may be able to plot maps of the rate of entropy increase imposed by physical environments in places over the world's surface.

O'Neill (1976) found small perturbation responses in systems with rapid turnover times, with larger quantities of consumers acting to maintain homeostatic control by adjusting consumption (see Fig. 13-21). Energy not used to maintain storage may stabilize by supporting a larger consumer control system.

ENERGY QUALITY AND INFORMATION

Tribus and McIrvine (1971) found that higher-quality communication equipment transmitted more bits of information per calorie of heat. This pattern is consistent with the energy chain spectra discussed in Chapter 15. The higher-quality flows have less actual energy, but more control actions of an informational nature. Bits per actual calorie used by Tribus and McIrvine may be a useful quality control index.

HISTORICAL NOTE

Concepts of entropy in thermodynamics relating it to logarithm of the molecular states was by Boltzmann (1886) (Lotka, 1922a,b). This concept was used for macroscopic items and items in communication under the term "information" by Shannon (1948), Shannon and Weaver (1949), and Wiener (1948) [see review by Tribus (1979)]. Extension of information to assemblages of species in the analysis of ecosystems was initiated by Margalef (1975b, 1958). The use of variance as a measure of information was due to Fisher (1921).

SUMMARY

In this chapter complexity was measured as the number of possibilities considering units, combinations, and connections. The information measures used in information theory were defined as the logarithm of the possibilities, and various formulas were introduced for representing information. Information is higher where types of unit are in equal numbers. At the molecular level the complexity of states is measured by the information content of these states, and this measure is entropy. Sometimes increased complexity is regarded as increased disorder. An alternative view regards the molecular complexity as useful structure in the same way that macroscopic complexity is regarded as orderly. Although a departure from tradition, the trend toward entropy increase can be viewed as a trend to molecular organization. See chapter 18.

Either in closed reversible situations or in open systems, the amount of complexity and information was found to be a logarithmic function of the energy. Variance was also related to energy. Spontaneous processes require an energy source that has either a lower or higher entropy content. The difference can drive a spontaneous process, causing an overall increase in entropy.

The changes accompanying a spontaneous process included an overall decrease in free energy, overall increase in entropy, but either an overall increase or decrease in entropy changes of state within the system. Spontaneous processes are the consequence of displacements from equilibrium. Structural storages spontaneously return to equilibrium in their depreciation. Steady states resemble equilibrium states enough so that approximate prediction can be made by calculating distributions with maximum entropy.

Larger units have longer time constants, have controlling functions, last longer, and carry more information, properties consistent with the role of these units as high-quality energy controls.

QUESTIONS AND PROBLEMS

1. What are some measures of complexity, considering the number of units, types of unit, and connections between units?

2. What is information in the special scientific sense of information theory? How is it defined in relation to measures of complexity given in question 1? How are possibilities related to probabilities?

3. What are ways of calculating information of units, types of unit, and connections between units?

4. What is a bit? What is a nit? How does one convert information expressed in one log base to that of another?

5. Do complexity measures and information measures provide a scale for uncertainty, certainty, or both?

6. What is the relationship between information and entropy?

7. How do the magnitudes compare between the microscopic entropy of molecules and that of the macroscopic entities of the biosphere that are made up of those molecules?

8. How is negentropy related to entropy? How is molecular entropy related to absolute zero?

9. According to classical views, what is happening to order as heat is added to a state at abso-

lute zero? Why is more and more heat required per unit entropy added as entropy is increased? Compare this to the addition of species and ecological states with increase in energy realm.

10. What is the range of bits per individual found in macroscopic systems? What ratios of entities have maximum entropy?

11. What ideas have been offered for complexity and information affecting stability, where stability is used in the sense of continued function following pathway losses?

12. If interunit organizational costs are in proportion to possible connections, how do energy needs vary with number of units to be organized? For systems of such type, how is energy related to number of units? This question could be answered for units or for types of units where only the types are interorganized, and the duplicates are redundant extensions of the one category.

13. To which is energy most related, the complexity or the informational measure of that complexity?

14. Give two contrasting views of what order may be as related to entropy and diversity. (In one view, molecular complexity is viewed differently from macroscopic complexity.)

15. Explain the changes in free energy, enthalpy, entropy of the systems of state, and environmental entropy when an organism eats food and reconstitutes part of those resources into its own structure as new order.

16. Define the following: affinity; Helmholtz free energy; exergy; Le Chatelier's principle; evenness; Sterling's approximation; N factorial; exergonic process; exothermic process; endothermic process; White noise.

17. How is energy per unit bit related to the concept of energy quality?

18. What properties of a climax rain forest indicate level of structure and ability to compete?

19. How is turnover time related to metabolism (Chapter 3)? How is size related to position in energy-quality hierarchy? How is information storage and inheritance related to size and time constant? How is DNA related to hierarchy of components within a biological cell? How is the DNA of ecosystems in successional ecosystems compared to more complex climax ecosystems?

20. Which most controls the timing of a system: a population of many units, each with short time constant; or one coupled unit with a long time constant?

21. Compare the return to equilibrium of a displacement. Use concepts of free energy, potential energy, entropy generation rate, and maximum power. Next, describe in similar terms the return to steady state of an open system that is displaced by a small free-energy deviation. Finally, consider a large displacement of a steady-state open system in which the potential energy of the displacement is sufficient to support autocatalytic process growth (see also Chapters 9 and 22).

22. To which of the conditions referred to in question 21 does the principle of minimizing entropy generation rate apply?

SUGGESTED READINGS

Abramson, N. (1963). *Information Theory and Coding*, McGraw-Hill, New York.

Ash, R. B. (1965). *Information Theory*, Interscience, New York.

Brillouin, L. (1962). *Science and Information Theory*, Academic Press, New York.

De Angelis, D. L. (1975). Stability and connectence in food web models. *Ecology* **56**, 238–243.

Evans, C. R. and A. D. J. Robertson (1968). *Cybernetics. Key Papers*, University Park Press, Baltimore.

Feinstein, A. (1958). *Foundations of Information Theory*, McGraw-Hill, New York.

Gatlin, I. I. (1972). *Information Theory and the Living System*, Columbia University Press, New York. 210 pp.

Goldman, S. (1953). *Information Theory*, Prentice-Hall, Englewood Cliffs, NJ.

Hancock, J. C. (1961). *An Introduction to the Principles of Communication Theory*, McGraw-Hill, New York.

Jantsch, E. (1981). A central aspect of dissipative self organization. In *Autopoiesis, a Theory of Living Organization*, M. Zeleny, Ed. North Holland, NY. pp 65–90.

Jantsch, E. ed., (1981). *The Evolutionary Vision*. Westview Press, Boulder. 200 pp.

Kelly, F. P. (1979). *Reversibility and Stochastic Networks*, Wiley, NY. 230 pp.

Kerner, E. H., Ed. (1972). *Gibbs Ensemble: Biological Ensemble*, Gordon and Breach, New York. 349 pp.

Kitaigorodskiy, A. E. (1967). *Order and Disorder in the World of Atoms*, Springer-Verlag, New York. 135 pp.

Levine, R. D. and M. Tribus (1978). *The Maximum Entropy Formalism*, MIT Press, Cambridge, MA. 498 pp.

Locker, A., Ed. (1973). *Biogenesis, Evolution, Homeostasis*, Springer-Verlag. 190 pp.

Margalef, R. (1956). Information theory in ecology. *Gen. Syst. Yearbook* **3,** 36–71.

Margalef, R. (1968). *Perspectives in Ecological Theory*, University of Chicago Press, Chicago. 111 pp.

Morowitz, H. J. (1968). *Energy Flow in Biology*, Academic Press, New York.

Morowitz, H. J. (1970). *Entropy for Biologists*, Academic Press, New York.

Nicolis, G. and I. Prigogine. (1977). *Self-Organization in Non-Equilibrium Systems*, Wiley, New York.

Pierce, J. R. (1967). *Waves and Messages*, Doubleday, New York.

Quastler, H. (1953). *Information Theory in Biology*, University of Illinois Press, Urbana.

Raisbeck, G. (1963). *Information Theory*, MIT Press, Cambridge, MA.

Reza, F. M. (1961). *An Introduction to Information Theory*, McGraw-Hill, New York.

Riedl, R. translated by R. P. S. Jefferies (1978). *Order in Living Organisms*, Wiley, New York. 311 pp.

Rothstein, J. (1958). *Communication, Organization, and Science*, Falcon Wings Press, Indian Hills, CO.

Sacher, G. A. (1978). Longevity and Aging in Vertebrate Evolaution. *Bioscience*, **28,** 497–501.

Schlegel, R. (1961). *Time and the Physical World*, Michigan State University Press. East Lansing, Mich. 211 pp.

Shannon, C. E. and W. Weaver (1949). *Mathematical Theory of Communication*, University of Illinois Press, Urbana.

Tribus, M. (1961). *Thermostatics and Thermodynamics*, Van Nostrand, Princeton, NJ.

Tribus, M. (1969). *Rational Descriptions, Decisions, and Designs*, Pergamon Press, Oxford.

Tribus, M. and E. McIrvine (1971). Energy and information. *Sci. Am.* **225,** 179–184.

Trincher, K. S. (1965). *Biology and Information* (English translation), Consultants Bureau, New York. 93 pp.

Von Foerster, H. and G. W. Zopf, Eds. (1962). *Principles of Self Organization*, Pergamon Press, New York.

Weinberg, G. M. (1975). *An Introduction to General Systems Thinking*, Wiley, New York. 279 pp.

CHAPTER EIGHTEEN

Spatial Distribution and Diversity

Systems of the earth form complex patterns of spatial distribution and diversity. To relate these distributions to theories, maps, graphs, and indices of many types are used, such as land-use maps, rank-order graphs, and species–individual graphs. Explanations of patterns are sometimes sought in stochastic processes and random distributions and sometimes in causal mechanisms and adaptative functions. In this chapter distribution and diversity are considered, especially in relation to distribution of energy sources and to the hierarchy of system structure. For other approaches see Pielou (1975, 1979).

SPATIAL HIERARCHY

Even when energy inflows are spatially uniform as with an even distribution of land surface, rainfall, and sunlight, uneven distributions of high-quality components develop. These are hierarchical, as extensive work in geography and ecology has shown. For example, Cristaller (1933) found towns and their surrounding agricultural support in hexagons (see Fig. 18-1). Smaller subordinate towns were related similarly to larger ones. These relations were called *central place hierarchies*. Berry et al. (1976) review studies of urban hierarchies, including accounts by Al-Muqaddasi in 986 A.D.

Similar patterns are found in ecosystems of the sea and the forest, where distributions of plant production and plankton are organized to be converging to centers of consumption. Many of these hierarchies are easily visualized with webs and chains of energy processing, introduced in Chapters 13–15. See Fig. 18-2 and examples by Bonner (1965), Sheldon et al. (1972), and Steele (1977) in Fig. 18-3*b*.

Spatial Hierarchy Generated by Energy-Quality Chains

As introduced with Figs. 15-1 and 15-2, the transformations of energy that are part of surviving systems are those that converge and concentrate spatially. As the embodied energy per calorie transformed increases, the transformed energy becomes more valuable—in effect, more concentrated—but in its action more widely feeding back, amplifying and cascading outward to control the system. Several manifestations of energy hierarchy are shown in Fig. 18-2 (see also Figs. 2-1 and 15-1). Figure 18-2 shows the concentration of energy into units of higher energy embodiment, being transformed in stages, forming a hierarchy. Whereas the actual size of the concentrated units increases, often with more actual calories, the territory occupied, controlled, and influenced by the high-quality units becomes much larger per unit as the number of units decreases higher and higher up the chain. Each unit has two size dimensions, the size of its structure at the core and the size of its territory.

The characteristic spatial distribution found where energies are planar and evenly distributed is hierarchical when viewed spatially, and tends to be exponential when viewed in a spectral summary graph (see Chapter 15). Many properties of observations and sampling can be readily explained by hierarchical energy transformation and the resulting distributions.

Greater Pulsing of Larger Units

Another hierarchical property of an adapted system is the pulsing caused by growth and renewal cycles

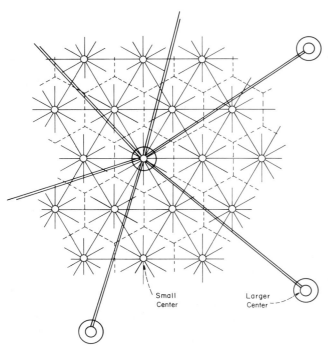

FIGURE 18-1. Concept of central place hierarchy where energy sources are spatially uniform following [Cristaller (1933)].

of the larger high-quality units toward which other units converge energies, as shown to the right in Fig. 18-2. Because of the diverging feedback of control actions by the pulsing at the top of the hierarchy, the rest of the pattern develops similar frequencies of variation. See pulsing models in Fig. 11-21 and theory of pulsing and hierarchy in Chapter 15. The higher the position in the hierarchy, the larger the spatial territory and the longer the time period of its cycle of growth and renewal. For example, Sacher and Staffeldt (1974) develop regression lines relating life span to size. In Fig. 18-3 Bonner (1965) plots size and generation time. Heron (1972) plots generation time and Fenchel (1974), metabolic rate and weight. McNaught (1979) plots components of oceanic plankton ecosystem on a useful plot of territory and temporal period. Sheldon et al. (1972) relate unit size to replacement time (Fig. 18-3b).

Greater Energy Flow Develops More Spatial Hierarchy and Complexity

Examination of many systems shows the principle that more energy flow develops more spatial structure. One example is the progression of fluid vortices that develop as one increases the temperature gradient through an oil that is located between two glass plates. The cells of convection are readily observed because flow affects the index of light refraction. As the energy applied increases, hexagonal cells develop; then forming row patterns that seem fairly regular; and with more energy developing, more complex hierarchies form that seem less regular (Bénard, 1900; Gmitro and Scriven, 1966).

Something similar occurs in farming and in deserts, where at lower levels of photosynthetic production there are regular rows, but when full photosynthetic potential develops as in the wild forest after a long period of succession has replaced monocultures, the hierarchical relationships are so complex that they no longer seem regular to the human eye.

Something in the evolution of aesthetics and understanding of value of nature follows this trend, where a simple pattern at lower energy seems neat and preferable to some, whereas the full natural complexity is recognized as value by others whose role is related to that complexity.

These concepts are sometimes counterintuitive. Some people think that the more vigorously one stirs a fluid, the more homogeneous it becomes, which is not true. The more energy put in, the more complex becomes the structure of turbulence and the longer the energy hierarchy of units of different quality.

Energy and Territories of Centers

With more energy flow, more total system activity can be supported with more high-quality centers. With higher energy flow density, less area is required to support each center and more centers develop with small territories. Smaller territories per center are observed in animal populations (Lack, 1966; Murray, 1979). Similarly, more towns and cities occur in areas with more energy flow.

High-energy density also allows emergence of new categories with larger areas with superhigh-quality centers to which the smaller territories are hierarchically converging. Support of larger carnivores and support of dominant national powers are examples.

DISTRIBUTIONS CONTROLLED BY SHAPE OF ENERGY SOURCE

The distribution of the structure developing for energy use is affected by the special shapes of the energy sources when they are not uniform. Whereas sun, rain, and some uniform land surfaces provide planar inputs, others are in the shape of lines,

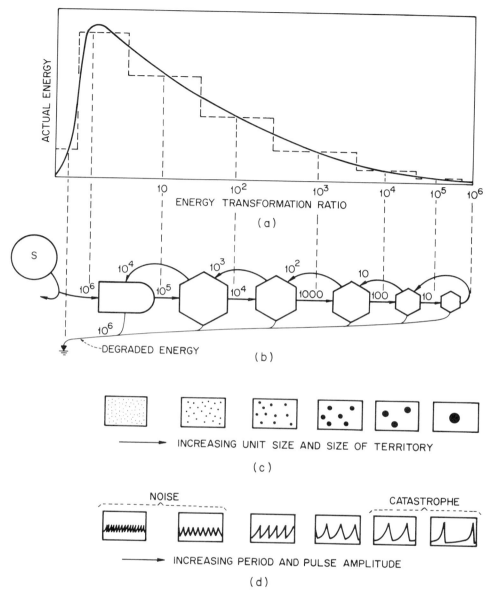

FIGURE 18-2. Concepts of size, territory, and period in relation to energy flow hierarchy: (a) energy quality spectrum; (b) energy transformation chain; (c) spatial patterns in an energy hierarchy; (d) temporal patterns in an energy hierarchy.

points, or other geometric forms. The surviving systems are those that orient their structure so as to feed back means to maximize the use of these special energies. If an energy source is sufficiently dominant, its spatial form may be imposed on the systems that result. Several spatial forms of energy sources and resulting systems are given in Fig. 18-4.

Point sources (Fig. 18-4c) include oasis springs in the desert, the presence of valuable minerals at a point outcrop, or the water available to a region from ice on a mountain peak. Energy in a point source tends to be concentrated and at higher quality so that it diverges outward in a control pattern, organizing lower-quality energy around it. Systems develop around the point.

Still other point sources are goods, services, and fuel arriving at a port, causing the rest of the area to be oriented around the source. The heat source from a volcanic center is a source to earth processes. A pipeline of fuel terminates in a low-fuel area. A point of debarkation is a source of human immigrants on a virgin continent. See Fig. 24-1.

Second in Fig. 18-4b is the line source such as the action of a river that is releasing energy evenly

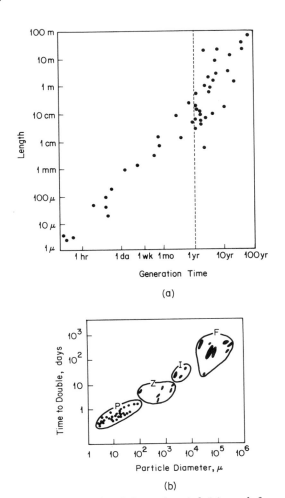

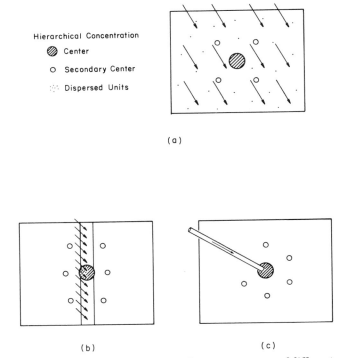

FIGURE 18-3. Graphs of size and period: (a) graph for organisms [from Bonner (1965) after May (1976)]; (b) marine organisms [from Steele (1977) after Sheldon et al. (1972)].

FIGURE 18-4. Spatial patterns for energy sources of different form: (a) planar, dilute source that converges to form hierarchy; (b) line source along which hierarchy is organized; (c) point source around which hierarchy becomes organized.

along its path. Here developments may be in a line. For example, line sources include wave energies caught at the shore, the energies of current action along the shores of a stream, or the energies flowing along a highway used by tourist facilities that develop along highways. The energy tends to organize spatial patterns on a line along which the system may converge energy to form a higher-order center.

Finally, and most common, is the planar source such as the sun, evenly falling rain, or uplift of land from below. Here energies are broadly distributed and developments are spread evenly over the earth surface, although there may be secondary concentrations due to hierarchical organization of higher-quality energies in centers scattered over the surface (Figs. 18-1 and 18-4a).

Quoting Launhardt and Palander, Isard (1956) summarizes geometric access of raw materials, markets, labor, and cost of transport affecting locational patterns of development. These are high-quality energy sources affecting distribution by their orientation. Berry et al. (1976) summarize Weber's theory for location of industrial plants as related to sources and users and mimimum transportation losses (Weber, 1929).

Interaction of More Than One Spatial Type of Source

However, as given in Fig. 14-14, most systems have several kinds of energy each of a different quality and with a mixture of area, line, and point sources interacting, with the high-quality ones amplifying the more abundant but lower-quality sources. For example, tidal marshes develop natural channel forms in the geomorphological interaction with tide and sediment that serve to catch further tidal energy, exerting its high-quality control action on lower-quality energy of photosynthesis of marsh plants. The spatial distribution depends on the spatial distribution of the energy signature.

Edges and Mixing Zones

Environmental systems develop discontinuities between air, water, and land and between subsystems in one medium. When allowed to find equilibrium, gas, liquid, and solids separate into phases. A system with phase distributions such as these at equilibrium do not have to use as much of their energy sources to maintain structure. The discontinuities provide organisms with more than one type of energy. After ecosystems are established, these recognizably sharp zones of transition are found to have more variety in energy signature. These zones are sometimes called *edges*. Examples are beaches on the sea, edges of the forest, air–sea interfaces, and passages between estuary and the sea. With more energy alternatives at the transition zone, there may be a wider variety of system adaptation as with birds on forest edges. On the other hand, with the sharply varying energy conditions in an estuarine pass, adaptation requires a different kind of diversity—not of types of species, but of special adaptations to changing chemistry, to turbulent motion, and to periods of interrupted function.

Patton (1975) provides a numerical index of abundance of edges for wildlife.

TEMPORAL CHARACTERISTICS OF ENERGY SOURCES

The inflow of energy from the external sources has temporal properties as well as spatial location. In the long run, the temporal characteristics of energy inflow can be represented by the frequency of repeating surges or waves. Higher-quality energies usually have larger, lower frequency.

The temporal properties are often represented by graphs of variance versus frequency, which are used in time series analysis as given in Fig. 15-21.

Spatial Dimensions for Utilizing Pulsing Energy Sources

In Fig. 13-15 the size of storages was related to the frequency of energy flow according to the concept of filters used in electronics. Utilization of an energy flow requires a receiving unit with chargeup characteristics of the same frequency as that of the incoming energy flow. In the process of adapting to energy sources, therefore, systems that survive are those developing the size dimensions necessary to develop appropriate frequency characteristics to use the energy. Unnecessarily large storages waste energy in depreciation; systems too small charge up so quickly that most of the pulse of incoming energy passes through without absorption.

Examples of the spatial dimensions developed by characteristic frequencies in the incoming energy sources are the size of marsh or desert plants in relation to hydroperiod and the size of plankton in relation to wave energies that generate turbulence (see Fig. 15-22). Larger dimensions render horizontal spatial dimensions more coarse.

Widely dispersed plankton and leaves have a time constant that can absorb the daily pulses of sunlight, whereas larger trunks, animals, and detritus pools have longer time constants able to utilize the longer seasonal pulses. These units are adaptively larger in polar regions or monsoon climates where seasonal interruptions are longer.

Lugo et al. (1978) relate the size of mangrove forests to the frequency of hurricanes.

Spatial Scan and Frequency Domain

When scans consist of spatial patterns of process and organization, data are converted to frequency form. Low frequencies occur with spatial patterns of larger dimension, and high frequencies occur with small structures passed over rapidly. To some extent where size is a measure of frequency of oscillations and variations, the spatial scan can be used as a temporal frequency analysis. For example, a spectrum from a spatial scan is sometimes used in oceanography for spectral analysis (see Fig. 15-22) (Platt, 1971). In these examples, high energy per unit is the low-frequency end and hierarchical energy taper explains why there are few.

Many examples of spatial scan analysis of terrestrial systems are given by Davis and McCullogh (1975).

Priorities in Maximization of Power and Greater Centralization During Growth

Power is maximized by feeding back high-quality energy production and storages to augment more energy inflow, with priority going to the pathway of feedback that augments energy most. With many trials and choices inherent in the variation in any system, the alternative with the greater feedback can maximize energy more. There are three alternatives: (1) feed back more energy to the growth of the means of accelerating energy from the source already in use, providing that source can be accelerated and is not source limited or requiring more energy to process than its net effect; (2) use the transformed high-quality production from one source to subsidize and bring in energy from a second source that may or may not be a net yielder of embodied energy; and (3) trade either high- or low-quality energy externally to outside systems for inflows with high amplifier action. See Fig. 14-10.

Almost by definition during growth, priorities go to alternative (1), which is to feed back energy into more of the same so long as it is a preferential yielder. As shown in the graph of Fig. 12-9a, a unit that is ahead in competitive growth or has a higher initial storage of high-quality energy will accelerate more rapidly than its competitor even though the percent growth rates are the same.

The role of initial seeding, dispersal, and initial population may be critical to establishment in a rapid growth regime. The spatial dimension of this is that centers of growth during initial growth periods can outgrow other centers in their vicinity, with the effect extending outward until the pathway friction and divergence effects neutralize the effect of the initial condition. However, after the energy source on which growth is based has been either used up or—in the case of renewable energies—drawn into use at a rate corresponding to the source rate of supply, the advantage of being ahead in growth disappears and a more decentralized hierarchy in which there is a better ratio of low-quality energy to matching feedback energy becomes preferable. Also, new structures require little maintenance at first but more later. The implications of these principles are that during growth hierarchies can be developed more centrally with more net depositions and exports.

Stated in other words by Toynbee, much of the history of human civilizations can be explained by alternation of centralization with energy innovations followed by decentralization as the innovation spreads. See Fig. 24-13.

MODELS OF CONTROL INFLUENCE WITH DISTANCE

Long used in geography is the *gravity model* relating centers and their surroundings. The control influence in an area around a center is proportional to the product interaction of the item with the center's mass diminished in inverse square with distance. Figure 18-5a gives the energy model for systems such as gravity, electrostatic charge, light emission, and other fields where influence decreases by inverse square because of the diluting effect of spreading influence outward in concentric spheres. The force or energy influence diminishes as the inverse square, because concentric spheres increase in proportion to the square of the area acting to dilute the intensity at any spot further out. In the diagram a heat sink is shown with diverging dilution $1/r^2$ because the divergence is an entropy increase and loss of potential energy.

Although the equations of the geographic gravity model and that for physical influences are similar in form, the energy diagrams show differences. The processes of spread from a population center are confined mainly to the planar surface and are transmitted primarily by transportation and communication processes with friction proportional to distance.

As shown in Fig. 18-5b, if the influence were channeled so that there was no geometric dilution effect and only friction, an influence curve should be inverse to the distance if friction were a resistance inverse to conductance (see Chapter 2). In Fig. 18-5c the influence is attenuated as a result of a geometric effect that is inverse to distance, but not inverse square, since it is not an expanding volume but only an expanding ring. Together, however, the friction and geometric effects produce an equation that is inverse square for an influence moving outward.

Where there is convergence from an area toward its hierarchical center, the concentration increases in proportion to distance but is diminished by frictional losses with distance. Models of spatial organization based on radius of travel of individuals are given by Doxiadis in Fig. 25-6. Figure 18-5d is an example of decreasing circulation with distance.

Stewart (1947) provided maps of population influence n/r^2. These were called a *population poten-*

FIGURE 18-5. Models of spatial influence of a center M: (a) influence decreasing by inverse square due to geometric dilution without frictional losses; example, a star; (b) influence confined to a linear conduit, the effect reciprocal to distance because of friction; example, road transport; (c) influence confined to a surface, the effect decreasing inversely with distance because of friction and inverse to distance because of geometric dilution; example, market demand; (d) example of decreasing influence with distance (Isard, 1956). United States Class I railroad shipments. Tonnage of all commodities, by distance shipped (25-mile zones), 1949.

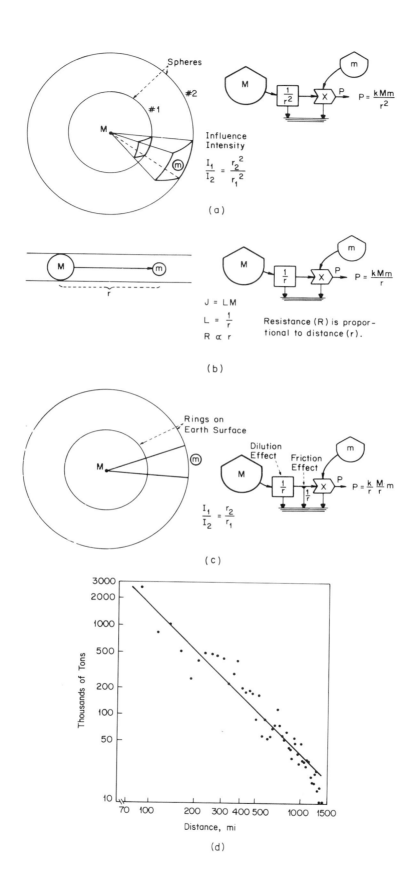

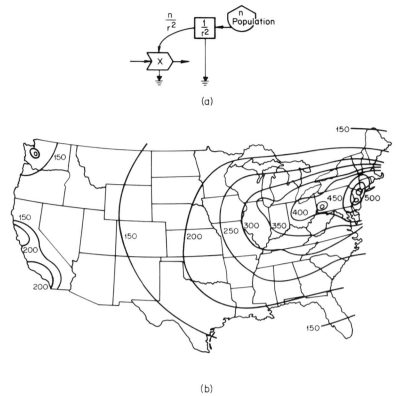

FIGURE 18-6. Spatial action of storage feeding outward and diminished with distance according to inverse square n/r^2, where n is population [Isard (1956) after Cox and Alderson (1950)]: (a) model of demand; (b) spatial distribution. Units are thousands of persons per mile.

tial. Since there are high-quality energies stored in people and in their associated structures, it may be correct to use the term "potential" in the sense of stored embodied energy (see Fig. 18-6).

The action of storage feedback modeled in Fig. 18-6a is one part of potential interactions. This was identified as demand in Chapter 8 (see Chapter 23 on economics for further discussion of demand). Spatial hierarchy of demand follows spatial hierarchy of units [see theory due to Lösch (1944, 1954)].

Clark (1951) found population density declining with distance according to a negative exponential function.

Energetics of the convergences of food in animal feeding as a function of distance was studied by Orians and Pearson (1979).

Area Required to Serve a Population

An equation was developed by Stephan (1977) for the area required to support a population density so as to minimize the time (and thus energy) spent in transportation to and from the center. The derivation is given in Fig. 18-7. Higher energies supporting higher population densities cause areas of support of the centers to be smaller.

M. T. Brown (1980) determined areas of influence for population centers, using matching energy of landscape required for interacting with energy concentrations of the center. The distance at which embodied energy density of the city equals that of the region is taken as the city's territorial influence.

Genetic Information and Cultural Flows

Among the highest quality flows of all are those with genetic content as with seeding and immigration from pools of organisms, cultures of humans, and inflows of know-how. If in sufficient quantity, these have major controlling action. They may be supplied in point, line, or planar fronts. They should be included in spatial models along with the other inflows. Most of this chapter gives examples of organization of spatial pattern and diversity due to

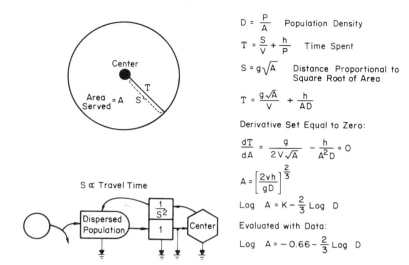

FIGURE 18-7. Model relating the area served by a center to the population density on the principle of minimizing time spent (i.e., saving energy) (Stephan, 1977), with population P, population density D, travel time to center T, velocity to center V, area served A, time cost of maintaining center h, mean distance to center S. The lower diagram contrasts divergent outflow with convergent inflow.

energy sources where high-quality information (genes, etc.) were tacitly assumed to be present. Often it is the genetic information that is the limiting energy source, which can control the spatial pattern, limit diversity, and by its absence cause selection for systems that put more energy into developing and conserving information.

The contours of spread of English sparrows from New York, moving into the new open-niche ecosystems of early America, are an example of the control exerted by genetic information and its role in each place as a driving force on the next place with a controlling action similar to the human population pressure diagrammed in Fig. 18-6b (Johnston et al., 1972). The spread of colonists in the earlier centuries was similar, carrying easily transported high-quality information interacting and using broadly distributed energy storages. Providing energy quality factors can be developed for various kinds of information, the embodied energy in information can be handled as with lower-quality flows in estimating potential production, limiting factors, growth trends, and spatial patterns.

A SPATIAL CLUSTER MODEL BASED ON MAXIMUM POWER SELECTION

Costanza (1977) developed a model of spatial organization based on the interchange between unit models scattered over the surface. In Fig. 18-8 the energy accumulated in storage is fed back either to more self-growth or exchanged with a neighboring unit for a limiting factor, whichever of the two gives the more gross production. Comparisons are made almost continuously between the alternatives by taking partial derivatives of alternative production strategies, selecting the one with maximum power transformation in production. When energy inflow is equally distributed, a set of models organizes its own central cluster of higher power because of the advantages of central position in exchanges. See Fig. 18-8c. The models' procedure of comparing partial derivatives is the same as adjusting prices according to marginal utility. By stating the model in energy terms, it applies to more systems than those involving humans and prices.

GAUSSIAN (NORMAL) SPATIAL DISTRIBUTION

Where there are spot concentrations of matter that carry concentrations of energy, the spatial distribution may fit the Gaussian normal distribution. The Gaussian curve is a negative quadratic exponential, and the spread of the curve is measured by the variance that is in the denominator of the exponent. Variance is the mean of the squared deviations from the mean. The square root of variance is the stan-

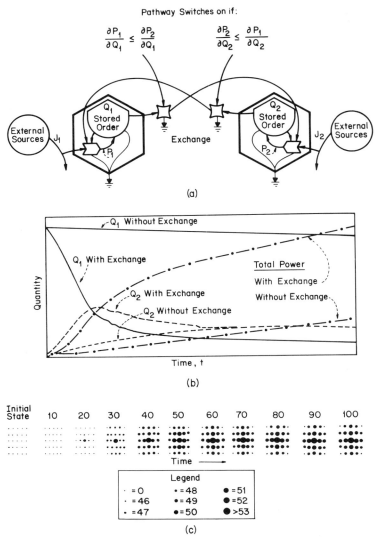

FIGURE 18-8. Model of energy exchange by Costanza (1977) according to marginal utility: (*a*) model for exchange by two units; (*b*) results of simulation of model in part *a*—situation when one of the cells (Q_1 in this case) represents a resource pool, operationally defined as a large storage that is "underutilized" locally relative to some other production area; (*c*) spatial pattern resulting from simulation of model in part *a* relating all units of a 5 × 5 spatial matrix.

dard deviation σ and is used as a scale unit to measure the width of the bell-shaped curve (see Fig. 18-9).

If energy and matter are concentrated in a bell shape as in the example of an isolated mountain or a concentration of waste from a point source, there are outward directed gradients that drive diffusion. Outward diffusion disperses the energy concentration. One model for describing a gradual spread of a concentration uses a Gaussian curve but increases the spread (standard deviation) with time according to the product of diffusion rate and time Dt. This model is used in spread of wastes from smoke stacks (Pasquill, 1974) or for the spread of eroding mountains (Scheidegger, 1970) (see Fig. 18-10).

Figure 18-9*b* shows the distribution of means of samples drawn from Fig. 18-9*a*. A sample of measurements is significantly different if its mean is two or three standard errors of the mean ($\sigma/\sqrt{n}$) separated from another. Even if the distribution of measurements in Fig. 18-9*a* were not Gaussian distributed, the means of samples probably would be.

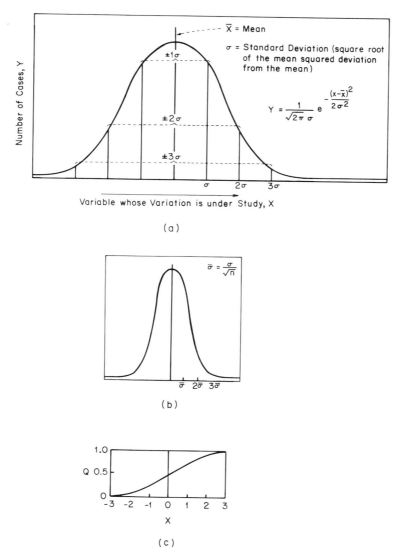

FIGURE 18-9. (a) Gaussian ("normal") distribution on which many statistical parameters are based and that is an hypothesis in many efforts to examine variation; σ is standard deviation (square root of the mean of the sum of squared deviations; (b) distribution of means of successive samples from part a; the standard deviation of this curve is called *standard error* of the mean ($\bar{\sigma}$); (c) cumulative form of normal distribution with cumulative population (integrated area under curve in (a) as a function of the parameter on the x axis expressed in units of standard deviation σ.

Spatial Dispersal

Populations disperse from points of concentration or from points of immigration, losing concentration through mortality and other energy losses. Such dispersal is part of the spatial hierarchy where influences are fed back from high-quality centers and is subject to geometric dilution (Fig. 18-6). Many models and simulations for dispersal use random trajectories (random walk) and other diffusion mechanisms. Concentric rings of influence result around the hierarchical centers. For example, Kitching (1971) used variance of normal distribution as variable controlling angles of directed movement of dispersing animals and mortality rates to generate curves of population with distance from

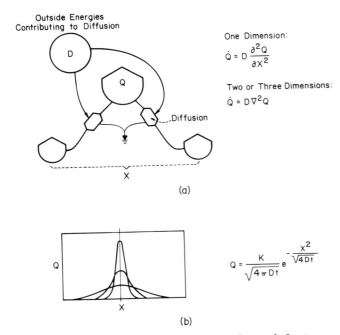

FIGURE 18-10. Diffusion model for spatial spread of a storage: (a) energy diagram and diffusion equation; (b) graphs of successive distribution of quantity when Gaussian equation in Fig. 18-9 has its standard deviation increasing with time t (Scheiddeger, 1970).

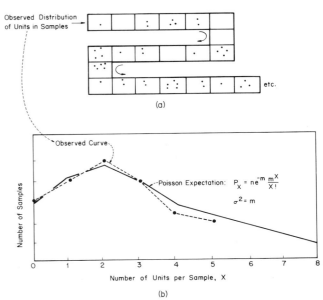

FIGURE 18-11. Poisson distribution for comparing observed frequency P_X of samples containing X units, where m is mean, n is the total number of samples, and σ^2 is variance.

centers. Steep curves resembling negative exponentials resulted. See Fig. 20-23 for more information on dispersion models.

Cumulative Graph of Normal Distribution

When a graph is constructed of the cumulative number in a normal distribution as a function of the variable on the X axis starting with the tail on the left as in Fig. 18-9, an S-shaped curve results (the cumulative form of the normal distribution in Fig. 18-9c).

POISSON FITTING TESTS RANDOM SPATIAL DISTRIBUTION

Where discrete units are distributed among spaces at random, a Poisson distribution results. As given in Fig. 18-11, such a distribution has some slight grouping and some empty spaces. The distribution is neither organized in simple rows nor clustered in hierarchies. By sampling quadrats and counting the number of units, one may compare the frequencies of plots with number of individuals, with 1, with 2, with 3, and so on, with that pattern expected by the Poisson distribution for the same number of total individuals. Then the observed and predicted curves can be compared for significance with χ^2 tests.

Where the distributions are Poisson, the variance is equal to the mean, and thus the two are related. If variance is proportional to main power flow, density (mean) is also proportional to power flow.

What is usually found is that natural distributions are more ordered than "random" Poisson, as might be expected from the kinds of hierarchical energy flow organization that exist in most systems.

Crowding Index

Deevey (1947), studying barnacles, used an index of crowding that includes the radius of the individual r and the probability of interaction n^2 of individuals per area n:

$$C = 2\pi r^2 n^2 \qquad (18\text{-}1)$$

GRAPHS OF COMMONNESS AND RARITY

In this chapter spatial distributions are derived from the spatial organization of energy flow hierarchy and the special effects of spatial patterns in en-

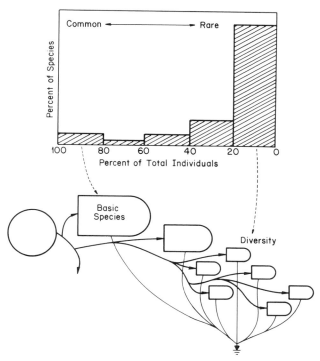

FIGURE 18-12. Raunkier (1934) distribution of abundant and scarce-plant species.

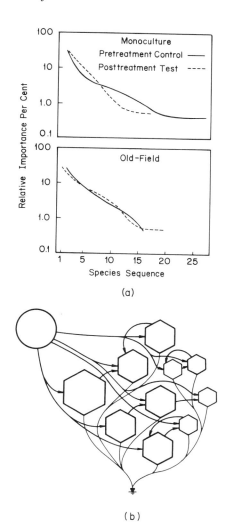

FIGURE 18-13. A rank-order graph that may reflect hierarchical relationships of the web of life adapting to energy flow: (*a*) insects in crop ecosystems (above) and in old fields before and after use of pesticide (Suttman and Barrett, 1979); (*b*) web of common and rare components.

ergy sources and previous storages. A large area of environmental research has been concerned with the distribution of commonness and rarity that have been represented with graphs, equations, and indices of various types. Most of these seem to be explained by the hierarchies that are developed by energy webs, which are sampled by those making measurements of numbers of units and numbers of types.

Number of Types as a Function of the Quantity of Each

One of the first graphical forms showing order in the relation of commonness and rarity is the graph of number of types that have number of individuals. Raunkier (1934) gave graphs of the number of families with genera, the number of genera with species, and the number of species with individuals. Figure 18-12 shows the typical hollow curve that shows many rare species dropping to very few, with many individuals each. The fact that natural systems have many rare types is related to the retention of variety so as to meet contingencies and maintain good self-design for maximum power.

The quantity of energy in basic functions on the left tapers to the right but supports high diversity of information. In the energy diagram (Fig. 18-12*b*) all rare species, including plants, may be regarded as part of higher-quality information storage held for use in adaptation and self-organization for contingencies.

Rank-Order Graphs

Data on species can be expressed in a graph of number or quantity of each type as a function of the rank in order of the most abundant. Figure 18-13 is a

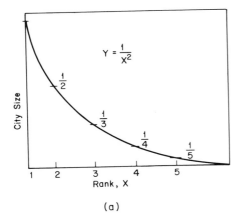

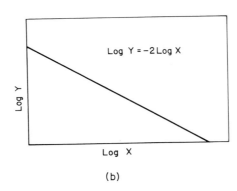

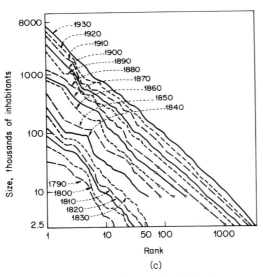

FIGURE 18-14. (a) Rank-order graph of city size compared by Zipf (1941) to an inverse-square series; (b) same on a log–log plot; (c) graph of log population versus log of rank for cities during time of increasing energy base (Zipf, 1919).

rank-order graph. When a rank-order graph is normalized so that the ordinate is a fraction of the total population of individuals (percent), then it expresses probabilities and can be called a *probability density function*.

Many distributions on rank-order plot seem to be exponentially distributed. That is, each succeeding rank is a constant fraction of the previous one. Such distributions are straight on a semilog graph with rank. More than one model has been offered for these. The hierarchy of energy transformation was used to generate distributions in Chapter 15, and these may account for the distribution of many common species and few rarer ones that represent higher embodied energy of special roles and maintenance for contingencies.

Rank-order graphs can also be expressed on log–log plot, which tends to straighten out many of the curves, suggesting functions involving products and powers.

Zipf (1919, 1941) used rank-order graphs, arranging cities by order of size. As shown in Fig. 18-14, the graph was described by an inverse power series, the size being proportional to the inverse square of the rank order. As given in Fig. 18-14c, the graph was straight on log–log plot. He found a similar pattern for frequency of work use and rank.

Pareto Distribution

Pareto (1927, 1971) fitted the steep hollow curve found for distribution of incomes and other aspects of society with an inverse power function

$$Y = 1 + \frac{b}{X^a} \qquad (18\text{-}2)$$

where Y is the number that has the value of X (Johnson, 1937). When exponent a is 2, the distribution is proportional to the inverse square as in Fig. 18-14. In the case of plots on log–log coordinates, the lines are straight with a slope corresponding to the exponent

$$\log Y = \log b - a \log X \qquad (18\text{-}3)$$

Cumulative Species Versus Individuals and Log of Individuals

A very convenient reference graph for data is the *cumulative number of types* found versus the *number of individuals* counted (see Fig. 18-15). The typical

Spatial Distribution and Diversity

shape is curved to the right. As first studied by Gleason (1922) on a semilog plot, the curve is usually fairly straight up to 1000 units and then usually curves upward more rapidly. Similar curves of increasing functions in human settlements with increasing realm were quoted by Haggett [(1966) from Stafford (1963)], Naroli and Von Bertalanffy (1956), and Odum (1971a). In the range of about 1000, the slope of the graph is a useful empirical diversity index easily described as "types per thousand." Or it may be expressed as types per unit log of individuals. On cycle log paper the slope can be expressed as species per cycle. A wide range of diversity was found by Saunders (1968) in sea bottom animals (from 15 to 85 species per thousand individuals).

Since the horizontal axis is in units of logarithm of the number of units, it is an information measure and the slope that is used as an index is species per unit information. Energy explanations of the curves were proposed with Fig. 17-5.

Cumulative Types as a Function of Area

Graphs of *cumulative species* versus *area sampled* were also made by Gleason (1922) with curves resulting such as those in Fig. 18-16 with slopes steep at first and gradually decreasing (Kilburn, 1966). However, the slopes never level and often turn upward as areas expand (Preston, 1962; Diamond, 1974). The most common species are found first since they are most numerous. Graphs of cumulative species versus area are similar to those of cumulative species versus number of individuals counted when the individuals are evenly spread over the landscape as in the case of trees. See Fig. 17-5 for an energy explanation, where more area represents more energy available, but species that can be added per used energy diminish as complexity of potential interactions requiring energy increases with area.

For very large areas, a double-log relation has been derived from data:

$$\log S = \log C + Z \log A,$$

where S is species and A is area (Gorman, 1979). The number of species on an island increases as the 3/4 power of the area (Diamond, 1974).

Log-Normal Distribution of Species

In Chapter 15 log-normal distribution used to fit skewed distributions seemed to be related to energy

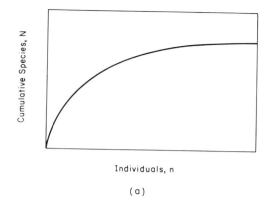

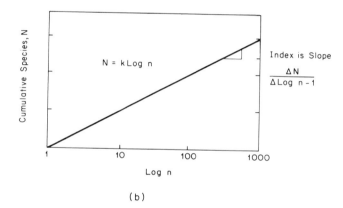

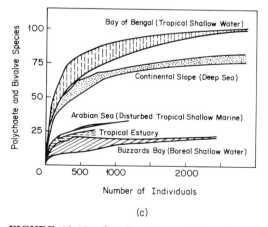

FIGURE 18-15. Graphs of cumulative types versus units counted showing common diversity index and semilog plot and tendency to curve beyond 1000: (*a*) linear plot; (*b*) semilog plot; (*c*) example from Saunders (1968).

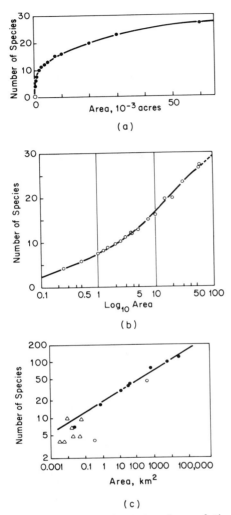

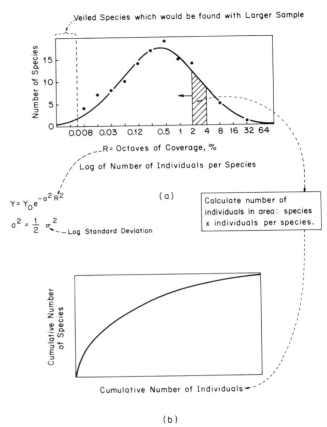

FIGURE 18-16. Example of graphs of cumulative species–cumulative area: (a) brachen–blueberry plant community in Michigan with both axes linear (Gleason, 1922); (b) semilog plot [Gleason (1922) after Vestal (1949)]; (c) log–log plot of data from islands (Diamond, 1974). Open symbols are islands recently depopulated by volcanic or tidal disaster.

FIGURE 18-17. Commonness and rarity graphed as a log normal distribution (Preston, 1962): (a) number of species plotted according to the log of the number of individuals in each; example from Whittaker (1965); (b) cumulative species as a function of cumulative individuals counted in the log-normal distribution starting on the right.

hierarchies. Species patterns may be another example.

Whereas the distribution of species on an abscissa of the quantity of individuals of each is a hollow curve sharply skewed to the right end where quantities are large, as in Fig. 18-17, a shape resembling part of the bell-shaped probability curve results when the horizontal axis is made logarithmic. This graph is a log-normal distribution (Cassie, 1962). An example is Fig. 18-17a. Preston (1962, 1980) introduced the plot to consider the number of species in small areas and samples as related to larger areas.

It was shown to at least one or two standard deviations that the graph was at least approximately log-normal near the center. Whether it was actually log-normal in the tails or not did not matter to the interesting manipulations that Preston then applied, if the results are considered semiquantitatively. A small sample from a large community seemed to be chopped off on the rare-species side, as might be expected with inadequate sampling; thus only a part of the log-normal hump-shaped curve showed. The rest was said to be veiled in the graph. One could approximate it by extending the graph to the left symmetrically. If larger samples are taken, the peak of the humps move to the right toward more individuals per species as more rare species are found.

A very interesting discovery was a different pattern for island, reef, or other isolated systems. A sample from the isolated community required less individuals in the total collections and counts to show a fully formed log-normal than a similar number from the continental area. It was possible to distinguish between a full sample of species-restricted island populations and a similar number taken as a subsample of a large system and hence not adequately sampled to include all species. These studies showed the inherent limits of small areas for supporting many species and raised questions about island fauna being depauperate simply because they never experienced species emigration. Also see evidence from the New Zealand area (Williams, 1981). Diversities of enclosed microcosms are less than those of their less isolated prototypes (Odum and Hoskin, 1957; Odum et al., 1963, 1970) even when more species have been frequently seeded.

The apparent superiority of a fauna–flora system on a larger continent to that on a smaller body when mixed may reflect the greater percentages of energy that can be used for evolutionary progress with a large system compared to a small system where more must be utilized for overcoming gene loss from stochastic instabilities.

Berry (1961) related the rank-size graph of cities to a log-normal distribution, graphing cities on coordinates of log size and a scale of cumulative probability, so that a straight line would indicate log-normal distribution. These cities were from countries that have hierarchy peaks in their signature of energy sources.

As suggested in Fig. 18-17 the area under the distribution curve may be integrated in steps from right to left. Since the area under the curve represents the number of individuals, this procedure plots cumulative individuals versus cumulative number of species. In this way species distribution graphs can be related to another kind of graph frequently used, namely, cumulative species versus cumulative individuals (introduced in Figs. 18-15 and 18-16).

For more on log normal distribution see Atchison and Brown (1966).

Comparison of Rank-Order Graphs to Cumulative-Type Graphs

When the number of types found in an area studied are graphed in order from most numerous to least numerous, the graph is also essentially a graph of area per individual from low values on the left to high values on the right.
The relationship between the species–area graph and the rank–frequency graph can be recognized as

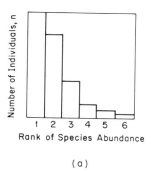

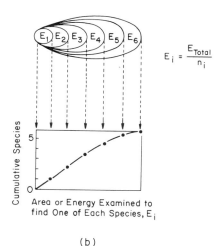

FIGURE 18-18. Relation between a rank-order graph and a cumulative-species graph: (a) rank-order graph; (b) graph of cumulative species with increasing area surveyed (n_i is number of individuals of type i); E is area or energy.

follows. As one samples a larger and larger area, one tends to first find the species with the greatest frequency and hence smallest area per individual. The numerous ones with the smallest area would be the first to be found as one starts a selection process with increasing area of sampling, if there were no statistical errors in sampling. When the area sampled is a little larger, equaling the area per individual of the second-ranked species, it appears in the cumulative list. One tends to find the third rank when the area is larger and equal to the area per individual of that species, and so on.

Thus the rank–frequency graph is a plot of number of individuals per area taken in order of most abundant. The cumulative species–area graph is on the average a plot of the species rank and area per individual. The rank–frequency graph can be converted into a cumulative species graph by plotting the reciprocal as shown in Fig. 18-18.

Dominance and Diversity

The term "dominance" in plant ecology refers to predominant structures of the forest and is sometimes defined as the basal area of trees. Other measures of quantity have also been used with the term "dominance." Observed gains in net accumulation of biomass (one type of net production) have also been used to indicate dominance. Dominance and numbers of individuals have been related to diversity.

Given in Fig. 18-19a are dominance–diversity graphs from Whittaker (1965, 1972). Because the ordinate is proportional to energy storage or flow, the dominance rank order is a kind of energy spectrum related to those given in Chapter 15, reflecting a hierarchy from the common on the left to scarce on the right.

Sometimes energy dominance, where energy is measured in calories, has been used to indicate what is important to a system. The common species on the left in Fig. 18-19 would be regarded as the important members according to this concept. However, in Chapter 14 scarce members of a system were often found to represent convergences of energy transformations that gave them particular amplifying abilities, making them much more important than their weight, numbers, or actual calories would indicate. Embodied energy is a better measure. A controlling member can dominate functions by control actions without being dominant in the sense of large quantity. Whether scarce members are always more valuable remains to be studied. In the sense of marginal utility and potential from the gene pool, the scarce members are potentially of great importance.

THEORETICAL MODELS FOR EXPLAINING COMMONNESS AND RARITY

Most of the graphs used to represent commonness and rarity can be transformed from one to another, with the same data being used to chart a different perspective with the use of a different set of coordinates. A number of theoretical models have been given to account for the observed graphs in terms of causal functions. For example, Whittaker (1965) compares four of these (see Fig. 18-20). These and others follow.

Resources Partition Model

One way of arriving at an exponential distribution was to visualize the most dominant common species taking some fraction of the total resource, the second species taking that fraction of the next, the third species a fraction of the remaining resource, and so on.

Random Partition Model

Sometimes called the *broken-stick model* (MacArthur, 1957; Hutchinson and MacArthur, 1959), the resources are apportioned among species by sections of a line broken at random. Pielou (1975), following MacArthur, shows that this basis generates a de-

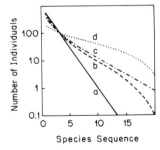

FIGURE 18-20. Rank-order graphs given by Whittaker (1965) with frequency on a logarithmic scale showing shapes of graphs according to four hypotheses. "Curves to fit dominance–diversity relations—four major hypotheses: (curve a) geometric series of Motomura (I), $c = 0.5$; (curve b) log-normal distribution of Preston (1962); (curve c) logarithmic series of Fisher et al. (1943); (curve d) random niche hypothesis of MacArthur (10). Numbers of individuals in the species, on the ordinate, are plotted against species number in the sequence of species from most to least abundant, on the abscissa. The curves are all computed for a hypothetical sample of 1000 individuals in 20 species."

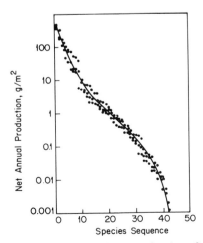

FIGURE 18-19. Rank order of net production of a plant community in the smoky mountains from Whittaker (1965).

clining exponential relation between number of individuals and rank order.

First studying distributions of occupations in industry (Sterndl, 1965) and comparing these with rank-order distributions of species, Cohen (1966) related these to the random division of available resources by competition as the basis for their similarities.

Theory of Hierarchical Dependence of Types on Individuals

Berry and Garrison (1958) showed the presence of specialized stores related to supporting population. Odum et al. (1960) suggested that if species were each based on some number of other species in larger numbers as in food chains and other kinds of dependence, the fractional change in individuals would be proportional to the species added. As given in Fig. 18-21, this makes species added a log function of the individuals accumulated. Conversely, the individuals added with a species are an exponential function of the species number. This model was compared to tables in human organizations and ratios of occupation specialties to total population. Each species is dependent on some number of inputs from one or more species on which it is dependent as in the food web and chain model. There is a tapering of numbers from the common basic species close to the energy source to the specialized dependent ones that require more and more energy for their justification.

The ones on the right are the high-quality ones, whether they are predators, parasites, or specialists using wastes. In any case, since they require a larger web for their support, they have a large energy base in energy flows of a different type, either directly or indirectly. Depending on the energy source, this model predicts a graph that is exponential unless modified by secondary energy sources of different quality. See Fig. 17-6.

May (1975) shows this model to be similar mathematically to a geometric series for apportioning niche spaces.

ENERGY HIERARCHY AND RANK ORDER

The hierarchical energy given in spectral forms in Chapter 15 can be used to derive models of rank order. Mandelbrot (Cherry, 1957b) provides a derivation of an inverse rank order based on an exponential energy hierarchy and a relation of energy to rank that is logarithmic. The derivation was made for messages where energy per message was proportional to length (see Fig. 18-22). Margalef (1958b) used the suggestion by Mandelbrot that rare types cost more:

$$P_z = \frac{P}{(z + B)^\gamma} \quad (18\text{-}4)$$

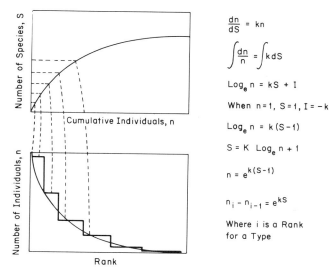

FIGURE 18-21. Hierarchical dependency model for adding species as individuals are added (Odum et al., 1960).

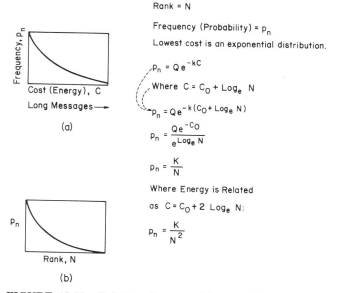

FIGURE 18-22. Relation of exponential energy hierarchy and rank order of messages after Mandelbrot (Cherry, 1957b, p. 210): (a) exponential energy graph; (b) rank-order graph.

where P_z is frequency of types in rank position Z and B, P, and γ are constants. A set of plankton data was fit to this distribution.

Kerner (1972) uses statistical mechanics to consider species.

Quadratic Energy Organization Model Expressed on Rank-Order Graph

In Fig. 17-5 a model was given for adding species according to energy, each with added species requiring energy according to the square of the number of species. A curve of species was constructed as a function of area. The reciprocal, area per species, can be generated from the same hypothesis as explained in Fig. 18-23. The result is a rank-order graph generated by the reciprocal square function. The energy theory makes the rank order a geometric progression on an inverse square. Note the urban (city) case in Fig. 18-14 given by Zipf (1941). This theory can be applied to graphs of cumulative species versus individuals also where individuals are of similar size and energy involvement.

Taxonomic Hierarchies

Willis (1922) found that graphs of number of genera having one, two, etc. species were hollow curves suggesting hierarchical relationships. Similar graphs were found for genera and families, families and orders, etc. Larger realms were involved for larger taxonomic categories involving longer times for evolutionary development. Mandelbrot (1956) compared the distribution to stable energy distributions of molecules as a model.

Taxonomic hierarchies may be regarded as an energy spectrum of evolution. The species is a smaller unit with faster time constant that processes energy necessary to generate the next higher taxonomic category, the genus, etc.

Logarithmic Series

Using data from insects attracted to light traps, Fisher et al. (1943) and Williams (1964) found some degree of fit of Fisher's logarithmic series to the distribution of species and individuals (Fig. 18-24). The frequency of species with a number of individuals as given by the successive terms of the logarithmic series written in the form as follows, in which α is

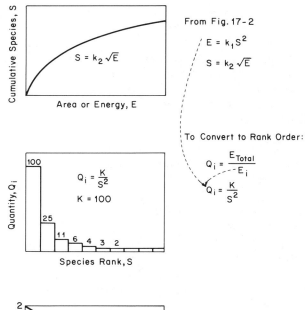

FIGURE 18-23. Diagrams showing a rank-order graph with an inverse square derived from the theoretical model of quadratic energy requirements for organization of types: (a) graph of cumulative species; (b) rank-order graph with quantity proportional to the inverse square of the rank order; (c) double-log plot of rank order. See previous applications to rain forests (Odum, 1964, 1971a; Odum and Pigeon, 1970).

described as a diversity index and k a constant in proportion to sample size:

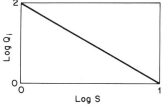

Number of species	αX,	$\dfrac{\alpha k^2}{2}$,	$\dfrac{\alpha k^3}{3}$,	$\dfrac{\alpha k^4}{4}$,	$\dfrac{\alpha k^5}{5}$,	etc.	(18-5)
Individuals	1	2	3	4	5	etc.	

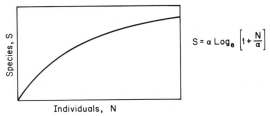

FIGURE 18-24. Species related to number of individuals according to logarithmic series.

In the logarithmic series model the total number of species is divided among rank classes in inverse proportion to the individuals. Thus there are the most species with one individual, less with 2, even less with 3, and so forth. Greig-Smith (1957, 1964) in comparing the pattern with Preston log-normal distribution, indicated that observed plant samples usually had fewer species with only one individual than some other classes.

Williams (1964), like other authors in later papers, stated that the pattern did not fit perfectly and thus questioned the theoretical basis. An advantage of the Williams formulation was a graph for test of statistical significance and confidence limits. Use of the index was made for marine plankton by Hulbert (1963).

DIVERSITY

The word "diversity" has been used in study of systems to refer to various aspects of the numbers and kinds of units and their properties examined with various functions and ways of studying distributions. Since diversity requires flows of energy, according to theory preferred, and is a stored manifestation of previous energy flow, indices of diversity can be used as an indicator of the state of a system in the balance between energy flows that develop diversity and those negative actions that may decrease diversity. Studies of pollution, for example, show decreases in diversity indices correlated with negative actions. Often high diversity requires large flows of energy relative to negative actions, and often time is required to develop diversity starting with low diversity states.

Among the reasons why more energy is required for maintaining variety than for maintaining units of one type is the energy or organization required to prevent competition, a property of living and nonliving autocatalytic units. Without organization, units of different type tend to exclude each other and become simple.

Among the indices of diversity used as overall indicators of ecosystem states and distributions are measures of information, slopes of curves of types and individuals, and coefficients of statistical distributions, as already covered in previous sections (Chapter 17). Some of these have theoretical importance in conceiving how systems operate; others are useful in data processing to establish statistics of inference; others seem descriptive for lack of theories of functional explanation. A summary list of some diversity indices is given in Table 18-1.

TABLE 18-1. Summary of Diversity Indices[a,b]

Description	Formula
Information per individual in groups	$\frac{1}{n} \log n_1! \, n_2! \cdots n_k!$
Information in assemblage of units with p the frequency of each (Shannon and Weaver, 1949)	$-\Sigma p_i \log p_i$
Information per individual not in groups (Brillouin, 1962; Margalef, 1975a,b)	$\frac{1}{n} \log \frac{n!}{n_1! \, n_2! \cdots n_k!}$ where $n = \Sigma_i^k n_i$
Information in identifying the species of every individual (Margalef, 1957a,b)	$\log n!$
Information in species S	$\log S!$
Information in potential species connections	$\log S^2$
Information in web use where p is probability of path use (MacArthur, 1955)	$-\Sigma p_i \log p_i$
Cumulative species per log individual	$\frac{S-1}{\log n}$
Log series index α (Fisher et al., 1943)	$S = \alpha \log_e \left(1 + \frac{n}{\alpha}\right)$
Total species in log-normal distribution (Preston, 1948, 1962) see Fig. 18-17.	$N = \int_{-\infty}^{+\infty} Y_o e^{-a^2 R^2} dR$
Probability of successive sampling (Simpson, 1949)	Σp_i^2
McIntosh (1967)	$\sqrt{\Sigma p_i^2}$
Ascendency, measure of network coherence (Ulanowicz, 1981)[c]	$T \sum_k \sum_j f_{kj} Q_k \log\left[\frac{f_{kj}}{\sum_i f_{ij} Q_i}\right]$

[a] See Chapter 17 for explanations.
[b] Symbols: species, S; individuals, n.
[c] T is total system throughput and f_{kj} is probability of throughput from k to j. Q_k is probability of flow through k.

To some, the terms "diversity" and "complexity" are synonyms. To some extent, diversity has been used more for variety of units and complexity more for considering relationships of units or the lack of it.

Margalef (1956) correlated and compared diversity indices. Single indices of diversity are often sought to appraise the conditions of a community in relation to its sources and stresses (Odum et al.,

1963d; Kaesler et al., 1978). Also, an index can be used as a state variable for models as in Fig. 16-16.

Diversity Indices Using Information

Information content as the logarithm of possibilities and various measures of systems with this concept was initially discussed in Chapter 17. Many of these were introduced by Margalef (1956) and included in Table 18-1. Ulanowicz (1981) uses information to measure coherence of flows in a network. See Table 18-1.

A theory was also given with Figs. 17-5 and 17-6 relating information as the logarithm of the energy flows or storages. Indices that do not use logs (noninformation indices) may be more directly related to the energy that is ultimately causal.

Diversity and Energy Budget

Energy was related to permutations of rain-forest species complexity (Odum, 1964) and energy required for diversity estimated as in Fig. 17-5. Connell and Orias (1964) related energy to diversity by using the model in Fig. 18-25.

The requirement that all the various functions of the system originate from the energy budget with economic division of the resource renders many quite different phenomena complementary when examined on an energy basis. One idea is that diversity of units is a low-priority use of energy that operates when other basic physiological functions have been met. Diversity of structure for physiology may have priority over diversity of species. Diversity, then, is an indicator of a good balance between energy sources and stresses. Increasing physiological requirements for adaptation require a decline in the variety of units and number of circuits that can be supported. A decline in the budget reduces the number of processes possible. An increase in deteriorating influences requiring maintenance energies diminishes other abilities. Note the role of stress in diverting energy storages or flows in Fig. 18-26, which summarizes these concepts.

Quadratic Minimodel of Diversity and Stability

In Fig. 18-26 diversity is represented by a storage state variable developed as a second priority after production, consumption, and recycle are developed.

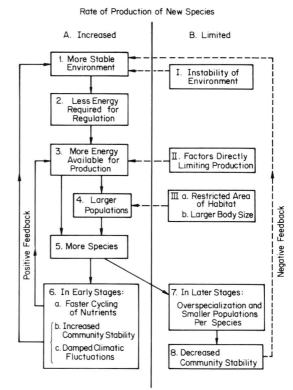

FIGURE 18-25. Model of role of diversity in ecosystems from Connell and Orias (1964).

As shown diversity requires a quadratic energy flow for its maintenance as explained for information with Fig. 17-5. A quadratic drain makes diversity a stabilizing autocatalytic process. The equation for n becomes logistic (see Fig. 9-8). Representing diver-

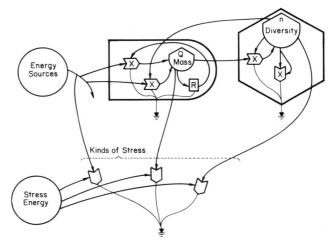

FIGURE 18-26. Minimodel of the role of diversity, where n is the number of types (see Fig. 16-16 for example). Stress energy can divert energy from various storages or flows with the same effect, a lower diversity.

sity as a storage is a way of aggregating hierarchical properties and their energy role. It is also a way of realizing that diversity is itself a regulatory mechanism that helps to maximize useful power and efficiency, as explained next. Terborgh (1974) used a quadratic extinction rate of species to represent interspecific competition.

As the number of species is increased, the energy required increases as the square of related function, and growth is stabilized, even with a constant-force energy supply. Evidence for a quadratic relation is given with Fig. 17-5.

The convergence of energies to support organized complexity utilizes energy to develop diversity, a high-quality energy, which has feedback roles to improve the energy flow of the supporting chain (see Fig. 18-26).

Square Root as a Diversity Index

Mehnhinick (1964) and E. P. Odum (1971) found a correlation between number of species and square root of the number of individuals. The quadratic theory of energy control of diversity (Figs. 17-5 and 18-26) may provide an explanation.

Two Contrasting Mechanisms of Developing Diversity from Energy Flows

Saunders (1968) presented graphs of commonness and rarity from the deep sea (Fig. 18-15c), showing very high diversity where physical conditions of temperature, salinity, and possibly other variables were very uniform. With these temperatures near freezing, energy loss due to thermal disordering is minimal. These results support the concept of diversity being high when there are few stresses to eliminate energy available to generate diversity. By this theory, diversity is the result of using energy to generate order. The diversity of some complex communities such as coral reefs and rain forests has obvious organizational relationships showing in the symbiotic behaviors, physically connected relationships, and energy-expensive adaptations for communication.

Whereas diversity is sometimes generated by stability that allows energy to be used for organizations, there are other situations where diversity has been related to fluctuating conditions where diversity is due to no one condition remaining long enough for any species to prevail in competition (Hutchinson, 1961).

Grenney et al. (1974) simulated two nutrient-limited, competing phytoplankton species with chemostat type of dilution and renewal flushing (see Fig. 12-25). With flushing conditions constant, one or the other species prevailed excluding the other; but when flushing conditions were alternated, there was coexistence with each species gaining during the period favorable to it.

These two concepts for developing diversity may not be conflicting. Each may be an adaptive system mechanism when conditions are appropriate. Where energies are in surges and potentially a stress to organized communities, systems that are adapted to noisy energy conditions can use the stress to achieve diversity by the second mechanism (unorganized competition) without spending energy for organizational relationships.

A demonstration of the way disturbance energy can maintain diversity was given by Kemp and Mitsch (1979), who simulated phytoplankton competition with recycle stressed by turbulent energy (see Fig. 18-27). Jacobs (1977) simulated diversity by varying food and predators.

Both mechanisms provide diversity adapted to appropriate energy flows and in the process maximize power. In the first case it is to a steady energy and in the second, to a variable one. Perhaps the model in Fig. 18-26 is correct for either. With stress, the quadratic organizational drain is eliminated. After adaptation, varying energy sources are regarded as part of the regular energy inflows and are no longer regarded as stresses.

Internal and External Diversity

When conditions of adaptation are severe, diversity of species is usually observed to be low. Species that prevail have energy utilized in special adaptations such as special biochemical systems, special organs, and special seasonal adaptive programs. Presumably, energy is being utilized for adapting to special conditions with less energy remaining for the special functions required to prevent competitive exclusion and otherwise organize species for cooperative coexistence. Measurements of internal physiological, morphological, and biochemical diversity are relatively few. H. P. Jeffries (1969, 1970, 1979) examined diversity of amino acids and fatty acids, finding variations and seasonal changes in biochemical in-

oping large detritus pools. See discussions of this controversy [Rosenweig quoted by MacArthur (1968)]. Steady, natural conditions of eutrophication often have high diversity (see Chapter 22).

SPATIAL MANIFESTATION OF GROWTH

Models of autocatalytic growth competition may be given spatial expression by simulating growth in a grid of spatial compartments. Space is consumed as a resource by species competing for larval access. Energy is unlimited, and other kinds of ecosystem regulation are not included. For example, see the spatial model of coral communities (Maguire and Porter, 1977) in Fig. 18-28. Like the single-

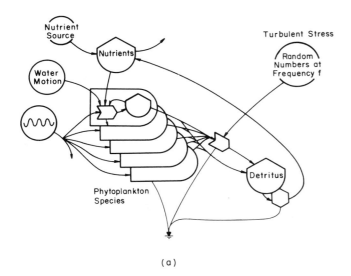

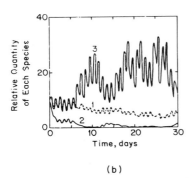

FIGURE 18-27. Model of stress maintaining diversity by arresting competitive exclusion (Kemp and Mitsch, 1979): (a) energy diagram; (b) example of simulation that maintains coexistence diversity by stress when period of stress is that of plankton. Pulses are sinusoidal.

dices. Open sea with more species diversity had less biochemical diversity.

Some questions about which case really is the deep sea were raised by Dayton and Hessler (1972) and which case applies in coral reefs by Connell (1973).

Eutrophication Surges

A surge of enrichment often generates increase of dominants, some competitive exclusion, and variation as suggested by Yount (1956) with studies of diatoms in Silver Springs. The presence of a diversity of rarer species is partly masked. The system may be stabilized in other ways such as in devel-

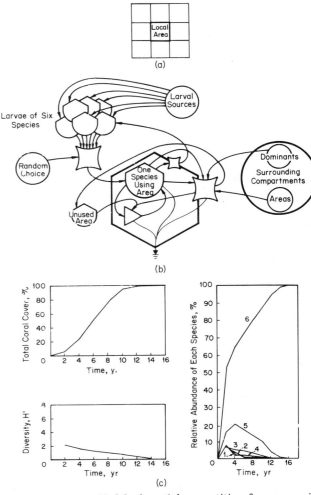

FIGURE 18-28. Model of spatial competition for community dominance involving six species (Maguire and Porter, 1977): (a) spatial blocks each with unit model given in part b; (c) simulation without interruptions, catastrophes, or other regulation.

compartment versions of these models (Chapter 9), species dominance tends to result unless catastrophic and randomness factors are added, arresting the succession in a stage preceding dominance.

Maps of Energy Distribution; Embodied Energy

Part of the basic data for recognizing spatial patterns are maps of distribution spatially of energy inflows such as maps of sunlight, rain, land uplift rates, tides, and winds. The energy stored from the past can also be represented with maps of soil storages, water storages, population, storage of economic assets, and so on.

The theory of maximum power suggests that energies are of different quality and that maximum power develops when high-quality energy flows and storages amplify the low-quality ones through interactive processes such as multiplicative production (see Chapters 8 and 14). Thus synthetic maps may combine separate maps of storage and flow according to models for combined energy use that maximize gross production and survival.

Planning by Combining Overlay Maps

Often associated with McHarg (1969) is the procedure of developing detailed maps of various properties of an area and then overlaying them so as to obtain a spatial pattern of properties that arise from combinations of the map characteristics. Maps may be of soil, slope, population, weather elements, proximity to valuable sites, or other geographic variables. The overlay procedure may be used, for example, to locate a highway where other values are less.

To combine maps is really to interact them according to some unit model. Among the models are adding, multiplying, or calculating the logistic value (sum of useful value inputs and subtraction of quadratic factor of the storage of economic development) (see also Figs. 14-15 and 21-4).

Mapping the output of a unit model of the interaction of the many properties can provide maps of predicted growth rate, density of development, maximum power, and other factors. The overlay modeling combination provides planners some predictions of spatial trends consistent with their ideas of relationships.

Models for relating energy signature to production were given in Fig. 14-15. Some of these can be adapted for weighting map overlays: (1) multiplication of maps (multiplication of the value at each spatial point by the value of every other map's value at that point; and then drawing isopleths for the product); (2) conversion of all maps into embodied energy (the energy of one type required to develop that energy flow or flow from storage) and adding these; and (3) addition of the lowest-quality energy flow to that part of higher-quality energy flows that are in excess of the low-quality energy. This recognizes that low-quality energy originally contributed to the higher-quality energy and should not be double-counted in an addition. See Fig. 14-15g.

Figure 18-29 gives an example of a synthetic map

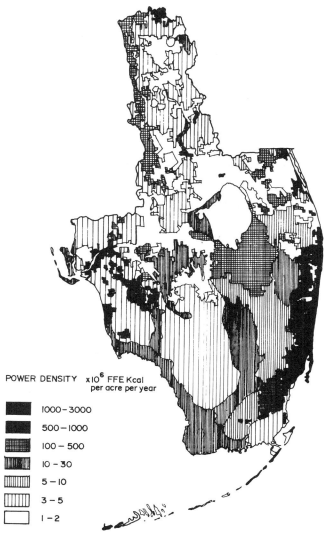

FIGURE 18-29. Maps of energy flow in units of equal energy quality from Costanza (1975). FFE is fossil fuel equivalents.

that utilizes the data of many maps of energy storage and flow to indicate areas of maximum production. Regan's (1977) statistical study correlating growth in 67 Florida counties with embodied energies helps validate these approaches.

CONTINUUM

As one moves across the landscape, the energy signature and frequency characteristics vary, requiring different combinations of structural adaptation to provide for the best energy interaction for maximum productivity and survival. In ecosystems this adaptation is often observed to be concomitant with changes in the ratios of species and other components so that when one inflow is less, such as water, species that conserve water become more prevalent, and so on. The changes are usually gradual, although some sharp changes do occur at forest edges and boundaries between phases of the biosphere (air, sea, land, discontinuities, etc.). The gradients of properties and gradients in structural adaptations are called a *continuum*, and many of the indices of distribution and diversity are plotted to show gradual changes and relate them to the spatial indices of environmental storage and inflow (Curtis, 1959; Kershaw, 1964). A different species may be dominant in different zones where combinations of factors are different (e.g., see Fig. 18-30).

A procedure called *ordination* has been used to determine indices along gradients where different species predominate (Kelsey et al., 1977). Continuum indices are formulated from the proportions of species so that areas with similar index will have similar species. These are also compared with gradients of environmental properties such as soil moisture (Austin, 1972; Bray and Curtis, 1957).

SPATIAL PATTERN FROM DIFFUSION OSCILLATION

Figure 9-23 gives a basic model of an autocatalytic growth balanced by diffusion that is dependent on the curvature of the distribution. Such processes may oscillate if constrained by a boundary. Patterns of banding are observed with this pattern in chemisty (Glandsdorf and Prigogine, 1971), plankton ecosystems, and morphogenesis (Ransom, 1981). Rosinsky (1977) calculates the steady-state field of oscillation, mapping lines that represent the nodes of oscillation that compare with observed patterns in insect embryos (see nodes in an eliptical shaped system in Fig. 18-31). It is possible to imagine a distribution of ecosystems affected by diffusion of species

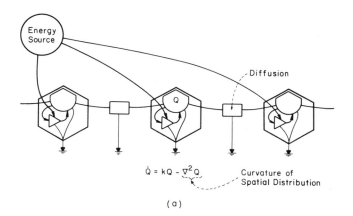

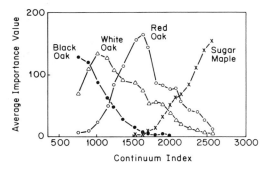

FIGURE 18-30. Distribution of dominant forest trees along a gradient of properties (McIntosh, 1963; after Curtis and McIntosh, 1951).

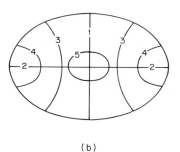

FIGURE 18-31. Spatial pattern of oscillations of autocatalytic growth with diffusion (see Fig. 9-23): (*a*) energy diagram of components in one dimension; (*b*) distribution of nodes that may occur in an elliptical shaped area depending on size. Numbers are nodes that occur at the same time. (Kauffman, 1977; Rosinsky, 1977).

or countries affected by diffusion of innovative culture. Simulations of models of populations involved the same principle, but with more factors (Stesen et al., 1977; Mimura and Yamaguti, 1982; Wroblewski and O'Brien 1976).

Dubois (1975, 1979), Parker (1976), and Hilborn (1979) studied propagation of ring waves and oscillations resulting from plankton patches in the sea following the models of outward diffusion balanced with Lotka–Volterra kinetics inside.

THREE-DIMENSIONAL SPATIAL FORMS

The flows of energy develop forms in three spatial dimensions. The exponential distribution of the atmosphere with height was indicated in Fig. 15-14 with the upper atmosphere low in quantity but at a high potential, of high quality, and able to control larger masses below. In the sea, the food chains based on energy received at the sea surface work downward in energy chains that are partly biological and partly through physical exchanges. High-quality units may be found at lower levels. Time constants are longer; units of flashing luminescence have longer periods with depth (Clarke and Hubbard, 1959). Populations serve the world, requiring more energy-support transformations and representing the convergence of more processes. The diversity and information involved is extensive. Whether they feed back to have control over the surface phenomena remains to be seen.

When energy concentrations are high in air and water, highly organized structures develop. Some of the three-dimensional forms observed in chemical reactions and in high-energy thunderstorms are shown in Fig. 18-32. In the latter the tornado is under the upcurrent core, and the severe downdrafts and hail are under the downdraft part of the core (Eagleman et al., 1975).

SUMMARY

Spatial characteristics of systems were related to energy signatures and the hierarchical energy webs and chains that develop as part of maximizing power. The spatial characteristics of initial storages, the spatial characteristics of input energies, and feedback from concentrated storage interacting with spatial resources can be related by means of models to hierarchies and to control the inverse square characteristic of influence and demand with distance. The hierarchy of structure and units in webs may account for the patterns of commonness and rarity observed with various kinds of graph: species frequency; rank order; log-normal; cumulative species with individuals; cumulative species with area; logarithmic series; and semilog and log–log plots of these.

Measures of diversity were summarized, including parameters of the various graphs for representing commonness and rarity and information theory measures of numerical properties. Energy was related as much to the actual complexities of structure as to the log of the complexities (information content) of the possibilities. An energy minimodel relates diversity to maximizing power, regulating stability, and stress.

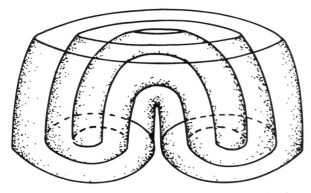

FIGURE 18-32. Three-dimensional structures developing where the flow of potential energy in fluids develops circulatory patterns. Forms observed in some chemical reaction solutions (Winfree, 1973) and on much larger scale of heavy thunderstorms (Eagleman et al., 1975).

QUESTIONS AND PROBLEMS

1. Describe patterns that result from hierarchy development when most of the energy sources enter on a broad area base (sun, rain, land, etc.).

2. What are some other spatial patterns associated with special spatial arrangements of energy sources?

3. By what distance function are units at one level related to the low-quality units of lower quality that are convergent and that receive services and recycle back? Explain. How does

this affect the ratio of units at one level serving another?

4. Can the area in support of a center be related by time of movement?

5. Explain how spatial patterns can result from the exchange of resources between spatial units with energy loss for distance and with priorities for growth or trading made according to marginal effect.

6. How does diffusion from a point develop a Gaussian spatial distribution with standard error increasing with time? What is the equation for a normal distribution?

7. What is a Poisson distribution? What are numerical tests of this distribution?

8. How can Monte-Carlo selection of numbers generate a pattern of diffusion spatially?

9. How is diversity distributed relative to the energy hierarchy?

10. What is a rank-order graph? What is the usual shape of the graph? Can you relate this to energy hierarchies?

11. What is a cumulative species graph? What is its shape on semilog coordinates? What diversity index can be taken from this graph? How is the cumulative-species graph related to the rank-order graph? From Chapter 17, how is energy related to cumulative declining rate of addition of species?

12. What is a log-normal distribution? How does a skewed distribution become more symmetrical on log-normal plot? If species abundance was inverse to their quality, what shape would result to rank order, cumulative species, and log-normal graphs?

13. Where energy requirement for addition of species increases as the square of the number, what is the relation of quantity to rank order? How is square root related as a diversity index?

14. How was the logarithmic series related to species diversity?

15. How is diversity related to energy priorities in newly developing systems?

16. How do maps of embodied energy help in theory of energy control of spatial pattern?

17. What is a continuum? What is a continuum index that can be used to represent distributions of species?

18. What kinds of shape develop in three dimensions where energy flows are intense?

SUGGESTED READINGS

Ashby, M. (1963). *Introduction to Plant Ecology*, Macmillan, New York. 235 pp.

Atchison, F. and J. A. C. Brown (1957). *The Log Normal Distribution*, Cambridge University Press, London.

Berry, B. J. L., E. C. Conkling, and D. M. Ray (1976). *The Geography of Economics Systems*, Prentice-Hall, Englewood Cliffs, NJ. 529 pp.

Cain, S. A. and G. M. deOliviera Castro (1959). *Manual of Vegetation Analysis,* Harper, New York. 328 pp.

Cody, M. L. and J. M. Diamond (1975). *Ecology and Evolution of Communities*, Belknap Press, Harvard University Press, Cambridge, MA.

Cormack, R. M. and J. K. Ord eds. (1979). *Spatial and Temporal Analysis in Ecology.* Statistical Ecology Series **8,** International Cooperative Publishing House, Burtonsville, MD.

Elton, C. (1927). *Animal Ecology,* Sidgwick and Jackson, London.

Gorman, M. L. (1979). *Island Ecology,* Wiley, New York.

Grassle, J. F., G. P. Patil, W. K. Smith, and C. Taillie (1979). *Ecological Diversity in Theory and Practice*, Statistical Ecology Series **6,** International Cooperative Publishing House, Burtonsville, Md.

Greig-Smith, P. (1964). *Quantitative Plant Ecology,* Butterworths, London.

Haggett, P. (1966). *Locational Analysis in Human Geography*, St. Martins Press, New York.

Haggett, P., A. D. Cliffs, and A. Frey (1977). *Locational Methods*, Halstead Press (Wiley), New York.

Hesse, R., W. C. Allee, and K. P. Schmidt (1931). *Ecological Animal Geography,* Wiley, New York.

Hutchinson, G. E. (1978). *An Introduction to Population Ecology*, Yale University Press, New Haven, CT.

Isard, W. (1960). *Methods of Regional Analysis—An Introduction to Regional Science*, Wiley, New York.

Isard, W. (1956). *Location and Space Economy*, MIT Press, Cambridge, MA.

Jardine, N. and R. Sibsen (1971). *Mathematical Taxonomy*, Wiley, New York. 285 pp.

Kelly, F. P. (1979). *Reversibility and Stochastic Networks,* Wiley, New York. 230 pp.

Kerner, E. H. (1972). *Gibbs Ensemble: Biological Ensemble, Applications of Statistical Mechanics to Ecological, Neurological, and Biological Methods,* Gordon and Breach, New York.

Kerr, S. R. (1974). Theory of size distribution in ecological communities. *J. Fisheries Research Board of Canada* **31,** 1859–1862.

Kershaw, N. A. (1964). *Quantitative and Dynamic Ecology*, American Elsevier, New York.

Kessell, S. R. (1979). *Gradient Modelling.* Springer-Verlag, New York.

Koch, A. L. (1966). The logarithm in Biology I. mechanisms generating the log-Normal distribution exactly. *J. Theoret. Biol.* **12**, 276–290.

Leigh, E. (1971). *Adaptation and Diversity*, Freeman, San Francisco. 288 pp.

MacArthur, R. H. (1972). *Geographical Ecology*, Harper, New York. 269 pp.

Mandelbrot, B. B. (1977). *Fractals*, Freeman, London.

Margalef, D. R. (1956). Information theory in ecology. *Gen. Syst. Yearbook* **3**, 36–71.

Mills, J. (1979). *Vegetation Dynamics*, Chapman and Hall, London. 80 pp.

Monin, A. S. and A. M. Yaglom (1965, 1981). *Statistical Fluid Mechanics*, M.I.T. Press, Cambridge, MA.

Mostow, G. D. (1975). *Mathematical Models in Cell Rearrangement*, Yale University Press, New Haven, CT.

Ord, J. K., G. P. Patil, and C. Taillie eds. (1979). *Statistical Distributions in Ecological Works*. Statistical Ecology Series, **4**, International Cooperative Publishing House, Burtonsville, Md.

Patil, G. P., E. C. Pielou, and W. E. Waters eds. (1971). *Spatial Patterns and Statistical Distributions*. Statistical Ecology Series, **1**, Pennsylvania State University Press, University Park, PA.

Pielou, E. C. (1975). *Ecological Diversity*, Wiley, New York.

Pielou, E. C. (1979). *Biogeography*, Wiley, New York. 352 pp.

Poole, R. W. (1974). *Quantitative Ecology*, McGraw-Hill, New York. 531 pp.

Scheidegger, A. E. (1970). *Theoretical Geomorphology*, Springer-Verlag, New York.

Seber, C. A. T. (1973). *The Estimation of Animal Abundance*. Hafner Press, New York. 506 pp.

Sheath, P. H. A. and R. R. Sokal (1973). *Numerical Taxonomy*, W. H. Freeman, San Francisco. 573 pp.

Simmons, I. G. (1979). *Biogeography*, Duxbury Press, North Scituate, MA.

Stonehouse, B. and C. Perrins, Eds. (1977). *Evolutionary Ecology*, Univ. Park Press, Baltimore, MD. 310 pp.

Stokes, A. W. Ed. (1974). *Territory*. Benchmark Papers in Animal Behavior, Dowden, Hutchinson and Ross, Stroudsburg, PA. 398 pp.

Southwood, T. R. E. (1966). *Ecological Methods*, Methuen, London. 311 pp.

Thompson, D. (1942). *On Growth and Form*, 2nd ed., Cambridge University Press, London.

Tribus, M. (1969). *Rational Descriptions, Decisions and Designs*, Pergamon Press, New York. 478 pp.

Udvardy, M. D. F. (1969). *Dynamic Zoogeography*. Van Nostrand Reinhold, New York. 445 pp.

Whittaker, A. H. and S. A. Levin, Eds. (1975). *Niche Theory and Applications*. Dowden, Hutchinson and Ross, Wiley, New York. 445 pp.

Williams, C. B. (1964). *Patterns in the Balance of Nature*, Academic Press, London.

Williamson, M. (1981). *Island Populations*. Oxford University Press, Oxyford, U.K. 286 pp.

Woldenberg, M. J. (1968). Hierarchical Systems: Cities, Rivers, Alpine glaciers, bovine liver, and trees. Ph.D. Dissertation Dept. of Geography, Columbia University, New York.

PART FOUR

SYSTEMS OF NATURE AND HUMANITY

CHAPTER NINETEEN

PRODUCERS

Producers are subsystem units that use several types of high-quality energy flow as interacting amplifiers to convert low-quality energy into higher-quality energy, usually yielding net production to other units. Green plants are examples of producers found in forests and aquatic environments. Green plants use nutrients, genetic inputs, biomass, and other high-quality flows to convert sunlight into organic matter that supports other members of ecosystems. Some bacteria produce organic matter from chemical fuels. Many industries that are primary in an urban economy are producers since they combine raw material, equipment, fuel energy, and labor inflows into high-quality products. Atmospheric processes interacting with landscapes generate geological production such as organization of geomorphology. Chapter 8 introduced various types of intersection. This chapter presents basic systems of production that combine intersections into energy transformation systems.

A typical pattern of interaction for a producer is given in Fig. 19-1. It has a large volume inflow of low-quality energy, interactions with higher-quality inflows that act as amplifiers, internal Michaelis–Menten loops, interaction with feedback from the structural storage of the unit, feedback from other systems of higher quality, exports to other units, and energy dispersal from processes and storages as required by the energy laws. This chapter concerns some of the primary producer units, their design and performance. Chapter 23 has economic producers.

MINIMODELS OF PRODUCTION

Aggregation Problems and Energy Limits

As in previous chapters, there are the problems of aggregation of detail, where models must be sufficiently simple to understand and simulate but correct in main essence of mechanism and performance. For example, the full detail of photosynthetic production involves chains of loops and storages and interactions whose aggregated whole is often simplified to models such as that in Fig. 19-1. In this chapter we consider both the simplified models and some more complex representations that have well-known details, comparing the performance of patterns observed in nature and those models that can generate maximum power.

There are two basic approaches: (1) an existing pattern is described and coefficients are fixed from measurements of these units; or (2) a model is constructed that can change coefficients and mechanisms as necessary to achieve the model's maximum power potential. In real systems this is done often by substituting species. When models are made according to energy potentials, there should be no internal limitations so that external energy is the only limit. Obviously, some simple systems operated by humans as in agriculture must be modeled by approach 1 for short-term results, whereas results in the long run where natural selective actions and adaptive mechanisms are operating should be of type 2.

Choice of Producer or Consumer Symbol

There is some production in all self-maintaining units. To some extent, the term "producer" is a relative one, depending on one's boundaries of view. A unit at the low-quality-energy end of one system may be at the high-quality-energy end of another. The producers are at the low-energy end of a system. They are sometimes called *primary producers*. Units processing low-quality energies close to sunlight and plant actions are generally enclosed in producer symbols, whereas those higher in the chain from sunlight are generally given consumer symbols. In general, the producers provide net energy produc-

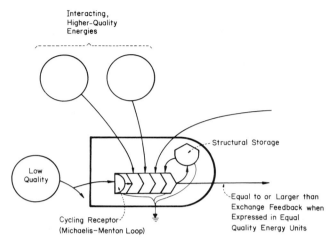

FIGURE 19-1. Features of a typical producer unit.

tion, whereas the consumers may often be net consumers of energy, returning less than they receive even when exchanges are expressed in units of the same quality. Both producers and consumers may have autocatalytic designs with positive rates of growth. Figure 19-2 shows producer and consumer units.

Early Models of Production

Rabinowitch (1945) reviews history of photosynthesis concepts. Early models of production in aquatic media originally in equation form are diagrammed in Fig. 19-3. These are energy transformations that lack the interaction of more than one type of energy that characterizes most production processes. Tamiya (1951), Lumry and Rieske (1959), and Lumry and Spikes (1957, Chapter 10) showed the kinetics of the basic photosynthetic process was of Michaelis–Menten type (Fig. 19-3c). Patten (1968) reviewed plankton production models.

Simplified Producer Models

The simplest models of producer units are given in Fig. 19-4. These are sometimes used where other aspects are constant. Each simplified design is included in a producer symbol to indicate that some simplification is used to represent a more complex producer unit for purposes of the modeling.

In Fig. 19-4a there is only the interaction between feedback energy flow and the basic low-quality source with storages and other sources omitted. Figure 19-4b includes the Michaelis–Menten loop, which contains the storage and material recycling component but no high-quality amplifiers from outside. In Fig. 19-4c there is no feedback from storage. Fig. 19-4b includes an internal nutrient cycle.

ECONOMIC SYSTEM PRODUCTION

Although most of the examples have been plants, economic production involves the interaction of limiting factors and similar designs. An example of an economic producer is given in Fig. 19-5.

Money Overlay for a Producer Unit

Where the producer system is part of the economic system, there are countercurrents of money for those flows that are involving purchases and sales with

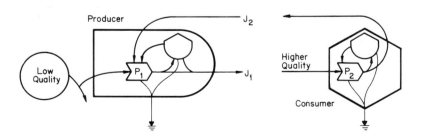

$J_1 \geq J_2$ when expressed in equal quality units.

$J_1 \gg J_2$ when expressed in actual energy units.

FIGURE 19-2. Diagram showing uses of producer and consumer symbols depending on quality of energy and net production as evaluated in energy-quality units, where P_1 is primary production in a producer and P_2 is secondary production in a consumer.

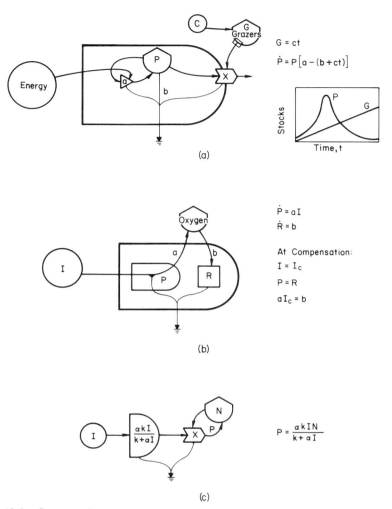

FIGURE 19-3. Some production models of historical interest: (*a*) plankton bloom (Fleming, 1939); (*b*) model due to Sverdrup et al. (1942); (*c*) Michaelis–Menten kinetics connected to light conversion in photosynthesis (Tamiya, 1951).

human beings involved. Some flows, however, are drawn from the environment external to the closed money loops and would not be evident if only the money flows were shown (e.g., see Fig. 19-5).

Cobb–Douglas Production Function

Introduced into economics is the production function with output proportional to the product of interacting commodities; with each component and exponent is arrived at with empirical regressions. For example, production models are often an interaction of labor and capital. This same function can be used with the ecosystem. The function also generates an asymptotic curve as one of its interacting flows becomes limited. An example of a Cobb–Douglas production function is given in Fig. 19-6.

PHOTOSYNTHETIC PRODUCTION

First Unit in Photosynthesis: Transforming Light

Figure 19-7 gives models of the first step in light uptake in photosynthesis and related photochemical systems basic to living food chains. Light interacts with a cycling material that is activated as the means of transforming light energy to chemical potential energy. Sometimes these models can be used as a summary of the more complex biochemical sys-

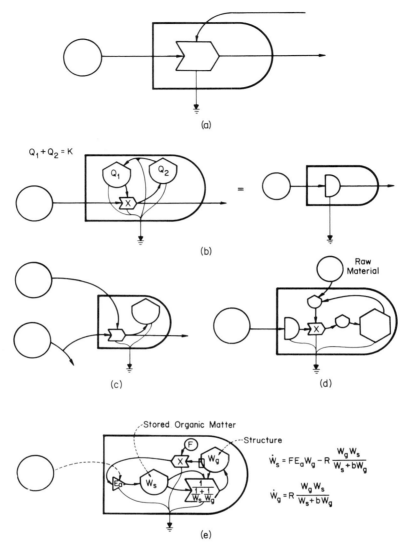

FIGURE 19-4. Simplified producer units: (a) one interaction only; (b) internal storage and recycle only; (c) interaction and storage only; (d) cycling receptor, respiring subunit, and recycle; (e) producer with resistance model due to Thornley (1976).

tems. All have a chlorophyll recycle and Michaelis–Menten kinetics (Lumry and Spikes, 1957).

Light is neither oxidizing nor reducing initially. Photons react with chlorophyll and separate plus from minus charges. Various kinds of photosynthesis were given by Arnon et al. (1961). In primitive photosynthesis the two types of charge react again under catalytic conditions in which their energy is transferred to form ATP (ATP, the energy flywheel of biochemical systems that is used to drive other processes). In this kind of photosynthesis there are no gaseous products or raw materials (see Fig. 19-7a).

In most photosynthesis, however, there is a reaction of water with the positive-charged activated chlorophyll releasing oxygen gas and returning the chlorophyll to a neutral receptive state ready for reaction with more light. The removal of an oxidized substance from the positive arm of the unit leaves the unit negative, and when this reacts with recycling biochemical substances such as ADP or triphosphopyridine nucleotide (TPN), organic matter synthesis is driven. In overall consideration the release of an oxidized substance leaves the living system reduced. Since the external medium of most life in the biosphere is oxidized, some means of keeping

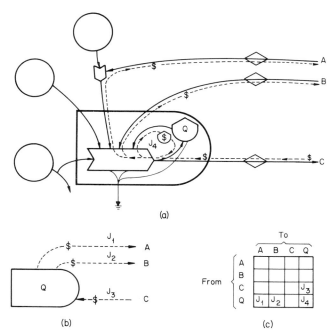

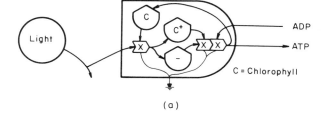

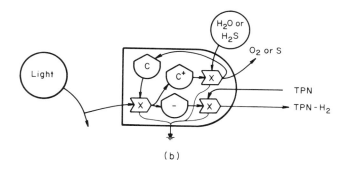

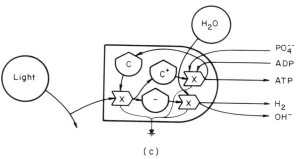

FIGURE 19-5. Producer unit within economic system with money counter currents: (a) detail; (b) representation as in economic input–output models; (c) same as part b with matrix form.

the interior reduced is required. The type of photosynthesis in Fig. 19-7b is thus the one adapted.

In Fig. 19-7c the negative area of the initial energy reception reacts with water or other substances so as to reduce the hydrogen from neutral to reduced state with hydrogen gas going off. This leaves the interior of the living system positive, and the synthesis of organic matter must be coupled to this drive. This kind of photosynthesis occurs in purple sulfur bacteria that are found in very reduced envi-

FIGURE 19-7. First processes transforming light energy in photosynthesis. (Odum, 1967a): (a) primitive types; (b) regular type that releases oxygen; (c) photosynthesis in reduced environment (*chromatium*).

ronments. It may be that keeping the interior of the living unit more oxidized than the medium is done by pumping out reduced substances (Odum, 1967a).

Reception of Light a Michaelis–Menten Loop

When pure energy inflows as in light or other wave transmissions, some recycling material is required to transform its energy into that of flowing matter. Examples of light transformation by separation of charge in chlorophyll and rhodopsin are given in Figs. 19-7 and 19-17. In Chapter 8 such loops were shown to be Michaelis–Menten loops. Apparently any system that can transform wave energy has this form.

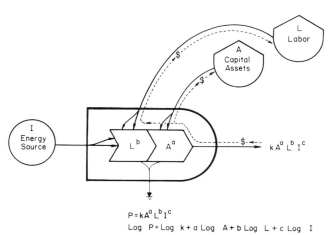

FIGURE 19-6. Production function given by Cobb and Douglas (1928). The energy source was originally omitted.

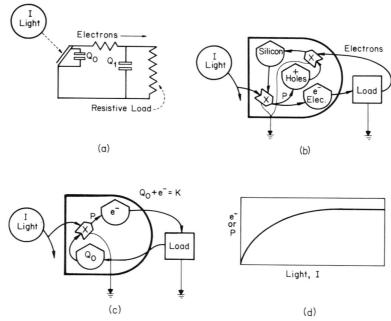

FIGURE 19-8. Photovoltaic Cells. (*a*) electrical circuit diagram; (*b*) energy diagram; (*c*) simplified energy diagram; (*d*) response to light when circuit has a load. Graph is that of Michaelis–Menten kinetics given in Chapter 10.

Analogous Photocells

Even simpler examples are the selenium and silicon solar cells used in photographic light meters and in conversion of light to electricity. As shown in Fig. 19-8*b*, there is a separation of negative-charged electrons from the silicon or selenium that becomes plus charged. After the electrons circulate through pathways provided, they return to their "holes," neutralizing the semiconductor and rendering it receptive to light again. The Michaelis–Menten kinetics apply.

A selenium cell is a good analog simulator of the photoreceptive process, also found in plants. The response of the meter is similar to that seen in the Michaelis–Menten graphs, although the curve is also close enough to a logarithmic one over part of its range to be used for calibration graphs. If the system is given no load, the output approaches linearity. Linear light meters are made by having a large cell area and small resistance in the device used to read the current.

Maximum Power Loading in Photosynthesis

In Chapter 8 the efficiency for maximum power was shown to be that at an intermediate loading, neither as efficient nor as fast as some other loadings. It was shown that the Michaelis–Menten loop design helps to maintain high levels of productivity by letting light go to the next units after maximum power is achieved in the upper units.

Billig and Plessner (1949) found maximum power with intermediate loading of photovoltaic cells, another example of optimum efficiency for maximum power (Fig. 7-21). Various loadings produce different power outputs in photosynthesis also. Biochemical studies have been made with isolated fragments of the photosynthetic apparatus including mainly the first stages. Figure 19-9 shows the design of those reaction studies by which photosynthesis is operated with some artificial oxidative and reducing substances supplied in place of the normal feedback loading; notice that there is an optimum loading that gives maximum power in the photosynthetic process. Adaptation of quantity of pigment to light intensity was considered in Fig. 10-10.

Inflows and Outflows of Photosynthesis

The light transformed in the first process is used to drive the synthesis of carbohydrate that uses an inflow of carbon dioxide. Figure 19-10*a* includes the interaction of the additional inflow. As indicated

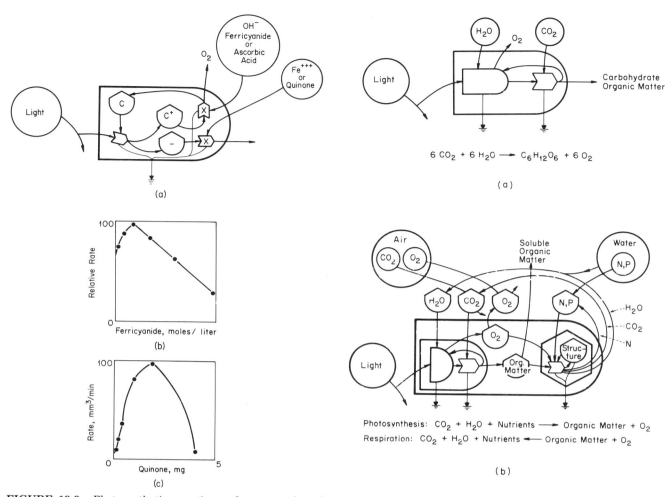

FIGURE 19-9. Photosynthetic reactions of macerated and isolated parts of chloroplasts of plants that carry out part of photosynthesis when oxidative and reducing reactants are supplied: (a) energy diagram; (b) power as a function of loading concentrations in experiments by Lumry and Spikes (1957) with saturating light; (c) experiment with quinone by Clendenning and Ehrmantraut (1951) [see Rabinowitch (1956)].

earlier, these models omit most of the storages and loops that are part of the actual biochemistry.

Next, and as shown in Fig. 19-10b, the organic matter is the fuel and source of material for building of structure and chemical substances used in the system. For this, other elements are needed such as nitrogen and phosphorus. Figures 19-10b and 19-11 show the incorporation of these substances as structure is made and stored. The structure is an inherent part of the whole system, although part of the feedback from structure is omitted from these models.

One may measure photosynthesis by measuring one or more of the inflows or outflows.

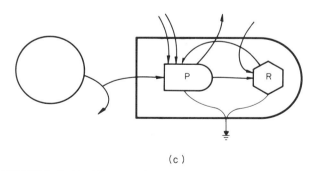

FIGURE 19-10. Photosynthetic production: (a) light transformation (regular photosynthesis) and the incorporation of carbon dioxide to produce carbohydrate organic matter; (b) same with addition of synthesis of protein and other organic structure by means of respiration using nutrients such as phosphorus and nitrogen as additional raw materials; (c) abbreviation of the model of plant producers showing P and R within producers.

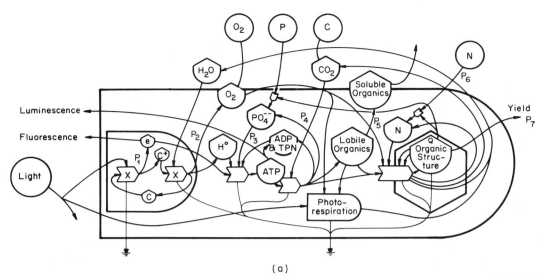

$106\ CO_2 + 90\ H_2O + 16\ NO_3 + 1\ PO_4 + \text{Minerals} + 1{,}300{,}000\ \text{Cal. sun} \rightarrow 154\ O_2 + 3258\ g\ \text{protoplasm}\ (106\ C, 180\ H, 46\ O, 16\ N, 1\ P, 815\ g\ \text{ash}) + 1{,}287{,}000\ \text{Cal! dispersed heat}$

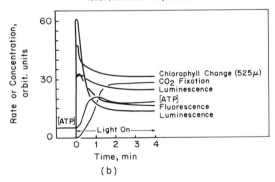

FIGURE 19-11. A model of photosynthesis by algae compartmentalized to include main inflows and outflows and their interactions: (a) energy diagram representing typical photosynthetic equation for production; (b) typical graph of observed charge up when light is turned on (Strehler, 1957). The terms $P_1 - P_7$ are exchanges with the outside medium that are used to estimate production rates. Degraded energy sinks include one for light reactions on the left and one for dark reactions on the right.

Respiration Units Within Producers

In the last process shown in Fig. 19-10b some organic matter built in the first part of the process is oxidized, generating energy for the synthesis of the structural elements such as proteins. This last part of the system is similar to respiration in an animal consumer that uses oxygen and food to operate its machinery with carbon dioxide and water as by-products. It is appropriately boxed in the hexagon group symbol as a consumer.

Respiration includes depreciation of structure and the work processes in keeping up with synthesis to replace depreciation. The overall equation for respiration (Fig. 19-10b) is the reverse of that for photosynthesis.

The respiration is necessary to operate the system at night when there is no energy flow coming in from photosynthesis. In some plants synthesis pathways are separate from respiration.

For some purposes, the summary model in Fig. 19-10c is used to represent photosynthesis. Notice that it has a Michaelis–Menten production unit and an autocatalytic respiration unit.

CHARACTERISTICS OF PRODUCTION AS A CHAIN

Figure 19-11 gives a model of photosynthesis with intermediate complexity showing main inflows, outflows, and more of the important storages within the system than shown in minimodels. The high-energy compound ATP is shown in its role as an energy transporter, where P_1 through P_7 are pathways that can be measured as indices of productivity. Those that originate closer to the light-energy source have stronger flows because energy is lost on passing downstream and because storages filter out energy surges. Thus, values for productivity measured with different inflows and outflows should not be exactly comparable.

Productivity has as many definitions as there are flows that can be measured or calculated. The best way of studying and reporting on a photosynthetic system is to give a diagram of the model believed pertinent with measurements identified in the diagram as flows and storage change rates.

RESPONSE TO LIGHT

The response of Michaelis–Menten units to increasing light is a characteristic asymptotic curve as derived in Fig. 10-5. The P–R unit as a whole is a Michaelis–Menten loop. The first production unit of photosynthesis is a Michaelis–Menten loop with recycle of chlorophyll as we showed in Fig. 19-7. As indicated in Fig. 11-25, a chain of Michaelis–Menten modules has an overall transfer function output like one such unit.

The graph of response of photosynthetic units to light is given in Fig. 19-12 with limiting asymptotes, "half-maximum" light intensity, and other expressions for describing the curve of light response.

Rapid Charge-Up

The first steps in photosynthesis have very small storages and reach quasi-steady state very rapidly when lights are turned on (see Fig. 19-11b). The first step in which light is transformed into a separation of charge can be measured by back fluorescence, a process in which the negatively charged electrons drop back into their holes, releasing energy in photons of energy lower than that absorbed.

Simplified P–R Model

In Fig. 19-13 the models given in Figs. 19-10 and 19-11 are simplified to two interactions, one representing photosynthesis and one representing respiration. This is the model given in Fig. 2-7 as representative of balanced aquaria or the entire biosphere. It is also used for plants since they have production and respiration within them. In Fig. 19-13a oxygen is an external storage variable, as is appropriate for plants in water. Figure 19-13b, however, shows land plants where the oxygen content is so great that it is hardly affected by the plants over short periods of time and is omitted. These models are Michaelis–Menten modules with appropriate kinetics.

Usual patterns of charging of storages and discharging of storages as light is turned on and off are given in Fig. 19-13c. Photosynthesis is maximum at the start of the light period since the raw material tank is fullest then. Respiration is maximum at the end of the light period since the organic matter and oxygen tanks are maximum at that time.

One typical approach for modeling of P–R units is given in Fig. 19-14, where the output of a P–R model is shown as the difference between P operating as a Michaelis–Menten module and R. The slope of the graph (derivative) can be graphed. It is similar in shape but has a quadratic denominator (Fig. 19-14c). The P–R model was the basis for back calculating production rates from diurnal measurements of oxygen and carbon dioxide (Odum and Hoskin, 1958; Odum and Wilson, 1962; Odum et al., 1963c).

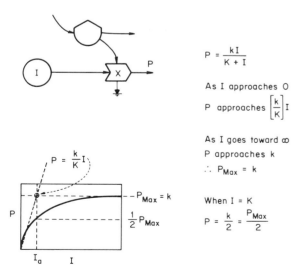

FIGURE 19-12. Light effect on production where other flows are limiting, either from outside or from recycle (see derivation in Fig. 8-10 or 10-5).

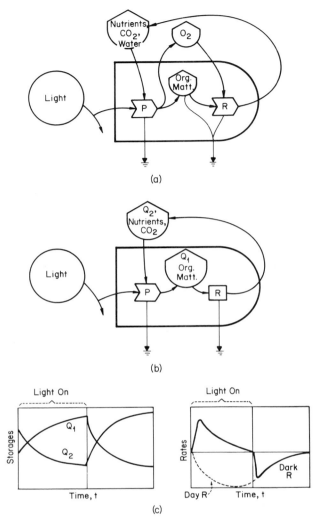

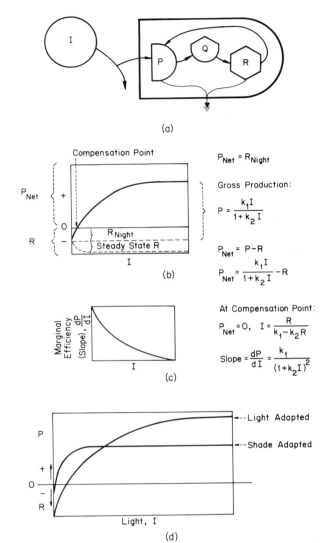

FIGURE 19-13. Photosynthesis–respiration $(P-R)$ models with closed cycle of materials: (a) with oxygen varying as in water; (b) with oxygen constant as in air; (c) graphs of organic matter Q_1 and nutrients Q_2; (d) graphs of rates of net production and respiration.

FIGURE 19-14. Response of photosynthesis and respiration units to variation in light intensity: (a) energy model; (b) Michaelis–Menten curve for gross production P and concurrent respiration A; (c) derivative indicating marginal effect of light; (d) light adaptation curves from Boysen–Jensen (1932).

Start and Stop Surges When Storage Units Are Small

When cells are very tiny, they are called *nanoplankton* and pass through the smallest plankton nets. Their storages are charged up to steady-state level in an hour or two to ensure that their production and consumption are equal thereafter. Then at night the storages are used up in an hour or so, and the respiration returns to a very low rate. A simple tank model was given in passive analog by Odum et al. (1963a), which was analyzed with control theory analysis by Milsum (1966) (see Figs. 3-18 and 3-19).

The model was used to account for the very low net production observed in waters with nanoplankton, especially when done with radiocarbon methods involving times as short as the time constant of the cells. The models showed that gross production was much larger than that inferred from net production data due to the artifacts of long-term measurements. Conversion and useful work based on photosynthesis was much larger than inferred from field measurements (Wilson, 1963). These theories were confirmed by measurements due to Pearl and Mackenzie (1977), showing net fixation of carbon showing only in the first light and large losses of carbon

TABLE 19-1. Equations for Response of Simple-Unit Production Systems to Increase in Light (I).

Name or Source	Equation for P	Reference to Model
Response curve with level asymptote:		
Blackman (1919) linear increase with limit (P_m)	kI and P_m	Fig. 8-14
External limiting factors of Monod (1942) model	$\dfrac{kI}{K+I}$	Figs. 8-10, 19-12
Michaelis–Menten (1913) model for internal limiting recycle	$\dfrac{kI}{K+I}$	Fig. 10-5
Smith (1936)	$P = \dfrac{I\sqrt{P_{\max}^2 - P^2}}{I_k}$	
Same, manipulated	$P = \dfrac{P_{\max} I}{I_k^2 + I}$	Fig. 19-12
Mitscherlich (1909) exponential	$k(1 - e^{-kI})$	Fig. 8-15
Cobb–Douglas (1928)	$k_I a_X b$	Fig. 19-6
Jassby and Platt (1976)	$P_m \tanh \dfrac{\alpha I}{P_m}$	Empirical fitting
Response curves with maximum:		
Steele (1965)	$\dfrac{I}{I_m} e^{(1-I/I_m)}$	Empirical
Parker (1974)	$P = kP_{\max} I e^{(1-kI)}$	
Platt and Gallegos (1980)	$P = P_s(1 - e^{-\alpha I/P_s}) e^{-\beta I/P_s}$	

in the first darkness. When time constants are very small, the pulses of net production and respiration given in Fig. 19-13 become very large, short bursts. See Fig. 3-19e.

P and R in Frequency Domain

In streams with algae, oxygen rises and falls minute to minute due to P–R balance as passing clouds vary the sunlight. Gallegos et al. (1977) transformed data on light and net production to frequency domain with Fourier transform algorithms. An equation for productivity in terms of frequency responses was derived so that net productivity could be calculated from characteristics of frequency functions. The cross-correlation between light and productivity increased as the period of light fluctuation became shorter. The productive response was higher and leading the light during pulsing. The model given in Fig. 3-19 was selected as appropriate. Gallegos et al. (1977) proposed a model [Eq. (19-1)] in time domain for production with a differential term adding productivity when conditions were pulsing:

$$P - R = g\left[I + \tau \frac{dI}{dt}\right] \quad (19-1)$$

Higher yield with pulsing light was shown in laboratory studies of algae because biochemical storages have time to develop larger concentrations (Rabinowitch, 1951) (see discussion of pulsing lags with Fig. 3-17 and p. 195).

Other Equations for Light Response

Many different equations have been written for the response of productive systems to varying light intensity. The observed responses are usually asymptotic curves such as that in Fig. 19-14b. Light as an external limiting factor was modeled in Fig. 8-8. Since production is often a function of the interactions of several inputs, the equation for light response is tantamount to the model for production. Equations for response of simpler production models to light are listed in Table 19-1, which contains cross-references to appropriate sections in this or other chapters.

Selection of a light response equation is not an *a priori* modeling step but should be an output of developing, simulating, and validating a model of the production subsystem. Combination of equations for components gives different results than determination of equations for a unified model (e.g., see Figs. 8-18 and 8-19).

Equations are included in Table 19-1 that a decreased output at high light intensity was observed for marine plankton cells adapted for intermediate light intensities with intermediate chlorophyll (Odum et al., 1958). Engqvist and Sjöberg (1980) provide a calculation method for use of Steele's curve.

MEASUREMENT OF PRODUCTION

The measurement of rates of energy transformation or storage from outside a system are done by measuring the rate of utilization of some raw material, the rate of release of some by-product, or the rate of storage accumulating inside the system. The items crossing the unit boundary in Fig. 19-11 are some of the items that are sometimes measured to estimate the processes in the system, or the storages inside (P, N, Q) are estimated in samples before and after a period of production. Since such measurements are only net differences between inflows, outflows, and feedbacks, the actual process must be calculated with the help of a model. Back fluorescence measures the first step (Samuelsson, 1978).

Measurements of nutrients (P,N) are the most difficult conceptually since they are pumped in by some cells when there is no production as an adaptation to low-nutrient environment. Also, their uptake is far downstream, producing measurements somewhat different from those of oxygen or carbon.

An important example is given in Fig. 19-15, where a radio tracer uptake is being used to estimate production. Figure 19-15b shows a situation in steady state without growth; Fig. 19-15c shows growth also. Tracer enters from the left and is stored, and some of it begins to come out again immediately. The catchup curve in Fig. 19-15b is a first-order process similar to that studied in Chapter 3. The procedure of estimating uptake of radio-tracer as a measure of the production uptake gives rates that are far to low. The uptake is not a measure of net production either since there is none in Fig. 19-15b. To determine the actual uptake rate, determination of several points on the graph is necessary. Since uptake is a chargeup exponential, the curve can be subtracted from the asymptote and plotted on semilog graph to produce a straight line to facilitate fitting of points. R. F. Wilson (1963) generated curves of this type with algal production experiments for varying time lengths.

In Fig. 19-15c, where there is growth, the uptake catchup curve approaches the growth curve, but the uptake rates over discrete time intervals are neither gross production nor growth (net gain of storage).

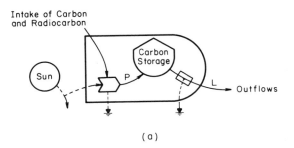

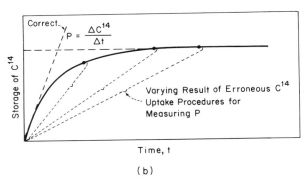

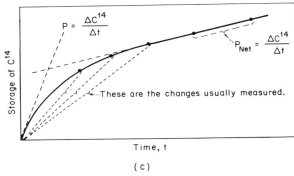

FIGURE 19-15. Meaning of rates of uptake of a radioactive tracer in production: radiocarbon uptake in bottles containing estuarine plankton, where P is gross production; and P_{net} is net storage rate $(P-L)$: (a) model; (b) catchup in nongrowing cells; (c) catchup in growing cells.

The smaller the time constant of the storage (small cells), the larger the errors and the lower the actual net growth rates. Many tracer production results are incorrect (Petersen, 1980).

Graphs of Rate of Change of Storage

One type of production is the *net production of growth*, which is defined as the rate of change of storage. The rate of change of storage is the slope of the graph of stocks with time. For example, Conover (1958) gave the rates of production of bottom grow-

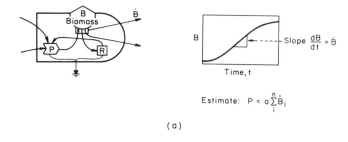

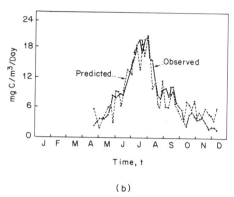

FIGURE 19-16. Model for relating rates of phytoplankton net growth to rates of radioactive carbon uptake (DeAmezaga et al., 1971). For each species, rates of net growth ($\dot{B}$) were calculated from slope of graphs of biomass with time. Species rates were then added to obtain net production N. (a) Definition of biomass based net productivity P; (b) correlation of calculated index P with observed rate of ^{14}C uptake of the whole phytoplankton community.

ing macrophytic algae with time by graphing the rate of change from graphs of biomass.

In Fig. 19-16 De Amezaga et al. (1971) obtain net growth rates from each of the dominant phytoplankton in lakes. These were correlated with the radioactive carbon uptake of the whole association with a good correlation, even with only five to nine of the species present. This was evidence that carbon tracer fixation was measuring the net growth and conversely that rates of change of data on stocks of dominants were comparable to process rates measured physiologically.

PHOTORESPIRATION

In ecosystems in which there are severe conditions such a low temperatures, high salinity, or toxins, respiration may be too slow to recycle with sufficient rapidity to keep photosynthesis and respiration in balance. In these and to a lesser extent in other systems there may be a pathway that connects light to drive the respiration process [see pathway marked (PhR) in Fig. 19-17a]. Figure 19-17b shows the same pathway, except with oxygen as a state variable (see Fig. 19-13 for $P-R$ models without photorespiration).

Evidence for photorespiration as an adaptive mechanism in systems with poor conditions for ordinary respiration were obtained in pink brine bacteria by Nixon (1969) and Odum et al. (1971) using the model in Fig. 19-17b. Physiological mechanisms of the proton pump in brine bacteria were shown by Stoeckenius (1978) and Osterhelt and Stoeckenius (1973). Figure 19-17c shows an enlargement of the photorespiration unit of brine bacteria. These pink organisms have a pigment rhodopsin (the same as in the eye), which receives photons just as chlorophyll does, except that it carries a negative charge when activated after releasing a positive charge (proton). Thereafter the rhodopsin reduces oxygen and is returned to the receptive state. The positive-charged protons react with organic matter accelerating respiration. Examination of the design of the photorespiration unit in Fig. 19-17c shows that, similar to the chlorophyll cycling photosynthetic process, the proton pumping recycle of rhodopsin is also a Michaelis–Menten loop with similar kinetics.

As shown in Fig. 19-17d, the effect of light on a unit with photorespiration and an accumulation of organic matter is to depress oxygen in the light at first until recycle is sufficient to stimulate photosynthetic components. Curves such as that in Fig. 19-17d were obtained in brine systems that had both photosynthetic algae and photorespiring brine bacteria *Halobacterium* (Nixon, 1969; Odum et al. 1971). A passive analog (Fig. 19-18c) was used to generate the observed curve (Fig. 19-18d). Talbert and Osmond (1976) discuss roles of photorespiration.

PHOTOELECTRIC ECOSYSTEM; BLUE–GREEN ALGAL MAT

In very shallow waters (<10 cm) of high fertility, the range of oxygen in the water is very high, reaching zero for much of the night. A blue–green algal mat develops on the bottom, which has oxidized upper surface and reduced lower surface, the two separated by 1–2 millimeters. In the photosynthetic process organic matter is developed and the oxygen is released to the upper surface, where it causes the

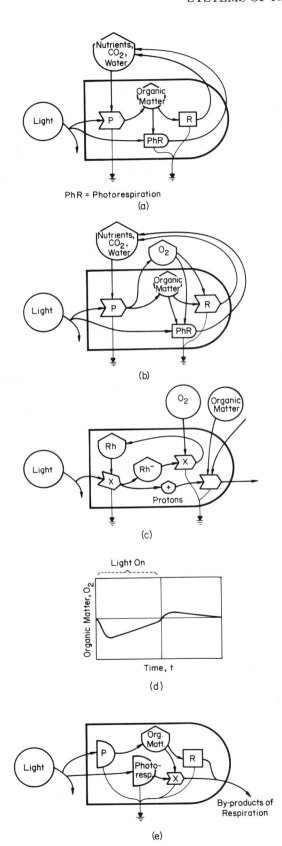
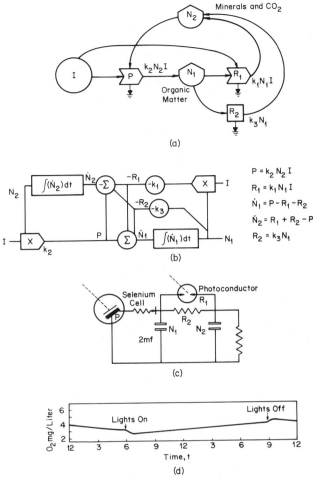

FIGURE 19-18. Models of production with photorespiration: (a) energy language; (b) block diagram; (c) passive electrical analog; (d) response to light (Nixon, 1969b; Nixon and Odum, 1970; Odum et al., 1971).

medium to be oxidized during the day, leaving the rest of the mat and undersediments reduced. The result is a gradient of oxidation–reduction potential of 0.5 volt over a short distance. It is a living photocell. The voltage apparently helps pull nutrient anions such as phosphates and nitrate from the regeneration zone under the mat through to the photosynthesizing cells in the upper lighted part of the

FIGURE 19-17. Photorespiration: (a) P–R model with photorespiration pathway added, with oxygen constant (Nixon et al., 1971); (b) same with oxygen varying; (c) photorespiration unit of brine bacteria with proton pump of Osterhelt and Stoeckenius (1973), where Rh is rhodopson; (d) response of P–R model with photorespiration to light on and off. (e) Photorespiration model from Thornley (1976).

Producers

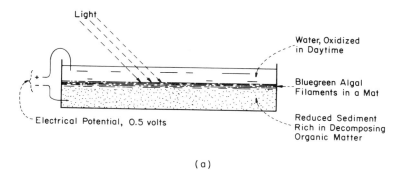

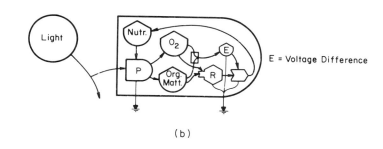

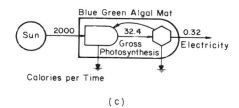

FIGURE 19-19. Photoelectric ecosystem, blue–green algal mat (Armstrong and Odum, 1964): (a) sketch of mat ecosystem isolated in a laboratory dish; (b) energy diagram; (c) aggregated diagram with energy flow values.

mat. Thus the energy that develops the electrical potential is fed back into the recycling process in the same way that the high-quality energies of a higher animal consumer organize and aid recycling. This system is shown in Fig. 19-19. Its design is similar to other producer recycle systems in this chapter with Michaelis–Menten response.

The energy diagram in Fig. 19-19b is aggregated further in Fig. 19-19c with energy values from Armstrong and Odum (1964). Here sunlight is transformed through photosynthesis first to organic matter and then to electricity. The ratio of conversion of low-quality light to high-quality electricity in this system includes self-maintenance of the structures involved (refer also to Figs. 7-19–7-21, where the power loading response is discussed).

LIGHT AND DEPTH WITHIN THE ECOSYSTEM

Light Penetration and Stratification of System Units

A familiar property of light penetration through an absorbing or deflecting medium is exponential extinction described by the Beer–Lambert law. It states that the percent change with distance is proportional to the quantity of material interfering with the passage of light (see Fig. 19-20). The material causing an extinction of light may be inert matter that diffracts, absorbs, or reflects light, or it may be living units that absorb the energy. Figure 19-20 shows the action of a series of units as light passes,

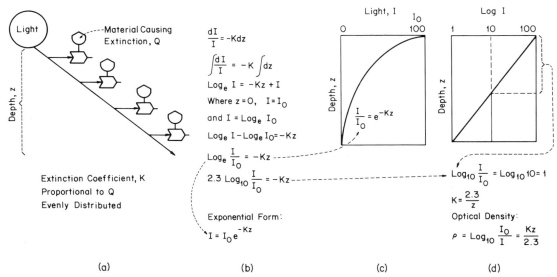

FIGURE 19-20. Extinction of light by Beer's law: (a) energy diagram; (b) equations; (c) exponential decrease of light with depth; (d) semilog graph.

with a fraction diverted at each stage. When the passage of light is graphed with depth exponential transmission curve results as predicted by Beer's law. When plotted on semilog paper, the line of absorption by Beer's law is straight and the slope is the light-extinction coefficient K.

Compensation Point

The compensation point is the light intensity at which photosynthesis is balanced by respiration. In Fig. 19-14b it is the point where the curve of production crosses the horizon of zero net production. Con-

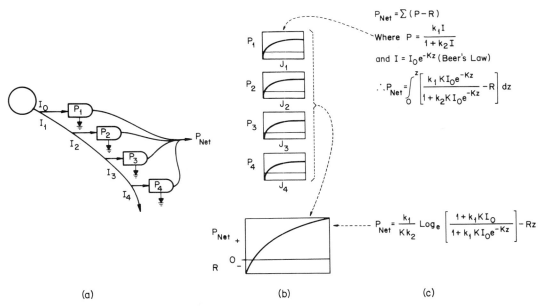

FIGURE 19-21. Production of a column as the sum of the productive units, each of which operates as a Michaelis–Menton loop and absorbing light according to Beer's law, a model given by Monsi and Saeki (1953): (a) energy diagram; (b) graphs of light response separately and integrated; (c) equations.

ditions that increase respiration raise the light intensity required to compensate. The model in Fig. 19-14a generates a curve with increasing light that has a compensation point as shown in Fig. 19-14b at which production is equal to respiration and there is zero net production.

Production of a Column of Photosynthetic Units

When there is a whole column of stacked photosynthetic units as with plankton in the sea or leaves in forest, light is used up as it is absorbed and deflected. Where the distribution of photosynthetic units is uniform, the light disappears according to Beer's law with an exponential curve with depth as in Fig. 19-20. Each photosynthetic unit manifests Michaelis–Menten kinetic response to light and some concurrent respiration as in Fig. 19-14. These two principles are combined in Fig. 19-21, where production minus respiration is integrated with depth by use of Beer's law for the distribution of light with depth. The result is a logarithmic curve for production with varying light intensity that rises but never levels as light is increased.

Benoit (1964) studied yield of *Chlorella* cultures at 40,000 foot-candles, four times daylight. Adapted cultures were not inhibited by high light intensity because deeper beds of cells and chlorophyll developed with proportionately higher photosynthesis per unit area of culture surface.

Light Attenuation

In many models the quantity of plankton, leaves, turbidity, or other quantity causes light extinction. Figure 19-22 shows two ways of indicating action of the quantity Q on the inflow of light. Both methods give sharply decreasing curves with increasing quantity of interfering substance. In Fig. 19-22a Beer's law is given with the present change in light proportional to the quantity. When integrated, an exponential interaction results. In Fig. 19-22b the unused light remainder I controls further interaction as already given in Fig. 8-8b.

Where the biomass is evenly distributed, the optical density in Beer's law calculated from vertical light graphs on a semilog plot (Fig. 19-20c) may be used as a biomass measurement technique (Westlake, 1964; Odum et al., 1963).

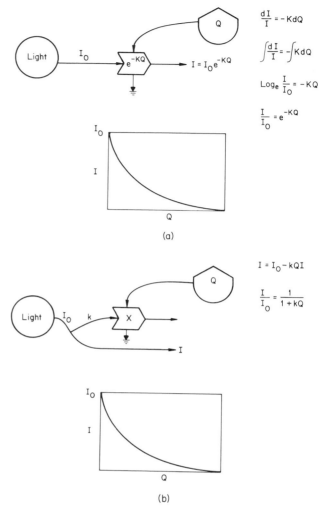

FIGURE 19-22. Two different models used for light attenuation as a function of quantity Q: (a) exponential attenuation; (b) remainder availability.

Net Production and Depth of Mixing

Where plankton are mixing well in a vertical layer as shown in Fig. 19-23, the depth of mixing becomes important to net production. If the mixing zone is not very deep, the plants remain in the light and production exceeds respiration. With deeper zones of mixing, the algal cells are in the dark for too long a period, respiration exceeds photosynthesis, and production of the entire layer is essentially stopped. Thus changes in depth of mixing because of changes in physical conditions or because of waters passing into shallow zones can cause dramatic bursts of net production. In Fig. 19-23 equations for net production are developed following Sverdrup (1953), Ragotzkie (1959), and Murphy (1962). The equation

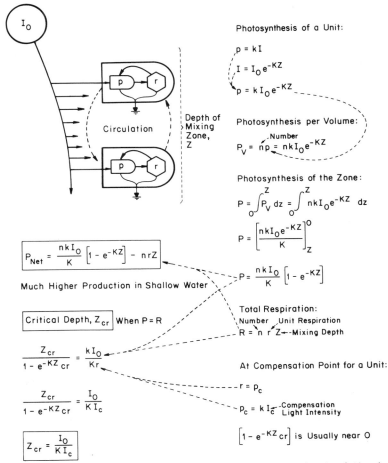

FIGURE 19-23. Net production and critical depth of production units circulating in a mixing layer (Sverdrup, 1953; Ragotzkie, 1959).

is an exponential one. Examples of curves showing the effect of depth are given in Fig. 19-24. Rawson (1953) discovered production decrease with depth of Canadian lakes.

Critical Depth

When the layer of mixing achieves a depth so that production and respiration are equal, this depth is called the *critical depth*. Mixing deeper than this stops net production. Critical depth (Fig. 19-23) is given in terms of the compensation point (I_c) of the component units that are circulating (Fig. 19-14).

SPECIAL PHYSIOLOGICAL ADAPTATIONS

Many adaptive mechanisms exist in plant cells for adapting to light and other interacting resources (Myers, 1946).

Nutrient Concentration System

Many algae living in low-nutrient environments have mechanisms for uptake of nutrients as they are available, even though photosynthesis is not operating (e.g., at night). Figure 19-25 gives a model for nutrient uptake as a separate mechanism.

Simulation of Photosynthesis That Couples Two Pigments

In Fig. 19-26 is an energy circuit translation of the model used by French and Fork (1961) to simulate the photosynthetic responses of algae with a two pigment system, each absorbing wavelengths differentially and then interacting to generate oxygen and organic matter. This was one of the early analog simulations in plant physiology. Use of more than one pigment is an adaptation for depth in waters in

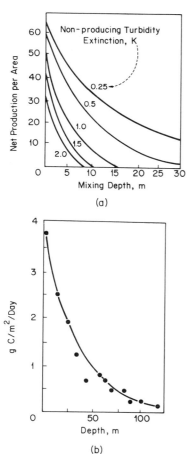

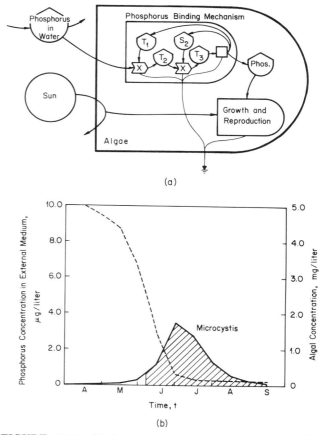

FIGURE 19-24. Graphs of net production and mixing depth: (a) redrawn from Murphy (1962) (see equation from Fig. 19-23); (b) Steeman-Nielsen (1963).

FIGURE 19-25. Model of phosphorus uptake by algae with growth later based on stored uptake (Bierman et al., 1975): (a) energy diagram, with growth details omitted; (b) simulation.

which only green wavelengths are penetrating deeply.

C_3 and C_4 Plants

Work on the biochemistry of pathways of plant photosynthesis shows many differences that provide alternative adaptations for different environmental conditions. One difference is commonly referred to as *C_3 and C_4 plants*. The abbreviated biochemical pathways are given in energy language in Fig. 19-27.

For fast production to compete and maximize power where nutrients are rich, C_4 is adapted. It has more net production but has less of a self-maintenance role to sustain structure when initial growth has maximized structure possible for that energy base. This is the one adapted to farming where quick yields and temporary growth are needed.

For self-maintenance, recycling, and maximizing power of gross production, which is ultimately the criterion of survival, the C_3 system is adaptive.

Ehleringer (1978) simulated net production of C_3 and C_4 grasses for different climatic conditions, finding that more tropical conditions generated C_3 with maximum growth and other conditions when roles were reversed. In the model the Michaelis–Menten response to light showed 20% less gross production with C_3 plants and a parabolic temperature curve.

Succulents

Many plants such as cactus with fleshy tissue operate carbon dioxide fixation at night (Gregory et al., 1954), carrying the process as far as malate. Then during the day, light is transformed and the cycles are completed, but the stomata remain closed, preventing water loss in the hot daytime. As shown in

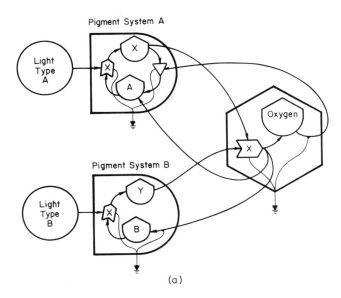

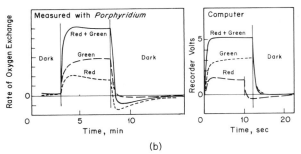

FIGURE 19-26. Model of photosynthesis with interaction of red and green light with two types of pigment receptor (French and Fork, 1961): (a) energy diagram; (b) comparison of simulation with observed responses of red algae to light and dark periods.

Fig. 19-28a, control of gas exchange by stomates and large storages of intermediates makes possible adaptation to desert conditions. This kind of metabolism is sometimes called *Crassulacean acid metabolism* (CAM) (Kluge and Ting, 1978).

Load Adaptation

In Fig. 10-10 the quantity of cycling receptor material in a Michaelis–Menten loop required to maximize power was found to increase hyperbolically as input energy levels were decreased. Cloudy regions develop intensely green plant cover to increase the capture of light energy, whereas sunny regions develop less chlorophyll in the top level since their input energy reception is sufficient to load the biochemical system on top, and by absorbing less, more energy passes below to cells that are shade adapted there with the development of high chlorophyll concentrations. In the mixing waters of the sea, plant cells are in a varying light regime and develop intermediate chlorophyll levels (Odum et al., 1958). A speculative model of the adaptation process is given in Fig. 19-29.

Boysen–Jensen (1932) show the production response curves to varying light to be different for the shade-adapted and light-adapted receptors as shown in Fig. 19-14d, each better adapted to transform more energy in its usual working zone.

Chlorophyll and Carotenoids and Maximum Power

Long recognized are the patterns of increased chlorophyll with rapid growth in high nutrients and the predominant yellow pigments and carotenoids in old cultures, leaves late in the season, and ecosystems that are more nearly in steady state. Odum and Hoskin (1957) switched the blue–green color of microcosms from green to yellow and back by control of nutrients, observing a high P/R ratio during the green period and similarity of P and R during the period of yellow preponderance.

Margalef (1960), discussing the aging of ecosystems, used the carotenoids:chlorophyll ratio (actually the ratio of absorbency of solvent extractions of pigments at 430 nanometers to 665 nanometers) as a measure of maturity and diversity with the suggestion that the carotenoids were part of the diversification absorbed in older plankton and microcosm systems that commenced with high nutrients and consumed the nutrients subsequently.

Odum et al. (1971) offered the theory that maximizing power required effective production and consumption–recycle components (in biology sometimes called *catabolism* and *anabolism*) and that when either of the two processes were limiting, the ecosystem was uncompetitive until mechanisms were organized to apply available energies to the process that limited the cycle and processing of energy. One mechanism for applying energy is through pigment regulation of solar application to either production with the use of chlorophyll or the photorecycling with the use of carotenoids. Nixon's dissertation showed the role of sunlight in the photorespiration cycle in brine bacteria, and physiological evidence of proton pumping by *Halobacteria* was given by Stoeckinius (1978). Kowallik (1967) showed carotenoid roles in other forms of photorespi-

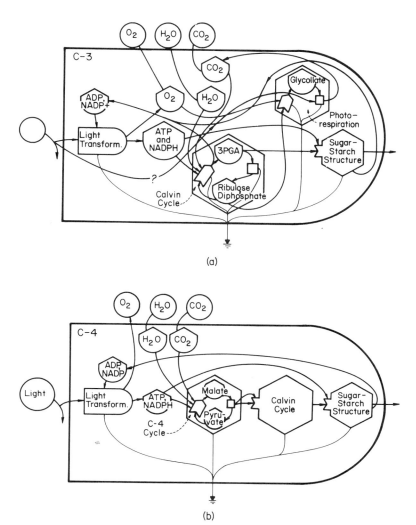

FIGURE 19-27. Energy diagram of types of plant photosynthesis, with regular respiration omitted: (a) CO_2 uptake through C_3 Calvin cycle with photorespiration–glycollate pathways providing higher gross production through recycle of some CO_2; (b) CO_2 uptake through C_4 cycle with higher net production but with CO_2 limiting gross production. For details, see Salisbury and Ross (1978).

ration in algae, and Eppley and Macias (1963) found a waste pond with photoconsumption a dominant process where organic matter was in excess.

Without showing all the known details of chlorophyll synthesis and carotenoids, Fig. 19-29 suggests the kind of switching control to be exacted if the theory of pigment control of power maximization is correct. When chlorophyll is in excess of need, the storage S_1 remains full and generates more carotenoids from synthesis. When carotenoids are in excess, nutrients stimulate chlorophyll from synthesis. This can happen when light is low or nutrients high. When light is high or nutrients low, the process is limited by recycle and carotenoids are generated.

AQUATIC PRODUCERS

Many of the examples of models of producers given so far represent aquatic producers that are controlled by light, nutrient limiting factors, and physical stirring (Falkowski, 1980) (see Fig. 19-30a). A popular model combines autocatalytic growth and eddy dispersion as in Fig. 9-23 with nutrient and other limitations (O'Brien, 1974; Wroblewski

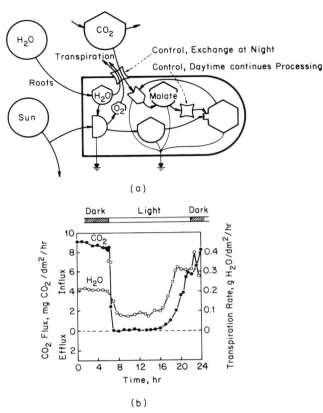

and O'Brien, 1976; Dubois, 1975). Spatial oscillation results. See Fig. 18-31.

TERRESTRIAL PRODUCERS

Terrestrial producers are more complex, with the development of leaves, trunks, roots, fruit, and so on. In addition to nutrients, there is the transpiration pull of wind and evaporation power of dry air. An example of a terrestrial plant producer model is given in Fig. 19-30b. Lommen et al. (1971) simulate leaf production with most of the properties given in Fig. 19-30b developing nomograms of net photosynthesis and transpiration as a function of light, heat, and water budgets. Cunningham and Reynolds (1978) provide a unit model for the growth of plant organs from the carbohydrate and other pools generated by the leaves from photosynthesis (see Fig. 6-26). A model for whole creosote bush trees connected the organs as a food chain to the pools including leaves, stems, buds, flowers, stems, and roots (see also the leaf production–heat balance model in Fig. 16-2).

A simulation of a terrestrial producer that draws structure from allocation of photosynthetic pools is given in Fig. 19-31. Jordan (1971) found more net organic storage at high latitudes where summer night respiration is short.

FIGURE 19-28. (a) Aggregated summary of delayed metabolism of succulent desert plants tissue fixing solar energy in day and taking CO_2 in at night; (b) diurnal record of gas exchange in a succulent (CO_2 fixation and transpiration rates of the CAM plant *Agave americana* during alternate light and dark periods). From Neales et al. (1968), given by Salisbury and Ross (1978).

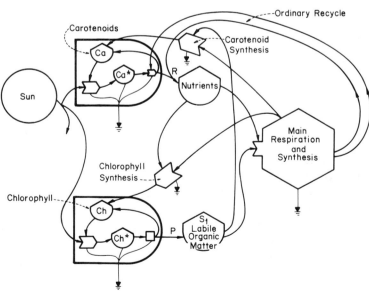

FIGURE 19-29. A model of production, photorespiration, and pigment synthesis that maximizes power by increasing the pigment concentration that is needed to keep P and R in balanced cycle of materials and that generates more chlorophyll at low light intensity (because R has a dark pathway).

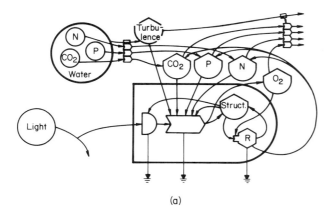

Seed Germination, Storage, and Dispersal for Survival

Seeds and other genetic propagules provide the high-quality means for transporting information to spread a successful system. Reproductives have a wide range of quality from many small seeds of successional species compared to the few, protected seeds of climax species (see Fig. 15-26).

Where conditions are varying, survival strategy for a species and its system role may require germination of only part of the seed stock in any one year so as to smooth out periods of poor and good yields. Cohen (1966) considers seed survival and production for various mortalities and probabilities of favorable yield energy (see Fig. 19-33). See Fields and Sharpe (1980).

PRODUCER–CONSUMERS IN ONE LIFE CYCLE

Figure 19-34 illustrates a life cycle where the units—of a moss that has green leafy producer on which another nonphotosynthetic generation is attached and derives energy doing work of reproducing—are formed successively as lower-quality producer; then as dependent consumer; and finally as high-quality concentration spore, dispersing them as feedback to begin the work of capture of low-quality energy again. Part of the high-quality work done is in manipulating the genes through the segregation and recombination process that uses energy to generate choices as part of the work of maintaining adaptation. The highest-quality components (with most kilocalories of solar energy per actual kilocalorie stored) are the concentrated spores prior to dispersal (see also Fig. 20-16).

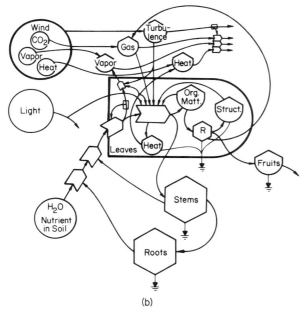

FIGURE 19-30. Typical producers: (*a*) aquatic algae; (*b*) terrestrial plant.

Tree Model

Figure 19-32 gives a model of a whole tree that includes many of the important processes and parts. A tree is a whole hierarchy from chloroplasts and roots converging to stem center. Depending on local energy signature, various parts are relatively more prominent or missing, and the species involved are appropriately selected. A model of a leaf response to light, heat, and water exchange is given in Fig. 16-2. Murphy and Gresham (1974) model the control action of stomata (the cell controlled pores through which gas exchange occurs).

Arctic Tundra Plants

In one model, Lawrence, et al. (1978) simulate a population of grasses in Arctic tundra with compartments that follow the life cycle from seed to seedlings to rhizome-based "tillers" (first year to sixth year) to inflorescence, flowers, and seeds. See Fig. 19-35. Sugars are generated from photosynthesis in proportion to light, which has biomass attenuation function. Sugars are allocated to the labile pool and then to rhizomes, leaves, and roots of each tiller. Empirical functions for allocation of sugars to different units and different times were derived from field

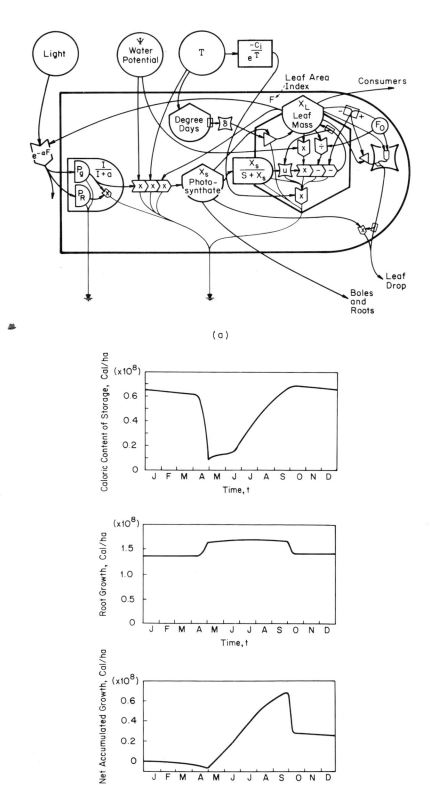

FIGURE 19-31. Model of terrestrial production (Shugart et al., 1974): (*a*) energy diagram; (*b*) simulations.

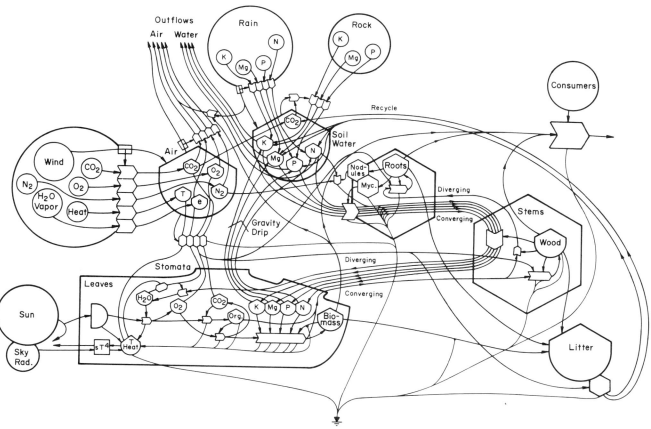

FIGURE 19-32. Model of the functions of a tree.

data. Lemming grazing was added in some simulations. Other simulations involving phosphorus and nitrogen used the same circle of compartments. Results of simulation were seasonal, and year-to-year growth patterns partially validated with field data.

This approach is simple in concept, with mostly empirical graphical functions guiding the flow of energy through the system. It is fairly complex in the total number of compartments. This kind of systems model is a synthesizer of experimental observations. The difficulty with using empirical relationships is that predictability is uncertain beyond the situation and time of the measurements. It may not be very satisfying for the objectives of understanding and representing the main essence of systems. It may be argued that ecological models don't need the detailed mechanisms of organism functions that might be required if the model were rendered deterministic from causes. However, it may also be argued that the detailed mechanisms are constrained to generate characteristic functions that achieve energy adaptation, and thus there are usable aggregated models that are causal.

SYNTHESIS SIMULATION OF EMPIRICAL MEASUREMENTS

A much-used technique in systems evaluation of production is simulation that combines measurements of relationships of the parts. The system is modeled to the extent of identifying the parts and pathways, but not in determining relationships from concepts. Instead, the measurements of relationships are graphed with or without statistical fitting and entered in computer memories as an empirical function. Then the simulation synthesis is run, computing states with time as flows and interactions are generated by empirical functions. If the resulting graphs of larger categories with time are validated by measured time graphs, consistency has been achieved between measurements of parts, processes, and larger categories. This approach was involved in much of the modeling of the U.S. International Biological Program. The thinking about relationships is done with the parts and the submodels connected. This approach appears inevitable if equations are written for parts before the whole model is

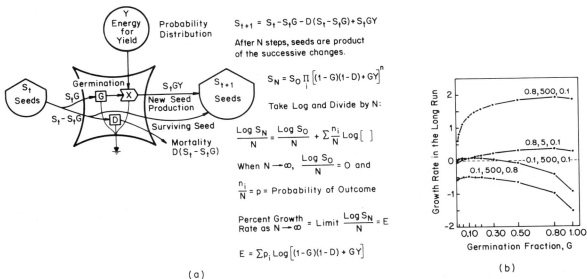

FIGURE 19-33. Seed germination G and growth for long-range adaptation of probabilities of environmental energy for yield Y and mortality D (Cohen, 1966, 1968): (a) energy diagram for discrete state transition for a unit of time t; (b) simulation.

completed. This is different from much of the approach in this book, where the model is considered first as a whole and parts are made consistent with the whole.

Examples of synthesis simulation of a producer system is given in several papers on tundra studies (Tiezen, 1978) (see Fig. 19-35). Van Dyne (1969) fits growth of roots, leaves, dead standing matter, and litter each to sine-wave seasonal functions and uses the set of resulting equations as a grassland systems model (Walters, 1971).

SUMMARY

In this chapter we considered models of primary production. Production functions involving interaction

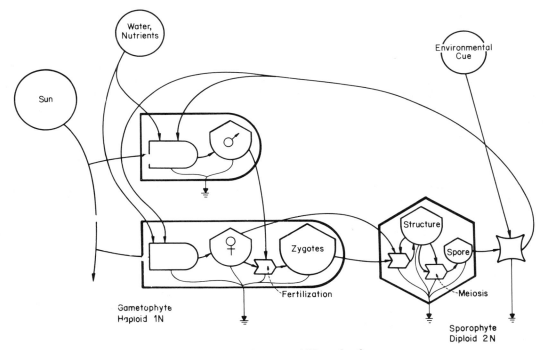

FIGURE 19-34. Energy diagram of life cycle of a moss.

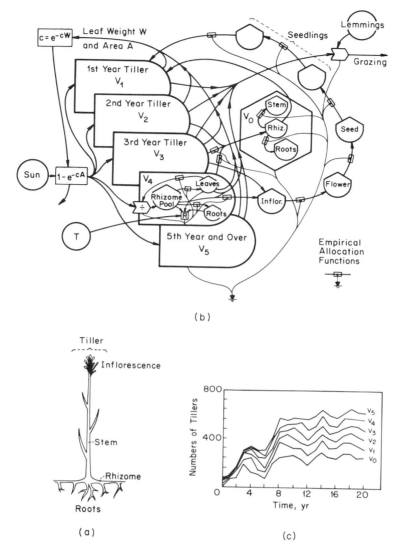

FIGURE 19-35. Producer model of Arctic grasses, using empirical functions that allocate photosynthate (Lawrence et al., 1978): (*a*) gross form; (*b*) energy diagram; (*c*) simulation of growth.

of light, nutrients, carbon dioxide, and internal structure in generating organic matter were compared. Methods for clarifying production concepts with models and converting estimates of storage accumulations to rates of gross production, respiration, and other flows were summarized.

The phenomena considered in this chapter included limiting factors, role of pigments, shade adaptation, depth effects, photorespiration, adaptations for net production, adaptations for land and water, adaptations for desert, and development of electric potential.

Simple production–respiration models account for many characteristic diurnal and seasonal patterns. Most models useful for understanding main concepts were simple, but a few more complex models were included from the literature.

QUESTIONS AND PROBLEMS

1. Draw minimodels representing production that include interactions of several qualities of energy, feedback from storage, and consumption that is part of producers for dark reactions.

2. Diagram details of the chlorophyll receptor cy-

cle, a rhodopsin photooxidation receptor, voltage developed by a blue–green algal mat, a C_4 specialist for net production, a double-pigment system, a nutrient pump, a leaf, and a photometer hardware cell.

3. Diagram the production process so as to show the external exchanges between environment and producer system that can be used to infer processes in the system.

4. Describe properties of producers with depth, including light penetration, shade adaptation (Chapter 10), compensation point, critical depth, and relation between production and respiration.

5. Can you graph the shape of uptake curves of radio-tracers of a raw material added to the medium of a producer? How does the curve differ during growth from uptake when the producer is at steady state?

6. Diagram a life history cycle of a plant and indicate the relation of stages to energy quality.

7. Give examples of producers with large lags in exchange of raw materials relative to the timing of light energy.

8. What expressions are used to model the role of light as a limiting factor? Compare two different ways of modeling the demand of energy from a flow of light so that the process becomes limited to the same extent that light becomes incorporated into production.

9. How does the rate of change of storage compare with net production indicated by tracer uptake?

10. Describe the Cobb–Douglas production function.

11. Define production in a general-system way so as to apply to systems other than those with green plants.

12. Discuss theories of receptor adaptation for maximizing resource use and recycling. Give examples from chlorophyll and carotenoid adaptations in ecosystems.

13. Is the first step of photosynthesis oxidization or reduction? Is oxygen always released in photosynthesis? Under what conditions does photosynthesis cause its immediate environment to be oxidized? Reduced? Discuss reasons.

SUGGESTED READINGS

Anderson, J. W. (1980). *Bioenergetics of Autotrophs and Heterotrophs,* Institute of Biology, Studies in Biology **126**, Edward Arnold, London, U.K.

Dame, R. F. (1979). *Marsh Estuarine Systems Simulation,* Belle W. Baruch Library in Marine Science No. 8. University of South Carolina Press, Columbia.

DeWit, C. T. (1978). *Simulation of Assimilation, Respiration, and Transpiration of Crop Surface,* Halsted Press, Wiley, New York. 146 pp.

Eckardt, F. E. (1968). *Functioning of Terrestrial Ecosystems at the Primary Production Level,* UNESCO, New York.

Falkowski, P. G. (1980). *Primary Productivity in the Sea,* Plenum Press, New York. 530 pp.

Goldberg, E. D., I. N. McCave, J. J. O'Brien, and J. H. Steele, Eds. (1977). *The Sea,* Vol. 6, *Marine Modelling,* Wiley, New York. 1048 pp.

Grime, J. D. (1979). *Plant Strategies and Vegetation Processes,* Wiley, New York.

Hall, C. A. S. and J. Day (1977). *Ecosystem Modelling in Theory and Practice,* Wiley, New York. 685 pp.

Halle, F., R. A. A. Oldeman, and P. B. Tomlinson (1977). *Tropical Trees and Forests,* Springer-Verlag, New York. 411 pp.

Hesketh, J. D. and J. W. Jones (1979). *Predicting Photosynthesis for Ecosystem Models,* CRC Press, Boca Raton, FL. Vol. 1, 273 pp; Vol 2, 279 pp.

Ivlev, V. S. (1945). The biological productivity of waters (translation by W. E. Ricker). *Adv. Mod. Biol.* **19**, 98–120.

Middlebrooks, E. J., D. H. Falkenborg, and T. E. Maloney, Eds. (1974). *Modeling the Eutrophication Process,* Ann Arbor Science, Ann Arbor, MI. 228 pp.

Lieth, H. F. H. (1978). Patterns of primary production in the biosphere. *Benchmark Papers in Ecology 8,* Wiley, New York.

Lieth, H. F. H. and R. H. Whittaker (1975). *Primary Productivity of the Biosphere, Ecological Studies,* No. 14, Springer-Verlag, New York.

Monteith, J. L. (1973). *Principles of Environmental Physics,* Edward Arnold, London.

Parsons, T. and T. Masaguki (1973). *Biological Oceanographic Processes,* Pergamon Press, New York. 186 pp.

Patten, B. C. (1968). Mathematical models of plankton production. *Internatl. Rev. Hydrobiol.* **53** (3), 357–408.

Petersen, B. J. (1980). Aquatic primary productivity and the ^{14}C-CO_2 method. A history of the productivity problem. *Ann. Rev. Ecol.,* **11**, 359–381.

Platt, T., K. I. Denman, and A. D. Jassby (1977). Modeling the productivity of phytoplankton. *In the Sea,* Vol. 6, E. D. Goldberg, Ed., Wiley, New York. 807–856 pp.

Stoekenius, W. (1976). The purple membrane of salt-loving bacteria. *Sci. American* **234**, (6), 38–46.

Symposium (1971). Factors that regulate the wax and wane of algal populations. Mitteilungen No. 9. International Association of Theoretical and Applied Limnology. 318 pp.

Thornley, J. H. M. (1976). *Mathematical Models in Plant Physiology, Monographs in Experimental Botany,* Vol. 8, Academic Press, New York. 318 pp.

Wiegert, R. (1976). Ecological energetics. *Benchmark Papers in Ecology 4,* Wiley, New York.

CHAPTER TWENTY

Consumers

In Chapter 20 we consider consumers and their models. Consumers occupy the higher-quality end of ecosystems, using more foods and fuels than they produce but contributing complex services that have high embodied energy values. Main components and processes of consumers are illustrated in Fig. 20-1, including feeding; recruitment and reproduction; mortality; dispersal; respiration and maintenance; waste disposal; use of auxiliary energies; and systems of inheritance, age structure, social organization, mass, numbers, and control. Whereas models of internal organs, cells, and biochemistry of organisms are becoming advanced and detailed, overview models of consumers and their roles in ecosystems require aggregation of most of the inside details. This chapter considers various ways of representing consumer functions and roles. Human consumers in economic systems are considered in Chapter 23.

MINIMODELS AND BASIC PATTERNS OF CONSUMPTION

Energy Intake

Intake of energy may be modeled in various ways as given in Fig. 20-2 and in Chapter 9. When modeling a single species, limits in its ability to demand energy must be considered (Holling, 1959). Thus forms such as those shown in Fig. 20-2f, 20-2h, 20-2i, or 20-3 may be used. However, when a consumer category is modeled for maximum power in which species substitutions can eliminate any limit other than thermodynamic ones, the simpler models may be appropriate (e.g., as in Fig. 20-2b, 20-2e, or 20-2g).

Increase in energy sources availability and concentration usually has the same kind of diminishing return for consumer production as given in Chapter 19 (e.g., see Fig. 8-10 and 19-12) for primary producers because of limiting rates of supply of interacting commodities or of internal limitations.

Modeling without Separation of Mass and Number

In many models of consumer populations the state variable for the stock is given either as mass B or as the number of individuals n, but both are not in the model. Mass is regarded on the average as proportional to numbers in these models and vice versa. In most of the models in Chapter 9, for example, there is only the one state variable. Those thinking in terms of populations use number, and those thinking of energy and material cycles often use biomass, but the equations are often the same as in the logistic, for example. In Fig. 20-4a the equations for biomass B and number n are made identical where the ratio m is constant.

Modeling Mass and Number Separately

Where more understanding of the details is of concern, biomass and number are separated with reproduction and mortality concerned with numbers and growth and respiration concerned with the mass (e.g., see Figs. 20-4b–20-4f). Since mass and number are coupled, modeling them both is necessary for understanding their relationships. Often, particularly with larger organisms, the scales of time are quite different for numbers that may be in years and mass growth and turnover in days or months. Since different organisms do these maneuvers in different ways and with different timing, the models for mass and number need not be the same. In Fig. 20-4b respiration–depreciation drain of mass B is separated from the mortality that drains n, a model suitable for a population where mortality is the

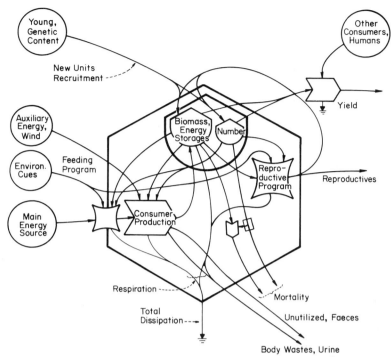

FIGURE 20-1. Energy diagram of the main sources and outflows for a typical population of consumer units.

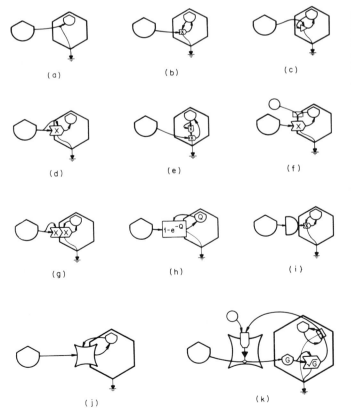

changeable and most interesting variable. In Fig. 20-4c, as in some marine populations, numbers are recruited from outside and biomass develops, depending on survival and food supply. In Fig. 20-4d biomass is used by numbers to generate new numbers through reproduction; numbers interacting with external energy sources control new biomass, in addition to which existing mass generates more by growth. Mortality removes biomass in proportion to mass per unit. In Fig. 20-4e growth of biomass is further separated from reproduction of number, although numbers draw energy from biomass and drain with mortality. In Fig. 20-4f numbers prevail, controlling food intake to a storage of mass per individual m, which serves as a control on reproduction.

In Fig. 20-5a number and mass are separated by using calculus as a language for showing a link between each rate and each storage. Such a relationship can be translated and energy sources added as

FIGURE 20-2. Some energy intake functions in consumer units: (a) linear donor; (b) autocatalytic; (c) consumer control; (d) quadratic donor; (e) quadratic consumer; (f) difference from threshold consumer; (g) donor and consumer pumping; (h) Watt–Ivlev function; (i) asymptotic limit; (j) logic control; (k) hunger threshold in rats (Booth and Toates, 1974).

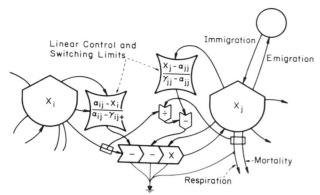

FIGURE 20-3. Energy diagram of consumer pathways and control mechnisms given by Wiegert (1974).

shown. Figure 20-5b diagrams a mechanism in which numbers per unit mass m, acting as a threshold, turn reproduction on or off. This mechanism is used in Fig. 20-5c to control both reproduction (when level is above a high threshold) and mortality when the mass per unit is below a given threshold.

A storage of number where there is a separate storage for mass has the digital information aspect. It does not have depreciation, but it has dispersion, which is the equivalent entropy increasing pathway that can be shown if it is considered sufficiently large. Generation of numbers requires an energy supply from the biomass or other source. By both recruitment and mortality, the numbers can be regarded as controlling the biomass and energies. In Fig. 20-4 numbers are shown downstream in higher-quality energy position with feedback roles.

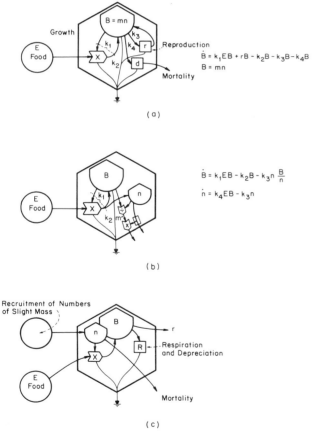

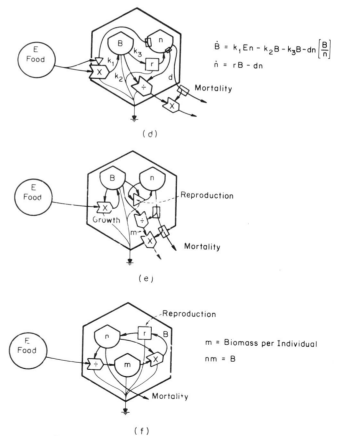

FIGURE 20-4. Population minimodels with mass B and number n. (a) With mass and number combined (mass per individual constant m); (b) mass and number produced together but numbers and mortality separated (mortality drains biomass); (c) numbers controlling growth of mass (mortality and respiration separate and recruitment of number external); (d) mass generating number and number controlling mass; (e) mass and number each autocatalytic separately with mortality draining mass; (f) mass per individual m as state variable generating mass and number.

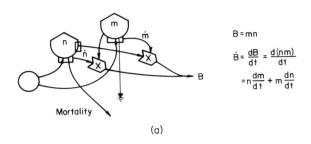
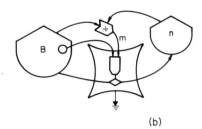
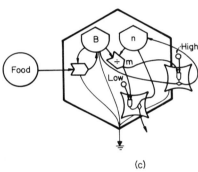

FIGURE 20-5. Population minimodels of mass and number, where mass B controls number n and number controls mass: (a) separation by calculus followed by energy circuit translation; (b) threshold level of mass per unit m controlling reproduction; c high threshold level of m controls reproduction, and low threshold controls mortality.

Numbers act as a high-quality control system on biomass. In Fig. 20-5c the ratio of biomass to numbers is a variable m, a condition index. Beyond a threshold, reproduction generating more numbers from mass is switched on. Breeding in the wood stork (Browder, 1976, 1978) is apparently one example (Fig. 20-6).

The division of mass into units of varying number and size provides many properties that can be organized to maximize power. Digital properties emerge, providing digital control mechanisms. By adjusting mass per unit, various sizes can be used; from the sizes can be regulated metabolism, turnovers, and abilities to respond to various frequencies of external energy inflows and stress.

TOP CONSUMER AS INDICATOR

Joan Browder (1976, 1978) modeled the wood stork populations (Fig. 20-6) with spreading and contracting waters acting as alternative accelerator of photosynthesis of plants followed by concentrations of small fishes to wood storks. Wood storks are modeled with breeding set off by threshold of body condition (correct correlate but not necessarily the mechanism). Populations at the apex of system energy convergence are indicators of the regime's success in concentrating energy quality. Often wildlife management has used species at the top as focus of protection and control and an index of success. In the wood stork example, the population serves as an index of effective water management for regional productivity. A diagram of energy convergence is Fig. 15-6.

The relationships of biomass to number in Figs. 20-4 and 20-5 are not the only ones that may be encountered. There are many functions for recruitment (new individuals incorporated from outside), reproduction, and mortality. Some of these are considered next.

RECRUITMENT OF NEW INDIVIDUALS

For models that have numbers (in place of or in addition to biomass), various functions are used for introduction of new individuals to replace those lost by mortality and dispersion. The incorporation of new individuals is sometimes called *recruitment* (Ricker, 1954; Beverton and Holt, 1957; Jones and Hall, 1973). Examples of recruitment are given in Fig. 20-7. See review of models by Steele (1979).

New individuals may be supplied by immigration, invasion, or other external functions as a source. The supply may be source limited, proportional to the gradient in concentration of individuals between inside and out, or a constant population force maintaining a concentration.

New individuals may be supplied by reproduction within the system, where the function varies with the kind of consumer. Some reproduction is proportional to the energy storage. Other reproduction is a constant number per unit of existing population. Some reproduction is in proportion to area occupied.

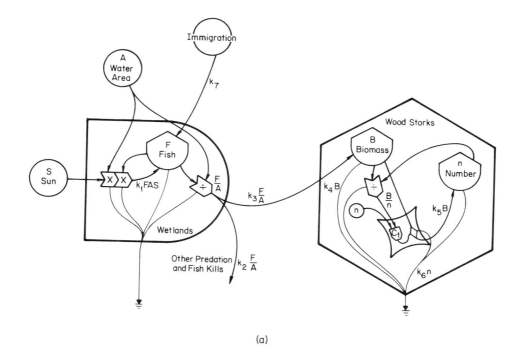

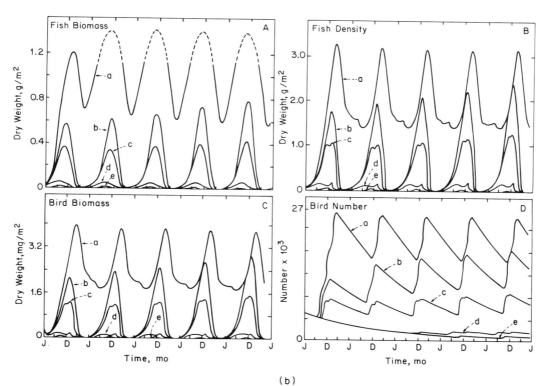

FIGURE 20-6. Wood storks as indicator of good energy use of hydroperiod (Browder, 1976): (*a*) energy model; (*b*) simulation; curves *a–e* represent increasing condition of channelization and drainage of the region. Fish density is fish biomass divided by water area that varies.

SYSTEMS OF NATURE AND HUMANITY

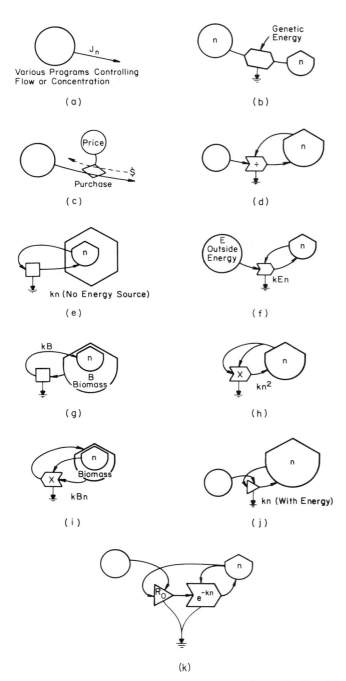

FIGURE 20-7. Some types of recruitment and reproduction: (a) externally controlled; (b) diffusion; (c) purchased; (d) based on energy per unit; (e) reproduction by use of stored energy only; (f) based on numbers and energy; (g) based on biomass storage; (h) cooperative interaction; (i) interaction of biomass and number; (j) based on numbers with unlimited energy; (k) reproduction with an exponential crowding term for cranes (Miller and Botkin, 1974).

See table of equations for dispersal and mortality of parasites (Kranz, 1974).

MORTALITY

The loss of numbers of the population due to death is mortality. The flow of mortality is sometimes described with life tables, sometimes with functions, and sometimes with graphs. There are many models for the mortality pathway or pathways, some appropriate for one situation and species and some for another. Some principal mortality models are given in Fig. 20-8.

One customary way of tabulating or graphing mortality characteristics is to show the time course of survival of a group of individuals born at the same time. Survivorship curves are included in Fig. 20-7, with the initial condition represented as 100%. The first group of decay functions were considered in Chapter 3.

In Fig. 20-8 a constant mortality (Fig. 20-8a) may occur when an external process such as a consumer is removing a set number per time. In Fig. 20-8b the mortality per individual is constant. For example, the probability of death due to encounter of some lethal influence such as gamma radiation is proportional to the number of units exposed. In Fig. 20-8c mortality is as the square of the number as in an epidemic where spread of disease is proportional to the interactions among units. Deevey (1947) used an index of crowding in barnacles with a quadratic term (Eq. 18-1). In Fig. 20-8d the mortality per individual is decreased by the number as in the examples of fish diluting toxic material by taking it up in a shared way. In Fig. 20-8e mortality on the average is proportional to number as in Fig. 20-8b, but the toxic effect is normally distributed, producing an envelope of mortality curves (Waller et al., 1971).

In Fig. 20-8f mortality depends on an accumulation with time that depends on the number. The integrated curve was generated by analog computer.

In Fig. 20-9 distribution of mortality with time is given by the Weibull distribution, which produces the various kinds of survivorship curves observed in populations depending on the coefficient c (Pinder et al., 1978).

Mortality Webs

A model of mortality of one population is often the model of consumption for the rest of the system. For

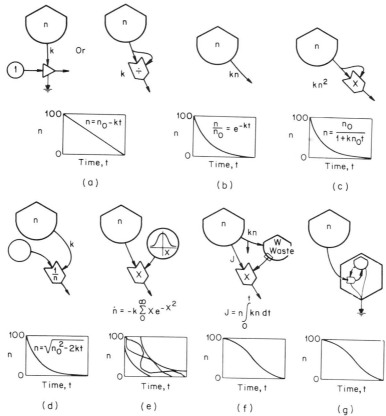

FIGURE 20-8. Types of mortality flow, each expressed with energy diagram and survivorship curve where the age class starts with 100%: (a) constant; (b) linear; (c) quadratic; (d) reciprocal; (e) normal distribution; (f) cumulative; (g) autocatalytic.

example, Chamberlin and Mitchell (1978) find bacterial mortality to increase with nutrients, decrease by eddy diffusion that prevents settling out, and increase by solar penetration.

Mortality Due to Energy Stress on a Declining Threshold

A model of energy stress that fits human mortality over most ages was given by Strehler and Mildvan (1960), Mildvan and Strehler (1960), and Strehler (1962), represented in Fig. 20-10. Death occurred when the combination of random stresses exceeded the individual's threshold for survival. Data on human functions showed linear decline with time due to age. Thus the threshold E_t was decreased as a linear function of time $-kt$. The probability p of exceeding the threshold was given by the exponential

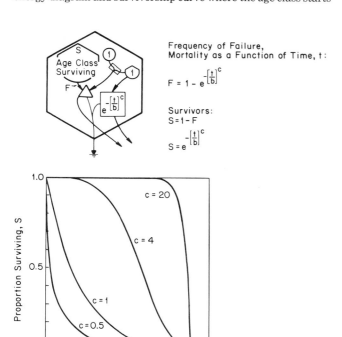

FIGURE 20-9. Mortality according to Weibull distribution (Pinder et al., 1978). Values of b are chosen so S approaches zero at age 8.

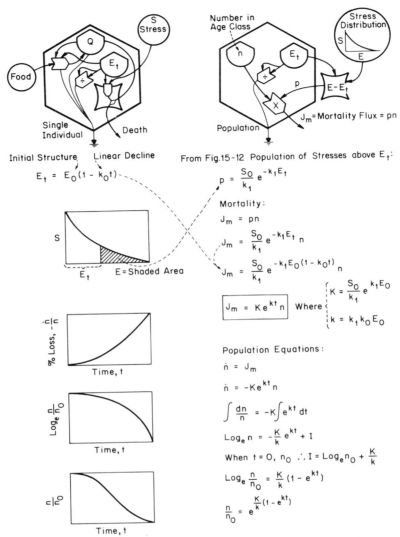

FIGURE 20-10. Energy stress theory of mortality and the declining Gompertz equation (Strehler, 1960).

energy distribution of the type given in Chapter 15, where the generality of such energy distribution is indicated. The ratio of stresses above the threshold to the total population of energy is as the exponential $\exp -k_1 E_t$. Combination of these equations produces an equation where the percent mortality loss is exponential, a function that is called the *declining Gompertz function*. It is S shaped on a linear plot.

Stress and Aging

Any external or internal stress that removes structure from the essential compartment E in the Strehler energy model for mortality accelerates the senescence process, in effect moving the threshold for mortality (E_t in Fig. 20-10) more rapidly toward the level where mortality becomes more probable. Ionizing radiation, chemical mutagens, and high temperatures can accelerate aging in this way by disordering essential and irreplaceable informational structure (see Chapter 17). Sensitivity is size dependent (Fig. 15-20).

Included in Fig. 20-8 is the Brody–Failla theory, in which the decline of vitality is not linear but is exponential (proportional to that remaining); the Simms–Jones theory, according to which the damage is proportional to the damage accumulated (autocatalytic) so that the vitality decline is positive

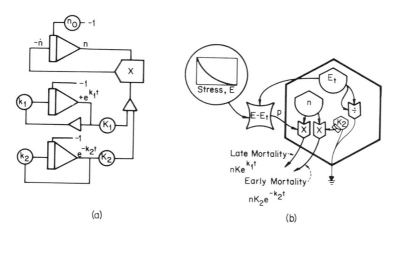

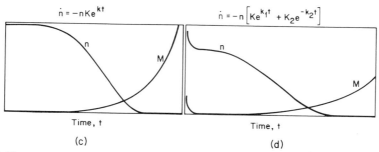

FIGURE 20-11. Analog simulation of survivorship curves by Kevin Henderson (Henderson, 1977) (M = mortality flux; n = survivors): (a) analog program; (b) energy diagram with Strehler equation (see Fig. 20-10); (c) simulation; (d) with Henderson's modification.

exponential; the Sacher theory, according to which the probabilities of function become inadequate as the mean population function declines linearly; and the Strehler–Mildvan theory, which states that as the physiologic functions decline linearly, the probability of a combined random stress exceeding a threshold ability to recover increases exponentially as in a Maxwell–Boltzmann distribution of kinetic energy among gas molecules (Fig. 15-12).

Mortality Declining and Then Rising

Recognizing that many populations have declining mortality rate in early years, Kevin Henderson added a negative exponential term to the positive mortality rate derived by Strehler. Both equations were simulated with analog program with the results given in Fig. 20-11, matching some human mortality curves well in early middle and late periods.

AGE GROUPS

As they age, populations often occupy successive roles in the ecosystem just as people occupy successive roles in their social system. Various ages serve different roles as if they were different species. Modeling of the successive stages, numbers, and roles can be done with different methods including graphs of age structures, flow charts, and transition matrices. Figure 20-12a shows a species composed of different subgroups called *age classes* or *cohorts* that move from one unit to the next in annual switching.

Age Class Graphs

Figure 20-12 contains graphs of age classes. Such graphs may show number, mass, embodied energy such as solar energy equivalents, and dollar value. The number is highest at the start during recruitment and decreases each year with mortality. Unit

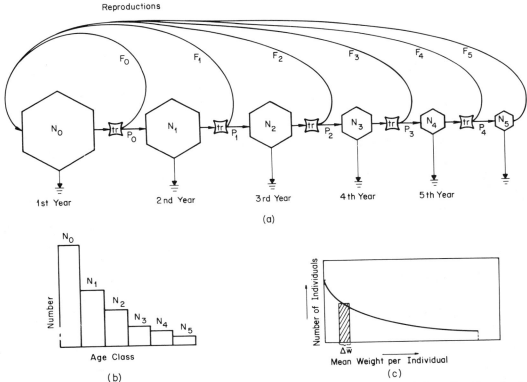

FIGURE 20-12. Age classes: (a) energy model of age class transformation; (b) bar graph of decreasing numbers with age; (c) use of area under age graph for estimation of net production of a population following Allen (1953).

mass, however, increases as each unit grows using energy sources, eventually reaching a maximum and declining as mortality effects exceed growth effects. According to one idea of harvest, the age of maximum weight is the age to harvest. However, alternative arguments are given for harvesting at age of maximum value. The value of the products increase as fish grow larger. The value can be measured in cumulative energy used in the development or by dollar value.

Matrix Model of Age Classes

The accounting and manipulation of a population, keeping track of age classes, fecundity rates, and survival rates is readily done in tabular form and can be represented as a vector–matrix model (Fig. 20-13) (Leslie, 1945). The numbers in age classes are represented by a column vector. An energy diagram of the same system was given in Fig. 20-12a. A vector representing the age distribution at one time can be converted to the vector representing the age distribution at the next time by adding natality and subtracting mortality. These are obtained by multiplying each age group by its fecundity and each age by its survival. The table of fecundities and survival factors is the transition matrix. Multiplication of the age class vector by the transition matrix produces the new vector of age classes at the later time.

There are several advantages in expressing the operation in vector notation. First, the operations can sometimes be summarized in short form with vector matrix notation. Repetitive data manipulations are readily done on computer because the matrix operations are part of the readily available software. Most important, there are operations and theorems from matrix algebra that can sometimes be used to recognize properties of the system not visualized in other ways.

Repetitive operation of the matrix on the age class vector eventually produces an age vector that is unchanging. This is called the *stable age distribution*. It is a vector that gives rise to itself. A matrix procedure that finds a vector that gives rise to itself is the process of finding eigenvectors. The process also generates single numbers (called *eigenvalues*) that, multiplied by the vector, have the same effect

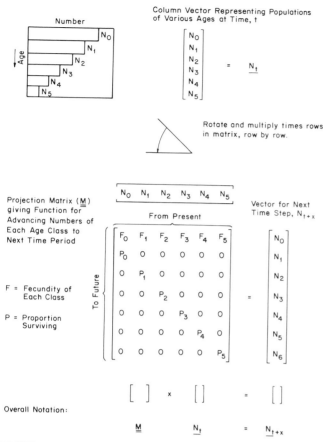

FIGURE 20-13. Leslie (1945) vector–matrix representation of age classes and their transition in time steps (Poole, 1974). Matrix multiplication rotates vector, multiplies by the transition functions, and adds results in the row to obtain new numbers.

as the whole matrix (see Fig. 13-23). Finding of the stable age distribution (eigenvector) also generates reproduction rates that, multiplied by the age class vector, generate themselves or a multiple. In this case the eigenvalues are a survival rate that, multiplied by stocks at each age and summed, gives the total survival (Usher, 1972).

COMPLEX LIFE CYCLES

Many consumer populations have complex stages in their life history that control the balance of births and deaths and their response to external environment. Figure 20-14 gives a model of rotifers by King and Paulick (1967). The model is diagrammed in the dynamics symbols given by Forrester. Figure 20-15 shows a translation of the same model into energy circuit language arranging steps from left to right

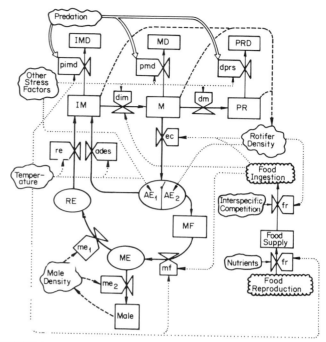

FIGURE 20-14. Diagram of life history of rotifers [by C. E. King from Paulik (1967)].

with the energy flow and life cycle, making a loop from eggs to immatures, matures, and then around to eggs again. Temperature is an accelerator on all the stages of the life cycle in the loop. Without stress, the cycle generates immatures from matures parthenogenetically (without sex process). With stress, the cycle switches to genetic eggs and a population of males that fertilize them to form resting eggs used at a later time. The energy spent on simple reproduction and that spent on genetic variety can be switched by control of these alternative loops in the life cycle.

Quality and Life Cycles: The Chicken and the Egg

Old paradoxes concern the question regarding which is first and more valuable: the chicken or the egg (Jacobsen, 1955). Consideration of energy quality in a model of the chicken and the egg in Fig. 20-16 clarifies the issue as being one of concentration. So long as the eggs are concentrated as a product of the chicken in and as a part of the chicken, they represent a transformation of, and thus a concentration of, the embodied energy of the chicken and its work, a high-quality potential. However, as soon as the

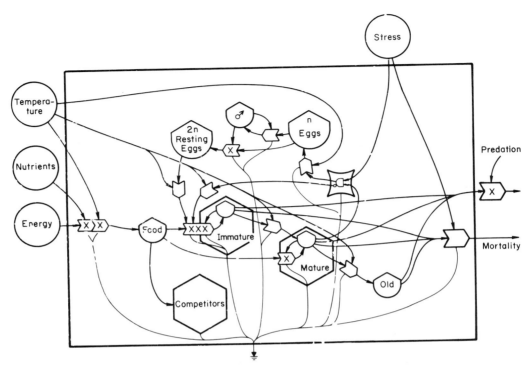

FIGURE 20-15. Energy diagram of rotifer life history given in Fig. 20-14.

egg is returned to the environment as a chick with a dispersed territory, it represents a feedback dispersion and dilution, returning high-quality seeding to a larger territory, making its concentration low again. In its growth and survival it concentrates energy again, ultimately moving to higher quality of the adult chicken and ultimately to the group of high-quality eggs, which attain maximum quality just before they are dispersed to become lowest quality in the cycle.

Other Life Cycles

The fantastic variety of life cycles and symbiotic relationships in the living biota are among the building blocks selected in self-organizational design work by the larger environmental systems. Usually these are taught according to taxonomic criteria and evolutionary similarities. Representation of the relationships in energy diagramming helps to show their energy significance, the kinds of kinetics they contribute, and commonalities and differences of structure and function.

The alternation between haploid (one gene set) and diploid (two gene sets) conditions that takes place in chromosome segregation and recombination also provides more DNA in the diploid stage. The generation with larger size, quality, and operation time between repair is the one that usually is diploid (see Chapter 17).

There may be some advantages in these diagrams for teaching basic biology. Various types of life cycle are given in Figs. 20-17, 10-31, and 19-34.

SYMBIOTIC RELATIONSHIPS

Although the energy quality theory (Chapter 14) implies that all units in systems are symbiotic (mutually contributing to the others' survival), many species (and other kinds of units) have evolved very close mutual relationships wherein pairs oper-

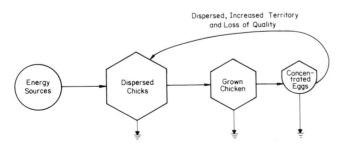

FIGURE 20-16. Energy quality in the concentration and dispersal cycle of organism's life cycle.

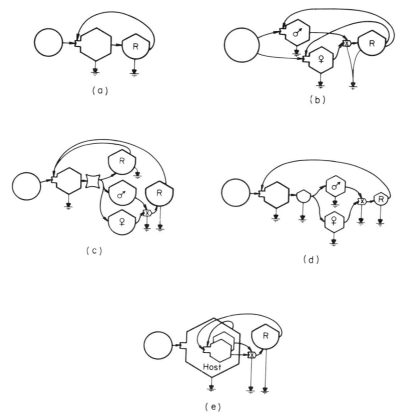

FIGURE 20-17. Kinds of consumer life cycle (δ, male; $\circ$, female; R, reproductives): (a) asexual division; (b) typical sexual life cycle; (c) sexual and asexual cycle; (d) alternation of generations; (e) parasite–host (see also Fig. 11-22).

ate almost as a single unit, each requiring the other. Some of the simplest relationships were modeled as part of the study of series and parallel pairs in Chapters 11 and 12. Some other more complex kinds of coupling are given in Fig. 20-18. Those concerned with producer–consumer symbiosis are given in Chapter 21.

SOCIAL INSECTS

As given in Fig. 20-19, social insect populations are differentiated into occupational subgroups, each of which is specialized for part of the work of the colony. The programming of roles is more permanent and less flexible than that of human beings. The model of termites shown in Fig. 20-19, by L. Burns with the biological advice of E. A. MacMahan, has workers, soldiers, and reproductives. The simulation shown in Fig. 20-19b produced an internal oscillation with a period that was entrained by a seasonal cue to breed. The model describes and relates some of the facts about the termite populations observed.

The differentiation of social insects into castes that are structurally and behavioristically different provides a hierarchy of increasing quality with many workers converging on a few reproductives.

Oster and Wilson (1978) develop a model of workers and queens drawing on an energy source (see Figs. 20-18e and 20-18f).

GENETIC SYSTEMS

Part of the main work of the life cycle is transmission of information to the future so that operations may continue as part of survival. Highly developed and specialized information processing systems occur in organisms. The genetic systems of organisms process genetic units so that recombinations can provide choice for selection maintaining adaptability. More choice is provided by disordering energies

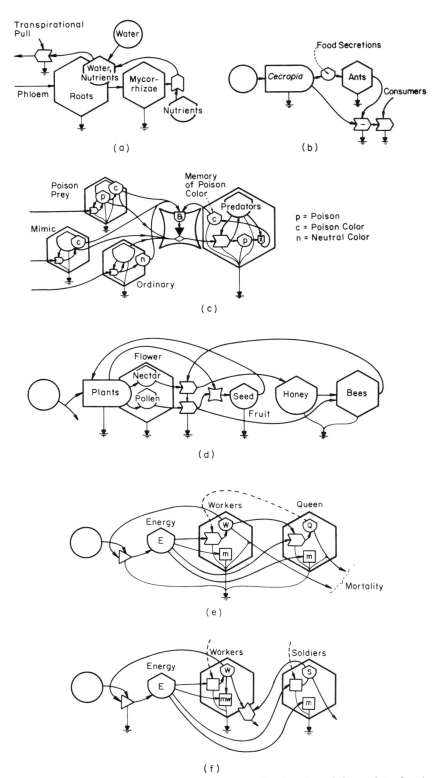

FIGURE 20-18. Kinds of consumer symbiosis: (a) Mycorrhizae concentrating nutrients on roots; (b) soldier protection as in *Cecropia* and ants; (c) mimic protected by poison in the prey that previously poisoned the predator leaving memory; (d) pollination; (e) relation of workers and queens (Oster and Wilson, 1978); (f) workers and soldiers (Oster and Wilson, 1978).

396

Consumers

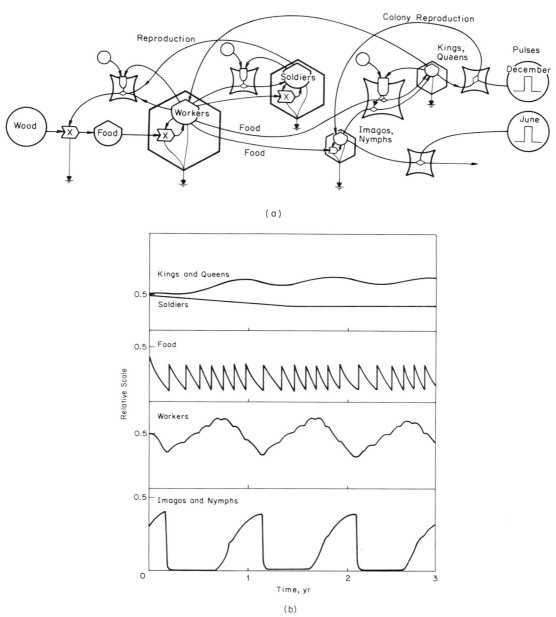

FIGURE 20-19. Model of relationships of caste systems in a species of termite (Burns, 1971): (*a*) energy diagram; (*b*) results of simulation of termite model in part *a* by Burns (1971).

of environment such as heat, radiation, and chemical stress. Examples of genetic systems are given in Fig. 20-20–Fig. 20-22. In usual genetics in each cell there are two units (called *alleles*) for each character, and the expression of that character depends on whether the two are the same or different. If different, one may dominate the action of the other or blend in its action. During the sexual process the two characters are separated when gametes are formed (as with eggs and sperm) and recombined with the fertilization that precedes development of the new organism.

Figure 20-20 shows the release of a population of genes X in reproductive segregation that recombine with genes for the same characteristic from outside, from which selection process generates the surviving genes. The main energy storages are the power supply. Outside mutation makes a few of the genes into alternative genes.

One of the basic principles for understanding the

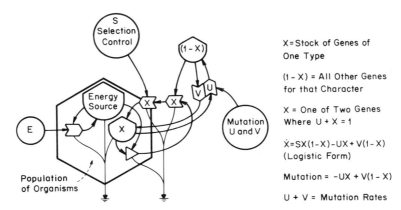

FIGURE 20-20. Energy diagram of population with genetic segregation, mutation, recombination, and selection (Levins, 1965).

behavior of genes in populations is the Hardy–Weinberg law, which states that the ratios of individuals that are double-recessive, heterozygous (mixed genes), and double dominant is the product of the probabilities where the probabilities are in proportion to the frequencies of the genes in the breeding population. This is illustrated in Fig. 20-21.

Using the model in Fig. 20-22, Chickilly (1976) varied the mortality for a population with sickle-celled blood. With malaria sickle cells survived; without malaria they were eliminated. When diagramming information processing systems, the power supply may be taken for granted as constant, although this may not be the natural state.

DISPERSAL

Dispersal is often coupled to energy sources such as wind or transport by other organisms. As a spreading process, inverse square attenuation may be involved (Fig. 18-6). Some models of dispersal driven by random energies are given in Fig. 20-23.

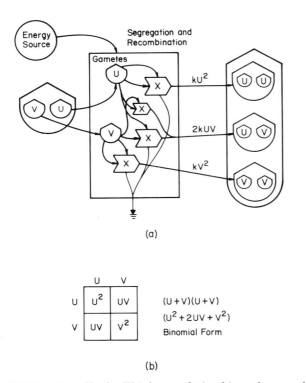

FIGURE 20-21. Hardy–Weinberg relationships where each individual has two genes; U and V are numbers of genes in populations. Energy diagram of Hardy-Weinberg Law for population recombination: (a) energy diagram; (b) genetic equation.

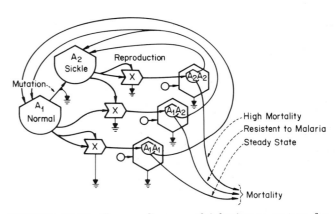

FIGURE 20-22. Energy diagram of inheritance system for sickle cell anemia (Chickilly, 1976). Model generates a steady state balance when survival of A_1A_2 is preferential because of malaria. Without malaria, sickle genes decrease and total human population increases.

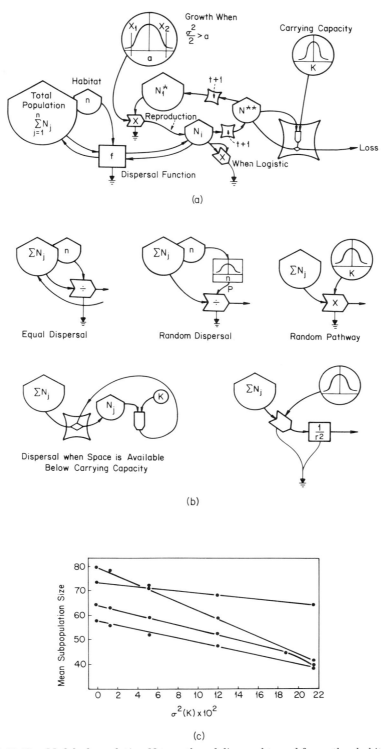

FIGURE 20-23. Model of population N_i growth and dispersal to and from other habitats ($\Sigma\, N_j$) (Roff, 1974): (*a*) model; (*b*) some types of dispersal; (*c*) effect of variance on population.

Dispersal and Energy Quality

As with the example of the chicken and the egg, high-quality reproductive units of systems feed back to their larger areas through dispersal as an informational pressure as part of the means for maintaining organizational control and spatial integration.

The access to pools of high-quality genetic information makes possible self-organizational processes that constitute succession and homeostasis of environmental systems. The addition of external population pressures stabilizes many models (W. H. B. Smith, 1976) (see Fig. 9-10).

One of the most serious stresses applied by humans that interferes with ecological systems is the elimination of free access of areas to dispersal exchange to which species were evolved and that may be required. Caribou areas in Alaska, for example, are being separated, and excesses from one area are not being allowed to compensate for shortages in another.

The same concept can be viewed as a hierarchical means by which each system can maximize power by contributing to the next-larger system, where the larger area permits some special kinds of high-quality service to be developed to feed back. An isolated system that does not exchange to the larger realm lacks some special feedbacks and develops less power for the same resources locally. Since high-quality flows provide the most economic means for transfer of embodied energy (less energy for transportation), exchange and participation in the surrounding larger realms is done in large measure with dispersal and migration. In human affairs this transfer is accomplished by trade. See Fig. 24-1 and 24-6.

Since life cycles have a trend toward high quality followed by a feedback dispersal to a lower and more dispersed energy base, the pattern of population migration can be related to concentration followed by dispersion as in bird migration.

Hall (1972) showed the generality of the movement of fish from low-energy parts of streams at their headwaters moving downstream with larger units occupying larger roles in higher-energy waters, but the return cycle is provided by a return migration up to the headwaters for dispersal of reproductives into the broad area of dispersed high quantity, low-quality energy. With salmon there is a double dispersal, one upstream and another into the sea.

Monroe (1967) finds dispersal in insects a means

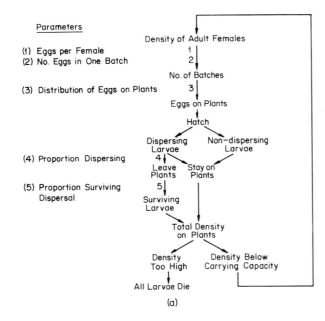

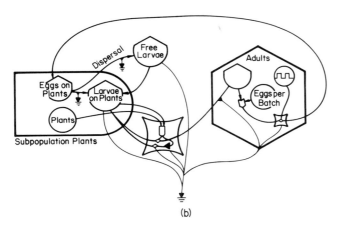

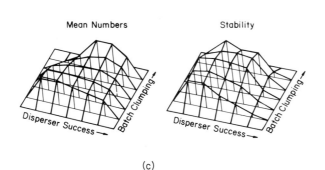

FIGURE 20-24. Stability of insects against their own overpopulation through dispersal and optimal clustering of eggs (Myers, 1976): (*a*) flow chart of model; (*b*) energy diagram; (*c*) population as a function of dispersal success and batch dumping. Overpopulation kills all larvae on plant.

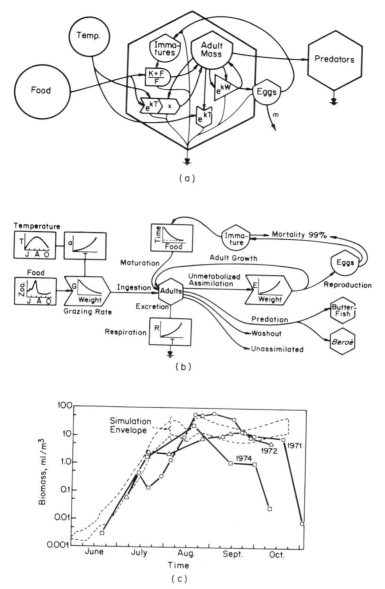

FIGURE 20-25. Simulation model of a Ctenophore jellyfish population (*Mnemiopsis leidigei*) dominating estuaries in summer (P. Kremer, 1976): (*a*) energy diagram; (*b*) diagram with graphs of relationships; (*c*) observed data and simulation envelope for different initial biomasses.

for population stabilization. In Fig. 20-24, Myers (1976) simulates insect dispersal and clustering of eggs using the ratio of mean to variance as an index of stability.

De Angelis et al. (1977) modeled seed dispersal as a transition matrix to step animals that carried seed from one point to another with attracters, and repellers, with probabilities of dropping seeds for each species.

Dispersal and Diversity

Maintenance of diversity was given as a balance between input dispersal and extinction of species in the single tank model in Fig. 3-20. However, participation of species in ecosystem reward loops may be as important. See Chapter 18.

Energy dispersal transportation means are sometimes those of the population, are often those of en-

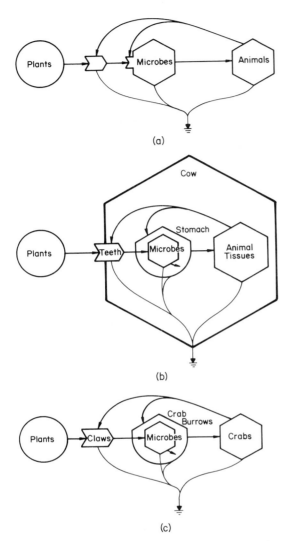

FIGURE 20-26. Diagrams illustrating role of animals as microbe managers: (a) concept; (b) cow with digestive microbes in ruminant stomach; (c) crabs in marsh or mangrove swamp.

vironmental winds and currents, and often make use of disorderly processes. Dispersal as an entropy increasing process (because it is divergent) can use disorderly energies, but since the result is a seeding of high-quality amplifiers to interact with large quantities of lower-quality energies, new order results. Dispersal is another example of hierarchical order–disorder cycle (see Fig. 17-10 and 17-12).

CONSUMPTION

Consumption serves as a drain pathway for storages in larger systems, and various minimodels in this chapter may be used to represent consumer metabolism. These may range from simple linear drains to the minimodels of Chapter 9 or those in Figs. 22-25 for the consumption of sewage in a stream. Mahloch (1974) summarizes some models used for BOD (biochemical oxygen demand) consumption.

Maximum Consumption

Systems of consumers can develop at whatever rate that limiting energy flows and interactants are supplied. Busch (1965) and Servizi et al. (1963) show that metabolism of sludge is proportional to the Gibbs free energy supplied. The efficiency of conversion was about 50% as expressed in heat calorie equivalents. Rapid metabolism requires rapid stirring and mixing of fuels and oxygen just as in the case of a hot fire. The cold fire of concentrated metabolism is as great as that of some fires releasing similar quantities of heat. Spontaneous combustion of compost is an example.

High-metabolism systems are of high quality and generally occur only where the main energy supplies are concentrated as in the waste outflows of our cities that are high energy because of the concentrations of fossil fuels. However, high-energy metabolic systems also occur at the convergence of food chains, in the contracting waters of desert areas, in the concentrations of schooling fish, and in fecal deposits of larger animals.

SIMULATION OF A CONSUMER POPULATION USING EMPIRICAL FUNCTIONS

Figure 20-25 gives an example of a consumer model that includes various combinations of basic functions for reproduction, mortality, predation, consumption, feeding, and so on. When built up from the separate processes, models may not be so readily visualized as a whole as with the minimodels that are generated from holistic consideration (Chapter 9) and subsequently modified to show particulars of population phenomena.

INTELLIGENT SYSTEMS

Complex, high-quality organisms have intelligent neural systems facilitating adaptation and participation by learning. Loose (1974) developed an en-

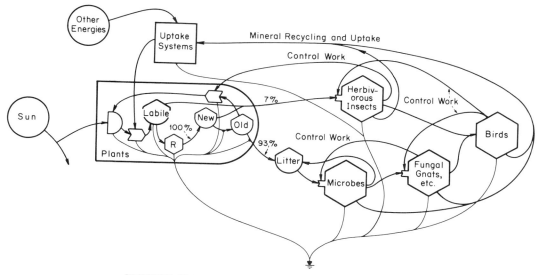

FIGURE 20-27. Convergence of grazing food chain with one step and decomposition food chains with two steps in rain forest; modified from Odum (1968). Programs of control work are shown.

ergy circuit model that was simulated, generating curves of human learning. The model has motivation storages with threshold switching actions (Fig. 13-30).

CONSUMER CONTROL ECOSYSTEMS

In chapters on ecosystem webs (Chapters 2 and 13) there was much discussion given on roles of consumers in control of other aspects of the ecosystems. Examples are given in the following paragraphs. Also see Holdgate and Woodman (1978).

Animals and Microbes

In most ecosystems there are microbes that have the variety and adaptability to provide chemical processes such as consumption and regeneration and animals that have the size and information to manage and control. The common pattern of symbiosis between animals and microbes, or between the large and the small, is shown in Fig. 20-26. Sometimes animals are called *microbe managers*. Two examples are given: in Fig. 20-26b the example of the cow and its symbiotic cellulose digesting bacteria in its special stomach; and in Fig. 20-26c the action of small crabs in cutting plant matter and placing it in underwater burrows in mangroves and marshes where decomposition by microbes can take place. In this analogy crab claws serve as the teeth of the marsh and the burrows serve as the stomach. Some aspect of this subsystem should be looked for in any ecosystem. Other works done by crabs in marsh control were measured by Montague (1980).

Guilds and Control

Classes of roles are called *guilds,* and species are assigned to these categories as one way of aggregating and visualizing main features of ecosystem function. One may often assume that a guild is in operation whereas knowledge to predict a particular species rarely exists.

Mobile herbivores eating directly on live plants and the plant parasites that are nonmobile organisms eating plants are usually in a small role, taking about 7% as the means for continued existence. They switch into epidemic mode and take large quantities in low-diversity environments and in some ecosystems that undergo self-regression.

Osmotrophs, those that utilize dissolved foods, are found in high-energy fluids, usually in microhabitats, but they can become dominant in sewage.

Detritivores consume the mixture of decomposed plant and animal matter, microbial bodies, and enzymes that are called *detritus*. This rich material constitutes the main energy storage in the food chain after plant biomass. Because of their size and programming, detritivores have mechanical roles that accelerate detritus processing.

The plant regulator:detritus regulator ratio is maintained by the convergence of these as common consumers of more versatile higher consumers such as birds and fish, and sometimes by the ratio of seed eaters and carnivores among the birds or in one species. Figure 20-27 shows the way convergence of detritivore and herbivore chains can be regulated by common consumers. There are two transformations through detritus and one through herbivores. Regulation of the herbivores and detritivores in similar ratio may keep the detritus chain 10 times larger than the herbivores. This plant matter is mainly consumed after it is released from production.

Kangas and Risser (1979) use guild concepts to compare human consumers in restaurants with animal consumers.

SUMMARY

Consumer subsystems have high-quality roles with programs in space and time, which their life histories perform. Consumer models may be of the consumer's internal body system, populations, or role in the larger environmental systems. Size, number, and mass become important; and the balance of consumer stocks include recruitment, reproduction, mortality, maintenance, age classes, genetic structure, and dispersal.

Some of the complexity provides means for controls that maximize power. Gompertz and Weibull models of survivorship result from declining energy thresholds for lethal stress. Age classes move populations up energy quality chains, and reproduction and dispersal distribute reproductives as part of the means of spatial organization and concentration of dilute energy. The hierarchical energy structure is found within the castes of social insects, the distribution of nerves and DNA. High levels of energy are required for maintaining the molecular information for high-quality units, and their roles in the ecosystem are observed to be very high quality in the sense of having control actions with large amplification. Models of complex consumer subsystems include social insects, complex life cycles, genetic transformations, and complex means of spatial dispersal and organization. Interesting as these systems are in themselves, even more important are the services they supply to the ecosystem as timers, programmers, and adaptive mechanisms.

QUESTIONS AND PROBLEMS

1. Give a general definition of a consumer in terms of its role in a larger system. What is its main useful output?

2. From Chapter 9, give some minimodels of consumers that have properties of growth, leveling, and regression.

3. Can you diagram the main features of an animal so far as its external exchanges? Its internal organs and main functions?

4. Draw minimodels that have mass and number as separate but interacting variables.

5. Enumerate and give equations for various recruitment–reproduction functions.

6. Enumerate and give equations for various mortality functions.

7. Explain models that have Gompertz and Weibull functions for mortality.

8. Give various functions for stress. Define stress in energy terms. Discuss conditions under which an energy flow is a stress or a resource.

9. Express life history stages as a network of discrete transfers. Represent this same network in matrix form, including recruitment, mortality, and transfer from one age group to another.

10. Select several animal and microbial life histories from biology references, and diagram in energy language with special effort to arrange according to energy quality with reasonable kinetic functions used in selection of symbols.

11. Diagram types of symbiosis.

12. Diagram the castes of a social insect system.

13. Diagram a population genetics system with storages for frequencies of alleles in diploid and haploid states. (If necessary, refer back to a general biology text to review genetic principles.)

14. Describe ways of introducing stochastic variability into pathways of dynamic models.

15. Enumerate various dispersal functions and give their equations.

16. Diagram the basic biochemistry of typical consumer organisms. Either use an aggregated model or diagram the full details of known

chemical reactions and loops as given in basic biology texts.

SUGGESTED READINGS

Anderson, R. M., B. W. Turner, and L. R. Taylor, Eds. (1980). *Population Dynamics, The 20th Symposium of British Ecological Society,* Halsted, Wiley, New York. 456 pp.

Bailey, N. T. J. (1957). *The Mathematical Theory of Epidemics.* Griff, London. 194 pp.

Bailey, N. T. J. (1975). *The Mathematical Theory of Infection Diseases and its Applications,* Hafner, NY.

Bartlett, M. S. (1960). *Stochastic Gradation Models,* Methuen, London.

Bazin, M. J. (1982). *Microbial Population Dynamics,* CRC Press, Boca Raton, FL. 216 pp.

Begon, M. and M. Mortimer (1981). *Population Ecology,* Sinauer Associates, Sunderland, MA. 216 pp.

Bodenheimer, F. S. (1938). *Problems of Animal Ecology Today,* Oxford University Press, London.

Brian, M. V. (1965). *Social Insect Populations,* Academic Press, New York. 135 pp.

Browning, T. O. (1963). *Animal Populations,* Science Today Series, Harper and Row, New York.

Chapman, D. G. and V. Gallucci, Eds. (1979). Quantitative Population Dynamics, *Statistical Ecology Series* **13**, International Cooperative Publ. House, Burtonsville, MD.

Cole, L. (1957). *Population Studies, Animal Ecology and Demography, Cold Spring Harbor Symposia,* Vol. 22. Biological Lab, Cold Spring Harbor, Long Island, New York.

Cox, D. R. (1962). *Renewal Theory,* Methuen, London.

Cushing, D. H. (1968). *Fisheries Biology,* University of Wisconsin Press, Madison.

Cushing, D. H. (1975). *Marine Ecology and Fisheries,* Cambridge University Press, London.

Dempster, J. P. (1975). *Animal Population Ecology,* Academic Press, New York.

Eberhardt, L. L. (1969). Similarity, allometry and food chains. *J. Theoret. Biol.* **24,** 43–55.

Freedman, H. I. (1980). *Deterministic Mathematical Models in Population Ecology,* M. Dekker, New York. 254 pp.

Fretwell, S. D. (1972). *Population in a Seasonal Environment,* Princeton University Press, Princeton, NJ.

Goel, N. S., S. C. Maitra, and E. W. Montroll (1971). *Nonlinear Models of Interacting Populations,* Academic Press, New York.

Goel, N. S. and N. Richter-Dyn (1974). *Stochastic Models in Biology,* Academic Press, New York. 269 pp.

Goh, B. S. (1980). *Management and Analysis of Biological Populations,* Elsevier, Amsterdam, Neth. 258 pp.

Gulland, J. A. (1977). *Fish Population Dynamics,* Wiley, New York.

Harris, C. J., Ed. (1979). *Mathematical Modelling of Turbulent Diffusion in the Environment,* Academic Press, New York.

Hochachka, P. W. and G. N. Somero (1873). *Strategies of Biochemical Adaptation,* Saunders, Philadelphia.

Horn, D. J., R. D. Mitchell, and G. R. Stairs (1979). *Analysis of Ecological Systems,* Ohio State University Press, Columbus. 312 pp.

Jeffers, J. N. R. ed. (1972). *Mathematical Models in Ecology,* Blackwell Scientific Publ., London. 398 pp.

Johnson, C. (1976). *Introduction to Natural Selection,* University Park Press, Baltimore.

Keyfitz, N. (1968). *Introduction to the Mathematics of Population,* Addison-Wesley, Reading, MA.

Kostitzin, V. A. (1934). *Symboise, Parasitisme et Evolution,* Hermann, Paris.

Krebs, C. J. (1972). *Ecology,* Harper, New York. 694 pp.

Levins, R. (1968). *Evolution in Changing Environments, Monographs in Population Biology,* No. 2, Princeton University Press, Princeton, NJ. 119 pp.

Lewontin, R. C. (1968). *Population Biology and Evolution,* Syracuse University Press, Syracuse, New York.

Lewontin, R. C. (1974). *The Genetic Basis of Evolutionary Change.* Columbia University Press, New York. 346 pp.

Locker, A. (1968). *Quantitative Biology of Metabolism,* Springer-Verlag, New York.

MacFadyen, A. (1963). *Animal Ecology. Aims and Methods,* Pitman, London.

McLaren I. A. (1971). *Natural Regulation of Animal Populations,* Atherton Press, 193 pp.

Manley, B. F. J. (1977). A model of dispersion experiments. *Oecologica,* **31,** 119–130.

Marler, P. and W. J. Hamilton, III (1966). *Mechanisms of Animal Behavior,* Wiley, New York. 771 pp.

Merrell, D. J. (1982). *Ecological Genetics,* Univ. of Minnesota Press, Minneapolis, MN.

Palmer, J. D. (1976). *An Introduction to Biological Rhythms,* Academic Press, New York. 375 pp.

Ransom, R. (1981). *Computers and Embryos: Models in Developmental Biology,* Wiley, New York. 224 pp.

Solomon, M. E. (1969). *Population Dynamics.* Edward Arnold, London. 60 pp.

Steele, J. H. (1977). *Fisheries Mathematics,* Academic Press, New York.

Strehler, B. L. (1962). *Time, Cells, and Aging,* Academic Press, New York.

Swift, M. J., O. W. Heal, and J. M. Anderson (1979). *Decomposition in Terrestrial Ecosystems,* University of California Press, Berkeley.

Van der Spoel, S. and A. C. Pierrot-Bults (1979). *Zoogeography and Diversity in Plankton,* Halsted, Wiley, New York. 410 pp.

Waltman, P. (1974). *Deterministic Threshold Models in the Theory of Epidemics,* Springer-Verlag, New York. 1974 pp.

Waters, T. F. Secondary production in inland waters. *Adva. Ecol. Res.* **10,** 1–176.

Weatherby, A. L. (1972). *Growth and Ecology of Fish Populations,* Academic Press, New York.

Winberg, G. C. and A. Duncan (1971). *Methods for the Estimation of Production of Aquatic Animals.* Academic Press, New York. 175 pp.

Zaret, T. M. (1980). *Predation and Freshwater Communities,* Yale University Press, New Haven. 208 pp.

Zlotin, R. I. and K. S. Khodashova (1980). *The Role of Animals in Biological Cycling of Forest-steppe Ecosystems,* Dowden, Hutchinson, and Ross, Stroudsburgh, PA.

CHAPTER TWENTY-ONE

Ecosystems

In this chapter we continue the consideration of ecosystems begun in Chapter 2. Models are used to analyze, overview, and generalize about typical ecosystems on land and in lakes, streams, and estuaries. Minimodels are used as controlled experiments and to simulate microcosms. Models of medium complexity are simulated to examine the real-world interplay of energy, materials, and information. Complex model diagrams are used to inventory components, recognize impacts, and synthesize knowledge.

Main components of ecosystems, the producers and consumers, were examined in Chapters 19 and 20, and the patterns of changing structure of ecosystems with time are considered in Chapter 22. Larger ecosystems of landscape dimensions that include humans are given in Chapters 23–25.

THE MANY FACETS OF THE JEWEL

Like the blind person touching the proverbial elephant in different places with different premises, the ecologists of this century feel the ecosystem in different contexts with different models of thought. Those concentrating on physiological relationships of organisms to environment see the system as the result of many autecologic relationships. Those studying competition, symbiosis, and strategy of individual adaptations see the system as a collection of struggling strategies dividing up a realm of niches. Those concentrating on overall performance criteria in relation to energy see the system as a homeostasis of production, recycle, and energy sources into which the components contribute or are eliminated because their subsystem has less with which to contribute. Those who concentrate on variation see the ecosystem as an interaction of statistical distributions in which the order of nature is visualized as emergence from disorder. Those studying variety and distribution see hierarchies and spatial relationships. Those concentrating on evolutionary history and genetic inheritance see dispersal and phylogenetic trajectory as determining what lives in a given place.

All these properties may be valid aspects of the ecosystem. Since all the approaches are interrelated, the ultimate ideal of systems modeling for some purposes may be to try to represent the various concepts simultaneously. This chapter has overview models that combine parts given in previous chapters.

FEATURES OF ECOSYSTEMS

Consider some of the main features of ecosystems that must be considered in modeling for understanding. Table 21-1 lists some of the main driving functions, components, and processes that may be included, depending on the system and the purposes. First, there is the energy signature to which each ecosystem may develop organization to maximize power. Figure 2-2 showed some of the sources and driving functions supporting ecosystems. The combination of physical factors affecting an ecosystem has been called a *biotope*.

As shown in Fig. 21-1, producers are coupled to consumers with a web that converges, develops quality, and recycles control services and materials. High-quality flows from outside or inside interact in production with lower-quality energy.

Figure 21-2 shows other features of ecosystems such as temporal programs, information–diversity storages, imports, exports, energy interactions in production, and reward loops.

Figure 21-3 is an example of an ecosystem with many of the features in Table 21-1. Boynton (1975) developed this model to simulate diurnal patterns of production as affected by large inputs from the river. Another example in Fig. 7-6 is a stream ecosystem

TABLE 21-1. Design Features of Ecosystems

A set of driving energy sources constituting an energy signature

A web of components that includes feedback loops

Convergence of successive energy transformations to form a chain of quality

Increase of time constant and spatial size along the quality chain that can absorb various frequencies of energy flow and filter variation

Pulsing of system controlled by oscillatory period of the terminal consumer phenomena

Storage of mass and information

Recycle of materials

Feedback control systems of a switching nature, including temporal and spatial programs

Interactions of energy flows of different quality so as to maximize power and reward those components that contribute

Parallel units that can adjust their relative loading for maximum power

Coupling of producers with consumers

---

that includes other energies in addition to those involved in food chains (Odum, 1955). The human interface has a money loop attracted by embodied energy of nature's work.

The typical webbed ecosystem patterns (in Fig. 21-1–21-3) are generally more stable than those with simple series, connected with multiplier interactions causing wide oscillations (see Fig. 11-28).

Characteristic, power maximizing designs may form by self-organizing selection of components from a large variety of species made available by surrounding systems. Different species associations may develop similar webs for various combinations of climate and ecologic factors. Each prevailing species helps to maximize production by its ability to use well each particular combination of inputs.

ECOSYSTEM MINIMODELS

Ecosystem minimodels attempt to be comprehensive in overall constraints such as energy, nutrient budgets, organic storages, diversity, and driving functions, but are kept simple in regard to the number of state variables and pathways. In minimodeling, there is an underlying faith that the system has adaptive characteristics that are determined by the dominant energy sources because of energy laws and the maximum power principle. According to this view, many details of species and spatial structure

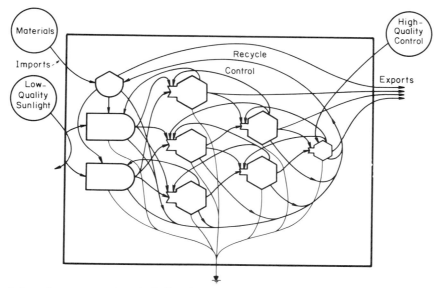

FIGURE 21-1. General form for an ecosystem, including imports, exports, heat sinks, units in parallel and series organized as a web with recycle controls, obligate interaction intersections in the production of most units, and a high diversity of unit population and pathways that operate some of the time.

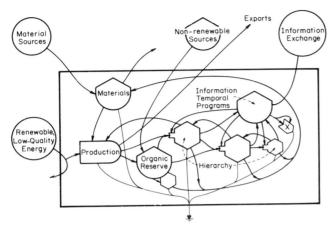

FIGURE 21-2. Characteristic features of ecosystems, including hierarchy, recycle, reserve storages, material cycles, renewable and nonrenewable energy, and information and diversity.

are mechanisms developed and selected during self-organizational periods to achieve the maximum potentials of the environmental resources. By modeling the overall functions by which energy, matter and numbers are operating when they are at their maximum potential, one models the result toward which the details are being self-organized. Thus work with minimodels can sometimes avoid the questions of species substitution and the detail of flip-flop switching of ecosystems. Minimodels have the same rewarding intellectual effect of treating fundamental and important overall mechanisms that simple models of populations do.

In this chapter some main principles are illustrated with minimodels showing how the external signature guides classifications and homeostasis of production and consumption. Chapter 22 considers these features in succession, climax, and regression.

Minimodels for Classifying Ecosystems

Ecosystems are frequently classified with minimodels of ecosystem functions, used consciously or unconsciously. Types of ecosystem with classification of traits by Naumann (1932) recognized different names for each type of predominant energy flow using the suffix "trophy." Odum et al. (1967, 1972) classified coastal marine ecosystems according to the most prominent driving influences (embodied energies). Many classifications of ecosystems are made by using only one or two of the flows of the signature. For example, Fig. 21-4 includes some minimodels that have been used to classify ecosystems. These classifications take the form of plotting factors on graphs or combining them in statistical or dynamic relationships that reflect the effect of the variables on production functions or such outputs as evapotranspiration to which production is often correlated. Use of the whole set of driving variables (Fig. 2-2) is more complete but more difficult in simple graphs or integrative functions short of full modeling of the web of relationships (see Fig. 14-15, where minimodels are given for combining multiple factors of the energy signature to estimate a system's maximum production).

Minimodels Relating Production and Consumption

The simplest overall minimodel for ecosystems is the production–consumption model [production-respiration ($P-R$) model] introduced in Fig. 2-7 and Fig. 19-13 (see Figs. 21-5 and 21-6). In simplest form for a balanced aquarium closed to matter, there is solar energy, driving production in proportion to nutrient storage and consumption and recycling in proportion to organic storage.

Sometimes single, isolated plants alone have enough consumption to balance their own production so that the $P-R$ model can apply to single, isolated producers as given in Fig. 19-10 (see also a microcosm with autolytic recycle in Fig. 21-6a). Usually, however, the consumption includes several components of the ecosystem web with animals and microorganisms.

More complex $P-R$ models are given successively in Fig. 21-5. In Fig. 21-5b external inflows and outflows of materials are given. In Fig. 21-6c the production separates oxygen, carbon dioxide, and organic matter, and these react again in consumers. This model is apppropriate in aquatic systems where oxygen and CO_2 are not held constant by wind flow. See also Fig. 2-6.

In these models organic matter of all components is aggregated, and species are assumed to develop as may be necessary to utilize the potentials that the model simulates. The energy limit is that of a renewable, flow-regulated source.

FIGURE 21-3. Estuarine ecosystem model for diurnal phenomena in Appalachicola Bay, Florida (Boynton, 1975): (a) evaluated diagram; (b) simulation results, diurnal changes in metabolism. Nitrogen inputs: 1, as scaled; 2, double; 3, 5 times.

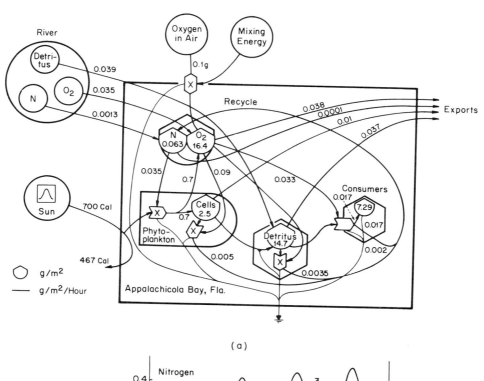

(a)

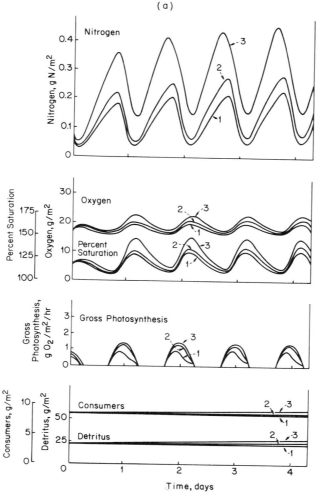

(b)

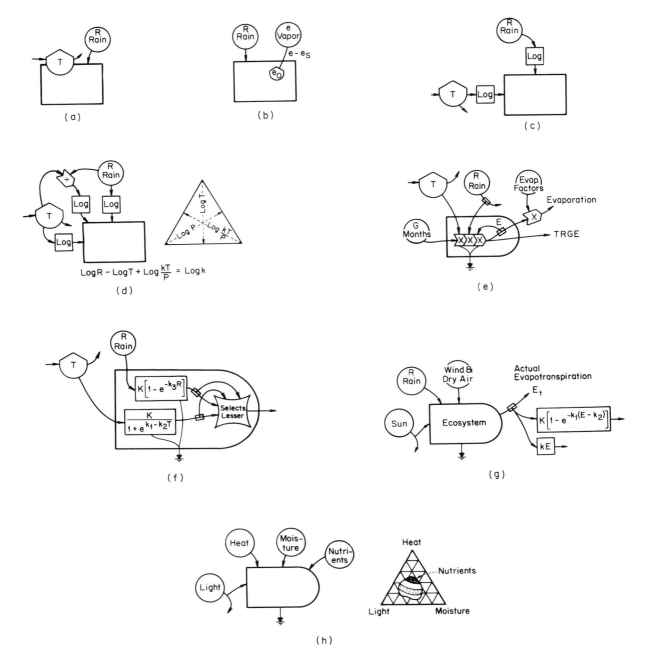

FIGURE 21-4. Energy diagrams of factors (sources and storages) and minimodels that have been used to classify and map ecosystems according to prominent driving functions, states, or outputs: (a) rainfall and temperature (McDougall, 1925); (b) rainfall and saturation deficit (Meyer, 1926); (c) log of temperature and rainfall (Totsuka, 1963); (d) log of temperature and rainfall and log of the ratio (Holderidge, 1947); (e) climate–vegetation–productivity index (Paterson, 1962); (f) selection of lesser of two exponential limiting functions, one for temperature and one for rainfall (Lieth, 1975); (g) evapotranspiration correlated with net production [Rosenzweig (1968), quoted by Macarthur (1972); Leith (1975)]; (h) graphical representation of ecosystem factors controlling forests (Bakuzis, 1969).

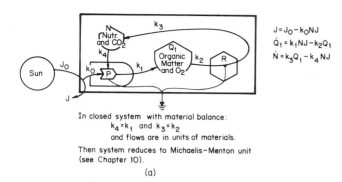
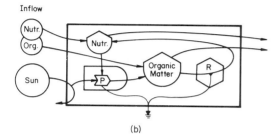
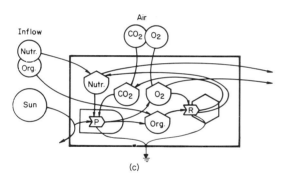
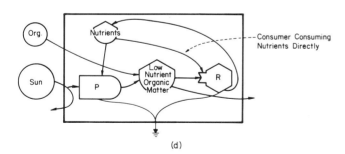
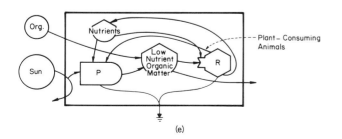

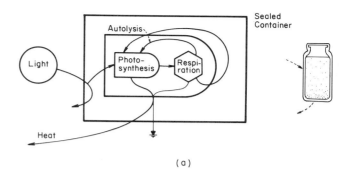
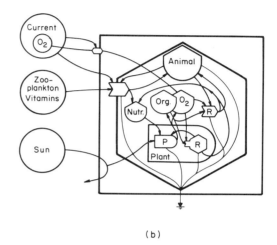

FIGURE 21-6. Minimodels of symbiosis: (*a*) unialgal microbe-free population in sealed bottle for 10 years; (*b*) coral with photosynthetic symbiotic zooxanthellae.

P–R Symbiosis

Symbiotic relationships are couplings of populations that collectively maximize power and favor survival by their collective organization. In a sense, all members of recurring ecosystems are symbiotic. The name is most appropriate when the coupled members are physically connected. In Figure 21-6*b* includes examples of *P–R* symbiosis with producer–consumer coupling. Examples are the zooxanthellae containing animals of coral reefs, the chlorella-containing hydra, the green algae containing fresh-

---

FIGURE 21-5. Some production consumption ecosystems (*P–R* models): (*a*) closed to matter; (*b*) nutrient and organic inflow–outflow included; (*c*) gases included separately; (*d*) competition for nutrients between plants and consumers where large organic matter inflows are deficient in nitrogen and phosphorus nutrients (Tenore, 1977); (*e*) same as in part *d* but with plants adapted to consume animals for nutrients.

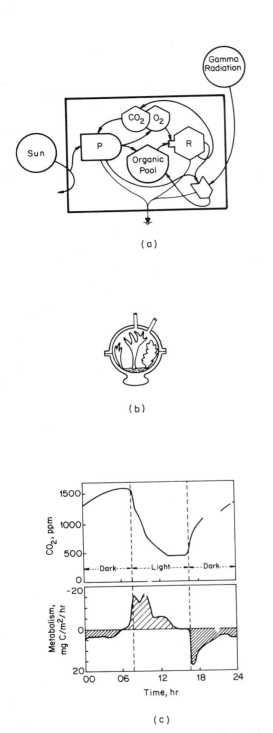
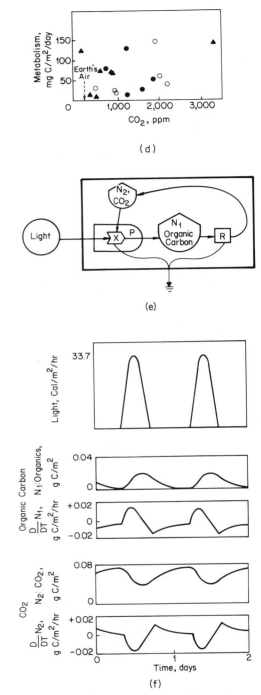

FIGURE 21-7. Terrestrial microcosms adapted to rain forest shade treated with 25,000 roentgens of γ irradiation (Odum and Lugo, 1970): (a) energy diagram; (b) sketch; (c) diurnal curve of carbon dioxide when given light with a square wave; (d) metabolism of balanced systems showing mean CO_2 levels maintained; (e) P–R model used to simulate microcosms; (f) simulation of microcosms (Burns, 1970).

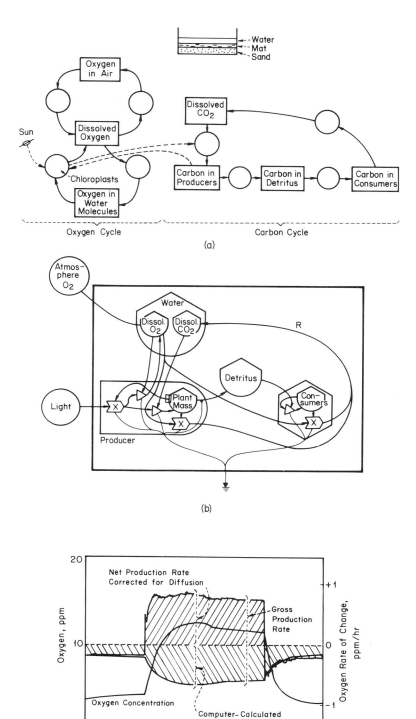

FIGURE 21-8. Sketch and model for simulating diurnal patterns in microcosms in Fig. 19-19 with blue–green algal mats (Sollins, 1970): (a) model in Forrester symbols emphasizing the complete material separation of oxygen in the production process and the oxygen circulating as part of the carbon cycle (dashed lines indicate nonmaterial interactions); (b) energy diagram; (c) observed pattern with special feature of computing day respiration as proportional to dissolved oxygen storage.

water sponges, and the lichen symbiosis of algae and fungus (Bliss and Hadley, 1964).

Self-Optimizing Combination of Production and Respiration with Series and Parallel Connections

Figure 21-5*d* shows photosynthetic production and consumer respiration connected through the presence of opportunistic species so that it operates either in sequence like any *P–R* model, in parallel when organic matter inputs are large, or with combinations. Scarce nutrients are used, preferably to amplify the larger pathway. This design generates a more stable production to changing inflows than do simple chains not webbed in this way. Examples are estuaries receiving organic matter from marsh grass (Thayer, 1971) or soil bacteria operating on compost that may compete with green plant roots.

Plants as Animal Consumers

Figure 21-5*e* shows plants consuming animals for their nutrients. For example, bogs have pitcher plants and venus flytraps that short-cut the general release of nutrients to the general environment (see also Fig. 21-6*b*, where corals catch zooplankton in a nutrient poor environment).

ECOLOGICAL MICROCOSMS

Ecological microcosms are small contained ecosystems that have the main features of larger ones. They are replicated by mixing, and without mixing initial duplicates drift apart. Since microcosms can be replicated somewhat, they can be used for experimental testing where statistical replications are essential, in a way not done cheaply with larger systems. Because ecological microcosms are small and somewhat simpler than other ecosystems, they can be modeled readily and understood in some of their overall aspect. They are interesting in themselves, too. Some examples of microcosms are given in Figs. 21-7, 21-8, and 19-18.

Unless special seeding is arranged, microcosms are isolated from exchange, trade, and embodied energy of the next-larger system. Therefore, their metabolism and energy flows will theoretically be less than those of the larger systems that they most resemble. With less metabolism, the diversity is likely to be less. In other words, when area is less, high-

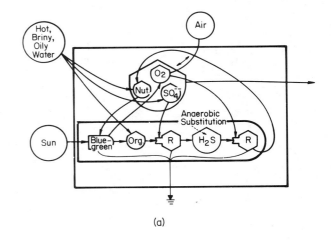

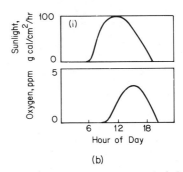

FIGURE 21-9. Delayed oxygen curve in briny, bleedwater lagoon in Texas with blue–green bottom mat, July 25, 1961 (Odum et al., 1963c): (*a*) energy diagram; (*b*) diurnal oxygen record.

quality energies of diversity are less, which is one of the reasons for lower metabolism.

What is the smallest ecosystem possible that processes energy flow in a steady or repeating pattern maintaining structure? Possibly a small population of algal cells of a species that undergoes autolysis (see Fig. 21-6*a*).

DIURNAL PATTERNS

The pulse of diurnal energies to ecosystems produces a wave of energy transformation through the energy chain with increase of productivity and chlorophyll first and later a maximum activity of consumers, animals, and respiration at the end of the day when energy pulse reaches their food. Patterns of response to energy pulses are given in Fig. 3-15 and Fig. 19-13. The response of the ecosystems to alternating light and dark is a general rise and fall of stored products and rates during day and night as given in Fig. 21-7. Simulation of a more complex model for an

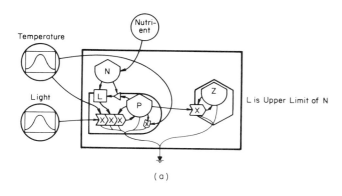

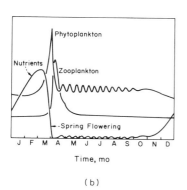

FIGURE 21-10. Seasonal simulation of zooplankton–phytoplankton–nutrient relationships from Davidson and Clymer (1966): (a) energy diagram; (b) simulation.

open estuary was given in Fig. 21-3, where typical diurnal curves were found. Some examples of the diurnal cycle of overall production and consumption of the $P-R$ model are given in Fig. 21-7. Microcosms have sharp diurnal patterns that reflect the alternation of new production in light and consumption in the dark. The thinner an ecosystem the more concentrated are changes. Thin aquatic films have slow diffusion rates, lacking turbulence (Odum, 1967a). Diurnal oxygen patterns in Fig. 21-8 result from blue-green algal ecosystems (Sollins, 1970).

Inverted diurnal patterns due to predominant photorespiration were given in Figs. 19-17 and 19-18. Some ecosystems develop storages that are utilized with a time lag, causing products to be released at opposite time of day than the usual pattern. Thin fertile, aquatic ecosystems that become anaerobic soon after dark are examples (see Fig. 21-9). Deficits are compensated for by saving anaerobic products until oxygen is available. Succulent plants such as cactus are terrestrial examples. In the succulents water is conserved since stomata can remain closed in the daytime (see Fig. 19-28).

Diurnal oxygen production was considered in the frequency domain by Gallegos et al. (1977), who found the ecosystem leading the light changes, an adaptation to high efficiency in pulsing light (see Chapter 19). Whereas most models of diurnal metabolism and exchanges have been computer simulated, Sugiura (1953, 1956) developed state equations for oxygen variation in the sea including reaeration and eddy diffusion in water.

SEASONAL PATTERNS

Because the external energy drives vary seasonally, ecosystems vary seasonally in their structures and adaptations for maximizing power. Producers generally increase and decrease their structures for light reception. Consumers are often adapted through life cycles, migration, or fast growths to fit their maximum demands on the pulse of energy passing up the energy chain following increase of main energy inflows (e.g., light or water).

Examples of simulations of seasonal patterns are given in Figs. 21-10 and 21-11, where rise and fall of sunlight, temperature, and physical stirring energy produce rises and falls in metabolic rates and stocks.

In some kinds of simplified laboratory culture algae bloom, and these stored resources cause a later pulse of consumer animals. However, in many real adapted systems in the wild, there is no lag and the consumers are adapted through their own storages to increase their populations to the same extent that there is a rise in plant productivity. Lead models of the class that have feedback multipliers (most models) can have their consumers in phase with producers or not with slight variations in coefficients.

For smaller components with rapid turnover times, seasonal patterns may include species substitution and new succession each year with matter and reproductives stored to help the previously selected members continue to be predominant, reappearing in the next year. Barker (1978) supplies a compendium of examples of migration.

OSCILLATIONS IN PLANKTON MODELS

When models of plankton communities with series relation such as that in Chapter 11 are simulated, the short time constants cause oscillations. Davidson and Clymer (1966) obtained a classical shape of plankton stock with spring bloom followed by oscillations during the quasisteady state of summer (Fig.

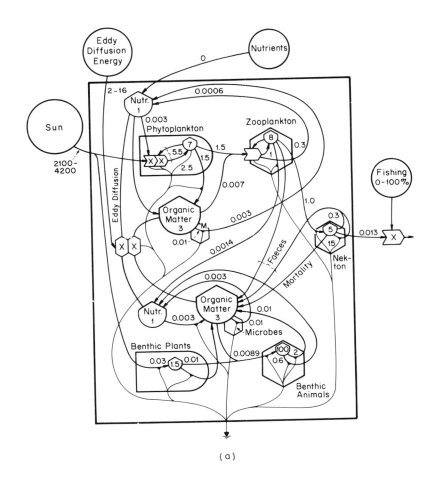

(a)

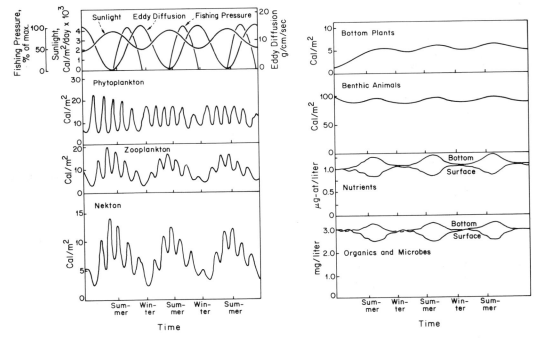

(b)

Ecosystems

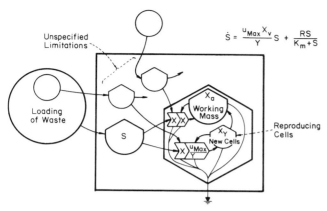

FIGURE 21-12. Diagram of equation given by Jones (1978) for steady-state activated sludge with two main rate processes. Energy diagram translates the equation and role of maintenance.

21-10) (Parker, 1976). McKellar and Odum (1972) simulated a plankton ecosystem of the Gulf shelf (Fig. 21-11). The oscillations are less than the time periods of spatial mixing frequencies. The stability effect of mixing can override any inherent oscillating tendency. The range of oscillation is also that of observed data.

The mixing energy substitutes for other energies that would be required to maintain ecosystem structure. By adapting to the mixing energy, the ecosystems do not have to use other energies in spatial and temporal stabilization required in other circumstances. See discussion of plankton diffusion model (Fig. 9-23), the critical size for a bloom, and the spatial oscillation fields that may result (Fig. 18-31). Wroblewski and O'Brien (1976) simulated models

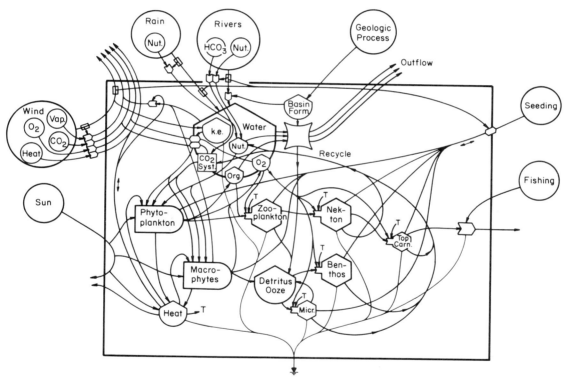

FIGURE 21-13. An unstratified lake ecosystem: k.e., kinetic energy; T, heat pathway. For details of CO_2 system, see Figure 26-15d.

FIGURE 21-11. Model of Gulf of Mexico shelf ecosystem simulated by McKellar and Odum (1972) (M = microbes): (*a*) energy diagram, where nutrients are in microgram-atoms per liter, living compartments calories per square meter, organic matter milligrams per liter; (*b*) simulation with seasonal variations of sources in the top graph.

with the diffusion effect included with other population regulatory mechanisms.

The plankton ecosystems tend to oscillate without diffusion, and models with only diffusion also oscillate. Entrainment is possible between the two. Oscillation is a form of persisting steady state that may maximize power; (see pp. 117, 195, and 444).

ECOSYSTEM WITH MAXIMUM METABOLISM

What are the upper limits to ecosystem metabolism, if any? Whereas the plant systems are generally limited by solar regimes, what systems develop when energy of light is greater than that usual on earth, such as in space? Benoit (1964) studied algal ecosystems at 40,000 foot-candles, four times the usual light levels on earth. The algae developed deeper beds of chlorophyll, each adapting to light and shade in the usual manner so that total productivity was much higher. High metabolic rates occur where complexity is large helping feedback materials as in rain forests (Odum and Pigeon, 1970) and coral reefs (Odum and Odum, 1955).

Activated sludge waste treatment is a process with a high degree of metabolism in a small space. As given by Jones (1978), maximum power occurs, not with maximum growth, but in steady state with only a few cells reproducing (see Fig. 21-12). Most of the energy goes into maintenance and processing, another example of climax. Like some other minimodels for succession and climax (pp. 216 and 468) there are two autocatalytic loops, one for rapid succession and the other for maintenance. See section on maximum consumption (p. 402).

LAKES

Composite diagrams of some of the models currently used for modeling lakes are given in Figs. 2-5 and 21-13. Dissolved oxygen is a necessary part of many aquatic models because of its large range of variation. Although large lakes may be modeled in spatial compartments, as in the marine environment, the smaller lakes can often be modeled in a single aggregated model since the boundaries of the basin cause circulations to make more of a single unit than in some situations in the sea.

Figure 21-14a shows a model of main features of production and consumption in a stratified lake or marine basin that develops a low oxygen zone at the bottom in water or sediment [see also Figs. 21-8 and 21-9 and Walters (1980)]. Dahl-Madsen and Gargas (1975) simulated a stratified saltwater fjord where some properties are similar to those in Fig. 21-14.

High-Energy Ecosystems, Eutrophy

Since the definitions by Thieneman (1928) and Nauman (1932), eutrophy has meant high productivity, usually due to high levels of nutrients concentrated in small volumes. The communities that develop in natural eutrophic systems sometimes include high diversity as in shallow grass flats at uniform salinity and at other times have lower diversity associated with sharply varying water levels or temperatures in very thin marine waters. Many of the heavily fertilized conditions associated with humans are new situations and tend to be of the latter type.

The increased variability and fluctuations observed in eutrophic waters may be another example of increased disorderly randomness generated by increased power flows, with the variability possibly helping to maintain adaptation to high-energy conditions. See Gilpin (1972).

The natural eutrophic lakes and the artificially produced eutrophy are characterized by low oxygen zones and periods, but this may also be adaptive since anaerobic digesters (Figs. 21-12 and 21-14) are a very compact mechanism for maximizing consumption and power flow, whether they are domesticated in sewage plants or in the sediments or hypolimnions of environments. Oxygen reaeration is maximized. Whereas the diversity of larger animals tends to be smaller, the diversity of the algae, microorganisms, and enzymes in the consumers may be as large as in the oligotrophic systems where species diversity tends to be high.

Lake Dimensions and Eutrophy

Shallow lakes have been known to be more eutrophic than deeper ones, other aspects being similar. Energy resources such as inflowing nutrients, sunlight, and organic matter are diluted because of their distribution in a greater volume. Curves of decreasing productivity with increased depth or area of the effective zone of production were given in Fig. 19-24. With curves from Rawson (1953) Hayes developed models of production and depth (Hayes 1957; Hayes and Anthony 1964; Vollenweider,

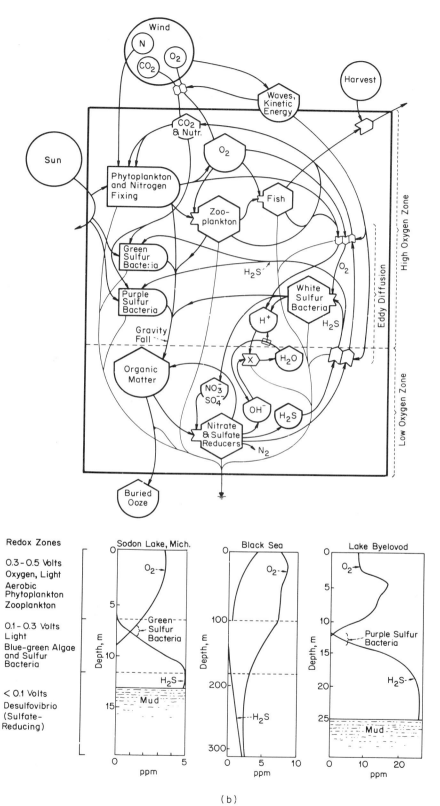

FIGURE 21-14. Stratified aquatic ecosystem including sulfur cycle: (*a*) energy diagram; (*b*) vertical distribution of properties in stratified, eutrophic, aquatic ecosystems: Sodon Lake (Newcombe and Slater, 1950); Black Sea (Caspers, 1957).

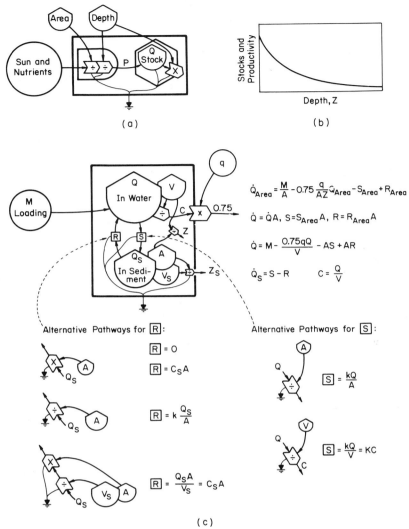

FIGURE 21-15. Nutrient loading in lakes: (a) model of diluting effect of volume on stocks and productivity; (b) decrease in productivity with depth in Canadian Lakes (Rawson, 1953); (c) models of nutrient loading effect from Vollenweider (1969) with alternatives from Welch et al. (1975) and Lorenzen (1974, 1978), where Q is quantity of nutrient stores, V is volume, S is sedimentation, R is recycle from sediment, A is area, and C is concentration Q/V.

1966). Other models relating production and nutrients to depth are given in Fig. 21-15, where type of exchange with sediment was found most important and apparently varying in different lakes. After eutrophication, decrease of inflow results in decreasing nutrients depending on the sediment exchange. The exchange with the bottom was the most sensitive item in eutrophication of Lake Okeechobee in Florida (Gayle, 1975) (see Fig. 21-16).

In small lakes where access to winds and waves is restricted, high nutrient loading generates floating plant cover such as on the water hyacinth lake modelled by Mitsch (1976).

Nutrient Loading

For purposes of appraisal and management of eutrophic lakes, rates of addition of nutrients per area, called *loading*, are related to area, volume, and sediments by using models so that nutrient levels can be predicted from proposed loading (loading models are given in Fig. 21-15). This is similar to the simple models used for classification in Fig. 21-4, subject to error when other driving functions are important. Reckshow (1979) reviews empirical models of phosphorus as related to loading. Also see Loehr et al. (1980).

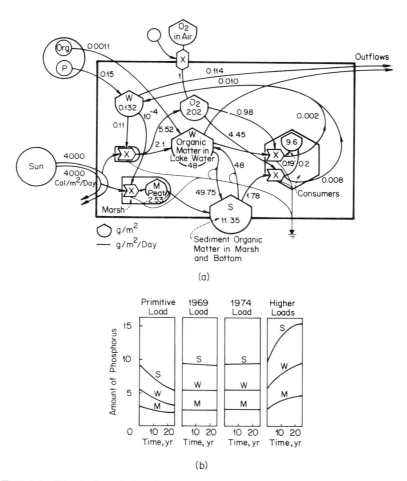

FIGURE 21-16. Simulation of phosphorus in Lake Okeechobee, Florida (Gayle, 1975): (*a*) energy diagram with stocks and material flows written (data are in g/m²); (*b*) simulation of eutrophication trends for alternative loading of inflows or phosphorus with initial condition of 1976, where S = sediment, gP/m²; M = marsh, gP/m²; W = total phosphorus in water column (parts per billion) × 10^{-1}. The simulation is made with simplifications holding constant those pathways not important to changes over a 25-year period.

System of Bottom Rooted Plants Recycling Sedimentary Nutrients in a Eutrophic Lake

Fontaine (1978, 1981 and Fontaine and Ewel, 1981), measuring metabolism in a culturally eutrophic lake in Florida, simulated a model showing the way rooted bottom plants (*Hydrilla*) help to maximize nutrient circulation and metabolism in a system whose sediments were rich as a result of earlier external waste inflows. Figure 21-17*a* gives a minimodel of the system that shows recycle pathways of benthic plants and white amur, the root organ storages, the ultimate nutrient dependence on inflow–flushout relations, and the annual consumers.

Fontaine simulated a more complex model, showing the gradual decline of eutrophication characteristics, biomass, and metabolism due to nutrient washout. Addition of herbivorous fish, white amur, was simulated. The plants were removed more quickly, causing a pulse in animals, but much phosphorus was left in the sediments, ultimately delaying the return to a more oligotrophic condition.

STREAMS

Streams, with their one-way flows, have a large inflow:outflow ratio and have special features. A generalized unit model of a stream is given in Fig. 21-18, which is a composite of many of the features included in stream models. Often the models used are applied to many sectors with simpler aspects to the model of each unit. Some of the characteristic

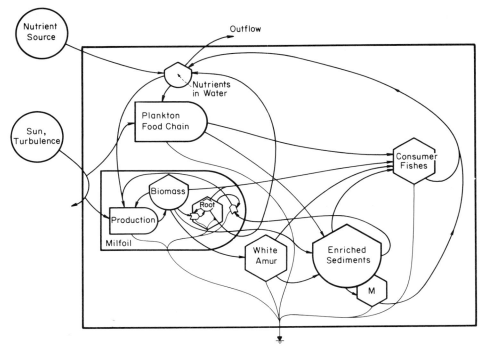

FIGURE 21-17. Hydrilla–white amur–phosphorus relations in Lake Conway, Florida (simplified from Fontaine, 1978; Fontaine and Ewel, 1981).

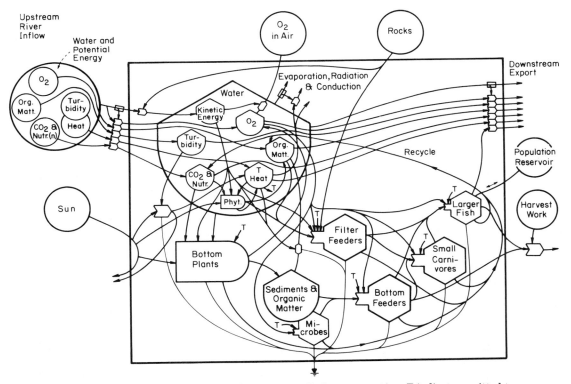

FIGURE 21-18. Generalized model for an unstratified stream section. T indicates omitted temperature pathway.

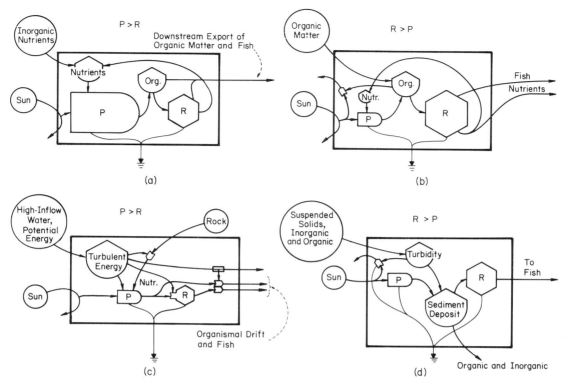

FIGURE 21-19. Minimodels summarizing some stream types based on dominant inflow: (a) spring run with inorganic nutrients; (b) forest stream with organic inflows; (c) turbulent mountain stream; (d) turbid, sluggish river in floodplain.

stream minimodels used for some purposes are given in Fig. 21-19. Much depends on depth and the role of the bottom in photosynthesis, respiration, or storage. An evaluated web for an artificial stream microcosm is given in Fig. 21-20.

Cummins (1977) summarized longitudinal patterns in many streams, finding P greater than R in the middle section only. Armstrong and Gloyna (1968) found a close fit of respiration and photosynthesis in artificial streams.

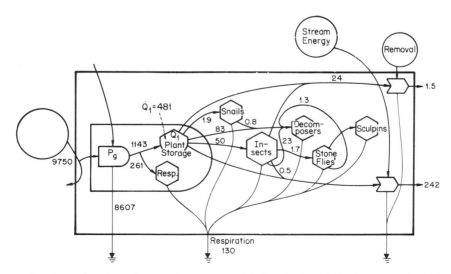

FIGURE 21-20. Energy flow in a laboratory stream microcosm (McIntire et al., 1964; Brocksen et al., 1968; Warren and Doudoroff, 1971). Numbers are actual kilocalories per square meter per 74 days.

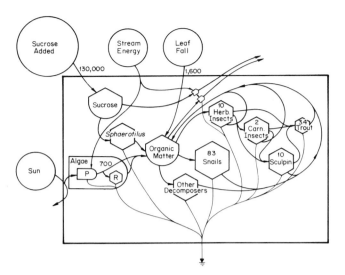

FIGURE 21-21. Diagram of stream riffles in Oregon with a source of sucrose dripping into the water to simulate wood sugar pollution from logging (Warren et al., 1964).

Diagrams of spring streams are given in Figs. 7-6 and Fig. 13-2a. Silver springs with fertile clear water and sun has high net production exported downstream. Root Spring receives forest leaves and is more heterotrophic. Karr and Schlosser (1978) found diversity of fish species proportional to but less than the diversity of habitat within the stream.

Summaries of stream ecology were by Hynes (1960), Warren and Doudoroff (1971), and Nemerow (1974).

Experimental Addition of High-Quality Fuel

A stream in Oregon was studied with drip infusion of sugar. Below this point the rocks were white with colonial bacteria. Similar loading occurs in the great rivers such as the Columbia River, as a result of the leaching of sugar from sap of freshly cut trees being floated to market. The system with sugar inflow is given in Fig. 21-21.

MARINE ECOSYSTEMS

Typical marine models are given in Figs. 21-22 and 21-23, one for open sea where vertical eddy diffusion is important (Riley et al., 1949) and one for an estuary where rivers, tidal exchange, and bottom are important. Notice the respective roles of waves, tides, winds, rivers, salinity relationships, phytoplankton, zooplankton, bottom animals, fish, pelagic animals, and economic harvests.

Because of large sizes and distances, often considered in marine models and appropriately because of

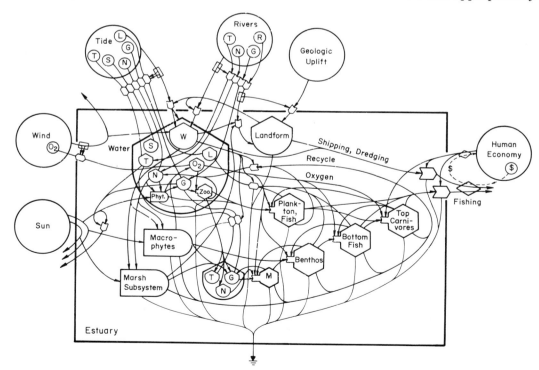

FIGURE 21-22. Overview model of an estuary: G, organic matter in water; L, larvae; M, microbes; N, nutrients including CO_2; O, oxygen; R, sand and gravel; S, salt; T, clay turbidity (inorganic); W, waves, current energy, and turbulence.

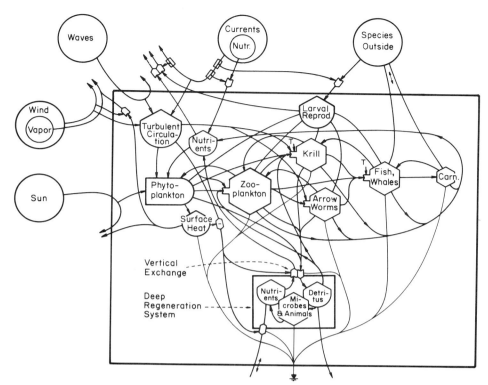

FIGURE 21-23. Upper thousand meters of open sea. Variables that are not varying significantly are not diagrammed separately.

large energies of water movement and mixing, modeling of marine systems often involves division of an area into sectors, each with its own model, exchanging by various functions with the next. Some marine models have had a few relationships modeled in many sectors, such as water and salt (Fig. 13-16). At the other extreme the unit model of a marine ecosystem in one sector has a very large number of possible components and relationships. Aggregated models have been used for whole estuaries or representative zones. Sometimes these are regarded as unit models, but in some cases they are simulated and checked with observed data in the aggregate (Jansson and Wulf, 1977). Vinogradov and Menshutkin (1977) use a new set of symbols to diagram and simulate a plankton ecosystem. Sjöberg (1977) finds nutrient cycling a stabilizing feature in pelagic models. See Fig. 10-8.

Figures 13-24 and 21-24 show some commonly simulated minimodels that are simple, with few equations, but become complex when applied to an area of many sectors by spatial disaggregation. A simulation of a simple estuarine model that generated realistic diurnal patterns was given in Fig. 21-3.

Figure 21-25 shows a unit model developed for the San Francisco Bay area by DiToro et al. (1971) that was able to generate seasonal and longitudinal patterns of nutrients and chlorophyll in various sectors.

Upwelling Models

Considerable focus of interest has been in modeling the most fertile parts of the sea, the upwelling areas. Models are given in Fig. 22-24 with an example of validation. Here there is an import of high nutrients interacting with sunlight and moving laterally with some aspects of patches that are under dispersive tendencies. These models have some of the features of stream models and longitudinal succession with high ratio of import:export.

Marine Microcosms

Figure 21-26 shows a sequence of five microecosystems that were operated by Copeland et al. (1972); Cooper and Copeland (1973) simulate the sequence of salinity zones in an estuary with river water flowing from the left and saltwater from the right. The

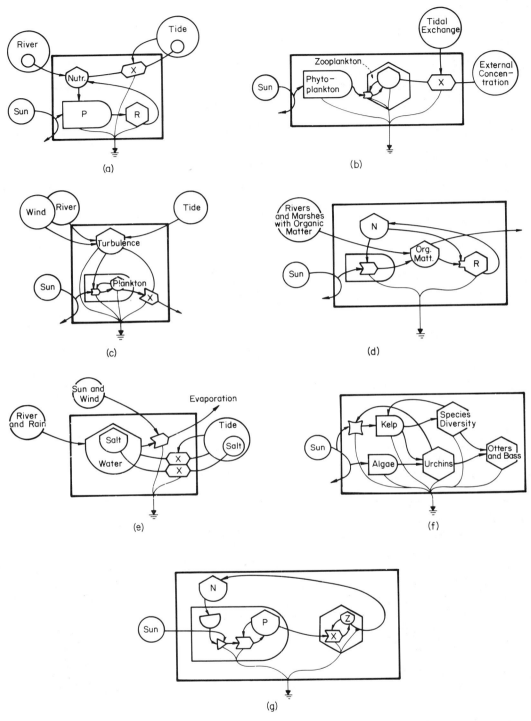

FIGURE 21-24. Minimodels of marine ecosystems: (*a*) eutrophication from river; (*b*) plankton in an estuary; (*c*) turbulence control of size and time constant; (*d*) autotrophic–heterotrophic competition; (*e*) salt balance; (*f*) competition of encrusting algae and kelp with urchins controlled by otters and sea bass, thus permitting kelp and high-diversity system to prevail (Estes et al., 1978); (*g*) plankton (O'Brien and Wroblewski, 1972).

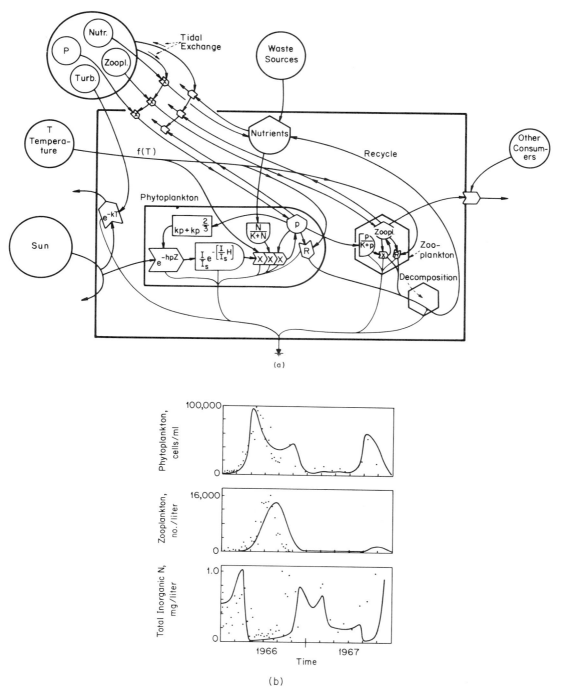

FIGURE 21-25. Model of estuarine production applied to the Sacramento-San Joaquin delta in California by DiToro et al. (1971): (a) energy diagram drawn from the equations of the authors; (b) results of simulation with observed data for the river at Mossdale in 1966–1967.

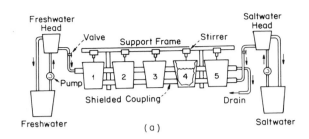

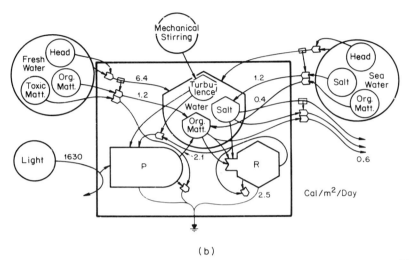

FIGURE 21-26. Estuarine microecosystems (Copeland et al., 1972). (a) apparatus; (b) evaluated model of one cell.

water has concentrations of organic matter and nutrients characteristic of some Texas bay situations. Diurnal production was proportional to nocturnal consumption. When the river water was turned off, simulating drought, productivity and diversity of the cells on the left decreased, but productivity of those on the right increased, associated with evaporative concentration. See also Fig. 21-8.

Littoral System

The ecosystem of attached organisms on the edge of water bodies is often very concentrated because it receives special convergence of energies of waves, currents, and interactions of land and water influences. Figure 21-27 is a simulation of a littoral algal ecosystem in a northern region with short summer.

Reefs

Reefs are aquatic ecosystems with high concentrations of life based on high physical energies of current and wave action. A model and simulation of an oyster reef system based on filtering organic matter is given in Fig. 16-16. Minimodels of coral reefs based on photosynthetic production are diagrammed in Fig. 21-6b. DiSalvo (1969) studied the role of microbial detritus in porous channels within the reef with current and filtering actions by contained animals (see Fig. 21-28). See also Fig. 13-31b.

TERRESTRIAL ECOSYSTEMS

Figure 21-29 is a composite energy diagram of the main features of many terrestrial ecosystem models. Included are dominant plants and trees, which are divided into leaves, trunks, limbs, roots, litter detritus, insects in several categories of consumer roles, and recycle of one or more critical nutrient materials. Inflows include the usual set of terrestrial driving functions the sun, wind, rain inflows and outflows, land uplift, genetic information in species, and the interactions with humans and their economy. The diagram in Fig. 21-29 probably has more than is needed for every problem and simulation,

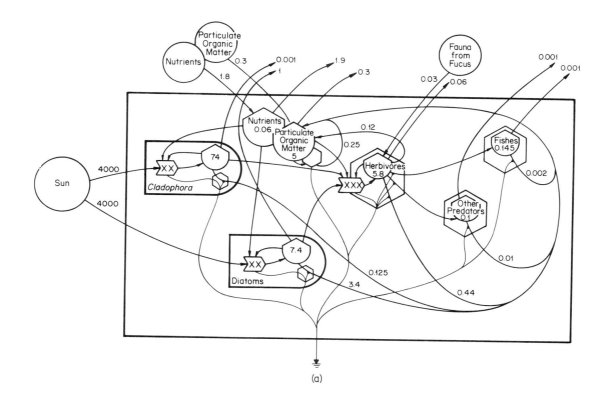

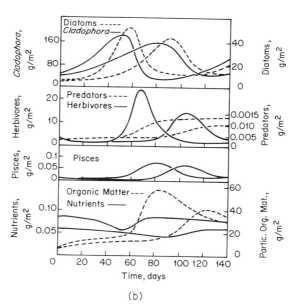

FIGURE 21-27. Attached algal ecosystem, *Cladophora* in Baltic Sea, modified from A. M. Jansson (1974): (*a*) energy diagram; (*b*) simulation in solid lines; simulation with 30% increase in nutrients in dashed lines.

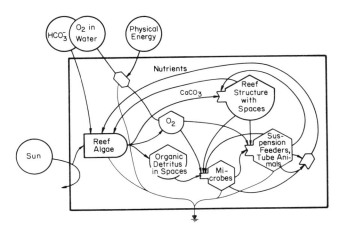

FIGURE 21-28. System of detritus formation and use in the spaces of coral reef. Modified from DiSalvo (1956).

but it may serve to help the scanning process in setting up new models. This model is of moderate complexity, and because it is drawn in a hierarchy of symbols within symbols, it can be remembered in the mind more or less intact after one has become used to the energy language and traces the relationships thinking functionally.

A terrestrial model in most places has to have a submodel for water, a submodel for nutrients, and large time constants with large storages of organic matter in woody structures and soils. Figure 21-30 presents minimodels for terrestrial ecosystems (see also Fig. 21-7 and Chapter 22).

Soil

Soil is part of the ecosystem that generates it from inputs of energy and materials from above interacting as a countercurrent with the slow movement of the geologic substrate from below, weathering in the process. Models of the soil subsystem deal with various mechanisms such as the holding and release of nutrients by base exchange in percolating fluids (Overman, 1975). A model of soil genesis has to include the whole ecosystem at least in an aggregated form so as to include the main driving interactions. The work of soil structural formation includes roots, microbes, microzoa, chemical processes, horizon formation, depositions, spatial organization, and so on. Hundred year periods are usually required (Jenny, 1980). Richards (1974) provides models of soil relations in energy circuit language.

When the above-ground ecosystem is removed and agriculture is substituted, the soil represents a valuable storage that usually degrades until it is returned to a complex ecosystem for restoration. Some agroecosystems of high diversity may maintain soil structure and storages after forming a new type.

Deserts

Ecosystems of the desert have low levels of water input, and in most cases it is intermittent. Figure 21-31 gives minimodels of desert conditions as suggested by Noy-Mier (1973). Figure 21-32 models the role of water in overview, with a pulse–reserve concept, and a pattern of increasing stability with size resulting from greater access to larger water storages (Noy-Mier, 1973).

Wetlands

Wetlands have the terrestrial energy signature of sun, wind, and radiation field, with the aquatic ecosystems signature of water and chemical conditions that together often, but not always, produce high productivity. Some general characteristics are given for a floodplain (Fig. 21-33), a cypress swamp (Fig. 21-34), a salt marsh (Fig. 13-2b, Figs. 21-35, and 21-36), and mangrove swamps (Figs. 21-37 and 21-38). Adaptations to hydrological flows and hydroperiod predominate, causing species diversity to be small in the plants but high in the terrestrial insects. A swamp simulation given in Fig. 21-34 shows the role of fire interacting with water-level management in controlling forest yield. A salt marsh simulation given in Fig. 21-36 provides an aggregated coarse view of the massive cycle of grass growth shifting to dead biomass and detritus. Burns and Taylor (1977) simulated nutrient uptake. A mangrove simulation given in Fig. 21-37 illustrates the response of a wetland to absorption of high nutrients and the role of hurricanes in setting back succession. The model in Fig. 21-38 shows the effect of shortage of freshwater or of exchange in causing water shortages and dwarfing of mangroves. Impact of massive herbicide spraying of mangroves in Vietnam was simulated showing shortage of seeding propagules delaying reforestation (Odum et al., 1974).

Rykiel (1978) and Littlejohn (1977) simulated water levels in swamps and superficial water tables. Klose and Deb (1978) simulated dissolved oxygen levels and Wildi (1978) the development of peat in bogs. Summers et al. (1980), using submodels for

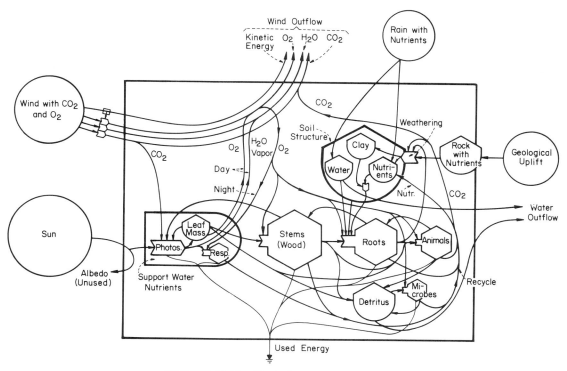

FIGURE 21-29. Typical terrestrial ecosystem.

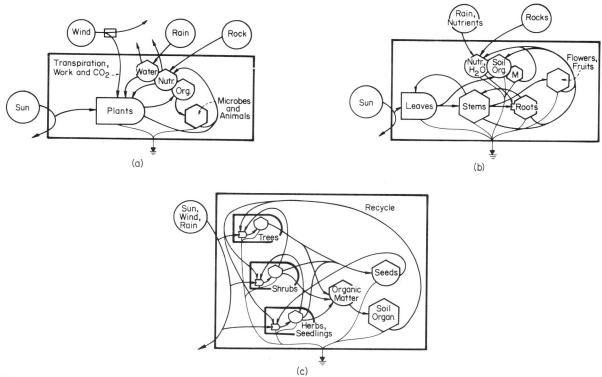

FIGURE 21-30. Minimodels of terrestrial ecosystem: (*a*) biota and soil; (*b*) aggregated according to main parts of trees; (*c*) seeds and stratification (*M*, microbes).

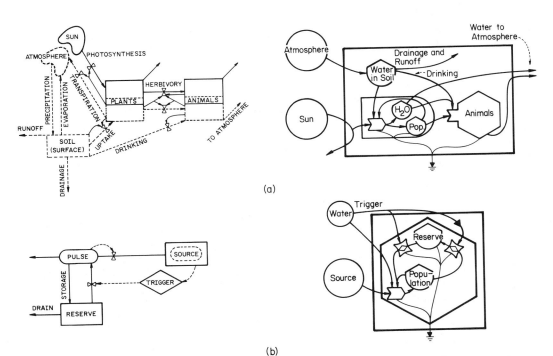

FIGURE 21-31. Models of desert ecosystems, including diagrams from Noy-Mier (1973) and equivalent energy diagrams: (a) ecosystem with water conservation pathways; (b) reserves released when water conditions favorable and reserve restored as drought returns.

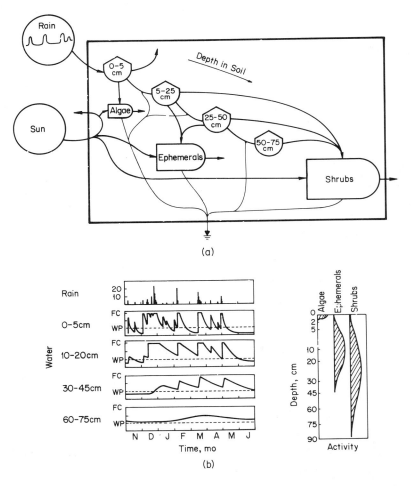

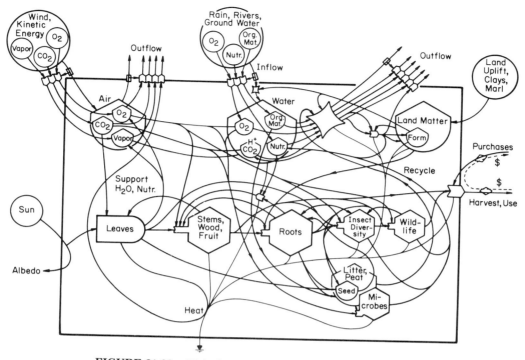

FIGURE 21-33. Main features of a floodplain swamp.

marsh, oyster reef, and estuarine water, generated realistic curves for production, respiration, and exchange in a South Carolina estuary. Both P and R were in phase in the marsh and system as a whole, with maximum metabolism in summer.

Wetlands often build peat storages in which nutrients are bound. Browder and Volk (1978) simulated the release of nutrients from the oxidation of peat as storages are relaxed. Sedlik (1976) simulated wetlands competition with an energy threshold model.

HISTORICAL MODELS

Figures 21-39–21-42 give some energy diagrams of models that have been important in generalizations about ecosystems. See also Fig. 19-3 for the first marine models. A milestone was the quantitative evaluation and hand simulation of marine plankton ecosystems by Riley (1947) (Fig. 21-40). For early ecosystem diagrams see Allee et al., (1949).

PASSIVE ELECTRICAL WEBS FOR ECOSYSTEMS

Passive electrical circuits that use resistors and capacitors (RC circuits) were given in Figs. 3-8, 3-19 and have been used to model ecosystems and make inexpensive simulators (Odum, 1960; Vogel and Ewel, 1972). Some examples are given in Figs. 21-43 and 21-44, including one with zener diodes that act as threshold pathways that flow when pressure exceeds a threshold.

COMPLEX ECOSYSTEM MODELS

For purposes of organizing data and concepts and sometimes for simulation, more complex models are developed—some that have more components and processes than can be shown on a single page. These have less value in helping conceptualize systems but

---

FIGURE 21-32. Hydroperiod adaptation in desert ecosystems after Noy-Mier (1973): (*a*) energy diagram showing relationships; (*b*) attenuation of water fluctuations through the soil storage system, lengthening hydroperiod available to larger producers that represent a convergence to higher quality and longer periods.

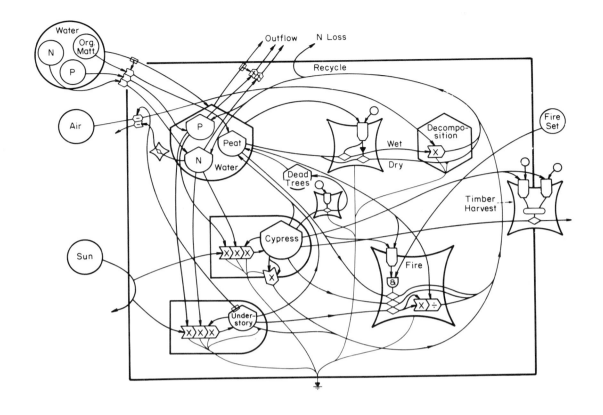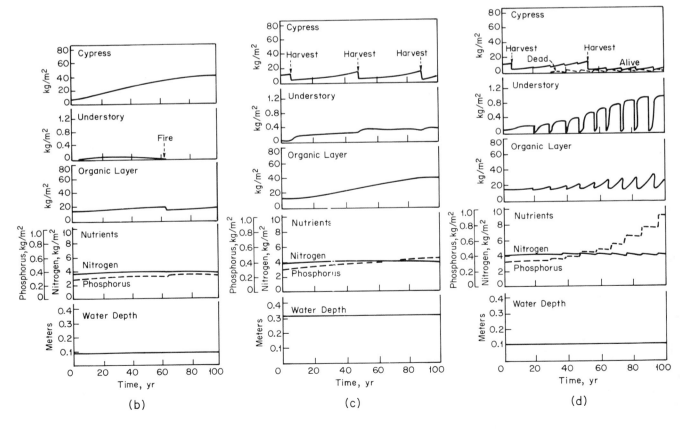

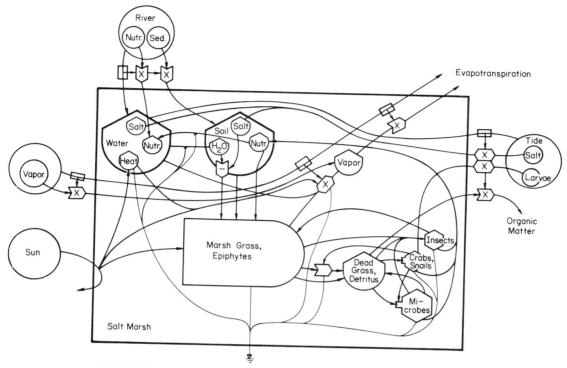

FIGURE 21-35. Main features of salt marsh ecosystem (S = salinity).

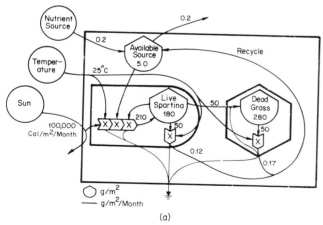

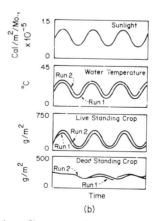

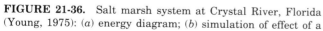

FIGURE 21-36. Salt marsh system at Crystal River, Florida (Young, 1975): (a) energy diagram; (b) simulation of effect of a thermal effluent (run 2) compared with lower temperature (run 1).

FIGURE 21-34. Cypress swamp model from Mitsch (1975): (a) energy diagram; (b) simulation with fire; (c) simulation with harvest; (d) simulation with fire and harvest.

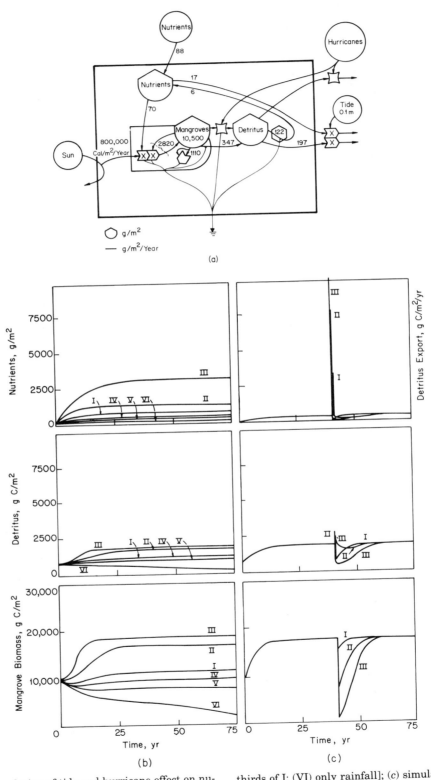

FIGURE 21-37. Simulation of tide and hurricane effect on nutrients of mangrove swamp in Florida (Sell, 1977): (a) energy diagram; (b) simulation of effect of nutrient inflow [(I) 87.7 g/m² · year; (II) doubled; (III) four times; (IV) one-third of I; (V) two-thirds of I; (VI) only rainfall]; (c) simulation of the effect of hurricanes [(I) 58 m/sec; (II) 90 m/sec; (III) 90 m/sec and severe tidal damage].

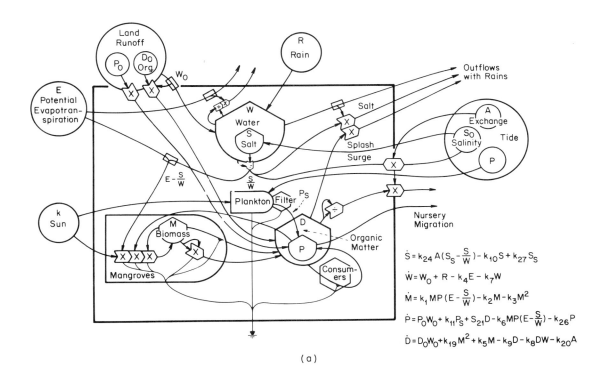
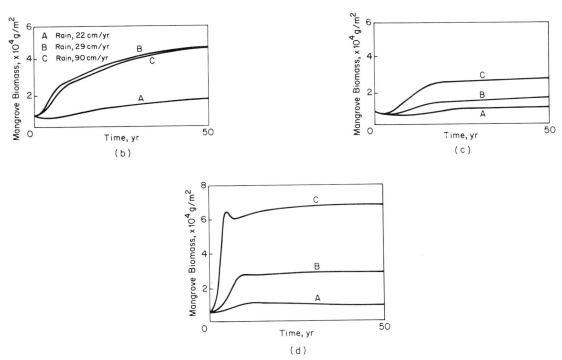

FIGURE 21-38. Simulation of salinity and nutrient exchange between mangrove swamp and the sea (Sell in Odum et al., 1977): (a) energy diagram and equations; (b) effect of freshwater source, land runoff, 3.8×10^6 g/m$^2 \cdot$ year with 0.065 ppm P; (c) effect of tidal exchange when freshwater sources are small; (d) increased growth with freshwater and nutrients.

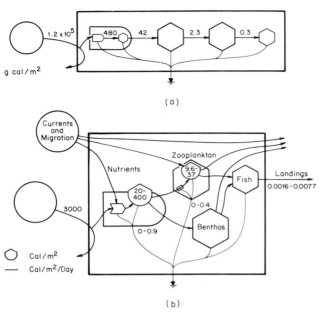

FIGURE 21-39. Historical models diagrammed: (*a*) Lake Mendota (Juday, 1940); (*b*) continental shelf, Georges Bank (Clarke, 1946, 1959) (see also cogwheel diagram in Fig. 15-24).

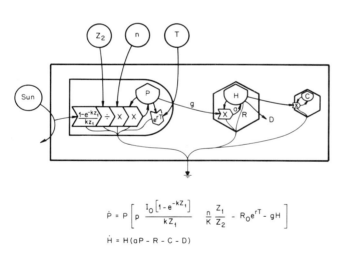

$$\dot{P} = P\left[p\, \frac{I_0\left[1-e^{-kZ_1}\right]}{kZ_1}\, \frac{n}{K}\frac{Z_1}{Z_2} - R_0 e^{rT} - gH \right]$$

$$\dot{H} = H(aP - R - C - D)$$

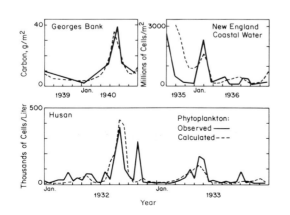

have the special uses of inventory, impact summary, and so on. Their usefulness is increased if they are drawn in hierarchy with new subunits, each with subsystems within. See some models of the U.S. International Biological Program. Detailed models with evaluations will help a person making a new model for that system. He can aggregate and reduce or add and change without the extensive work of developing a new one entirely from the start. Complex diagrams that have been evaluated and partially simulated contain understanding of ecosystem relationships obtained from those who have worked with such systems for many years. A model becomes large when everything in mind is included without any discriminating effort to determine what has a large effect in the case under consideration.

Complex models ordinarily may not be as useful for simulating overall performance since the effect of most of the detail becomes too small at the larger scale of total aggregation. The human recognition of the lack of utility of these models is the fact that they are not much used after development, that even those developing them may not understand them adequately, and that they are not as thoroughly studied for sensitivities and bugs. At least they require large expenditures of money and time compared to the extra understanding achieved. However, if work is cumulative, with many runs over a period of years, complex models may develop an ability to simulate varied situations within the range of variation of driving functions. However, support of simulation studies of ecosystems for extended periods has not yet been adequate to test this aspiration. See Chapter 27.

HIERARCHICAL MODEL STRUCTURE

Recognizing the hierarchical nature of ecosystems, Overton (1975) organized computer programs in systems and subsystems so as to maintain similar levels of detail for each level of modeling. See application to streams (McIntire and Colby, 1978).

SUMMARY

Typical ecological systems are given as models in Chapter 21 with their characteristic oscillations and diurnal and seasonal patterns. Included are repre-

FIGURE 21-40. Simulation model of marine plankton (Riley, 1946, 1947).

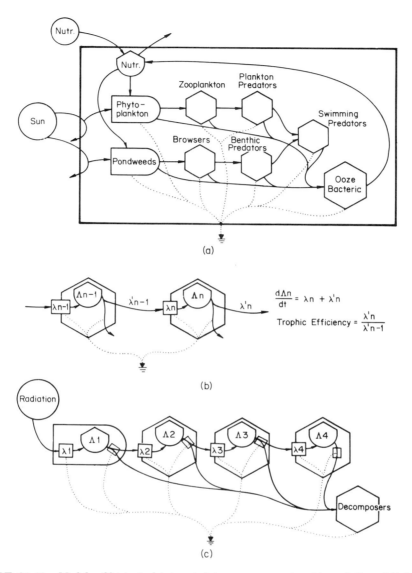

FIGURE 21-41. Models of historical interest: (a) rearrangement and translation of Cedar Bog model (Lindeman, 1941) simulated by Williams (1971); (b) trophic level equation (Hutchinson, 1942); (c) trophic levels (Lindeman, 1942). Heat sinks were added (dotted lines).

sentative models of streams, ponds, lakes, swamps, forests, deserts, reefs, estuaries, open-sea ecosystems, and historically important concepts.

Responding to the varied signature of energy inflows driving their self-design, ecosystems of the many habitats of the biosphere simultaneously exhibit almost unlimited variety of detail but similarity in the topology of patterns conforming to the basic web of energy quality with loops of recycle and control and with temporal programs and storages related to temporal programs of external driving patterns. Models readily generate diurnal patterns, and gross aspects of seasonal change but are less proven over longer periods. Many principles are well illustrated by living models called *microcosms*, especially where they are also theoretically modeled and simulated. The historical trend has been toward modeling with greater complexity, but some of the role of models as interfaces between reality and the human mind is lost as the models become too complex to visualize or manipulate easily.

Differences between land, lakes, and marine environments are mainly the result of substitution of driving functions. The aquatic systems, with high

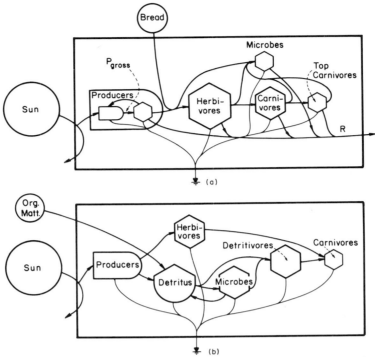

FIGURE 21-42. Historical models: (a) Silver Springs (Odum, 1955); (b) detritus dominant food chains (E. P. Odum, 1962). A similar one included materials recycle (Odum, 1962).

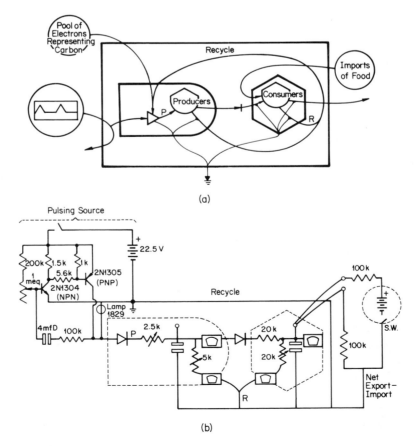

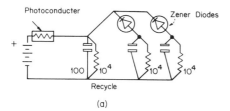

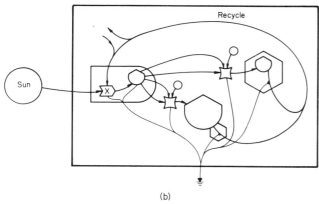

FIGURE 21-44. Passive analog circuit for ecosystems after Vogel and Ewel (1972) (see also Figs. 3-8 and 21-43): (a) electrical circuit; (b) equivalent energy diagram.

flows of high-quality mechanical energy, adapt to these by developing matching high frequencies of storage. Terrestrial systems responding to large frequencies in rainfall inputs develop long-range adaptive storages. By the time convergence of energy reaches top consumers, storages and frequencies are similar.

Systems modeling that does not provide mechanisms for substitution and self-organization of components is incomplete. The longer-range changes involving evolution, self-organization, and system learning are considered in Chapter 22.

QUESTIONS AND PROBLEMS

1. From Chapter 2, review the basic production–consumption minimodel for ecosystems. Discuss modifications to the diagrams to represent autotrophic net production, heterotrophic net consumption, multiple input–outputs of both nutrients and organic matter, a system with nutrient limitation but massive organic matter storages, a model that keeps track of nutrient, oxygen and carbon dioxide separately, and a model with water representing desert ecosystems.

2. What were some of the first ecosystem models that simulated seasonal patterns? Can you diagram these? What are some seasonal patterns of principle components of ecosystems (draw graphs).

3. Explain usual diurnal curves of oxygen, carbon dioxide, or organic matter stocks in terms of minimodels. Use other models to explain some less common situations in which there are large half-day lags, so that curves of gas exchanges are out of phase with usual responses to light (include an aquatic and a land case).

4. Discuss animal microbial relationships and energy quality.

5. What are some ecosystem properties that develop in natural eutrophy? Include aspects of aerobic and anaerobic zonation, diversity, role of larger animals, total power flow, net production, and variation.

6. Draw a typical diagram for each of the following environments: stream, pond, ocean system, upwelling zone, fresh-water wetland, and marine wetland. In each case identify the aspect of the energy signature that makes that kind of system special and predominates the internal special adaptations.

7. Describe the effect of increasing depth on the concentration of functions and storages, on total metabolism, on the proportion of larger animals, on net production, and on diversity per unit metabolism.

8. Discuss the virtues and disadvantages of minimodels, mesomodels, and macromodels (high complexity) for understanding, inventory and impact purposes, simulation, management, and prediction.

9. Give an example of a microcosm for each class of environments listed in question 6.

FIGURE 21-43. Pulsing ecosystem simulator (Odum and Baltzer, 1962; Odum, 1962a): (a) equivalent energy diagram; time constants were varied by substituting capacitors and adjusting variable resistors; (b) electrical network. Electrical current (energy flow) and population potential N were read in various positions with oscilloscope.

SUGGESTED READINGS

Anderson, J. M. (1981). *Ecology for Environmental Sciences*, Wiley, New York. 150 pp.

Arnold, G. and C. T. De Wit, Eds. (1976). *Critical Evaluation of*

Systems Analysis in Ecosystems, Centre for Agricult. Pub. and Documentation, Wageningen, Neth. 108 pp.

Bakuzis, E. V. (1974). *Foundations of Forest Ecosystems*, College of Forestry, University of Minnesota, St. Paul, MN.

Barnes, R. S. K. and K. Mann, Eds. (1980). *Fundamentals of Aquatic Ecosystems*, Blackwell-Mossy, St. Louis, MO. 240 pp.

Bliss, L. C., J. B. Cragg, O. W. Heal, and J. J. Moore (1980). *Tundra Ecosystems, A Comparative Analysis*, Cambridge Univ. Press, New York. 670 pp.

Brown, J., P. C. Miller, L. L. Tieszen, and F. Bunnell (1980). *An Arctic Ecosystem*, Dowden, Hutchinson and Ross, Stroudsburg, PA. 571 pp.

Canale, R. P. Ed. (1976). *Modeling Biochemical Processes in Aquatic Ecosystems*, Ann Arbor Science, Ann Arbor, MI.

Cobb, J. S. and M. M. Hardin (1976). *Marine Ecology, Selected Readings*. University Park Press, Baltimore.

Deininger, R. A. (1973). *Models for Environmental Pollution Control*, Ann Arbor Science, Ann Arbor, MI.

Ford, R. F. and W. E. Hazen (1972). *Readings in Aquatic Ecology*. Saunders, Philadelphia. 398 pp.

Giesy, J. P., Ed. (1978). *Microcosms in Ecological Research*, Conf 781101 Technical Information Center, U.S. Dept. of Energy, Oak Ridge, TN.

Goldberg, E. D., I. N. McCave, J. J. Obrien, and J. H. Steele (1977). *Marine Modeling, The Sea*, Vol. 6., Wiley, New York. 1048 pp.

Haldin, S., Ed. (1979). *Comparison of Forest Water and Energy Exchange Models*, Developments in Agricultural and Managed-Forest Ecology, **9**, Elsevier, Amsterdam, Neth. 258 pp.

Hamilton, P. and K. B. Macdonald (1980). *Estuarine and Wetland Processes*, Plenum Press, New York.

Hazen, W. E. (1964). *Readings in Population and Community Ecology*, Saunders, Philadelphia.

Heal, O. W. and D. F. Perkins (1978). *Production Ecology of British Moors and Montane Grasslands, Ecology Studies*, Vol. 27, Springer-Verlag, New York.

Innis, G., Ed. (1975). *New Directions in the Analysis of Ecological Systems*, Parts I and II, Simulations Councils, La Jolla, CA.

James, A. (1978). *Mathematical Models in Water Pollution Control*, Wiley, New York.

Jeffers, J. N. R. (1972). *Mathematical Models in Ecology*, Blackwell Scientific Publishers, Oxford. 398 pp.

Jorgensen, S. E., Ed. (1979). *State of the Art in Ecological Modelling*, Pergammon, New York. 891 pp.

Kerfoot, W. C. (1980). *Evolution and Ecology of Zooplankton Communities*. University Press of New England, Hanover, NH.

Kessell, S. R. (1979). *Gradient Modelling*, Springer-Verlag, New York.

Kinné, O., Ed. (1978). *Marine Ecology*, Vol. IV, *Dynamics*, Wiley, New York, 746 pp.

Kormondy, E. J. and J. F. McCormick (1981). *Handbook of Contemporary Developments in World Ecology*, Greenwood Press, Westport, CT. 766 pp.

Kraus, E. B., Ed. (1975). *Modelling and Prediction of the Upper Layers of the Ocean*, Pergamon Press, New York.

Kremer, J. N. and S. W. Nixon (1978). *A Coastal Marine Ecosystem, Simulation and Analysis*, Springer-Verlag, New York. 217 pp.

Le Cren, E. D. and R. H. Lowe-McConnell, Eds. (1979). *The Functioning of Freshwater Ecosystems*, Cambridge University Press, New York. 624 pp.

Levin, S. A. (1974). *Ecosystem Analysis and Production*, Siam Institute for Mathematics and Society. 337 pp.

Lock, M. A. and D. D. Williams (1981). *Perspectives in Running Water Ecology*, Plenum Press, New York. 448 pp.

Longhurst, A. R. (1981). *Analysis of Marine Ecosystems*. Academic Press, New York. 741 pp.

Lugo, A. E. and S. C. Snedaker, Eds. (1971). *Readings on Ecological Systems*, MSS Educational Publishing Company, New York. 353 pp.

Mann, K. H. (1982). *Ecology of Coastal Waters*. University of California Press, Berkeley, CA. 322 pp.

Margalef, R. (1948). General concepts of population dynamics and food links. In *Marine Ecology*, Vol. IV, O. Kinne, Ed., Wiley, New York, pp. 617–704.

Middlebrooks, E. J., D. H. Falkenborg, and T. E. Maloney (1974). *Modeling the Eutrophication Process*, Ann Arbor Science, Ann Arbor, MI. 218 pp.

Rich, L. G. (1973). *Environmental Systems Engineering*, McGraw-Hill, New York.

Scavia, D. and A. Robertson (1979). *Perspectives in Lake Ecosystem Modeling*, Ann Arbor Science, Ann Arbor, MI. 326 pp.

Smith, J. M. (1974). *Models in Ecology*, Cambridge University Press, London.

Steele, J. M. (1974). *The Structure of Marine Ecosystems*, Harvard University Press, Cambridge, MA. 128 pp.

Sundermann, J. and K. Holz (1980). Mathematical modeling of estuarine physics. Springer-Verlag, Berlin.

Thibodeaux, L. J. (1979). *Chemodynamics*, Wiley, New York.

Thomann, R. V. (1972). *Systems Analysis and Water Quality Management*, McGraw-Hill, New York. 286 pp.

Velz, C. J. (1970). *Applied Stream Sanitation*, Wiley-Interscience, New York.

Whittaker, R. H. (1970, 1975). *Communities and Ecosystems*, 2nd ed., Macmillan, New York. 385 pp.

Williams, W. D. (1981). *Salt Lakes*, W. Junk, Hague, Belg. 44 pp.

CHAPTER TWENTY-TWO

Succession

The self-organizational process by which ecosystems develop structure and processes from available energies is called *succession*. The name includes the time dimension of ecosystems. Apparently changes in structure improve the system's adaptation to utilize resources. Programs of succession become part of the informational structure of systems themselves, providing mechanisms of adaptation to external changes and variations. Different mechanisms are involved in different realms of size and energy hierarchies. Many models describe systems well, and similar short-term processes and variations but do not have features that can substitute components and change structure as real systems are observed to do. In this chapter patterns of self-organization are given with models concerned with succession and oscillations of ecosystems. Succession differs from other changes with time mainly in degree, involving more change in structure, program, and adaptation. The larger systems and the longer periods of time concerned with the biosphere are considered in Chapter 26.

Succession involves the whole system. For example, the cleared ground—receiving light, water, nutrients, immigration of seeding and living animals and utilizing existing storages—develops a sequence of stages. Where similar sequences have occurred frequently, a cluster of components develops that constitutes a stored and ready program of succession that is readily released when the ground is again cleared. Although the terms used to describe the process are different for changes that take place in other kinds of system and on different time scales, the self-organizational processes are similar, and much of what happens can be explained by the survival criterion of contributing maximum power at each level of size. In general, the process of organizing new programs of succession involves offering of choices, followed by selection by the next-larger system for those patterns that feed back work toward maximizing power input and useful transformation. Biological evolution, succession, and learning are all essentially similar, differing mainly in their details and time scale. In ecological succession the smallest components with rapid turnovers evolve, the medium-sized ones disperse choices from which competition and system reinforcement makes selection, and the larger ones may reorganize their memory patterns with regular learning. The land surfaces are also organized, feeding back their structure in maximizing work contributions of vegetation, water, winds, waves, and so on.

SUCCESSION, CLIMAX, AND RETROGRESSION

Succession includes self-organization and change by which ecosystems become established and sometimes retrogress. Succession is regarded as terminated when a steady pattern is reached or when the system returns to a less organized state to begin succession again. Climax is the crest of growth or the high plateau if one is attained, as in Fig. 22-1. *Retrogression* (or regression) is a period when assets decrease for any reason. Growth may be asymptotic as shown in Fig. 22-1a unless the initial conditions have higher quantities of resources and organisms than the condition that is ongoing or repeating. Other causes of overshoot are given below. The presure of genetic dispersal provides a large choice of organisms whose populations, if they prevail, contribute to the process of transforming the various energy inflows of various qualities of the energy signature into storages, patterns, numerical relations, diversities, and other dimensions of organization that maximize the flow of power through producers, consumers, and the hierarchical webs and spatial cells. Some of the successional processes known or postulated are given in Fig. 22-1. Some have long

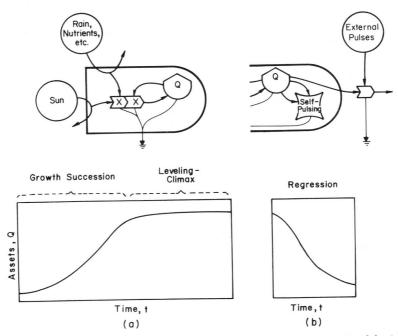

FIGURE 22-1. Basic concepts of succession expressed with simulation minimodels: (a) succession and climax; (b) regression (retrogression) by self-pulsing and/or by pulsing of external larger-scale systems that have a longer period.

programmatic steps; others seem to proceed immediately to climax.

There are more kinds of initial condition than there are climaxes, where the term "climax" is used for any state that follows a period of rapid growth in mass, structure, and/or diversity. Succession may be measured by growth curves of the main parameters of the ecosystem, such as live biomass, nutrient storage, total organic matter storages, diversity, web parameters, total metabolism, total energy receptor quantities, and so on. Many of the growth models presented in Chapters 3–13 can be used as minimodels of successional increase of the whole ecosystems.

Figure 21-1a shows a simple minimodel that generates some of the classical idea of succession such as fast growth and a climax steady state. Sequences of stages (seres) are part of the classical concept of succession in ecosystems, each stage a prerequisite for the next. Some sequences are diagrammed in Fig. 22-2.

A New Stage Using Storages of Predecessor

Storages of structure developed and maintained by one stage in succession are available as an energy and material resource to stages that follow. The extinction of some large mammals by an agrarian society as it replaced hunting and gathering humans may be an example. Extinction of giant birds in New Zealand by Maoris attaining agriculture is another example. A current example may be the consumption of forests and soils by a fuel-based economic culture (see also the use of mineral deposits by evolving humanity in Fig. 26-5b).

Pulsing

In Chapters 11 and 15 and subsequent chapters the hypothesis was considered that the maximum power principle causes systems to store and pulse their feedback services to the system of which they are a part, thus maximizing their roles in useful work and control at that size scale. The convergence of many small, short pulses to larger and longer pulses occurs with steps along the energy-quality chain so that each unit absorbs and evens out those pulses converging to it but imposes on those below it in the chain the period of its own pulse. Period and amplitude of pulsing thus increases along the energy chain, with the uppermost ones constituting such large surges as to seem like disasters in disruption of the smaller ones. During pulsed energy feedback a "frenzy system" tends to be operated, which is

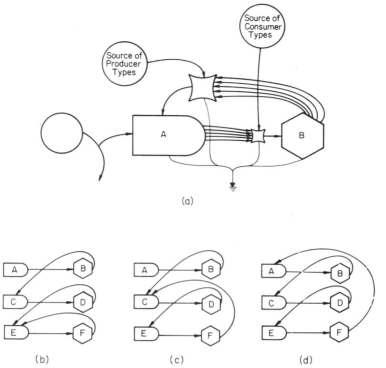

FIGURE 22-2. Succession of producers and consumers: (a) diagram in which changing flows initiate substituted components; (b–d) substitution sequences; (b) last stage is climax; (c) cycles to an intermediate stage; (d) cycles to start.

transient but important. See Alexander's model minimodel for the pulse delivery system (Fig. 11-21).

From such theoretical considerations, the many observed patterns of pulsing and oscillation, and the many observed mechanisms for pulsing and oscillation observed and modeled, it is apparent that absolutely constant patterns are not the rule, although some long periods of little change with hundreds of turnover times have been observed in various special cases. The steady productivity at Silver Springs, Florida is one example (Odum, 1955; Knight, 1980). Another is the steady phosphorescence of dinoflagellates in Bahia Fosphorecente in Puerto Rico for many years and the steady zooplankton populations there (Coker and Gonzalez, 1960).

Catastrophe Pulsing—a Question of Scale

Many observational studies find catastrophes such as floods, wind storms, earthquakes, and externally started fire, waves of human colonization, and so forth, a principal influence in ecosystems [e.g., see review by Bormann and Likens (1979b)]. Rather than regard these as capricious external influences, the hierarchy concept explains large pulses of large systems as the intermittent feedback of high-quality energy. The pulse seems unusually catastrophic in local effect, whereas it is part of regular pulsing of the next-larger systems. It becomes a part of the

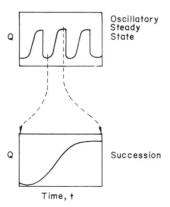

FIGURE 22-3. Succession as a part of an oscillating type of climax when examined on a large scale of time or space (Q = structure).

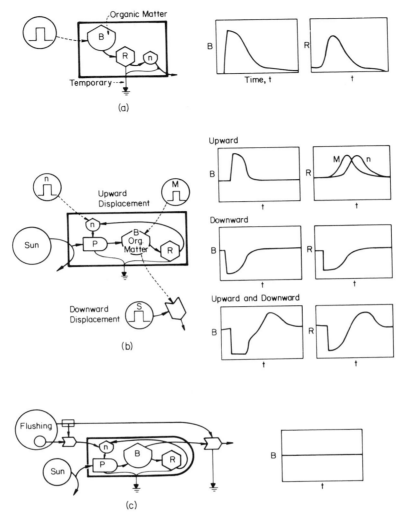

FIGURE 22-4. Minimodels of displacement and succession: (a) displacement of equilibrium; (b) upward and downward displacement of steady state; (c) steady state controlled by physical drain. P, production; B, biomass; n, nutrients; S, stress; R, respiration; M, migration or import.

normal understanding of succession to expect the pulses that are on a period appropriate to the next-sized system (see time–space relations in Chapter 14). When the larger system is examined, the source of pulsing is found as a high-quality component at the end of the energy chain that spreads its concentrated influence from spot to spot. In the sea there are physical eddies of nearly all sizes, providing each smaller system with its pulsing reorganization the energy of which adapted systems may convert to use.

Examples of top consumers controlling whole ecosystems include: marsh by crabs (Montague, 1980) and pools by small fish (Hurlbert et al., 1972). Gyllenberg (1977) models animals in tundra.

Stabilization of Climax by Disordering Energy

If the pulsing is supplied externally from a physical drain such as a turbulence or an advection, control of the lower parts of the energy chain is supplied according to the pattern of the external energy control. If it is steady, a very steady state can be observed. Simple minimodels of production and removal were given earlier, with chemostat examples in Chapter 11, the turbulence and exponential model in Fig. 9-23b, and the example for Silver Springs in Fig. 7-6 (see p. 520 also).

Whether more power can be drawn from the system by pulsing the external control is still an open question (see pp. 117, 195, and 573).

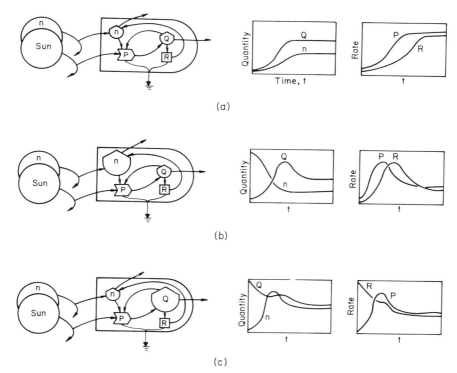

FIGURE 22-5. Kinds of biomass succession as related to ratio of initial storage conditions and renewable energy flows: (a) steady renewable, initial storages small; (b) nutrient n originally larger than at steady state, biomass Q initially small; (c) biomass Q originally larger than at steady state; nutrients n initially small.

Scale of Succession or Climax

From the foregoing description of pulsing and hierarchy, it is apparent that considered in space or in time, every steady pattern is composed of pulsing components. More space means a realm of phenomena of larger period. In a forest trees grow and fall, each of which has periods of growth and regression, but when considered over longer periods, average out as steady. Probably the term "climax" is equivalent in definition to "steady state." In Fig. 22-3 succession and climax on a short time scale are shown part of a system with a pulsing steady state of larger period.

Displacement of Equilibrium and Steady State

Homeostatic open systems, which probably includes all long surviving systems, when displaced return to their steady pattern (or repeating pattern) (Burton, 1939). Closed equilibrium systems when displaced return to their equilibrium state. Displacement of equilibrium is tantamount to adding a temporary energy storage source (see p. 312).

Succession often involves depletion of a temporary energy storage toward equilibrium plus displacement of a steady repeating pattern of an open energy flow. As with the minimodel in Figs. 9-6 and Fig. 13-9b, we can relate storage, steady sources, and displacement to succession and climax simply with the minimodels in Fig. 22-4. More elaborate alternatives are given in Fig. 22-5. Butterfield and Purdy (1931) grew bacteria and a bacteria-consuming protozoan *Colpidium* in pure batch culture separately and together. After an initial surge of both, the combined populations were stabilized at a low level.

SUCCESSIONAL PRINCIPLES AND MINIMODELS

Regimes as Related to Energy Source Patterns

The kinds of successional, climax, or regression sequences found are readily generated by minimodels that have appropriate initial storages and energy

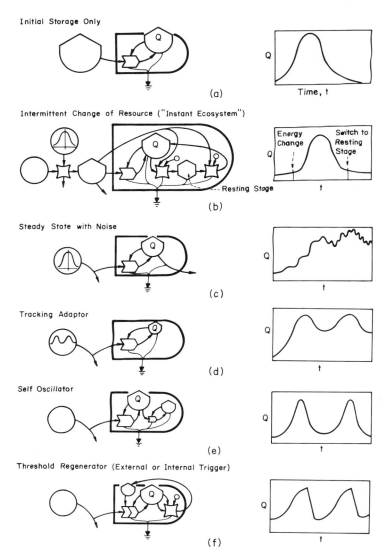

FIGURE 22-6. Types of climax and their minimodels. Q is structure.

inflows from forcing functions. In Fig. 22-5 some of the minimodels are given with their characteristic patterns and the kinds of ecosystem that are examples.

There are many kinds of climax state (postgrowth states). There are some where renewable energy flows are unchanging and the ecosystems have been proven unchanged for many turnover times. Others have been shown to be very variable in a stochastic manner. Still others undergo bursts of growth and dormancy in response to energy regimes that are intermittent, regular, and irregular (types of climax are given in Fig. 22-6, and species accumulations for minimodels are given in Fig. 22-7).

Succession often starts with depauperate area so that most parameters increase as resources are accumulated and organized. Since productivity and other metabolism increases with time and in these conditions, succession is often thought of as a progression from a low nutrient, low productivity condition called *oligotrophy* to a more fertile and productive condition called *eutrophy*. However, this is not a general rule, since many successions proceed from initially rich productive conditions to less fertile conditions where the initial condition of fertility is due to an initial storage rather than to renewable resources. The shape of the growth or decline may be estimated from initial state as related to final state

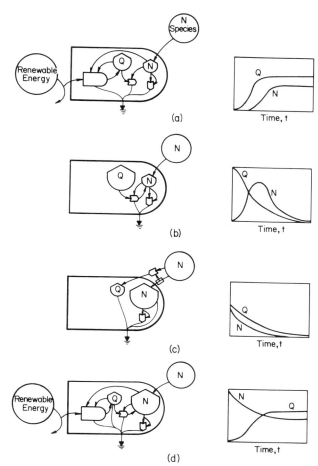

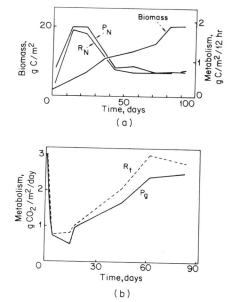

FIGURE 22-7. Patterns of biomass Q and species diversity N in relation to sources and initial conditions: (a) initial mass and diversity small, renewable energy; (b) large biomass initially, no renewable energy; (c) species variety initially seeded larger than supportable without other energy source than dispersal; (d) initially more species than at steady state; biomass initially small.

FIGURE 22-8. Production and respiration during successional changes: (a) succession in microcosm following large initial change of nutrients (Cooke, 1967), where P_N is net daytime production and R_N is net nighttime respiration; (b) adaptive reorganization following a reduction in level of light energy (Copeland, 1965). P_g is gross production and R_t is total 24-hour respiration.

as in Figs. 22-4 and 22-5. See also E. P. Odum (1971a, page 271). A maximum (overshoot) between growth and a more steady pattern results when initial storages are higher than steady-state storages past the end of growth. An example of overshoot following large initial storage of resources is given by Cooke (1967) in Fig. 22-8. An example of noisy climax was the day to day productivities in an Indian pond (Sitaramiah, 1961).

An example of decline during succession after initial seeding is given from a microcosm study in Fig. 22-9. Periphyton from Silver Springs were initially introduced but the microcosm sustained a reduced level of diversity and production.

An example of successional sequence with greater metabolism in early succession than later with R greater than P in final conditions is the plankton in rice paddy fields during rice cultivation (Kurusawa, 1956).

An example of a successional trend with initial concentrations of organic matter but no input at climax is the decomposition microcosms given by Renn (1937) and von Brand and Rakestraw (1941). This system trends toward the final stability of a closed system. A closed system is a limiting case for the open system.

Effect of Materials for Recycling on Succession

Illustrated with simulation of a $P-R$ model in Fig. 22-10 is the limitation of material shortages that interfere with levels of structure and metabolism that may be obtained. Note the higher levels and later arrival at climax when the recycling pool is larger.

Adapting to Intermittent Regimes

Where resources for ecosystems are intermittent as in the desert, adapted ecosystems hold seeds, roots,

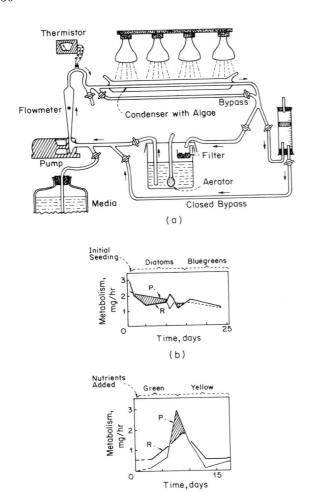

FIGURE 22-9. Successional responses of a periphyton microcosm to introduced quantities (Odum and Hoskin, 1957): (a) microcosm apparatus; (b) succession with declining metabolism and diversity; (c) response to nutrient pulse, first green, shifting to a carotenoid-rich recycling regime.

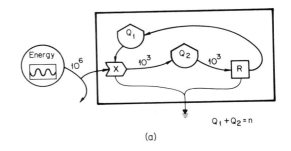

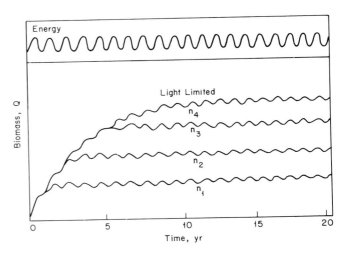

FIGURE 22-10. Simulation of effect of increasing material supply where main energy source is unchanging: (a) energy diagram; (b) simulation results with increasing nutrients N_1 to N_4. $N = Q_1 + kQ_2$.

spores, and so on in reserve until environmental cues are received that turn loose rapid growth and reproduction. As the intermittent resource begins to be used up, cues are received that cause the adapted system to switch energies into storage reserves and reproductive bodies that can remain dormant for a long period until the next curve of resource. Noy-Mier (1973) provides some minimodel diagrams of desert functions. Some are translated into energy language in Figs. 21-31 and 21-32.

Maximum power implies ecosystems develop mechanisms for their own pulsing. However, where external energies are periodic (e.g., diurnal and seasonal) the pulsing is already available for systems of the size and time scale to take advantage at that period.

Many ecosystems have a resting aspect when dry and spring into life and fresh growth when wet. So-called *instant ecosystems* include the blue–green algal mats of estuaries (Fig. 21-8), the dried muds of desert playas, the sediments of pools on granite outcrops, and the lichens that cover rocks in tropics and arctic. Bliss and Hadley (1964) found lichens accelerating photosynthesis with P greater than R when wet and becoming more of a consumer when dry. Lugo (1969) studied succession in artificial outcrops and found a high P/R ratio initially.

Self-Organization of Parallel Components

Some of the automatic self-organizing mechanisms of having species in parallel collectively contribut-

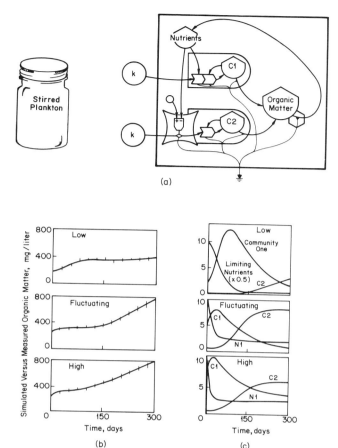

FIGURE 22-11. Simulation of succession in bottle microcosms (Kelly, 1971) in three regimes, low temperature, high temperature, and daily fluctuating temperature: (a) simulation model; (b) graphs of organic matter concentration during succession, where bars represent range of value in replications, and the line is from the simulation; (c) curves of nutrients and two population groups (C_1 and C_2) generated by the model and validated semiquantitatively.

were removed, the remaining ones automatically divided up the realm with larger proportions for each, but there was a decline in total production as some of the advantage of division of labor was lost.

Substitution of Components During Succession

Typically, succession, as the name suggests, has substitution of species as time passes, each preparing the system for the next. An example of the substitution of species is given from Kelly (1971) in Figs. 22-11 and 16-5. Microcosms developed one characteristic set of algae and consumers during its early self-organizational period and then flipped to another set as initial nutrients declined. Three sets of microcosms were included, each with a different temperature regime (low, high, and fluctuating daily). A minimodel in Fig. 22-11a was adjusted to generate the observed curves of organic matter accumulation in the three regimes (Fig. 22-11b). Thus calibrated, the model generated curves for the effect of increased nutrients in Fig. 22-11b. A model of the effect of temperature regime was also simulated in Fig. 16-5.

Transition Matrix for Succession

Waggoner and Stephens (1970), Horn (1975) and Horn et al. (1979) represent the seedling frequencies under trees in a transition probability matrix. This matrix constitutes a model for successional changes. The diagonal represents the reproduction rate continuing the same species. A climax is found as a matrix that reproduces itself.

Maximum Power as Self-Design Principle

The theory of maximum power control of self-organization suggests survival of those combinations of components that contribute most to the collective power of their system, thus making their habitat most favorable for each. Maximization of power involves an increasing energy flow rate during successions that start with little initial storages, but power maximization may involve a rise and then lowering of power levels where stored and renewable resources are being drawn in.

The role of natural selection in system self-

ing to the metabolism and mineral cycle were suggested in relation to Fig. 12-5. Species are reinforced that divide up and optimize use of resources so that collectively they maximize production. A good demonstration of actions of population models to self-organize within an ecosystem model was given by O'Neill and Giddings (1979). A model of phytoplankton production included 10 species with variations in response to light and nutrients. Under the competition arrangement each species developed different times of growth and decline, contributing to an overall seasonal pattern such as that in more aggregated models. See for example Fig. 12-25. When species

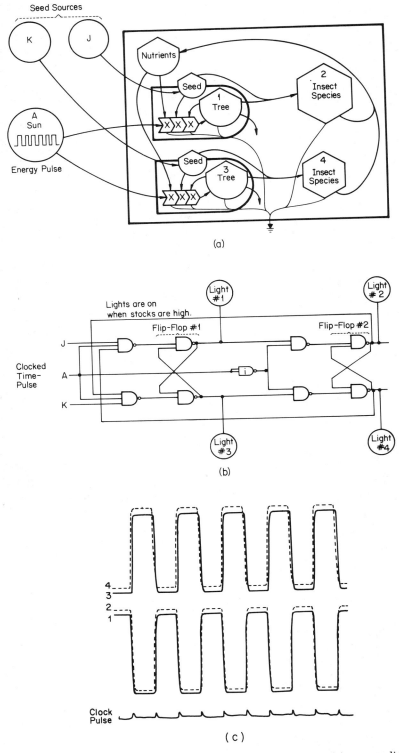

FIGURE 22-12. Plant–insect pairs constituting a flip-flop oscillator: (*a*) energy diagram; (*b*) logic circuit with use of NAND gates for master–slave flip-flop analogous to parallel chains in part *a*; (*c*) graph of simulation of *b*.

452

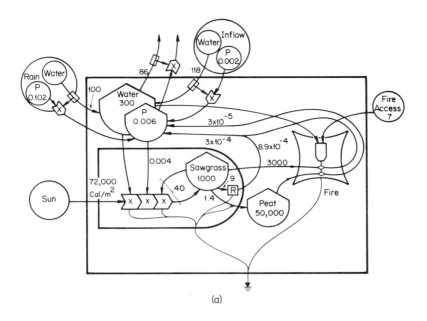

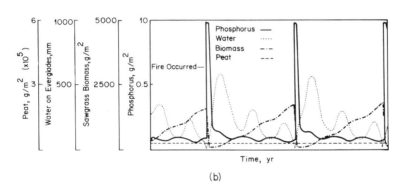

FIGURE 22-13. System of water and grass in Everglades; period of oscillation of fire oscillators dependent on level of energy resources: (a) energy diagram with magnitudes of storage and flows per month; (b) simulation which burned every third year (Bayley and Odum, 1976).

organization was clearly stated by Tansley (1935): "If organization of the possible elements of a system does not result, no system forms or an incipient system breaks up. There is in fact a kind of natural selection of incipient systems and those which can attain the most stable equilibrium survive the longest."

Much confusion followed Wynne-Edwards' (1962) book on particular mechanisms of "group selection." Part of the problem was semantic. By system selection, Lotka, Tansley, and theories in this volume refer to self-organizational choices that retain units and mechanisms that contribute to the systems resources for meeting contingencies. Darwinian selfish selection is regarded as secondary priority for survival.

OSCILLATING PATTERNS

Circular Succession: Flip-Flop Oscillation

Circular succession is a changing climax where one state produces another until eventually the first state occurs. One mechanism of repeating pulses is circular succession. Sequences of actions that are programmatic are included in Fig. 22-2. An example that has logic aspects is modeled in Fig. 22-12. A system of parallel food chains consists of two producer–consumer pairs with the use of basic energy supplies in parallel. Both the producers and the consumers have negative effects on the competing energy budgets; thus whichever is started remains

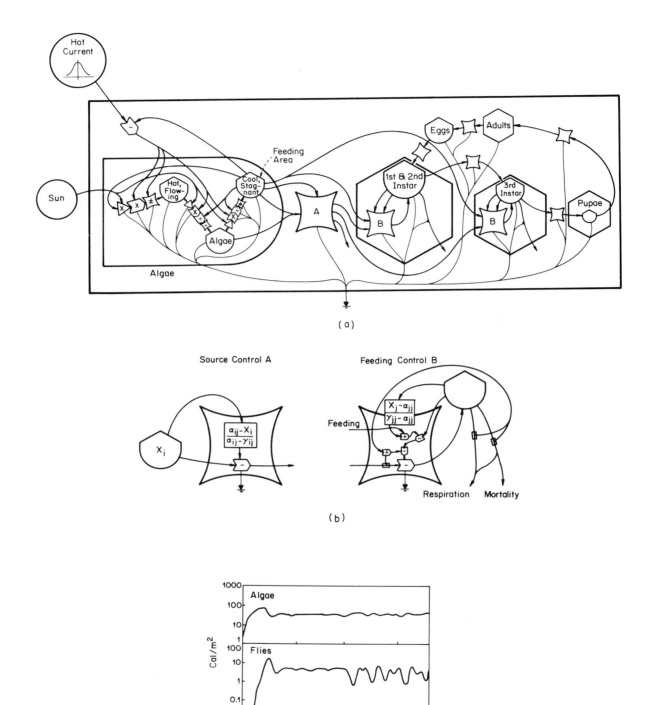

FIGURE 22-14. Model of algae and flies in Yellowstone Hot Springs (Wiegert, 1975): (*a*) energy translation of model; (*b*) detail of energy controls by and to populations; (*c*) example of simulation with oscillatory characteristics observed in the spring environment.

454

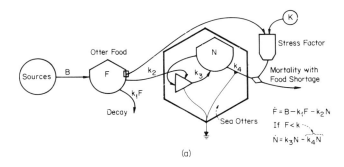

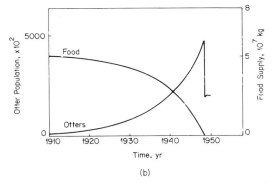

FIGURE 22-15. Model of sea otters on Amchitka Island with consumer-controlled pulsing (Raines et al., 1971): (a) energy diagram; (b) simulation.

dominant as long as energy sources are continuous. The growth of the dominant plant sets off a growth of its specialized consumer that cuts off the transmission of the reproductives to start the new cycle of growth if there should be an interruption such as restart in a new season. In effect, the consumer's action sets up conditions for replacement with the alternate parallel food chain in the next growing cycle. The process repeats and the next interruption generates the original. Such a double-flip-flop system provides an oscillating dominant.

Patterns of this kind have been studied by Janzen (1970, 1971). Specialized insect consumers of rain forest tree species by their numbers cut off self-reproduction in that vicinity so that the next trees that grow are a different species. These studies show at least one mechanism for the d'Aubreville theory of mosaic growth in rainforests, according to which each species is replaced by another rather than itself. The overall effect is increased diversity and possibly greater stability against large damaging epidemics. A logic circuit often described as the *master–slave* flip-flop, is shown in Fig. 22-12b. The first flip-flop is roughly comparable to the competing tree species, and the second flip-flop is roughly comparable to the insect consumers. The electrical output is given in Fig. 22-12c; NAND gates were used because this was the building block of the logic demonstrator used.

The two inputs represent immigration of seeds necessary to start the system and reseed new growth after interruption. The clocked time pulse at A represents the input energy that is interrupted in pulses that represent seasons. In the tree example, the pulse represents growth, decline, and falling of a tree repeated again with a new tree and new pulse of growth and decline. The electrical system starts with light 1 on representing the dominance of tree species 1. Then as the energy pulse declines, the downstream flip-flop is enabled with light 2 going on, representing excessive growths of this consumer predominating with decline of the tree. The rise of a new pulse of energy and the cross stimulus of the consumer stimulating the new growth of the opposite species causes the reverse flip-flop with light 3 going on. Then, as in the first cycle, the light 4 goes on as the downstream flip-flop is kicked with the declining phase of the pulse. The rising pulse downstream as energy arrives represents the dominance of the second consumer insect species, indicated with light 4. This model is readily patched on analog equipment or microcomputer and its performance studied with a manual operated energy pulse at A. Since competing pairs of species have the general property of a flip-flop, this class of digital action is to be expected in competition phenomena, even when the competition is harnessed as a component in a larger system to form mechanisms of complex structure maintenance.

The model and the phenomena it resembles may be regarded as a circular succession. A good question is when circular succession is an advantage over simple noncircular succession. Circular succession is one kind of pulsing mechanism.

Pulsing Consumer Regression and Regeneration

Figure 27-4 models a commonly recurring ecosystem mode in which production is fairly continuous but consumption is sharply pulsing. By storing moderately high quality energy from production and delivering its feedback actions and recycle in a short pulse, the energy-quality amplifier effect is much higher, at least during the period of the rapid consumption and work feedback. The development of

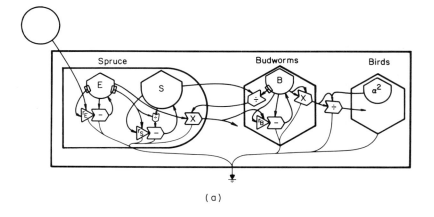

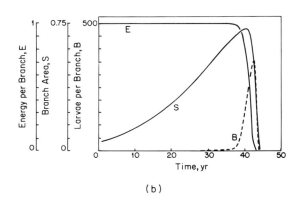

FIGURE 22-16. Model of spruce budworm epidemics with fast consumer (Ludwig et al., 1978): (a) energy diagram; (b) simulation.

higher-quality action may make such systems surpass those that do not accumulate and pulse. The theory explained in Chapter 15 is that more control is exerted for the same calorie flow by that system with the highest-quality energy.

The mechanism for the pulsing of the consumer action may vary. Self-organization among multiplicity of species and component systems may feed back a reward loop to any design that has a pulsing mechanism. The following are examples of pulsing systems that may illustrate these principles and be represented by the model in Figs. 11-21 and 27-4: recurring fire climax, forest and grassland; recurring red tide fish kills; epidemics and epizootics; and predator–prey interactions (see Figs. 11-17 and 11-18).

The principle may also apply to other systems such as earthquakes based on accumulated strains and actions of governments in feedback to population. Systems with short time constants can utilize aspects of the seasonal cycle as mechanisms of pulsing under certain conditions. Bayley and Odum (1976) (Fig. 22-13) provide an example with simulation of Everglades with fire. Mueller-Dombois (1980) describes natural dieback cycles in mountain rainforests of Hawaii involving storages in soils.

The oscillating buildup of blue–green algal mats in hot springs is followed by action of fly larvae in eating the structure of the mat, causing them to break off and start over. The flies are not able to reproduce in large quantity until the mat saturates again, providing a barrier to hot water (see Fig. 22-14 as studied by Wiegert).

Raines et al. (1971) model the role of sea otters in control actions of kelp as a pulsing system (see Fig. 22-15). Ludwig et al. (1978) studied the oscillation of spruce budworms according to the model in Fig. 22-16. Casti (1979) uses this example to illustrate catastrophe surfaces. An equilibrium equation derived from the differential equations is found to have the form of a cubic equation with a folded-shape surface that gives a discontinuity. A crash or catastrophe may occur when energy levels are high. The important question is why such a systems design is selected for in the systems self-organization. For this answer, we return to energetic reasons for pulsing patterns.

Alexander (1978) found that the sequence model in Fig. 11-21a simulated a series of growth–pulse use systems including an earthquake, forest fire, and flood damage. A variation of this model is given in Fig. 27-4.

Miller et al. (1975), simulating a tundra ecosystem, found a prey–predator type of pulsing emerging from their simulations of field data. Bormann and Likens (1979a) found northern forests with self renewing cycles.

Fox (1978) finds fire and precipitation involved in the pulsing of snowshoe–hare population oscillation. The pulsing of plant and animals are apparently coupled, although it may be incorrect to try to assign cause to only one part of a pulsing ecosystem.

Mattson and Addy (1975) collect evidence for herbivorous insects contributing to forest productivity by pulsing consumption of canopies, helping to develop vertical structure.

REMOVAL AND STRESS

Climax and Removal Times

The balance between production and removal was modeled in Fig. 11-12 and 11-14 on chemostats. This is an example of the general principle of homeostasis. The amount of structure is a balance between the rate of production and the rates of removal and the losses due to thermal disordering depreciation. Reduction of removal allows more structure to de-

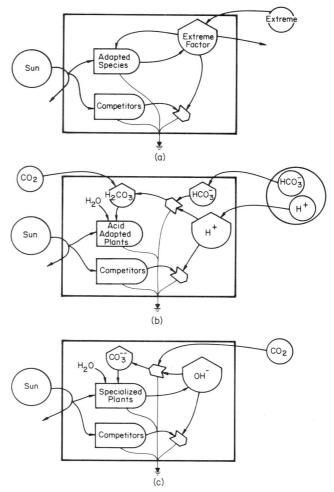

FIGURE 22-17. Adaptation of ecosystems to use extreme factor and help maintain it extreme: (a) adaptation to an extreme makes it a source to stress competition and reinforce itself; (b) acid ecosystem; (c) alkaline ecosystem.

velop, but power yield is maximized at intermediate removal rates (see population steady states in Daphnia studied by Slobodkin in Fig. 11-11). Yield maximization leads to survival, where the yield generates protective feedback services that are supplied by the outside user of yield. Without the feedback, power is maximized by use of potential yield to increase internal structure capable of improving efficiency instead. In other words continuing systems require coupling of feedback loops as in Fig. 15-1.

Stress and System Adaptation

A stress can be defined as a high-energy effect produced by a factor with high-amplifier effect and probably one with high embodied energy. The system that maximizes power and control is one that utilizes the energy flows that would be stresses in such a way as to contribute to survival. For example, by adapting to toxicity, a species has diverted the potential stress into an energy action against competitors.

When the stresses come in surges as part of frequency and space patterns of larger systems of the biosphere, they can adapt by developing that structure and a regeneration cycle that uses the stress energy on either the production side or the consumption side. The extrinsic pulses provide high-quality control energy for regressive recycle, enabling the system that is organized in this way to prevail.

When there is a general pattern of high-quality energy that has its own hierarchy of high- and low-quality energy—in the same manner that waves, eddies, and turbulence have in the sea—the means for convergent and divergence coupling of dispersed production to converged consumers is already established, requiring only that the species have some of the correct time constants and weights to be transported on the one hand and some relative motion to aid individual processes on the other.

In our terminology adapted systems do not have stresses, although the action of an energy surge may be regarded as a stress on components of the system. Irradiation studies with ionizing radiation have been useful in studying ecosystem responses to a fairly general stress of high quality. Marshall (1962, 1965), for example, found lower population levels maintained while the disordering actions were taking place. See Chapter 17.

Minimodel with Flip-Flop Adaptation Action Reinforcing an Extreme

Many ecosystems that are adapted to some extreme quality such as acidity, high alkalinity, aridity, swampiness, or another special property requiring special adaptations may feed energy into reinforcing that property so that it serves to favor the adapted species, excluding those more general species that, lacking the special adaptations, could prevail in ordinary conditions. If the environmental tendencies are toward the extreme property, the ecosystem may feed back energy to maintain that property. If conditions move away from the extreme, it may change with a flip-flop action such as species substitution capable of reinforcing a new condition that locks the system into the new state. Cypress swamps tend to conserve water; arid communities tend to maintain a radiation and heat balance that favors their continuation (see Chapter 21). In Fig. 22-17 carbonate-using plants reinforce high pH. In acid photosynthesis affects pH little.

Stress and Oscillation

Response to toxicity of a chemical element was studied by Von Voris et al. (1978) (see Fig. 22-18). Here microcosms with low frequency oscillations indicating high quality structure were more stable when stressed.

MAXIMUM BIOMASS, MAXIMUM POWER, AND SUCCESSION

The increase in biomass is one of the products of the net production early in succession. Then, as respiratory requirements accumulate for the maintenance of more structure, the net production is less [see Boysen-Jensen (1932) and many others]. One view of this is that later more biomass can be maintained for less production (Margalef, 1963; E. P. Odum, 1969).

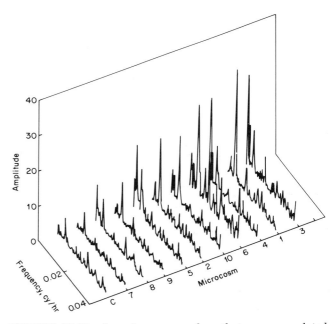

FIGURE 22-18. Low frequency pulses that were correlated with stable retention of calcium in microcosms with cadmium stress (Van Voris et al., 1980).

With the maximum power theory in mind, a different perspective results. As structure is accumulated, biomass is added in those situations where biomass facilitates power maximization. There is, however, a curve of diminishing returns in the addition of biomass and other structure as the energy required for maintenance of the structure becomes equal to the gross production and other energy incomes that can be induced (see Fig. 22-19). The key measures in succession, therefore, may be the gross production and total respiratory metabolism (not net production or biomass). Total metabolism is a measure of the total structure and of the dissipation rate for the low-entropy structure being maintained. Where organic inputs are small, gross production is a measure of maximum power. Increases in these total measures are observed in succession to climax. If the maximum power theory holds, observed patterns of power pulsing and regression may provide means for increasing total power flow in the long run, accelerating consumer metabolism during downturn but setting the stage for greater production and consumption later.

Some systems have large organic imports and/or exports. The equation for steady-state balance is $P + I = R + E + f$ (Odum, 1956) (see Fig. 22-19). Total power maximization requires maximization of R from both production and import income. The export is not a loss if in the surrounding systems it facilitates power maximization in the larger-scale view. For example, the export may be part of exchange that increases I. Succession can go toward more eutrophic or less eutrophic conditions, depending on the relative fertility of the initial stocks versus those maintained in the long run. The long run may be a fluctuating one. The value of P may be either greater or less than R in succession or at climax depending on the balance of P and I.

SPECIES AND DIVERSITY

Species-Information Access

Dispersal of species is an important and often limiting driving function (dispersal mechanisms are given in Fig. 20-23). Continual replacement is a major source of stability, especially for scarcer members. If the initial state is underseeded with species, its diversity may increase (see Fig. 3-20b). If the diversity is overseeded, it may decrease. The role of species-information access may be separated from other aspects of succession by studies of colonization

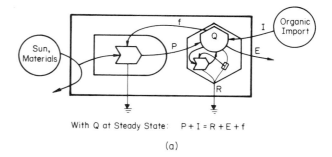

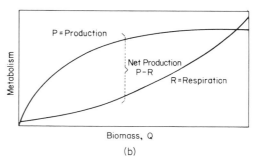

FIGURE 22-19. Relationships of production, respiration, and biomass: (a) energy diagram; when respiration is measured with the gas exchange methods, f is incorporated in R; when flows are in Calories, feedback f is small in actual calories but high in embodied energy; (b) effect of biomass Q on gross production P, structural energy requirements R, and net production $P-R$.

in beakers that do not supply appreciable energy for growth and development (Maguire, 1971) (see Fig. 22-20). However, if energies are available, more species may become part of a self-maintaining system, but some species are competitively excluded. Minimodels with species as a storage are given in Fig. 22-7.

Using the colonization model in Fig. 22-21, Shugart and Hett (1975) find faster turnover of species in succession than in later stages and plot the species replacement rate on semilog plot, using its slope b as an index of succession.

Maturity Concept

The concept of maturity as a more organized and diversified system used by Margalef (1968) applies only to situations that start with low quantities of resource materials and species, so that time causes increases in these in reaching the steady pattern. If, however, initial conditions have higher quantities of nutrients and species than the starting condition, the succession is a decrease of these characteristics.

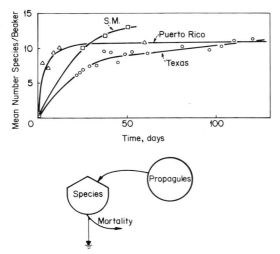

FIGURE 22-20. Colonization curves of microscopic animals in beakers (Maguire, 1971).

Maturity then has less than the starting state. Examples are the decline in production of formerly fertilized fish ponds or of Lake Washington after nutrient removal.

Use of the P/R ratio as a measure of maturity is valid only for ecosystems for negligible inflows of nutrients or organic matter. The ratio of carotenoid to chlorophyll is given as an index of P/R status in Fig. 19-29.

The controversy over the relation of stress energy and diversity was considered in Chapters 17 and 18.

Apparently diversity is a means to utilize stress energies in some conditions. Whether high diversity of coral reefs is because uniform conditions allow energy for organization or because coral reefs are frequently subjected to varying conditions causing diversity (Talbot et al., 1976) remains to be proven.

LONGITUDINAL SUCCESSION

Longitudinal succession is the progression of changes that occur with time and space as a result of the system being embedded in a moving medium. Examples are the increase in plankton community moving downstream from emergence of a clear spring (Odum, 1957) and the succession of communities that develop downstream from the influx of pollution into a stream. If longitudinal succession is in a steady state, distance represents time following the initial intersection of energy sources interacting in productive processes. The steady pattern in space is a climax, but the sequence of climaxes along a line represent increased time for development [Fig. 22-22 shows longitudinal sequence in a stream; Fig. 22-23 shows longitudinal succession in refinery waste (Copeland, 1963)].

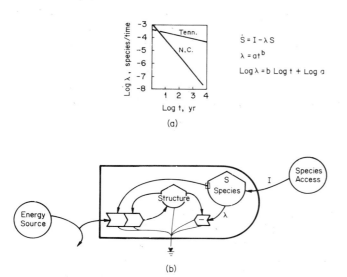

FIGURE 22-21. Decline of extinction rate with succession (Shugart and Hett, 1975): (a) logarithmic graph used to obtain rate of change of the extinction flux; upland forest examples; (b) a model with declining rate of species extinction with succession.

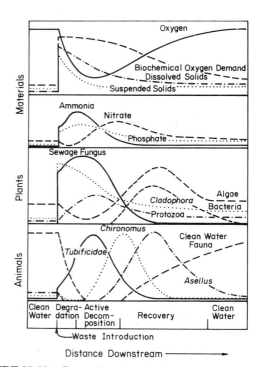

FIGURE 22-22. Examples of longitudinal succession in stream receiving waste [from Hynes (1960), after Warren and Doudoroff (1971)]. For model appropriate to this longitudinal pattern, see Fig. 21-18.

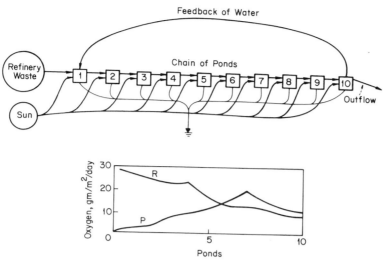

FIGURE 22-23. Example of longitudinal succession in a chain of refinery waste ponds (Martin, 1973).

Going downstream in an energy transformation stream, there is convergence and development of higher quality (Fig. 18-1). For example, as they grow larger, fish move downstream to larger realms, sending them back as an organizing pulse to the small upstream environs (Hall, 1972).

Upwelling Models

There has been considerable focus of interest in modeling the most fertile parts of the sea, the upwelling areas. The progression of plankton community succession that develops in the upwelling outflows off the Peruvian coast starts with nutrients, generates rich phytoplankton, anchovies and later a diverse fishery, great bird populations, and immense guano deposits on the nesting islands. A simulation model is given in Fig. 22-24 with validation data. Here there is an import of high quantities of nutrients interacting with sunlight and moving laterally with some aspects of patches that are under dispersive tendencies. See also Wroblewski (1977). These models have some of the features of stream models and longitudinal succession with a high import : storage ratio.

Oxygen Sag Curve

One of the first minimodels to be used in a practical way is the oxygen sag model for estimating the depletion and restoration of dissolved oxygen in water after an initial slug of organic matter is added, such as that from a sewage outfall. Measured in units called *biochemical oxygen demand* (BOD), the organic matter combines with oxygen in its biological oxidation by microbes that are usually ubiquitously present and adapt rapidly to utilize the potentials for consumption. The oxygen is used up rapidly but is partially being restored by reaeration (diffusion) in from atmosphere in proportion to the difference in partial pressures. The lower the oxygen level in the water, the more rapid the diffusion. The lower oxygen levels cause depression in the rates of oxidation, also. Ultimately, as the initial slug of organic matter is depleted, the oxygen level rises again as diffusion exceeds respiration.

Since the organic slug is added at a point along a stream or estuary, the "sag" curve in that batch with time also represents the sag with distance away from the source.

Figure 22-25a shows the simplest form of the oxygen sag model. Figure 22-25b shows a more complex version that includes plant production and some storage and autocatalysis in development in consumer stock. The second model should be compared with that in Fig. 21-5; it is a version of the basic P–R model. Wang et al. (1978) simulate a model of this type, comparing it with one by Rich (1973).

ARRESTED SUCCESSION

If structure that may be maintained depends on the conditions tending to disperse the system, the

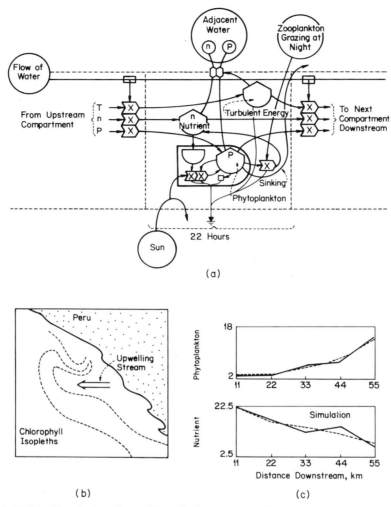

FIGURE 22-24. Simulation of upwelling plankton system off Peru (Walsh and Dugdale, 1970): (a) energy diagram of one of the chain of compartments in the model; (b) spatial location; (c) simulation (dashed line) and data along the current flow.

climax depends on both the energy sources and pumped outflows. When the development of structure at climax is limited by strong outflows, the succession is in an *arrested* state compared to what would develop without the drains. Eliminating such outflows often provides surprises as to ecosystems that develop after their protection.

For example, the irradiated (ionizing gamma rays) forest at Brookhaven developed a state of low diversity in steady state (Woodwell, 1967). Many pollution situations resemble early successional states. Removal of grazing animals allows forests to develop. Nixon (1969) found delicate spatial structures developing in brine microcosms when the usual pattern of increasing salinity was arrested by maintaining brine salinity constant. For any ecosystem, more structure can be developed by decreasing the disordering energy flows so that washout is less frequent.

Another example is the flushing of sewage ponds at varying turnover times (Oswald et al., 1953). When flushed every day, a steady state develops in which populations are arrested in rapid net growth stages in what would ordinarily be regarded as early succession. If ponds are flushed out every month, the systems are in a steady state with respiration as high as production and most energy in structural maintenance.

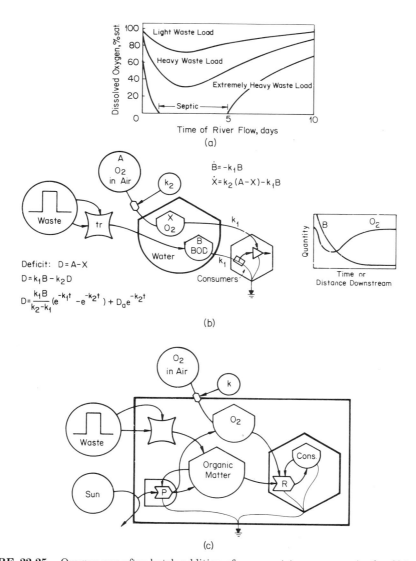

FIGURE 22-25. Oxygen sag after batch addition of sewage: (*a*) oxygen sag in the Ohio River from Whipple et al. (1927) quoted by Warren and Doudoroff (1971); (*b*) simple model (Streeter and Phelps, 1925); (*c*) more accurate model which is a variation of *P–R* model.

SUCCESSION IN AQUATIC MICROCOSMS

Modeling of succession in aquatic ecosystems is illustrated by examples in Fig. 22-26, where a sequence of photosynthesis of microbial processes followed annual load of bird droppings in tide pools in springtime. The system can be compared to sewage with successional change as storages diminish.

Figure 22-27 shows the classical hay infusion sequence described by Woodruff (1912). Other examples are given by Kurihare (1957) for bamboo microcosms with mosquitoes and by Maguire (1971) without mosquitoes. Similar successional patterns were given for marine fouling surfaces by Redfield and Deevey (1952). See other microcosms in Fig. 22-28.

Gnotobiotic Ecosystems

Gnotobiotic cultures are free of any organisms other than the one under culture. Considerable skill in handling is required to start and maintain such cultures available for experimental study. An equivalent culture at the level of the ecosystem contains a web of organisms that were each added from a pure

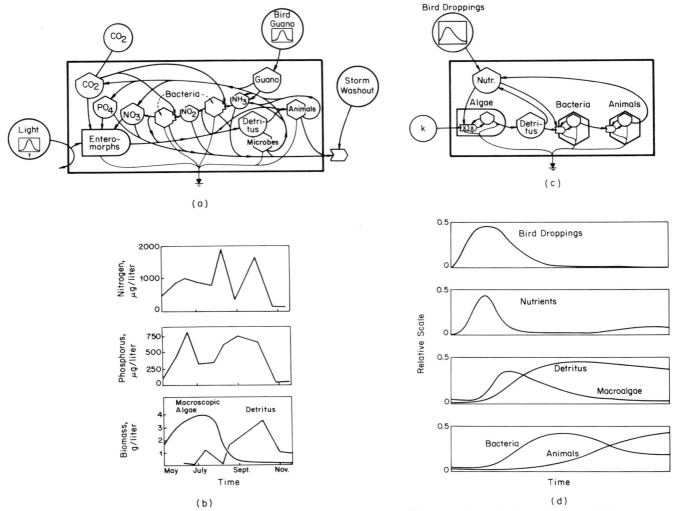

FIGURE 22-26. Tide pools in Sweden (Ganning and Wulff, 1969, 1970): (a) system; (b) graphs of succession; (c) simulation model (Wulff, 1970); (d) results of simulating model in part c.

culture or isolation; thus the culture is completely defined in regard to species with the only microorganisms being the ones added.

In Figs. 22-29 and 22-30 are two examples of gnotobiotic ecosystem experiments. In one, by Nixon (1969a), the species isolated from a brine ecosystem were recombined. In the other, by Taub (1968, 1969) and Taub and McKenzie (1973), species originating from a species culture bank and that did not already have working relationships were used. In both cases the larger consumer that was added became extinct after a period of growth, and the ultimate steady state was at a lower metabolic rate, with recycling not as good as growth on initial resources. Possible relationships are diagrammed. Larger components may require larger realms or external population immigration as needed.

ROLE OF BIOLOGICAL EVOLUTION

Self-organization utilizes multiple inputs of information from which the choices and selections are made by the system. New combinations and choices are generated by the system from which later selection is made. Many complex programs of information are generated and carried by the living organisms in which mechanisms of genetics have been part of developing and selecting information available from the past for guiding the choices and selection in the future. Models of genetic action were included in Fig. 20-20–Fig. 20-22. The sequence of changes in life that result from the eternal repetition of self-organization, self-maintenance, and survival of systems is called *evolution*. Succession is sometimes regarded as the patterns of life resulting

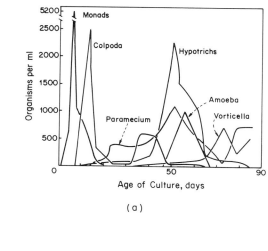

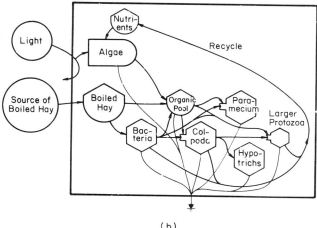

FIGURE 22-27. Succession in aquatic microcosms: (a) hay infusion succession [Woodruff (1912), Woodruff and Fine (1910) as quoted by McIntosh (1963)]; (b) model of components involved.

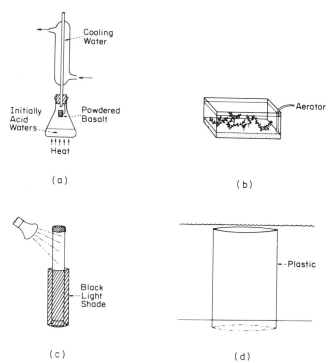

FIGURE 22-28. Microcosms used to study principles of ecosystem succession: (a) simulation of seawater development by Goldschmidt reaction (see Fig. 26-4e) (Odum and Abbott, 1972); (b) replicated aquarium microcosms used by Beyers (1962, 1963a, b) to study temperature adaptation and ecosystem adaptation after overgrazing; (c) tube model of plankton; (d) plastic cylinder within waters for study of diurnal metabolism (Hood, 1962) and effect of isolation (Lund, 1972) (see also Figs. 16-5, 19-18, 19-19, 21-6, 21-7, 21-8, 22-11, 22-18, 22-26, 22-27, 22-29, 22-30).

from the trajectories of evolution, but the reverse may be true that evolution is the life resulting from the trajectories of succession. Van Valen (1976) discusses the energy basis for evolution. See also Chapter 17.

The gnotobiotic studies in Figs. 22-29 and 22-30 show the action of self-organization using whatever species are available. The species survives if it is used by the ecosystem. Evolution is the sequence of changes that emerge from the continued repetition of system self-organization with species adaptation.

Lewontin (1966) describes genetic mechanisms and their immediate selection as a process of local action and short time period with no long-term memory. Evolution of larger organisms and long-term patterns are part of the larger scales of time and space considered in Chapter 26.

The very rapid selection and reconstitution of the tropical forest ecosystem at El Verde after γ irradiation is an example where genetic damage was observed to delay succession but only by part of a year as compared with a cut control (Odum, 1970) (see model of regenerative action in Fig. 22-31a, which is a special case of disorder system given in Fig. 17-10).

Figure 22-31b gives an example of the self-organizational process acting on the informational storages used in succession. Cooke et al. (1968) used ionizing radiation to disorder and change the information available to the microbes of a microecosystem. The effect was a delay in succession in which time was required to reorganize.

Biological survival in evolution requires a species to find one ecosystem after another with which it may be a successful part, changing as needed to make the system continuously competitive to ensure that it has continuity of existence as a carrier of ecosystem information from the past to the future.

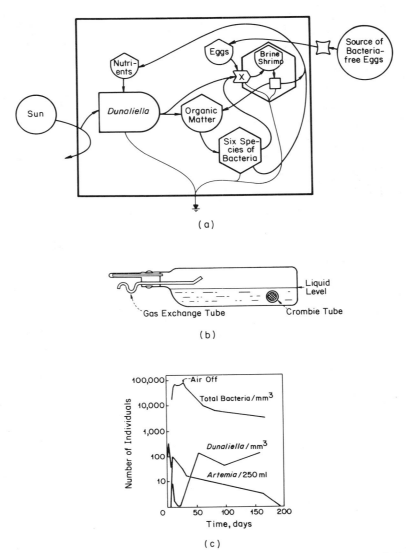

FIGURE 22-29. Defined resynthesis of brine ecosystem after components were isolated into pure cultures (Nixon, 1969a); (*a*) energy diagram; (*b*) microcosm; (*c*) succession graph.

Margalef (1961) correctly compared evolution to information in a channel being transmitted from past to the future. The ecosystem provides the channel.

McCormick and Platt (1962) and Murphy (1970) found disordering of radiation to be a delay and arresting agent in succession in terrestrial microcosms representing granite outcrop communities.

PLANKTON SUCCESSION

Fairly steady state plankton associations do occur, as shown by R. E. Coker and others for the Bahia Fosforecente in Puerto Rico before the bay was disturbed by industrial developments (Odum et al., 1959). See p. 445.

Succession in plankton communities in temperate and higher latitudes is rapid because of the small sizes and time constants so that it can track the changing seasonal variables while making some storages and progressions that represent increases after minimum states in winter.

Lassiter and Kearns (1974) studied the kinds of species substitution possible with a production model with nutrients, temperature, organic matter, and recycle and six species as competing alternatives with different properties (Fig. 22-32). After winter, there was an initial large bloom and net

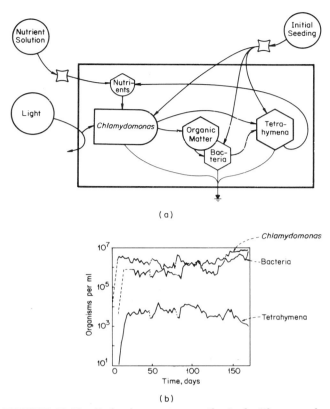

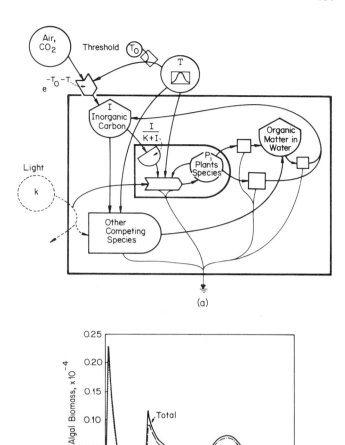

FIGURE 22-30. Defined ecosystem synthesized with pure cultures, including algae (*Chlamydomonas*), protozoa (*Tetrahymena*), and bacteria (Taub, 1968, 1969): (*a*) energy diagram; (*b*) observed curves.

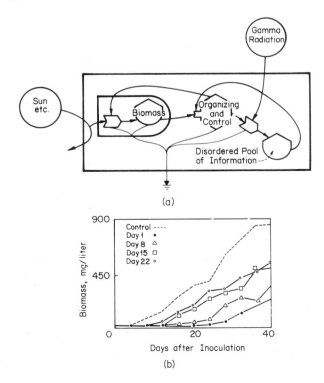

FIGURE 22-32. Model of plankton succession (Lassiter and Kearns, 1974): (*a*) energy diagram; (*b*) simulation that resembles seasonal patterns in temperate lakes.

gain based on high initial nutrients, which was replaced by species that could maximize production more under lower nutrient and recycle conditions at higher temperatures. This is like the diatom to blue–green sequence in many lakes and is comparable to sequence on land of net producing weeds followed by longer-lived associations of recycling types. Concepts of niche are hardly necessary as the web automatically brings different species into domi-

FIGURE 22-31. Genetic disruption and reconstitution: (*a*) model based on rainforest irradiation (Odum, 1970); (*b*) effect of damaging radiation on succession and recovery from bioregenerative systems (Cooke, 1967; Cooke, Beyers and Odum, 1968). Graphs of growth are given after varying periods of delay following irradiation.

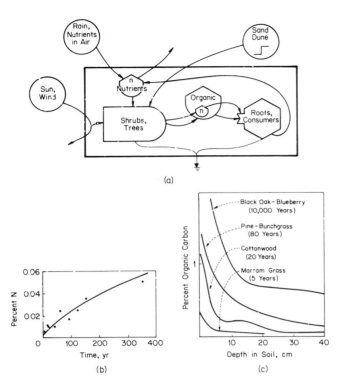

FIGURE 22-33. Examples of succession, forest on sand dunes of Great Lakes (Cowles, 1899; Olson, 1958): (*a*) suggested minimodel; (*b*, *c*) data on accumulations from Olson (1958).

nance, maximizing total power and retaining the other species as minor until conditions are favorable. Yet as the model shows, the system is organized by its recycle (also see Fig. 12-5).

SOIL AND TERRESTRIAL ECOSYSTEM MODELS

The accumulation of high-quality matter and structure by ecosystems on land is partly in the form of soil and partly in the vegetational structure of roots, boles, limbs, leaves, microbes, and animals. Early demonstrations of the timing of storages of organic matter buildup were given by Cowles (1899) and Olson (1958). Figure 22-33 gives a model summarizing the concepts involved in explanation. An early model by Jenny (1941) and simulated by Neel and Olson (1962) is given in Fig. 5-19.

The model in Fig. 22-34*a*, by its distribution of time constants, automatically transfers energy from developing the short-term rapidly charging autocatalytic loop first to the longer time cycle that develops structure more slowly, displacing the short-term successional species. Simulated by Burns, the

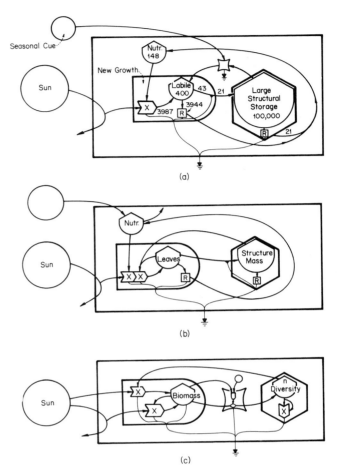

FIGURE 22-34. Ecosystem successional minimodels: (*a*) high-pass and low-pass filter (Burns, 1970); (*b*) fast cover and added productivity with development of long-range structure (Regan, 1977); (*c*) excess production into diversity with added productivity (Odum and Peterson, 1972).

model reproduces some overall characteristics of tropical forest energy use.

In Fig. 22-34*b* a fast time constant storage covers the ground with a leaf cover in three years, after which longer-lasting structure develops and gradually replaces the early successional storage. A more elaborate successional model on this principle was developed by Regan (1977) and evaluated with data from Florida (Fig. 22-34*b*).

In the model shown in Fig. 22-34*c* energy is transferred to diversity after initial production is established. Gross productivity increases through the extra energy tapped by the diversity.

The model of ecosystem development shown in Fig. 22-35 includes early development of plant cover and nutrient storage followed by later development of vertical structure and differentiation of sun and

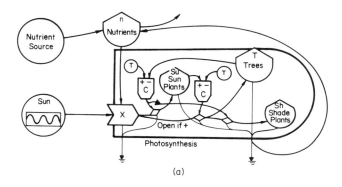

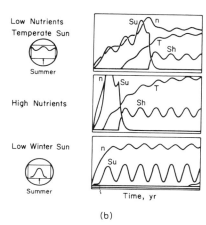

FIGURE 22-35. Minimodel of succession and results of simulation with solar radiation patterns: (*a*) energy diagram; (*b*) simulation for temperate, eutrophic, and arctic conditions (Odum, 1976a).

shade forms. Substitutions are controlled by switches that represent the kind of action that sometimes control species substitutions. This model, when simulated with only short seasonal periods of sunlight, fails to develop a vertical structure and resembles tundra. Kangas (1981) simulated succession on disturbed land after phosphate strip mining, using a simple model with substitution of plants with tall ones shading the short ones and litter controlling germination (see Fig. 22-36).

Data on terrestrial succession at Hubbard Brook are given in Fig. 22-37. Summary diagrams are given interpreting principal processes as described.

P–R Models with High- and Low-Pass Filters

Many ecosystems have large storages of organic matter in living and dead biomass of trunks and soil organic matter. These storages respond to long-term pulses, whereas small labile storages such as those in leaves can respond to rapid pulses so as to utilize the pulses for energy. The fast and slow response abilities of the real forest are included by having a pair of storages, one rapid and one slow time constant. For example, Fig. 22-34*a* shows a minimodel used to represent a tropical forest in overall aspect by having one rapid pathway and one slow one, thus allowing energy pulses of more than one frequency to be used (see also p. 216 and p. 232).

Models of Combined Tree Populations

Translated and presented in energy language form, the model of forest tree composition shown in Fig. 22-38*a* was adapted from one by Botkin et al. (1972a,b) and Shugart et al. (1977) [see review by Shugart and West (1980)]. The approach here is to iterate year by year the composite behavior of the individual trees. At each iteration, increments of growth or seeding are added to each individual as known from species studies. The known characteristics of major dominant tree species in their growth and mortality responses to the environment conditions, including their own shading are calculated for each year, their interaction with light shading effects, nutrients, and other competition calculated, and the new state of the populations indicated.

This is a moderate-sized unit model, but when iterated for many species and age classes, becomes a very large model. It was run for simulated times of hundreds of years to allow study of the consequences of the concepts held by the modelers. Since no observed records span such a long period, and because there are natural climatic excursions beyond that included in these long runs, there is no way to check the long-range characteristics, but the shorter-range results produce the kind of species growth and substitution observed. The individual relationships are the kind already covered in other chapters. Emmanuel et al. (1978) found an oscillation pattern as canopy opened in oscillatory cycle followed by regrowth of new recruitment (see Fig. 22-38*b*). Time series analysis showed frequency peaks at 50, 100, and 250 years. Removal of chestnut caused a large oscillation simulating the chestnut blight action 50 years ago.

Choices are provided by the species with their individual, genetically controlled characteristics, but the control loops of total energy constraint in shading by tall tree leaf area and biomass crowding provide reward loops and selection.

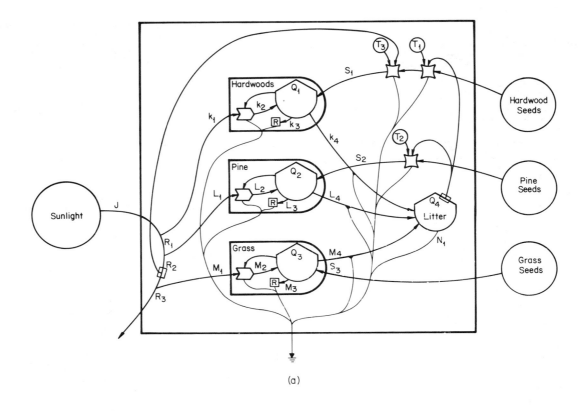

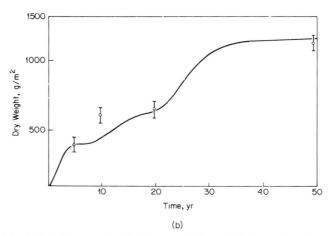

FIGURE 22-36. Simulation model of old-field succession in the southeastern U.S. Q_1 = hardwood biomass (Kangas, 1982): (a) model driven by shading effects of different stages of succession and by switching action of germination of seed inputs as a function of litter and light levels; (b) simulation and data.

The model translates data on tree species such as growth, age, and mortality into stand characteristics. The data have hidden relationships about adaptation such as the energy required for longevity and insect relations. If oscillatory characteristics maximize power and self-organization for survival, the observed properties of the species are those already selected for by higher overall production accompanying that composition of species and characteristics. The model is attractive to managers as it may indicate some effects of species planting, trimming, cutting, and so on.

What may be missing are other regulatory feedbacks from stand to recruitment and availability of alternative species so that new designs and species associations can organize for new conditions.

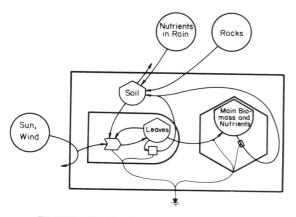 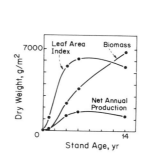 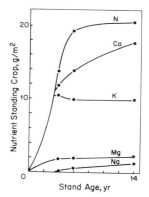

FIGURE 22-37. Succession following forest cutting at Hubbard Brook, New Hampshire (Marks and Bormann, 1972).

Grassland Model

Gutierrez and Fey (1980) provide a grassland succession model developed by cascading many coefficients in the Dynamo tradition. The energy diagram in Fig. 22-39a shows production and consumption of the nutrients and a complex production function with stages for vigor and diversity, both feeding back to augment production.

There are several switches M that select minima of alternate inputs, such as selecting the flow needed for maintenance or the minimum flow available, whichever is smaller.

A representative simulation is given in Fig. 22-39b showing succession followed by a steady state, sometimes an oscillating one as shown. Most of the grassland models of the International Biological Program were more complex (Innis, 1978).

LANDFORM AS PART OF ECOSYSTEM STORAGE

Succession of ecosystems usually involves long-term storages of structure and information in the form of the land, in basins, floodplain deposits, peat accumulation, networks of tributaries or distributaries, and so on. These are among the features requiring the most embodied energy, converging work in space and over long periods of time. An example of landform development as part of long-range succession is given in Fig. 22-40 of a cypress swamp basin pond in karst landscape where acid percolation develops the basin. Calculations (Odum, 1977) suggest times of approximately 4550 years to develop the swamp pond basin one meter deep. Kangas (1982) evaluated models for energy basis of combined ecosystems–landforms. See simulation by Gilliland (1976).

Ultimately, ecological succession is controlled by the larger systems of the biosphere. For example, Randerson (1979), using dynamics symbols, simulates marsh growth and sediment capture with limitation by emergence beyond the level of tidal submergence. See Doornkamp and Cuchlaine (1971). Peat structure is a main feature of bog models by Clymo (1978) and Jones and Gore (1978).

SUMMARY

In this chapter changes of system structure with time are considered under the general heading of succession. Succession includes both growth and retrogression, organizational changes, new self-design and repeating programs of adaptation to recurring environmental change. Early succession is dominated by trends driven by initial storages, but later succession adapts to the renewable or recurring sources. Every successional pattern is part of a larger steady state when longer time periods or larger spaces are considered. Similarly, every steady state is made up of component items in succession in space or in time. In its broadest senses, succession includes the return of equilibria from displacement but more often concerns the pattern of structure and operation of open systems as they develop structure to maximize resources and continuation of patterns. Patterns of succession in flowing streams develop series in space called longitudinal succession that resemble sequences in time.

Many of the main features of succession can be understood with overall minimodels such as the $P-$

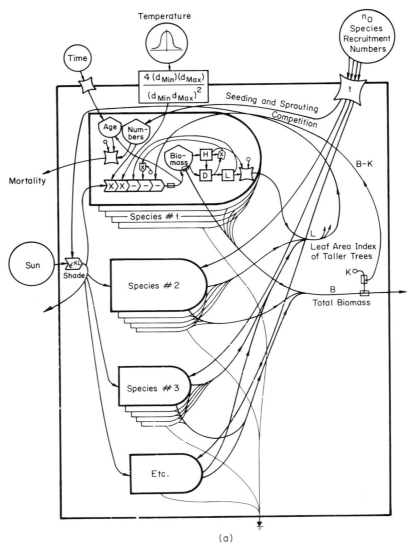

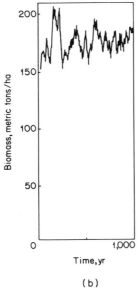

FIGURE 22-38. Forest succession model of type given by Botkin et al. (1972a, b) and Shugart and West (1977); diagrammed from Emanuel et al. (1978) (D, diameter; H, height; L, leaf area index): (a) energy diagram of some principal features; (b) simulation of Appalachian forest develops small pulses.

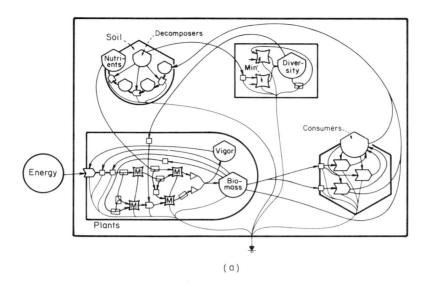

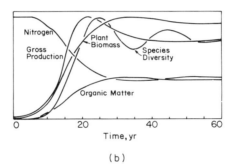

FIGURE 22-39. Simulation model of grassland (Gutierrez and Fey, 1980): (a) energy diagram representation of original Dynamo program; control boxes with M select one of two pathways that is least; small squares are graphical or tabular functions in the program; (b) one of the simulations.

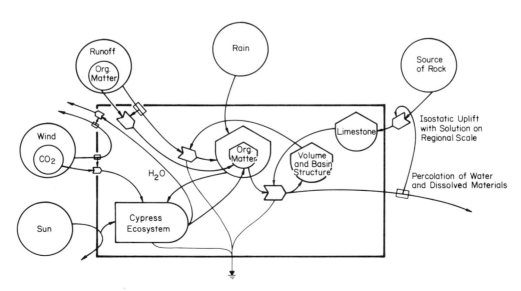

FIGURE 22-40. Model of cypress swamp including ecosystem work in generating its own basin through acid percolation.

R models, but more complex models that keep track of individual trees evaluated according to observed data on their performance generate similarities. Fluctuations and oscillations are more general than very steady patterns, but the repeating patterns may be good steady states and climaxes when viewed over longer periods and over larger spaces.

Many of the best perspectives on succession are provided by study of microcosms, where holistic measurements have been possible. Successional priorities are set by relative effect in utilization of resources to generate more structure for gaining more resources. In order of priority and first appearance comes adaptation to physiological requirements, growth of mass, development of storages, exchange with external systems, diversification, and temporal programs.

QUESTIONS AND PROBLEMS

1. What is the classical concept of ecosystem succession? Consider the hierarchical organization of nature and explain the way a steady overall pattern can result from combination of pulsing members, each growing and being replaced. Consider how a long-range steady pattern can be made up of regularly or irregularly varying patterns. Explain how a repeating pattern, when examined over a short period, would be judged to be in growth, climax, or regression.

2. Using minimodels (see also Chapter 11), diagram and describe simple mechanisms for some common features of succession, including increase in gross production, net production maximized during only part of succession, overshoot with some initial conditions, late development of diversity, increased cycling, increased feedback of metabolic work to maximize power, and substitution of species specialists.

3. What is a temporary, frenzied catastrophic consumer pulse system and its role in a long-range repeating climax? On what level of perspective is this regarded as a disaster? Give reasons from maximum power and energy quality theory for pulsing of higher consumer controls that cause the whole system to pulse. Give several mechanisms (minimodels) for pulsing (see also Chapter 11), and give real-world examples.

4. Give examples of microcosm studies of succession in various types of environment; illustrate with minimodels of the performance observed.

5. What is a gnotobiotic microcosm? How are species and their genetic information a driving force? Discuss their embodied energy and energy quality. How does import–export exchange of genetic information affect stability of models with oscillatory tendencies?

6. What are circular successional patterns? Compare changes in predominant stocks in circular pattern with pathways on a phase plane graph used for population studies.

7. How is oscillation in time related to banding in space? Discuss attributes affecting survival of alternatives of oscillatory type of order and stochastic type of variation. Consider stability, maximum power, and ease of long-range programming.

8. How is biomass a limiting factor? Discuss the energy advantages and disadvantages of storages and the circumstances that make a large storage adaptive.

9. What are the energy requirements for maintaining information as high-quality control components? Compare energy for generating new information, energy for replication, energy for correction of error, energy for new adaptation, and energy of spatial and temporal transmission. Give examples from human society and its written and educational informational storages. Give examples from species formation, extinction, and dispersal in ecosystems. Give examples from the biosphere and its geological memories.

10. Describe the P/R ratio with various initial and steady-state patterns, including streams, brines, tide pools, waste disposal, and stored waters.

12. What is longitudinal succession? Give examples. Try diagramming a sequence of models connected by the longitudinal flow.

13. Discuss adaptations to strong physical and chemical stress and dispersal energies. How are ecosystems adapted to use such energies toward stability, maximum power, survival, and reduction of the energies of maintenance and defense?

14. Consider both interunit and intraunit diversity as a means for adaptation to energy signature and the relative priorities.
15. What are some historical demonstrations of ecosystem change usually given in teaching about succession?

SUGGESTED READINGS

Ashby, M. (1963). *Introduction to Plant Ecology*, Macmillan, London.

Barnett, G. W. and R. Rosenberg (1981). *Stress Effects on Natural Ecosystems*, Wiley, New York.

Bazzaz, F. A. (1979). The physiological ecology of plant succession. *Ann. Rev. Ecol. and Syst.* 10, 351–371.

Bormann, F. H. and G. E. Likens (1979). *Pattern and Process in a Forested Ecosystem*, Springer-Verlag, New York. 253 pp.

Botkin, D. D. and R. S. Miller (1974). Complex ecosystem models and Prediction. *Amer. Scientist* 62, 757–768.

Bradshaw, A. D. and M. J. Chadwick (1980). The restoration of land. Univ. of California Press, Berkeley, CA. 317 pp.

Brody, S. (1945). *Bioenergetics and Growth*, Reinhold, New York.

Cairns, J. (1980). *The Recovery Process in Damaged Ecosystems*, Ann Arbor Science, Ann Arbor, MI.

Canale, R. P., Ed. (1976). Modeling Biochemical processes in Aquatic ecosystems, Ann Arbor Science, Ann Arbor MI. 387 pp.

Cody, M. L. and J. M. Diamond (1975). *Ecology and Evolution of Communities*. Belknap Press, Harvard University, Cambridge, MA. 545 pp.

Etherington, J. R. (1975). *Environment and Plant Ecology*, Wiley, New York.

Giesy, J. P. Ed. (1980). *Microcosms in Ecological Research*, CONF-781101, Symposium Series No. 52., Department of Energy.

Gorham, E., P. M. Vitousek, and W. R. Reiners (1979). The regulation of chemical budgets over the course of terrestrial ecosystem succession. *Ann. Rev. Ecol. Syst.* 10, 53–84.

Grime, J. P. (1979). *Plant Strategies and Vegetation Processes*, Wiley, New York. 222 pp.

Holdgate, M. W. and M. J. Woodman (1978). *The Breakdown and Restoration of Ecosystems*, Plenum Press, New York. 496 pp.

Horn, H. S. (1974). The ecology of secondary succession. *Ann. Rev. Ecol. and Syst.* 5, 25–37.

Lassiter, R. R. (1979). Microcosms as ecosystems for testing ecological models. *Ecol. Modelling* 7, 127–162.

Lugo, A. E. and S. C. Snedaker (1971). *Readings on Ecological Systems*. MSS Educational Publishing Co., New York.

McIntosh, R. P. (1980). The relationship between succession and the recovery process in ecosystems. In *The Recovery Process in Damaged Ecosystems*, J. Cairns, Ed., Ann Arbor Science, Ann Arbor, MI. pp. 11–62.

Miles, J. (1979). *Vegetation Dynamics*, Chapman and Hall, London. 80 pp.

Nikolskei, G. V. (1969). *Theory of Fish Population Dynamics*, Oliver and Boyd, Edinburgh. 323 pp.

Odum, E. P. (1969). The strategy of ecosystem development. *Science* 164, 262.

Odum, E. P. (1971). *Fundamentals of Ecology*, Saunders, Philadelphia.

Risser, P. G., E. C. Birney, H. D. Blocker, S. W. May, W. J. Parton, and J. A. Wiens (1981). *The True Prairie Ecosystem*, Academic Press, New York. 544 pp.

Rich, L. G. (1973). *Environmental Systems Engineering*, McGraw-Hill, New York.

Russell, G. S., Ed. (1975). *Ecological Modelling*, Resources for the Future, Washington, DC.

Scavia, D. and A. Robertson (1979). *Lake Ecosystem Modeling*, Ann Arbor Science, Ann Arbor, MI.

Swartzman, G. L. and G. M. Van Dyne (1972). An ecologically based simulation optimization approach to natural resource planning. *Ann. Rev. Ecol. Syst.* 3, 347–398.

Swift, M. J., O. W. Heal, and J. M. Anderson (1979). *Decomposition in Terrestrial Ecosystems*, University of California Press, Los Angeles.

Taub, R. B. (1974). Closed ecological systems. *Ann. Rev. of Ecol. and Systematics* 5, 139–160.

Trofymow, J. A., J. Gurnsey, and D. C. Coleman (1980). A gnotobiotic plant microcosm. In *Plant and Soil* 55, 167–170.

Udvardy, M. D. F. (1969). *Dynamic Zoogeography*, Van Nostrand Reinhold, New York. 445 pp.

Van Dobben, W. H. and R. H. Lowe-McConnell (1975). *Unifying Concepts in Ecology*, W. Junk B. V. Publishers, Hague, Netherlands. 300 pp.

Velz, C. J. (1970). *Applied Stream Sanitation*, Wiley-Interscience, New York. 618 pp.

Warren, C. E. and P. Doudoroff (1971). *Biology and Water Pollution Control*, Saunders, Philadelphia.

West, D. C., H. T. Shugart, and D. B. Botkin (1981). *Forest Succession*, Springer-Verlag, New York. 516 pp.

Wiegert, R. E. (1976). *Ecological energetics. Benchmark Papers in Ecology 4*, Wiley, New York.

CHAPTER TWENTY-THREE

Economic Systems and the Nation

Where humans are part of systems, there are economic transactions and flows of money. Economic behavior of human beings causes money, a symbolic form of information, to flow in countercurrents to the flow of commodities bearing energy. Because of commonalities among systems, there are many similarities between systems of economics and those of ecology. In this chapter some of the mechanisms of economics and environment are combined in energy language, keeping money and the wealth it represents separate but coupled to energy flows with price mechanisms. The relationships between money and energy are only beginning to be understood. Comparing and unifying these fields is currently a fertile scientific frontier (Odum, 1971; Hannon, 1973a,b; Nijkamp, 1977; Rapport and Turner, 1977; Hirshleifer, 1977; Slesser, 1978; Fluck and Baird, 1980; Bernstein, 1981).

MONEY CIRCULATION

As illustrated in the aggregated model of the United States in Fig. 23-1, the money countercurrent forms closed loops where pathways are controlled by human beings but does not flow along the pathways that are part of the nonhuman components of the life-support system of the biosphere. Circulation of money keeps a system better adapted to maximize power than without the economic pathways and programs. Selection for maximum power may lead to optimum rate of circulation of money and maintains power-yielding behavioral mechanisms.

Externalities

Figure 23-2 gives some simple configurations of money and energy. Whereas the money is in closed loops, the energy comes in through a source and goes out in used, degraded form through the heat sinks. The basis for the buying power of the money is the useful flow of energy. The amount of work the money will buy depends on the energy flowing from the externalities. The way externalities generate values that are successively used and transformed in centers of human life was appropriately diagrammed by Collens (1876).

Simple Model of Money

Given in energy language in Fig. 23-2a is the simple model for the relation of money flow to energy flow as given in economics texts, except that emphasis is placed here on the ultimate source of work from the external source.

Rate of Spending

In general, money is spent in proportion to its storage. In common language, one spends money as rapidly as one receives it. If one saves it in a bank, it is being spent for that person and maintained in circulation. In the United States money circulates about four times per year in comparison to its storage at any given time. If there is steady state, capital dollar storages M_1 and M_2 are constant and flow $J_1 = J_2$.

Inflation

If money is added and energy input is unchanged, as in Fig. 23-2b, the flows of energy per dollar decrease; this is called *inflation*.

Economic Systems and the Nation

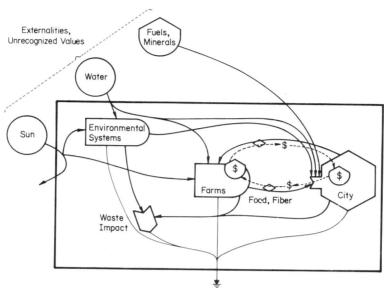

FIGURE 23-1. Systems of humanity and nature showing relation of circulation of money to externalities that ultimately determine value of the money, renewable resources, nonrenewable resources, and environmental impacts.

Inflation as Growth Accelerator

Behavior that feeds money back into new investments more rapidly accelerates production and growth when there are unutilized energy flows to permit expansion. Behavior that creates money such as borrowing is reinforced during periods when growth can occur. Thus the money supply M may grow if real assets also grow.

Inflation Due to Decreasing Energy Flow

If the flow of energy into a system is retarded, for example, by being rationed from the source, and if the money circulating is the same, the ratio of money flow to energy flow is reduced and there is inflation.

Macroscopic Model Determining Prices

If one has a macroscopic minimodel of an entire economy and can model the energy flow and the money flow, the ratio of the two is the price as shown in Fig. 23-2c. Since each flow is dependent on recycle from the others, the model is self-regulating.

Transaction Heat Sink

There is work in a transaction involving complex human thinking, passage of paper, accounting, transmission of information, and so on. The energy used in the transaction is the energy cost of doing business. The transaction symbol shown in Fig. 23-3a has a heat sink where the energy flow is explicitly separated and evaluated. Figure 23-3b, however, omits the heat sink, and according to the convention of the notation, the energy used in doing the transaction is implied to be part of the heat sinks at either end or both.

Notation for Prices

Where the transaction is believed to generate price as in the model in Fig. 23-2c, price is indicated by a barbed pathway flowing from the transaction symbol (Fig. 23-3c). However, where the price is determined from outside the transaction and imposed on it, the price is indicated on the pathway by a barb showing direction of control of the pathway as in Fig. 23-3d.

Model of an Economic Sector

For each unit in an economic system flows, storages, and concepts may be identified and related with the help of the energy diagram in Figs. 6-24 and 23-4. Money saved generates capital that is reinvested, generating capital assets. The energy basis includes externalities of fuel and renewable inflows plus the

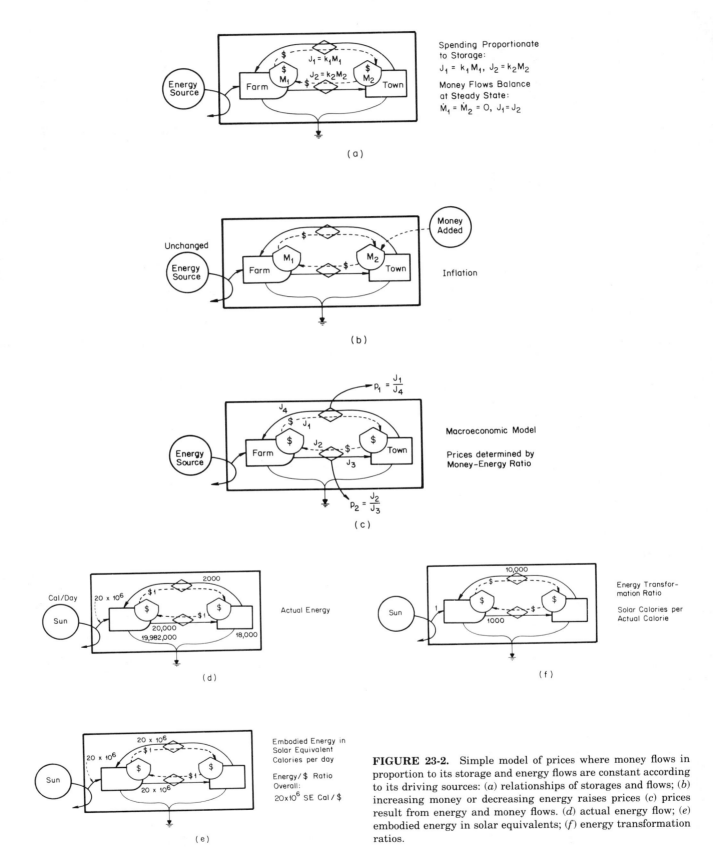

FIGURE 23-2. Simple model of prices where money flows in proportion to its storage and energy flows are constant according to its driving sources: (a) relationships of storages and flows; (b) increasing money or decreasing energy raises prices (c) prices result from energy and money flows. (d) actual energy flow; (e) embodied energy in solar equivalents; (f) energy transformation ratios.

Economic Systems and the Nation

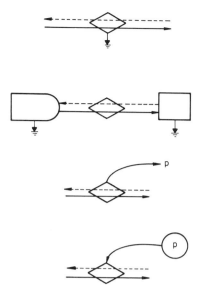

FIGURE 23-3. Economic notations in energy language: (a) energy used in business transaction isolated; (b) energy used in transaction not isolated but remains in heat sinks at the ends of the exchange; (c) price generated as the ratio of money flow to energy flow; (d) price generated externally and controls the transaction.

high-quality energies that are purchased with income generated by the production.

Some have claimed that energy flows and economic flows are alternative analyses giving the same result. As shown in Fig. 23-1 and 23-4, some pathways have dollar flow and energy flow values that bear constant relation to each other in some situations, but many pathways are not represented by money, especially the sources, sinks, and life-support system. Figure 23-5 shows that money is not exactly proportional to any other energy flows and that a full understanding requires consideration of both.

Balance of Money Flows

In Fig. 23-2a the flow of money from town to farm must equal money from farm to town at steady state. A balance of payments is required. Money must be kept moving to maintain maximum flow of energy possible. Accumulation of money at one place increases the price when it is used to purchase and decreases the prices in its return pathway until the flows are again balanced and not accumulating dollars at any point. Where money is accumulated, its buying power is less. There is a local inflation.

ENERGY RELATION TO MONEY

The simple model in Fig. 23-2d is shown with actual calories dispersing into degraded heat with successive transformations. The embodied solar energy equivalents are given in Fig. 23-2e. The energy transformation ratio in Fig. 23-2f is the ratio of embodied energy in Fig. 23-2e to actual energy in Fig.

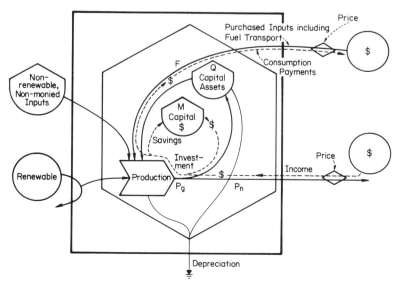

FIGURE 23-4. Relationship between production, capital, capital assets, fuel, renewable environmental energies, income, consumption, savings and investment. Profit is rate of increase of M; P_g is gross production; P_n is net production (one kind); and $(P_n - F)$ is net production net energy if in embodied energy of one type).

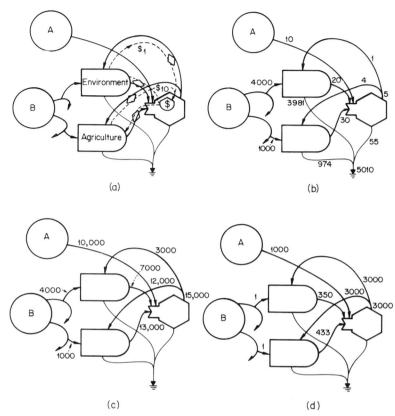

FIGURE 23-5. Web with flows of money and energy; (a) money flow; (b) actual energy flow in calories per time; (c) embodied energy in units of energy type B; calculated by using energy transformation ratio for relating type A to B, $F_{A/B} = 1000$; (d) energy transformation ratios; number in part c divided by those in b.

23-2d. Notice that the ratio of money flowing to actual calories flowing increases as one passes around the energy chain. Embodied energy flow and money flow are both constant around the loop in this simple case. The money flow is just as important to the left as to the right, recognizing the symbiosis of the two arms of the loop. In Fig. 14-3 we indicated that a closed loop with equal embodied energy was equal value in the sense of energy required.

The flow of money is an indication of the evaluation of the effect of the energy flow by the unit receiving the energy. Whereas embodied energy measures energy required, money spent tends to measure energy effect to the extent that the money flows are free to adjust as in the free market. In Fig. 23-2e the overall ratio of energy flow to the dollar flow through the final sector (consumer on the right) is 20 million solar equivalent calories per dollar. The dollar : embodied energy ratio is useful in practice to calculate embodied energy in feedbacks.

The energy transformation ratios in Fig. 23-3f are a measure of the concentration of embodied energy and are an intensive measure of energy quality (Fig. 14-1).

Money in a Web

Figure 23-5 shows a more complex system, with two sources and money circulating as a counter current in more than one closed loop. Source A is a high-quality externality such as oil with high embodied energy per actual calorie, but because it is concentrated and free, little money is paid to anyone to get it. The upper loop is priced low because it is perceived to be less limiting than the lower one. Circulation of money also flows more in one loop than the other because more of the human controlled services are involved in one than the other.

The embodied energy within the closed loops are different because the feedback from the consumer sector is divided into two parts, with more embodied energy going to the lower sector than the upper one. (See distinction between division of an energy flow

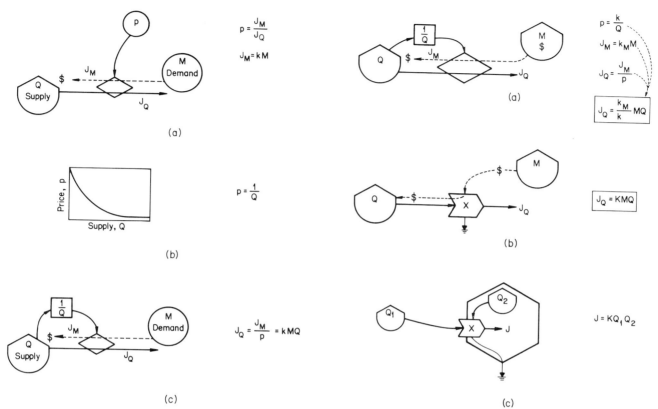

FIGURE 23-6. Flows of energy (commodities carry embodied energy) as a function of demand depends on price function (demand is fixed): (a) constant fixed price; (b) graphs of price function used in part c; (c) price inverse to supply.

FIGURE 23-7. A model of demand J_M dependant on money with price inverse to supply as a product function: a diagram and equations; (b) alternative notation; (c) analogy with predator demand for prey without money.

of one quality and two by-products of flows as explained in Fig. 14-11.) Feedback energy is proportioned by money spent for it.

Energy : Money Ratio

As shown in both Figs. 23-2 and 23-5, the ratio of actual energy to dollar flow decreases as one passes from source to consumer services along the pathways of the energy web. There is little wonder that the public has trouble understanding the energy basis for economics when the actual energy decreases as the accompanying dollar value increases.

The ratio of embodied energy (expressed in equivalent calories of one type) to dollars also varies within the web depending on the proportion of embodied energy entering directly as an externality and the proportion contributed as a feedback from the high-quality consumers. The embodied energy ratio is often highest where a rich resource is being processed into the economy.

Although the energy:dollar ratios vary along the pathways of the web, the overall ratio of total embodied energy to dollars circulating through the most valuable consumer sector provides a useful index of the economy as a whole and a means for evaluation of the feedbacks of high-quality goods and services from the terminal sectors. In Fig. 23-5 the overall embodied energy ratio to dollars is 15,000 calories of energy of quality B divided by $11 circulating through the final sector. The embodied energy:dollar ratio, therefore, is 1365 B cal/$.

The distribution of energy:dollar ratios is a useful parameter for relating the perceived value of a flow and the real contribution of energy to the economy by that flow. See, for example, the study of the energy:dollar ratios in Götland (Jansson and Zucchetto, 1978). The system of evaluation of energy and embodied energy just considered involves the decision to assign within closed loops the embodied energy of all flows contributing to that loop, a procedure already discussed in Chapter 14, where some other concepts of embodied energy were also described (Fig. 14-17).

DEMAND AND PRICE

The flow of money directed toward the purchase of an energy flow is the demand (see flow J_m in Fig. 23-6). Consider several models for the way energy flows in response to demand in systems controlled by economic mechanisms. In Fig. 23-6a the price is externally fixed and the flow of money of demand generates a flow of energy from supply J_Q. The quantity of supply in this case has no effect on the flow since the price is fixed until supply runs out.

In Fig. 23-6b and Fig. 23-6c, the price is inverse to the supply. As supply decreases, the price is adjusted upward according to a hyperbola that is often observed in price–supply curves.

Demand Dependent on Money; Product Function

In Fig. 23-6c and 23-7 the demand is a function of the supply of money held by the purchaser. In Fig. 23-7a demand is shown in proportion to local money available; price is inverse to supply Q. The resulting equation is a product of money and supply. In some models this relationship has been abbreviated with the equivalent notation in Fig. 23-7b.

For comparison, examine the prey–predator interaction model in Fig. 23-7c, which is also a product function. In a sense a predator effort to obtain food is a demand except its work is directed instead of its money (see Fig. 11-17).

Demand Inverse to Supply

Another model of flow in response to demand is given in Fig. 23-8. Here demand by the purchaser is inverse to the perceived supply. For example, behavior of the buyer might be to increase buying when supply is getting scarce, anticipating later market manipulations. As in Fig. 23-7, price is inverse to supply. The resulting equation is in proportion to the money because the effects of supply cancel. Lower prices cost less but attract more spending. In abbreviated notation this model is given in Fig. 23-8b.

Demand Proportional to Supply but Inverse to Price

If, however, demand is proportional to supply as in the case of buying when items are abundant and is

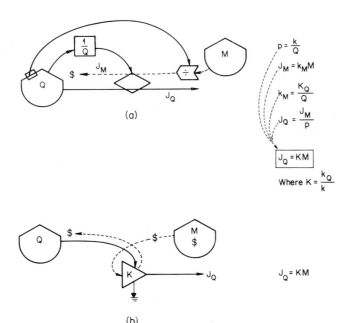

FIGURE 23-8. A model with spending inverse to supply and price inverse to supply, causing cancellation of supply effects: (a) model and equation; (b) alternative notation.

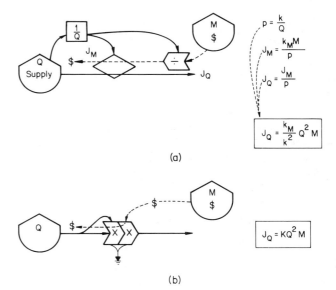

FIGURE 23-9. Demand inverse to price and price inverse to supply: (a) model and equations; (b) alternative notation.

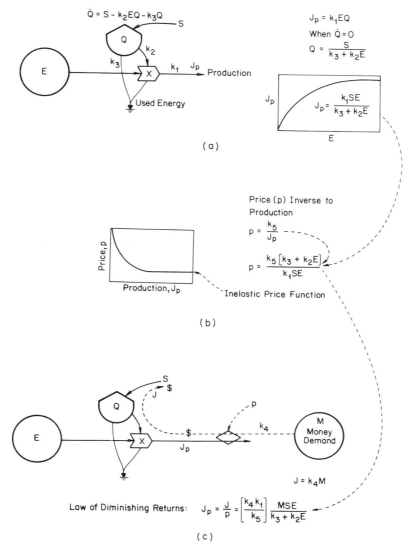

FIGURE 23-10. Model for the law of diminishing returns based on price inverse to production and production based on limiting supply S to an interactive process (this model has inelastic price relation to supply): (a) without money; (b) graph of inelastic price function; (c) model with money and law of diminishing returns.

inverse to price, a quadratic equation results (Fig. 23-9). Two of these relationships were used to simulate a producer–consumer economy by Bayley and Odum (1976), who found the model to be very stable.

Alternative notation is given in Fig. 23-9b. Alternative notations that are more nearly straight translations of equations are easier for the mathematical oriented person to visualize, but the notation that shows how the prices and spending are controlled gives more specific information about mechanisms.

Price with Limited Response; Elasticity

Prices that respond sharply to supply as in Fig. 23-6b are said to be *elastic*. Other prices respond less to supply and are inelastic. Figure 23-10 gives an energy design in which supply is the product of E and Q, where Q is supplied by the limiting flow S. The flow of the production of supply has the form of a limiting factor as indicated by the derivation given of production (see more information on limiting factors without money in Fig. 8-10). In Fig. 23-10 price

484   SYSTEMS OF NATURE AND HUMANITY

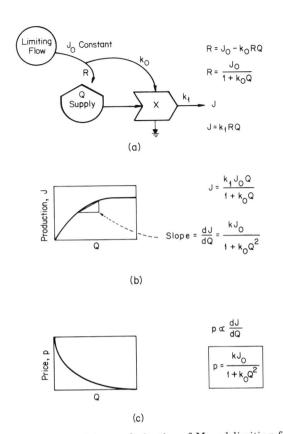

FIGURE 23-11. Price as derivative of Monod limiting factor function used to model economic law of diminishing returns (a) Limiting-factor concept; (b) slope is derivative; (c) price proportional to slope as the marginal utility.

is inverse to the supply, and since the supply has a limiting factor curve, the price curve is the same in inverse form. It levels off at a price well above zero. The price is less elastic.

With this price function, and with demand proportional to money available, the productivity equation is the same as the limiting-factor curve, except multiplied by money supply M.

LAW OF DIMINISHING RETURNS, MARGINAL EFFECT

The limiting factor curve given in Fig. 23-10a and in Chapter 8 is called the *law of diminishing returns* in economics. When quantity Q is small, the effect on production is large, but as Q increases, the effect diminishes. In Chapter 8 a Michaelis–Menten equation was obtained similar to that in Fig. 23-10, relating production to quantity where something was

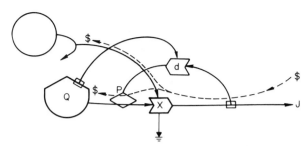

FIGURE 23-12. Effect of price p in stimulating money flow toward quantity Q in short supply. Price p is the derivative d of production J with respect to supply Q.

supplied at a limiting rate or recycling at a limiting rate. Others have used an exponential equation (Fig. 19-6). Different mathematical functions have been used in economics for the same basic process (see Chapter 8 for its use outside economics).

In Fig. 23-11 the amplifier action of quantity Q on production is the slope of the limiting-factor curve; it is a derivative (see equation in Fig. 23-11b, which is a reciprocal quadratic). See also Fig. 10-5. In economics this slope has traditionally been called the *marginal* effect, measuring the effect of the last bit of quantity Q added. Figure 23-11b shows one of a number of models that can generate more productive flow as supply Q is varied (see derivations for this function in Fig. 8-10).

Marginal Utility Theory of Price

Where behavior becomes programmed to value that which has the most effect on useful production J, prices are inverse to the slope. In other words, prices would rise as quantity Q becomes scarce, becomes more limiting, is having greater effect per unit, and thus is perceived as more valuable. The marginal utility and prices rise with scarcity. In Fig. 23-11c price is made proportional to the slope (the derivative) of the limiting-factor curve according to the theory of marginal utility. The derivative curve with supply superficially resembles the simple inverse in Fig. 23-6, but its effect is sharper because of the quadratic term in the denominator.

Flow of Money To Alleviate Shortage

When supply is limited, the marginal utility (derivative) increases. If the price is set according to the

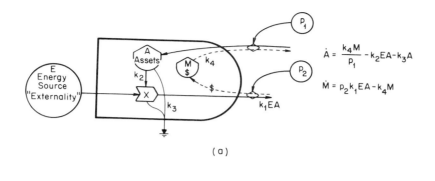

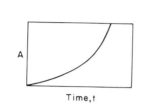

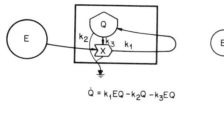 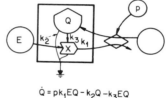

FIGURE 23-13. Exponential growth of economic production when energy sources are unlimited (constant force) and prices constant: (*a*) model; (*b*) growth; (*c*) simpler versions that also have exponential growth.

utility, it increases, causing more money to flow along the pathway that is in short supply. These relationships are diagrammed in Fig. 23-12. If people buy more when supply is large, their action will generate a demand proportional to marginal utility.

Linear Programming

The automatic mechansisms for eliminating shortages generates a web that tends to maximize production is a free market economy. A procedure for allocating commodities among more than one production process is linear programming. See Fig. 12-4.

SOURCES, PRODUCTION AND GROWTH

Exponential Growth

As long as externality effects are constant, economic models generate exponential or faster growth. A constant force from outside the economy means an unlimited energy source (see Figs. 9-3 and 9-11).

Kinetics of growth of economic production are given in Fig. 23-13, with equations as given in previous diagrams. If the time constant of money is small, it tends to develop a steady-state balance between income and outgo of money. Combining the equations produces a differential equation recognizable as similar to two tank web autocatalytic given in Fig. 10-8c.

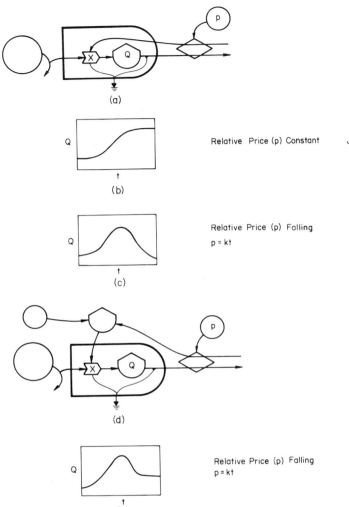

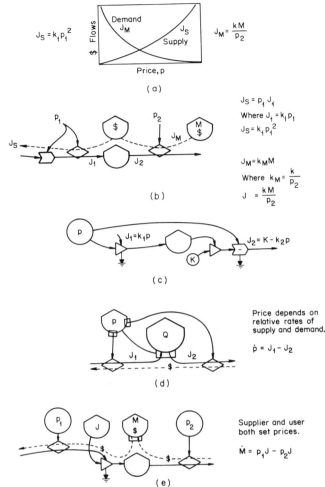

FIGURE 23-14. Minimodels of growth and exchange where main energy source is flow limited: (a) model of production requiring interaction (matching) of source and trade input; (b) graph of growth in part a with relative price constant; (c) graph of growth in a with relative price rising; (d) model with source and inputs usable separately or together; (e) graph of model in d with rise in relative price.

FIGURE 23-15. Models for coupling supply and demand: (a) graph of supply and demand; (b) price based on quantities; (c) linear (Brewer, 1976); (d) Walrus model based on price due to relative rate of supply; (e) model with two prices.

Source-Limited Growth

Since most energy sources are not constant pressure, the action of externalities are given by a limiting function (Fig. 9-5). As available flow is incorporated, there is no more energy for further growth, and growth stops. The minimodels in Fig. 23-14 have a sigmoid growth pattern with relative price constant. These are good overview models for isolated effects of international trade. When relative price p is allowed to fall, as it is doing in the 1980s with declining worldwide net energy of fuels, growth crests and decreases. With the first model (Fig. 23-14a) assets approach zero as p decreases, whereas the second model (Fig. 23-14b) provides for use of renewable resources without much trade and assets level at a lower level than reached in the growth period. See also Fig. 13-9b. The diagram in Fig. 23-14 is probably identical with the Domar growth model, which is derived by setting investment proportional to output.

Supply and Demand

Production is coupled to consumption by balancing the rate of supply to equal the demand. Figure 23-15

Economic Systems and the Nation

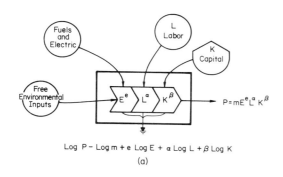

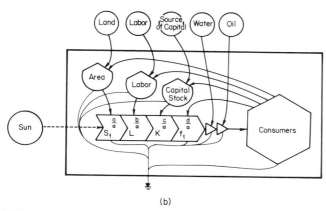

FIGURE 23-16. Model of economic production with use of Cobb–Douglas function: (a) example of inclusion of energy in production function of Finnish paper industry (Cunningham, 1974); (b) translation of model of Japan (Sugayima and Shimazu, 1972).

gives some alternative models proposed to represent supply and demand coupling. Mechanisms may vary in different systems.

Graphs of rate of use decline as a function of price, and graphs of rate of supply increase as a function of price. Where these lines cross, the rates are equal and the processes are matched. Mechanisms for maintaining a match maximize flows and power and are observed to be self-regulating. When the price is lower than match point, supply production drops, stock drops, and price rises; when price is higher than match point, demand falls, stock increases, and prices fall.

Models of Consumer Strategy and Scarcity

Textbook relationships of supply and demand, consumer strategy for scarce resources, cost and benefit, and so on have been applied to relationships of organisms in ecosystems without humans, and many close parallels are found. See the review by Rapport and Turner (1977).

Other Production Functions; Cobb–Douglas Model

The models of the production function used so far in this chapter are those of limiting-factor concepts as used in many sciences (Chapter 8), sometimes called *Monod equations*. Figures 21-10–21-13 show these kinetics to be consistent with traditional economic concepts of diminishing returns and marginal utility.

The traditional production functions used in economics are products with exponents. For example, in the Cobb–Douglas function (see Fig. 19-6) production is the product of labor, capital assets, and so on, each raised by one power. The log of the production is the sum of the log terms for each interacting quantity, a property making it easy to fit data on graphs to obtain coefficients.

Cunningham (1974), fitting data on forest industry production to the Cobb–Douglas model, includes an energy category and obtains as good a correlation as with capital and labor only. (Energy calculated was that separate from that embodied in labor and capital.) Energy used was correlated with capital (see Fig. 23-16a).

A model by Sugayima and Shimazu (1972) of the economy of Japan is diagrammed in energy language in Fig. 23-16b. The main variables in productions were modeled as labor, capital stocks, and land area.

ECONOMIC WEBS AND SPECTRA

The sectors of the economy are organized like other natural systems with branches and feedback loops. Figure 23-17 shows a moderately aggregated model for the United States in which lower quality items are on the left and higher quality ones, requiring more support of embodied energy equivalents, on the right. One can arrange occupations and industries in a power spectrum with low-quality external energies on the left, fossil fuel inputs and electricity in the middle, and human services on the right. Because of the high inflow of moderately high quality fuel, there are proportionately more medium and high-quality sectors in the economy than in solar-based ones.

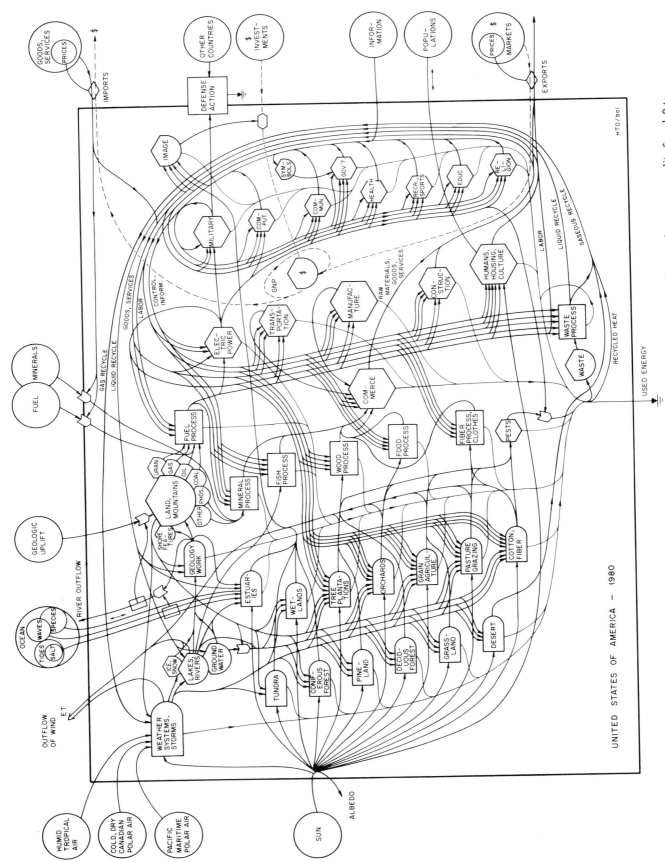

FIGURE 23-17. Aggregated model of the economy of the United States with sectors arranged in order of increasing energy quality from left to right (Odum and Odum, 1980). Courtesy McGraw-Hill Book Company.

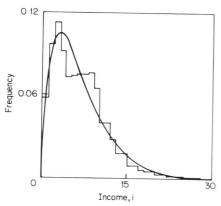

FIGURE 23-18. Distribution of income in Sweden in 1964 with an energy-based theoretical curve given by Peterson and Von Foerster (1971).

INCOME DISTRIBUTION

The tapering hierarchy of the energy chain generates a tapering distribution of units (Chapter 15). As Fig. 23-2d shows, the money:energy ratio is highest in feedback service from the right.

Within a population of human occupations there are many with smaller embodied energy and a few with larger embodied energies (Charpentier, 1976) (see Fig. 23-18, which shows an observed distribution of income). Pareto (1927, 1971) used an inverse-square function as a model for these distributions, which may be related to energy theory as in Fig. 18-14. Where a hierarchy is based on convergence from low-quality solar energy sources, a declining exponential may be appropriate, as discussed in Fig. 15-3. Also see Mandelbrot (1960).

However, in this century, with predominantly higher-quality available energies entering in the middle zone of the spectrum, an income distribution with a middle-class peak bulge is to be expected (Atkinson, 1976). See, for example, simulation by Brown (1980) in Fig. 11-24 and Fig. 15-10c. Peterson and Von Foerster (1971) develop an energy-based equation for value that generated an observed income distribution curve given in Fig. 23-18.

NATIONAL MINIMODELS

Simplified Models of the United States

The model in Fig. 23-19 has renewable and nonrenewable energies separated and the economy as four sectors, three of which are directly involved with dollar–human economy and one as part of externality production. The flow of money through the consumer unit on the right is for human settlements and government and is called the *gross national product* (GNP) (see Fig. 23-17). The money flowing through from right to left represents the flow of real work and service from left to right. Division of the

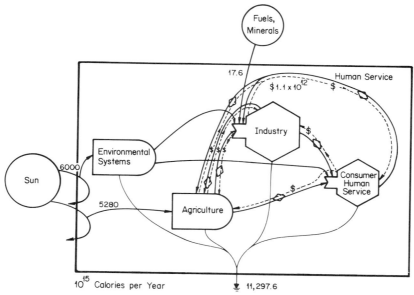

FIGURE 23-19. Aggregated model of the economy of the United States in 1974 showing renewable (sun) energy and fuel base that converge in value in labor. Values are flows of actual energy (Odum, 1978).

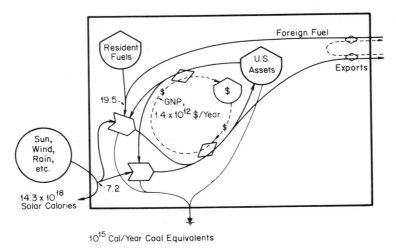

FIGURE 23-20. Aggregated energy diagram of the system of the United States, including the circulation of money, renewable sources, nonrenewable sources, and sources from foreign trade, all expressed in coal equivalent calories of embodied energy.

direct solar energies by 2000 converts them to approximate embodied energy in coal equivalents, showing the basis for the U.S. economy to be 75% from fuels and the rest from renewable sources (tentative, pending revision of conversion factors).

Figure 23-20 gives an even more aggregated model of the U.S. economy, showing the interplay of renewable, nonrenewable, and externally purchased energy. This diagram can be used to estimate the impact on inflation of a change in relative price of foreign oil. The percent change in expected money circulation compared to percent change in energy flow provides the inflation rate due to this effect. Models of this type stop growing and retrogress as energy becomes limiting (Alexander et al., 1976). The model illustrates the calculation of dollar : energy ratios for a given year. The renewable energy is converted to coal equivalents and added to fuel consumption.

National Income, Expenditures, and Government

Given in Fig. 23-21a from Schultze (1971) is a closed-loop model of the flows of money in the United States. It is aggregated to show separately the government, the households (final demand), and the main flow of production and maintenance of capital assets. This diagram has been combined with energy relationships in a translation in Fig. 23-21b. On the left is the production function that is an interaction of labor, capital assets, and government control. The output of the production is the gross national product which fans out to maintain assets, government, and households, which return their services in a loop. The money flows are countercurrents to the flows of work. Cross-connections given by Schultze include transfer payments such as welfare, several tax flows, and return of savings to investment. The role of government is as a high-quality control sector.

Energy Embodied in Labor

In some approaches to energy analysis, the fact that human service is at the end of the energy chain constitutes the basis for the anthropocentric concept that the purpose of the economy is to supply the consumer. However, as shown in Fig. 23-17, the human consumers provide the means for feeding back the main controls to make the system do complex things. The most energy-intensive item in embodied energy is human labor, in which the embodied calorie/calorie values are very high. Since most of the rest of the economy converges on the human services, it is a reasonable approximation to estimate the energy for labor by multiplying by the average energy:dollar ratio with energy expressed in equivalents of the same quality. The energy used in this ratio should include the environmental free inputs as well as the fuels free or purchased (see discussion of Fig. 23-5).

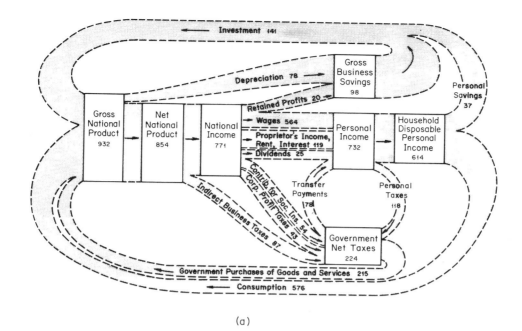

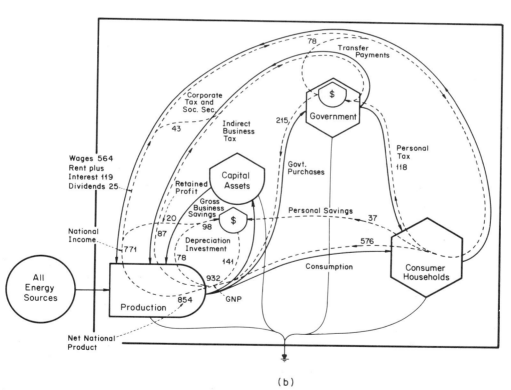

FIGURE 23-21. Income expenditure flows given by Schultze (1971): (a) circulation of money with flows for 1969 in billions of dollars per year; (b) energy diagram of part a with energy source added.

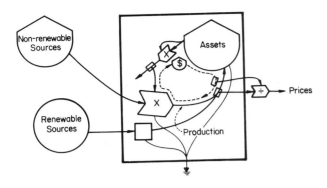

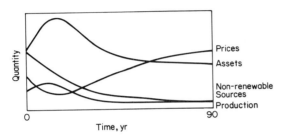

FIGURE 23-22. Simulation of a world minimodel with cooperative autocatalytic use of non renewable resources.

Aggregated Model of World Economy

Since the national economy is driven by the larger world economy through its control of export-import prices, an aggregated minimodel is included here to show the way declining nonrenewable resources increase relative prices to the nation.

Figure 23-22 is an aggregated model of a world economy showing renewable resources, nonrenewable sources, productivity, and the circulation of money. Overall inflation index of prices is taken as the money:energy ratio generated by the model as energy of the nonrenewable storages is used up. The model illustrates the dependence of productivity on external richness of sources and on the feedback of such assets of technology reaching a maximum before the assets reach a maximum. Prices of energy reach a minimum as productivity is maximum and then as productivity declines inflation due to changes in energy accelerates and later diminishes again.

Oscillation

In Fig. 11-21 and Chapter 22 alternate models with pulsing consumer actions are considered as possibly prevalent because of their role in maximizing power. The pulsing minimodel in Fig. 27-4 may be suitable. These build up storages which are then consumed in a burst of consumer activity. Cycles of economic activity show a spectrum of oscillations of many frequencies (Granger, 1964).

Forrester (1976) finds evidence for a 50-year cycle of storage and discharge of the capital assets of the human economic system, referred to as the *Kondratieff cycle* after the Russian exile who suggested it. This is in addition to the three-year cycle of stock and spending of the business cycle. These oscillations seem to be the pulsing of consumers possibly to be expected from the energy theories and minimodel simulations. Taylor (1979) gives a power spectrum of U.S. business cycles with larger amplitudes with lower frequency.

ECONOMIC COMPETITION

The phenomena of competition were considered in Chapter 12 for units without money. Figure 23-23 gives models of economic competition in which parallel units may compete for resources or for markets (income from the market facilitates energy access). The high information levels of human-guided competition produces complex digital actions, switching purchases wholly from one pathway to another as profit, and thus power is maximized.

ENERGY ANALYSIS OF ECONOMIC ALTERNATIVES

Calculating Net Energy Contribution to Economic Activity

To calculate the net input of an external sector to the main economy, one evaluates the yield relative to the feedback, both expressed in energy units of equivalent type (see Fig. 14-8). In Fig. 23-24a the yield is readily expressed by multiplying heat calories in Y by the appropriate energy transformation ratio. The feedback from the main economy usually involves a mix of total economic activity, especially labor, which is a convergence of all the embodied energy. Thus the money flow is converted to dollar flow by use of a total dollar:total energy flow ratio for the appropriate year. This is added to embodied energies of fuels, minerals, or other raw products that are purchased to obtain F. Net energy yield ratios for energy sources important in the

Economic Systems and the Nation

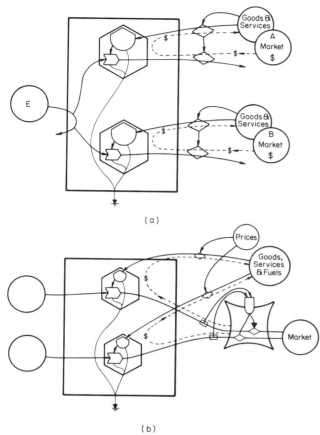

FIGURE 23-23. Models of economic competition (Odum, 1976a) (*a*) competition for externality *E*, where the system with the highest level of production from *E* wins; (*b*) digital selection of competition for market, where sales go to system with greatest yield where costs are similar.

United States were mainly 3 to 12 (Odum et al., 1976b). For detailed explanation, see Gilliland (1975) and her treatment of geothermal energy.

A source that has a higher net energy than all others is a primary source, and its net energies are fed back by the system into further growth, into alternative sources, or into efficiency mechanisms such as specialization and diversity. Thus the activities that are receiving feedback of energy from the primary sources need not have net energy from their own external sources to make a contribution. See Fig. 14-10.

Investment Ratio as an Energy Cost–Benefit Method

In Chapter 14 the investment ratio was defined as the ratio of high-quality feedback F of energy attracted from outside a system to the lower quality inflow I of energy into the system (Fig. 14-8). The investment ratio is defined as F/I in Fig. 23-24*b*. It operates for human-operated systems through market competition and money. The flow of energy on the left provides the attraction by which energy inflows from the outside economy are connected into a productive interaction. The output of the production flows to the right generating an inflow of money through sales, services, and attracting mechanisms that bring people and their money (Fig. 23-24*b*). The money attracted travels around the loop and out to buy continued inflow of outside embodied energy.

If the ratio of feedback of invested energy (i.e., that incoming from the main economy) F is small relative to the resident inflows I, where both are compared in embodied energy equivalents of the same energy type, the system is receiving more free subsidies than competitors, and its sales are made at lower prices because its costs around the loop are less. It captures the market and receives more flows, probably growing.

When the ratio of invested energy F to resident energy I is large, almost all the inputs must be paid for, prices rise, and the system competes poorly and grows no further. The investment ratio of competitors serves as an index as to whether a proposed development will maximize power and thus survive as an economic activity. The calculation can be made with energies and thus without problems of data subject to inflationary change.

Evaluating an Externality with Embodied Energy

As shown in Fig. 23-25, the use of a rich externality such as a mineral, a fuel, or another product of natural systems (e.g., wood) involves processing of the inflow through successive stages in which more and more of the economy's work is used to interact with the flow, increasing its value. These flows are called *value added*. The ultimate flow of money is the one shown on the right part of the dollar loops, and not the left-hand one, which is only the first price paid for the raw product. The ultimate value to the economy may be estimated by accounting for the average energy : dollar ratio of the economy and the proportion that the inflowing embodied energy is of the total embodied energy of the economy. This is the proportion of the GNP ultimately caused by the externality, even though the pathways of action may be indirect and unrecognized.

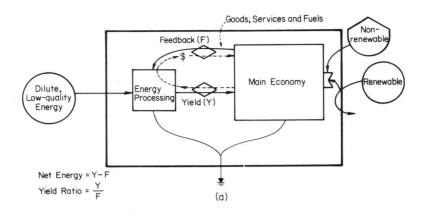

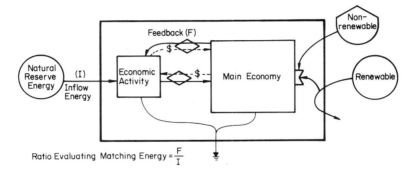

FIGURE 23-24. Ratios of energy flows for energy analysis; calculated in calorie equivalents of the same quality: (a) net energy evaluation of a source, using net energy yield ratio; (b) investment ratio evaluation of the contribution of environment to making an activity economic.

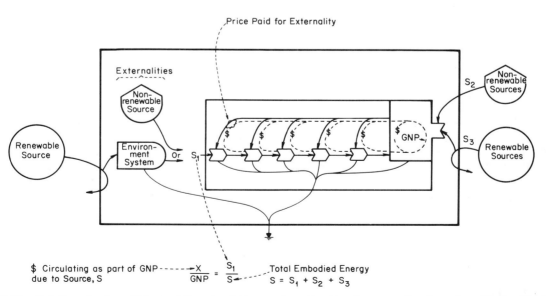

FIGURE 23-25. Relation of externalities to dollar circulation in an economy. The ultimate contribution of the environmental sources are much greater than the first price paid at point of entry of the inputs. The calculation of value in dollars per year is made by estimating the proportion the externality is of the total flow of embodied energy. This proportion of the GNP is due to the source evaluated (Odum, 1981).

Environmental Impact Diagrams

The moderately complex energy diagrams that are first drawn to organize thoughts and information about a system of humanity and environment serve well as impact statements for those who have learned the energy symbols. The diagrams can be used to key in tables, charts, and discussions of impact. Examples which can serve this role are Figs. 25-8, 25-9, 25-12, and 25-13. There are many more in a South Florida study (Odum and Brown, 1975) and popular summary (Browder et al., 1975). Leopold (1971) suggested a matrix tabulation of interconnections and intersections as a way to summarize and inventory environmental impact. Odum (1972b) suggested diagrams.

Energy Investment Ratio for Judging Environmental Loading

On p. 493 the investment ratio was compared with that of an alternative investment ratio prevailing in successive economic activity. The difference may be taken as that measure of the impetus for investment. In Fig. 23-20 the U.S. ratio of feedback to renewable energy is 2.7:1. Comparison to this ratio indicates how economic a proposition may be.

The ratio also measures environmental load. The higher is the investment ratio, the greater is the load on the environmental resource.

M. J. Lavine suggested that the feedback investment from storage (F_2/I in Fig. 13-7d) may be an important index of appropriate energy matching also.

Coupling of Sale Prices to Cost Prices

For an economic activity to prevail, it must charge a price for its outputs that will generate money to pay for those of its costs that require money. In energy notation, this is given in Fig. 23-26. Money flow J_m is the same for purchases and sales. At steady state, money flow includes all costs, such as the support of the humans involved, replacement of equipment and buildings, operation, and maintenance.

In addition, if the economic activity is embedded in a situation where energy resources support expansion, the activity must grow as much as competitors or more or have its relative position inhibited. Thus there must be some growth in assets and with it the money that corresponds to those assets, usually called *profit*. However, if the sector of competitors is not growing, operation without any profit is

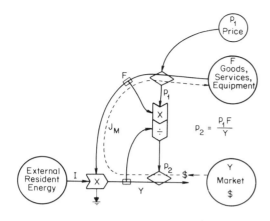

FIGURE 23-26. Diagram showing sale price as a function of purchase price plus contribution of externalities (Zucchetto, 1975b). To include effect of external energy, express Y in units of contributing energy flows that are not by-products of each other.

As the externality becomes scarce, the price rises as the externality has a decreasing net effect on production.

If one follows the production all the way to the final consumers, one finds the ultimate money flow and feedback energy that the externality has induced in matching interactions. For example, Boynton (1975) studied the processing of oysters and found the final ratio of feedback to inflow energies when all were calculated as embodied energy of one type to be 2.1, whereas the overall investment ratio in the United States at that time was 2.4.

Burr (1977) evaluated direct dollar yield and energy basis for the contributions of cypress swamps in Collier County, Florida where wetlands are prominent. Energy measurements give much higher estimations of environmental effects on dollar economy. Odum et al. (1982) include an Appendix: manual of environmental evaluation with embodied energy method. Alexander (1979) evaluates a nonrenewable source, coal.

Shabman and Batie (1978) criticize these methods, and a rebuttal followed (Odum and Odum, 1979).

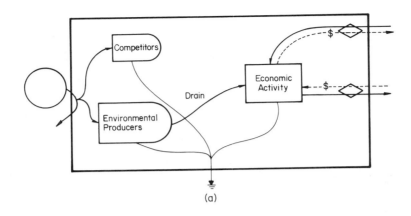

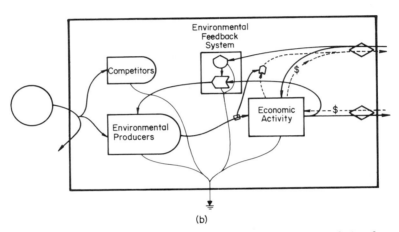

FIGURE 23-27. Comparison of economic couplings: (a) use is a drain and competitors increase and economic activity decreases; (b) feedback maintains chain of use competitive and economic activity stable.

competitive if the payments include necessary innovations for efficiency or progress possible within the same budget.

In Fig. 23-26 the adjustment of price on the lower arm is made according to the upper arm on the principle that the total money flows must have continuity; outflows must equal inflows. The resulting formula for price p_2 is common in economics.

Including External Energy in Price Coupling

Figure 23-26 the traditional formula for adjusting prices omits any reference to the ultimate causal flow for the whole process, the energy inflows from the left that are free. If these input supply services I are not bought, the price P_2 that can be set is lower. This relationship is shown in the lower equation in Fig. 23-26. Price is lower because renewable energy flow I is included.

Feedback Loop for Coupling Economic Activity to Environment

When an economic activity is connected to a natural production system, the economic activity drains some of the product (Fig. 23-27a). Examples are fisheries, forestry, and wildlife. The chain that produces the items of economic use is thus stressed, and its position as a competitor for energy inflow is decreased. As given in Fig. 23-27b, the theory of good symbiotic coupling requires a feedback loop that amplifies the chain that is used as much as the use is a drain. Agriculture and oyster culture are examples of where that feedback exists; fisheries based on the open sea, and cutting of our largest trees are

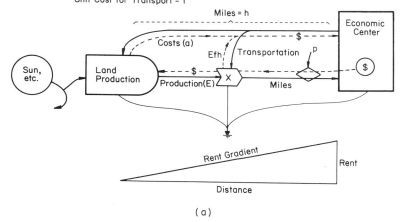

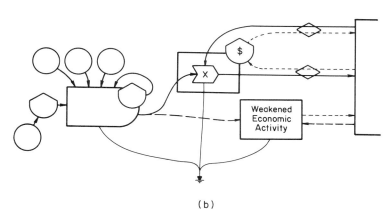

FIGURE 23-28. Energy flows that form the basis for land value: (a) model of land rent given by Von Thunen in 1826, quoted by Berry et al. (1976); (b) land coupled directly to economic activity, taking free subsidy away from the economic system elsewhere, but with earnings now yielding tax money.

examples where this feedback loop rarely exists and the ultimate effect is loss of the economic activity.

Value of Land, Tax

The model in Fig. 23-28 shows the energy basis of land and its economic interaction. Land is formed and renewed by uplift, a slow, high-quality flow driven by earth cycles. A process of soil differentiation results from interaction of rain, wind, sun, and vegetation. Finally an economic interaction develops dependent on proximity to sources of fossil fuel and fossil fuel-based investment energies with which to interact. The first two processes produce land and are free externalities. The proximity to economic activity centers is recognized by our tax system, which causes a flow of money according to potential for interactive work with the economy. The tax forces the land to be coupled to the economy even at the expense of diverting some of its formerly free services to the economy. Figure 23-28 shows the situation with and without the economic coupling. As the diagram shows, land rights are really the means for using most of the renewable energy flows. Tax policies can accelerate economic development, even to the detriment of the existing economy (Fig. 23-28b).

A model for land rent was given by Von Thunen (1826) that recognizes the proximity to economic centers. Rents decreased with distance, based on increasing transportation cost required (see Fig. 23-28a). In the model in Fig. 23-28 the value of land is

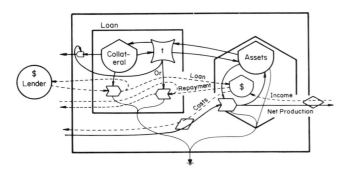

FIGURE 23-29. Energy diagram of a loan system.

represented as an interaction of the soil processes (externality) from the left and the proximity factor from the nearest economic center on the right. The force of the latter declines in inverse square with distance from the economic source. Money flow generated that recognizes this pumping action as land tax or land rent declines with distance reflecting the energy effect.

Loans

Loans with pathways of repayment and interest are shown in Fig. 23-29. The loan increases the money supply circulating since both the borrower and the lender have paper with which they can buy things as a result. One has the money, and the other has the note.

Cost–Benefit Analysis

When assets have been accumulated from previous growth and when capital money has been accumulated, there are alternatives for feeding these funds and assets back into interaction with additional resources (energy sources) to generate more flow of energy, more money, and more vitality. Figure 23-24 shows the overall interplay of money and energy associated with new endeavors; Fig. 23-30 gives more detail.

The economic cost–benefit procedure as shown in Fig. 23-30 consists of a comparison of the proposed investment alternative to start a new endeavor with the simple alternative, one of investing in some standard interest-yielding procedure such as a bank deposit. If the new endeavor causes more money to circulate in the new activity than in the comparative investment, the project is regarded as a good one. The rate of earnings of the comparative investment is called the *discount rate*. The *benefit:cost ratio* is the ratio between the circulating money and

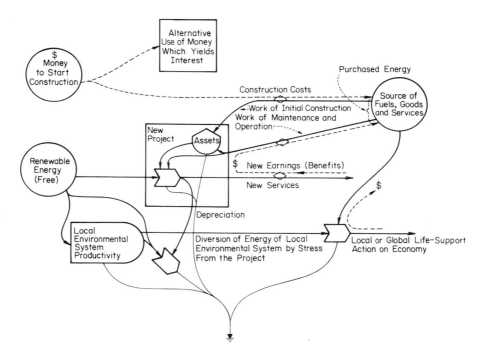

FIGURE 23-30. Diagram showing energy and economic cost–benefit analysis.

the sum of the money used at the start plus what it would have earned elsewhere.

As shown in Fig. 23-30, the money initially is used to develop the capital assets and start operations. This is prorated over the time interval used for the calculation. The flow of output goods, services, attractions, and so on, brings in new money that is subsequently used to defray the cost of inputs that sustain operation, maintenance, repayment of debts, and so on.

Shown in Fig. 23-30 are the diverted and stressed systems of the environment that were previously using the same energy sources to feed into the economy indirectly and directly. Some accounting of these losses is required.

A main difficulty with many cost benefit procedures is that the output production of the system depends on having good input of energies, and nothing in usual procedure measures this directly. Implicit assumptions used are that the process will yield as much economic flows as similar situations have in the past from which coefficients are obtained. If the environmental energies are less, if inputs based on the outside fossil fuels that come in with purchases are at higher prices, or if the environmental resources have already been crowded so that they no longer give as much free subsidy as previously, the economic yields of the new system are less than expected. Another difficulty is that the free pathways of life support from the environment in maintaining air, water, soils, groundwaters, landforms, and so on in usable form are not easily measured and are often very much underestimated.

Hannon (1982) discusses energy discounting analagous to money discounting. Both assume conditions of growth, which is the only time when it can be assumed that a storage invested will generate a net increase.

Economic evaluation of environment is now a very large literature with complex methods offered to get around the difficulties mentioned here and others (Nijkamp, 1977; James et al., 1978).

Cost Benefit of Ecological Adaptations

Using energy instead of money costs of biological structures and adaptations have been compared with yields fostered by structures. For example, Solbrig and Orians (1977) determine payback times in the same ways as is done for technology. See reviews by Cody (1974) and Bernstein (1982).

INPUT–OUTPUT MATRIX—GNP

In Figs. 6-16 and 13-26 flow networks were expressed as matrices. The intersections of the matrix table represent the pathways connecting the units. These can be expressed in units of material, energy, and so forth. In Fig. 23-31 the economic input–output matrix is given as represented by Leontief (1965, 1966, 1971). The figure has a steady-state economy and thus does not have changes with time due to growth or decline. Note that the energy flows in one direction (Fig. 23-31a) and the money flows in an opposite countercurrent (Fig. 23-31b). If, as done in economics, the external inflows and energy outflows from the money circles are omitted, a closed

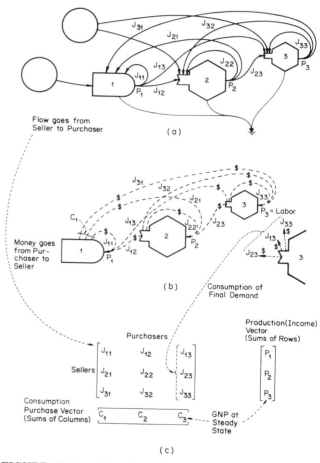

FIGURE 23-31. Network and matrix representation of a steady-state economic system: (a) energy flows; (b) money flows isolated; (c) input–output matrix of money flows per time. The term P_3 includes capital, taxes, and wages; C_1 and C_2 are values added by feedbacks. In some usages final demand flows are labeled Y; thus J_{13} is Y_1, $J_{23} = Y_2$, and $J_{33} = Y_3$. The sum of P values is sometimes called *total demand*.

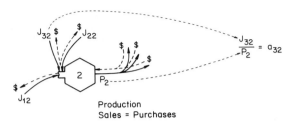

Input-Output coefficient is defined as the fraction of production (P) (sales) for a purchase pathway.

FIGURE 23-32. Diagram of input–output coefficients.

network results (Fig. 23-31b). The matrix describes the flows of money per time along each pathway. The money inflows (goods production output, money sales) are shown on the horizontal row and the money outflows (for purchases) are shown on the vertical column, but the commodities represented are reversed—inflowing on the vertical and outflowing on the horizontal row. (This results in a different custom for inflow and outflow from some other initiatives in ecology where money is not involved.) Notation for subscripts is different from those used in Chapter 13. Here J_{13} means flow from unit 1 to unit 3 and dollar flow from unit 3 to unit 1.

The flows through the end of the economic train (highest quality consumer) to humans are called the *final demand* and the dollar flow through this sector is called the *gross national product* (GNP). Note that flows forming closed loops but not passing through the last sector are not part of the GNP.

Roughly speaking, the GNP is the total flow through people. Since people control all other economic sectors by their feedbacks, the GNP measures the total system also, since feedbacks must amplify in proportion to their cost to be competitive.

A more precise definition of GNP as used in economics includes all the money that goes to the final-demand sector. A *sector* is a compartment of the ecosystem—a unit in Fig. 23-31b, for example. The final demand includes flows to replace depreciation, to account for net growth, imports and exports.

Input–Output Coefficients and Economic Structure

For each sector the fraction that an input flow represents of that sectors' output production is the input–output coefficient of that inflow. Note inflow J_{32} in Fig. 23-32. Its ratio to the output production P_2 is a_{32}, the coefficient that describes their flow relationships. During growth of the whole system, these ratios change more slowly than the flows. See also Fig. 13-26 where input–output coefficients for mineral cycles were used without dollars being involved. The matrix of these coefficients is a quantitative representation of the *economic structure*. After these coefficients have been obtained from one set of data, they can be used to model various situations for purposes of extrapolation and theory.

Matrix Notation

As shown in both the diagram and the matrix in Fig. 23-31, final demand is the flow through the human consumers at the end of the energy chain J_{13}, J_{23}. It is the total productive outputs of various sectors minus the part of the production that circulates to other parts of the system without going to the final consumer compartment. Final demand is represented in the last column in the input–output matrix. In the steady-state diagram, consumer receipts equal expenditures. In other words, the sum of the last column and the sum of the last row are equal. Equations for final demand are given in Fig. 23-33 in terms of the productions P and input–output coefficients a. They are manipulated there so they can be expressed in matrix form using an identity matrix (1's on diagonals; 0's elsewhere) and finally at the bottom the very compact form of the system of equations in matrix notation.

Energy Coefficients for the Input–Output Matrix of Dollar Flows

The relationships of energy and dollar flows are given in Fig. 23-34 using notations for assigning embodied energy with matrix procedures. The output X_j of sector j is described as the sum of interunit sales $a_{ji}X_i$ and final demand Y_j. Energy:dollar ratios are calculated for each pathway so that a matrix of ϵ's is available for tracing the energy required for various sectors. Where embodied energy is called energy intensity (Herendeen, 1981), these ratios are sometimes called *energy intensity coefficients*. In Fig. 23-34 the energy balance (steady-state flow) in each sector is the sum of the inputs from outside the economy E_j and from other sectors $\epsilon_i X_{ij}$ combining to produce the embodied energy in the productive output of the sector.

The energy intensity coefficients are calculated by evaluating the external inputs E's in energy

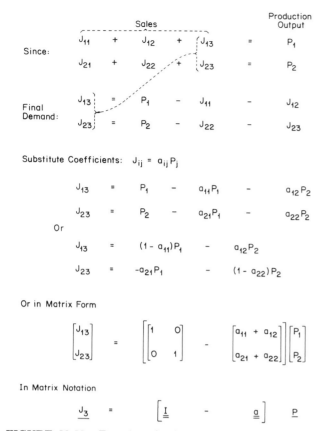

FIGURE 23-33. Equations for final demand of the system in Fig. 23-31 in terms of production P and the structure of input–output coefficients a.

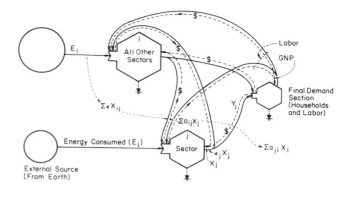

FIGURE 23-34. Diagram of economy for showing energy analysis methodology of Herendeen (1974, 1981) and Bullard et al. (1976). Expression and terms are theirs; X_j is production output of sector j and Y_j is final demand flow from sector j.

units of one quality such as coal equivalents, evaluating the intrasector dollar flows of the economy, and then determining the ϵ as the unknown variable by a matrix inversion. This has the effect of assigning the external energies to pathways so that they are additive and weighted according to the dollar flows. See Krenz (1976), Bullard et al. (1976), and Herendeen (1973, 1981). As described in Fig. 14-17 this is a different concept from the one that gives all pathways the embodied energy required to generate the flow. It is an arbitrary assignment of inflowing energies to pathways. Applying the method to the U. S. economy, Costanza (1980) included an estimate of environmental energy and fuels. Also see debate about the method (Huettner and Costanza, 1982).

If data on the pathways are in some units other than dollars, the procedure still works, but the coefficients are ratios of energy to the carrying medium.

Hannon (1973, 1976) and Herendeen (1981) applied the procedure to Silver Springs data on food webs that were in actual energies. The energy intensity coefficients were then ratios of assigned embodied energy to actual energy. Costanza and Neill (1981), also using Silver Springs data and the method in Fig. 14-16c, determined embodied solar calories per calorie for the various flows.

An Error in Estimating Embodied Energy with Input–Output Matrix

In estimating embodied energy from fuels and environment reaching the terminal consumer sector 3 for the whole economy in Fig. 23-31b, it would be incorrect to add in the embodied energy of the feedback loop of human service J_{33} since this would involve counting of the same energy twice. The human service is a downstream byproduct of the input energy chain.

However, when calculations are made of the embodied energy in one small sector, it is necessary to evaluate the labor feedbacks as shown in Fig. 23-34. Consider the evaluation of the embodied energy of sector j. One must include the direct energy used (E_j)

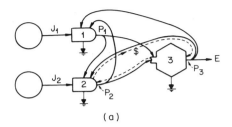

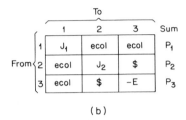

FIGURE 23-35. Mixture of economic and ecologic flows in input–output matrix used to calculate environmental production coefficients (Isard, 1968): (a) system with two sectors connected by dollar flows and sector number 1 connected only by externalities "ecologic commodities"; (b) matrix representation; (c) input–output coefficients as defined in Fig. 23-32. The four coefficients, a_{12}, a_{13}, a_{21}, and a_{31} are ecologic–economic coefficients.

plus that from other sectors (sum of embodied energy from "all other sectors i"), which includes direct inputs to those sectors E_i. These are usually included in energy analysis procedures. As shown in Fig. 23-34, one must also include that energy that comes in as E_i and goes to the final demand sector and then back to sector j as labor. This pathway is not usually included, and its omission causes the energy embodied in many high-quality (high embodied energy) sectors to be underestimated.

Ecologic–Economic Input–Output Coefficients

Isard (1968) represented marine food chains as an input–output matrix and combined ecological and economic flows as one matrix developing interprocess coefficients for relating environmental flows to dollar flow data (Isard, 1968; Isard and Langford, 1971; Isard, 1969). Selected ecological *commodities* were chosen. The coefficients relating dollar flow to other flows from environment are measures of the environment's basis for the economy. The concept is demonstrated in Fig. 23-35 with a mixed ecologic–economic network. In the examples given by Isard, no effort was made to include the total environmental part of the network. Only selected inputs were included. More complete inputs are feasible with the use of holistic energy diagrams evaluated in embodied energy.

MAXIMIZATION OF PROFIT

If profit is the excess of income over costs, it is the net rate of increase of money in a tank as shown M in Fig. 23-4. If pathways are in steady state, there is no profit, but money flows, energy flows, and systems survive. During rapid growth, however, maximum power depends on maximization of growth where energy sources can deliver more energy than is being demanded. During this period there is increase in real assets, and thus there can be an increase in money (real buying power). During these times maximization of both profit and power have similar effect and are proportional. At steady state maximization of the steady flows of energy produces competitive survival, but in this case profit is not a measure of survival conditions.

The relationships of profit to energy from a renewable resource are often indicated with a logistic model of population growth (fisheries, wildlife, trees, etc.) that is being harvested. From Fig. 11-13, yield is a parabolic function of effort; and if prices are constant, income from sales is a parabolic function also. Costs, on the other hand, may rise with effort, and if prices are constant, the curve is linear. Profit is the difference between income and costs, and the break-even point is indicated at the intersection of the curves in Fig. 23-36. Maximum profit results when the second derivative is set to zero and is at a much lower population.

One difficulty with this theory is that harvest without feedback stimulation work (see Fig. 23-27) causes the harvested species to lose out to competitors and the species to be much less than its carrying capacity or to become extinct. Clark (1973) gives other reasons for the fishing to be in excess of the density for maximum profit or break-even point. In time of expanding energy, feedback from richer portions of the economic system can receive loans and

Economic Systems and the Nation

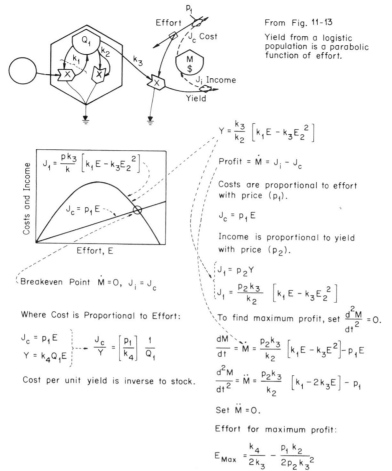

FIGURE 23-36. Financial break even point where harvest is from a logistic population [Clark 1971]. However, the population would be subject to decrease due to competition of other species unless there is feedback from the user of the yield to improve the energy sources or efficiency of the recruitment. Effort to maximize profit is half of the break-even level.

investments for harvest based on good investment ratios at first, even though the fish stocks may be depleted by the end of the life of the boats, loans, and investments. Better matching of time constants is needed.

Optimum Efficiency for Maximum Profit

Wesley (1974) provides a derivation showing maximum profit at an intermediate loading and intermediate efficiency for generalized heat engines that use purchased heat to generate work for sale. The system is given in Fig. 23-37a along with his derivations, where T is the temperature of the source; T_1 is the temperature at the working site of the production transformation, a temperature that is lower due to losses in diffusion; and T_0 is the heat sink temperature. Heat differences do work, causing sale of transformed work at price A. The profit is the money remaining after the income is spent on the processing of heat at price B. The maximum profit is found by varying efficiency (loading) to find the maximum. The derivative of the expression for profit flow J_P is set to zero to find the maximum. Then the equations are given in terms of efficiencies of the transformation and of the overall process. The graph in Fig. 23-37b (Wesley, 1974) shows the maximum profit curve with a maximum.

The equation for maximum profit J_P was transformed into an equation for maximum power, showing an equivalence between net energy and profit so that maximization of profit in this case causes maximization of power (see relationship between power and loading in Fig. 7-19).

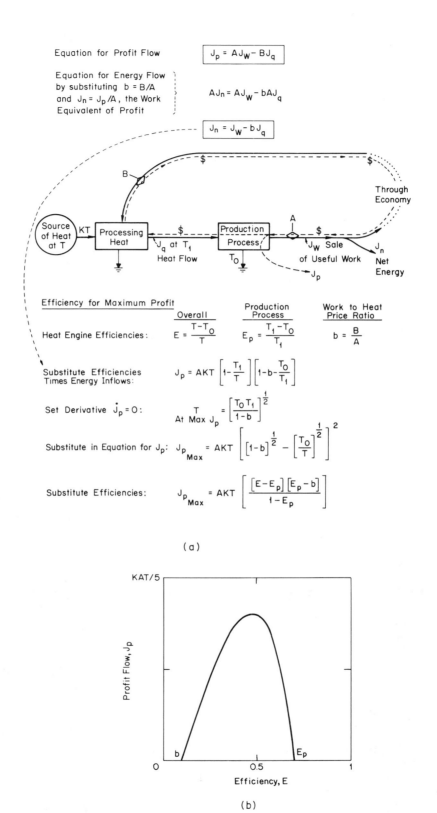

FIGURE 23-37. Diagram of profit–heat engine relations given by Wesley (1974): (*a*) energy diagram; (*b*) graph of profit rate as a function of efficiency for a heat engine driving work for sale. Derivations of power-efficiency relations of heat engines were given by Odum and Pinkerton (1955), Tribus (1961), Andresen et al. (1971), Curzon and Ahlborn (1975), and Fairen and Ross (1981).

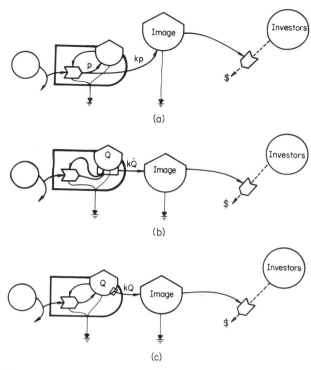

FIGURE 23-38. Alternative concepts of image, as a collective storage of information about value to attract interests. Image changed by (a) production, (b) rate of change, and (c) sensing storage levels.

IMAGE AND HUMAN DEMAND

The response of humans using behavioral responses to generate demand is part of the economic mechanism, although it is an open question as to the extent to which the mean human behavior is tracking economic success. A population has a collective view of value and opportunity that may guide its attraction to an area to a market or to an enterprise. The collective "image" is a sort of sum of stored impressions that decay with forgetting and mortality. Figure 23-38 gives several theories of image. In Fig. 23-38a image input is from production; in Fig. 23-38b inflow is proportional to rate of change of assets; and in Fig. 23-38c it is proportional to storage. Which concept of image controls which process? Image as a state variable is used in Figs. 25-8 and 25-9.

ENERGY AND ECONOMIC VITALITY

By providing economic theory with an underpinning of energy systems, realities of the environmental control of the economy may become apparent. Many tenets of economics turn out to be true only for special periods of fast growth. Growth is not a permanent state nor are profits prevalent at all times. Economic vitality does not require growth when resource bases cannot be expanded. Public policies may be reversed for climax, retrogression, and periods of environmental net production. For elaboration on the subject see Odum (1973), Odum and Odum (1976, 1981), Passet (1979), and Van Raag and Lugo (1974).

SUMMARY

The web of the system of human economy has organizational patterns similar to that of the nonhuman ecosystems with multiple sources, interactive production processes, convergence of sectors to a final consumer sector, and feedbacks of high-quality services forming necessary and controlling closed loops. The economic web has a countercurrent cycle of money over most of the pathways that are human regulated, and competitive pressures have caused the development of human microbehavioral response to increased prices with scarcity that acts to tune the system to maximize power more rapidly than without the monied mechanisms. In microeconomic situations prices are a driving function on energy, but in macroeconomic situations energy flows in a relation to money flows determine prices. Energy diagrams and input–output matrices are alternate ways of representing the network, each facilitating appropriate perspectives and calculations.

Models of limiting factors and production transferred from ecological theory are mathematically different but generate graphs similar to those traditionally used in economics. The role of externalities supplying energy to the closed loops of the economic system is shown by minimodels. The minimodels guide calculation of net energies to evaluate primary energy sources, investment ratio to evaluate secondary sources, and sensitivities to evaluate consumer efficiencies. Energy analysis of systems that explicitly diagram energy and money flows separately provide easier ways of teaching economic concepts, especially the relations of externalities in controlling growth and predicting what will be economically competitive under conditions of changing net energy and environmental impact.

Overview models of the United States include both the network of the wild environment and the web of the economy, but overviews are generated by aggregated minimodels that show the main effect of changing external resources.

QUESTIONS AND PROBLEMS

1. Draw a typical web of humanity and nature that has several economic sectors, including a final-demand sector. Show the money circulation. Indicate the gross economic product and relate the energy basis for vital economic circulation.

2. Which changes in the externalities can contribute to inflation?

3. Diagram the money flows within an autocatalytic minimodel. If the external energy source is one that supplies a constant pressure available for use according to the systems demand, what kind of growth pattern results? For an increase in real dollars, what must also increase?

4. Give several kinds of relationship that control prices for a transaction. For each, indicate the equation for production that results. Describe two methods for diagramming transactions that are the product of working capital of the purchaser and the storage of commodity. Diagram the situation where the purchaser anticipates business advantage by spending more on purchases when prices are low?

5. Compare the effect of a limiting energy resource on an interactive process that does not involve humans and money with one that involves human behavior spending inverse to production.

6. Give alternate equations for the curve of diminishing returns and for the curve of price dependent on marginal effect. Explain how price responses to marginal utility have the effect of removing limits to maximizing power.

7. Give models that relate supply price and demand price.

8. Compare Cobb–Douglas and Monod hyperbola models for economic production.

9. Explain how the quality of energy sources affects the spectrum of income distribution.

10. Show the role of government in the economy by drawing an appropriate web that has production sectors, government, and consumer households. Identify GNP in this diagram.

11. What pattern (i.e., shape of curve) of growth and regression results with simulations of minimodels of the total economy drawing on nonrenewable and renewable energy resources and foreign exchange that draws on energy sources there?

12. Diagram net energy of an external energy source and explain why price of that resource is not proportional to the ultimate value of that resource's effect within the economy. In the same diagram, indicate the way the prices for sale are adjusted to prices of the high quality feedback from the main economy.

13. Explain which work loops are necessary from the monied economy to the external webs to ensure that the economic activity remains competitive. How does the investment ratio estimate economic–environmental loading?

14. What part of the embodied energy of land is evaluated by the distance-dependent "land rent"? Does this value represent externality input or feedback utility?

15. Use a diagram to explain cost–benefit methods that compare discounted returns with the money circulation that develops on a new external resource. How can the investment ratio help predict whether economic circulation started will be competitive?

16. Draw an economic web and write its equivalent input–output matrix. What are the input–output coefficients? In times of change, which may be changing most, the coefficients or the values of the flows?

17. Where environmental and energy sources join the web, coefficients may be related to dollar flows through that sector as energy coefficients. How may input–output coefficients be used to identify the energy externalities contributing to the production of each sector? What embodied energy is omitted in this procedure because the input–output matrix coefficients do not usually include the final-demand sector?

18. What input–output coefficients can be used to relate externalities to the monied flows within the combined web of humanity and nature?

19. Plot a graph of a parabola of environmental

production yield and effort (Chapter 11). Also draw a dollar–cost curve and estimate the breakeven point. Is this higher than the classical optimum catch? Is the optimum catch competitive when the web of environmental competion is considered?

20. Recall criteria for maximum power during growth and during steady state. How is total profit during growth and steady state? Under what conditions does maximizing profit maximize power? How is efficiency related to power (Chapter 7)? To profit growth?

21. Give some alternative models for image.

SUGGESTED READINGS

Allen, R. C. D. (1968). *Macroeconomic Theory*, St. Martin's Press, New York. 420 pp.

Ayres, R. U. (1978). *Resources, Environment, and Economics*, Wiley-Interscience, New York. 207 pp.

Bahm, D. and A. V. Kneese (1971). *Economics of Environment*, Macmillan, New York.

Banks, F. E. (1976). *The Economics of Natural Resources*, Plenum Press, New York. 267 pp.

Baumol, W. J. (1961). *Economic Theory and Operations Analysis*, Prentice-Hall, Englewood Cliffs, NJ. 438 pp.

Baumol, W. J. (1970). *Economic Dynamics*, 3rd Ed., Macmillan, London. 472 pp.

Bernstein, B. B. (1981). Ecology and Economics. *Ann. Rev. Ecol. Syst.* **12**, 309–330.

Berry, B. J., E. C. Conkling, and D. M. Ray (1976). *The Geography of Economic Systems*, Prentice-Hall, Englewood Cliffs, NJ.

Bullard, C. W., III, P. S. Penner, and D. A. Pilati (1976). *Energy Analysis Handbook*, CAC Document No. 214, Center for Advanced Computation, Univ. of Illinois, Urbana–Champaign.

Clark, C. W. (1976). *Mathematical Bioeconomics*, Wiley-Interscience, New York. 352 pp.

Cottrell, F. (1955). *Energy and Society*, McGraw-Hill, New York.

Dahlberg, A. O. (1962). *Money in Motion*, University Press, New York.

Fisher, A. C. and F. M. Peterson (1976). The environment in economics: A survey. *J. Econ. Lit.* **14**, 1–33.

Georgescu-Roegen, N. (1971). *The Entropy Law and the Economic Process*, Harvard University Press, Cambridge, MA. 457 pp.

Georgescu-Roegen, N. (1976). *Energy and Economic Myths*, Pergamon Press, New York. 380 pp.

Granger, C. W. J. (1964). *Spectral Analysis of Economic Time Series*, Princeton University Press, Princeton, NJ.

Hall, C. A. S. and J. Day (1977). *Ecosystem Models in Theory and Practice*, Wiley, New York. 685 pp.

Hamberg, D. (1971). *Models of Economic Growth*, Harper, New York. 246 pp.

Herendeen, R. A. and C. W. Bullard, III (1967). *Energy Costs of Goods and Services*, 1963 and 1967, CAC Document No. 140, Center for Advanced Computation, University of Illinois at Urbana–Champaign. 380 pp.

Hirschleifer, J. (1977). Economics from a Biological Viewpoint. *J. of Law and Economics*, 2152, 1–52.

House, P. W., C. D. Swinburn, and J. McLeod (1977). *Large Scale Models for Policy Evaluation*, Wiley, New York.

Howe, C. W. (1979). *Natural Resource Economics*, Wiley, New York. 350 pp.

Hune, C. W. (1979). *National Resource Economics*. 350 pp.

Kavanagh, R., Ed. (1980). *Energy Systems Analysis*, Reidel, Dordrecht, Holland. 678 pp.

Klein, L. R. (1976). *Econometric Model Performance*, University of Pennsylvania Press, Philadelphia.

Krenz, J. H. (1976). *Energy Conversion and Utilization*, Allyn and Bacon, Boston.

Leontief, W. (1966). *Input–Output Economics*, Oxford University Press, New York.

Lucas, R. E. (1981). *Studies in Business Cycle Theory*, MIT Press, Cambridge, MA.

Mathews, R. (1958). *The Business Cycle*, University of Chicago Press.

Mishran, E. J. (1976). *Cost Benefit Analysis*, Praeger, New York. 454 pp.

Mitsch, W. J., R. K. Radade, R. W. Bosserman, and J. A. Dillon, Jr, Eds. (1982). *Energetics and Systems*, Ann Arbor Press, Ann Arbor, MI. 132 pp.

Morishima, M. (1964). *Equilibrium, Stability, and Growth*, Clarendon Press, Oxford.

Naylor, T. H., Ed. (1980). *Simulation in Business Planning and Decision Making*, Society for Computer Simulation, La Jolla, CA.

Nijkamp, P. (1977). *Theory and Application of Environmental Economics*, North-Holland, Amsterdam. 332 pp.

O'Sullivan, P. (1981). *Geographical Economics*, Halsted, Wiley, New York. 190 pp.

Passet, Rene (1979). *L'Economique et le Vivant*, Pargot, Paris. 287 pp.

Rapport, D. J. and J. E. Turner (1977). Economic Models in Ecology. *Science*, **195**, 367–373.

Roberts, F. S. (1975). *Energy Mathematics and Models*, Proceedings of SIMS Conference, Society for Industrial and Applied Mathematics.

Searl, M. F., Ed. (1973). *Energy Modeling*, Resources for the Future. 436 pp.

Soddy, F. (1933). *Wealth, Vertical Wealth, and Debt*, 2nd ed., Dalton, New York.

Townsend, C. R. and P. Calow, Eds. (1981). *Physiological Ecology: An Evolutionary Approach to Resource Use*, Sinaur Associates, Sunderland, MA. 410 pp.

Vogely, W. A. et al. (1973). *Energy Modeling*, IPC Science and Technology Press, Guildford, Surrey, England. 170 pp.

Wilson, A. G., P. H. Rees, and C. M. Leigh. *Models of Cities and Regions*, Wiley, New York. 536 pp.

CHAPTER TWENTY-FOUR

Ecosystems with Humans

Much of the earth is occupied by humanity, either as part of ecosystems or interfacing as users and controllers. Where humans comprise a major part, new kinds of systems evolve with human culture at the hierarchical center. Information processing, social structure, symbolism, money, political power, and war become important components along with the vegetation, consumer organisms, and the inanimate work of the biosphere. This chapter considers the ecosystems that contain humans and those ecosystems developing under use by humanity. First characteristics are given for times of low energy with hunting and gathering systems and with agrarian systems. Then use of ecosystems is considered during high-energy times with various kinds of interfaced relationship between humanity and nature.

PATTERN OF HUMANS IN ECOSYSTEMS

Figure 24-1a gives a general pattern of humans at the hierarchical, controlling apex of ecosystems *within* which they are a part. Figure 24-1b shows the pattern of interface of humans located spatially *outside* of an ecosystem drawing convergent energies and exerting feedback controls. In either case the ecosystem's environmental components have generally smaller dimensions in space and time than does the controlling human culture; thus the system is subject to pulsing by the human users with frequency set by the culture.

Notice that there are feedbacks of embodied high-quality energies in the form of amplifier actions, nutrient controls, selection, pulsing consumption, and many other controlling actions. Surviving systems are observed to have a good closed loop of service from the environmental components to humanity and humanity back to the landscape ecosystems. In a local spot or for a short time period, there may be excessive consumption. As discussed in Chapter 15 and 22, periods of severe consumption may be adaptive; however, our knowledge of this possibility is inadequate.

Survival probably requires closed-loop actions. Systems without the feedback services from humanity are drained compared to alternatives and are displaced. Public fisheries and larger animals, for example, are often drained without much mutual feedback, and these tend to disappear. The energies go into other organisms that have a better feedback from their consumers.

Culture

Apparently, humans evolved from earlier animal stages with a fundamentally greater ability to be reprogrammed by environmental conditions. This ability allowed humans to become the ecosystem's computer program, readily organized and programmed with behavior to feed back control, maximize power, and compete. The human species gradually moved along the energy-quality scale—to the right in energy diagrams. Humans became the main programming entities for new kinds of ecosystem such as systems of hunting and gathering, cattle grazing, and fishing given in this chapter. The more complex ecosystems of humanity that are organized on a larger scale because of higher-energy flows are considered later in Chapter 25 on cities and regions.

Emerging with the new ability to program were new mechanisms of controls, such as economic motivations, family, government, and religion. The social structure and programs are called *culture* (White, 1943, 1959).

Nature and Culture

Wheeler (1928) and Allee (1938) showed the roots of many of the institutions in the animal populations

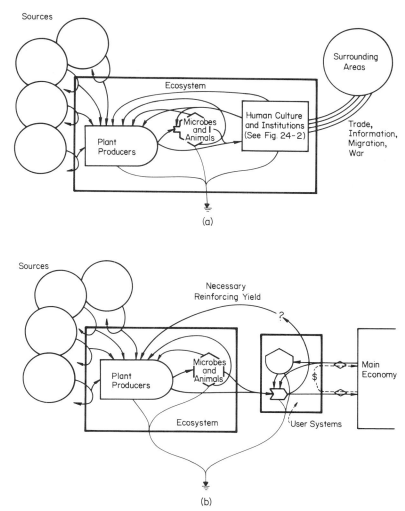

FIGURE 24-1. Patterns of humanity and ecosystems: (*a*) humans within ecosystem; (*b*) human use of ecosystems.

such as cooperation and pecking order, and many in social biology (Wilson, 1975) have shown inherited components of behavior such as hormones and belligerence. These inherited algorithms are necessarily included, but usually overridden or controlled by the learning and cultural patterns in the newly reprogrammed behavior patterns. Hence the higher-quality control is by flexible human cultural systems. However, inherited elements are necessary parts of any culture program, and when there is pathological disruption of culture, the inherent elements can express more primitive responses. In adapted systems, overriding control by the larger systems generally puts the genetic inheritance under cultural control. For example, biological programs of reproduction are generally managed by cultural inheritance.

Culture as a Controlling Program

Culture is the shared program of behavior of a population that organizes the individual, the group, and the environment into a high-quality system adapted to its energy regime and providing service to its territory. Figure 24-2 shows some of the categories that make up the cultural system. Each function has institutional structure that operates with individuals to aid in operation. These include government, religious institutions, family, war, feeding, and housing. Programs are established during training of the population, but control of the system flows from the common shared concept of culture to the human populations and then to the ecosystems. In the concept of hierarchy, social structure is a convergent component of higher quality than the population

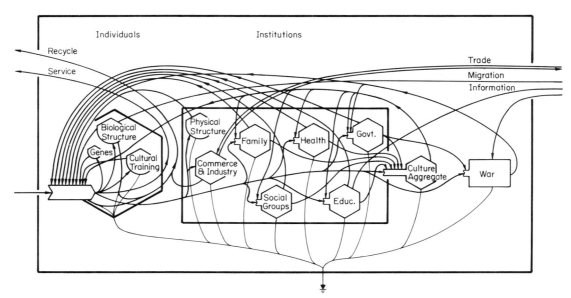

FIGURE 24-2. Details of hierarchical relations of social systems, including individuals, institutions, and cultures (see connections in Fig. 24-1).

with longer time constants and larger scale of action. As part of culture, Adams (1975) identifies a hierarchy of political units with increasing size along the chain of energy processing. An attempt is made in Fig. 24-2 to arrange components of culture on a quality hierarchy scale from left to right.

Chain Reaction of Cultural Information and Energy

Figure 24-3 relates high-quality information as a processer of lower-quality basic energy sources and vice versa, the use of basic energy to generate and maintain information. There may be selection necessary to keep information tuned to sources. This minimodel may represent the role of cultural humanity and its resource base. Comparison with designs in Fig. 10-25 show the system is a chain reaction capable of explosive growth when energy source is adequate and quenching not too large. For example, intensive oil energy and religious programming in Iran helped to produce an explosive revolution in 1979.

CULTURAL EVOLUTION OF SYSTEMS OF INCREASING ENERGY

The systems of humanity and nature can be considered in their historical sequence starting with humanity as a first consumer in the ecosystem. The historical progress in energy diagrams are illustrated in Fig. 24-4. First tribal cultures such as pygmies (Turnbull, 1963) were based on hunting and gathering, but with some large measure of control of the forest or fields through management of cutting, fire, and migration (Fig. 24-4a).

Then, with the development of agricultural cultivation and animal husbandry, larger parts of the energy realms were used and the practice of consuming larger storages began, introducing the pulsing patterns in which the culture moved like a wave over the energy storage positions or increased and decreased with the storages on which they were dependent. Land was rotated between farming use and wildlife restoration.

Increased energy base provided a larger hierarchy of concentrated high-quality centers so that vil-

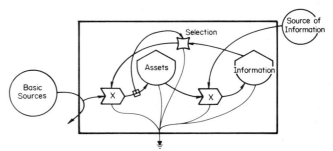

FIGURE 24-3. Interaction of basic energy source and information. Compare with configuration of explosive chain reaction in Chapter 10 (Fig. 10-25).

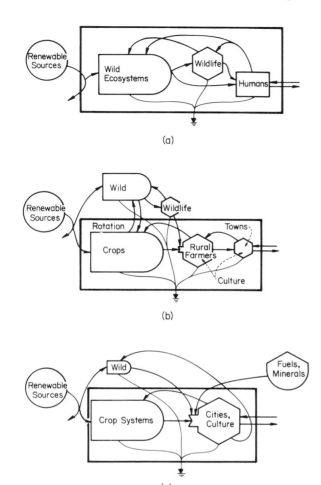

FIGURE 24-4. Minimodels of stages of energy development in human history: (a) hunting and gathering; (b) agrarian; (c) urban culture with fuels and minerals. The human subsystem is shown within the rectangle first as part of the wildsystem.

lages evolved into cities along with greater areas in cultivation Fig. 24-4b. Notice decline in size of the wild sector.

Then, with industrial revolution, mineral fuel energies were added that were even more concentrated, supporting larger cities and populations (Fig. 24-4c). In modern times many questions are raised as to the future return to use of the renewable energies and as to whether the old systems should be used or whether some low energy technology will replace simple agriculture.

Systems in Low-Energy Times

In times of lower energy before the present, most of the resources to an ecosystem were the renewable energy flows. Even the special high-quality items obtained by trade were mostly from renewable sources. Figure 24-5 gives an example of a cultural ecosystem based on an energy signature of land, water, and marine resources: the Maori system in New Zealand, which developed several hundred years before European colonization. Notice typical web with convergence to political leaders and religious symbols at the right. A large part of diet was from marine fisheries and birds and later the polynesian potato, kumara. "Flax" from the marshes supplied clothing and roofing.

Cultural Systems According to Environment

Some of the variety of cultural diversity is given in Figs. 24-5–24-9, including life based on marine, freshwater, and terrestrial regimes. As in the case of ecosystems discussed in Chapter 21, dominant structures can often be identified as an adaptation to use the special types of dominant sources in the energy signature. For example Lee (1968) relates Bushmen to water regimes.

Trade and the Surrounding System

Even with the simplest hunting and gathering system, the smaller human systems were part of and subordinate to larger realms through trade for critical amplifier items. For example, All (1977) uses plus–minus diagrams to describe the important role of obsidian for arrows and hunting (see Fig. 24-6). The system shown has some agricultural plots as a supplement.

Human Conquests and War as Energy Controlling Mechanisms

In Chapters 14, 15, 18, and 22 it was theorized that high-quality systems maintain control and maximize power by converging energy and feeding back service with pulsed use. Human cultures at the top of the hierarchy, engaging in converging energy and feeding its action back in rapid use over a short time sometimes utilize war and conquest. The spread of conquest over a large area has energy similarities to fire, epidemic locusts, and the movement of some

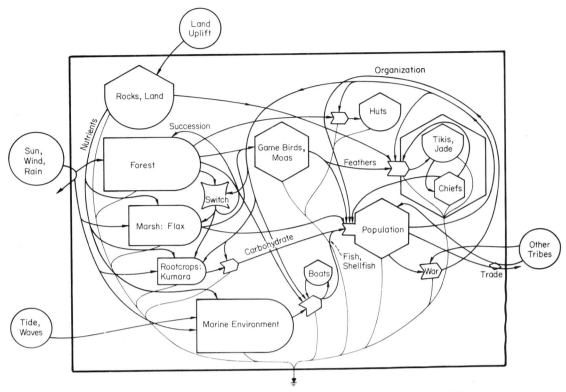

FIGURE 24-5. Early Maori culture in New Zealand (Odum et al., 1981).

weather systems. Pulsing models, like Fig. 27-4 with a frenzied consuming aspect, dependent on the available energies may be appropriate. Alcock (1972) suggests an oscillation between periods of defense build-up and war use of storages.

When energy levels are less, the scale of storage and pulsed consumption is smaller and the war actions are less severe and may be consummated in many centers, each feeding back to defend a territory protecting its energy base to the extent of the energy for this spatial action.

The cycles of peace and war in low-energy society are illustrated by Rappaport's (1971) "pigs for the ancestors." Figure 24-8 gives a model of the war mechanism simulated by Christine Padoch that generates a pulsing regime of alternative peace and war related to the energy base, the accumulation of storages, and the ritualistic release of those storages into territorial testing and protection as part of the process of maintaining order, hierarchy, and recycling. After each pulse the pattern starts over.

The linking of spatial realms to temporal cycles explains much about human culture and why it is

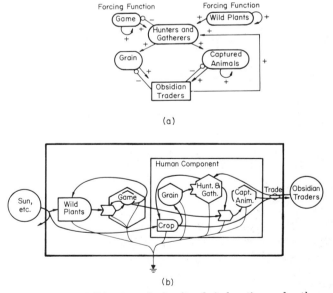

FIGURE 24-6. Diagram of role of trade in hunting and gathering society. Diagramming in energy language allows kinds and hierarchy of embodied energy to show. Obsidian is a high-quality amplifier and easily transported. (a) Plus–minus diagram [All (1977) after Jacobs (1969)]; (b) energy diagram.

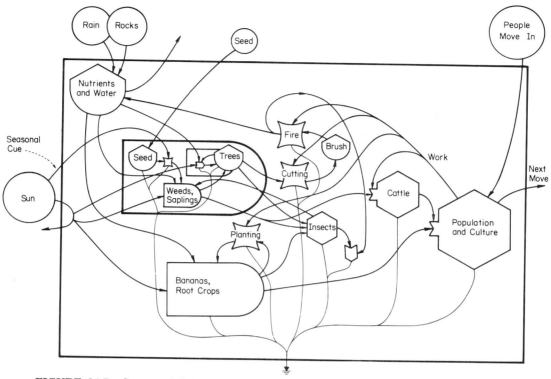

FIGURE 24-7. System of shifting tropical agriculture. Population moves after weed competition, nutrient shortage, and insects reduce yields.

competitively selected with alternating patterns—maximizing power of the whole system in the long run.

AGROECOSYSTEMS

Shifting Agriculture

Figure 24-7 shows the shifting agricultural system, which is one of the low-energy agricultural systems that emerged from hunting and gathering as garden plots increased. The shifting agriculture has the property of pulsing land use that may maximize power. The period of alternation is programmed from the human top. The rotation also provides a way for fertilizer needs and pest problems to be met by the action of the wild components in the fallow period. This system concentrates nutrients from rain and rocks weathering. The human tribal culture receives the convergence of many small pulsing plots. Modern agriculture also loses soils (Lucas, 1976) ultimately requiring rotation to a complex ecosystem for restoration (Jenny, 1980).

Dooryard Garden Ecosystem

Another low-energy ecosystem is the dooryard garden pattern found among the Choco Indians of Darien and in Indonesia (see Fig. 24-9). A large variety of food trees operates on solar energy. A recycle and protein source is provided by pigs and chickens under elevated houses where wastes are involved in recycle.

Seed and Agricultural Intensification

As shown in Fig. 14-10, one energy source may be fed back to develop another. Often this involves the use of capital remnants of one system to accelerate development of a new pattern. The evolution of more intensive agriculture from hunting and gathering may have occurred easily by gradual swap of emphasis from hunting as major and cultivation as minor to the reverse role as diagrammed by All (1977). The minor crop unit in Fig. 24-6 grows so as to dominate the system. Increasing centralization of larger-scale

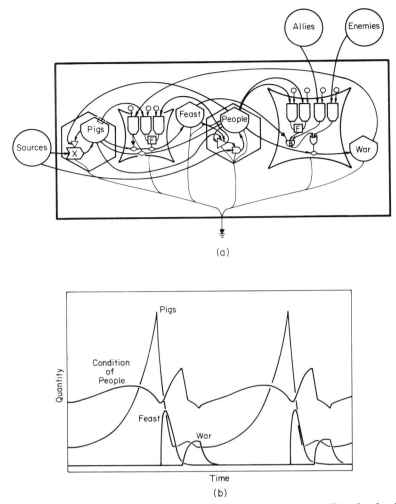

FIGURE 24-8. Model of New Guinea population in relation to war, or "Pigs for the Ancestors" (Rappaport, 1971): (a) simulation model (Padoch, 1973); (b) a simulation result.

hierarchical patterns contributed to change by control action of trade.

Seeds as valuable control mechanisms were important in organizing early economies. The planting cycle provides a way of controlling the pulsing of a productive system. Figure 24-10 shows the kind of fully agrarian human agroecosystem that developed in the monsoon climates of the world where agrarian cultures developed high complexity and quality without much nonrenewable fuels. The pattern of sacred cows in agriculture in India as shown by Harris (1965) has typical webs and feedbacks. Figure 24-11 gives a simulation model of the sacred cow agriculture that was simulated on analog computer with the use of flip-flops and track store units for the generation and successive planting of seed. This helps one visualize the seed planting as a logic control pulsing system consistent with use of digital action for high-quality control. See Fig. 5-18.

Roles of Centralized Public Structures

As total energy involved under human control increased, more complex high-quality political action and symbolic units were possible at the hierarchical convergence. Public works such as pyramids may provide one means for integrating a culture, maintaining a pulse. In times of high energy these may be in the form of symbolic monuments and other structures with large embodied energy that help in the orientation of individual patriotic and religious programming. Figure 24-12a includes a diagram of the Egyptian pyramids, an example of public works

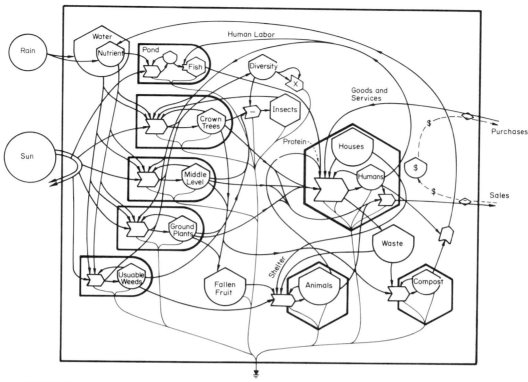

FIGURE 24-9. Dooryard garden system based on description by O. Soemarwoto (1974, 1976) as observed in Java.

that absorbs energy and help in the organization of a society (Mendolssohn, 1971). The public works may have some similar roles to war and substitute for it. Since the shape of the structures may be symbolic rather than functional, they can be individualistic, making the culture different. Heizer (1966) gives energy calculations of public works with human bodily forces. The modern pattern in Fig. 24-12b is not so different.

INNOVATION AND DOMINANCE

With more energies, higher levels of hierarchical dominance were possible with larger spatial domi-

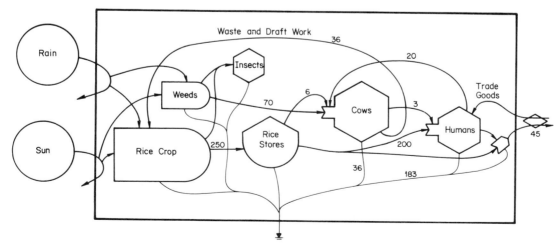

FIGURE 24-10. Diagram of Indian agriculture and the sacred cow adapted to sharply varying monsoon climate based on Harris (1965).

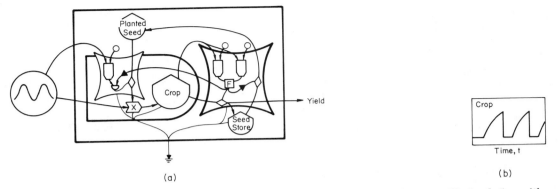

FIGURE 24-11. Crop harvest and replanting: (*a*) energy diagram; (*b*) simulation with use of track–store analog units for seed storages.

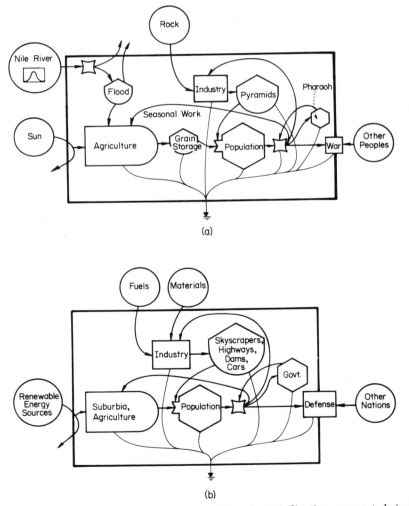

FIGURE 24-12. Construction work as an energy modulation: (*a*) pyramids as organizers in ancient Egypt (symbolic, religious); (*b*) modern civilization aggregated similarly for comparison.

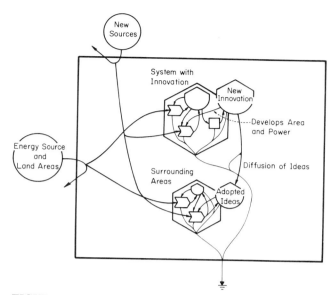

FIGURE 24-13. Role of innovation in dominance after Toynbee (1935–1939). System with innovation gains area and power but loses dominance as innovations diffuse to surrounding systems.

nance for regimes and longer time periods for oscillation. With some new innovation in resource or information, a unit that was on the same level with others could move into ascendancy to a higher level of convergence and control. Many examples of historical uses of centers of culture economic life and military power could be cited (Greeks, Romans, Mayans, etc.). Where based on innovation, these roles were temporary.

Following theories proposed by Toynbee (1935–1939), Fig. 24-13 shows how innovation helps one system become a dominant controller but loses it as the information of the innovation diffuses to adjoining competing systems.

SIMULATION OF HISTORY

Some of the models of this chapter are simulations of the main transformations of history. This is certainly one of the least developed scientific endeavors, possibly inhibited by the widely prevalent belief that history is not deterministic. However, many deterministic ideas about "force" and energy are available for simulation (Adams, 1943; Tygart, 1960). Rashevsky (1968) modeled many mechanisms of human participation, such as decay of prejudice,

transitions based on change in storage levels, and diffusion models of group exchanges, and a form of energy determinism was implied in relating cultural development to shoreline development. Taagepera (1968) fit several historical civilizations to the logistic model.

In Fig. 24-14a is a model of historical action by Farned (1974), who considered the role of combat in Greek life. Notice the order–disorder cycle (social structure and dissonance) with a civil war loop operating with a threshold. Military is increased either by greed sensor of expansion of resources through conquest, or by fear sensor on external military action.

In the more complex model in Fig. 24-14b, a system of tension and war includes military advantage loops affecting available resources and autocatalytic effect of tension. Simulations included some with erratic oscillation (see war simulation in Fig. 25-27 also).

HIGH-ENERGY TIMES

The energies of the highly developed human agrarian systems were used to develop the systems of fuel and mineral use. Soil and wood resources were used rapidly in a nonrenewable way to gain the assets that were able to develop coal, oil, and other sources. As shown in Fig. 24-4c, the fuels and minerals of high quality were incorporated in the hierarchical centers increasing the prevalence of urban systems. However, power was maximized by interacting these as amplifiers of more dilute energies of the suburban and rural environments with new kinds of interfaces.

Mineral Processing System

With the rising use of fuels and minerals, some human ecosystems became adapted for processing of nonrenewable resources (see Fig. 24-15). Such systems undergo an ascendant period and die away as the nonrenewable resource is exhausted. The system can stay in existence, only by moving to new sites as in shifting agriculture. See also Fig. 9-6 and Fig. 13-9. Georgescu-Roegen (1971) emphasizes the rising entropy in the loss of concentration of mineral resources as main factor in economic decline. See Fig. 26-21.

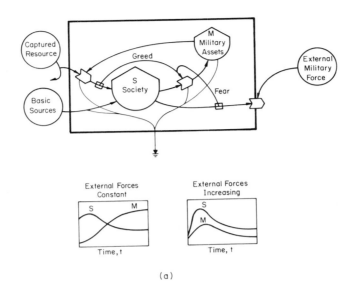

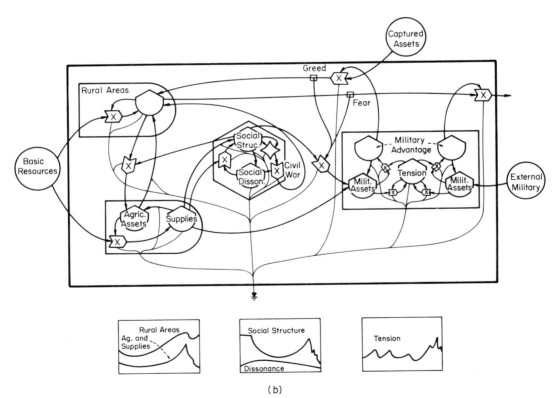

FIGURE 24-14. Simulation of military roles in social structure and access to resources (Farned, 1974): (a) military assets as a resource amplifier; (b) military tension system that controls interfaces with outside societies.

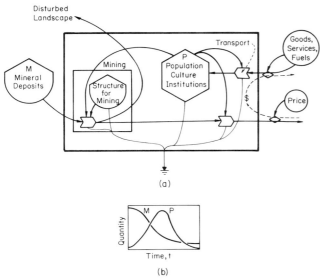

FIGURE 24-15. Temporary system based on trading of mineral deposits: (*a*) energy diagram; (*b*) simulation.

Increasing Emphasis on Cash Crop

Figure 24-16 diagrams coffee production, an example of a cash crop. The lower story of coffee trees is partially shaded by an overstory of trees that help maintain mineral cycles, soil quality, and optimum light intensities. Sales from the plantation provide funds for most other needs of the people involved.

As greater energies become available through trade for fuels or for goods and services based on fuels, agriculture becomes based increasingly on inputs from sales of crops and less on the environmental energies of sun, wind, rain, and soil. Cash crops began to replace diverse farms.

High Intensity Agroecosystems

One of the main ways of interfacing urban high-quality energy with rural low-quality energy was with technological inputs to agriculture, fertilizer, pesticide, fuel-based farm machinery, electric devices, and new genetic stocks that had been bred to route energies formerly used in self-maintenance into yield.

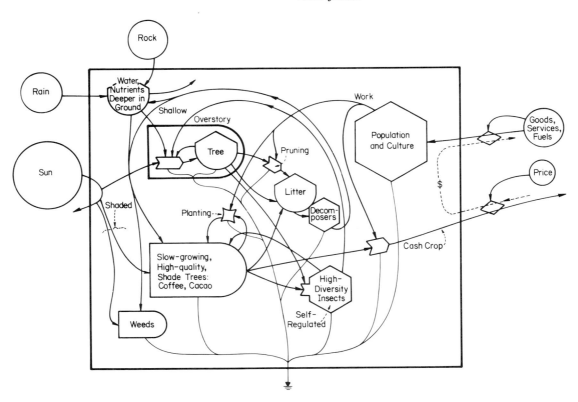

FIGURE 24-16. Overstory agriculture with shade trees and exported cash crop in a simplified ecosystem (Zevallos and Alvim, 1967).

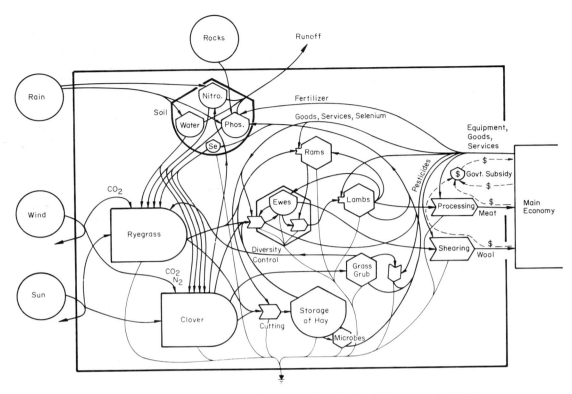

FIGURE 24-17. Sheep pastoral system of New Zealand (Odum et al., 1981).

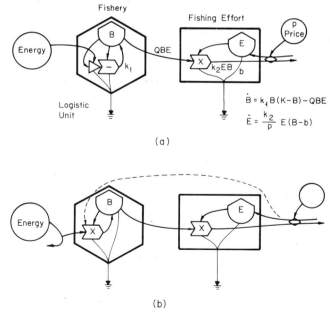

FIGURE 24-18. Minimodels of yields and sale: (*a*) fishing effort constitutes a predator where growth depends on prey–predator relation, where resource population is stabilized by logistic model (Schaefer, 1954); (*b*) same with resource population stabilized by limitations of energy flow at the source. Note dashed feedback line, which is required if fishing food chain is to be kept biologically competitive.

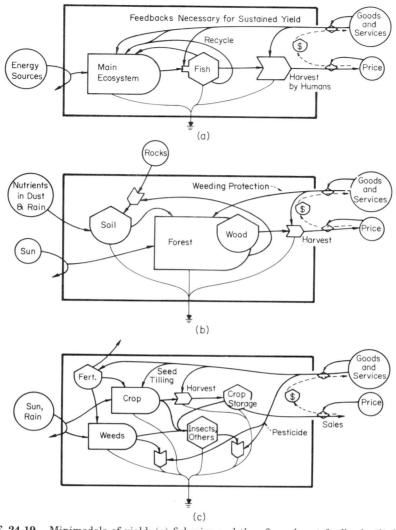

FIGURE 24-19. Minimodels of yield: (a) fisheries and the often absent feedback; (b) forestry, with seeding and nutrients required; (c) agriculture with pesticide roles.

The New Zealand sheep paddock system in Fig. 24-17 is an example of a yield system that uses a substantial embodied energy of nonrenewable fuel sources to develop high yields per area. A sheep system was evaluated for Sweden by Zucchetto and Jansson (1978). Klopatek (1977) summarizes high levels of embodied fuel-based energy supporting agriculture in Oklahoma. Principles of effective yield systems are considered next.

YIELD AND USE

Especially during high-energy times based on fuels, various new demands on the ecosystems developed, usually generating yields for economic activity, such as for cash crops. A famous minimodel for economic exploitation by Schaefer (1954) is diagrammed in Fig. 24-18a. It shows the role of price, a prey–predator coupling, and a resource system stabilized by logistic mechanism. Its response to increasing demand is similar to that in Fig. 11-13 and Fig. 23-36. A more accurate minimodel is given with a flow-limited energy source in Fig. 24-18b.

Note that the demand and yields decline with decrease of relative price. The relative price is not affected by simple inflation. It does decrease where inflation is driven by reducing net energy of the fuel and mineral sources as deposits with easy access are depleted.

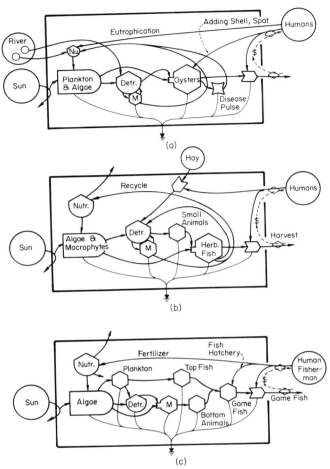

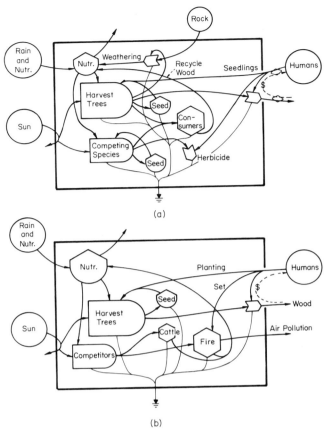

FIGURE 24-20. Minimodels of types of aquatic yield systems; ecosystems in which harvest is by humans: (*a*) oyster culture; (*b*) carp pond; (*c*) game fish pond (M = microbes).

FIGURE 24-21. Minimodels of terrestrial systems used for harvest: (*a*) forestry production of wood; (*b*) fire-managed forestry.

Yield Minimodels

Human systems usually contain subsystems that generate food and fiber from environmental energies. For purposes of understanding the basis for yields in agriculture, fisheries, forestry, and wildlife systems, ecosystem minimodels can be aggregated to emphasize the total external input resources that are the basis for yield and are necessary for the yield to be a sustained one. These models can be driven by annual and year-to-year variations in resources. Some yield overview minimodels are given in Figs. 24-19–24-21. Another in Fig. 24-11 was given earlier illustrating the use of track–store units.

Since a drain such as yield will tend to depress the part of the ecosystem that is desirable, a sustained yield requires an energy feedback amplifier that reinforces the yield pathways. These are present in Fig. 24-19 and Fig. 24-20.

The seasonal alternation of planting and harvest entrains the pulsing cycle of storage and use that may be inherent in human society and may maximize yield. See de Klerk and Gatto (1981).

Successional Position of Yield Systems

Species that constitute successional systems tend to have more net production than those in climax. Consequently, the successional species are the main ones domesticated for basic food production. Some products that have more embodied energy require a more complex yield system, more like those of a climax. Hart (1980) arranges tropical crops on a scale of position in the scale of ecosystem complexity that goes with succession. Grains are early but tree crops, coffee, and cacao are later. Compare simpler yield systems in Fig. 24-19 with those in Figs. 24-9

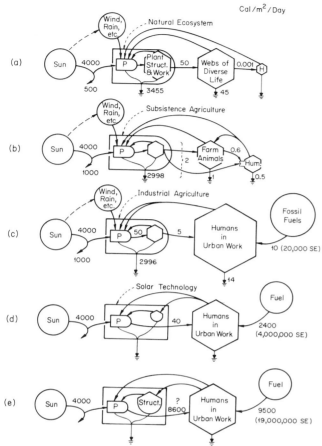

FIGURE 24-22. Comparison of human uses of solar energy (Odum, 1978). Numbers are actual calories except as indicated. Diagrams are arranged in order of increasing concentration of human and technological activity, from one person per square mile at the top to a solar tower converting agricultural waste at the bottom. (a) Hunting and gathering as part of the ecosystem; (b) subsistence agriculture; (c) industrial agriculture; (d) solar cells; (e) solar tower that uses heat to turn farm wastes to fertilizer.

and 24-16, where more structure and respiration operate.

Net Energy

Yield systems based on renewable energy develop net energy temporarily that is used to develop feedbacks of high-quality inputs of equal embodied energy after which there is no net energy. When outside additional energies are attracted, the yield ratio decreases as the yield increases. Industrial agriculture seldom yields net energy. Examples follow.

Figure 24-22 compares solar conversion. Hunting and gathering and the subsistence agriculture have small net energy yields on renewable energy basis. In Figs. 24-22c–24-22e fuel-based inputs are used to augment yield, but these are not net energy yielding. In these an interaction of solar- and fuel-based inputs is used to obtain more output.

External Energy Control of Yield of Agroecosystems

The models of agricultural systems show productivity in providing net yield beyond parasites, predators, and competing organisms as dependent on the energy supplied in high-quality control form from outside. In Fig. 24-23 Mitchiner et al. (1975) show that the pesticide that can be used to maintain an economic balance is a function of the prices, which are in turn dependent on the levels of fossil fuel energy available to the economy in which the yield system is embedded.

In Fig. 24-24a Gutierrez (1978) gives models for pastures in Florida in which the amount of intensive improvements that are economic, and thus the yields, are dependent on the energy costs. Fig. 24-24b is a simulation when energy costs were low. Simulations show lesser control energies used with rising energy costs. The successful pasture then becomes one with more diversity, less management, less yield, and much less cost. See grazing models of Goodall (1967).

In Fig. 24-25 Gutierrez also shows the embodied energy of feedbacks to Florida agriculture compared to the solar energy contribution also expressed in embodied coal equivalents CE. The least intensive rangeland agriculture at the top has a low investment ratio (11.5/268) compared to citrus fruit at the bottom with a high ratio (3.64/0.76). DeBellevue (1976) found winter crops in Hendry County, Florida with 25/1, indicating these products are luxuries.

OPTIMIZATION

Systems, in their self-organizing adjustments to survival and maximizing power, may optimize their allocations of resources and feedbacks. When humans are part of the system, the humans become the instrument for optimization and the models they generate are one of the means (Rosen, 1967). For example, linear programming models are used to

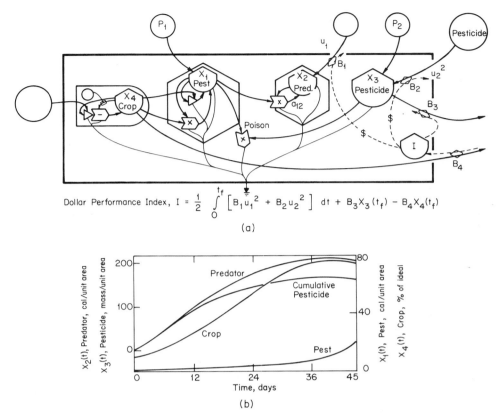

FIGURE 24-23. Model of pest regulation and cost given by Mitchiner, et al. (1975): (a) diagram of equations given; (b) simulation when effectiveness of money spent is maximized with dollar performance index.

prevent uneven use of resources (see Fig. 12-4). Fig. 24-24 has self-optimizing features adapting to decreasing availability of fossil fuels.

Humans often view yield and resulting dollar profit as the property to be maximized. This may be tantamount to maximizing power since more external embodied energy may be purchased with greater yields.

Swartzman and Van Dyne (1972) review optimization models for yield systems. An Australian range management model is given with experiences in generating maximum yields from sheep where other consumers such as kangaroos are present. Transfer of sheep from drought areas served as a principal mechanism of optimization.

INTERFACES WITH TECHNOLOGY

Still fairly new are the many contacts and uses of environment by still changing technological systems. Technology has special uses of environmental products, special waste materials returned to environment, and special land uses by humans whose main role is in technology. Solar energy uses may be arranged on a scale according to the investment ratio of high-quality inputs. In Fig. 24-22 subsistence agriculture with small net energy is at the top, grading into extremely intensive processes where the embodied energy calculation from the sun is a minor fraction of the energy used. The sun in this frame of reference provides a means for conservation of other inputs, giving them matching action.

Intermediate Technology

During present times of high energy, the primitive agroecosystems and land uses are not generally competitive since they do not make use of the available fuels, minerals, goods, and services (whose high embodied energy is from use of fuels and minerals).

However, considerable exploration and testing is under way of lower-quality energy technology that uses a higher percentage of environmental energy

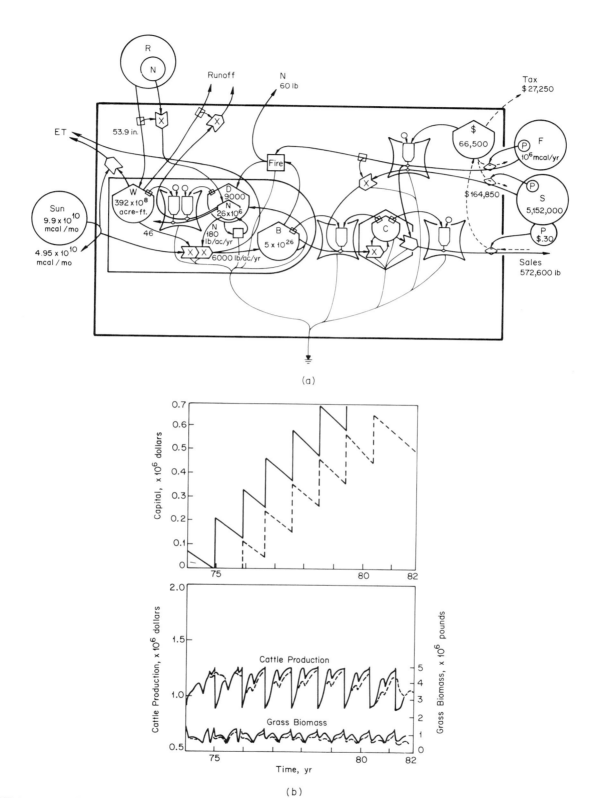

FIGURE 24-24. Model of improved pasture evaluated and simulatd by Gutierrez (1978): (a) model with values, with F = fuels, supplies, and labor; supplemental feeds; B = grass biomass; C = cattle biomass; N = nutrients; D = organic matter; R = rainfall; W = water; $ = capital; P = price; T = taxes; numbers represent 15,000-acre system and 2000 cattle in Kissimmee River basin of Florida; (b) typical simulation when prices of energy and derived inputs permit profit; Dashed line is for less nutrients.

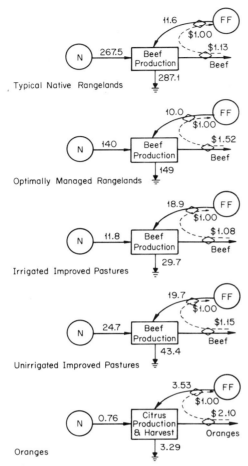

FIGURE 24-25. Energy summary of some examples of Florida agriculture (Gutierrez, 1978). Energy units are embodied coal equivalent calories. Investment ratio increases from top to bottom.

but with some high-quality energy from the modern economy. Burnett (1978) studied modern examples of appropriate technology agriculture with lower investment ratios but with some technological inputs (e.g., note the high degree of self-sufficiency in Fig. 24-26).

Technoecosystems

Some formerly wild components of ecosystems may be incorporated into technological systems as hybrids of living units and hardware homeostatically coupled. These might be called *technoecosystems*. For example, in Fig. 24-27 an electromicrobial system that uses *Hydrogenomonas* converts electric power into organic matter and oxygen, the same products as in photosynthesis, using electrolysis to produce hydrogen, followed by living chemosynthesis.

Shimazu et al. (1972) simulated a model of the *Hydrogenomonas* system coupled to human function and found the system fairly stable to the pulsing byproducts of human activity. Reservations about supporting humans without a large area and full hierarchy were given earlier (Odum, 1963).

An example of an ecosystem controlling its own technology was arranged by Beyers (1974) (see Fig. 24-28). Rises of pH are a product of photosynthesis and sensed with pH electrode, and the signal is fed back to allow for selection of changes correlated with more energy conversion. Thus the ecosystem gradually turned the system to full light, maximizing photosynthetic power.

Role of Computers

Norbert Wiener wrote philosophically of the symbiosis emerging between systems of technology and human functions in which part of the control actions and information would be performed by computers. Wiener asked whether computers, if they could become sufficiently intelligent to assume roles now performed by humans, could have self-consciousness, developed souls, and be admitted to church (Wiener, 1964). A system in which there is a combination of human and hardware components is drawn in Fig. 24-29. So far the human mind apparently has more embodied energy and control ability.

Suburbia

Another interface between fuel-based urban technology and the renewable energy resources evolved through development of suburban settlements around the cities where feedbacks of technology from the city interacted with land, sun, wind, and water, generating a high standard of human living.

Grassy Lawn Interface

A subsystem dominant in urban culture is the grassy lawn, which is managed something like a grain crop with fertilizing, harvesting, insect control, and so on. An energy analysis provided by Falk

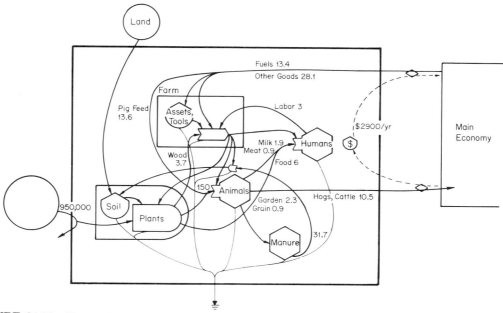

FIGURE 24-26. Energy flows on Taylor farm, Arkansas (Burnett, 1978). Flows are million calories per year.

(1976) is given in Fig. 24-30. The main output to the human systems is apparently primarily aesthetic and organizational and secondarily as a maintainer of landscape and soils, since other ecosystems can maintain and develop landscape values with less expensive inputs from the economy. This environmental interface system may be displaced with alternatives such as gardens, orchards, and vineyards as energy sources become scarce.

The aesthetic ideal of neatness and order may have been adaptive on the frontier, but an opposite aesthetic ideal may be developing now, one that favors complex vegetation that appears disorderly because of the many levels of complexity in its orderly pattern but is better as a waste absorber, rebuilder of soils, and interface with urban life.

Parks

A systems model and quantitative energy overlay was made of Everglades National Park, including several types of ecosystem, with the management, tourists, and surrounding economy affected (De-Bellevue et al., 1979). The ratio of attracted outside energy was smaller in relation to environmental resource than typical investments. This might be interpreted as evidence that the Everglades National Park is underutilized. The numerical ratio for this park needs comparison with other more used parks.

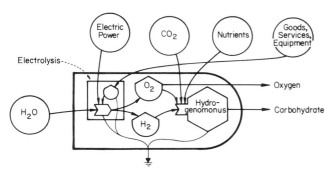

FIGURE 24-27. Electromicrobial system that converts electric power into organic matter and oxygen with aid of *Hydrogenomonas* bacteria; example of a system studied for technological production and support of humans in space (Bongers and Medici, 1966).

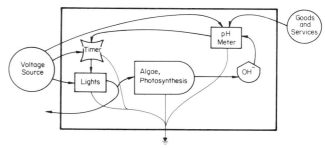

FIGURE 24-28. Biofeedback control of power source by algae with use of pH (Beyers, 1974).

528   SYSTEMS OF NATURE AND HUMANITY

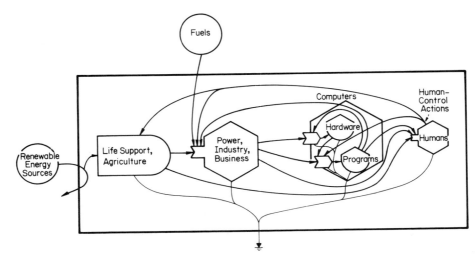

FIGURE 24-29. Diagram of humans and computers with control pathways.

Toxic Actions

The long-term pattern of survival of ecosystems in the biosphere substitutes adapted systems for those that would be stressed by energy flows, even those that tend to be harmful to life. Many substances are useful at low concentrations and become toxic at higher concentrations. The toxic substances become part of cycles that develop combinations of tolerance and binding and burying mechanisms that through natural selection supply ecosystems for the varying conditions of the earth. For example, ecosystems near volcanic emissions are restricted in variety and are minimal, with much of the productive energies being used for special adaptation or replacement. In some extremes the surviving process systems are nearly without life.

Figure 24-31 gives a minimodel to indicate main properties of a toxic substance that is involved in a self-reproducing system. There is an inflow, an outflow, and a depreciation–elimination process. Some toxic action is external, causing loss of structure and thus indirectly of process. Some toxic substance is incorporated into storage of structure. While thus stored, it causes additional depreciation and recycle. Some toxic action contributes to production beneficially in the low range where it is recycled as structure is recycled. Schomer (1976) simulated recovery of Escambia Bay, Florida.

The effects of toxicity can be read by monitoring

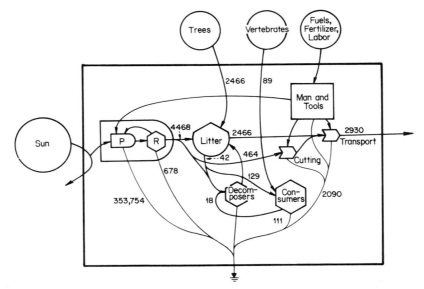

FIGURE 24-30. Energy flow in a suburban lawn (Data from Falk, 1976). Flows are actual calories per square meter per year.

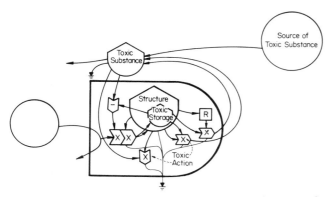

FIGURE 24-31. Minimodel of main interaction of a toxic substance with a self-reproducing entity [see cadmium example in Knight (1980)].

one or more aspects of the system such as the structure, the productive process, the rate of respiratory process, the rate of destructive removal, and the rate of accumulation. The model helps to show the relationships of these various system properties to each other and to toxic actions.

The greater the amplifier action of the toxin, the less of it is required to have effects that are either harmful or that can be used in support of the system. For example, adaptation to a toxin provides a system an easy means of eliminating competition from units not so adapted. An interesting theory is offered that toxic action is proportional to embodied energy in the same way as other active systems that have experienced competition are predicted to have energy effects related to embodied energy required to develop their use.

Knight (1980) developed models for action of cadmium showing its very high embodied energy, its high sensitivity of amplifier actions, and its role as a controller of ecosystems with and without top consumer animals. His generalized curve of chemical action was hump-shaped tapering at high concentrations.

Thomann (1979) used size of organisms as a variable in models of retention of polychlorinated biphenyls (PCBs). Mitsch et al. (1978) found pollution stress greater on organisms with higher energy quality (higher energy transformation ratios).

Environmental Overload

When the levels of embodied energy of purchased inputs are high compared to the matching environmental resource flow, there may be an overload of environment. The investment ratio (Fig. 14-8) is higher than in other places. The system is not economic because the environmental contribution is less than in competing activities. Because there is a mismatch of high- and low-quality energies, and because of destruction of the environmental system, the output is less than it could be with the same energy sources applied elsewhere.

Hornbach and Fitz (1977) gave a matrix of ecosystem relations, using plus or minus signs to relate cattle in the Sahel of Africa and its role in environmental deterioration. Antonini et al. (1975) simulated development trends, using energy language models for landscapes in the Dominican Republic under impacts of deforestation and dam building. Reduced deforestation results in extended life and capacity of dam, and reduced erosion causes increase in land productivity.

Waste Interface

Another interface between the urban culture and the environmental energies is the waste release of urban and industrial society. Whereas the source sink concept was prevalent, as humanity became so predominant in the landscape, it became necessary for recycle mechanisms to develop. Once again the wild ecosystems were needed for general waste recycling and restoration of soils, forest stocks, and so on. Figure 24-32 contrasts the technological efforts with the wild ecosystem interface method. The environmental technology (Fig. 24-32a) disposes waste, but in some cases they make more waste stress than they ameliorate. In contrast, Fig. 24-32c uses a renewable swamp ecosystem as an interface to receive and recycle waste, for forest growth, and to conserve wastes—all on solar energy with much less cost of goods and sources and fuels. The investment ratio (economic activity divided by environmental input with both in units of embodied energy of the same type) with the swamp is generally about the same as for the total United States.

Conscious interfacing to encourage self design is ecological engineering (Odum, 1976a). The interface principle used in Florida wetlands and elsewhere may be a general mechanism for fitting human society to environments (Odum et al., 1977a; Ewel and Odum, 1982).

For ways of using environmental systems representations as environmental impact summaries see

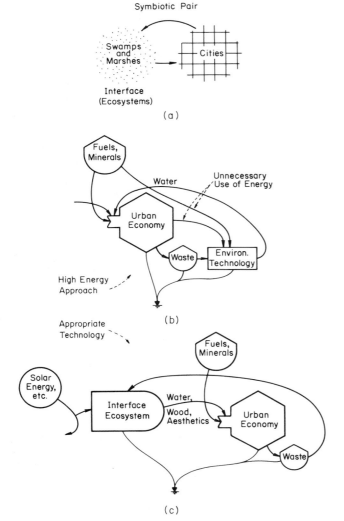

FIGURE 24-32. Interface ecosystem concept with swamp as a symbiotic partner of the economic system: (a) spatial sketch; (b) technological waste treatment with use of fuel-based inputs; (c) treatment by solar energy-based interface ecosystem.

environmental evaluation in Chapter 23. Martin (1973) used an index of environmental performance:

$$\text{Perf.} = (\text{index of ecosystem})^2 + (\text{control costs})^2 \quad (24\text{-}1)$$

Urban Recycle with an Agrarian Economy

Ultimately, urban wastes may be recycled to food production systems as in preindustrial times. Stanhill (1977) provides energy and nutrient analysis of a very intensive system of agriculture part of the Paris urban system in 1882 (Fig. 25-22). Wastes from 96,000 horses that served the city constituted the basis of high-intensity, high-yield crops. Without fossil fuels, an urban economy requires a tight symbiotic relationship between farm products converging for city use and wastes dispersing as recycle (Fig. 23-2).

SUMMARY

Ecological systems that include humans as controlling consumers range from primitive tribal consumer interfaces through early agroecosystems to the increasingly urban patterns of the twentieth century that so utilize and crowd their external resource base as to reach thresholds beyond which they are no longer competitive. Models help to show the ways in which the systems of nature and culture may control the progress of history. Also in this chapter models were given for various environmental utilization interface systems from the past and the present. Many modern systems are really based on high levels of direct and indirect fossil fuel use, and these environmental use systems may be expected to give way to older ones that are more symbiotic. Included are some questions of net energy among proposed systems of interfaces between nature and humanity such as the relative productivity of green plant (chlorophyll) solar base versus solar technology. The positions of humans and their culture in the energy diagrams were arranged according to hierarchical position postulated in the scale of embodied energy.

QUESTIONS AND PROBLEMS

1. Using simple models, indicate the role of humans in systems of humanity in nature, including a hunting and gathering economy, a renewable energy agrarian economy, and a fossil fuel-based economy.

2. How are the main aspects of culture and the life of human individuals arranged to the large system of environmental support?

3. What are mechanisms and possible adaptive roles of war affecting temporal patterns and spatial patterns? What is the effect of adding rich fossil fuels to such a system?

4. How does rising relative price of energy affect pressure of human effort on stocks generated in the external environments?

5. What are the main sectors of a relatively self-sufficient economy as judged by a low investment ratio? Which investment ratios are characteristic of modern agriculture, forestry, fisheries, and the grassy suburban lawn?

6. What is the *Hydrogenomonas* system? How does it compare with photosynthesis?

7. How do high-quality symbols and functions organize and control social systems? What kinds of pulse are generated by cultural control mechanisms? What events in history may be interpreted in terms of pulsing oscillators? Diagram energy relations for different historical stages of U.S. history.

8. What properties of human systems make the combined systems of humans and nature competitive in maximizing power and long-range survival?

9. Describe the environmental impacts of modern human systems requiring system reorganization. How are these diagrammed? How can wilderness be retained and contribute to maximum power where most externalities are being matched by mined flows of high-quality energy flow?

10. What properties of culture and social organization may be different during climax and regression as compared to recent period of growth?

SUGGESTED READINGS

Barrett, G. W. and R. Rosenberg (1981). *Stress Effects on Natural Ecosystems,* Wiley, New York. 304 pp.

Bennett, R. J. and R. J. chorley (1978). *Environment Systems,* Methuen, London. 624 pp.

Boulding, K. E. (1978). *Ecodynamics: A New Theory of Societal Evolution,* Sage Publ., Beverly Hills, CA. 368 pp.

Brebbia, C. A. (1976). *Mathematical Models for Environmental Problems,* Wiley, New York.

Chadwick, M. J. and G. T. Goodman, Eds. (1975). *The Ecology of Resource Degradation and Renewal.* Blackwell, Oxford, U.K. 480 pp.

Charnes, A. and W. R. Lynn, Ed. (1975). *Mathematical Analysis of Decision Problems in Ecology, Lecture Notes in Biomathematics,* No. 5, Springer-Verlag, New York.

Chorley, R. J. and P. Haggett (1967). *Integrated Methods in Geography,* Methuen, London.

Clapham, E. B. (1981). *Human Ecosystems.* Macmillan, New York. 416 pp.

Clapham, W. B., Jr. (1981). *Human Ecosystems,* Macmillan, New York. 416 pp.

DeSouza, G. R. *System Methods for Socioeconomik and Environmental Impact Analysis,* Lexington, Heath, Lexington, Mass.

Domenico, P. A. (1972). *Concepts and Models in Groundwater Hydrology,* McGraw-Hill, New York.

Eschenroeder, A., E. Irvine, A. Lloyd, C. Tashure, and K. Tran (1980). Computer Simulation Models for Assessment of Toxic Substances, in *Hazard Assessment of Toxic Chemicals,* Ann Arbor Science, R. Hague, Ed., Ann Arbor, MI, pp. 323–368.

Haldin, S., Ed. (1979). Comparison of Forest Water and Energy Exchange Models, in *Developments in Agricultural and Managed-Forest Ecology,* vol. 9. Elsevier, Amsterdam, Neth. 258 pp.

Harris, C. J. (1979). *Mathematical Modelling of Turbulent Diffusion in the Environment,* Academic Press, New York. 500 pp.

Harris, M. (1979). *Culture, People and Nature,* 3rd ed., Harper, New York.

Howe, C. W. (1979). *Natural Resource Economics,* Wiley, New York. 350 pp.

Iskander, I. K., Ed. (1981). *Modeling Wastewater Renovation,* Wiley, New York. 801 pp.

Keith, A. (1948). *A New Theory of Human Evolution,* Watts, London.

Lee, R. B. (1979). *The Kung San: Men, Women, and Work in a Foraging Society,* Cambridge, Univ. Press, Cambridge, U.K.

Moran, E. F. (1979). *Human Adaptability, An Introduction to Ecological Anthropology,* Duxbury Press, North Scituate, MA. 404 pp.

Pimentel, D. (1979). *Food, Energy, and Society,* Halstead Press (Wiley), New York.

Rosenzweig, M. L. (1971). *And Replenish the Earth,* Harper, New York. 304 pp.

Simmons, I. G. (1981). *The Ecology of Natural Resources,* 2nd Ed., Wiley, New York. 436 pp.

Stanhill, G. (1974). Energy and agriculture. *Agroecosystems* 1, 205–217.

Steinhart, J. S. and C. E. Steinhart (1974). *Energy: Sources, Use, and Role in Human Affairs,* Duxbury Press, North Scituate, MA.

Tygart, F. J. (1960). *Theory and Process of History,* University of California Press, Berkeley.

Van Dyne, G. M., Ed. (1969). *The Ecosystem Concept in Natural Resource Management,* Academic Press, New York.

Vogely, W. A., Ed. (1975). *Mineral Materials Modeling,* Resources for the Future, Washington, DC. 404 pp.

Waintz, W. and P. Wolff (1971). *Breakthroughs in Geography,* New American Library, New York.

Welch, E. B. and T. Lindell (1980). *Ecological Effects of Waste Water,* Cambridge Univ. Press, New York. 349 pp.

Wheeler, J. O. and P. O. Muller (1981). *Economic Geography,* Wiley, New York.

White, L. H. (1959). *The Evolution of Culture,* McGraw-Hill, New York.

Wilson, A. G. (1981). *Geography and the Environment* (1981). Wiley, New York. 296 pp.

CHAPTER TWENTY-FIVE

Cities and Regions

Regional ecosystems are an organized mosaic of land-forming geologic and meteorologic processes, wild ecosystems, human-dominated ecosystems, human economy and institutions, and hierarchical urban centers. Regional systems have hierarchical spatial organization, interplay of many kinds of energy of the climatic signature, and emergent properties of landscape on a larger scale. As introduced in Chapter 15, the landscape hierarchy may be visualized with a hierarchical model given in Fig. 2-1. Continuing the use of energy principles for spatial organization given in Chapter 18 and the study of human-used ecosystems in Chapter 24, this chapter uses models to consider the organization and development of regions and their cities.

Because the parts are so visible to human beings, who are themselves parts within the landscape systems, there is a natural tendency to consider the systems as the result of human behavioral, economic, and environmental actions only. However, if we are consistent in our approach to all systems and believe that such overall design criteria as maximum power are operating, we can aggregate overview models by using the principles of energy organization already given. The hypothesis implied is that human behavior has become programmed genetically, through cultural inheritance and through contemporary learning to accomplish the operations necessary to produce the modular patterns that all systems develop. Porteous (1977) related individual participation in urban systems, finding varying perceptions of the complex system by individuals.

On the other hand, those less convinced of the predominance of the overall energy laws on human and environmental behavior prefer to model the observed mechanisms first and then aggregate these parts to find the overall models. Perhaps both may give the same answer and there may be little inconsistency between having principles of the parts and those of the whole. In any case, it is appropriate to present various models of cities, regions, and states, translated as much as it is possible to the common energy language of this book. Wilson (1974), and Wilson et al. (1977) summarize models of cities and regions, whereas Greenberger et al. (1977) review history of models of public policy. See also Shugart et al. (1981).

OVERVIEW MODELS OF CITIES AND REGIONS

Energy Basis in Simplest View

Figure 25-1 offers a very simple view of the systems of the landscape. The renewable resources interact with stored resources so as to maintain assets against depreciation and process costs, sending out enough products and attractions to bring in, by exchange, outside energy in the form of fuels, materials, goods, services, and information.

The resources in the simple model are the same as those in the larger ones. The ultimate outcome of such a system is rapid growth on stored resources, building up to a higher peak and returning to a lower steady state, while operating on renewable resources and the available outside exchanges. The outside exchanges also rise to a peak and decrease, making the final steady state a lower one. By "steady state" here we do not necessarily mean a very steady state, since the long-range steady state may be a repetitive or pulsing one.

A simulation minimodel for this simplest system is given in Fig. 25-1. Here the stored resources are defined as the net resources, the energy yield over and beyond energy used in transporting the energy into the system. By this definition, specification of the function by which the declining resources become less available may be postponed. Thus any nec-

Cities and Regions

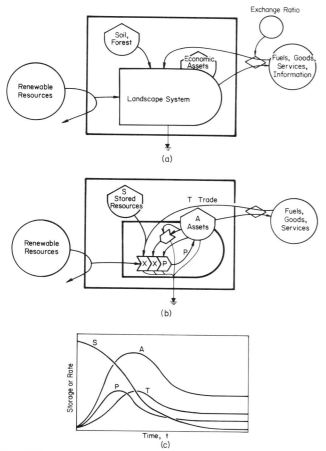

FIGURE 25-1. Main features of a landscape system: (a) overview; (b) simulation model; (c) pattern of growth and steady state as S is depleted.

essary items must come from the foreign exchange based ultimately on renewable resources. The drains include regular, linear depreciation, and quadratic costs such as crowding, diversity, cooperation, and conflict. The quadratic term is proportionately larger for larger units. Figure 25-1c shows a typical result obtained by using numbers reasonable for a regional system. With the quadratic pathway, the model has some logistic properties.

Examples of simulations of regional systems that generate curves of this shape are those by Zucchetto (1975a,b), Brown (1980), Alexander (1978), and Sipe (1978a,b).

The model grows and then levels, either because the pumping from renewable resources is equal to the rate of supply or because the quadratic terms of increasing numbers and assets have equaled the production. The quadratic term as given in this model accelerates growth and thus is competitive during growth.

An Economic Overview

The model in Fig. 25-1 has been elaborated in Fig. 25-2 to show the flows of income and payments for incoming fuels, goods, and services. The renewable resources and the stored resources interact with the feedback of high-quality assets and with trade imports to generate production that maintain assets against depreciation and process costs and also gen-

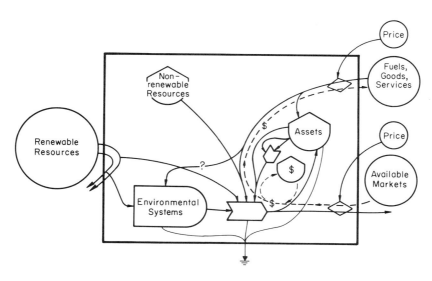

FIGURE 25-2. Generalized model of a city or regional economy with exchange of money with the outside economy.

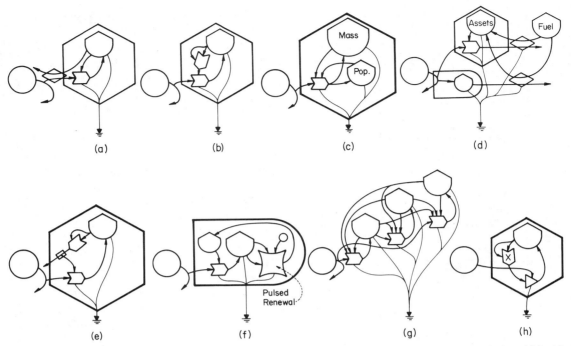

FIGURE 25-3. Minimodels of cities and regions: (a) exchange for renewable resources; (b) cooperative use of renewable resources; (c) population and physical assets on renewable resources; (d) renewable resources exchanged for fuel; (e) energy drain proportional to a square of population; (f) institutionalized pulsing system; (g) three sectors interorganized; (h) quadratic growth on unlimited energy.

erates foreign products for sale or attractions for foreign funds to investment. Notice that both sales and imports depend on prices that may be determined by the larger system in which the landscape is embedded. If the energies available abroad are decreasing in availability or if more energy is going back into their mining, the prices of imports rise more rapidly than those of sales. In other words, more goes out for what is imported. The effect of changing energy exchange ratio is a decrease in effect of imports on growth rate or on steady state.

The model in Fig. 25-2 also separates the environment as a subsystem, showing the way some environmental yields are drawn into the economic production and the way other environmental energies are diverted away from the environmental subsystem initially existing. The pathway marked with a question mark is weakly developed in many modern patterns. It is the feedback from the urban economy necessary to maintain the environmental subsystem healthy enough to provide a good symbiotic input to the economy. In the absence of this path by conscious design, it may be developing nevertheless as the surviving natural ecosystems tend to be those that can better utilize human waste recycle.

Minimodels

Many of the autocatalytic models from earlier chapters can be used to represent or simulate cities and regions from overview. Some of these are given in Fig. 25-3. Quadratic acceleration (Figs. 25-3b, 25-3e, and 25-3h) comes from interactive efficiency of city clustering.

Von Foerster et al. (1960) found percent rate of U.S. growth before 1973 quadratic. When translated, his model is that of Fig. 25-3h. This is one of several minimodels for overviews of cities, regimes, or counties (see also Fig. 9-13).

As in logistic models, drains are also quadratic. The effect of clustering may be neutralized if the quadratic feedback does not find energies sufficient to justify high drain. Baumol (1967) finds external costs such as pollution increasing as the square of population.

HIERARCHY OF LANDSCAPE SYSTEMS

The landscapes of nature are hierarchically organized, as already described in Chapter 15. Examples

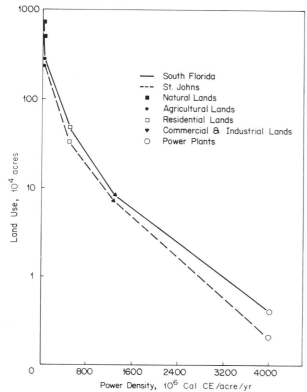

FIGURE 25-4. Distribution of land use areas as a function of embodied energy density (Brown, 1980).

are converging stream networks, desert arroyos, delta distributions, and convergences of plate movements to form mountains. Water is often the mechanism for converging solar energies into landscape organizational work.

Water flows organize landscape into a chain of land or water area uses that are hierarchical with increasing energy quality (see Chapter 15). For example, streams converge from the broad surfaces of rainfall to tributaries and finally to high-energy and -volume rivers. These land areas represent convergence of energy resource and quality to which human developments have always responded in their coupling.

Another example is the convergence of water-bearing energies from upland ecosystems to relatively smaller areas of swamps that serve as filters and recharge actions much as consumers serve producers. M. T. Brown (1980), in Fig. 25-4, provides a frequency diagram of land uses with hierarchical relation of common ecosystems converging to less common but more intensive land uses that have more embodied energy on the right.

For the large scale of the human-dominated landscape, Fig. 25-5 may be appropriate, showing the relationship of the city as the concentrated core and controlling unit on the larger units being supported. Modeling a city can be done either in the context of its larger supporting area or, if isolated, with great care to identify the many energy inflows and outflows, their magnitudes, and programs of supply.

HIERARCHY OF DOXIADIS KINETIC FIELDS

Within the city the hierarchy of convergence from lower-quality energy to more concentrated functions is expressed in the organization of transportation, communication, and movement. As depicted in Fig. 25-6, Doxiadis (1968), under the name *ekistics*, shows many circles representing walking distance for people on foot. Areas of such organization become tied together into larger circles that are the radius of easy movement by energy augmented transportation such as vehicles. Finally, these areas of size and organization are further interconnected by major arteries of mass transport. As function coverges from the broadly distributed to the concentrated, the energy embodied in values of the transported or the communication increases, justifying greater energies in the means of transport. With motion as an organizational principle, Doxiadis finds the hierarchy of organization that can be represented as another case of energy convergence (see Chapter 15). In the evolution and growth of cities, which accompanied expansion of the energy base in fossil fuels, the chain lengthened, with more intensive central organizations based on greater convergence. Many phenomena such as skyscrapers, information concentrations, and increased processing costs resulted (see Fig. 24-12).

A More Detailed Regional View

Figure 25-7 is a more detailed model which is like the models in Fig. 25-1 and 25-2 but further disaggregated so as to separate agricultural components, tourism, manufacturing, population, labor, households, transportation, and health. Because the main driving functions are unchanged, the overall behavior of the model is generally similar, with rapid growth followed by decreases and leveling brought on by disappearance of stored resources that were used up and return to renewable resources within and without the system. For different regions, some

536   SYSTEMS OF NATURE AND HUMANITY

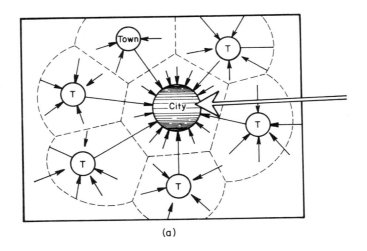

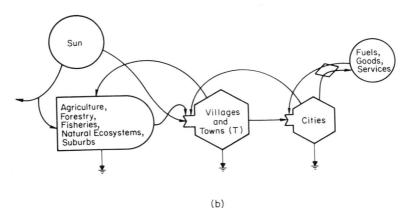

FIGURE 25-5. Hierarchy of landscape with solar energies converging as they interact with feedbacks from towns and cities, which have high-quality energy flows of fuels, goods, and services: (a) spatial pattern showing converging pathways and high-quality fuels, electricity, and goods and services incoming from the right (arrow); Feedbacks are omitted. (b) energy diagram with converging pathways and diverging feedbacks.

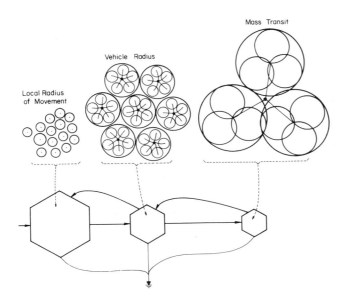

features would be more important and others omitted. A model in a marine area would have more marine resources, such as tide, for example, as driving functions and as subsystems.

County Models

County models are given in Figs. 25-8 and 25-9. Notice the use of image as a connection between resource production and human adaptation. Boynton et al. (1975) and Boynton (1975) simulated the

FIGURE 25-6. Hierarchical patterns within a city in which radius of movement is drawn as given by Doxiadis (1968). Many people with small movements support a few where functions are of larger dimension.

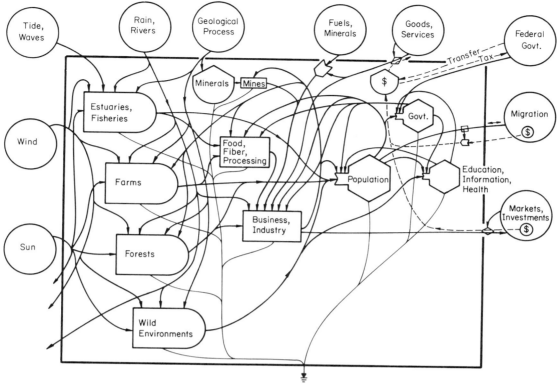

FIGURE 25-7. Main features of regional systems of landscape and human settlement.

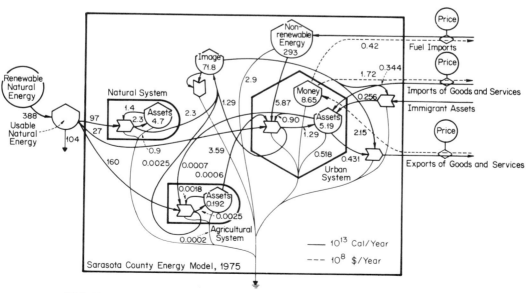

FIGURE 25-8. Evaluated model of Sarasota County, Florida (Sipe, 1978b).

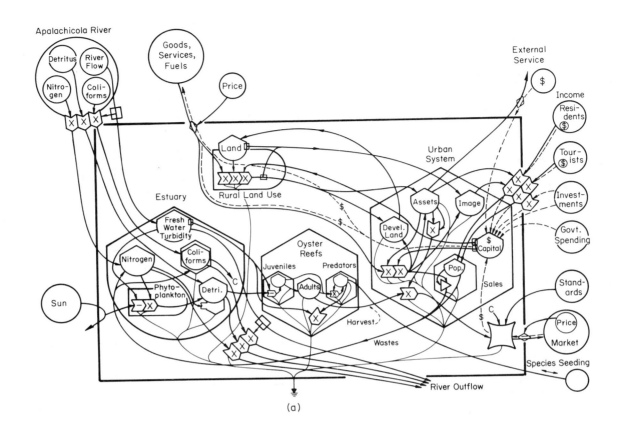

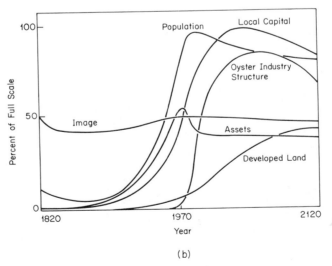

FIGURE 25-9. Simulation model of a coastal county with an oyster industry, Franklin County, Florida [modified from Boynton (1975)]: (a) energy diagram; (b) simulation of growth from pioneer times developing an energy-limited plateau by using the model in part a.

economic system of a coastal county in Florida that has had a stable economy based on the oyster industry of a river-fed estuary. The simulations included ups and downs in oyster economy with rising stress of housing developments. Brown (1975) simulated urban growth in Lee County Florida, an area attracting retirees.

CENTRAL PLACES AND SUPPORTING AREAS

Starting originally with agrarian economy, the energy hierarchy converging processed enough energy to generate cities that are intense self-regulating systems high in energy flow. When fuel use was added, intensity of the city increased further to levels on the order of 4000 kcal/(m² · day) coal equivalents of embodied energy in New York City. Cities are centers of communication, information processing, and control operations by which they serve their region.

Cities as High-Quality Consumers

When attention is focused on a city alone, its base in the landscape should be remembered. The rate of the larger supporting area shows up as a high percentage of outside–inside flows. The higher the quality position in the hierarchy, the larger the city, the larger the supporting area, and the larger percent of exchange support.

One of the difficult questions in city and regional planning is predicting which cities will emerge from one level in hierarchy to a higher level as a center of an even larger area.

Determining the Support Area of a City

Given a geographic unit such as a city, county, or state, its position in the hierarchy of regional support is measurable by the import–export exchange with surrounding areas. The units that are at the center of regional hierarchies have larger support area and larger percent of their economy in import–export exchange with their surroundings. Understanding a model of a geographic unit requires an evaluation of the import–export trade. One way to estimate these exchanges is to determine the "basic–nonbasic" portions of the economy.

Basic–Nonbasic Part of an Economy

The external dollar flow that is used to purchase fuels and embodied energy from outside may be estimated from the employment in each sector of the economy that is larger or smaller in percentage than the average. That fraction of workers in a sector larger than average is the fraction of that sector involved in exports. External–internal ratios have been called *basic–nonbasic analyses* (Ullmann and Cacey, 1962; Heilbrun, 1974). External percentage increases with concentration from 6% for the total United States, with about 10–15% for states, 20% for smaller cities, and 35% for large cities. The larger external exchange reflects the larger area that is the external basis of the more concentrated zone studied.

Economic Density and Exchange

Brown (1980) made calculations of basic–nonbasic parts of the economy for many counties, states, and countries and developed graphs similar to that in Fig. 25-10, relating exchange to the economic density. The more concentrated economic areas are more in the center of hierarchies and have higher dollar exhanges across the boundary. This graph can be used where direct statistics are not available for estimating external energy sources in the embodied energy of trade.

Another manifestation of the hierarchical organization and trade is the radius of movement of people,

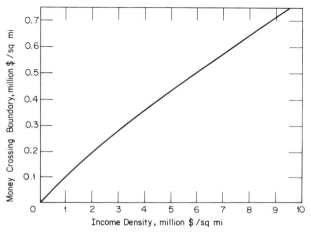

FIGURE 25-10. Percent of the total economic circulation that is crossing the boundaries as input–output exchange as a function of economic density (Brown, 1980).

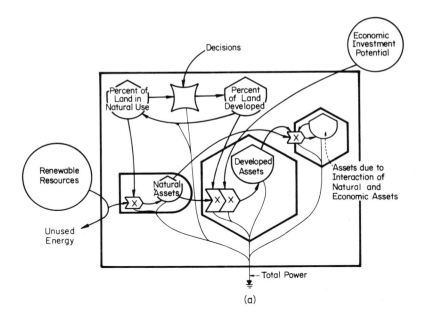

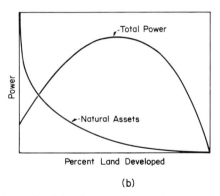

FIGURE 25-11. Model of optimum land development for maximum power [modified from Odum and Odum (1972)]: (a) energy diagram; (b) steady states for different percentages of developed land.

which is larger for those involved in higher levels of hierarchy as they conduct their functions of converging resources and diverging services to the larger area of regional support (see Fig. 25-6).

Wages and City Role

Hock (1976) relates city size to its demand curve for high-wage labor. Ultimately, however, the demand for such labor depends on the hierarchical position of that city in the regional system, with the higher positions having highest dollar flows and attracting the most labor.

Investment Ratio for Relating Regional Support

The investment ratio introduced in Fig. 14-8 relates the high-quality energies usually processed through hierarchical centers to the matching lower-quality energies usually spatially distributed as necessary support over the landscape (Odum et al., 1976a). The average ratio in a region is a guideline as to which activities may be feasible because they have good matching and which activities may not be feasible because they utilize too much high-quality energy without good matching of lower-quality energy. The ratio provides a means for determining when a proposed economic activity may be successfully eco-

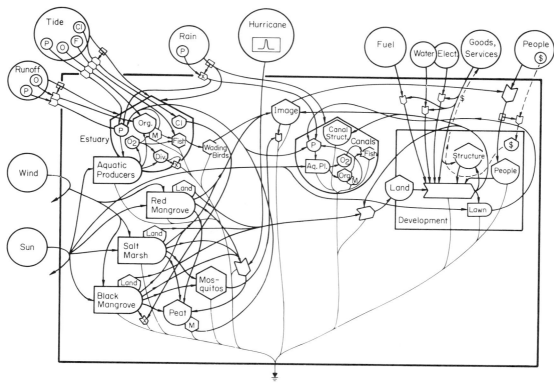

FIGURE 25-12. Impact diagram of housing development in mangrove swamp in south Florida [modified from Steller, (1976)].

nomic (see Fig. 23-24) and for determining, from the flow of high-quality energy, how far out into the region a city is drawing and converging matching low-quality support. See, for example, radii of influence of Florida cities estimated by Brown (1980). For other approaches to carrying capacity see Bishop et al. (1974).

Optimum Density for Maximum Economic Vitality

Total economic vitality depends on using environmental values well, attracting outside flows that require direct use of local energy flows and the interaction of the development and natural values. A model shows that an optimum density maximizes the economy. Models of the type in Fig. 25-11 have been simulated in various forms with data for several regions (Odum et al., 1972; Odum and Odum, 1976). The environment-based production on the left is the attraction for the investment of external energy during development.

The principle of high-quality energy maximizing power by matching and amplifying energy is illustrated in Fig. 25-11. As land (and its renewable energies) is moved from environmental production to urban use, the matching goes through a maximum that may be the concentration that maximizes total power.

MODELS OF ENVIRONMENTALLY LIMITED ECONOMIC DEVELOPMENT

In Chapter 14 the concept of investment attractions was introduced. Regular renewable resources provide the basis for attracting and coupling additional energies from outside that take the form of economic developments usually based on fossil fuels. An example of a system of development involving the resident energy flows and those developing to match it is the estuarine finger canal and housing system in Florida. See Steller (1976). Figure 25-12, shows the interplay of marine resources and the housing development with its flows of money and purchased energy, goods and services, and so on. This model was one done to evaluate the environmental impact at

Marco Island, Florida and was accompanied by energy calculations of housing density, suggesting that proposed plans would constitute an overdevelopment and ultimate loss of resident and attracted value. Some of this development was denied by the permitting process.

A simulation model was developed by Sell (1977) that has economic development dependent on matching of environmental energy and purchased economy (see Fig. 25-13). It develops an overshoot because of lags in time of construction. Favorable environment stimulates more construction than the optimum. Development with intermediate density housing generated more economic value than densities that excluded environment.

Water in Regional Models

Analogous to the ATP in biological systems are the several energies of water that are flexibly used by humanity and nature as a way of converting solar energy into more concentrated work and organizational processes of the landscape.

Water is a moderately high quality energy generated by the solar heat engines that feeds back to control and organize the landscapes. Because of its controlling influence and high quality, water is not only important to human enterprises but is a principal means of management by which humans control the landscape. Thus regional models often are based on systems analysis of water. Often as in deserts, it is a prevailing value (see Fig. 21-31). Water has already been found in a predominant role in ecosystem models.

Regional models can be organized around the water systems. Figure 25-14a gives the typical water model that may be considered whether isolated or included within a regional model (Fig. 25-14b). There are flows of rain, run-in (runoff from surroundings) percolation into groundwater, returns from groundwater, evaporation from surfaces and ponds, and especially transpiration of plant covers driven by sun and wind energies. The wind drying is a combination of advection of dry air and turbulent energy maintaining steep gradients in diffusion shells. Water flow over and through ground vegetation is linear, but water flow through canals is of higher order, and quadratic flows are sometimes appropriate. Digital switching is involved where there are overflows of constraining barriers such as weirs and dams. See Domenico (1972).

Competition for Water and Maximum Power

The use of water in renewable sectors so as to attract and also supply economic developments is given in Fig. 25-15. The competition for water generates maximum power if feedbacks develop means for that system to generate more power per unit water to reinforce the best pathways. However, maximizing power of the larger system requires that water maximize the utilization of renewable energies so that they can attract external energies obtainable by trade and exchange. Water has a high-amplifier action on urban developments, but the urban development may depend on the renewable energy use as attractant. Particularly in desert regions, the diversion of water anticipating urban development may undermine the region's ability to attract and hold that urban development. Also, water is lost to evapotranspiration in reservoirs used for regulating water and hydroelectric use.

S. Brown (1978) simulated a model like that in Fig. 25-14 for the Green Swamp region near Tampa, Florida, finding that swamp ditching would lose a large percentage of the available water for economic development because of the increased evapotranspiration and loss of storage.

GROWTH PATTERNS

Succession

Succession of cities and regional systems has phenomena similar to those found in ecosystems: growth, homeostasis, retrogression, renewal and oscillation. The model in Fig. 25-16 has three of the features of power maximizing: (1) total mass, which grows exponentially; (2) diversity, which adds some new energy input; and (3) trade, which is developed for additional energy.

Dansereau (1976), considering land use map and the development of Montreal, arranged stages in order as wilderness first, agriculture next, industrial next, and finally housing and populations. These made sense in terms of the sequence of increasing energy maximization as more and more energy is converged and the quality chain is lengthened. Housing and populations in modern urban systems have very high embodied energy (see Fig. 25-5).

Pulsing consumption and renewal is familiar in cities with cycles of growth and urban renewal. Appropriate simulations of pulsing minimodels are given in previous Chapters 11 and 22.

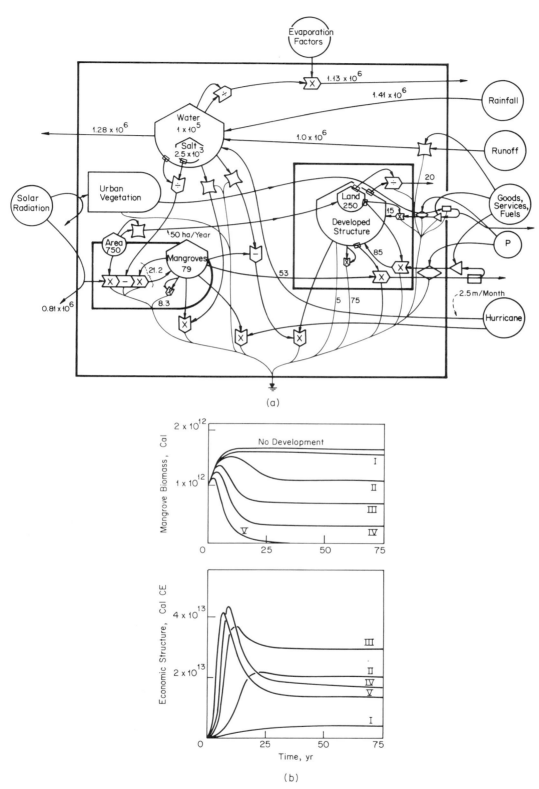

FIGURE 25-13. (a) Simulation model for economic development as a matching process to coastal wetlands in Florida (Sell, 1977), with sunlight in kcal/m^2 · year; others in g/m^2 × 10^{10}; (b) simulation of economic growth (Sell, 1977) for increasing densities of housing, I–IV.

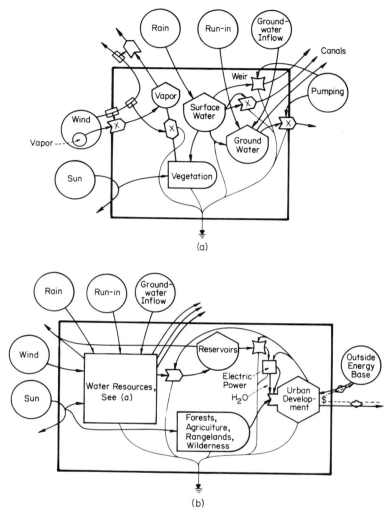

FIGURE 25-14. Typical regional water model: (a) water flows alone; (b) as part of regional model.

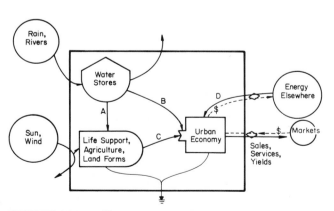

FIGURE 25-15. Alternative use of water for agriculture and environment or to supply urban economy. Moving water flow from A to B may not encourage urban economy if the attraction at (C) is generating new investments at D.

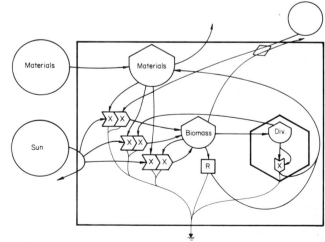

FIGURE 25-16. A model of succession with growth feedback, diversifications, and external exchange.

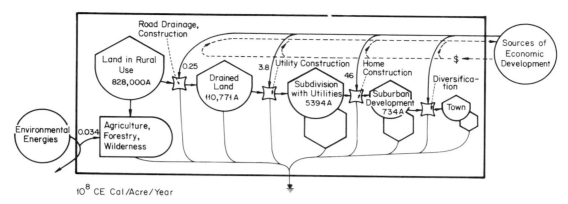

FIGURE 25-17. Hierarchy of land consumption in successive development of land in Golden Gate, Collier County, Florida (modified from Miller, 1980). Embodied energies are 10^8 coal equivalents per acre per year in 1975 (A = acres). Flows are calories of coal equivalents CE per acre.

Land Use Transformation

Holling (1970, 1971) conceived of development as the processing of land like food through a food chain, with successive transformations. Miller (1980) found development of land in Florida in an energy transformation sequence with an order of magnitude increase in embodied energy at each stage as developers put in more goods and services. As shown in Fig. 25-17, stages in development involve roads and drainage, then utilities, then homes, and later owners add values and diversification. Land is processed like any other environmental source, reaching its full involvement in the economy after many steps so that the initial price is no measure of the proportional contribution to the economy. Also see Fig. 23-28.

Antonini et al. (1974) evaluate a matrix for the transformation of land uses in the Dominican Republic and examine models for explaining these observed transformations.

Succession with Minerals from Sources and Recycle

In the model in Fig. 25-18 minerals are mined as auxiliary energy sources with feedback from main system assets, but recycle is also provided, with the energy derived from the main storage of assets. As particular minerals become limiting from the outside sources, the recycle pathways become more important. The presence of rich mineral sources substitutes for the energy required to recycle, indicating by its effect the energy equivalence of the minerals. This is another example of one source eliminating limitations of another so that only overall energy is limiting after organization is adapted (see Fig. 13-8 for other models of webs with more than one input commodity).

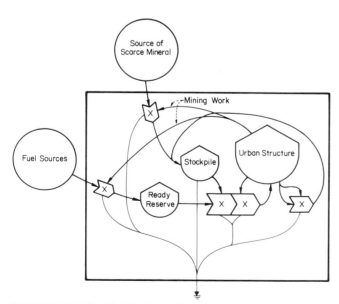

FIGURE 25-18. Model of use of critical materials with energy sources facilitating work of mining and recycling. For simulation, see Odum (1976a).

MODEL FOR A STATE

A model for Florida is given in Fig. 25-19 showing main income sources used to develop auxiliary energy. History of Florida is illustrated by Figs. 25-20a–25-20c, which shows an Indian stage (Fig. 25-

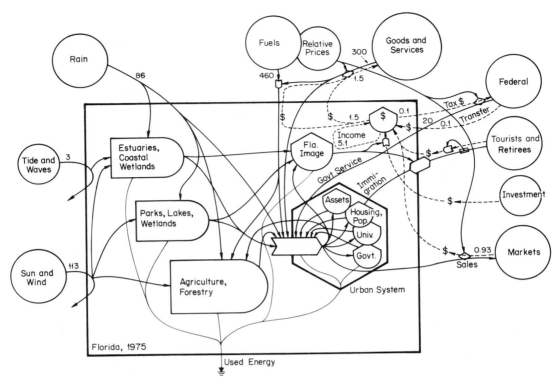

FIGURE 25-19. Evaluated model of Florida in 1975. Flows are 10^{12} kilocalories of coal equivalents per year. Dollar flows (dashed lines) are 10^{10} \$ per year.

20a), a colonial stage (Fig. 25-20b), a period of exploiting original stores of timber and soil (Fig. 25-20c), and a modern period of luxury services to an urban nation. History is the record of observed succession, and its models may often be similar.

SIMULATION OF CITY MODELS

At the convergence of the regional systems are the cities receiving high energy through convergence and as centers of exchange and trade from other regions. Hanga and Amle (1976) describe mass and metabolism of cities as more concentrated than other systems. When energies were mainly environmental, the cities were less concentrated since so much energy was dispersed in the inherent process of concentrating energy through the hierarchy. With rich fuels and minerals, the cities developed to larger proportion, with richest energies moving outward from the cities to be matched by lower-quality energies.

In overview, the cities are consumers with autocatalytic mechanisms as indicated in the minimodels in Fig. 25-3. As consumers, they are generally driven by the energy sources and productive processes elsewhere. Like other consumers, they have pulses, long periods, major controlling actions, and their basis in the feedback loops to and from their support regions.

Modeling cities separate from their regions may be too small a scope to be very predictive, but much effort has been expended in city models, some of which are given here for comparison and contract.

In Fig. 25-21 is given a general model for the components of a city and its many driving functions from outside. Compare this with the more primitive city given in Fig. 25-22, where transportation and communication is that of horse and buggy. An energy analysis of the agriculture supporting Paris in the nineteenth century was given by Stanhill (1977).

A Simulation

A simulation model of Miami is given by Zucchetto (1975a,b), which was simplified from the more general one by aggregating various functions but re-

Cities and Regions 547

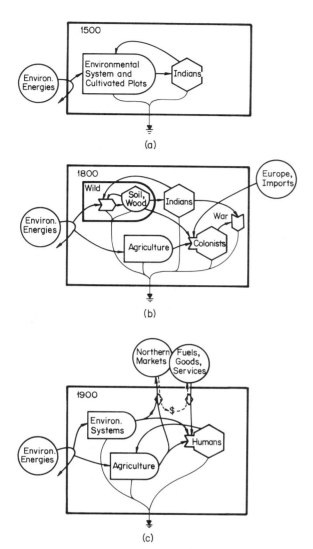

FIGURE 25-20. Minimodels of Florida history (Odum and Brown, 1976b; Sipe, 1978a): (a) Indian as part of ecosystem; (b) colonial period; high-quality items from Europe and energy of soil and wood storages used to displace Indians; (c) economy based on trade of wood, oranges, beef, and other goods. For 1975, see Fig. 25-19.

taining the main driving functions (see Fig. 25-23). The simulations made in 1974 related the city to its support capacity in Florida and to rising relative prices of foreign resources and decreasing tourist resources from U.S. citizens because of declining net energy. These simulations suggested an approaching limit to growth. What actually happened was a change in status of Miami from a regional center to an international center utilizing the rising energy usages of Venezuela, Brazil, and many other countries that funnel U.S. trade through a bilingual city.

Urban Models that Extrapolate Growth

Some urban models assume economic or population growth. Various algorithms generate consequences in labor, housing, spatial expansion, and other variables. A summary of these types of model was given by Goldberg (1977). Steinitz et al. (1977), for example, simulate spread of growth in New England with a model that translates an assumption about overall regional economic growth into locations of new development.

Urban Dynamics of Forrester

Forrester (1969) documents a model of urban structure consisting of three stages of business development, three levels of housing, and three levels of labor (managers, labor, and underemployed). The diagram in Fig. 25-24 shows Forrester's summary. Each gate (valve) shown is controlled by 5–20 coefficients representing information sensors of the nine levels, or additional stored properties such as tax attractiveness perceptions, land occupancy, and various ratios. Twelve pages are required to diagram all the pathways. Figure 25-24b gives an aggregated energetic view of this model. The implied energy sources are included in dotted lines.

A literal translation to energy language would involve substituting tanks for squares, interaction gates for the Forrester gate, and a detailed translation of the program equations, with the addition of sources for inputs from outside. Coefficients are recombined with the resource pools. In a simpler model, that of the world economy in Fig. 26-23, such a translation has been attempted with attempts to position and arrange according to structural associations and energy quality.

The 1969 urban model is too complex a diagram for one page, and there is the question as to whether models are useful when so little is detailed to such degree. Only three sectors of the city are included, environment is omitted, and main driving functions from outside are omitted (e.g., net energy effects). In a general way, attractiveness perceptions attract labor, thus stimulating business and housing and tax expenditures and improving attractiveness. The process is autocatalytic but slows down as housing and business depreciates, reducing jobs and attractiveness, and some people leave.

With simulation (Fig. 25-24c), the model generates a steady state following exponential growth

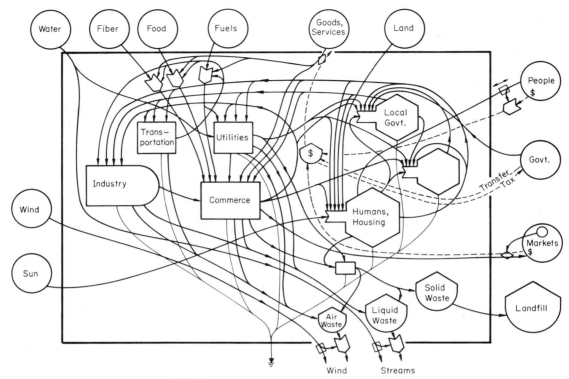

FIGURE 25-21. Representative city system.

mainly because many coefficients determined empirically have optima and thus are quadratic. In other words, the system is an extremely disaggregated logistic type of model limited by interactions rather than by sources that were regarded as infinite pools.

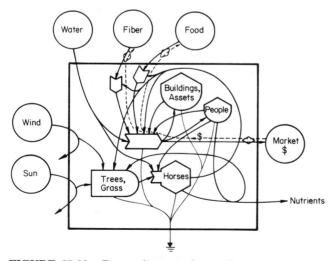

FIGURE 25-22. Energy diagram of a city (Paris in 1850) based on a mainly agrarian regional economy (Stanhill, 1974).

Employment Model

Batty (1976) reviews a series of models applied to cities in which changes in population and land use are generated iteratively from the employment, with additional employment generated by the population and land developed. The closed-loop aspect is a form of exponential growth depending on an economic function for supplying employment from outside. The type of model is diagrammed in Fig. 25-25. These models are linear, and some are represented in matrix form. The population and the employment are located spatially by the distribution of working trips that are allocated in relation to centers at the start using gravity model. The model does not generate growth predictions but is a means for extending predetermined economic trends to extrapolate consequent trends in population and land use.

SIMULATION OF REGIONAL MODELS

Spatial Distribution Models

The interactions and trade between zones can be represented as a systems model, using matrix or net-

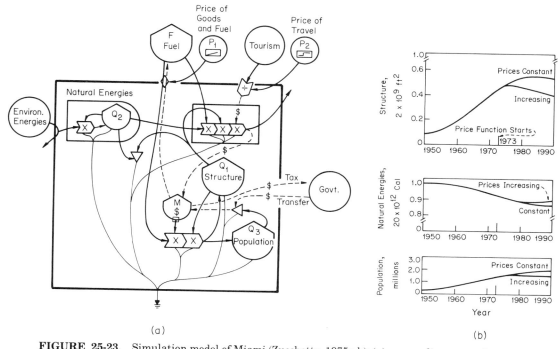

FIGURE 25-23. Simulation model of Miami (Zucchetto, 1975a,b): (*a*) energy diagram, redrawn; (*b*) simulation showing effect of rise in prices.

work diagram to represent relationships. Following examples given by Masser (1972), Fig. 25-26 shows a simple matrix, an equivalent energy diagram, and two alternate unit equations. The first equation has trips between sectors according to the gravity model, and the second has an exponential distance–product model. For an example involving trips between zones, the model expresses the distribution of transportation and job interactions as a function of storages in each zone.

A model of spatial aggregation according to marginal effect of storages on growth or with trade was given by Costanza (1979) in Fig. 18-8. It was simulated for south Florida, generating the most growth in areas that later developed urban centers at Miami and Fort Myers.

Simulation of Region at War

Some aspects of war are given in the simulation model in Fig. 25-27. Commitment of resources of one side is proportional to that of the other when it is possible. Assets are disordered so that the normal state of order is maintained at a lower level during war. People, materials, and lands disordered by the war build up as a pool stimulating sufficient regrowth to balance destruction. If and when some steady state is achieved, the war has seemed to reestablish influence boundaries according to energies available. A more complex model of Vietnam at war was simulated by Brown (1977) studying the impact of destruction of coastal mangroves by herbicide. The model suggested an 11% effect on the general economy. See also Milstein and Mitchell (1968). Voevodsky (1971), fitting data on wars, found army strengths S to be a function of the casualties C:

$$S = aC^B \quad (25\text{-}1)$$

He also showed commitment of armies followed a chargeup curve.

Using game theory matrices, Richardson (1960) analyzed the relationships of countries. The distribution spectrum of wars showed the hierarchical pattern we emphasized in Chapter 15. There were many small quarrels but most deaths were in the large ones.

SUMMARY

Models of cities and regions, those "ecosystems" of larger dimension that include humanity, nature,

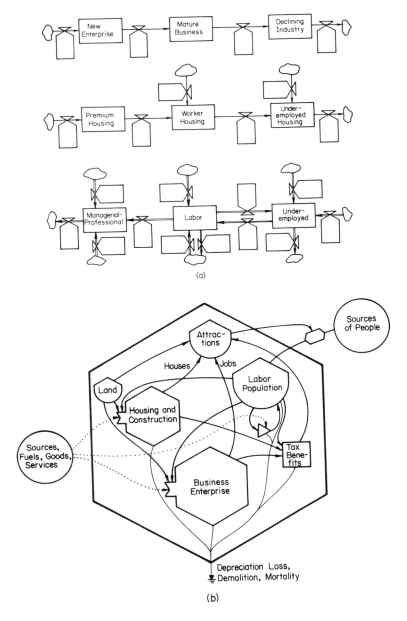
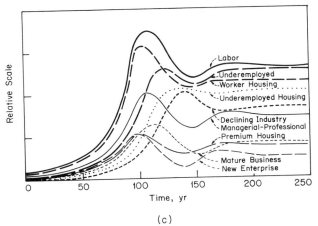

FIGURE 25-24. Urban model given by Forrester (1969): (a) summary; (b) aggregated energy translation; (c) example of simulation.

economic systems, and land use, were given in this chapter. Aggregated properties and spatial patterns were considered as the interaction of environmental and driving functions, previous storages, and the self-organizing development of energy webs that maximize power. Models given for cities, regions, and states included some from ecological traditions and some in the literature with other precedents. See also Lee (1973).

Autocatalytic features foster growth when resources are available for increased utilization, but the models level and undergo regression when external resources decrease in flow or concentration or when relative prices of exchanges increase as with the current trends of decreasing net energy of mineral–fuel sources. Models are given that estimate carrying capacity as that level of development that maximizes competitive economy by utilizing free ex-

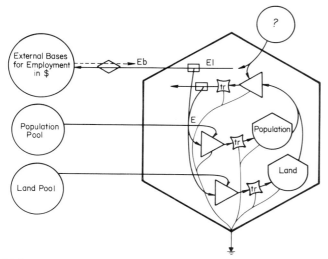

FIGURE 25-25. Energy diagram of one of the linear economic-driven employment–population–land models reviewed by Batty (1976). Energy sources and sinks have been added.

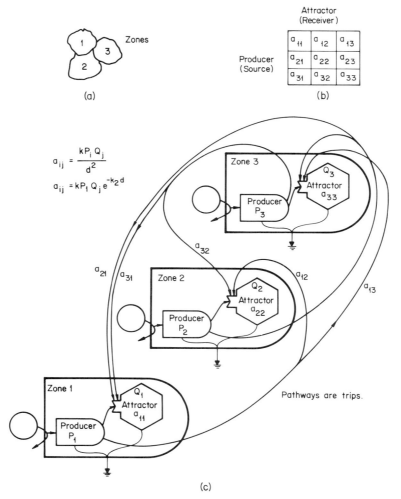

FIGURE 25-26. Spatial distribution matrix for representing interactions between zones from Masser (1972): (a) zones; (b) matrix; (c) unit equations and energy translation. For example, trips may be represented from one zone to another.

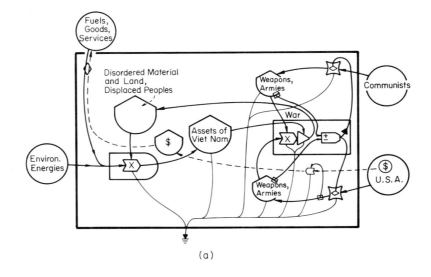

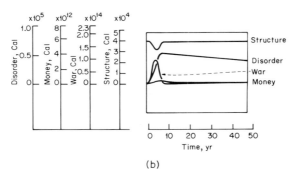

FIGURE 25-27. Simulation model of Vietnam war zone in 1970, where support by each warring side is adjusted in proportion to that supplied to the other: (*a*) energy diagram; (*b*) simulation with steady-state war developing as a prelude to cessation of hostilities (Swallows and Odum, 1976).

ternalities well in the attraction of purchased additional resources.

Many of the very complex models generated for cities, states, and regions are faulty as they lack resource limitations and have many coefficients that are empirically determined that are really resource-dependent variables. Complex models have rarely been received with confidence from others.

QUESTIONS AND PROBLEMS

1. Explain the role a city has in its region regarding spatial pattern, energy quality, matching of energy, and service. Then use minimodels to show the structure and mechanics within a city and its major sectors and pathways.

2. What ratios of external dollar flow to internal circulating dollars are characteristic of cities? Considering the investment ratio concept, what matching rural energy is required to buffer the city functions to make a competitive system? Calculate the area necessary to provide 2000 kilocalories of sunlight for each kilocalorie of fossil fuel back of the export–import dollar flows (use 10,000 kilocalories/dollar to estimate embodied energy in external dollar exchange flow).

3. Give two mechanisms that tend to limit growth of cities, even during times of constant price and availability of external fuels and goods and services.

4. Which properties of cities are self-interactive, involving communication and information so as to generate quadratic energy drain from storages? Which minimodels generate maximum productivity at intermediate density?

5. Diagram the typical urban models that generate new jobs as an initial economic growth premise affecting population, housing, services, de-

mands, and so on. How are such models made useful in spatial extension of growth predictions?

6. Draw a diagram for a typical city. Draw a diagram for your state. Evaluate the main energy sources, and calculate the investment ratio to estimate relative degree of development.

7. What pattern occurs in development based on nonrenewable resources such as mining? Considering models that interface renewable environmental attractions and economic development, how can overshoot occur with more development than can be sustained in steady state? Where does this phenomenon occur?

8. What is the relationship between transportation and hierarchical spatial organization? What is the effect of increasing or decreasing energy availability on the pattern of spatial order? Considering the hypothesis that high-quality energy is more readily transported, what are the tendencies as to who and what is regularly transported in city zones?

9. Diagram models for water in the landscape.

SUGGESTED READINGS

Aguilar, R. J. (1973). *System Analysis and Design,* Prentice-Hall, Englewood Cliffs, NJ.

Bathersby, A. (1964, 1970). *Network Analysis for Planning and Scheduling,* Wiley, New York.

Bohm, P. and A. V. Kneese (1971). *Economics of Environment,* Macmillan, St. Martins Press.

Bradshaw, A. D. and M. J. Chadwick (1980). *The Restoration of Land,* Univ. of California Press, Berkeley, CA. 317 pp.

Brown, C. B. (1977). Analytical methods for environmental assessment and decision making. In *Regional Environmental Systems*. Report NSF/ENV 76-04273. Department of Civil Engineering, University of Washington, Seattle. pp. 180–206.

Butler, G. C., Ed. (1978). *Principles of Ecotoxicology,* Wiley, New York. 350 pp.

Chadwick, G. (1971, 1978). *A Systems View of Planning,* 2nd ed., Pergamon Press, New York. 432 pp.

Christian, K. R., M. Freer, J. R. Donnelly, J. L. Davidson, and J. S. Armstrong (1978). *Simulation of Grazing Systems,* Pudoc, Wageningen, Neth. 121 pp.

Dickey, J. W. and T. M. Watts (1978). *Analytical Techniques in Urban and Regional Planning,* McGraw-Hill, New York. 542 pp.

Douglas, I. (1981). The city as an ecosystem. In *Progress in Physical Geography,* Edward Arnold, London. Sections 5,3.

Foell, W. K. (1979). *Management of Energy—Environment Systems,* International Series on Applied Systems Analysis, Wiley, New York.

Gazis, D. C. (1974). *Traffic Science.* Wiley, New York.

Haggett, P. and R. J. Chorley (1969). *Network Analysis in Geography,* St. Martin's Press, New York.

Harary, F., R. Z. Norman, and D. Cartwright (1965). *Structural Models,* Wiley, New York.

Havlick, S. W. (1974). *The Urban Organism,* Macmillan, New York. 515 pp.

Hillier, F. S. and G. J. Lieberman (1967, 1974). *Operations Research,* 2nd ed., Holden-Day, San Francisco.

Holling, C. S. and M. A. Goldbert (1971). Ecology and planning, *J. of the American Institute of Planners,* 37(4):221–230.

Isard, W. (1960). *Methods of Regional Analyses—An Introduction to Regional Science,* Wiley, New York.

Isard, W. and P. Llossatos (1979). *Spatial Dynamics and Optimal Space Time Development,* North Holland, Amsterdam, Neth. 434 pp.

Lee, C. (1973). *Models in Planning,* Pergamon Press, New York.

Martin, L. and L. March (1972). *Urban Space and Structures,* Cambridge Urban and Architectural Studies, Cambridge University Press, London. 272 pp.

Mercer, R. L. (1962). *A Communications Theory of Urban Growth,* MIT Press, Cambridge, MA.

Naveh, Zev (1982). Landscape ecology as an emerging brand of human ecosystem science, in *Advances in Ecol. Research* **12**, 190–237.

Pugh, R. E. (1977). *Evaluation of Policy Simulation Models,* Information Resources Press, Washington, D.C. 350 pp.

Richardson, L. F. (1960). *A Mathematical Study of the Causes and Origins of War.* Pillsbury.

Richardson, H. W. (1973). *Regional Growth Theory,* Wiley, New York. 264 pp.

Richardson, H. W. (1973). *The Economics of Urban Size,* Lexington Press, Lexington, MA.

Rummel, R. J. (1979). *Understanding Conflict and War,* Sage Publications, Beverly Hills, London.

Sage, A. P. (1977). *Methodology for Large-Scale Systems,* McGraw-Hill, New York. 442 pp.

Simon, H. A. (1957). *Models of Man,* Wiley, New York.

Smith, K. J. (1975). *Finite Mathematics—A Discrete Approach,* Scott, Forsman, Dallas, Texas.

Stearns, F. W. and T. Montagg (1974). *Urban Ecosystem,* Dowden, Hutchinson and Ross, Wiley, New York.

Sudo, F. M. and J. R. Ziegler (1978). *Golden Age of Theoretical Ecology—1923–1940.* Lecture Notes in Biomathematics, No. 22. Springer-Verlag, New York. 491 pp.

Webber, M. M., J. W. Dyckman, A. Z. Guttenberg, W. L. C. Wheaton, and C. B. Wurster (1964). *Explorations into Urban Structure,* University of Pennsylvania Press, Philadelphia.

Wilson, A. G., Ed. (1972). *Patterns and Processes in Urban and Regional Systems,* Pion, London. 324 pp.

Wilson, A. G. (1974). *Urban and Regional Models in Geography and Planning,* Wiley, New York.

Wolman, A. (1965). The metabolism of cities. *Sci. Am.* **213**(3),179–188.

CHAPTER TWENTY-SIX

World Patterns

The largest ecosystem is the biosphere of the whole planet. Mainly closed to matter, the biosphere runs on the flow of sunlight, which drives the cycles of the atmosphere, the oceans, much of the geologic systems, and the fabric of life and humanity. These parts interact to generate world productivity as suggested in Fig. 26-1. Various models have been made to consider global systems, usually considering one phase at a time such as the atmosphere, the carbon cycle, or the economy. Models of the earth are given in this chapter, including overall macroscopic minimodels, representative models of elemental cycles, models of phases of the biosphere, and models of the world human presence.

MODELS OF THE BIOSPHERE

Because there are fewer external driving functions, modeling of the biosphere ought to be simpler than that of regions. The interplay of air, land, and water is so intimate that models ought to consider all the phases as they interact as a single system. However, simple global models are less often attempted, possibly because the sciences of the earth have been developing separately and because so much detail is known to those doing modeling.

Attempts to represent the biosphere to show main phases of air, sea, and land and the relative hierarchical positions are given in Figs. 26-1–26-3. Figure 26-1 shows each phase to be a part of the ordered structures being maintained by production and cycles. Here geologic production such as land forming is included with biological production of organic matter.

Figure 26-2 further disaggregates the biosphere into producers and consumers. Figure 26-3 is further divided and is also evaluated to show the declining actual enery and rising embodied energy as energy flows from left to right. Land building, volcanoes, glaciers, and human endeavors are examples of the high-quality controlling and pulsing phenomena, which feed back controls that dominate the biosphere. The worldwide human innovations, conquests, and frenzied consumption is a burst of energy that impacts the whole biosphere with its programs (Bertine and Goldberg, 1971).

The basic production processes include the convergence of the sun and wind in interacting with the sea to generate rains and cyclic salts in the air; interacting with the land; and building soils, vegetation, and oxygen. Then the products of production are converged by the water cycle and other pathways to generate the sedimentary and igneous rock cycles used in urban activity under human control. The consumers recycle water, carbon dioxide, and acid volatiles. The acid volatiles and the igneous rocks, spreading outward as a recycle process, interact with the atmosphere again in the production of soils and sediments. The production is spread over the earth's surface, whereas the consumer–recycle processes are concentrated, individually more prominent, pulsing more, and hierarchically arranged in centers, feeding back their products to the whole biosphere again.

In the overview, biological, geologic, and human processes are operating somewhat in parallel, with the human role being a late and more efficient means for recycle that was done by less organized life and inorganic processes earlier.

In Fig. 26-3 a characteristic ecological web results, with feedbacks, recycling, and convergence of energy in quality from the sun to the intensive and pulsing consumer energies to the right. Embodied energy is given in Table 14-1 on the basis of Fig. 26-3. The biosphere uses every by-product so that waste is only a relative term. Consequently, the energy transformation ratios between the sun and the various world processes may be the most efficient possible at maximum power. Presumably, the system

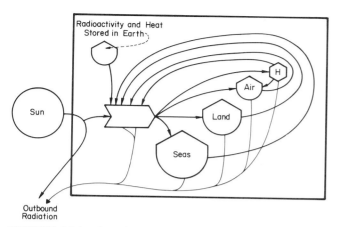

FIGURE 26-1. Interdependant phases of the biosphere in which structures of air, seas, and earth are maintained by interactive cycles driven by solar energy (H = human activity).

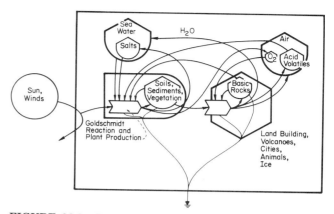

FIGURE 26-2. Production–consumption model for earth cycles including Goldschmidt reaction of acid volatiles and basic rock to generate soils, sediments, and seawater, followed by consumption of redox and other energies to generate recycle.

that organizes so as to use all by-products is more efficient than one that does each operation alone with by-products unused. For world biological productivity estimates are given by Reichle et al. (1975a) and Lieth and Whittaker (1975).

Steady-State Proportions

In Chapter 15 similarities between steady states and equilibria were discussed with suggestion that maximum power was favored by distribution of materials in states similar to the one that would be calculated for closed equilibria (see Fig. 15-7). The prevalence of water in the sea and the calculations of equilibria for salts in the sea by Sillén (1967) were given as examples (Fig. 26-4b).

Hierarchical Pulsing

If hierarchical concepts are correct, the biosphere is exposed to pulsing rhythms that have time periods increasing with size. On earth the pulses of humans, ice, and volcanic–orographic phenomena have the largest embodied energy and largest dimensions in space and time. The pulses from the astronomical surroundings, however, are of even larger dimension and period. Mechanisms by which sunspot cycle pulses affect earth weather are being sought. Pulses of very high quality (high energy) radiation of solar wind of charged particles, may have control actions acting first on ionosphere. See Alexander (1979), Markson and Muir (1980), and Evans (1982).

Macroscopic Minimodels of Processes of the Biosphere

Macroscopic mimimodels of aspects of the biosphere are shown in Figs. 26-4 and 26-5. Figure 26-4a shows the hydrological cycle; Fig. 26-4c shows the sediment–igneous rock relationship in simplest form; and Fig. 26-4d illustrates the cycle of salt that accompanies the action of wind on the sea followed by return of salt to the sea in rivers. Rubey's (1951) concept of steady acretion to the biosphere of water and other elements from the earth through hot springs and volcanoes is shown in Fig. 26-4e, and the Goldschmidt reaction, by which acid volatiles react with basic igneous rocks to generate soils, sediments, and seawater is shown in Fig. 26-4f.

Figure 26-5a shows a minimodel that aggregates the biosphere as producing for parallel consumers, including animal–microbes, thermal geologic activity, and human technology, all of which have some common properties in recycling "nutrients."

In Fig. 26-5b capture of old reserves by new pathways is shown. The expanding system of technological humans is replacing part of the earth cycle and is gaining a boost initially by using the storages of minerals that were a necessary part of the cycle prior to human use.

PHASES OF THE EARTH ATMOSPHERE

Within the atmosphere, hierarchies of cloud systems are recognized. Figure 26-6a relates for levels of convergence often recognized in tropical meteorology.

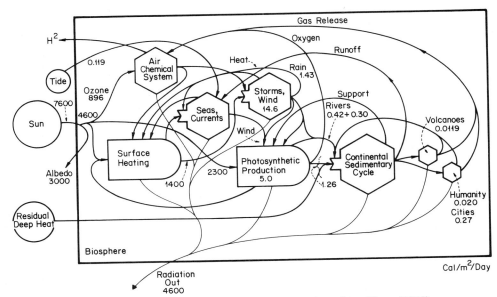

FIGURE 26-3. Web of energy flow in the biosphere from Odum (1978).

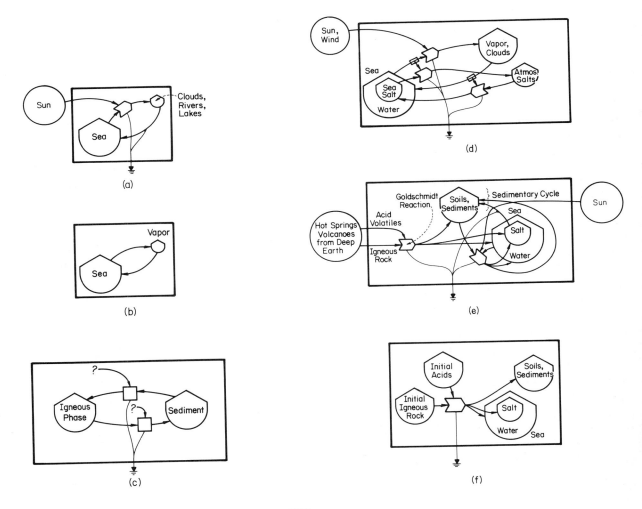

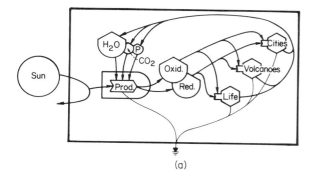

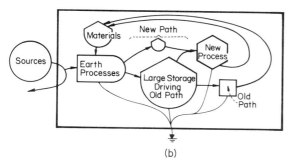

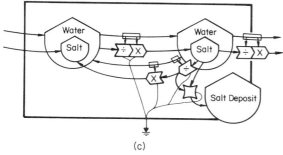

FIGURE 26-5. Minimodels of earth processes: (*a*) model of biosphere production as an oxidation–reduction separation (Odum, 1972); (*b*) model of a new pathway of use absorbing storages necessary to older pathways; (*c*) model of salt deposition (Briggs and Pollock, 1967).

Figure 26-6*b* diagrams the production and consumption process of atmospheric momentum from earth heating with some representative values. The broad heating of the earth surface is converged into potential energy and kinetic energy of storms that feed back their pulsing and concentrated services and re-

FIGURE 26-4. Minimodels of earth states and evolution: (*a*) steady-state hydrological cycle; (*b*) equilibrium model of ocean; (*c*) earth cycle (Harbaugh and Bonham-Carter, 1970); (*d*) model of cyclic salt (Conway, 1943; Livingston, 1963b, 1964); (*e*) model of steady accretion to the biosphere (Rubey, 1951); (*f*) Goldschmidt (1933) model of differentiation of the biosphere.

cycling of air by means of the general circulation (Lorenz, 1955) (Fig. 26-6*b*).

A word diagram of the climatic parameters needed for an aggregated climatic model was given by Kellogg and Schneider (1974). Instead, most meteorological modeling has been by iterative solution of equations of motion and energy relations of many points distributed in a spatial grid. This approach uses enormous computer time without much higher-level aggregation hierarchies or control structures.

A more detailed overview of the atmospheric heat and water budget system is given in Fig. 26-7, which shows main processes by which solar heating drives the earth heat and vapor engine, generating ice in poles and glaciers. Sergin (1979, 1980) simulated ice age oscillations with a model something like Fig. 26-7. Developing ice interacts with areas of ocean by isostatic land adjustment. Clouds of snow and ice increase albedo. Ice age pulses develop with sharp temperature gradients and storms that utilize stored energy. This is followed in the simulation by longer interglacial periods (see Fig. 26-7*b*).

Chemically, the great reservoir of oxygen and nitrogen mass acts as stabilizing storage and reference for many of the chemical and living phenomena. These are maintained by the production cycles of the biosphere. Subsystems of atmospheric chemical action interact directly with solar energy and human air pollution.

Oceanic Systems

The ocean subsystem is driven by atmospheric winds and thermal gradients developed by differential heating from the sun and atmosphere. A very tentative hypothesis as to the hierarchy within the sea's physical system considers energy convergence from weak thermal gradients to strong currents and finally sea ice formation. Whereas there is nothing inherently high quality about any particular physical state such as ice, its position in the hierarchy depends on whether it receives a convergence of embodied energy required to generate a scarce quantity that has the ability in its feedback actions to amplify the whole system.

Chemical modeling in the sea is often aggregated according to main water mass compartments that are somewhat distinct. See, for example, the diagram of compartments used by Broeker (1974) in Fig. 26-8.

The chemical system of the sea is mainly driven

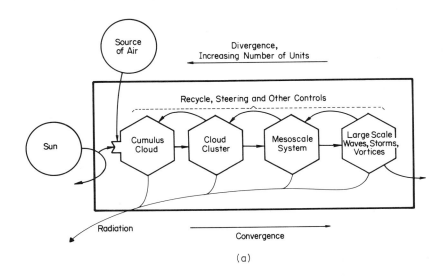

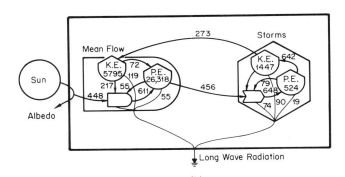

FIGURE 26-6. Model of atmospheric hierarchy: (a) hierarchical concept of tropical convection when vertical organization predominates over latitudinal organization; (b) production and consumption of atmospheric kinetic energy (Riehl, 1979). Numbers are from Charney and Philips (1953) after Hess (1959).

by the sedimentary and igneous cycles of the whole earth. Cycles of elements pass from sea to sediment or air and back through main and rivers to the sea again as outlined in Figs. 26-2 and 26-3 and the minimodels of Fig. 26-4. Figure 26-9 has a minimodel of the circulation of mass presented earlier to suggest the ocean could be quasisteady state. See Odum (1951), Barth (1961), Livingston (1963a), and MacKenzie and Garrels (1966).

Some of the main chemical entities that are part of the open, quasisteady-state chemical homeostasis in the sea are shown in Fig. 26-10. It has the sedimentary–orographic cycle in more detail and includes main elemental pathways into and out of the ocean. Many chemical reactions are fast relative to the geochemical circulation so that many aspects remain close to equilibrium, although the system is an open steady state. The CO_2 equilibrium within the sea is an example. The minimodel in Fig. 26-5c represents the formation of hypersaline deposits of sulfates and salt.

Earth Terrestrial System

The hierarchy of earth system—judging by spatial distribution, intensity and period of pulsing, and energy transformation ratios—moves from areas of broad slow sediment deposition in the sea to land, converging as part of plate movements to zones of volcanoes, earthquakes, and heat concentration (see Figs. 26-3 and 26-10). Figure 26-10 includes the main kinds of sediment formation and reincorporation in rock.

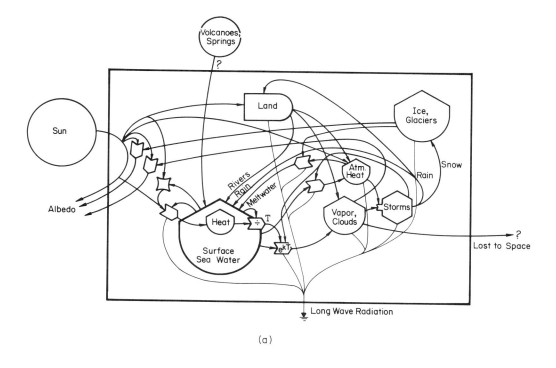

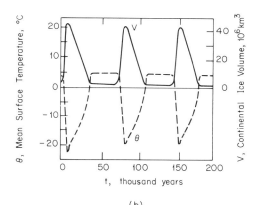

FIGURE 26-7. Models of atmosphere important to large-scale and long-range oscillations: (a) a model of atmospheric heat and water, flows and compartments; (b) oscillations of mean surface temperature (potential temperature) and ice volume on continents by Sergin (1980) from simulations of an overview model like that in part a for northern hemisphere.

Included in Fig. 26-10 and detailed in Fig. 26-4f is the Goldschmidt reaction, by which volatile acids interact with basic rocks to reform soils and sediments plus the seawater oxidation, and reduction is generated simultaneously by production being separated into gas and solids. Consumption converges the oxidized and reduced flows for intensive reaction. The surface of the earth becomes the zone of control. Reduced materials are generated where energy converges. Volcanoes and metamorphic convergence concentrate sulfides.

Figure 26-3 gives the earth's crust model aggregated with the widespread dilute weather and ocean circulation developing and converging rain to generate erosion and sediments, including biological actions concentrating organic matter and radioactivity. Next, convergence of the sediments generates sedimentary rock and sedimentary mountain uplift and land. Finally, further convergence gives high temperatures, heat, magma, and igneous processes. The terminal end has pulsing of volcanic eruptions, earthquakes, and surges that feed back work.

560 SYSTEMS OF NATURE AND HUMANITY

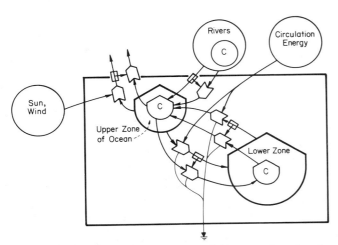

FIGURE 26-8. Compartments for modeling constituents c in the sea after Broeker (1974).

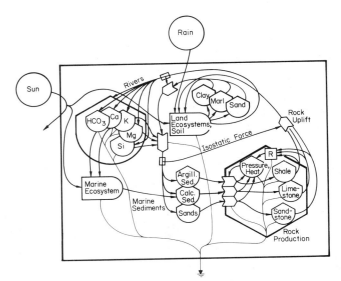

FIGURE 26-10. Models of earth sedimentary cycle modified from Odum (1972) (R = radioactive heat generation).

The chemical constitution of the seas is homeostatically regulated as a balance between river and rain inflows and five main processes of removal from the sea (water evaporation and salt recycle, brine deposition, marl deposition, shale deposition, and sand bed deposits). The behavior of the major elements of the sea are shown as part of this homeostasis in Fig. 26-10. Figure 26-15d shows the pH–CO_2 equilibrium by which acidity and limestone deposition are regulated.

The much advertised industrial release of acid volatiles is presumably matched by equal moles of base in solids as ash and solid wastes of the civilization's manufacturing. Ultimately, the recombination of the two may help neutralize the pollution hazards feared from carbon dioxide, sulfur oxides, nitrogen oxides, and so on.

EARTH EVOLUTION

The evolutionary development of the earth is a long-range succession in which producers develop consumers that feed back different products. These generate different production processes that generate different consumers, and so forth, with each round of unit substitution becoming possibly closer to a steady state, although the steady pattern may in-

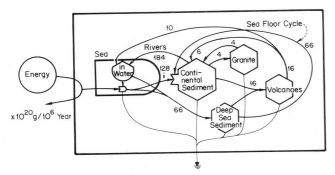

FIGURE 26-9. Model of world mass flow in steady state after Odum (1950, 1951).

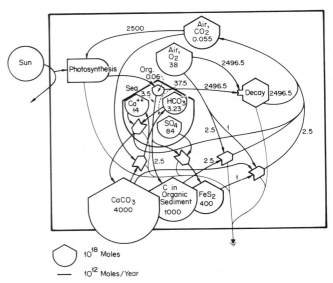

FIGURE 26-11. Energy diagram of global model of carbon by Garrels et al. (1976).

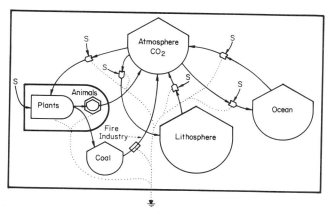

FIGURE 26-12. Carbon cycle after Lotka (1925) (S = implied energy source).

volve a pulsing surge in the higher-quality elements (Nicholls, 1965).

Types of succession are given in the models in Fig. 22-2 in which the development of a production component generates conditions for development of a consumer unit. The consumer unit then generates feedbacks, causing a new producer unit, which, in turn, causes yet a different consumer unit, and so on, changing in sequence until either a matching pair is developed as a steady state; the sequence repeats an older unit, thus forming a time continuous loop; or a pulse consumer starts the sytem from the beginning. The substitutions may be by development from a repertoire of seeded variety, from production from existing established system, or by evolution of new types.

This theory has some production and consumption from the start, although at each stage the diversity, fine tuning, and efficiencies of harnessing energies and building order may have increased and may still be increasing. This model has solar energy and cycles initially and later condenses, in a miniaturized way, continents, volcanoes, life, humans, and so on. This theory does not commence as oxidized or reduced, as only land or water, as igneous or sedimentary, but has primordial beginnings of all operating within a crude cycle. A model for evolution of life in successive steps in a circular process was given in the model in Fig. 10-31b. Some other views on evolution are given in Fig. 26-4.

A type of producer–consumer model is given by Garrels et al. (1976) (see Fig. 26-11). The increase in oxygen over geologic time was shown to reduce the levels of organic matter that are buried without oxidation compared to that of much earlier times. Odum and Lugo (1970) showed with experimental microcosms that steady-state levels of carbon dioxide could vary from 350 to 3000 ppm depending on living components. See simulation of these microcosms by Burns (1970) in Fig. 21-7c. A fairly steady state biosphere is visualized with homeostatic cycles for most elements.

Step Surges with Innovative Substitutions

In Fig. 10-4 storages upstream from pathway bottlenecks were shown to be larger in inverse relation to the pathway conductivity. Mineral accumulations

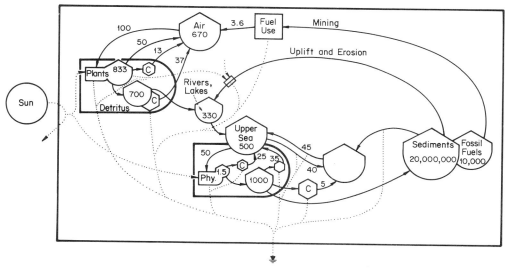

FIGURE 26-13. Carbon cycle interpreted from Reiners (1973) (C = consumers; dotted lines indicate driving energy flows).

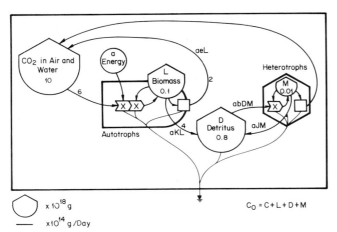

FIGURE 26-14. World carbon cycle model by Wesley (1974).

were required to develop cycles. However, when a new system for processing and recycling replaces the old one that can process more power, it can replace the former pathway by competition (see Fig. 26-5b). By processing more, it can develop a steady state with less storage. In the transition the new system can thus draw down the old storage to the new steady-state level by using the previous stored resource for a pulse of growth that helps its transition and evolution. The energy flow then declines to a lower level. So may pass the Twentieth Century.

BIOGEOCHEMICAL CYCLES

The biogeochemical approach to world systems takes one chemical element at a time, showing its storages, pathways, and recirculation. Cycles are one kind of subsystem, a way to represent data and

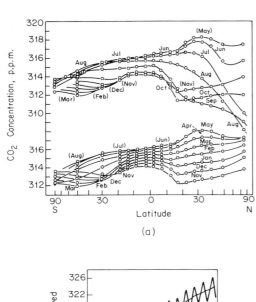

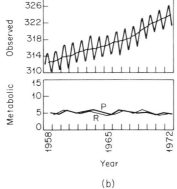

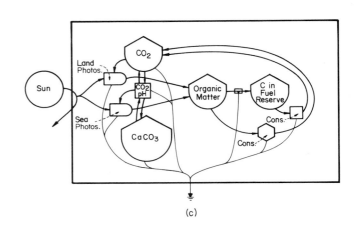

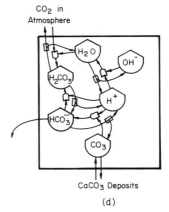

FIGURE 26-15. Carbon dioxide variation in the atmosphere as a function of the balance of P and R: (a) monthly pattern for various latitudes (Bolin and Keeling, 1963); (b) estimation of P and R by subtracting cultural contributions and oceanic absorption (Hall et al., 1975); (c) model of carbon with terrestrial and marine photosynthesis; (d) detail of CO_2 quasiequilibrium in the sea; enlargement of box in part c.

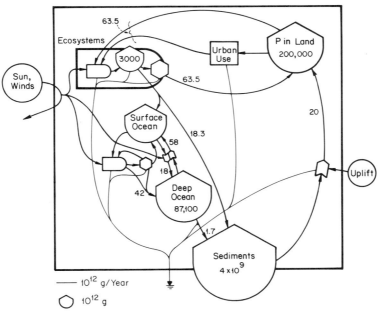

FIGURE 26-16. Biogeochemical cycle of phosphorus by Garrels et al. (1975) given in form of energy diagram.

organize knowledge, but they can be deceptive. Since every cycle is interacting in causal ways with other cycles, one cannot use a cycle model to predict or simulate without including all the interactive drives. The cycle diagrams may distract thinking away from the natural units of system organization.

Some models of single chemical cycles are given with and without driving functions in Figs. 26-12–26-18. These include a selection from the literature, including some that have been simulated. For an introduction, see Holland (1978) and Bowen (1979).

Carbon Cycle

Because it constitutes half of the living fractions, is a main component for the fossil fuels, and is being changed by human activity, the carbon cycle has been used to gain an overview of the biospheric processes. The carbon cycle is driven by the production–consumption processes running on solar energy and may be represented most simply by the P–R model with values for the biosphere as in Fig. 26-5a.

One of the earliest carbon cycles considered with

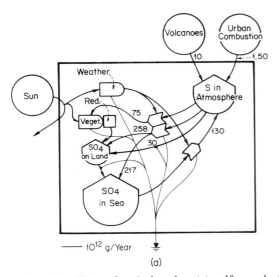

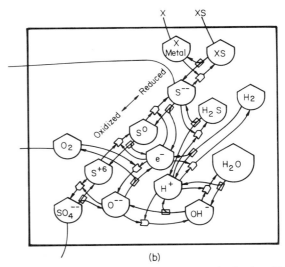

FIGURE 26-17. Biogeochemical cycles: (a) sulfur cycle (Kellogg et al., 1972) [volcanic contribution from Stoiber and Jepsen (1973)], evaluated energy diagram; (b) details of redox box (Red.).

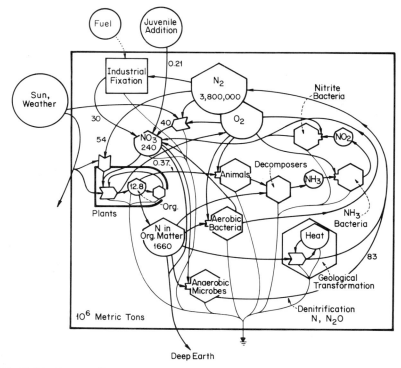

FIGURE 26-18. Energy diagram of nitrogen cycle modified from Delwiche (1970). Land and ocean compartments for the same quantity were aggregated.

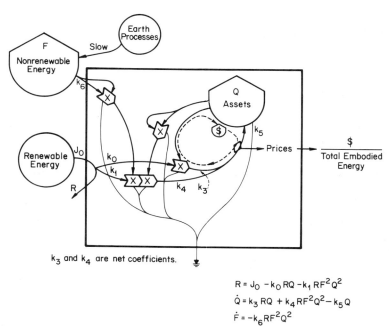

FIGURE 26-19. Minimodel of world economic growth based on use of stored and renewable energy resource.

kinetic considerations was given by Lotka (1925) as translated in Fig. 26-12. Energy drives have been added, although they are inherent in the coefficients of the pathways when they are omitted.

As a summary of a symposium on carbon cycles, Reiners (1973) gave a partially evaluated world cycle to summarize estimates of others (see Fig. 26-13). Compare this carbon cycle with the general model of the biosphere in Fig. 26-2, noting that most of the phases are represented.

Seasonal Simulation of Carbon Cycles

The diurnal simulations of carbon due to the alternation of production and consumption were given in Chapters 19 and 21.

The rise and fall of carbon dioxide with seasons, presumably driven mainly by photosynthesis and respiration balances seasonal changes in interchange with the sea and seasonal increases in carbon fuel consumption, are observed in Fig. 26-15a. Hall et al. (1975) accounted for the seasonal observed patterns of carbon dioxide shown in Fig. 26-15b with a $P-R$ model with the use of world figures.

Figure 26-15c gives a type of $P-R$ model for the carbon dioxide balance of the atmosphere to represent main factors. The underwater photosynthesis is coupled to calcium carbonate production because the bicarbonate state in the sea used for photosynthesis raises the pH, helping to drive calcium carbonate deposition tendencies in skeleton-forming organisms. Seasonal variations in oceanic heating may drive CO_2 into the air in warm seasons. Cutting and burning forests generates CO_2, but immediately afterward net growths fix CO_2 into wood and soils again. The pathway from CO_2 reacting with alkaline wastes, volcanic rocks, desert deposits, and rocks in general weathering may be important (see Fig. 26-15).

Concern over increasing CO_2 causing a greenhouse heating effect on the entire earth is controversial. Evidence suggests net cooling during the use of carbon dioxide since 1940 (Agee, 1980). Sunspot cycles were correlated. However, sea level trends have been up. Higher tropical temperatures may increase clouds and snow in high latitudes.

Wesley (1974) gives a series of overview minimodels of the biosphere using the carbon cycle. Energy translations of his more complex one is given in Fig. 26-14. His mathematical analysis of the model using reasonable figures for carbon quantities suggested some damped oscillatory characteristics with very short periods (3–10 years). Human activity was included in the heterotrophic consumers. This model omits longer period carbon cycles involving flow through inorganic cycles of limestone and deeper earth.

Shimazu and Urabe (1967) simulate earth chemical cycles in steady state. A carbon model was also used to simulate past events. Onset of photosynthesis lowered available carbon sharply and with time lag produced oscillation. Extensive worldwide measurements and new models are in progress (Bolin et al., 1979).

It is not clear, however, that answers about future carbon states can be obtained by narrowing focus on carbon alone except as part of an aggregated world model that has geologic, meteorologic, and anthroprogenic interactions and driving functions.

Other Chemical Cycles

The literature on the quantitative facts about biogeochemical cycles is now voluminous. Some examples of other cycle models are given in Fig. 26-16 (phosphorus), Fig. 26-17 (sulfur) and 26-18 (nitrogen). These are translated in the energy language from other authors by using their pathways and numbers but adding the driving energy sources and sinks and arranging components by quality. This helps to show the way each element fits into the main functional units of the biosphere system. Gilliland (1976) simulated phosphate deposit formation in Florida over a 25 million year period.

Linear Simulations

Illustrating one approach to global modeling, Wycoff and Mulholland (1979) simulated a model for the world nitrogen cycle by using donor-controlled linear pathways. Inputs were given variation with a Gaussian distribution, and impacts by human civilization were studied. A linear model in effect sets the interacting driving forces that control each pathway constant. The study of displacements of world cycles with linear pathways may be valid for low-energy displacements, but not where displacement energy is sufficient to change the linear coefficients. Study of linear carbon models showed the percent change in storages to be greater than percent change in pathways (Gardner et al., 1980a). Testing of sensitivity of linear global models (Gowdy, et al., 1975) of systems that are not linear should underestimate the correct sensitivities. See Bledsoe (1976) for comparisons of linear–nonlinear simulations.

AGGREGATED WORLD SYSTEM OF HUMANITY

Now worldly in scope, organization, and impacts, the human system has emerged to program many aspects of the biosphere, consciously and without intent (see sector H in Fig. 26-1). The major patterns of energy support and growth can be represented with world minimodels. Figure 26-19 shows the characteristic pattern with renewable and nonrenewable sources. As already given in Fig. 22-5, this class of model grows to a peak and returns to a lower level as the nonrenewable part of the energy is mainly used up. Figure 26-20 gives one of the simplest minimodels for representing the world pattern; this minimodel is sometimes useful as a function generator to drive other models where world assets are main causal actions. A variation of this model given in Fig. 23-22 was used to show the relation of prices of energy to growth and the inflation that increases as productivity declines. One property of this class of models is that the peak arrives sooner if the nonrenewable energy storage is larger, a somewhat counterintuitive property. Alexander et al. (1976)

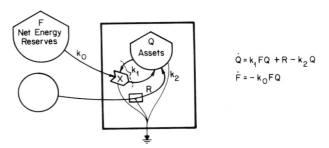

FIGURE 26-20. Simple minimodel for world economic development used as a driving function for some regional models. For evaluation, both inflows to assets are given as net energies expressed in embodied energies of the same quality.

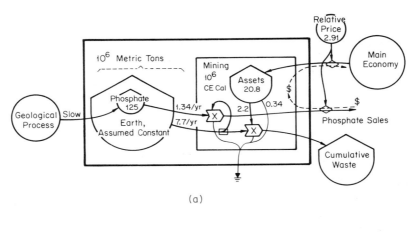

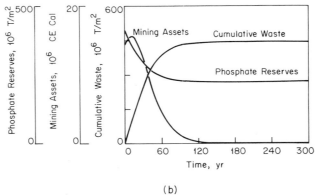

FIGURE 26-21. Model of mining system that includes increasing feedback effort as Florida phosphate reserves decline in net energy (Leibowitz, 1980): (a) energy Diagram; (b) simulation.

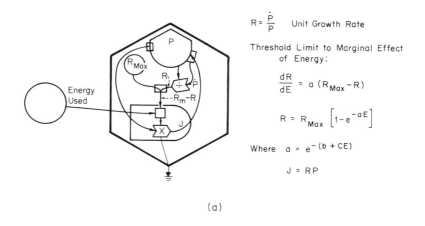

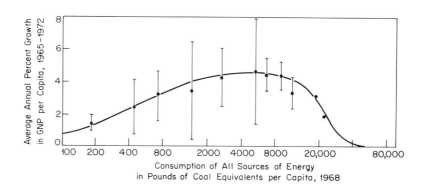

FIGURE 26-22. Relation of per capita growth and energy activity (Watt et al., 1977): (a) energy diagram and derivation of exponential limit function based on threshold for unit growth (see Fig. 8-15); (b) simulation curve and rate data.

simulated a more complex version utilizing embodied energy to evaluate storages and flows.

The world minimodel in Fig. 26-19 has the feature of matching interaction of lower- and higher-quality energy. An alternate system of pumping from nonrenewable energy also parallels so that a renewable system can readily replace the nonrenewable one as energies decline. This model also has a quadratic contribution from the nonrenewable sources. This may be appropriate since energy required to go deeper in mining (e.g., removal of overburden) increases as the square of the depth.

Leibowitz (1980), recognizing the economic behavior that increases prices and demand as a source begins to be limiting, provided a model (Fig. 26-21) that he simulated for phosphate mining but that may be applicable more generally. It increases effort and greatly increases environmental impact as the source availability declines. As fuels become scarce and unavailable to help concentrate scarce materials, these become limiting to the economic system (Georgescu-Roegen, 1971; Park and Freeman, 1975).

Inhibitory Effects of High Energy

The theory that high-quality fuel and fuel-based goods and services need to interact and match with low-quality energy to maximize power implies that continued additions of fuels would eventually cause limitations to develop from the renewable energies (see discussions in Chapters 14 and 23).

Watt et al. (1977) approached the question of too much energy with a different model given in Fig. 26-22. This is a variation of the exponential food limitation model for animal populations given in Figs. 8-15 and 9-16. The simulations were validated with data on annual percent growth. Also see Krenz (1976).

Barney (1980) reviewed results of world simulation models by E. Pesto and M. D. Mesarovic, W. Leontiev, H. Linnemann, and others. In all, growth continued past the year 2000 with a worsening ratio of food supplies to population.

World Dynamics Model

Starting from considerations of population, service, developments of industry, wastes, and resources, Forrester and Meadows evaluated and simulated models for the economy of the world. Figure 26-23a gives one of their models simulated in Dynamo, and in Fig. 26-23b this has been translated into energy circuit language with care to aggregate sectors in hierarchical form to provide more immediate understanding and perspective from inspection. The energy sources have been given their predominant role. Considered in the aggregate, it is not so different from other autocatalytic models with multiple storages and interactions. That model has subsystems for agriculture, industry, human populations, human service, and wastes. Included is a demographic population submodel, many coefficients of the effect of one sector on another. These coefficients are given as dashed lines in the Dynamics notation (see Fig. 6-25). In energy language, these pathways are regular flows of high-quality energy embodied in control actions, information, and other feedback influences.

In translating Forrester's model, some features were necessarily added that were implied but not specifically given. These are in dotted lines. Heat sinks were added. This model was built up from the parts by using empirical coefficients. If some of these are energy dependent, they may change with energy level in the real system but not in the model.

However, when it is aggregated in the translating with some grouping to fit conventions about autocatalytic loops, energy-quality chains, and depreciations, the resulting model has similarities when compared with the ones in Figs. 26-19 and 26-20, which were derived from energy principles and overviews of driving function.

International Trade and World Maximum Power

From an analysis of New Zealand, Odum and Odum (1980) argue that embodied energy of trade is the measure of its ultimate effect on the economy receiving that trade. It has been suggested that the flow of dollars in a country's GNP is due to the embodied energy received in imports relative to exports. Embodied energy rather than balance of money payments is a measure of effect on the economy.

When a country through balance of money payments causes more embodied energy to flow from an underdeveloped resource area to a developed urban center, the world system may be compelled to maximize power, even if the effect on the country supplying resources is to make it subordinate and with a lower economic standard of living.

When energy levels in the world are sufficiently high, the underdeveloped area cannot break off its balance of payments mechanism and impose trade on an embodied energy principle because the central dominant countries in the world hierarchy have enough feedback power through military, economic sanctions, intellectual influence, and other control mechanisms to maintain the economy as part of the world pattern, even though it subordinates the less developed area. As energies decline, however, and as transportation and military spheres of influence decline, more and more countries can become self-sufficient again, with a higher relative standard of living. Maximization of world power will be done with less centralized hierarchy.

A simulation model for New Zealand suggested that decreasing foreign trade of cash crops of wool, wood, meat, and other goods would increase the economic vitality because of the poor ratio of embodied energy presently offered to New Zealand by its current trade pattern (Odum and Odum, 1980). This model was similar to the trade minimodel in Fig. 23-14.

Modern Defense and War

Models of conflict and defense in simplest form were given in Figs. 12-23 and 12-24 in abstract and in more complexity for simpler cultural systems in Figs. 24-8 and 24-14. Models are given for war in modern states based on fuels and minerals in Fig. 23-27. As the diagram shows, the war phenomenon draws energy from both sides, creates a surge into the disorder–order cycle, accelerates demands on energy storages and sources, causes spatial boundaries to adjust to relative energy availabilities of the two sides, and causes extensive reorganizations. If the overall effect of the war is destruction of more energy storages than it causes to be replaced by tapping new sources through innovation, extensive inflation results. War in high-energy times is much

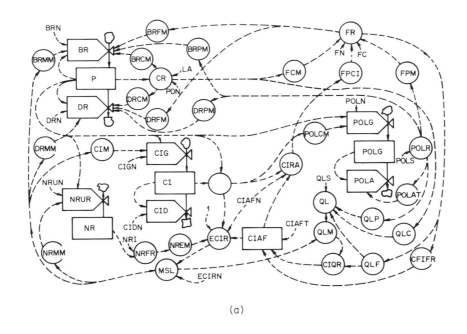

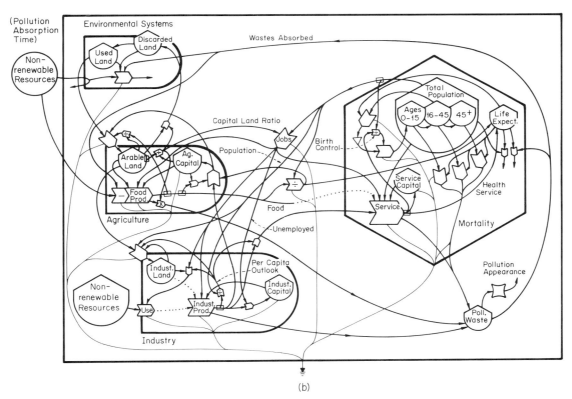

FIGURE 26-23. World model integrating population, natural resources, capital investment, agricultural assets, and pollution (Forrester, 1971; Meadows et al., 1972): (*a*) diagram with dynamics symbols; (*b*) energy diagram translation with some pathways added for energy sources and continuity (dotted lines).

more destructive and the radius of influence is very dependent on energy for transport. War in low-energy times was more like birds defending territories with small amounts of energy given to spatial organization, whereas modern high energy war is possibly a detriment to maximum power adaptations.

Simulation of war models is given in Fig. 25-27. As shown by simulation, war may go into steady state with order and disordering balanced and advantages to combatant sides equalized, after which truces may take place, when further war may no longer be increasing total power. Wars may provide means for developing excessive consumption as part of pulsing cycles. Surges in productive growth and destructive consumption and recycle may maximize power in some regimes.

SUMMARY

Ways of overviewing the world with models are given in Chapter 26. These include minimodels of energy flow, biogeochemical cycles, and overviews of the world economy and war. Models are also given for the main phases of the biosphere, earth, air, and oceans. Some of the world models of single elements are given, although this kind of representation is incomplete since most of the pathways of a single element are driven by coupling energy relations with other processes and cycles.

The carbon cycle is coupled to atmosphere, oceans, and geological cycles and constitutes nearly half of the mass of the living network. It has been simulated for studying the biosphere's state and future. Hierarchical models portray components of the biosphere with energy converging to high-quality controlling units such as storms, oceanic vortices, volcanoes, glaciers, rare mineral deposits, genetic information, culture, and technology.

The worldwide systems of trade, economic growth, and war has pulsing mechanisms that may develop spatial organization for maximum power.

QUESTIONS AND PROBLEMS

1. How is the biosphere interorganized with product and by-product webs between earth, air, seas, and living cover? Draw minimodels and mesomodels of the biosphere involving the main phases and organized according to energy quality.

2. In a more classical manner, draw separate models for water, earth, cyclic salt, sediments, and the Goldschmidt volcanic cycle. Indicate which changes in the flow of these systems are currently observed due to human activity.

3. With diagrams, discuss the carbon cycle of the earth and its link with the various phases of the biosphere and the unanswered questions about its future.

4. Which models summarize in overview the world economy as controlled and limited by resources of the biosphere? What are the main features of different initiatives in world modeling?

5. What are the pulsing mechanisms of the biosphere that are high-quality controls, recycling and feeding back actions over broad areas? Consider all phases of the biosphere. In what way are these oscillations formative of order?

6. Give the theories regarding the evolution of the earth's sedimentary and volcanic storages. How were oxidized and reduced states affected by solar driven photochemistry?

7. Diagram main features of cycles of nitrogen, sulfur, hydrogen, and oxygen.

8. What are the main processes of atmospheric chemistry and their interaction with earth releases of the human industrial system and volcanoes?

9. What kinds of closed system are possibly consistent with energetics hierarchies and natural selection?

10. Simulate world minimodels given in the chapter.

SUGGESTED READINGS

Ahrens, L. H. (1965). *Distribution of the Elements in Our Planet*, McGraw-Hill, New York. 110 pp.

Andersen, N. R. and A. Malahoff, Eds. (1977). The Fate of Fossil Fuel CO_2 in the Oceans, *Marine Science* **6.** Plenum Press, New York. 749 pp.

Barney, G. O. (1980). *The Global 2000 Report to the President*, U.S. Government: Council on Environmental Quality and Department of State, Pergamon, New York. 766 pp.

Barth, T. F. W. (1952, 1962). *Theoretical Petrology*, 2nd ed. Wiley, New York.

Barth, T. F. W. (1961). Abundances of the elements, area average, and geochemical cycles. *Geochem. Cosmochim. Acta* **23**, 1–8.

Bolin, B., E. T. Degens, S. Kempe, and P. Ketner (1979). *The Global Carbon Cycle*, Scientific Committee on Problems of the Environment (SCOPE), No. 13, Wiley, New York.

American Meteorological Society (1981). *Fifth Conference on Numerical Weather Prediction*, American Meteorological Society, Boston, Mass. 315 pp.

Baumgartner, A. and E. Reichel (1975). *The World Water Balance*, Elsevier, Amsterdam.

Bolin, B. (1981). *Carbon Cycle Modelling*, ICSU Scope 16, Wiley, New York. 390 pp.

Bolin, B. and R. J. Charlsen (1976). On the role of tropospheric sulfur cycle in the shortwave radiation climate of the earth. *Ambio* **5**, 47–54.

Brancazio, P. (1964). *The Origin and Evolution of Atmospheres and Oceans*, Wiley, New York.

Brown, H. J. M. (1979). *Environmental Chemistry of the Elements*, Academic Press, New York.

Campbell, I. M. (1977). *Energy and the Atmosphere*, Wiley, New York. 378 pp.

Churchmen, C. W. and R. O. Mason (1976). *World Modeling Priority*, North Holland, Amsterdam. 163 pp.

Clark, J. and S. Cole with R. Cornow and M. Hopkins (1975). *Global Simulation Models, A Comprehensive Study*, Wiley, New York.

Curran, S. C. and J. C. Curran (1979). *Energy and Human Needs*, Wiley, New York. 330 pp.

Dodd, J. R. and R. J. Stanton (1981). *Paleoecology, Concepts and Application*. Wiley, New York. 558 pp.

Doornkamp, J. C. and C. A. M. King (1971). *Numerical Analysis in Geomorphology*, St. Martins, New York. 372 pp.

Drever, J. I. (1982). *The Geochemistry of Natural Waters, Prentice Hall*, Englewood Cliffs, NJ. 388 pp.

Dyrrssen, D. and D. Jugner (1972). *The Changing Chemistry of the Ocean*, Wiley, New York.

Fairbridge, R. W. (1967). *Encyclopedia of Atmospheric Sciences*, Reinhold, New York.

Fortescue, J. A. C. (1980). *Environmental Geochemistry*, Springer-Verlag, New York. 347 pp.

Goldberg, E. D., I. N. McCave, J. J. O'Brien, and J. H. Steele (1977). *The Sea*, Vol. 6, *Marine Modeling*, Wiley, New York.

Hamberg, D. (1971). *Models of Economic Growth*, Harper, New York.

Holland, H. D. (1978). *The Chemistry of the Atmosphere and Ocean*, Wiley, New York. 351 pp.

Jorgensen, S. E. and I. Johnsen (1981). *Principles of Environmental Science and Technology,* Elsevier, Amsterdam, Netherlands. 516 pp.

Kostitzin, V. A. (1935). Evolution de l'atmosphere. *Act. Sci. Indust.*, p. 271, Hermann, Paris.

Kraus, E. B., Ed. (1977). *Modelling and Prediction of the Upper Layers of the Ocean*, Pergamon Press, New York. 321 pp.

Likens, G. E., Ed. (1981). *Some Perspectives of the Major Biogeochemical Cycles*. Wiley, New York. 192 pp.

Mieghem, J. V. (1973). *Atmospheric Energetics*, Clarendon Press, Oxford.

Miller, D. H. (1981). *Energy at the Surface of the Earth*, Academic Press, New York. 516 pp.

Monin, A. S., V. M. Kamenkovich, and V. G. Kort (1977). *Variability of the Oceans* (English editing by John J. Lumley), Wiley, New York.

Park, C. F. and M. C. Freeman (1975). *Earthbound, Minerals, Energy and Man's Future*, Freeman, San Francisco. 279 pp.

Pearman, G. E., Ed. (1980). *Carbon Dioxide and Climate*, Australian Academy of Science, Canberra, Australia. 217 pp.

Rasool, S. I. (1973). *Chemistry of the Lower Atmosphere*. Plenum Press, New York.

Ridley, B. K. (1979). *The Physical Environment*, Wiley, New York. 236 pp.

Riehl, H. (1979). *Climate and Weather in the Tropics*, Academic press, New York. 611 pp.

Shen, H. W., Ed. (1979). *Modeling of Rivers*, Wiley, New York.

Siegel, F. R. (1974). *Applied Geochemistry*, Wiley-Interscience, New York. 351 pp.

Siegel, F. R. (1979). *Review of Research on Modern Problems in Geochemistry*, No. 16, UNESCO. 290 pp.

Singer, S. F., Ed. (1975). *The Changing Global Environment*, Reidel, Dordrecht, Holland. 423 pp.

Stumm, W. and J. J. Morgan (1970). *Aquatic Chemistry*, Wiley, New York.

Thornes, J. B. and D. Brunsden (1977). *Geomorphology and Time*, Wiley, New York.

Van Mieghem, J. (1973). *Atmospheric Energetics*, Clarenden, Oxford, U.K. 306 pp.

Watt, D. E. F. (1981). *Understanding the Environment*, Allyn Bacon, Rockleigh, NJ. 448 pp.

Wylie, P. J. (1971). *The Dynamic Earth*, Wiley, New York. 416 pp.

# CHAPTER TWENTY-SEVEN

Summary: The Unity of Systems

Examination of systems of many sizes with energy models suggests that there is a hierarchy of systems within systems, each with similar designs, differing mainly in scales of time and space. The possibility of generalizing and teaching the nature of the universe in this way with relatively fewer principles is most exciting for a time when the energy to support the world's summit of knowledge may decline. Reorganization of concepts around the fewer principles of design and function may shorten the educational process and allow more to be learned with less, to paraphrase Buckminster Fuller's favorite motto.

Existence of common designs and similar patterns with time provides a starting place for modeling with a unified theory of systems. Variation and error are manifestations of smaller-scale systems affecting consideration of larger systems. Even Gaussian distributions may be derived from hierarchical order rather than vice versa (see Chapters 15 and 16). Use of energy language to represent many different approaches to systems facilitates the realization of a converging consensus about systems ultimately due to the commonalities in their nature.

In this final chapter a recapitulation of some unified views of systems is given with remarks about their modeling.

COMMON PATTERNS

Figure 27-1 provides the common design of the general system that was recognized in varying degrees at each level of size from molecules to stars. As potential energy flows from sources to sink, self-organization for maximum power generates stored potential energy of low-entropy structure that is either hotter, colder, more organized, or otherwise different and able to feed back special work. The differential in states between the structure and its surrounding constitutes a low-entropy energy availability. Cycles of materials pass from disordered parts to organized wholes and back.

Universal patterns of hierarchy were found with energy transformation in stages through webs developing different energy quality, measurable in units of embodied energy. Position in the hierarchy was represented in the form of model diagrams, by spectra, and by the increasing time periods of pulsing of feedback control actions (see Fig. 27-2). Compare these concepts with other generalization initiatives.

"Living Systems" Hierarchical Web Due to J. G. Miller

Miller (1955, 1978), drawing from backgrounds and experience with psychology, communication, and biology, developed a general plan of systems with 19 types of subsystem. This has been applied to many examples, including living cells, organisms, human organization, cities, and countries. Emphasis in the compartmentalization is on processing of information. In Fig. 27-3 energy diagramming of Miller's categories is attempted, placing the information and high-quality units to the right, coupling informational and lower-quality energy and matter pathways. The resulting translation shows the similarity of hierarchical concepts to others given in this book. In general, information units use matter and energy, and matter and energy units are controlled by information units. If correctly done, the diagram suggests the differential equations for a general simulation model.

Whereas Miller did not apply his categories to ecosystems, the names seem equally appropriate there. For example, roots are ingestors, biomass is matter energy storage, top carnivores are deciders, seed pools are information storages, reefs and trunks are supporter structures, migration is output

Summary: The Unity of Systems

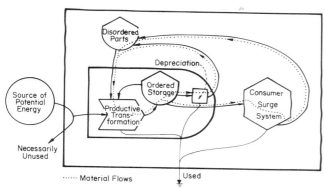

FIGURE 27-1. General pattern of production, stored structure, and consumer recycle.

transduction, songbirds operating territories are encoders and decoders, and landscape drainage organization are distributors. The word "living systems" used for this comparative analysis is inappropriate because nonliving systems have similar structures, as suggested in Chapter 26.

Principles of Energy Design

Many such as Makridakis (1977) find tendencies toward evolving order as pervasive as the dispersion tendencies of the second law (Blum, 1951). The second law and maximum power principle turn out to be coupled principles. Energy is degraded faster by developing evolutionary structural order capable of better feedbacks to maximize power flow. Conversely, systems may develop structure readily by

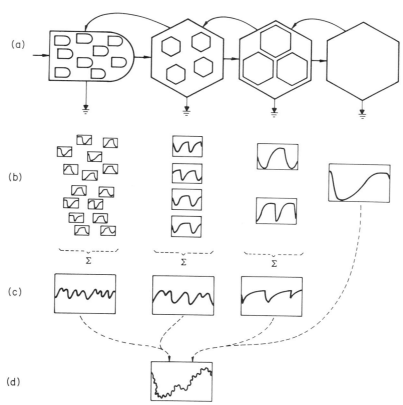

FIGURE 27-2. Schematic explanation of the convergence of pulses of smaller sizes to units with larger sizes and lower-frequency pulses: (a) energy diagram of hierarchy; (b) graphs of intrinsic pulses of component units without the noise generated by smaller pulses of lower-level units and without the catastrophic feedback pulses; (c) effect of pulses of component units in adding noise to the overall pattern; (d) combined time series graph of storage with time containing "noise" from components pulsing on short periods and large-scale, long-period pulsing.

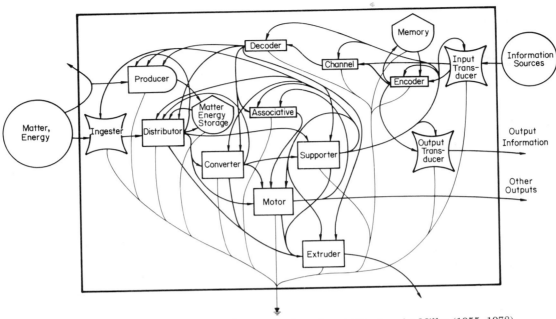

FIGURE 27-3. Energy diagram of general plan of systems given by Miller (1955, 1978).

using (and degrading) energy sources most rapidly. One orders to disorder, and vice versa.

Oscillation is an inherent property of systems varying with scale of size dimension. Time requires a storage that fills and is refilled in order to measure it; in other words, time exists relative to oscillators so that there are as many time scales as there are size scales. Since energy transformation can concentrate in space or by accumulations in one place in time, there is an energy equivalence of time and space. Prigogine (1978a,b), following Wiener (1948), finds oscillation a mechanism by which molecular phenomena develop order on a larger scale; for example, chemical binding is an oscillation of reactions in space. This is probably equivalent to the principle of selection through maximizing power with pulsing.

A pattern of pulsing seems to occur in growth and replacement of units of all sizes, but convergence in the chain allows the pulses at each level to be averaged at the next, absorbed in the convergence to a larger storage. The larger units have larger pulses. Because the higher-quality energies have longer pulses, they have the greatest control actions (see Fig. 27-2 and Fig. 18-2).

Alexander (1978) identified disasters with the pulsing of the higher-quality units of the earth. Their effects were arranged in order of their energy quality. Humans in their biological and cultural evolution have advanced up the quality chain just as fish move up their chain with age. The human activities also pulse. The largest human pulse may be the fossil fuel surge and pulse of the twentieth century.

Figure 27-4 gives a world model with pulsing consumption in which there is a pulse of frenzied consumption following long periods of accumulation. Each pulse restarts production and these may be recurring. Will there be frenzied consumption during the regression of civilization as or if energy resources decline? Richardson and Odum (1982) find pulsing maximizes power over a medium range of energy availability.

Fowler (1977) uses a Gaussian distribution with time to predict the shape of the pulse of world fuel consumption during decline. In the model the probabilities of finding oil were normally distributed around the time of maximum discovery.

Statistical Relationships to Structure

Rather than starting with a statistical normal distribution and from this deriving energy patterns and the structure of systems, the formulations in Chapters 14–18 suggest the reverse view may be preferable, namely, that energy flows under maximum power selection generate exponential energy distributions that cause Gaussian distributions of those properties that are related to energy as the square. Other kinds of skewed distribution result when energy is a linear or logarithmic function. This makes Gaussian distributed molecular veloci-

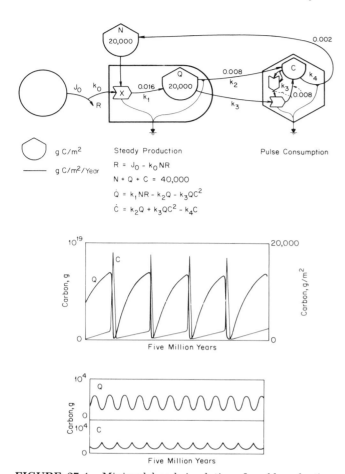

FIGURE 27-4. Minimodel and simulation of world production and pulsing consumption. (Odum, 1981, 1982). The long period simulation was calibrated with Q as world fuel reserves, C as world economic capital assets, and a land uplift erosion rate of 4 cm per 1000 years.

ties evidence of structural relationships rather than of randomness in the sense of indeterminacy.

The reverse view from the traditional one regards the increase of molecular complexity with temperature in the same way as increase of ecosystem complexity accompanies greater energy budgets, both as increased structure rather than increased randomness. This view makes entropy a measure of molecular structure in micro and macro realms, and the generating of entropy a contribution to structure. In this we have the early statement of Boltzmann (1905):

> The general struggle for existence of animal beings is therefore not a struggle for raw materials—these for organisms are air, water, and soil, all abundantly available—nor for energy which exists in plenty in any body or form of heat (albeit unfortunately not transformable),

but a struggle for entropy which becomes available through the transition of energy from the hot sun to the cold earth. . . . the products of the chemical kitchen (of plants) constitute the object of the animal world.

Dynamic pulsing at each level of space and time may be perceived as statistical variation at the next level, and as pulsing is converged and integrated (Fig. 27-2), it may generate a Gaussian error distribution that resembles indeterminate randomness. However determined, variation is a reality that needs adding to dynamic models for many purposes (Goel and Richter-Dyn, 1974; Pielou, 1977).

Equilibrium hierarchies were similar to steady-state hierarchies. The higher entropy of state of matter at higher temperatures can be interpreted as evidence of molecular structure rather than disorder (Chapter 18). Shimazu (1978) summarizes several models of world futures developed by the SMLES Research Group of Nagoya, discussing the alternative concepts of minimizing entropy, maximizing power, or maximizing efficiency, suggesting that each has its time of relevance.

If, as proposed in Chapter 15, selection for maximum power flow occurs within the microstates of a closed system also developing equilibrium hierarchies (Figs. 15-7 and 15-8), the second law may be regarded as an incomplete component pathway of a larger equilibrium system (see Fig. 27-5a). Web chains converge energy to higher quality and concentration and cycle distributive service back to dispersed and higher entropy state. The hierarchy and cycles from structure to disarray are found in both open and closed systems. Useful storages include variations in distributions of energy that are fed back in works of generating choice.

The Universe

The pattern of the universe is hierarchical. Note in Fig. 27-5b the hierarchical distribution of cosmic rays reflecting the hierarchical structure of processes, many low-energy units and a few large, pulsing ones. See also Fig. 27-5c, where the distribution of chemical elements is hierarchical. The heavier elements formed in higher-energy processes are less abundant (Ahrens, 1965). Figure 27-5, like the molecular chain in Fig. 15-8b, may represent an equilibrium hierarchy when viewed on a large enough scale. Low energy photons are shown converging to form a hierarchy of units that decrease in number but increase in quality. The feedback is according to the second law, possibly recycling energy in the way the models on smaller scale recycle matter (Fig. 27-1). Natural selection may be a Maxwell demon that allows energy to wander until it converges to the point where it is selected by being used. The hierarchical pattern may survive in competition with others because power in energy exchanges is maximized and because pulses can develop and control. The pulsing of parts of the universe may produce expanding and contracting oscillations. The pulsing of quasars may share a role with Brownian motion at a vastly smaller scale (see stellar model in Fig. 9-26).

Size, Time, and Embodied Energy

Since the larger sizes have larger time constants in their production and consumption pulses, there is a correlation between size and time, a property shown in many fields. See, for example, size and replacement of organisms in Figs. 15-9 and 18-3. Examples of size and time correlations in the sea are given by McNaught (1979) and Steele (1977). The speculative graph in Fig. 27-6 relates size and turnover time for many orders of magnitude in the hierarchy of nature. The upper right may include universe pulsing (Bondi, 1952).

Similarly, a graph of embodied energy and size may be drawn representing the convergence of energy flows at one level on those at another level. The connection between these variables should allow information on one to be used to calculate another. Parameters for dimensions of space and time should provide predictions of structure and function beyond those established by empirical measurement. The pulses of larger systems that appear as catastrophes in comparison to smaller sizes may become more understandable. In Fig. 27-6 the position of human activity has moved from small sizes to very large ones as its energy base has increased. Humanity qualifies for catastrophe status.

Time and Hierarchy

Time emerges with storage, and the scale of time that is pertinent to a phenomenon is the scale of the time constant of the storages. We can either correlate hierarchy with a scale of longer time pulses with size or, alternatively, consider the pulse time to be appropriately the same at each size scale but change the time scale with size. Whitrow (1980) reviews the many concepts of time [see also Thornes and Brunsden (1977) and Winfree (1980)].

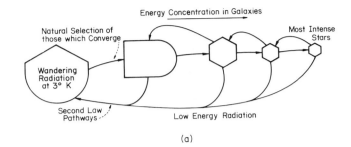

(a)

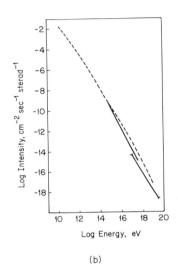

(b)

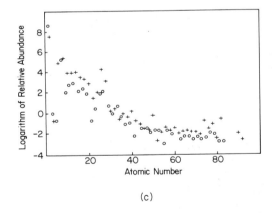

(c)

FIGURE 27-5. Energy hierarchy in the universe: (*a*) energy diagram; (*b*) cosmic rays indicating energy hierarchy [Yakovlev (1965) in Fairbridge (1978)]; (*c*) elemental abundance as a power spectrum (Mason, 1952).

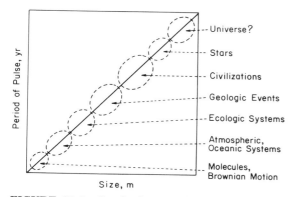

FIGURE 27-6. Graph of time of pulsing replacement.

The proper timing of organized systems requires structure and energy use in control. However, the inherent tendencies for subsystems to pulse and oscillate provides low energy means for timing controls to which other functions can be set or entrained. See the extensive field of biological clocks and their models (Brady, 1979; Pavlidis, 1973). A hierarchy may be tapped at different places to get different frequencies. See Fig. 15-24.

There may be other large-scale oscillations whose regularity of space and time dimensions are still mostly speculative in the orogenic, ice age, climatic and astronomic regimes [see summary by Gribbin (1978)].

The pulsing model of production and consumption shown in Fig. 27-4, when evaluated for production and consumption of organic matter of soils and forests, produces oscillations in simulation. However, the period of oscillation is very sensitive to coefficients, and the relation of frequency to maximizing power is an important open question. Large rates of consumption may decrease pulse amplitude and increase frequency.

Stability

Considerable interest exists in stability of ecosystems with several concepts often included in this subject (Paine, 1966; Holling, 1973; Leigh, 1971; Pielou, 1975; Orians, 1975). See review by Van Voris et al. (1980). Aspects of stability are: return acceleration after displacement; constancy of sources and constancy of system; resistance to change; resilience of the change; regularity; variability; speed of reorganization following change and alternate pathways. Most of the theory is for linear systems whose use is limited to unimportant changes. See Chapter 3 and 13. Stable sources operate diverse systems and to some extent vice versa.

Stability in the sense of survival was related to maximum power designs (Lotka, 1945). The point of view employed in this book relates patterns with time to hierarchy with a spectrum of pulses to which the system continually adapts as part of survival. The systems may be generally adapted to absorb perturbations smaller than their space–time realm and capable of self-reorganization, adapting to those perturbations of larger dimension as in Fig. 27-7.

Homeostasis of the Earth

The holistic concepts that the biosphere is homeostatically self-regulating are old views that receive restatements and clearer understanding as science of environmental systems progresses. For example, see:

Fechner (1872)	Stability principle
Henderson (1913)	Fitness of environment
Lotka (1922a,b)	Self-regulating material cycles; maximum power principle
Smuts (1926)	Holism
Kostitzin (1935)	Evolution of atmosphere
Vernadsky (1944)	Noosphere
Hutchinson (1948)	Teleological mechanisms
Odum (1951, 1955, 1970)	Selection of earth cycles for stability and for maximum power; evolutionary control of climate
Lovelock (1972)	Gaia hypothesis that the biosphere is self-organized for life

Simulation of suitable overview models of the biosphere is a worthy future objective. Some given in Chapter 26 may suggest new approaches.

Future of the Humanity–Nature System

If the theories about large-scale patterns having large-scale pulses are valid, the feedback of large oscillations of pulsing production and consumption may be expected in the future as they were in the past. The current consumer frenzy of human technological culture may be well adapted for the next oscillation downturn possibly beginning now. Catton

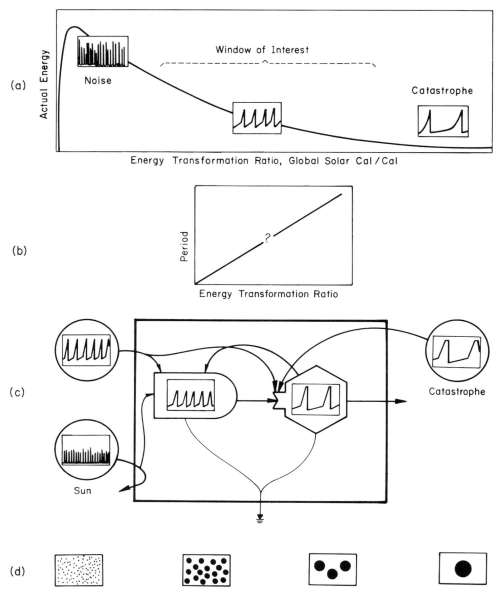

FIGURE 27-7. Frequency of pulses as a function of position in the hierarchy of energy quality. Pulses smaller than the system of concern are considered as noise, and larger pulses of systems larger than that of concern are regarded as catastrophes.

(1978) suggests a "bloom crash" model is a pertinent one for predicting human future (Odum and Odum, 1980).

In Fig. 27-4 the production-consumption model was simulated for the longer window of interest, which includes geological accumulation of mineral and fuel resources near the earths surface followed by frenzied consumption by a pulsing regenerator such as human civilization. The model was calibrated with estimated fuel reserves and world capital assets. The model shows the sharp pulse which is commonly called King Hubbert's blip. The model implies that something like the present civilization can repeat again in due time. A cogent strategy may carry forward critical information evolved in one period to better operate the next coming. This is not a bad theme for our present culture and its devotion to "progress."

SUMMARIZING REMARKS ON MODELING

Although this book contains many pages of models, it is only a sample of the rich literature. See bibliographies (O'Neill et al., 1977; Schultz, et al., 1976). Frenkel and Goodall (1978) provide a classification of models by type, size, and content.

Aspirations

Whereas the knowledge of relationships now in systems formulations and models is extensive, the shared use of this knowledge is scant. Few people can, or often are, given means to read other people's theoretical formulations. Because energy diagrams contain so much and rigorously demand so much acuity of those using them, the language could provide a means for eliminating the "tower of Babel" that now exists in the dispersed literature. Would it be a collective advance if everyone presented full models showing hierarchical relations, mechanistic relations, overall patterns, emergent properties, several levels of aggregation, energetics, kinetics, economics, and so on simultaneously? Diagrams could have the constraint of the real world such as material balance, energy conservation, energy hierarchical transformations, maximum power designs, and assumptions about linearities or homogeneities.

There may be a false security in science that relies on measurement and experiments without simulation. Only by testing models quantitatively can one know if explanatory theories are consistent with observation. Has science without modeling become second rate?

Although usage of the words vary, some modellers describe quantification with three terms: *Calibration* adjusts numerical responses of a part of the system. *Verification* confirms numerical response of the whole model to observed whole system behavior. *Validation* confirms agreement of model behavior with different data sets from that used in calibration.

A challenging aspiration is to model history, public policy and the future in overview. Some neglect these as indeterminate or too complex.

If the ultimate purpose of models is to provide humanity with a way to conceive of the greater complexity of the real world within the mind's capacity, the most valuable models will be those sufficiently simple to be understood mentally but containing information that is important for explanation, prediction, and control.

Is it possible to teach systems in grammar schools as well as in public discussions by use of the common visual language, emphasizing or deemphasizing the implied quantitative rigor, depending on the backgrounds of the listeners?

Systems Education

Our attempts to introduce these approaches to education were made with an introductory college book (Odum and Odum, 1976, 1981) with a compendium (Hall and Day, 1977), and at high school level with a booklet (Odum, Scott, and Odum, 1981). Attempts to introduce systems as a different approach in high school education have been tried before (Engineering Concepts Curriculum, 1971; Schaefer, 1972), but perhaps too many examples may have been of springs and balls instead of macroscopic minimodels of our future.

Recapitulation on Modeling

The steps in understanding a system and, therefore, to solve problems and predict patterns concerned with systems are summarized as follows:

1. Assemble information about the real system; assemble people knowledgeable about the system.

2. Together list the sources, the components that have storage, and the processes and interactions.

3. Make an energy language diagram by arranging sources by energy quality, arranging components within the system boundary also by energy-quality hierarchy, and connecting the pathways with interactions and intersections that are known mechanisms (Fig. 2-9).

4. Draw, redraw, and make full detailed systems diagrams when inventory and scanning of all knowledge is desirable; this is usually a good first step in a new situation.

5. Simplify and aggregate, retaining for emphasis, the main storages of concern, the interactions that are varying, and the parts of the system that are concerned with questions and problems.

6. Place numerical evaluations of flows and storages on a copy of the diagram.

7. Make another diagram with embodied energy values, all of one type of energy (e.g., sun or coal equivalents) where energy analysis is being done for such purposes as estimated value.

8. Translate aggregated diagrams into differential equations for simulation.

9. Run simulations with families of curves to clarify relationships, sensitivities, and possible futures.

10. Compare performance of each mechanism and of whole models to the performances of the real world, making changes if there are contradictions.

11. Look for phenomena, generalized designs, and features of the real world or of the models that may be new generalities. Use work of others by diagramming their equations, programs, and diagrams in other languages to find what is similar and different in their approaches and how this may be real and important.

12. Find ways to reorganize knowledge around the relatively fewer types of systems design that are found in the universe of systems of many levels of size.

For another approach, see Goodall (1972), who gives procedures for modeling with emphasis on use of data and range of generalization.

Less Than the Best Approaches

The following modeling approaches may have special-purpose use but are less than we might desire for giving the human mind theoretical formulations for understanding nature and humanity.

1. Simulation of the combined effects of mechanisms separately established and evaluated.

2. Trying to maximize the detail of a model so that there are 30, 100, or even 1000 state variables.

3. Applying one kind of mathematical analysis to a real system because it is satisfying in its elegance.

4. Theorizing and modeling at only one level of hierarchy, such as molecules or populations.

5. Assuming Gaussian randomness as a primary fact without relating variation to dynamic relationships.

6. Assuming dynamic relationships without recognizing the importance of disordering variation.

7. Modeling without building survival energetics into self-organization.

8. Modeling only one aspect of systems, such as material cycles, living components, human features, dollars, or water pipes.

9. Modeling with digital relations only or modeling with continuous functions only.

10. Modeling without spatial considerations or modeling spatial exchanges without emergent hierarchical functions.

11. Modeling with constants that change or are incorrectly formulated; these are unrecognized variables and systems components.

Pitfalls

Controversy continues as to what constitutes good modeling. Table 27-1, from Galomb (1968), contains a list of pitfalls. See also Berlinski (1976), who attacks uses of models that are verbal concepts without mathematical content, fallacies in complex dynamics models that don't establish mechanisms, and the gap between concepts of automata and real self-duplicating organisms. Gardner et al. (1980b) studied the effect of error by propagating the errors through models with Monte-Carlo simulation. In some models the percent error thus generated was

TABLE 27-1. Some Do's and Don'ts of Mathematical Modeling from Galomb (1968).

"Don't believe the 33rd-order consequences of a first-order model

Don't extrapolate beyond the region of fit

Don't apply any model until you understand the simplifying assumptions on which it is based, and can test their applicability

Don't believe that the model is the reality

Don't distort reality to fit the model

Don't limit yourself to a single model; more than one may be useful for understanding different aspects of the same phenomenon

Don't retain a discredited model

Don't fall in love with your model

Don't reject data that are in conflict with the model; use them to refute, modify, or improve the model

Don't apply the terminology of subject A to the problems of subject B if it is to the enrichment of neither

Don't expect that by having named a demon, you have destroyed it

Don't use terminology and notation merely to impress or confuse the uninitiated, but rather to enhance insight and to facilitate computation."

greater than initiated; in other models the errors were damped.

Optimization

There are many systems approaches for computer optimization and for finding designs and coefficients that maximize or minimize some property. Little effort was made to include these. Real systems are self-optimizing through choice generation and selection for maximum power designs. Hence the models we develop for real systems, if successful, will have the self-optimizing mechanisms and not require more conscious search and selection procedures. The characteristic hierarchical webs with their feedback amplifiers to upstream production may possibly have some of this property.

Where models are used in real time to guide management, questions arise as to representativeness of data. DePalma et al. (1979) used the Kalman filter method to smooth out errors in data. Statistical measures are used at each time step to transform observed data to better estimates for models.

Complex Models

Most of this book was concerned with minimodels and medium-sized models, "mesomodels" that can be diagrammed on one page and evaluated and simulated in a week or two. In this emphasis we are following the literature, which by and large may talk about what the large complex models are and do but don't really use them, present them in any way that a reader can follow them, or very often publish them. Raw computer programs of complex models are generally too complex for readers unless they plan to spend weeks working with the model. The complex models have taken much of the money for modeling in the last decade and in the International Biological Program before that, but it is not the complex models that are emerging. The permutations may be too large to help the human mind or help management where money is short. No doubt there are exceptions, but the burden of proof is on those who run them.

There is the possibility that the main use of models is to help humans visualize nature, and thus the models should be medium or small in order to be helpful. Aside from this there are questions about whether the big ones are really ever debugged or whether anyone understands them. Increasingly, workers question the stringing together of parts which are understood but whose whole is not.

There is an ideal represented well by the late George van Dyne to have a model (such as of grassland) that would be universal for any continent requiring only small substitutions to be used anywhere. If this is possible, it may not be by increasing the complexity of having everything in it. Main categories may suffice in which fine tuning is done after local knowledge and data are available.

There is some faith that the further one gets away from the macroscopic world of man and nature which everyone can see is complex, the simpler it gets—going to molecules or going up in size dimension to the stars. This is probably nonsense and a function of our human psychological perspective. The percent accuracy with which systems organize themselves and the percent accuracy that we normally need to understand a model and use it for prediction is probably the same for every level of size. Learning to deal with our macroscopic world around us requires the macroscope (Odum, 1971) which this book tries to provide with the energy systems way of thinking.

Where a model is simple in its unit set of equations but replicated spatially many times, the situation is different. It has large computer cost but is not complex. Relatively little information is required to describe it. Examples are the oceanic and atmospheric spatial models. Some complex models are receiving so much study and work that they may prove to be exceptions or even the disproof of the opinions given above. This author started to diagram some of these and found that it would require a funded research project, since it could not be done accurately without engaging those who wrote and operated the model. Perhaps it is time to do this for everyone's benefit.

Research Approaches

In the early days of ecology the science was criticized for naming things rather than being quantitative. Looking back over recent decades, this weakness has continued in a more sophisticated form. There has been a preoccupation with making such general terms as competition, diversity, order, stability, niche, and so on, quantitative according to one particular meaning. This causes confusion and a flurry of papers and symposia that are basically semantic because the topic word refers to many related concepts. It does stimulate work, but is this the right direction? Hopefully, the energy diagram formulations bypass the need for names and eliminate the semantic confusion so that effort can go for measurements, evaluations, simulations, analyses, synthesis, and principles.

Because the energy diagramming of models of different branches of science show so much basic similarity, unity and better communication may be fostered between scientific subcultures whose jargons are different. The disagreements between population and larger systems ecologists seem to disappear when the models of Chapters 6–13 are found within those of Chapters 19–26.

Much systems research has been directed at automatic methods of analyzing unknown systems by study of its black box responses (Bekey, 1970). This is an alternative way of taking a system apart and rebuilding a model with understanding, the theme of this book.

Many visualize systems as a series of state changes with time steps visualized as a matrix of successive operations. Thinking this way, it is easy to overlook the energy constraints and pathways by which the state changes are carried out and matter supplied. Perhaps diagrams are closer to the heterogeneity of the real world. However, we need many languages to see it all.

OPEN QUESTIONS

So many unanswered questions remain. So little of the great theoretical knowledge of humanity has been unified. What kinds of oscillation maximize power? What system designs are predictable from energy combinations of flows of different qualities and frequencies? What are energy transformation ratios for all the chemical substances, units of the earth, and components of human culture? If energy quality becomes very large, its actual energy may become unmeasurably small. Where does this leave science in relation to the highest-quality control phenomena that may exist? What is required to predict the substitution of types and species in longer time successions? What special accelerator consumer units are likely to develop, that reduces the present structure of human patterns in the downturn of our energy surge? How much memory storage is needed for maximization of power? Is there hierarchy of quality and structure in the subatomic realms? What minimodels summarize the biogeologic evolution of the earth and its atmosphere and oceans? How is the abundant knowledge of behaviors of humans and other organisms consistent with the theory for the role of these units in the pulsing control loops? What kinds of minimodel can be developed that are true in mechanism and correct in role? Will knowledge of nature's control loops permit the low energy, delicate management of the ecosystems of the biosphere by humans?

Finis

Finally, it is hoped that the modeling of ecological and general systems is helping to elucidate complex relations. Does the use of diagrammatic language unify, synthesize and teach systems knowledge? Does the unity of general systems facilitate communication among specialties and between abstraction and reality? Out great challenge is to use systems models to understand the pulsing changes in our biosphere and the role humanity can play. This mission is worth our best.

SUGGESTED READINGS

Arnold, G. and C. T. de Wit (Eds.) (1976). *Critical Evaluation of Systems Analysis in Ecosystems Research and Management,* Centre for Agricultural Publications and Documentation, Wageningen, Holland. 108 pp.

Arnold, G. W. and C. T. de Wit (1976). *Critical Evaluation of Systems Analysis in Ecosystem Research and Management,* Centre for Agric. Publ. and Documentation, Wageningen, Neth.

Bozic, S. M. (1980). *Digital and Kalman Filtering,* Wiley, New York. 157 pp.

Brown, M. T. and H. T. Odum, Eds. (1981). *Research Needs for a Basic Science of the System of Humanity and Nature,* Results of a NSF Workshop, Center for Wetlands, University of Fla. Gainesville. 154 pp.

Catton, W. R. (1980). *Overshoot,* University of Illinois Press, Urbana, IL. 297 pp.

Checkland, P. (1981). *System Thinking, System Practice,* Wiley, New York. 330 pp.

de Rosnery, J. (1979). *The Macroscop: A New World Scientific System.* 247 pp.

Ecology and the Environment (1976). A dissertation bibliography, University Microfilms, Ann Arbor, MI.

Gilmore, R. (1981). *Catastrophe Theory for Scientists and Engineers,* Wiley, New York. 666 pp.

Goh, B. S. (1980). *Management and Analysis of Biological Populations,* Elsevier, Amsterdam, Neth. 258 pp.

Graham, D. and V. D. Freeman (1976). A macroeconomic model displaying growth and cycles. *J. of Dynamic Systems, Measurement and Control* **98,** 63–67.

Gunther, R. (1973). Physiological Time and its evolution. In

Biogenesis, Evolution, Homeostasis, A. E. Locker, Ed., Springer-Verlag, New York. pp. 127–133.

Hollings, C. S. (Ed.) (1978). Adaptive Environmental Assessment and Management, In: *International Institute of Applied Systems Analysis,* Vol. 3. Wiley, New York. 377 pp.

Hughes, B. B. (1980). *World Modeling,* Lexington Books, Lexington, MA.

Iberall, A. S. (1972). *Towards a General Science of Viable Systems,* McGraw-Hill, New York. 414 pp.

Innes, G. S. and R. V. O'Neill, Eds. (1979). *System Analysis of Ecosystems,* Statistical Ecology Series, **9,** Cooperative Publ. House, Burtonsville, MD.

Jantsch, E. (1980). *The Self Organizing Universe,* Pergamon, New York. 343 pp.

Jorgensen, S. E. (1979). State of the art in ecology and modeling. *Proceedings of the Conference on Ecological Modeling,* Copenhagen, International Society for Ecological Modeling, Copenhagen.

Kormondy, E. J. and J. F. McCormick (1981). *Handbook of Contemporary Developments in World Ecology,* Greenwood Press, Westport, CT. 766 pp.

Lasker, G. E., Ed. (1981). *Applied Systems and Cybernetics Vols 1-6,* Pergamon, New York.

Lovelock, J. E. (1979). GAIA, Oxford University Press, Oxford. 157 pp.

Margalef, R. (1980). *La Biosfera entre La Termodynamica Y el Juego,* Ediciones Omega, Barcelona, Spain. 236 pp.

McNaught, D. C. (1979). Considerations of scale in modeling large aquatic ecosystems. In *Lake Ecosystem Modeling,* ed. by D. Scavia and A. Robertson, Ann Arbor Science, Ann Arbor, MI. pp. 3–24.

Melzak, Z. A. (1973, 1976). *Mathematical Ideas, Modeling, and Applications,* Vols. 1 and 2, Wiley, New York.

Nisbet, R. M. and W. S. C. Gurney (1982). *Modelling Fluctuating Populations,* Wiley, New York. 416 pp.

Odum, H. T. and E. C. Odum (1981). *Energy Basis of Man and Nature,* 2nd Ed., McGraw Hill, New York. 336 pp.

O'Neill, R. V., N. Ferguson, and J. A. Watt (1977). *A Bibliography of Mathematical Modeling in Ecology,* Oak Ridge National Laboratory, EDFB/IBP-7515.

Pielou, E. C. (1981). The usefulness of Ecological Models: A Stock Taking, *Quart. Rev. of Biol.* **56,** 17–31.

Rosen, R. (1967). *Optimality Principles in Biology,* Plenum Press, New York.

Schultz, V., L. L. Eberhardt, J. M. Thomas, and M. I. Cochran (1976). *A Bibliography of Quantitative Ecology,* Dowden, Hutchinson and Ross, Stroudsburg, PA. 362 pp.

Tillman, F. A., C. Hwang, and W. Juo (1980). *Optimization of Systems Reliability,* M. Deckker, New York. 328 pp.

Weinberg, G. M. and D. Weinberg (1979). *On the Design of Stable Systems,* Wiley, New York.

Wilson, C. and J. Grant (1980). *The Book of Time.* Westbridge David and Charles, North Pomfret, VT. 320 pp.

References*

Adams, B. (1943). *The Law of Civilization and Decay,* Knopf, New York.

Adams, R. N. (1975). *Energy and Structure, A Theory of Sound Power,* University of Texas Press, Austin. 353 pp.

Agee, E. M. (1980). Present climatic cooling and a proposed causative mechanism. *Bull. Am. Meteorol. Soc.* **61,** 1356–1367.

Aggarwal, Y. P. and L. R. Sykes (1978). Earthquakes, faults, and nuclear power plants in southern New York and northern New Jersey. *Science* **200,** 425–429.

Ahern, J. E. (1980). *The Exergy Method of Energy Systems Analysis.* Wiley, New York. 295 pp.

Ahrens, L. H. (1965). *Distribution of the Elements in Our Planet,* McGraw-Hill, New York. 110 pp.

Alcock, N. Z. (1972). *The War Disease.* Canadian Peace Research Institute, Oakville, Ontario, Canada.

Alexander, J. F. (1978). Energy Basis of Disasters and the Cycle of Order and Disorder, Ph.D. dissertation. Environmental Engineering Sciences, University of Florida, Gainesville. 232 pp.

Alexander, J. F. (1979a). *A Global Systems Ecology Model Changing Energy Use, Futures,* Vol. II, Pergamon Press, New York. pp. 443–456.

Alexander, J. F. (1979b). An energetics analysis of coal quality, in *Coal Burning Issues,* University of Florida Press, Gainesville, pp. 49–70.

Alexander, J. F., N. G. Sipe, and H. T. Odum (1976). A United States energy model economically driven by a global growth simulation. *Proceedings of the 7th Annual Pittsburgh Conference on Simulation and Modelling.* Publ. Rept. Instrument Society of America, Pittsburgh, PA.

All, T. F. H. (1977). Neolithic urban primacy. The case against the invention of agriculture. *J. Theor. Biol.* **66,** 169–180.

Allee, W. C., A. E. Emerson, O. Park, T. Park, and K. P. Schmidt (1949). *Principles of Animal Ecology.* Saunders, Philadelphia.

Allee, W. C. (1938, 1951). *Cooperation Among Animals,* Henry-Schuman, New York. 233 pp.

Allen, K. R. (1953). A method for computing the optimum size limit for a fishery. *Nature.* **172,** 210.

Anderson, R. M. and R. M. May (1978). Regulation and stability of host–parasite interaction and its effect on stability studied by a simulation model. II. Destabilizing process. *J. Animal Ecol.* **47,** 249–268.

Andresen, B., P. Salamon, and R. S. Berry (1971). Thermodynamics in finite time, extremes for imperfect heat engines. *J. Chem. Phys.* **66,** 1571–1577.

Andrews, J. F. (1968). A dynamic model of anaerobic digestion process. Publ. Environmental Systems. Engineering Department, Clemson University.

Antonini, G. A., K. C. Ewel, and J. J. Ewel (1974). Ecological Modelling of a Tropical Watershed: A Guide to Regional Planning, in *Spatial Aspects of Development,* B. Hoyle, Ed., Wiley, New York, pp. 51–74.

Antonini, G. A., K. C. Ewel, and H. M. Tupper (1975). *Population and Energy Analysis of Resource Utilization in the Dominican Republic,* University of Florida Press. 167 pp.

Armstrong, J. T. (1960). The dynamics of *Daphnia pulox* populations and *Dugesia tigrina* populations as modified by immigration. Ph.D. dissertation. Department of Zoology, University of Michigan, Ann Arbor. 102 pp.

Armstrong, N. and H. T. Odum (1964). Photoelectric ecosystem. *Science* **143,** 256–258.

Armstrong, N. E., and E. F. Gloyna (1968). Ecological aspects of stream pollution. In *Advances in Water Quality Improvement,* Water Resources Symposium, No. 1, E. F. Gloyna and W. W. Eckenfelder, Jr., Eds. 513 pp.; pp. 83–95.

Arnheim, R. 1971. *Entropy and Art,* University of California Press, Berkeley.

Arnon, D. I., M. Losada, M. Nozaki, and K. Tagawa (1961). Photoproduction of hydrogen and photofixation of nitrogen and a unified concept of photosynthesis. *Nature* **190** (4776), 601–603.

Arrhenius, S. (1889). *Z. Physiol. Chem.* **4,** 226 pp.

Ashby, W. R. (1963). *An Introduction to Cybernetics,* Wiley, New York. 295 pp.

Assmann, E. (1970). (Translated by S. H. Cordiner.) *Principles of Forest Yield Study,* Pergamon Press, New York. 506 pp.

Atchison, J. and J. A. C. Brown (1957). *The Log-Normal Distribution,* Cambridge University Press, London.

Atherton, D. P. (1981). *Stability of Non-Linear Systems,* Research Studies Press, Wiley, New York. 231 pp.

Atkinson, A. B. (1976). *The Personal Distribution of Incomes,* Westview Press, Boulder, CO. 349 pp.

Austin, M. P. (1972). Models and analysis of descriptive vegetation data. In *Mathematical Models in Ecology,* J. N. R. Jeffries, Ed., Blackwell Scientific Publishers, Oxford. pp. 61–86.

Bailey, N. T. J. (1967). *The Mathematical Approach to Biology and Medicine,* Wiley, New York. 296 pp.

Ballentine, T. (1976). A net energy analysis of surface mined coal from the northern great plains. Master's of engineering thesis. University of Florida, Gainesville. 148 pp.

Bakuzis, E. V. (1969). Forestry viewed in an ecosystem perspective. In *The Ecosystem Concept in Natural Resources Management,* G. M. Van Dyne, Ed., Academic Press, New York. pp. 189–258.

Bakuzis, E. V. (1974). *Foundations of Forest Ecosystems,* College of Forestry, Univ. of Minnesota, St. Paul, MN.

---

* These are cited references. See also suggested readings at the end of each chapter.

Barker, R. R. (1978). *Evaluating Ecology of Animal Migration*, Hodder and Stoughton, London, U.K.

Barney, G. O. (1980). *The Global 2000 Report to the President*, U.S. Government, 766 pp. Council on Environmental Quality, and Department of State, Pergamon, New York.

Barth, T. F. W. (1961). Abundance of the elements areal averages, and geochemical cycles. *Geochem. Cosmochem. Acta* **23**, 1–8.

Bartlett, M. S. (1960). *Stochastic Population Models in Ecology and Epidemiology*, Methuen, London. 90 pp.

Bartlett, M. S. and R. W. Hiorns (1973). *The Mathematical Theory of the Dynamics of Biological Populations*, Academic Press, London.

Batty, M. (1976). *Urban Modelling*, Cambridge University Press, London.

Baule, B. (1917). Zu Mitscherlichs Gesetz der Physiologischen Beziehunger. *Landw. Jahrb.* **51**, 363–385.

Baumol, W. J. (1967). Macroeconomics of unbalanced growth. *Am. Econ. Rev.* **57**, 415–426.

Bayley, S. and H. T. Odum (1976). Simulation of interrelations of the Everglades Marsh, peat, fire, water, and phosphorus. *Ecol. Model.* **2**, 169–188.

Bekey, G. A. (1970). System identification—an introduction and a survey. In *Simulation*, 151–166.

Bellman, R. (1960). *Introduction to Matrix Analysis*, McGraw-Hill. 327 pp.

Bénard, H. (1900). *Rev. gen. Sci Pure Applied* **11**, 1261.

Bennett, A. W. (1974). *Introduction to Computer Simulation*, West Publishing Company, St. Paul, MN.

Benoit, R. J. (1964). Mass culture of microalgae for photosynthetic gas exchange. In *Algae and Man*, D. F. Jackson, Ed. Plenum Press, New York. 434 pp.; pp. 413–425.

Berlinski, D. (1976). *On Systems Analysis*, MIT Press, Cambridge, MA. 186 pp.

Bernstein, B. B. (1981). Ecology and Economics. *Ann. Rev. Ecol. Syst.* **12**, 309–330.

Berry, B. J. L. and W. L. Garrison. (1958). Alternative explanations of urban rank-size relationships. *Ann. Assoc. Am. Geophys.* **48**, 83–91.

Berry, B. J. L., E. C. Conkling, and D. M. Ray (1976). *The Geography of Economic Systems*, Prentice-Hall, Englewood Cliffs, NJ. 529 pp.

Berry, B. J. L. (1961). City size distributions and economic development. *Econ. Devel. Cult. Change* **9**, 573–588.

Bertine, K. K. and E. Goldberg (1971). Fossil fuel combustion and the major sedimentary cycle. *Science* **173**, 233–235.

Beverton, R. J. H. and S. J. Holt (1957). Dynamics of fish populations. *Fish. Invest. Series II* **19**, 1–533.

Beyers, R. J. (1962). Relationship between temperature and the metabolism of experimental ecosystems. *Science* **136**, 980–982.

Beyers, R. J. (1963a). The metabolism of twelve aquatic laboratory microecosystems. *Ecol. Monogr.* **33**(4), 281–306.

Beyers, R. J. (1963b). A characteristic diurnal metabolic pattern in balanced microcosms. *Publ. Inst. Marine Science, Texas* **9**, 19–27.

Beyers, R. J. (1974). Report of Savannah River Laboratory, University of Georgia, Aiken, SC.

Bierman, U. J., F. H. Verhoff, T. L. Paulson, and M. W. Tenney (1975). Multidynamic modes of algal growth and species competition in eutrophic lakes. In *Modeling the Eutrophication Process*, 89–107. E. J. Middlebrooks, D. H. Falkenborg, and T. E. Maloney, Eds., Ann Arbor Science, Ann Arbor, MI.

Biggs, N. L., E. Keith Lloyd, and R. J. Wilson (1976). *Graph Theory*, Clarendon Press, Oxford.

Billig, E. and K. W. Plessner (1949). The efficiency of the selenium barrier–photocell when used as a converter of light into electrical energy. *Phil. Mag.* **40**, 568–572.

Bishop, A. B., H. H. Fullerton, A. B. Crawford, M. D. Chambers, and M. McKee (1974). Carrying capacity in regional environmental management. Environmental Protection Agency. EPA 600/5-74-021. 170 pp.

Blackman, F. F. (1905). Optima and living factors. *Ann. Bot.* **19**, 281–295.

Blackman, R. B. and J. W. Tukey (1958). *The Measurement of Power Spectra*, Dover Publication Reprint, New York. 190 pp.

Blackman, V. H. (1919). The compound interest law and plant growth. *Ann. Bot.* **33**, 353–360.

Blalock, H. M. (1971). *Causal Models in the Social Sciences*, Aldine-Albertson, Chicago. 515 pp.

Bledsoe, L. J. (1976). Linear and nonlinear approaches for ecosystem-dynamic modeling, pp. 283–298. In *Systems Analysis and Simulation in Ecology*, B. P. Patten, Ed., Academic Press, New York.

Blinchikoff, J. J. and A. L. Zverev (1976). *Filtering in the Time and Frequency Domains*, Wiley, New York. 494 pp.

Bliss, L. C. and E. B. Hadley (1964). Photosynthesis and respiration of alpine lichen. *Am. J. Botany* **51**, 870–874.

Blum, H. F. (1951). *Times Arrow and Evolution*, Princeton University Press, Princeton, NJ. 222 pp.

Bode, H. W. (1945). *Network Analysis and Feedback Amplifier Design*, Van Nostrand, New York.

Bogner, R. E. and A. G. Constantinides (1975). *Introduction to Digital Filtering*, Wiley, New York. 198 pp.

Bolin, B., E. T. Degens, S. Kempe, and P. Ketner (1979). *The Global Carbon Cycle*, Scientific Committee on Problems of the Environment (SCOPE) 13, Wiley, New York.

Bolin, B. and C. D. Keeling (1963). Large scale atmospheric mixing as deduced from the seasonal and meridional variations of carbon dioxide. *J. Geophysical Res.* **68**, 3899–3920.

Bolt, R., W. L. Koltun, O. H. Levine (1965). Doctoral feedback in higher education. *Science* **148**, 918.

Boltzmann, L. (1896). *Vorlesungen über Gastheorie*. 2 vol., Leipzig.

Boltzmann, L. (1886). Der Zweite hauptsatz der mechanischen Warmtheorie. *Almanach der K. Acad. wiss. mechanische, Wien* **36**, 225–259.

Boltzmann, L. (1905). The Second Law of Thermodynamics. *Populare Schriften*, Essay No. 3 (address to Imperial Academy of Science in 1886). Reprinted in English in *Theoretical Physics and Philosophical Problems, Selected Writings of L. Boltzmann*. D. Reidel, Dordrecht, Holland.

Bondi, H. (1952). *Cosmology*, Cambridge, University Press, London. 179 pp.

Bongers, L. and J. C. Medici (1966). Chemosynthetic metabolism of Hydrogeomonas. In *Bioregenerative Systems*, NASA

Scientific and Technical Information Division. NASA SP-165. pp. 9–18.

Bonner, J. T. (1965). Size and cycle—an essay on the structure of biology. Princeton University Press, Princeton, NJ.

Booth, D. A. and F. M. Toates (1974). A physiological control theory of food intake in the rat: Mark I. *Bull. Psychonomic Society* **3**, 442–444.

Bormann, F. H. and G. E. Likens. (1979a). *Pattern and Process in a Forested Ecosystem*, Springer-Verlag, New York.

Bormann, F. H. and G. E. Likens (1979b). Catastrophic disturbance and the steady state in northern hardwood forests. *Am. Sci.* **67**, 660–669.

Bosserman, R. W. (1980). Complexity measures for assessment of environmental impact in ecosystem networks, in *Proc. Pittsburg Conference on Modeling and Simulation*, Pittsburg, PA. pp. 811–820.

Bosserman, R. W. (1981). Effects of structural changes of matter on energy models of ecosystems, in *Energy and Ecological Modelling*, W. J. Mitsch, R. W. Bosserman, and J. M. Klopatek, Eds., Elsevier, Amsterdam, pp. 617–625.

Botkin, D. B., J. F. Janak, and J. R. Wallis (1972a). Some ecological consequences of a computer model of forest growth. *J. Ecol.* **60**, 849–873.

Botkin, D. B., J. F. Janak, and J. R. Wallis (1972b). Rationale, limitations and assumptions of a northeastern forest growth simulator. *IBM J. Res. Devel.* **16**, 101–116.

Boustead, I. and G. F. Hancock (1979). *Handbook of Industrial Energy Analysis*, Wiley, New York.

Bowdon, R. (1965). A method of linear causal analysis. *Am. Sociol. Rev.* **30**, 365–555.

Bowen, H. J. M. (1979). *Environmental Chemistry of the Elements*, Academic Press, New York.

Bowman, K. O., K. Hutcheson, E. P. Odum, and L. R. Shenton (1971). Comments on the distribution of indices. In *Statistical Ecology Series*. Vol. 3, pp. 316–365.

Box, G. E. P. and G. M. Jenkins (1976). *Time Series Analysis, Forestry and Control*, Holden-Day, San Francisco. 575 pp.

Boynton, W. (1975). Energy basis of a coastal region: Franklin County and Appalachicola Bay, Florida. Ph.D. dissertation. Environmental Engineering Sciences, University of Florida, Gainesville.

Boynton, W., D. Hawkins, and D. Gray (1975). Trends in growth effecting oyster economy in Appalachicola Bay and Franklin County. Report of Project R/EA-3 Sea Grant. University of Florida. 45 pp.

Boysen-Jensen, P. (1932). *Die Stoffproduktion der Planzen*, Gustav Fischer, Jena, Germany.

Brady, J. (1979). *Biological Clocks*, Edward Arnold, London.

Bray, J. R. and J. T. Curtis (1957). An ordination of the upland forest communities of southern Wisconsin. *Ecol. Monogr.* **27**, 325–349.

Brewer, J. W. (1976). Bond graphs of microeconomic systems. American Society of Mechanical Engineers 75-WA/Aut-8. Houston, TX.

Breymeyer, A. I. and G. M. Van Dyne (1980). *Grasslands, Systems Analysis and Man*, International Biological Programme 1-950. Cambridge University Press, London.

Briggs, L. L. and H. N. Pollock (1967). Digital model of evaporite sedimentation. *Science* **155**, 453–456.

Brillouin, L. (1962). *Science and Information Theory*, 2nd ed., Academic Press, New York. 351 pp.

Brocksen, R. W., G. E. Davis, and C. E. Warren (1968). Competition, food consumption, and production of sculpins and trout in laboratory stream communities. *J. Wildlife Manage.* **32** (1), 51–75.

Brody, S. (1945). *Bioenergetics and Growth*, Reinhold, New York. 1023 pp.

Broeker, W. S. (1974). *Chemical Oceanography*, Harcourt Brace Jovanovich, New York. 214 pp.

Brookes, L. G. (1974). Toward the all electric economy. In *Energy from Systems to Scarcity*, K. A. D. Inglis, Ed., Applied Science Publishers, Essex, U.K.

Browder, J. A. (1976). Water, wetlands, and wood storks in Southwest Florida. Ph.D. dissertation. Department of Environmental Engineering Science, University of Florida, Gainesville. 249 pp.

Browder, J. A. (1978). A modeling study of water, wetlands and wood storks. Wading Birds, Research Report No. 7. pp. 325–346. National Audubon Society.

Browder, J. A. and B. G. Volk (1978). Systems model of carbon transformations in soil subsidence. *Ecol. Model.* **5**, 269–292.

Browder, J. A., C. Littlejohn, and D. Young (1975). South Florida seeking a balance of man and nature. Division of State Planning, Florida Department of Administration, Tallahassee.

Brower, L. P., J. Zardt Brower, and C. T. Collins (1963). Experimental studies of mimicry 7. *Zoologica* **48**, 65–83.

Brown, M. (1975). Lee County—An Area of Accelerating Growth, Division of State Planning, Tallahassee, FL.

Brown, M. (1977). War, peace and the computer: simulation of disordering and ordering energies in South Vietnam. In *Ecosystem Modeling in Theory and Practice*, C. S. Hall and J. W. Day, Eds., pp. 394–417.

Brown, M. T. (1980). Energy basis for hierarchies in urban and regional landscapes. Ph.D. dissertation. Department of Environmental Engineering Sciences, University of Florida, Gainesville. 360 pp.

Brown, S. L. (1978). A comparison of cypress ecosystems in the landscape of Florida. Ph.D. dissertation. Department of Environmental Engineering Sciences, University of Florida, Gainesville.

Bryant, J. O. (1969). Frequency and transital response analysis of water and wastewater treatment systems. Application of Systems Analysis in Sanitary Engineering Workshop. Clemson University and American Association of Professors in Sanitary Engineering.

Bullard, C. W., III, P. S. Penner, and D. A. Pilati (1976). Energy analysis handbook for combining process and input-output analysis. CAC Document No. 214. Center for Advanced Computation. University of Illinois at Urbana–Champaign, Urbana. ERDA 77-61 YCB, 93.

Bungay, H. R. (1966). Analog simulation of interacting microbial cultures. Preprint 22B, Symposium on Chemical Engineering. American Institute of Chemical Engineering.

Bungay, H. R. (1967). Dynamic analysis of microbic systems. In *Water and Sewage Works*. pp. 196–198.

Burg, J. P. (1975). *Maximum Entropy Spectral Analysis*, Ph.D. Thesis, Stanford University. 123 pp.

Burnett, M. S. (1978). Energy analysis of intermediate technology agricultural systems. M.S. thesis. Environmental Engineering Sciences, University of Florida, Gainesville.

Burns, L. (1970). Analog Simulation of a Rain Forest with High-Low Pass Filters and a Programmatic Spring Pulse. Appendix p. I-284, In *A Tropical Rainforest*, H. T. Odum and R. F. Pigeon, Eds., Div. of Tech. Information, U.S. Atomic Energy Commission, Oak Ridge, TN.

Burns, L. A. and B. R. Taylor, III (1977). *A Model of Nutrient Uptake in Marsh Ecosystems*, American Society of Civil Engineers, San Francisco, CA.

Burr, D. Y. (1977). Value and water budget of cypress wetlands in Collier County, Florida. Master's thesis. Department of Environmental Engineering Science, University of Florida, Gainesville. 112 pp.

Burton, A. C. (1939). The properties of the steady state compared to those of equilibrium as shown in characteristic biological behavior. *J. Cell. Comp. Physiol.* **14**, 327–349.

Busch, A. W. (1965). Energy, total carbon, and oxygen demand. Proc. Twentieth Industrial Waste Conference, *Eng. Bull. Purdue University*, Series No. 8, 457–469.

Butterfield, C. T. and W. C. Purdy (1931). Some interrelationships of plankton and bacteria in natural purification of polluted water. *Ind. Eng. Chem.* **23**, 213–218.

Cain, S. A. (1930). Concerning certain phytosociological concepts. *Ecol. Monogr.* **2**, 475–508.

Cain, S. A. and G. M. de Oliveira Castro (1960). *Manual of Vegetation Analysis*, Harper, New York.

Cairns, J., K. L. Dickson, and E. E. Herricks (Eds.) (1980). *The Recovery and Restoration of Damaged Ecosystems*, University of Virginia Press, Charlottesville. 167 pp.

Campbell, G. S. (1972). *An Introduction to Environmental Biophysics*, Heidelberg Science Library, Springer. 159 pp.

Campbell, I. M. (1977). *Energy and the Atmosphere*, Wiley, New York.

Canale, R. P. (1969). Predator prey relationships in a model of activated process. *Biotechnol. Bioeng.* **11**, 887–907.

Cannon, W. B. (1939). *The Wisdom of the Body*, Norton, New York.

Canoy, M. (1970). DNA in ecological systems. Ph.D. dissertation. Zoology Department, University of North Carolina, Chapel Hill.

Capellos, C. and B. H. J. Belski (1980). *Kinetic Systems*. Kueger, Huntington, New York.

Caplan, S. R. (1966). The degree of coupling and its relation to efficiency of energy conversion in multiple flow systems. *J. Theor. Biol.* **10**, 209–235.

Carnot, S. (1824). Reflections sur la puissance matrice du fue, Paris. Translated 1960, American Society of Mechanical Engineers, New York. 107 pp.

Caspers, H. (1957). Black Sea and Sea of Azov. In *Treatise of Marine Ecology and Paleoecology*, Vol. 1, J. W. Hedgpeth, Ed., pp. 801–890, Geological Society of America, Memoir No. 67, Waverly Press, Baltimore, MD.

Cassie, R. M. (1962). Frequency distribution modes in the ecology of plankton and other organisms. *J. Animal Ecology* **31**, 65–92.

Casti, J. L. (1979). *Connectivity, Complexity, and Catastrophe in Large-Scale Systems*, Wiley, New York.

Caswell, H., H. E. Koenig, J. A. Rech, and W. R. Todd (1973). An introduction to systems science for ecologists. In *Systems Analysis and Simulation in Ecology*, Vol. 2, B. Patten, Ed., Academic Press, New York. pp. 2–80.

Catton, W. R. (1978). Carrying capacity, overshoot and the quality of life. In *Major Social Issues*, J. M. Yinger and S. J. Culler, Eds., Free Press, Macmillan, London. pp. 230–251.

Chamberlin, C. F. and R. Mitchell (1978). A decay model for enteric bacteria in natural waters. *Water Pollution Microbiol.* **2**, 325–347.

Chance, B. (1959). Electron transfer in biological systems. *Prof. Inst. Radio Eng.* **47**, 1821–1841.

Chance, B., J. G. Brainerd, F. A. Cajori, and G. A. Millikan (1940). The kinetics of the enzyme–substrate compound of peroxidase and their relation to the Michaelis theory. *Science* **92**, 445.

Chance, B., D. S. Greenstein, J. Higgins, and C. C. Yerd (1952). The mechanism of catalase action. Part II. *Arch. Biochem. Biophys.* **37** (2), 322, 339.

Chance, B., R. W. Estabrook, and J. R. Williamson, Eds. (1965). *Control of Energy Metabolism*, Academic Press, New York. 441 pp.

Chance, B., K. Pye, and J. Higgins (1967). *IEEE Spectrum* **4**, 79–86.

Chapman, R. N. (1931). *Animal Ecology*, McGraw-Hill, New York.

Charney, J. G. and N. A. Phillips (1953). Numerical integration of the quasigeostrophic equations for biotropic and simple baroclime flows. *J. Meteorol.* **10**, 71–99.

Charpentier, J. P. (1976). Toward a better understanding of energy consumption. The distribution of per capita energy consumption in the world. In *Energy*, Vol. 1, Pergamon Press, Oxford. pp. 320–334.

Chaston, I. (1971). *Mathematics for Ecologists*. Butterworths, London. 132 pp.

Chen, C. W. (1970). Concepts and utilities of ecologic model. *J. Sanit. Eng. Div. ASCE* **96**, 1085–1097.

Cherry, C., Ed. (1957a). *Information Theory*, Third London Symposium, Academic Press, New York.

Cherry, C. (1957, 1966). *On Human Communication*, 2nd ed., MIT Press, Cambridge, MA.

Cherry, C., Ed. (1961). *Information Theory*, Fourth London Symposium, Butterworths, London.

Chickilly, A. A. (1976). Conditions under which a stable polymorphism can be maintained. Class Report in Env. 561. H. T. Odum, instructor.

Childers, D. G., Ed. (1978). *Modern Spectrum Analysis*, IEEE Press, IBC, New York. 334 pp.

Clark, C. (1951). Urban population densities, *J. of Royal Statistical Soc.*, Series A, **114**, 490–496.

Clark, C. W. (1973). The economics of overexploitation. *Science* **181**, 630–634.

Clark, J. P. (1971). The second derivative and population modeling. *Ecology* **42**, 710–723.

Clarke, G. L. (1946). Dynamics of production in a marine area. *Ecol. Monogr.* **16**, 321–335.

Clarke, G. L. (1959). *Elements of Ecology*, Wiley, New York.

Clarke, G. L. and C. J. Hubbard (1959). Quantitative records of the luminescent flashing of oceanic animals at great depth. *Limnol. and Oceanogr.* **4**, 163–179.

Cleland, W. W. (1963). The kinetics of enzyme catalyzed reactions with two or more substrates or products: I. Nomenclature and equations. *Biochem. and Biophysica Acta* **67**, 104–137.

Clendenning, K. A. and H. C. Ehrmantraut (1951). Cited in *Photosynthesis,* Vol. 2, Part 2, E. T. Rabinowitch, Ed., Interscience, New York.

Clymo, R. S. (1978). A model of Peat Bog Growth, in Production Ecology of *British Moors and Montane Grasslands,* O. W. Heal and D. F. Perkins, Eds., Springer-Verlag, New York. pp. 187–223.

Cobb, C. W. and P. H. Douglas (1928). A theory of production. *American Economic Review,* supplement, pp. 139–165.

Cody, M. L. (1974). Optimization in Ecology. *Science* **183**, 1156–1164

Cohen, D. (1966). Optimizing reproduction in a randomly varying environment. *Theor. Biol.* **12**, 119–129.

Cohen, D. (1968). A general model of optimal reproduction in a randomly varying environment. *Ecology* **109**, 219–228.

Cohen, J. E. (1966). *A Model of Simple Competition,* Harvard University Press, Cambridge, MA. 138 pp.

Cohen, J. E. (1978). Food webs and niche space. In *Monographs in Population Biology,* Vol. II. Princeton University Press. 189 pp.

Coker, R. E. and J. G. Gonzalez (1960). Limnetic copepod populations of Bahia Fosforescente and adjacent waters, Puerto Rico. *J. of Elisha Mitchell Sci. Soc.* **76**(1), 8–28.

Cole, L. C. (1957). Sketches of general and comparative demography. *Cold Spring Harbor Symp. Quant. Biol.* **22**, 1–15.

Collens, T. W. (1876). Monogram of Political Economy, reproduced in *Seven American Utopias* by D. Hayden, Mass. Inst. Tech. Press, Cambridge, MA.

Connell, J. H. and E. Orias (1964). The ecological regulation of species diversity. *Am. Nat.* **98**, 399–414.

Connell, J. H. (1973). Population ecology of reef building corals. pp. 204–245. In *Biology and Geology of Coral Reefs,* Vol. 2, O. A. Jones and R. Endean, Eds., Academic Press, New York. pp. 204–245.

Conover, J. T. (1958). Seasonal growth of benthic marine plants as related to environmental factors in an estuary. *Publ. Inst. Marine Sci., Univ. Tex.* **5**, 97–147.

Constantinides, A., J. L. Spencer, and E. L. Gaden (1970). Optimization of batch fermentation processes. I. Development of mathematical models for batch penecillin fermentation. *Biotechnol. Bioeng.* **12**, 803–830.

Conway, E. J. (1942, 1943). The chemical evolution of the ocean. *Proc. Roy. Irish Acad. Sect. B* **48**, 119–159; 161–212.

Cook, E. (1976). *Man, Energy, Society.* Freeman, San Francisco. 478 pp.

Cooke, G. D. (1967). The pattern of autotrophic succession in laboratory microcosms. *Bioscience* **17**, 717–721.

Cooke, G. D., R. J. Beyers, and E. P. Odum (1968). The case for the multi-species ecological system with special reference to succession and stability in Bioregenerative studies. National Aeronautics and Space Administration, NASA special publication **165**, 129–139.

Coon, C. S. (1954, 1961). *The Story of Man,* Alfred A. Knopf, New York.

Cooper, D. C. and B. J. Copeland (1973). Responses of continuous series estuarine microecosystem to point source input variations. *Ecol. Monogr.* **43**, 213–236.

Cooper, R. B. (1972). *Introduction to Queueing Theory,* MacMillan, New York. 276 pp.

Copeland, B. J. (1963). Oxygen relationships in oil refinery effluent holding ponds. Ph.D. thesis. Oklahoma State University, Stillwater. 110 pp.

Copeland, B. J. (1965). Evidence for regulation of community metabolism in a marine ecosystem. *Ecology* **46**, 503–564.

Copeland, B. J., H. T. Odum, and D. C. Cooper (1972). Water quality for preservation of estuarine ecology. In *Water Resource Symposium,* No. 5, E. F. Glyna and W. S. Butcher, Eds., Center for Research in Water Resources, University of Texas, Austin. pp. 107–126.

Costanza, R. (1975). The spatial distribution of land use subsystems, incoming energy, and energy use in south Florida. M.A. paper. Department of Architecture, University of Florida, Gainesville.

Costanza, R. (1977). Energy models of exchange with simulation of the role of exchange in spatial organization. In *Energy Analysis of Models of the United States,* H. T. Odum and J. Alexander, Eds., Annual Report to the Department of Energy, Department of Environmental Engineering Science, University of Florida, Gainesville. pp. 96–112.

Costanza, R. (1979). Embodied energy basis for economic–ecologic systems. Ph.D. dissertation. Department of Environmental Engineering Sciences, University of Florida, Gainesville. 250 pp.

Costanza, R. (1980). Embodied energy and economic evaluation. *Science* **210**, 1219–1224.

Costanza, R. and C. Neill (1981a). The energy embodied in the products of ecological systems: a linear programming approach, in *Energy and Ecological Modelling,* W. F. Mitsch, R. W. Bosserman, and J. M. Klopatek, Eds., Elsevier, Amsterdam. pp. 661–670.

Costanza, R. and C. Neill (1981b). Energy embodied in the products of the biosphere, in *Energy and Ecological Modelling,* W. J. Mitsch, R. W. Bosserman, and J. M. Klopatek, Eds., Elsevier, Amsterdam. pp. 745–755.

Cottrell, F. (1955). *Energy and Society,* McGraw-Hill, NY.

Cowles, H. C. (1899). The ecological relations of the vegetation of the sand dunes of Lake Michigan. *Bot. Gaz.* **27**, 95–117, 167–202, 281–308, 361–391.

Cox, R. and W. Alderson Eds. (1950). *Theory in Marketing,* Richard D. Irwin, Inc., Homewood, IL.

Cristaller, W. (1933, 1966). *Central Places in Southern Germany.* Translated by C. W. Baskin, Prentice-Hall, Englewood Cliffs, NJ.

Culling, W. E. H. (1960). Analytical theory of erosion. *J. Geol.* **69**, 336–344.

Cumming, F. P. and R. J. Beyers (1970). Further laboratory studies of forest floor microcosms. In *A Tropical Rainforest.* H. T. Odum and R. F. Pigeon, Eds., Division of Technical Information. U.S. Atomic Energy Commission, Oak Ridge, TN. pp. 154–156.

Cummins, K. W. (1977). From headwater streams to rivers. *Am. Biol. Teacher,* 305–312.

Cunningham, G. L. and J. L. Reynolds (1978). A simulation model of primary production and carbon allocation in the creosote bush [*Larrea tridentata* (DC) Dov.]. *Ecology* **59**, 37–52.

Cunningham, J. P. (1974). An energetic model linking forest industry and ecosystems. *Communicationes Instituti Forestatis Fenniae* (Helsinki, Finland) **79** (3), 1–51.

Curtis, J. J. and R. McIntosh (1951). An upland forest continuum in the prairie forest border region of Wisconsin. *Ecology,* **32**, 476–496.

Curtis, J. T. (1959). *The Vegetation of Wisconsin.* University of Wisconsin Press, Madison. 657 pp.

Curzon, F. L. and B. Ahlborn (1975). Efficiency of a Carnot engine at maximum power output. *Amer. J. of Physics* **43**, 22–24.

Cushing, J. M. (1977). Integrodifferential equations and delay models in population dynamics. In *Lecture Notes in Biomathematics,* No. 20. Springer-Verlag, New York.

Czarnowski, M. S. (1961). *Dynamics of Even-Aged Forest Stands,* Louisiana State University Press, Baton Rouge. 132 pp.

Dahl-Madsen, K. I. and E. Gargas (1975). A preliminary eutrophication model of shallow fjords. *Progr. Water Technol.* **7**, 469–481.

Dainton, F. S. (1956, 1966). *Chain Reactions, Methuen's Monograph,* Wiley, New York.

Dansereau, P. (1973). *Inscape and Landscape,* CBC Learning Systems, Toronto, Canada.

Dansereau, P. (1976). An ecological grading of human settlements with special reference to the urban habitat. In *Science for Better Environment,* Proceedings of the International Congress on Human Environment, HESC, Science Council of Japan. pp. 234–243.

Davidson, R. S. and A. B. Clymer (1966). The desirability and applicability of simulating ecosystems. *Ann. NY Acad. Sci.* **128**, 790–794.

Davis, J. C. and M. J. McCullogh (1975). *Display and Analysis of Spatial Data.* Wiley, New York. 377 pp.

Dayton, K. and R. R. Hessler (1972). Role of biological disturbance in maintaining diversity in the deep sea. *Deep Sea Res.* **19**, 199–208.

De Amezaga, E., C. R. Goldman, and E. A. Stull (1971). Primary productivity and rate of change of biomass of various species of phytoplankton in Castle Lake, California. 18th Congress of International Association of Limnology. *Verh. Int. Verein. Limnol.* **18**, Part 3, 1768–1775.

DeAngelis, D. L., E. W. Stiles, W. C. Johnson, D. M. Sharpe, and R. K. Schreiber (1977). A model for the dispersal of seeds by animals. EDFB/EBP-77/5, Oak Ridge National Laboratory. 40 pp.

Debach, P. and H. T. Smith (1941). Effect of host density on the rate of reproduction of entomophagous parasites. *J. Econ. Entomol.* **34**, 741–745.

Debach, P. and R. A. Sundby (1963). Competitive displacement between ecological homologues. *Hilgarden* **34**, 105–166.

DeBellevue, E. B. (1976). Energy basis for an agriculture region. M.S. thesis. Department of Environmental Engineering Sciences, University of Florida, Gainesville.

DeBellevue, E., H. T. Odum, J. Browder, and G. Gardner (1979). Energy Analysis of the Everglades National Park. In *Proceedings of the First Conference on Scientific Research in the National Parks,* Vol. 1, R. M. Linn, Ed., National Park Service. pp. 31–43.

Deevey, E. S. (1947). Life tables for natural populations of animals. *Quart. Per. Biol.* **22**, 283–314.

DeGroot, S. R. (1952). *Thermodynamics of Irreversible Processes.* North-Holland, Amsterdam.

DeGroot, S. R. and P. Mazur (1962). *Non-equilibrium Thermodynamics.* North-Holland, Amsterdam.

de Klerk, P. and M. Gatto (1981). Some remarks on periodic harvesting of a fish population. *Math. Biosciences* **56**, 47–69.

Delwiche, C. C. (1970). Nitrogen cycle. *Sci. Am.* **223**, 136–146.

Denbigh, K. G. (1951). *Thermodynamics of the Steady State.* Methuen, London.

Denbigh, K. E. (1952). Entropy creation in open reaction systems. *Transact. Faraday Soc.* **48**, 389–394.

DePalma, L. M., R. P. Canale, and W. F. Powers (1979). A minimum cost surveillance plan for water-quality trend detection in Lake Michigan. In *Perspectives in Lake Ecosystem Modeling,* D. Scavia and A. Robertson, Eds., Ann Arbor Science. pp. 223–246.

Desmaris, A. P. and A. Vazquez (1970). Upper canopy crown closure at El Verde. pp. B123–B134 in *A Tropical Rainforest,* H. T. Odum and R. Pigeon, Eds., Div. of Tech. Information, U.S. Atomic Energy Commission, Oak Ridge, TN. 1600 pp.

De Wit, C. T. (1960). On competition. *Versl. Landbouwk. Onderzoek,* No. 668, Wageningen, Netherlands.

De Wit, C. T. and J. Goudriaan (1974). Simulation of ecological processes. Centre for Agricultural Publishing and Documentation. Wageningen, Netherlands.

Diamond, J. M. (1974). Colonization of exploded volcanic islands by birds: Their supertramp strategy. *Science* **184**, 803–806.

DiSalvo, L. H. (1969). Regenerative functions and microbial ecology of coral reefs. Ph.D. dissertation. Department of Zoology, University of North Carolina, Chapel Hill.

Di Toro, D. M. (1969). Maximum entropy in estuaries. *J. Hydraulics Division ASCE* **95**, 1247–1271.

DiToro, D. D., D. J. O'Connor, and R. V. Thomann (1971). A dynamic model of the phytoplankton population in the Sacramento–San Joaquin Delta. In *Nonequilibrium Systems in Natural Water Chemistry, Advances in Chemistry,* No. 106, American Chemical Society. Chapter 5, pp. 133–180.

Domenico, P. A. (1972). Concepts and models in groundwater hydrology. McGraw-Hill, New York. 405 pp.

Doornkamp, J. C. and A. M. K. Cuchlaine (1971). *Numerical Analysis in Geomorphology,* St. Martin's Press, New York. 372 pp.

Doxiadis, C. A. (1968). Man's movement and his city. *Science* **162**, 326–332.

Drake, J. F., J. L. Jost, and A. G. Frederickson (1968). The food chain. In *Bioregenerative Systems,* Scientific and Technical Information Division, National Aeronautics and Space Administration. NASA SP-165. pp. 87–95.

Dubois, D. M. (1973). Aspect mathematique de l'invariant en cybernetique. *Cybernetica* **3**, 161–176.

Dubois, D. M. (1975a). The influence of the quality of water on

REFERENCES

ecological systems. pp. 535–543. In *Computer Simulation of Water Resource Systems.* G. C. Vansteenkiste, Ed. North-Holland, Amsterdam.

Dubois, D. M. (1975b). A model of patchiness for prey–predator plankton populations. *Ecol. Model.* **1**, 67–80.

Dubois, D. M. (1979). State of the art of predator–prey cycles. In *State of the Art in Ecological Modelling. Environmental Sciences and Application,* Vol. 7. Pergamon Press, New York. pp. 163–217.

Dubois, D. M. and E. Schoffeniels (1974). A molecular model of action potentials. *Proc. Nat. Acad. Sci. (USA)* **71**, 2858–2862.

Duckham, A. N., J. G. Jones, and E. H. Roberts (1976). An approach to the planning and administration of human food chains and nutrient cycles. In *Food Production and Consumption,* North-Holland, Amsterdam. pp. 461–517.

Dwyer, R., S. W. Nixon, C. Oviatt, K. J. Perez, and T. J. Smayda (1978). Frequency response of a marine ecosystem subjected to time varying inputs. In *Energy and Environmental Systems.* J. H. Thorpe and J. W. G. Gibbon, Eds., Division of Technical Information, Department of Energy. pp. 19–38.

Eagleman, J. R. (1976). *The Visualization of Climate,* Lexington Books, London. 227 pp.

Eagleman, J. R., V. U. Muirhead, and N. Williams (1975). *Thunderstorms, Tornadoes and Building Damage,* Lexington Books, Heath and Company, Lexington, MA. 317 pp.

Ehleringer, J. R. (1978). Implications of quantum yield difference on the distributions of C3 and C4 grasses. *Oecologia* **31**, 255–267.

Ehrlich, G. G. and K. V. Slack (1969). Update and assimilation of nitrogen in microecological systems. *Am. Soc. Testing Mater. Spec. Tech. Publ.* **448**, 11–23.

Eigen, M. and P. Schuster (1979). *The Hypercycle,* Springer-Verlag, Berlin. 92 pp.

Einstein, A. (1905). On the electrodynamics of moving bodies (translation). In *The Principle of Relativity,* H. A. Lorentz, A. Einstein, H. Minkowski, and H. Weyl. Dover, New York. 216 pp.; pp. 35–69.

Electronic Associates, Inc. (1971). *Miniac Reference Handbook.* Pub. No. 00-800, 2071-0.

Electronic Associates, Inc. (1965). A host–parasite program. *Instrum. Control. Syst.* **38**, 129.

Electronic Associates (undated). Analog computer investigation of Van der Pol's equation. Training Department, Problem No. 2.

Elliott, J. M. (1977). *Some Methods for the Statistical Analysis of Samples of Benthic Invertebrates,* 2nd ed., Freshwater Biological Association Scientific Publication No. 25.

Elton, C. (1926). *Animal Ecology,* Sidgewick and Jackson, London. 209 pp.

Elton, C. (1953). *The Ecology of Animals,* Methuen, London. 87 pp.

Elton, C. S. (1958). *The Ecology of Invasions by Animals and Plants,* Methuen, London. 181 pp.

Emanuel, W. R., D. C. West, and H. H. Shugart (1978). Special analysis of forest model time series. *Ecol. Model.* **4**, 313–326.

Emlen, J. M. (1973). *Ecology: An Evolutionary Approach,* Addison-Wesley, Reading, MA.

Engineering Concepts Curriculum (1971). *The Man Made World,* McGraw-Hill, New York. 628 pp.

Engqvist, A. and S. Sjöberg (1980). An analytical integration method of computing diurnal primary production for Steele's light response curve. In *Ecological Modelling* **8**, 219–232. Elsevier, Amsterdam.

Epply, R. W. and F. M. Macias (1963). Role of the alga *Chlamydomones mundra* in anaerobic waste stabilization lagoons. *Limnol. Oceanogr.* **8**, 411–416.

Estes, J. A., N. S. Smith, and J. E. Palmisano (1978). Sea otter predation and community organization in the western Aleutian Islands, Alaska. *Ecology* **59**, 822–823.

Evans, J. V. (1982). The Sun's influence on the earth's Atmosphere and interplanetary space. *Science* **216**, 467–474.

Evans, R. (1969). A proof that essergy is the only consistent measure of potential work. Ph.D. dissertation. College of Engineering, Dartmouth.

Ewel, J. (1971). Experiments in arresting succession with cutting and herbicides in five tropical environments. Ph.D. dissertation. Department of Botany, University of North Carolina, Chapel Hill. 239 pp.

Ewel, K. C. and H. T. Odum, Eds. (1982). *Cypress Swamps,* Univ. of Florida Press, Gainesville, Fla. (in press).

Eyring, H. and D. W. Urry (1965). Thermodynamics and chemical kinetics. In *Theoretical and Mathematical Biology,* T. H. Waterman and H. T. Morowitz, Eds. Blaisdell, New York. pp. 57–95.

Fairbridge, R. W. (1967). *Encyclopedia of Atmospheric Sciences and Astrogeology.* Reinhold, New York, 1200 pp.

Fairen, V. and J. Ross 1981 On the efficiency of thermal engines with power output. *J. Chem Physics,* **75** (11), 5490–5496.

Fairen, V., N. D. Hatee, and J. Ross (1982). Thermodynamic processes, time scales, and entropy production. *J. Physical Chem.* **86**, 70–73.

Falk, J. H. (1976). Energetics of a suburban lawn ecosystem. *Ecology* **57**, 141–150.

Falkowski, P. G. (Ed.) (1980). *Primary Productivity in the Sea,* Plenum Press, New York. 530 pp.

Farned, J. (1974). Simple models for the study of history. Report for research course EES 670. 34 pp.

Fechner, G. T. (1872), quoted by S. J. Holmes (1948). The stability principle. *Quart. Rev. Biol.* **23**, 324.

Fenchel, T. (1974). Intrinsic rate of natural increase: The relationships with body size. *Oecologia* **14**, 317–326.

Fenn, W. O. (1924). A quantitative comparison between energy liberated and the work performed by the isolated Sartorius muscle. *J. Physiol.* **58**, 12.

Fields, P. E. and D. M. Sharpe (1980). Seedfall: A model of dispersal of tree seeds by wind. Oak Ridge National Laboratory. EDFB/IBP-78/2.

Finn, J. T. (1976). Measures of ecosystem structure and function derived from analyses of flows. *J. Theor. Biol.* **56**, 363–389.

Fisher, R. A. (1921). *Transact. Roy. Soc. Lond. Ser. A* **222**, 309–368.

Fisher, R. A., A. S. Corbet, and C. B. Williams (1943). The relation between the number of individuals in a random sample of an animal population. *J. Animal Ecol.* **12**, 42–48.

Fleming, R. H. (1939). The control of diatom populations by grazing. *J. Conseil Explor. Mer.* **14**, 210–227.

Fluck, R. C. and C. D. Baird (1980). *Agricultural Energetics*, AVI, Westport, CT.

Fontaine, T. D. (1978). Community metabolism patterns and a simulation model of a lake in central Florida. Ph.D. dissertation. Department of Environmental Engineering Sciences, University of Florida, Gainesville. 415 pp.

Fontaine, T. D. (1981). A self-designing model for testing hypotheses of ecosystem development. In *Progress in Ecological Engineering and Management by Mathematical Modeling. J. Int. Soc. Ecol. Model.* Copenhagen, Denmark. pp. 281–290.

Fontaine, T. D. and K. C. Ewel (1981). Metabolism of a Florida lake ecosystem. *Limnol. Oceanogr.* **26**, 754–763.

Forrester, J. W. (1961). *Industrial Dynamics*, MIT Press, Cambridge.

Forrester, J. W. (1969). *Urban Dynamics*, MIT Press, Cambridge.

Forrester, J. W. (1971). *World Dynamics*, Wright Allen Press, Cambridge.

Forrester, J. W. (1976). Business structure, economic cycles, and national policy. *Futures* (London) (June).

Fowler, N. (1977). The longevity of petroleum resources. *Energy* **2**, 189–195.

Fox, J. F. (1978). Forest fires and snowshoe hare–Canada lynx cycle. *Oecologia* **31**, 349–374.

Frankena, F. (1978). *Energy Analysis/Energy Accounting.* Vance Bibliographies, Box 229, Monticello, IL.

French, C. S. and D. C. Fork (1961). Computer simulations for photosynthesis rates from a two pigment model. *Biophys. J.* **1**, 669–681.

Frenkel, F. N. and D. W. Goodall Eds. (1978). *Simulation Modelling of Environmental Problems.* Wiley, New York. 112 pp.

Fukuda, N. M. and M. Sugita (1961). Mathematical analysis of metabolism using an analogue computer. I. Isotope levels of iodine metabolism in the thyroid gland. *J. Theor. Biol.* **1**, 440–454.

Fuller, B. (1969). Letter to Doxiadis. *Main Currents* **25** (4), 1–12.

Gages, D. C. (1974). *Traffic Science*, Wiley, New York.

Gallagher, R. G. (1968). *Information Theory and Reliable Communication*, Wiley, New York.

Gallegos, C. L., G. M. Hornberger, and M. G. Kelley (1977). A model of river benthic algal photosynthesis in response to rapid changes in light. *Limnol. Oceanogr.* **22**, 226–233.

Gallo, N. and S. Rinaldi (1977). Stability analyses of predator prey models via the Liapunov method. *Bull. Math. Biol.* **39**, 339–372.

Gallopin, G. C. (1971). A generalized model of a resource-population system Part I and II. *Oecologia* **7**, 382–413.

Gallucci, V. F. (1973). On the principles of thermodynamics in ecology. In *Annual Review of Ecology and Systematics*, Vol. 4, R. F. Johnston, P. W. Frank, and C. D. Michener, Eds., Annual Reviews, Palo Alto, CA. pp. 329–357.

Galomb, S. W. (1968). Mathematical models, uses and limitations. *Astronautics and Aeronautics*, 57–58.

Ganning, B. and F. Wulff (1969). The effects of bird droppings on chemical and biological dynamics in brackish water rockpools. *Oikos* **20**, 274–286.

Ganning, B. and F. Wulff (1970). Measurement of community metabolism in some brackish water rockpools by means of diel oxygen curves. *Oikos* **21**, 292–298.

Gardner, R. H., J. B. Mankin, and W. R. Emanuel (1980a). A comparison of three carbon models. *Ecol. Model.* **8**, 313–332.

Gardner, R. H., R. V. O'Neill, J. B. Mankin, and D. Kumar (1980b). Comparative error analysis of six predator–prey models. *Ecology* **6**, 323–332.

Garfinkel, D. (1962a). Digital computer simulation of ecological systems. *Nature* **194**, 856–857.

Garfinkel, D. (1962b). Computer simulation of steady state glutomate metabolism in rat brain. *J. Theor. Biol.* **3**, 412–422.

Garfinkel, D. (1962c). Simulation of ecological systems. In *Computers in Biomedical Research*, Vol. II, R. W. Stacy and B. Waxman, Eds. Academic Press, New York. Chapter 8, pp. 205–216.

Garfinkel, D. (1965). Computer simulation in biochemistry and ecology. In *Theoretical and Mathematical Biology*, T. H. Waterman and H. T. Morowitz, Eds., Blaisdell, New York. pp. 292–310.

Garfinkel, D. and R. Sack (1964). Digital computer simulation of an ecological system based on a modified mass action law. *Ecology* **45**, 502–507.

Garrels, R. M., F. T. Mackenzie, and C. Hunt (1975). Chemical cycles and the global environment. Kaufman, Los Altos, CA.

Garrels, R. M., A. Lerman, F. T. Mackenzie (1976). Controls of atmospheric O_2 and CO_2 past, present and future. *Am. Sci.* **64**, 306–315.

Gates, D. M. (1965). Energy, plants and ecology. *Ecology* **46**, 1–13.

Gates, D. M. (1980). *Biophysical Ecology*, Springer-Verlag, New York. 611 pp.

Gatlin, L. L. (1972). *Information Theory and the Living System*. Columbia University Press, New York. 210 pp.

Gause, G. F. (1934). *The Struggle for Existence*, Hofner (reprinted 1964), New York.

Gause, G. F. and A. A. Witt (1935). Behavior of mixes populations and the problem of natural selection. *Am. Natur.* **69**, 596–609.

Gayle, T. L. (1975). Systems models for understanding eutrophication in Lake Okeechobee. M.S. thesis. Department of Environmental Engineering Sciences, University of Florida. 119 pp.

Georgescu-Rogen, N. (1971). *The Entropy Law and the Economic Process.* Harvard University Press, Cambridge. 457 pp.

Gibbs, J. W. (1873). *On the Equilibrium of Heterogeneous Substances. Collected Works*, Vol. I, Yale University Press (reprinted 1928).

Gibbs, J. W. (1901). *Elementary Principles in Statistical Mechanics, Collected Works*, Vol. II, Yale University Press (reprinted 1948).

Gilpin, M. E. (1972). Enriched predator-prey systems theoretical stability. *Science* **177**, 902–904.

Gilliland, M. W. (1973). Man's impact on the phosphorus cycle in Florida. Ph.D. dissertation. Environmental Engineering Sciences, University of Florida, Gainesville.

Gilliland, M. W. (1975). Systems models for phosphorus management in Florida. In *Mineral Cycling in Southeastern Ecosystems,* F. G. Howell, J. B. Gentry and M. H. Smith, Eds. pp. 179–208. Energy Research and Development Administration.

Gilliland, M. W. (1976). A geochemical model for evaluating theories on the genesis of Florida's sedimentary phosphate deposits. *Math. Geol.* **8,** 219–242.

Gilliland, M. W. (1978). Energy analysis and public policy. *Science* **189,** 1051–1056.

Glansdorf, P. and I. Prigogine (1971). *Structure Stability and Fluctuations,* Wiley-Interscience, Chichester, U.K.

Gleason, H. A. (1922). On the relation between species and area. *Ecology* **3,** 158–162.

Gödel, K. (1931). Uber formal unentaherdbare satz der principia mulhematica und verwamaler systeme I. *Monalshefte fur Malhematik und Physik* **38,** 173–198.

Goel, N. S. and N. Richter-Dyn (1974). *Stochastic Models in Biology,* Academic Press, New York. 269 pp.

Goel, N. S., S. C. Muitra, and E. W. Montroll (1971). *Nonlinear Models of Interacting Populations* (Reviews of Modern Physics), Academic Press, New York. 145 pp.

Goldberg, E. D., I. N. McCave, J. J. O'Brien, and J. H. Steele (1977). *The Sea,* Vol. 6, *Marine Modeling,* Wiley, New York.

Goldberg, M. A. (1977). Simulating cities: Process product and prognosis. *AIP J.* **43,** 148–156.

Goldberg, M. A., C. S. Holling, and R. F. Kelley (1971). Vancouver regional simulation study. *Proceedings of the Igu Symposium in Urban Geography,* Gleerup Publishers, Lund, Sweden.

Goldman, S. (1953). *Information Theory,* Prentice-Hall, Englewood Cliffs, NJ.

Goldschmidt, V. M. (1933). Grundlagen der quantitativen geochemie. *Fontsch. Mineral. Krist. Petrog.* **17,** 112–156.

Gompertz, B. (1825). On the nature of the functional expression of the law of human mortality and on a new model of determining life contingencies. *Phil. Trans. Roy. Soc. Lond.* **115,** 513–585.

Goodall, D. W. (1967). Simulating the Grazing Situation. In *Concepts and Models of Biomathematics,* F. Heinmets, Ed., Marcel Dekker, New York. pp. 211–236.

Goodall, D. W. (1972). Building and testing ecosystem models. In *Mathematical Models in Ecology,* J. N. R. Jeffers, Ed., Blackwell, Oxford. pp. 173–194.

Goodwin, B. C. (1963). *Temporal Organization in Cells,* Academic Press, New York. 163 pp.

Gore, A. J. P. (1972). A field experiment, a small computer and model simulation. pp. 309–325. In *Mathematical Models in Ecology,* J. N. R. Jeffers, Ed., Blackwell Scientific Publ., Oxford, U.K., 398 pp.

Gorman, M. L. (1979). *Island Ecology. Outline Studies in Ecology,* Chapman and Hall, London.

Gowdy, C. M., R. J. Mulholland, and W. R. Emanuel (1975). Modelling the global carbon cycle. *Internatl. J. Systems Sci.* **6,** 965–976.

Granger, C. W. J. (1964). *Spectral Analysis of Economic Time Series,* Princeton University Press, Princeton, NJ.

Gray, I. E. (1954). Comparative study of the gill area of marine fishes. *Biol. Bull.* **107,** 219–225.

Greenberger, M., M. A. Crenson, and B. L. Crissey (1976). *Models in the Policy Process,* Russell Sage Foundation, New York. 353 pp.

Greenstein, C. H. (1978). *Dictionary of Logical Terms and Symbols,* Van Nostrand-Reinhold, New York.

Gregory, F. G., I. Spear, and K. V. Thiman (1954). Interrelation between carbon-dioxide metabolism and photoperiodism in Kalovchoe. *Plant Physiol.* **29,** 220–229.

Greig-Smith, P. (1964, 1957). *Quantitative Plant Ecology,* Butterworths, London.

Grenney, W. L., D. A. Bella, and H. C. Curl (1974). Effects of intracellular nutrient pools on growth dynamics of phytoplankton. *J. Water Pollution Control* **46,** 1751–1759.

Gribbin, J. (1978). *Climatic Change,* Cambridge University Press, London.

Groden, T. W. (1977). Modeling temperature and light adaptation of phytoplankton. Report No. 2, Center for Ecology Modeling. Rensselaer Polytechnic Institute, Troy, NY.

Grodins, F. S. (1963). *Control Theory and Biological Systems,* Columbia University Press, New York. 205 pp.

Guardabassi, G. (1976). The optimal periodic control problem. *J. A,* **17** (2), 75–83.

Guardabassi, G., A. Locatelli, and S. Rinaldi (1974). The status of periodic optimization of dynamical systems. *J. Opt. Theory Appl.* **13,** 1; **14,** 1–20.

Gutierrez, R. J. (1978). Energy analysis and computer simulations of pastures in Florida. M.S. thesis. Department of Environmental Engineering Sciences, University of Florida, Gainesville. 187 pp.

Gutierrez, L. T. and W. R. Fey (1980). *Ecosystem Succession,* MIT Press, Cambridge, MA.

Gyllenberg, G. (1977). Systems analysis in invertebrates, in *Fennoscandian Tundra Ecosystems,* Ecol. Studies No. 17, E. E. Wielgolski, Ed, Springer-Verlag, New York. pp. 267–272.

Hagerstrand, T. (1967). On Monte Carlo simulation of diffusion. *Northwest. Stud. Geogr.* **13,** 1–32.

Haggett, P. (1966). *Locational Analysis in Human Geography,* St. Martin's Press, New York. 339 pp.

Haken, H. (1977a). *Synergetics, A Workshop,* Springer-Verlag, Berlin. 276 pp.

Haken, H. (1977b). *Synergetics, An Introduction.* Springer-Verlag, New York. 370 pp.

Haldane, J. B. S. (1930). *Enzymes,* Longmans, London.

Haldane, J. B. S. (1957). The cost of natural selection. *J. Genet.* **55,** 511–524.

Halfon, E. (1979). *Theoretical Systems in Ecology,* Academic Press, New York. 515 pp.

Hall, C. A. S. (1972). Migration and metabolism in a temperate stream ecosystem. *Ecology* **53,** 585–604.

Hall, C. A. S. and J. Day (1977). *Ecosystem Models in Theory and Practice,* Wiley, New York. 685 pp.

Hall, C. A. S., C. A. Erdahl, and D. E. Wartenberg (1975). A fifteen year record of biotic metabolism in the northern hemisphere. *Nature* **255,** 136–138.

Hamilton, J. and H. W. Peng (1944). *Proc. Roy. Irish Acad. A* **49,** 197 pp.

Hamilton, P. and K. B. Macdonald (1980). *Estuarine and Wetland*

Processes. Marine Science No. 11, Plenum Press, New York. 683 pp.

Hanga, T. and Y. Amle (1976). A study on the metabolism of cities. In *Science for a Better Environment,* HSEC, Science Council of Japan. pp. 228–233.

Hannon, B. (1973a). Marginal product pricing in the ecosystem. *J. Theor. Biol.* **41,** 252–267.

Hannon, B. (1973b). An energy standard of value. *Ann. Am. Acad. Polit. Soc. Sci.* **410,** 139–153.

Hannon, B. (1976). The structure of ecosystems. *J. Theor. Biol.* **56,** 535–546.

Hannon, B. (1982). Energy Discounting, in *Energetics and Systems,* W. J. Mitsch, R. K. Ragade, R. W. Bosserman, and J. A. Dillon, Jr., Eds., Ann Arbor Science, Ann Arbor, Mich. pp. 73–93.

Harary, F. (1961). Who eats whom. *Yearbook Ben. Syst.* **6,** 41–49.

Harary, F. (1969). *Graph Theory,* Addison-Wesley, Reading, MA.

Harbaugh, J. W. and E. Bonham-Carter (1970). *Computer Simulation in Geology,* Wiley-Interscience, New York. 574 pp.

Harbaugh, J. W. and D. F. Merriam (1968). *Computer Applications in Stratigraphic Analysis,* Wiley, New York. 282 pp.

Hardin, G. (1960). The competitive exclusion principle. *Science* **131,** 1292–1297.

Hardin, G. (1963). The cybernetics of competition. University of Chicago Press, Chicago.

Hare, V. C., Jr. (1967). *Systems Analysis, A Diagonostic Approach,* Harcourt Brace, New York. 543 pp.

Harris, J. P. W. (1974). The kinetics of polyphagy. In *Ecological Stability,* M. B. Usher and M. H. Williamson, Eds. Chapman and Hall, London.

Harris, M. (1965). The myth of the sacred cow. pp. 217–228. In *Man, Culture, and Animals,* A. Leeds and Andrew P. Vayda, Eds. Publ. No. 78 of the American Association for the Advancement of Science, Washington, DC.

Harrison, H. L. (1970). System study of DDT transport. *Science* **170,** 303–508.

Hart, R. D. (1980). A natural ecosystem analog approach to the design of a successional crop system for tropical forest environment. *Tropical Succession, Biotropica,* Vol. 12 supplement.

Harte, J. and D. Levy (1975). On the vulnerability of ecosystems disturbed by man. In *Unifying Concepts in Ecology,* W. H. van Dobben and R. H. Lowe-McConnell, Eds., Center for Agricultural Publishing Documentation, Wageningen. 302 pp.

Harwell, M. A., W. P. Cropper, and H. L. Ragsdale (1977). Nutrient recycling and stability, a reevaluation. *Ecology* **58,** 660–666.

Hassell, N. P. and R. May (1973). Stability in insect host–parasite models. *J. Animal Ecology* **42,** 693–726.

Hatanaka, M. A. and M. Takahashi (1960a). Experimental study of utilization of food by young anchovy. *Engraulis Japonica Temminck. Tohoku J. Agr. Res.* **11,** 161–170.

Hatanaka, M. A. and M. Takahashi (1960b). Studies on the amounts of the anchovy consumed by the mackerel. *Tohoku J. Agr. Res.* **11,** 83–100.

Hayes, F. R. (1957). On the variation in bottom fauna and fish yield in relation to trophic level and lake dimensions. *J. Fish. Res. Board Canada* **14,** 1–32.

Hayes, F. R. and E. H. Anthony (1964). Productive capacity of North American lakes as related to the quantity and the trophic level of fish, the lake dimensions and the water chemistry. *Transact. Am. Fish. Soc.* **93,** 53–57.

Heilbrun, J. (1974). *Urban Economics and Public Policy,* St. Martin's Press, New York.

Heinmetz, F. (1966). *Analysis of Normal and Abnormal Cell Growth,* Plenum Press, New York. 288 pp.

Heinmetz, F. and A. Herschman (1962). Theoretical analysis of models for enzyme synthesis. In *Biological Prototypes and Synthetic Systems,* Vol. I, E. E. Bernard and M. R. Kane, Eds., Plenum Press, New York. 395 pp.; pp. 60–70.

Heizer, R. F. (1966). Ancient heavy transport, methods and achievements. *Science* **153,** 821–830.

Henderson, K. (1977). *Senility and Mortality.* Manuscript. 14 pp.

Henderson, L. J. (1913). *The Fitness of the Environment,* Macmillan, New York.

Henderson, L. J. (1917). *The Order of Nature,* Harvard University Press, Cambridge, MA.

Herbert, D. (1958). Continuous culture of microorganisms; some theoretical aspects. In *Continuous Cultivation of Microorganisms, a Symposium,* I. Malek, Ed., Czechoslovak Academy, Prague. 234 pp.; pp. 45–61.

Herbert, D. (1962) Multistage continuous culture. *In Continuous Cultivation of Micro-organisms,* I. Malek, K. Beran and J. Hospodka, Eds., Academic Press, New York. pp 23–53.

Herbert, D., R. Elsworth, and R. C. Telling (1956). The continuous culture of bacteria: A theoretical and experimental study. *J. Gen. Microbiol.* **14,** 601.

Herendeen, R. A. (1973). An energy input–output matrix for the U.S. *User's Guide,* Center for Advanced Computation, University of Illinois, Urbana.

Herendeen, R. A. (1974). Use of input–output analysis to determine the energy cost of goods and services. In *Energy: Demand Conservation and Distribution Problems,* M. S. Macrakis, Ed., MIT Press, Cambridge, MA.

Herendeen, R. A. (1981). Energy intensities in ecological and economic systems. *J. Theor. Biol.* **91,** 607–620.

Heron, A. C. (1972). Population ecology of a colonizing species: the plagic tunicate *Thalia democraleia. Oecologia* **10,** 269–312.

Hess, S. L. (1959). *Introduction to Theoretical Meteorology,* Holt, Rinehart, Winston, New York. 361 pp.

Hesse, R., W. C. Allee, and K. P. Schmidt (1951). *Ecological Animal Geography,* Wiley, New York. 591.9 H58e.

Hilborn, R. (1979). Some long term dynamics of predator–prey models with diffusion. *Ecol. Model.* **6,** 23–30.

Hill, T. L. (1977). *Free Energy Transduction in Biology.* Academic Press, New York.

Hinde, R. A. (1960). Energy models of motivation. In *Models and Analogues in Biology,* Society of Experimental Biology Symposium, Vol. 14, pp. 199–213.

Hintze, J. O. (1959, 1975). *Turbulence,* 2nd ed., McGraw-Hill, New York.

Hiramatsu, T. and Y. Shimazu (1970). Simulation of drainage network system and concept of entropy in geomorphology. *J. Physics Earth* **18** (2), 181–191.

REFERENCES

Hirschlerfer, J. (1977). Economics from a Biologist Viewpoint. *J. of Law and Economics,* 1977, 1–52.

Hirschlerfer, J. (1977). Economics from a Biologist Viewpoint. *J. of Law and Economics,* pp. 1–52.

Hobbie, J. E. and P. Rublee (1977). Radioisotope studies of heterotrophic bacteria in aquatic ecosystems. In *Aquatic Microbial Communities,* J. Cairns, Ed. Garland, New York. pp. 441–476.

Hock, I. (1976). City size effects, trends and policies. *Science* **193,** 856–863.

Hohn, F. E. (1965). *Elementary Matrix Algebra,* Macmillan, New York. 395 pp.

Holderige, L. R. (1947). Determination of world plant formations from simple climatic data. *Science* **105,** 367–368.

Holdgate, M. W. and M. J. Woodman (1978). *The Breakdown and Restoration of Ecosystems,* Plenum Press, New York. 496 pp.

Holland, H. D. (1978). *The Chemistry of the Atmosphere and Ocean,* Wiley, New York. 351 pp.

Holling, C. S. (1959). Some characteristics of simple types of predation and parasitism. *Canad. Entomol.* **91,** 385.

Holling, C. S. (1965). The functional response of predators to prey density and its role in mimicry and population regulation. *Mem. Entomol. Soc. Can.* **45,** 5–60.

Holling, C. S. (1970). Stability in ecological and social systems. Symposium Brookhaven National Laboratory. BNL 50175 (C-56). pp. 128–141.

Holling, C. S. (1973). Resilience and stability in ecological systems. *Ann. Rev. Ecol. Systems* **4,** 1–23.

Holling, C. S. and S. Ewing (1971). Blind man's bluff: exploring the response space generated by realistic ecological simulation models. *Proc. Internatl. Symp. Statist. Ecol. Yale Univ. Press* **2,** 207–229.

Homer, M. (1974). An acclimation model for small fishes. In *Progress Report to Florida Power Corporation, April.* H. T. Odum, H. N. McKellar, W. Smith, M. Lehman, D. Young, M. Kemp, M. Homer, and T. Gayle, Eds. pp. 78–92.

Hood, D. (1962). Field experiments on chemical changes in sea off Galveston, Texas; unpublished work.

Horn, H. S. (1968). Regulation of animal numbers: A model counter-example. *Ecology* **49,** 776–778.

Horn, H. S. (1975). Markovian properties of forest succession. In *Ecology and Evolution in Communities,* M. L. Cody and J. M. Diamond, Eds., Belknap Press, Harvard University Press, Cambridge, MA. pp. 196–211.

Horn, D. J., R. D. Mitchell, and G. R. Stairs (1979). *Analysis of Ecological Systems,* Ohio State University Press. 307 pp.

Hornbach, D. J. and R. Fitz (1977). A qualitative model of the quality of rangeland of the Sahelian ecosystem. In *Problem Analysis in Science and Engineering,* F. H. Branin and K. Huseyin, Eds. pp. 199–228.

Hornbeck, D. A. (1979). Metabolism of Salt Marshes and Their Role in the Economy of a Coastal Community. M.S. thesis. Department of Environmental Engineering Science, University of Florida, Gainesville. 175 pp.

Horton, R. E. (1945). Hydrophysical approach to quantitative morphology. *Bull. Geol. Soc. Am.* **56,** 275–370.

Huettner, D. A. and R. Costanza, (1982). Economic Values and Embodied Energy, *Science* **216,** 1141–1143.

Hughes, B. B. (1980). *World Modeling,* Lexington Books, Heath, Lexington, MA. 240 pp.

Hughes, J. L. (1968). *Computer Lab Workbook,* Digital Equipment Corporation, Boston, MA. 169 pp.

Hulbert, E. M. (1963). Diversity of phytoplankton populations in oceanic, coastal and estuarine regions. *J. Mar. Res.* **21,** 81–93.

Hunding, A. (1974). Limit cycles in enzyme systems with nonlinear negative feedback. *Biophys. Struct. Mechnism* **1,** 47–54.

Hurlbert, S. H., J. Zedler, and D. Fairbanks (1972). Ecosystem alteration by mosquitofish (*Gaubisa affus*) predation. *Science* **175,** 639–644.

Hutchings, M. J. and C. S. J. Budd (1981). Plant competition and its course through time, *Bioscience* **31,** 640–645.

Hutchinson, G. E. (1944). Limnological studies in Conn. VII. A critical examination of the supposed relationship between phytoplankton periodicity and chemical changes in lake water. *Ecology* **25,** 3–26.

Hutchinson, G. E. (1947). A note on the theory of competition between two social species. *Ecology* **28,** 319–321.

Hutchinson, G. E. (1948). Circular causal systems in ecology. *Ann. NY Acad. Sci.* **50,** 221–246.

Hutchinson, G. E. (1953). Concept of pattern in ecology. *Proc. Phila. Acad. Sci.* **105,** 1–12.

Hutchinson, G. E. (1954). Theoretical notes on oscillatory populations. *J. Wild. Manage.* **18,** 107–109.

Hutchinson, G. E. (1959). Homage to Santa Rosalia or why are there so many kinds of animals. *Am. Natur.* **93** (870), 155–160.

Hutchinson, G. E. (1959). Il concetto Moderno di Nicchia Ecologica. *Mem. Ist. Ital. Idrobio.* **11,** 9–22.

Hutchinson, G. E. (1961). The paradox of the plankton. *Am. Natur.* **95,** 137–145.

Hutchinson, G. E. (1965). *The Ecological Theater and the Evolutionary Play,* Yale University Press, New Haven, CT.

Hutchinson, G. E. (1978). *An Introduction to Population Ecology,* Yale University Press, New Haven, CT. 255 pp.

Hutchinson, G. E. and R. MacArthur (1959). A theoretical ecological model of size distributions among species of animals. *Am. Natur.* **93,** 117–126.

Huxley, J. S. (1932). *Problems of Relative Growth,* Methuen, London.

Hynes, H. B. N. (1960). *The Biology of Polluted Water,* Liverpool University Press, Liverpool, U.K. 202 pp.

IFIAS (1974). *Energy Analysis,* Report No. 6, International Federation of Institutes of Advanced Study, Stockholm, Sweden.

Ikusima, I. (1962). Biology of duckweeds with special reference to their growth. Parts I, II, and III. *Physiol. Ecol.* **10,** 130–164; **11,** 84–102, **11,** 120–137.

Innis, G. S. (1978). *Grassland Simulation Model, Ecology Studies,* No. 26, Springer-Verlag, New York.

Isard, W. (1956). *Location and Space Economy.* MIT Press, Cambridge, MA. 350 pp.

Isard, W. (1968). Some notes on the linkage of the ecologic and economic systems. *Pap. Proc. Reg. Sci. Assoc.* **22,** 84–96.

Isard, W. (1969). Methods of regional analysis. MIT Press, Cambridge, MA.

Isard, W. and T. W. Langford (1971). *Regional Input–Output Study*, MIT Press, Cambridge, MA.

Ivlev, V. S. (1939). Transformation of energy by aquatic animals. *Internatl. Rev. ges. Hydrobiol. u. Hydrogr.* **38**, 449–458.

Ivlev, V. S. (1961). *Experimental Ecology on the Feeding of Fish*, Yale University Press, New Haven, CT.

Jacobs, J. (1969). *The Economy of Cities*, Random House, New York.

Jacobs, J. (1977). Coexistence of similar zooplankton species by differential adaptation to reproduction and escape in an environment with fluctuating food and enemy densities. I. A model. *Oecologia* **29**, 233–247.

Jacobsen, H. (1955). Information, reproduction, and the origin of life. *Am. Sci.* **43**, 119–147.

Jacobsen, H. (1958). On models of reproduction. *Am. Sci.* **46**, 255–284.

Jacobsen, H. (1959). The informational content of mechanisms and circuits. *Inform. Control* **2**, 285–296.

Jaeger, J. C. (1949). *An Introduction to the Laplace Transformation*, Wiley, New York. 156 pp.

James, D. E., H. M. A. Jansen, and J. B. Opshoor (1978). *Economic Approaches to Environmental Problems*, Elsevier, Amsterdam.

Jannasch, H. W. (1962). Studies on the ecology of a marine spirillum in the chemostat. In *First International Symposium on Marine Microbiology*, C. H. Oppenheimer, Ed., Thomas, Springfield, IL, Chapter 51, pp. 558–565.

Jannasch, H. W. (1969). Current concepts in aquatic microbiology. *Verh. Intern. Verein. Limnol.* **17**, 25–39.

Jansson, A. M. (1974). Community structure, modelling and simulation of the *Cladoptra* ecosystem in the Baltic Sea. Contributions from the ASKO Laboratory. University of Stockholm, Sweden, No. 5. 130 pp.

Jansson, A. M. and J. Zucchetto (1978). Energy, economic and ecological relationships for Gotland, Sweden, a regional systems study. *Ecol. Bull.* **28**, 154.

Jansson, B. O. and F. Wulff (1977). Ecosystem analysis of a shallow sound in the northern Baltic—A joint study by the ASKO group units from ASKO Lab. No. 18. University of Stockholm, Sweden. pp. 1–169.

Jantsch, E. (1980). *The Self Organizing Universe*, Pergamon Press, Oxford.

Janzen, D. H. (1970). Herbivores and the number of tree species in a tropical forest. *Am. Natur.* **104**, 501–528.

Janzen, D. H. (1971). Seed predation by animals. *Ann. Rev. Ecol. Syst.* **2**, 465–492.

Jassby, A. D. and T. Platt (1976). Mathematical formulation of the relationship between photosynthesis and light for phytoplankton. *Limnol. Oceanogr.*

Jaynes, E. T. (1978). Where do we stand on maximum entropy. In *Maximum Entropy Formalism*, R. D. Levine and M. Tribus, Eds., MIT Press, Cambridge, MA. pp. 15–118.

Jeffers, J. N. R. (1978). *An Introduction to Systems Analysis with Ecological Applications*, Edward Arnold, London. 197 pp.

Jeffries, C. (1979). Stability of holistic ecosystem models. In *Theoretical Systems Ecology*, E. Halfon, Ed. Academic Press, New York, Chapter 20. pp. 489–504.

Jeffries, H. P. (1969). Seasonal composition of temperate plankton communities: Free amino acids. *Limnol. Oceanogr.* **14**, 41–52.

Jeffries, H. P. (1970). Seasonal composition of temperate plankton communities. *Limnol. Oceanogr.* **15**, 419–426.

Jeffries, H. P. (1979). Biochemical correlates of seasonal change in marine communities. *Am. Natur.* **113**, 643–658.

Jenny, H. (1930). An equation of state for soil nitrogen. *J. Phys. Chem.* **34**, 1053–1057.

Jenny, H. (1941). *Factors of Soil Formation, a System of Quantitative Pedology*, McGraw-Hill, New York. 281 pp.

Jenny, H. (1980). *The Soil Resource*. Springer-Verlag, New York.

Jensen, A. L. (1975). Comparison of logistic equations for population growth. *Biometrics* **31**, 853–862.

Jobin, W. R. and A. T. Ippen (1964). Ecological design of irrigation canals for snail control. *Science* **145**, 1325–1326.

Johnson, F. H., H. Eyring, and R. W. Williams (1942). The nature of enzyme inhibition in bacterial lumnescence: sulfanilamide, urethane, temperature and pressure. *J. Cell. Comp. Physiol.* **20**, 247–268.

Johnson, F. H. H. Eyring, and M. J. Polissar (1954). *The Kinetic Basis of Molecular Biology*, Wiley, New York. 874 pp.

Johnson, N. O. (1937). The Pareto Law. *Rev. Econ. Stat.* **19**, 20–26.

Johnsson. O. H. (1952). En studs flyltnings och födelseortsfäalt. *Svensk Geografiska Arshok*. **28**, 115–122.

Johnston, R. F., D. M. Niles, and S. A. Rohwer (1972). Hermon Bumpus and natural selection in the house sparrow. Passer Domesticus. *Evolution* **26**, 20–31.

Jones, F. and W. B. Hall (1973). A simulation model for studying the population dynamics of some fish species. In *The Mathematical Theory of the Dynamics of Biological Population*, M. S. Bartlett and R. W. Hiores, Eds., Academic Press, London.

Jones, G. L. (1978). Mathematical model for bacterial growth and substrate utilization in the activated sludge process. In *Mathematical Models in Water Pollution Control*, A. Jones, Ed., Wiley, New York, Chapter 12.

Jones, H. E. and A. J. P. Gore (1978). Simulation of production and decay in blanket bog, in *Production Ecology of British Moors and Montane Grasslands*, O. W. Heal and D. F. Perkins, Eds., Springer-Verlag, New York. pp. 160–233.

Jones, R. W. (1973). *Principles of Biological Regulation*. Academic Press, New York. 359 pp.

Jordan, C. F. (1971). A world pattern in plant energetics. *American Scientist* **59**, 425–433.

Jorgensen, S. E. and H. Mejer (1977). Ecological buffer capacity. *Ecol. Model.* **3**, 39–61.

Jorgensen, S. E. and H. Mejer (1979). A holistic approach to ecological modeling. *Ecol. Model.* **7**, 169–189.

Juday, C. (1940). The annual energy budget of an inland lake. *Ecology*, **21** (4), 438–450.

Judson, D. (1968). Erosion of the land, or what's happening to our continent. *Am. Sci.* **56**, 356–374.

Kaesler, R. L., E. E. Herricks, and J. S. Crossman (1978). Use of indices of diversity and hierarchical diversity in stream surveys. Special Technical Publication 652, American Society of Testing and Materials, Philadelphia, PA.

Kangas, P. (1982). Energy and Landforms. Ph.D. dissertation.

Environmental Engineering Sciences, University of Florida, Gainesville.

Kangas, P. C. and P. G. Risser (1979). Species packing in the fast food restaurant guild. *Bull. Ecol. Soc. of America* **60**, 143–148.

Kansky, K. J. (1963). Structure of transport network relationships between network geometry and regional characteristics. University of Chicago, Department of Geography, Research Papers, 84.

Kaplan, H. S. and L. E. Moses (1964). Biological complexity and radiosensitivity. *Science* **145** (3627), 21–25.

Karnopp, D. and R. Rosenberg (1975). *System Dynamics,* Wiley, New York.

Karplus, W. J. and J. R. Adler (1956). Atmospheric turbulent diffusion from infinite line sources: an electric analog source solution. *J. Meteorol.* **13**, 583–586.

Karr, J. R. and I. J. Schlosser (1978). Water resource and the land–water interface. *Science* **201**, 229–234.

Katchalsky, A. and O. Kedem (1962). Thermodynamics of flow processes in biological systems. *Biophys. J.* **2**, 53–78.

Kauffman, S. A. (1977). Chemical patterns, compartments and a binary energetic code in Drosophila. *Amer. Zoologist* **17**, 631–668.

Kellogg, W. W., R. D. Cadle, E. R. Allen, A. L. Lazrus, and E. A. Martell (1972). The sulfur cycle. *Science* **175**, 587–596.

Kellogg, W. W. and H. Schneider (1974). Climate stabilization: for better or worse? *Science* **186**, 1163–1172.

Kelly, R. A. (1971). The effects of fluctuating temperature on the metabolism of laboratory freshwater microcosm. M.A. thesis. Department of Zoology. University of North Carolina, Chapel Hill. 205 pp.

Kelsey, C. T., F. G. Goff, and D. Fields (1977). Theory and analysis of vegetation pattern. Oak Ridge National Laboratory. EDFB/IBP-76/3. Environmental Sciences Division Publication No. 922.

Kemeny, J. G., J. L. Snell, and G. L. Thompson (1966). *Introduction to Finite Mathematics,* Prentice-Hall, Englewood Cliffs, NJ. 465 pp.

Kemp, W. M. (1977). Energy analysis and ecological evaluation of a coastal power plant. Ph.D. dissertation. Department of Environmental Engineering Science, University of Florida, Gainesville. 560 pp.

Kemp, W. M. and W. J. Mitsch (1979). Turbulence and phytoplankton diversity—a general model of paradox of plankton. *Ecol. Model.* **7**, 201–272.

Kemp, W. M., W. H. B. Smith, H. N. McKellar, M. E. Lehman, M. Homer, D. L. Young, and H. T. Odum (1977). Energy cost-benefit analysis applied to power plants near Crystal River, Florida. In *Ecosystems Modelling in Theory and Practice,* C. A. S. Hall and J. Day, Eds., Wiley, New York. pp. 508–543.

Kendigh, D. C. (1941). Length of day and energy requirements for gonad development and egg-laying in birds. *Ecology* **22**, 237–248.

Kermach, W. O. and A. G. McKendrick (1927). A contribution to the Mathematical Theory of Epidemics. *Proc. Roy. Soc. Lond. Ser. A* **115**, 700–721.

Kerner, E. H. (1967). A statistical mechanics and of interacting biological species. *Bull. Math. Biophys.* **19**, 121–147.

Kerner, E. H. (1972). A statistical mechanism of integrating biological species, in *Gibbs Ensemble: Biological Ensemble,* E. H. Kerner, Ed., Gordon and Breach, NY., pp. 67–145.

Kershaw, K. A. (1964). *Quantitative and Dynamic Ecology,* American Elsevier, New York. 170 pp.

Kessler, W. J. (1950). Reciprocal aspects of transient and steady-state concepts. American Institute of Electrical Engineering, Technical Paper 59.

Kierstead, H. and L. B. Slobodkin (1953). The size of water masses containing plankton blooms. *J. Mar. Res.* **12** (11), 141–147.

Kilburn, P. D. (1966). Analysis of the species–area relation. *Ecology* **47** (5), 831–843.

Kilham, P. (1971). A hypothesis concerning silica and the freshwater planktonic diatoms. *Limnol. Oceanogr.* **16**, 10–18.

King, C. E. and G. J. Paulik (1967). Dynamic models and the simulation of ecological systems. *J. Theor. Biol.* **16**, 251–267.

King, E. L. and C. Altman (1956). A schematic method of deriving rate laws for enzyme catalyzed reactions. *J. Phys. Chem.* **60**, 1375.

Kitching, R. (1971). A simple simulation model of dispersal of animals among units of discrete habitats. *Oecologia* **7**, 95–116.

Kittel, C. (1969). *Thermal Physics,* Wiley, New York. 418 pp.

Klir, G. J. (1969). *An Approach to General Systems Theory,* Van Nostrand Reinhold, New York.

Klir, J. and M. Valach (1965). *Cybernetic Modelling,* ILIFFE Books, Van Nostrand. Princeton, NJ. 437 pp.

Klopatek, J. M. (1977). Energetics of land use in three Oklahoma counties. Ph.D. thesis. University of Oklahoma, Norman.

Klose, P. N. and A. K. Deb (1978). Unsteady state BOD simulation model for a cypress swamp. Flood control reservoir. Preprint paper, 39th Meeting of American Society of Limnology and Oceanography, Weston Environmental Consultants.

Kluge, M. and I. P. Ting (1978). *Drassulacean Acid Metabolism,* Ecological Studies 30, Springer-Verlag, New York. 209 pp.

Knight, R. L. (1980). Energy basis of control in aquatic ecosystems. Ph.D. dissertation. Department of Environmental Engineering Science, University of Florida, Gainesville. 198 pp.

Koenig, H. E. (1972). Principles of ecosystem design and management. *IEEE Transact. Systems Man Cybernet.* **SMC-2** (4).

Koenig, H. E., Y. Takad, H. K. Kesevan, and H. G. Hedges (1967). *Analysis of Discrete Physical Systems,* McGraw-Hill, New York. 447 pp.

Kolmogorov, A. M. (1941). The local structure of turbulence in an incompressible viscous liquid for very high reynolds numbers. *Prod. (Dokl) Acad. Sci. USSR* **30**, 4.

Kostitzin, A. (1939). *Mathematical Biology.* Harrap, London.

Kostitzin, V. A. (1935). Evolution de l'atmosphere. *Act. Sci. Indust.* Hermann, Paris. 271.

Kowallik, W. (1967). Chlorophyll-independent photochemistry in algae in energy conversion by the photosynthetic apparatus. *Brookhaven Symp. Biol.* **19**, 467–477.

Kranz, J. (1974). *Epidemics of Plant Diseases, Ecology Studies,* No. 13. Springer-Verlag, New York.

Kremer, J. N. and S. W. Nixon (1978). *A Coastal Marine Ecosystem,* Springer-Verlag, New York. 217 pp.

Kremer, P. (1976). *Population Dynamics and Ecological Energetics of a Pulsed Zooplankton Predator, the Ctenophore Mnemiopsis leidyi. Estuarine Processes,* Vol. 1, M. Wiley, Ed., Academic Press, New York.

Krenz, J. H. (1976). *Energy Conversion and Utilization,* Allyn and Bacon, New York.

Krigman, A. (1967). Direct simulation of physical systems. *Instrum. Control Syst.* **40,** 119–121.

Krol, J. G. (1969). Simplified simulation of the Elovich equation. *Instrum. Control Syst.* **42,** 108–109.

Kurihare, Y. (1957). Synecological analysis of the biotic community in microcosm. II. Studies in the relations of dipteran larvae to protozoa in bamboo containers. *Sci. Rep. Tohuku Univ.,* Ser. 4, **23,** 139–142.

Kurasawa, H. (1956). The weekly succession in the standing crop of Plankton and zoobenthos in the paddy field. *Miscellaneous Reports of the Research Institute for Natural Resources.* pp. 41–42; 86–98.

Labine, P. A. and D. H. Wilson (1973). A teaching model of population interactions in an algae–daphnia–predator system. *Bioscience* **23,** 162–167.

Lanchester, F. W. (1956). Mathematics in warfare. In *World of Mathematics,* Vol. 4, Simon and Schuster, New York, pp. 2138–2157.

Lack, D. (1966). *Population Studies of Birds,* Clarendon Press, Oxford, U.K.

Lane, P. (1978). Zooplankton niches and the community structure controversy. *Science* **200,** 458–460.

Lane, P. and D. C. McNaught (1970). A mathematical analysis of the niches of Lake Michigan zooplankton. *Proc. 13th Conf. Great Lakes Res.* 47–57.

Langbein, W. B. and L. B. Leopold (1964). Quasi-equilibrium states in channel morphology. *Am. J. Sci.* **262,** 782–794.

Langmuir, I. (1921). The mechanism of the catalytic action of platinum in the reactions $2CO + O_2 = 2CO_2$ and $2H_2 + O_2 = 2H_2O$. *Transact. Faraday Soc.* **17,** 621–654.

Lassiter, R. R. and P. K. Kearns (1974). Phytoplankton population changes and nutrient fluctuations in a simple aquatic ecosystem model. In *Modeling the Eutrophic Process,* E. J. Middlebrooks, D. H. Falkenborg, and T. E. Maloney, Eds., Ann Arbor Publishers, Ann Arbor, MI. pp. 131–138.

Laszlo, E. (1972). *The Systems View of the World.* George Braziller, New York.

Lavine, M. J., T. Butler, and A. H. Meyburg (1979). Energy analysis manual for environmental benefit–cost analysis of transportation actions. Report to Transportation Research Board Center for Environmental Research. Project 20-11B. Cornell University, Ithaca, NY.

Lawrence, B. A., M. C. Lewis, and P. C. Miller (1978). A simulation model of population processes of Arctic Tundra graminoids. In *Vegetation and Production Ecology of an Alaskan Arctic Tundra, Ecological Studies,* No. 29, L. L. Tiezen, Ed., Springer-Verlag, New York. pp. 599–619.

Lee, C. L. (1973). *Models in Planning, Urban and Regional Planning Series,* Vol. 4. Pergamon Press, New York. 142 pp.

Lee, R. B. (1968). Bushman subsistence; an input–output analysis in human ecology. In *An Anthropological Reader,* A. P. Vayda, Ed., Natural History Press, New York.

Lefever, R. (1968). *J. Chem. Phys.* **4a** (11), 4977.

Lehman, M. (1974). Oyster reefs at Crystal River, Florida and their adaptation to thermal plumes. Master's thesis. Department of Environmental Engineering Science, University of Florida, Gainesville.

Leibowitz, S. (1980). The embodied energy in U.S. soils. In *Energy Basis for the U.S. Report to Department of Energy,* H. T. Odum, J. Alexander, F. Wang, et al., Eds., Contract EY-76-S-05-4398. Department of Environmental Engineering Sciences, University of Florida, Gainesville.

Leigh, E. (1971). *Adaptation and Diversity.* Freeman, San Francisco. 298 pp.

Leontief, W. W. (1965). The structure of the U.S. economy. *Sci. Am.* **212,** 25–35.

Leontief, W. W. (1966). *Input–Output Economics,* Oxford University Press, New York. 257 pp.

Leontief, W. W. (1971). Input–output economics. *Sci. Am.* **185,** 15–21.

Leopold, A. C. (1975). Aging, senescence and turnover in plants. *Bioscience* **25,** 659–662.

Leopold, L. B. and K. S. Davis (1966). *Water,* Time Books, New York.

Leopold, L. B. and W. B. Langbein (1962). The concept of entropy in landscape evolution. U.S. Geol. Survey Prof. Paper 500-A, 20 pp.

Leopold, L. B. and J. P. Miller (1956). Ephemeral streams, hydraulic factors and the relation to drainage net. *Prof. paper U.S.G.S.* 282A, 1–38.

Leopold, L. B., M. C. Wolman, and J. P. Miller (1964). *Fluvial Processes in Geomorphology,* Freeman, San Francisco.

Leopold, L. B., F. E. Clarke, B. B. Hanshaw, and J. R. Baisley (1971). A procedure for evaluating environmental impact. U.S. Department of Interior Geological Survey Circular 645. 13 pp.

Leslie, P. H. (1945). On the use of matrices in certain population mathematics. *Biometrika* **33,** 183–212.

Leslie, P. H. (1948). Some further notes on the use of matrices in population mathematics. *Biometrika* **24,** 213–245.

Leslie, P. H. and J. C. Gower (1960). The properties of the stochastic model for the predator–prey type of interaction between two species. *Biometrika* **47,** 219–234.

Levenspiel, O. and N. deNevers (1974). The osmotic pump. *Science* **183,** 157–160.

Levin, S. (1975). *Lecture Notes in Biomathematics,* Springer-Verlag, New York.

Levin, S. A. (1974). Ecosystem analysis and prediction. Proceedings of a SIMS conference held July 1974.

Levine, R. D. and M. Tribus (1979). *The Maximum Entropy Formalism,* MIT Press, Cambridge, MA. 498 pp.

Levine, S. (1966). Enzyme amplifier kinetics. *Science* **15,** 651–653.

Levins, E. R. (1976). Applications of discrete and continuous network theory to linear population models. *Ecology* **57,** 33–47.

Levins, R. (1965). Genetic consequences of natural selection. In *Theoretical and Mathematical Biology,* T. H. Waterman and H. J. Morowitz, Eds., Blaisdell, New York. pp. 371–386.

Levins, R. (1968). *Evolution in Changing Environments, Monographs in Population Ecology,* Vol. 2. Princeton University Press, Princeton, NJ.

Levins, R. (1975a). Problems of signed digraphs in ecological theory. In *Ecosystem Analysis and Prediction. Proceedings of the Conference on Ecosystems.* SIAM Institute for Mathematics and Society, Philadelphia. pp. 264–276.

Levins, R. (1975b). Evolution in communities near equilibrium. In *Ecology and Evolution in Communities,* M. L. Cody and J. M. Diamond, Eds., Belknap Press, Harvard University Press, Cambridge, MA. pp. 16–51.

Lewontin, R. C. (1966). Is nature probable or capricious. *Bioscience* **16,** 25–27.

Liebig, J. (1840). *Chemistry in Its Application to Agriculture,* Taylor and Watson, London.

Lieth, H. (1963). The role of vegetation in the carbon dioxide content of the atmosphere. *J. Geophys. Res.* **68,** 3887–3898.

Lieth, H. (1975). Modeling the primary productivity of the world. Chap. 12. In *Primary Productivity of the Biosphere,* H. Lieth and R. H. Whittaker, Eds., Ecological Studies No. 14. Springer-Verlag, New York. pp. 238–263.

Lieth, H. and R. H. Whittaker, Eds. (1975). *Primary Productivity of the Biosphere.* Ecological Studies No. 14, Springer-Verlag, New York.

Limburg, K. (1981). Energy basis for ecosystem design. M.S. thesis. Environmental Engineering Sciences, University of Florida, Gainesville.

Lindeman, R. L. (1941). Seasonal food-cycle dynamics in a senescent lake. *Amer. Midl. Natur.* **26,** 636–673.

Lindeman, R. L. (1942). The trophic–dynamic aspect of ecology. *Ecology* **23,** 399–418.

Lineweaver, H. and D. J. Burk (1934). The development of enzyme association constants. *J. Am. Chem. Soc.* **56,** 658–666.

Linschitz, H. (1953). Information and physical entropy. In *Information Theory in Biology,* H. Quastler, Ed., University of Illinois Press, Urbana. pp. 14–20.

Littlejohn, C. (1977). An analysis of the role of natural wetlands in regional water management. In *Ecosystem Modeling in Theory and Practice,* pp. 451–476, Wiley, New York.

Livingston, D. A. (1963a). Chemical composition of rivers and lakes. In *Data of Geochemistry,* M. Fleischer, Ed., U.S. Government Printing Office, Washington, DC. Chapter G, pp. 1–64.

Livingston, D. A. (1963b). The sodium cycle and the age of the ocean. *Geochemica et Cosmochemica Acta* **27,** 1055–1069.

Livingston, D. A. (1964). The sodium cycle of the hydrosphere. *Verh. Internat. Verein. Limnol.* **15,** 289–292.

Lloyd, M. and R. J. Ghelardi (1964). A table for calculating the equilability component of species diversity. *J. Anim. Ecol.* **33,** 217–225.

Loehr, R. C., C. S. Martin, and W. Rast, Eds. (1980). *Phorphorus Management Strategies for Lakes,* Ann Arbor Science, Ann Arbor, Mich. 300 pp.

Lomeen, P. W., C. R. Schwintzer, C. S. Yocum, and D. M. Gates (1971). A model describing photosynthesis in terms of gas diffusion and enzyme kinetics. *Planta* **98,** 195–220.

Loose, K. D. (1974). Using an energy systems approach in modeling achievement motivation and some related sociological concepts. Ph.D. dissertation. Department of Education, University of Florida, Gainesville. 107 pp.

Lorenz, E. N. (1955). Available potential energy and the maintenance of general circulation. *Tellus* **7,** 157–167.

Lorenz, K. Z. (1950). The comparative method in studying innate behavior patterns. *Soc. Exp. Biol. Symp.* **4,** 221–268.

Lorenzen, M. W. (1974). Predicting the effects of nutrient diversion on lake recovery. In *Modeling the Eutrophication Process,* E. J. Middlebrooks, D. H. Falkenborg, and T. E. Maloney, Eds., Ann Arbor Science, Ann Arbor, MI. pp. 205–210.

Lorenzen, M. W. (1978). Phosphorus models and eutrophication. In *Water Pollution Microbiology,* R. Mitchell, Ed., Wiley, New York. pp. 31–50.

Lösch, A. (1944, 1954). *The Economics of Location* (translated by W. H. Sagglom and W. F. Stapler), Yale University Press, New Haven, CT.

Lotka, A. J. (1922a). Contribution to the energetics of evolution. *Proc. Natl. Acad. Sci.* **8,** 147–155.

Lotka, A. J. (1922b). Natural selection as a physical principle. *Proc. Natl. Acad. Sci.* **8,** 151–154.

Lotka, A. J. (1925). *Elements of Mathematical Biology,* Dover, New York.

Lotka, A. J. (1945). The law of evolution as a maximal principle. *Human Biol.* **17,** 167.

Lovelock, J. E. (1972). Gaia as seen through the atmosphere. *Atmosph. Envir.* **6,** 579–580.

Lovelock, J. E. (1979) GAIA, Oxford Univ. Press, Oxford, 157 pp.

Lowry, T. S. (1964). A model of metropolis. RM 4035-RI, Rand Corporation, Santa Monica, CA.

Lucas, R. (1976). Soil organic matter modification. Report to National Science Foundation. Design and Management of Rural Ecosystems. DMRE-76-7. Michigan State University.

Ludwig, D., D. D. Jones, and C. S. Holling (1978). Quantitative analysis of insect outbreak systems: The spruce budworm and forest. *J. Anim. Ecol.* **47,** 315–332.

Lugo, A. E. (1969). Energy, water and carbon budgets of a granite outcrop community. Ph.D. dissertation. Botany Department, University of North Carolina, Chapel Hill.

Lugo, A. E. (1978). Stress and ecosystems. In *Energy and Environmental Stress in Aquatic Systems,* J. H. Thorp and J. W. G. Gibbons, Eds., Division of Technical Information, Department of Energy. pp. 62–101.

Luikov, A. V. and Yu. A. Mikhailov (1965). *Theory of Energy and Mass Transfer* (translation by J. S. Dunn and J. A. Weightman), Pergamon Press, New York. 392 pp.

Lumry, R. and J. S. Rieske (1959). The mechanism of photochemical activity of isolated chloroplasts. V. Interpretation of the rate parameters. *Plant Physiol.* **34,** 301–305.

Lumry, R. and J. D. Spikes (1957). Chemical–kinetic studies of the Hill reaction. In *Research in Photosynthesis,* H. Gaffron, A. H. Brown, C. S. French, R. Livingston, E. T. Rabinowitch, B. L. Strehler, and N. E. Tolbert, Eds., Interscience, New York. 524 pp.; pp. 373–398.

Lumry, R. and J. D. Spikes (1960). Collection and utilization of energy in photosynthesis. *Rad. Res.* (Suppl.) **2,** 539–577.

Lund, J. W. G. (1972). Preliminary observations on the use of large experimental tubes in lakes. *Verh. Inst. Verein Theor. Angeio. Limnol.* **18,** 71–77.

Lynch, E. P. (1980). *Applied Symbolic Logic,* Wiley, New York.

MacArthur, R. (1955). Fluctuations of annual populations and a measure of community stability. *Ecology* **26,** 533–536.

MacArthur, R. and J. Connell (1966). *The Biology of Populations*, Wiley, New York. 200 pp.

MacArthur, R. H. (1957). On the relative abundance of bird species. *Proc. Natl. Acad. Sci. Wash.* **54**, 293–295.

MacArthur, R. H. and E. O. Wilson (1967). *Theory of Island Biogeography*, Princeton University Press, Princeton, NJ.

MacArthur, R. H. (1968). In *Population Biology and Evolution*, R. C. Lewontin, Ed., Syracuse University Press, Syracuse, NY. pp. 159–176.

MacArthur, R. H. (1972). *Geographical Ecology*, Harper, New York. 269 pp.

MacCready, P. B. (1953). Structure of atmospheric turbulence. *J. Meteorol.* **10**, 434–449.

MacKenzie, F. T. and R. M. Garrels (1966). Chemical mass balance between rivers and oceans. *Am. J. Sci.* **264**, 507–525.

Mackey, M. C. and L. Glass (1977). Oscillation and chaos in physiological control systems. Science **197**, 287–289.

MacNichol, E. F. (1959). An analog computer to simulate systems of coupled bimolecular reactions. *Proc. Inst. Radio Eng.* **47**, 1816–1820.

Maguire, B. (1971). Phytotelmata: Biota and community structure determination in plant-held waters. *Ann. Rev. Ecol. System.* **2**, 439–463.

Maguire, L. A. and J. W. Porter (1977). A spatial model of growth and competition strategies in coral communities. *Ecol. Model.* **3**, 249–271.

Mahloch, J. L. (1974). Comparative analysis of modeling techniques for coliform organisms in streams. *Appl. Microbiol.* **27**, 340–345.

Makridakis, S. (1977). The second law of systems. *Inform. J. Gen. Syst.* **4**, 1–12.

Malek, I. (1958). *Continuous Cultivation of Microorganisms—A Symposium*, Publishing House of the Czechoslovak Academy of Sciences, Prague. 234 pp.

Malek, I. and Z. Fencl, Eds. (1966). *Theoretical and Methodological Basis of Continuous Culture of Micro-organisms*, Publishing House of the Czechoslovak Academy of Sciences, Prague; Academic Press, New York. 655 pp.

Malkevitch, J. and W. Meyer (1974). *Graphs, Models, and Finite Mathematics*. Prentice-Hall, Englewood Cliffs, NJ. 515 pp.

Mandelbrot, B. (1953). On informational theory of the statistical structure of languages. In *Communication Theory*, W. Jackson, Ed., pp. 486–502, Academic Press, New York.

Mandelbrot, B. (1956). On the language of taxonomy, an outline of a thermostatistical theory of systems of categories with Willis (Natural) structure. Third Information Theory Symposium, C. Cherry, Ed., Butterswoths, London. pp. 135–145.

Mandelbrot, B. (1960). The Pareto–Levy law and the distribution of income. *Internatl. Econ. Rev.* **1**, 79.

Mandelbrot, B. B. (1977). *Fractals*, Freeman, London.

Margalef, D. R. (1956). Information theory in ecology. *Mem. R. Acad. Cienc. Artes Barcelona* **23**, 373–449 [translation in *Soc. Gen. Syst. Yearbook* **3**, 36–71 (1958)].

Margalef, R. (1957a). Nuevos aspectos del problema de la suspension en los organismos planctonicos. *Investacion Pesquero* **7**, 105–116.

Margalef, R. (1957b). Le teorie de la informacion en ecologia. *Mem. R. Acad. Cienc. Artes Barcelona* **32** (13), 373–449.

Margalef, R. (1958a). Information theory in ecology. *Yearbook Soc. Gen. Syst. Theory* **3**, 36–61.

Margalef, R. (1958b). Temporal succession and spatial heterogeneity in phytoplankton. In *Perspectives in Marine Biology*, A. A. Buaazti-Traverso, Ed., University of California Press, Berkeley. 621 pp.; pp. 323–349.

Margalef, R. (1960). Recientes progresos en el estudio de las communidades vegetales por medio de la extraccion de pigmentos. *Bole. R. Soc. Esp. Historia Natural* **63**, 291–300.

Margalef, R. (1961). Communication of structure in planktonic populations. *J. Limnol. Oceanogr.* **6**, 124–128.

Margalef, R. (1962). Modelos fisicos simplificados de poblaciones de organismos. *Mem. R. Acad. Cienc. Artes Barcelona* **34**, 1–66, 83–146 (1962).

Margalef, R. (1963). On certain unifying principles in ecology. *Am. Natur.* **97**, 357–374.

Margalef, R. (1968). *Perspectives in Ecological Theory*, University of Chicago Press. 111 pp.

Margalef, R. (1974). *Ecologia*, Ediciones Omega, SA, Barcelona. 951 pp.

Margalef, R. (1978). General concepts of population dynamics and food links. In *Marine Ecology*, Vol. IV, *Dynamics*. O. Kinne, Ed., Wiley, New York, pp. 617–704.

Marks, P. L. and F. H. Bormann (1972). Revegetation following forest cutting: Mechanisms for return to state nutrient cycling. *Science* **176**, 914–915.

Markson, R. and M. Muir (1980). Solar wind control of the Earth's electric field. *Science* **208**, 979–989.

Marshall, J. S. (1962). The effects of continuous gamma radiation on the intrinsic rate of natural increase of *Daphnia pulex*. *Ecology* **43**, 598–607.

Marshall, J. S. (1965). Population dynamics of Daphnia as modified by radiation stress. In *Radiation Effects on Natural Populations*, G. A. Sacher, Ed., Report of Colloquium, Argonne National Laboratory, Chicago. 61 pp.; pp. 43–44.

Martin, G. D. (1973). Optimal control of an oil refinery waste treatment facility: A total ecosystem approach. M.S. thesis. Oklahoma State University. 91 pp.

Martini (1921). Berechnungen und beobachungen zur Epidemiologie de Malaria, Gente, Hamburg.

Martino, R. L. (1965). *Allocating and Scheduling Resources, Project Management and Control*, Vol. III, American Management Association, New York.

Marx, Karl (1867, 1977). *Das Capital*. Translated by Ben Fowkes, Vintage Books, Random House, New York.

Mason, B. (1952, 1982). *Principles of Geochemistry*, 4th ed. Wiley, New York.

Mason, H. L. and J. H. Langenheim (1957). Language and the concept of environment. *Ecology* **38**, 325–340.

Mason, S. J. (1953). Feedback theory: Some properties of signal flow graphs. *Proc. Inst. Radio Eng.* **4L**, 1144–1156.

Masser, I. (1972). *Analytical Models for Urban and Regional Planning*, David and Charles, Newton Abbott, Devon, England.

Mattson, W. J. and N. D. Addy (1975). Phytophagous insects as regulators of forest primary production. *Science* **190**, 515–522.

Maxwell, J. C. (1877). *Matter and Motion*, Macmillan, New York (reprinted 1975) 162 pp.

Maxwell, J. C. (1892). *Treatise on Electricity and Magnetism,* Vol. I, 3rd ed., Oxford University Press, London. Articles 283–284, pp. 407–408.

May, R. M. (1973). *Stability and Complexity in Model Ecosystems,* Princeton University Press, Princeton, NJ.

May, R. M. (1974). Biological populations with non-overlapping generations, stable points, stable cycles, and chaos. *Science* **186,** 645–647.

May, R. M. (1975). Patterns of species abundance and diversity. In *Ecology and Evolution in Communities,* M. L. Cody and J. M. Diamond, Eds., Harvard University Press, Cambridge, MA. pp. 81–121.

May, R. M., Ed. (1976). *Theoretical Ecology: Principles and Applications,* Oxford Blackwell Scientific Publishers, London.

May, R. M. (1978). Factors controlling the stability and breakdown of ecosystems. In *The Breakdown and Restoration of Ecosystems,* M. W. Holdgate and M. J. Woodman, Eds., Plenum Press, New York. 496 pp.; pp. 11–22.

McCormick, F. and R. B. Platt (1962). Effects of ionizing radiation on a natural plant community. *Rad. Bot.* **2,** 161–188.

McCulloch, W. S. and W. Pitts (1943). A logical calculus of the ideas immanent in nervous activity. *Bull. Math. Biophys.* **3,** 115–133.

McDougall, E. (1925). The moisture belt of North America. *Ecology* **6,** 325–332.

McHarg, I. L. (1969). *Design with Nature,* American Museum of Natural History. Natural History Press, Garden City, NY. 197 pp.

McIntire, C. D. and J. A. Colby (1978). A hierarchical model of lotic ecosystems. *Ecological Monographs* **48,** 167–190.

McIntire, C. D., R. L. Garrison, H. K. Phinney, and C. E. Warren (1964). Primary production in laboratory streams. *Limnol. Oceanogr.* **9,** 92–102.

McIntosh, R. P. (1963). Ecosystem evolution and relational patterns of living organisms. *Am. Sci.* **51,** 246–268.

McIntosh, R. P. (1967). The continuum concept of vegetation. *Bot. Rev.* **33,** 130–187.

McIntosh, R. P. (1980). The relationship between succession and the recovery process in ecosystems. In *The Recovery Process in Damaged Ecosystems,* J. Cairns, Ed., Ann Arbor Science, Ann Arbor, MI. pp. 11–62.

McKellar, H. N. (1975). Metabolism and models of estuarine bay ecosystems affected by a coastal power plant. Ph.D. dissertation. Department of Environmental Engineering Sciences, University of Florida, Gainesville. 269 pp.

McKellar, H. N. (1977). Metabolism and model of an estuarine bay ecosystem affected by a coastal power plant. *Ecol. Model.* **3,** 85–118.

McKellar, H. and H. T. Odum (1972). A model of the Gulf shelf ecosystem. Report to National Science Foundation.

McLeod, J. (1968). *Simulation,* McGraw-Hill, New York.

McMahon, T. (1973). Size and shape in biology. *Science* **179,** 1201–1204.

McNaught, D. C. (1979). Considerations of scale in modeling large aquatic ecosystems. In *Lake Ecosystem Modeling.* D. Scavia and A. Robertson, Eds., Ann Arbor Science, Ann Arbor, MI. pp. 2–24.

Meadows, D. H., D. L. Meadows, J. Randers, and W. W. Behrens, III (1972). *The Limits to Growth: A Report for the Club of Rome's Project on the Predicament of Mankind.* A Potomac Associates Book, Universe Books, New York. Graphics by Potomac Associates.

Mehnhinick, E. F. (1964). A comparison of some species-individuals diversity indices applied to species of field insects. *Ecology* **45,** 859–861.

Mendelssohn, K. (1971). A scientist looks at the pyramids. *Am. Sci.* **59,** 210–220.

Menshutkin, V. V. (1973). A mathematical modeling of lake ecological system. *Verh. Internatl. Limnol.* **18,** 1843–1850.

Mesarovic, M. (1979). Practical applications of global modelling. In *Global and Large Scale Systems Models,* A. V. Balakrishnan and M. Thoma, Eds., Lecture Notes on Control and Information Sciences, pp. 42–55, Springer-Verlag, Berlin. 232 pp.

Meyer, A. (1926). Uber einige zusammenhange zwischen kilima and boden in Europa. *Chem. Erde* **2,** 209–347.

Meyer, H. A. (1933). *Schweiz Zectschift Forest* 84.

Michaelis, L. and M. L. Menten (1913). Die kinetik der invertinwirkung. *Biochemische Zietschrift* **49,** 333–369.

Mihram, G. A. (1971). *Simulations: Statistical Foundations and Methodology,* Academic Press, New York.

Mikolaj, P. G. (1972). Environmental applications of the Weibull distribution function: Oil pollution. *Science* **176,** 1019–1021.

Mildvan, A. S. and B. Strehler (1960). A critique of theories of mortality. In *The Biology of Aging,* B. Strehler, J. D. Ebert, H. B. Glass and N. W. Shock, Eds., American Institute of Biological Sciences, Washington, DC. pp. 216–235.

Milhorn, H. T. (1966). *The Application of Control Theory to Physiological Systems.* Saunders, Philadelphia. 386 pp.

Miller, J. G. (1955). Toward a general theory for the behavioral sciences. *Am. Psychol.* **10,** 513–531.

Miller, J. G. (1978). *Living Systems,* McGraw-Hill, New York.

Miller, M. A. (1980). Energy basis for housing systems in Collier County. M.S. thesis. Department of Environmental Engineering Sciences, University of Florida, Gainesville. 160 pp.

Miller, P. C., B. D. Collier, and F. L. Bunnell (1975). Development of ecosystem modelling in the U.S. IBP Tundra Biome. In *Systems Analysis and Simulations in Ecology,* Vol. 3, Patten, B. C. Ed., Academic Press. pp. 95–115.

Miller, R. S. and D. B. Botkin (1974). Endangered species. Models and predictions. *Am. Sci.* **62,** 172–181.

Miller, S. L. (1953). A production of amino acids under possible primitive conditions. *Science* **117,** 528–529.

Mimura, M. and M. Yamaguti (1982). Pattern formation in irregular interacting and diffusing systems in population biology. *Adv. Biophys.* **15,** 19–65.

Milstein, J. S. and W. C. Mitchell (1968). Dynamics of the Vietnam conflict. A quantitative analysis and predictive computer simulation papers. *Peace Res. Soc. Internatl.* **10,** 163–213.

Milsum, J. H. (1966). *Biological Control Systems Analysis,* McGraw-Hill, New York. 466 pp.

Milsum, J. H. (1968). Positive feedback. *A General System Approach to Positive Negative Feedback and Mutual Causality,* Pergamon Press, New York. 169 pp.

Mitchiner, J. L., W. J. Kennish, and J. W. Brewer (1975). Application of optimal control and optimal regulator theory to the "integrated" control of insect pests. *IEEE Transact. Syst. Man Cybernet.,* SMC-5, 111–116.

Mitsch, W. J. (1975). Systems analysis of nutrient disposal in cypress wetlands and lake ecosystems in Florida. Ph.D. dissertation. Department of Environmental Engineering Science, University of Florida, Gainesville.

Mitsch, W. J. (1976). Ecosystem modeling of water hyacinth management in Lake Alice, Florida. *Ecological Modelling* **2**, 69–89.

Mitsch, W. J., M. A. McPartlin, and R. D. Letterman (1978). Energetic evaluation of a stream ecosystem affected by coal mine drainage. *Verh. Internatl. Verein Limnol.* **20**, 1388–1395.

Mitscherlich, L. A. (1909). Das gesetz des Minimums und das Gesetz des absnehmenden Bodenertrages. *Landw. Jahrbucher.* **38**, 537–552.

Monin, A. S., V. M. Kumenkovich, and V. G. Kort (1976). *Variability of the Oceans,* Wiley, New York. 240 pp.

Monod, J. (1942). *Recherches du la Croissance des Cultures Bacteriennes,* Hermann et Cie, Paris.

Monroe, J. (1967). The exploitation and conservation of resources by populations of insects. *J. Anim. Ecol.* **36**, 531–547.

Monsi, M. and T. Saeki (1953). Uber den Lichtfaktor in den pflanzengesellschaften und seine becheutung fur die stoffproduktion. *Jap. J. Bot.* **14**, 22–56.

Montague, C. L. (1980). The net influence of the mud fiddler crab, Uca pugnax on carbon flow through a Georgia salt marsh: The importance of work by macroorganisms to the metabolism of ecosystems. Ph.D. dissertation. Department of Zoology, University of Georgia, Athens. 157 pp.

Monteith, J. L. (1963). Gas exchange in plant communities. In *Environmental Control of Plant Growth,* L. T. Evans, Ed., Academic Press, New York. pp. 95–112.

Morowitz, H. J. (1966). Physical background of cycles in Biological systems. *J. Theor. Biol.* **13**, 60–62.

Morowitz, H. J. (1967). Biological self-duplicating systems. In *Theoretical Biology,* Vol. I, Academic Press.

Morowitz, H. J. (1968). *Energy Flow in Biology,* Academic Press, New York. 179 pp.

Morowitz, H. J. (1978). *Foundations of Bioenergetics,* Academic Press, New York.

Morrill, R. L. (1970). *Spatial Organization of Society,* Duxbury Press, North Scituate, MA. 251 pp.

Mueller, P., T. Martin, and F. Putzrath (1962). General principles of operations in neuron nets with application to acoustical pattern recognition. In *Biological Prototypes and Synthetic Systems,* E. E. Bernard and M. R. Kare, Eds., Plenum Press. 397 pp.; pp. 192–212.

Mueller-Dombois, D. (1980). The Ohiá Dieback Phenomenon in the Hawaiian Rainforest. In *The Recovery Process in Damaged Ecosystems,* J. Cairns, Ed., Ann Arbor Science, Ann Arbor, Mich. pp. 153–161.

Murphy, C. E. and C. A. Gresham (1974). A model of leaf stomatal response to light and water stress. In *Summer Computer Simulation Conference,* Simulations Council, La Jolla, CA. pp. 550–556.

Murphy, G. I. (1962). Effect of mixing depth and turbidity on the productivity of freshwater impoundments. *Transact. Am. Fish. Soc.* **91**, 69–76.

Murphy, P. E. (1970). Ecological effects of acute beta irradiation from simulated fallout particles upon a natural community. Ph.D. dissertation. Botany Department, University of North Carolina, Chapel Hill.

Murray, B. G. (1979). *Population Dynamics Alternative Models,* Academic Press, New York. 211 pp.

Myers, J. (1946). Culture conditions and the development of the photosynthetic mechanism II and III. Influence of light intensity on cellular characteristics of chlorella. *J. Gen. Physiol.* **29**, 419–429; 429–440.

Myers, J. H. (1976). Distribution and dispersal resource depletion. A simulation model. *Oecologia* **23**, 259–270.

Nagel, E. J. and J. R. Newnan (1958). *Godel's Proof,* New York University Press, 118 pp.

Naroli, R. S. and L. Von Bertalanffy (1956). The principle of allemetry in biology and the social sciences. *Yearbook Gen. Syst. Theory* **1**, 76–89.

Nash, L. K. (1974). *Elements of Statistical Thermodynamics,* Addison-Wesley, Reading, MA.

National Academy of Science (1963). *Ocean Wave Spectra,* Prentice-Hall, Englewood Cliffs, NJ. 357 pp.

Naumann, E. (1932). *Grundzuge der regionalen Limnologie in Die Binnengewasser,* Vol. 11, Schweizerbarsche Verlagsbuchhandlung, Stuttgart.

Neales, T. F., A. A. Patterson, and V. J. Hartney (1968). Physiological adaptation of drought in the carbon assimilation and water loss of xerophytes. *Nature* **219**, 469–472.

Neel, R. B. and J. S. Olson (1962). Use of analog computers for simulating the movement of isotopes in ecological systems. Oak Ridge National Laboratory. ORNL-3172 UC-48. 119 pp.

Nemerow, N. L. (1974). *Scientific Stream Pollution Analysis,* McGraw-Hill, New York.

Newcombe, C. L., and J. V. Slater (1950). Environmental factors of Sodon Lake a dichotermic lake in southeastern Michigan. *Ecol. Monograph* **20**, 207–227.

Nicholls, G. D. (1965). The geochemical history of the oceans. In *Chemical Oceanography,* Vol. 2, J. P. Riley and G. Skirrow, Eds., Academic Press, New York. pp. 277–294.

Nicholson, A. J. (1954). An outline of the dynamics of animal populations. *Aust. J. of Zoology* **2**, 9–65.

Nicholson, A. J. and V. A. Bailey (1935). The balance of animal populations. *Proc. Zool. Soc. Lond.* (3), 551–598.

Nicolis, G. and I. Prigogine (1977). *Self-Organization in Nonequilibrium Systems,* Wiley, New York.

Nihoul, J. C. J. (Ed.) (1975). *Modelling of Marine Systems,* American Elsevier, New York.

Nijkamp, P. (1977). *Theory and Application of Environmental Economics,* North Holland, Amsterdam. 332 pp.

Nixon, S. W. (1969a). Synthetic microcosm. *Limnol. Oceanogr.* **14**, 142–145.

Nixon, S. W. (1969b). Characteristics of some hypersaline ecosystems. Ph.D. dissertation. Department of Botany, University of North Carolina, Chapel Hill.

Nixon, S. W. and H. T. Odum (1970). A model for photoregeneration in brines. Department of Environmental Sciences and Engineering, University of North Carolina, *ESE Notes,* Vol. 7, pp. 1–3.

Novick, A. and L. Szilard (1950). Description of the chemostat. *Science* **112**, 715–716.

Noy-Meir, I. (1973). Desert ecosystems environment and producers. In *Annual Reviews of Ecology and Systematics,* R. F.

Johnston, P. W. Frank, and C. D. Michener, Eds., Annual Reviews, Vol. 4, Palo Alto, CA. pp. 25–47.

O'Brien, J. J. (1974). The dynamics of nutrient limitation of phytoplankton algae: A model reconsidered. *Ecology* **55**, 135–141.

Odum, E. C., G. Scott, and H. T. Odum (1981a). *Energy and Environment in New Zealand*, Joint Centre for Environmental Science, 180 pp.

Odum, E. P. (1962). Relationships between structure and function in the ecosystem. *Jap. J. Ecol.* **12**, 108–118.

Odum, E. P. (1963). *Ecology*, Holt, Rinehart and Winston, New York. 152 pp.

Odum, E. P. with collaboration of H. T. Odum (1953, 1959). *Fundamentals of Ecology*, 1st and 2nd eds., Saunders, Philadelphia.

Odum, E. P. (1969). Strategy of ecosystems development. *Science* **164**, 262–269.

Odum, E. P. (1971). *Fundamentals of Ecology*, 3rd ed., Saunders, Philadelphia.

Odum, E. P. and H. T. Odum (1972). Natural areas as necessary components of man's total environment. In *Transactions of the 37th North American Wildlife Resources Conference*, Wildlife Management Institute, Washington, DC. pp. 178–189.

Odum, E. P. and H. T. Odum (1979). A rebuttal. *Coastal Zone Management J.* **5**, 239–241.

Odum, H. T. (1950). The biogeochemistry of strontium. Ph.D. dissertation. Yale University, New Haven, CT. 373 pp.

Odum, H. T. (1951). Stability of the world strontium cycle. *Science* **114**, 407–411.

Odum, H. T. (1955). Trophic structure and productivity of Silver Springs, Florida. *Ecol. Monogr.* **27**, 55–112.

Odum, H. T. (1956). Primary production in flowing waters. *Limnol. Oceanogr.* **1**, 102–117.

Odum, H. T. (1957). Primary production measurements in eleven Florida springs and a marine turtle grass community. *Limnol. Oceanogr.* **2**, 85–97.

Odum, H. T. (1960). Ecological potential and analogue circuits for the ecosystem. *Am. Sci.* **48**, 1–8.

Odum, H. T. (1962a). The use of a network energy simulator to synthesize systems and develop analogous theory: The ecosystem example. In *Proceedings of the Cullowhee Conference on Training in Biomathematics*, H. L. Lucas, Ed., Institute of Statistics, North Carolina State University. 390 pp.; pp. 291–297.

Odum, H. T. (1962b). Man in the ecosystem. Proceedings of the Lockwood Conference on the Suburban Forest and Ecology. *Bull. Conn. Agr. Sta.* **652**, 57–75.

Odum, H. T. (1963). Limits of Remote Ecosystems Containing Man. *American Biology Teacher* **35**, 429–434.

Odum, H. T. (1964). An energy basis for species diversity relations. In *Annual Report, The Rainforest Project*, Puerto Rico Nuclear Center, Rio Piedras. Report No. 34, 82 pp.; pp. 77–80. See also p. I-249 (Odum and Pigeon, 1970).

Odum, H. T. (1967a). Biological circuits and the marine systems of Texas. In *Pollution and Marine Ecology*, T. A. Olson and F. J. Burgess, Eds., Interscience, New York. pp. 99–157.

Odum, H. T. (1967b). Energetics of world food production. In *President's Science Advisory Committee, Report of Problems of World Food Supply*, Vol. 3, White House, Washington, DC. pp. 55–94.

Odum, H. T. (1968). Work circuits and systems stress. In *Mineral Cycling and Productivity of Forests*, H. Young, Ed., University of Maine. pp. 81–146.

Odum (1970a) p. I-222 in Odum and Pigeon (1970).

Odum, H. T. (1970b). Energy value of water. *Proceedings of the 19th Southern Water Resources and Pollution Control Conference*, Duke University, Durham, NC.

Odum, H. T. (1971a). *Environment, Power and Society*, Wiley-Interscience, New York. 336 pp.

Odum, H. T. (1971b). An energy circuit language for ecological and social systems, its physical basis. In *Systems Analysis and Simulation in Ecology*, Vol. 2, B. Patten, Ed. Academic Press, New York. pp. 139–211.

Odum, H. T. (1972). Chemical cycles with energy circuit models. In *Nobel Symposium No. 20*, Wiley, New York. pp. 223–257.

Odum, H. T. (1972b). Use of energy diagrams for environmental impact statements, in *Tools for Coastal Management*, Marine Technology Society, Washington, D.C., pp. 197–212.

Odum, H. T. (1973). Energy, ecology and economics. *Ambio* **2** (6), 220–227.

Odum, H. T. (1974). Energy cost benefit models for evaluating thermal plumes in thermal ecology. In *Proceedings of Symposium of the Savannah River Laboratory, Aiken, SC*, W. Gibbons and B. Sharitz, Eds., pp. 628–649, Division of Technical Information, U.S. Atomic Energy Commission.

Odum, H. T. (1975a). Combining energy laws and corollaries of the maximum power principle with visual systems mathematics. In *Ecosystem Analysis and Prediction, Proceedings of the Conference on Ecosystems*, SIAM Institute for Mathematics and Society. pp. 239–263.

Odum, H. T. (1975b). Marine ecosystems with energy circuit diagrams. In *Modelling of Marine Systems*, J. C. J. Nihoul, Ed., American Elsevier, New York. Chapter 6, pp. 127–151, and frontispiece.

Odum, H. T. (1976a). Macroscopic minimodels of man and nature. In *Systems Analysis and Simulation in Ecology*, Vol. 4, B. Patten, Ed., Academic Press, New York. pp. 249–280.

Odum, H. T. (1976b). Energy quality and carrying capacity of the earth. (Response at Prize Ceremony, Institute de la Vie, Paris.) *Trop. Ecol.* **16** (1), 1–8.

Odum, H. T. (1976c). Net benefits to society from alternate energy investments. In *Transactions of the 41st North American Wildlife and Natural Resource Conference*. Wildlife Management Institute, Washington, DC. pp. 327–338.

Odum, H. T. (1977). Value of wetlands as domestic ecosystems. In *National Wetland Protection Symposium*, J. H. Montanari and J. A. Kusler, Eds., U.S. Fish and Wildlife Service, FWS/Obs-78/97, U.S. Department of Interior. pp. 9–18.

Odum, H. T. (1978). Energy analysis, energy quality and environment. In *Energy Analysis, a New Public Policy Tool, AAAS Selected Symposium*, Vol. 9, M. Gilliland, Ed., Westview Press. pp. 55–87.

Odum, H. T. (1979). Energy quality control of ecosystem design. pp. 221–235. In *Marsh–Estuarine Systems Simulation*, P. F. Dame, Ed., Belle W. Baruch Library in Marine Science, No. 8. University of South Carolina Press. 260 pp.

Odum, H. T. (1981). Energy, economy, and environmental hierarchy. *Proceedings of the 3rd International Conference on Environment*, Vol. 4, Ministry of Environment. Paris.

Odum, H. T. (1982). Pulsing, Power, and Hierarchy. In *Energetics and Systems*, W. J. Mitsch, R. K. Ragade, R. W. Bosserman, and J. A. Dillon, Jr., Eds., Ann Arbor Science, Ann Arbor, Mich. pp. 33–59.

Odum, H. T. and W. Abbott (1972). p. 230 in Odum (1972).

Odum, H. T. and W. C. Allee. (1953). A note on the stable point of populations with both intraspecific cooperation and disoperation. *Ecology* **35**, 45–97.

Odum, H. T. and J. O. Baltzer (1962). A hardware demonstration. Annual Meeting Ecological Society of America.

Odum, H. T. and M. T. Brown, Eds. (1975). Carrying capacity for man and nature in South Florida. Report to Department of Interior National Park Service. Center for Wetlands, University of Florida, Gainesville.

Odum, H. T. and M. Brown (1976). Carrying capacity for man and nature in South Florida. Final report to U.S. Department of Interior and State of Florida Division of State Planning. Contract No. CX000130057.

Odum, H. T. and S. Brown (1976b). Energy: The Dynamo Word. In *Born of the Sun*, J. E. Gill and B. R. Read, Eds., Worth International Communications Corporation, Hollywood, FL. 192 pp.; pp. 146–151.

Odum, H. T. and C. M. Hoskin (1957). Metabolism of a laboratory stream microcosm. *Publ. Inst. Mar. Sci. Univ. Tex.* **4**, 115–133.

Odum, H. T. and C. M. Hoskin (1958). Comparative studies of the metabolism of marine waters. *Publ. Inst. Mar. Sci. Univ. Tex.* **5**, 16–46.

Odum, H. T. and A. Lugo (1970). Metabolism of forest floor microcosms. In *A Tropical Rainforest* (Odum and Pigeon, 1970). pp. 35–54.

Odum, H. T. and E. P. Odum (1955). Trophic structure and productivity of a windward coral reef community on Eniwetok Atoll. *Ecol. Monogr.* **25**, 291–320.

Odum, H. T. and E. C. Odum (1976, 1980). *Energy Basis for Man and Nature*, 2nd ed., McGraw-Hill, New York. 297 pp.

Odum, H. T. and E. C. Odum (1980). Energy system of New Zealand and the use of embodied energy for evaluating benefits of international trade. In *Proceedings of Energy Modeling Symposium, November 1979*. Technical Publication No. 7. New Zealand Ministry of Energy, Wellington, NZ. pp. 106–167.

Odum, H. T. and L. Peterson (1972). Relationship of energy and complexity in planning. *Arch. Design* **43**, 624–629.

Odum, H. T. and R. F. Pigeon, Eds. (1970). *A Tropical Rain Forest*, Division of Technical Information TID 24270, Atomic Energy Commission. Clearinghouse for Federal Scientific and Technical Information, Springfield, VA. 1660 pp.

Odum, H. T. and R. C. Pinkerton (1955). Time's speed regulator: The optimum efficiency for maximum power output in physical and biological systems. *Am. Sci.* **43**, 321–343.

Odum, H. T. and N. Vick (1962). The paradox that film ecosystems are anaerobic basins (abstract). In *Proceedings of the First National Coastal and Shallow Water Research Conference*, D. S. Gorsline, Ed., Tallahassee, FL. 897 pp.; p. 493.

Odum, H. T. and R. F. Wilson (1962). Further studies on reaeration and metabolism of Texas bays, 1958–1960. *Publ. Inst. Mar. Sci. Univ. Tex.* **8**, 23–55.

Odum, H. T., W. McConnell, and W. Abbott (1958). The chlorophyll "A" of communities. *Publ. Inst. Mar. Sci. Univ. Tex.* **5**, 65–96.

Odum, H. T., P. R. Burkholder, and J. Rivero (1959). Measurements of productivity of turtle grass flats, reefs, and the *Bahia fosfolescente* of southern Puerto Rico. *Publ. Inst. Mar. Sci. Univ. Tex.* **6**, 159–170.

Odum, H. T., J. Cantlon, and L. Kornicker (1960). An organizational hierarchy postulate for the interpretation of species–individual distributions, species entropy, ecosystem evolution, and the meaning of a species–variety index. *Ecology* **41**, 395–399.

Odum, H. T., R. J. Beyers, and N. E. Armstrong (1963a). Consequences of small storage capacity in nannoplankton pertinent to measurement of primary production in tropical waters. *J. Mar. Res.* **21** (3), 191–198.

Odum, H. T., B. J. Copeland, and R. Z. Brown (1963b). Direct and optical assay of leaf mass of the lower Montane Rain Forest of Puerto Rico. *Proc. Natl. Acad. Sci. (USA)* **49**, 429–434.

Odum, H. T., R. Cuzon, R. J. Beyers, and C. Allbaugh (1963c). Diurnal metabolism, total phosphorus, Ohle anomaly, and zooplankton diversity of abnormal marine ecosystems of Texas. *Publ. Inst. Mar. Sci. Univ. Tex.* **9**, 404–453.

Odum, H. T., W. L. Siler, R. J. Beyers, and N. Armstrong. (1963d). Experiments with engineering of marine ecosystems. *Publ. Inst. Mar. Sci. Univ. Tex.* **9**, 373–403.

Odum, H. T., B. J. Copeland, and E. McMahan (1967, 1972). *Coastal Ecosystems of the United States*, Vols. 1–4, Conservation Foundation.

Odum, H. T., S. Nixon, and L. DiSalvo (1971). Characteristics of photoregenerative systems. In *The Structure and Function of Freshwater Microbial Communities*, J. Cairns, Ed., American Microbiology Society, Virginia Polytechnic, Blacksburg. 301 pp.; pp. 1–29.

Odum, H. T., C. Littlejohn, and W. C. Huber (1972). An environmental evaluation of the Gordon River of Naples, Florida, and the impact of development plans. Report to the County Commissioners of Collier County, Florida.

Odum, H. T., M. Sell, M. Brown, J. Zucchetto, C. Swallows, J. Browder, T. Ahlstrom, and L. Peterson (1974). Models of herbicide, mangroves, and war in Vietnam. In *The Effects of Herbicides in South Vietnam*, Part B, *Working Papers*, National Academy of Sciences. February.

Odum, H. T., M. Brown, and R. Costanza (1976a). Developing a steady state for man and land: Energy procedures for regional planning. In *Science for Better Environment. Proceedings of the International Congress on the Human Environment (HESC), Kyoto,* Asahi Evening News, CPO Box 555, Tokyo, Japan. pp. 343–361.

Odum, H. T., C. Kylstra, J. Alexander, N. Sipe, and P. Lem (1976b). Net energy analysis of alternatives for the United States. In *Middle and Long-Term Energy Policies and Alternatives. Hearings of Subcommittee on Energy and Power, 94th Congress,* Serial No. 94-63. U.S. Government Printing Office, Washington, DC. pp. 258–302.

Odum, H. T., K. C. Ewel, W. J. Mitsch, and J. W. Ordway (1977a). Recycling treated sewage through cypress wetlands in Florida. In *Wastewater Renovation and Reuse*, F. D'Itri, Ed., Marcel Dekker, New York. pp. 35–68.

Odum, H. T., W. M. Kemp, M. Sell, W. Boynton, and M. Lehman

(1977). Energy analysis and coupling of man and estuaries. *Envir. Manage.* **1,** 297–315.

Odum, H. T., J. F. Alexander, F. Wang, M. Brown, M. Burnett, R. Costanza, P. Kangas, D. Swaney, S. Leibowitz, and S. Lemlich (1979). Energy Basis of the United States. Report to Department of Energy, Contract FY-76-S-05-4398, Systems Ecology and Energy Analysis Group, Environmental Engineering Sciences, Univ. of Florida, Gainesville.

Odum, H. T., M. J. Lavine, F. C. Wang, M. A. Miller, J. F. Alexander, Jr., and T. Butler (1981b). A manual for using energy analysis for plant siting. Report to the Nuclear Regulatory Commission, Contract NRC-04-77-123 Mod. 3. Nuclear Regulatory Commission, in press. 242 pp.

Oesterhelt, D. and W. Stoeckenius (1973). Functions of a new photoreceptor membrane, *Proc. Natl. Acad. Sci. (USA)* **70,** 2853–2859.

Olson, J. S. (1958). Rates of succession and soil changes on southern Lake Michigan sand dunes. *Bot. Gaz.* **119,** 125–170.

Olson, J. S. (1963a). Energy storage and the balance of producers and decomposers in ecological systems. *Ecology* **44,** 322–331.

Olson, J. S. (1963b). Analog computer models for movement of nuclides through ecosystems. In *Radioecology,* V. Shultz and A. W. Klement, Eds., Reinhold, New York. 746 pp.; pp. 121–126.

Olson, J. S. (1964). Gross and net production of terrestrial vegetation. In *British Ecology Society Jubilee Symposium Supplement.* Blackwell Scientific Publications, Oxford. 244 pp.; pp. 99–118.

O'Manique, J. (1969). *Energy in Evolution,* Humanities Press, New York.

O'Neill, R. V. (1969). Indirect estimation of energy fluxes in animal food webs. *J. Theor. Biol.* **22,** 284–290.

O'Neill, R. V. (1971). Examples of ecological transfer matrices. Oak Ridge National Lab. ORNL-IBP-71-3. 26 pp.

O'Neill, R. V. (1976). Ecosystem persistence and heterotrophic regulation. *Ecology* **57,** 1244–1253.

O'Neill, R. V. and O. W. Burke (1971). *A Simple Systems Model for DDT and DDE Movement in the Human Food Chain,* Oak Ridge National Laboratories Ecology Sciences Division, Publication No. 415, ORNL-IBP-71-9. 18 pp.

O'Neill, R. V. and J. M. Giddings (1979). Population interactions and ecosystem function: Phytoplankton competition and community production. In *Systems Analysis of Ecosystems,* G. S. Innis and R. V. O'Neill, Eds., International Cooperative Publishing House, Fairland, MD. pp. 103–123.

O'Neill, R. V., N. Ferguson, and J. A. Watts (1977). *A Bibliography of Mathematical Modeling in Ecology,* Oak Ridge National Laboratories, EDFB/IBP-75/5. 77 pp.

Onsager, L. (1931). Reciprocal relations in irreversible processes. *Phys. Rev.* **37,** 405; **38,** 2265.

Oparin, A. I. (1962). *Life, Its Nature, Origin, and Development,* Academic Press, New York.

Orians, G. H. (1975). Diversity, stability and maturity in natural ecosystems. In *Unifying Concepts in Ecology,* W. H. Van Dobben and R. H. Lowe-McConnell, Eds., Center for Agricultural Publishing and Documentation, Wageningen, Netherlands. pp. 139–150.

Orians, G. H. and N. E. Pearson (1979). On the Theory of Central Place Foraging. In *Analysis of Ecological Systems,* D. J. Horn, R. D. Mitchell and G. R. Stairs, Eds., Ohio State University Press. pp. 155–177.

Oster, G. F. and D. Auslander (1971). Topological representations of thermodynamics systems. I. Basic Concepts. *J. Franklin Inst.* **292,** 1–16.

Oster, G. F. and C. A. Desoer (1971). Tellegren's theorem and thermodynamic inequalities. *J. Theor. Biol.* **32,** 219–241.

Oster, G. T. and E. O. Wilson (1978). *Caste and Ecology in the Social Insects,* Princeton University Press, Princeton, NJ.

Osterhelt, D. and W. Stoekenius (1973). Functions of a new photoreceptor membrane. *Proc. Natl. Acad. Sciences* **70,** 2853–2857.

Ostwald, W. (1892). Lehrbadr der allgemeinen. *Chemie* **2,** 37.

Ostwald, W. (1907). The modern theory of energetics. *Monist* **17,** 511.

Oswald, W. J., H. B. Gotaas, H. F. Ludwig, and V. Lynch (1953). Algal symbiosis in oxidation ponds. *Sewage Ind. Wastes* **25,** 684–691.

Overman, A. R. (1975). Effluent irrigation as a physiochemical hydrodynamic problem. *Fla. Sci.* **38,** 207–214.

Overton, W. S. (1975). The ecosystem modeling approach in the coniferous forest biome, in *Systems Analysis and Simulations in Ecology,* Vol. 3, B. Patten, Ed., Academic Press, New York. pp. 117–138.

Ozmidov, R. V. (1965). Energy distribution between oceanic motions of different scales. *Fisika Atmosfer. Okean. Izv. Akad. Nauk* USSR **1,** 439–448. English Translation, American Geophysical Union. *Izvestiya, Atmospheric and Oceanic Physics* **1,** 257–261.

Padoch, C. (1972). An analog simulation of a simplified model of a New Guinea people, their pigs, ritual feasts and warfare. Report to Atomic Energy Commission, Contract At-(40-1)-4156, Environmental Engineering, University of Florida. pp. 649–658.

Paine, R. T. (1966). Food web complexity and species diversity. *Am. Natur.* **100,** 65–75.

Palmer, J. D. (1976). *An Introduction to Biological Rhythms,* Academic Press, New York. 375 pp.

Pamatmat, M. M. (1982). Heat production by sediment: ecological significance. *Science* **215,** 395–397.

Paperna, I. (1964). Metazoan parasite fauna of Israel inland water fishes. *Bamidgeh* **16,** 3–66.

Pareto, V. (1897). Cours de L'economie Politique. Paris.

Pareto, V. (1927, 1971). *Manual of Political Economy* (translated by A. Schwier), Augustus M. Kelley, Macmillan, New York.

Park, C. F. and M. C. Freeman (1975). *Earthbound, Minerals, Energy and Man's Future,* Freeman, San Francisco. 279 pp.

Park, R. A. (1978). *A Model for Simulating Lake Ecosystems,* Report No. 3, Center for Ecological Modeling, Rensselaer Polytechnic Institute. Troy, NY. 19 pp.

Parker, R. A. (1974). Empirical functions relating metabolic processes in aquatic systems to environmental variables. *J. Fish Res. Board Can.* **31,** 1550–1552.

Parker, R. A. (1975). The influence of environmental driving variables on the dynamics of an aquatic ecosystem model. *Ver. Internatl. Soc. Limnol.* **19,** 47–55.

Parker, R. A. (1976). The influence of eddy diffusion and advection in plankton population systems. *Internatl. J. Syst. Sci.* **7,** 957–962.

Parrish, W. (1933). *An Outline of Technocracy.* Rinehart, New York.

Parsons, T. R. and B. Harrison (1981). Energy utilization and evolution. *J. Social Biol. Struct.* **4,** 1–5.

Pasquill, F. (1974). *Atmospheric Diffusion,* 2nd ed., Wiley, New York. 429 pp.

Passet, R. (1979). *L'Economique et le Vivant,* Pagot, Paris. 287 pp.

Paterson, S. S. (1962). Der CVP-Index als. ausdruck fur forstliche produktionspotentiate. In *Die Staffproduction der Pflanzondecke,* H. Lieth, Ed., Fischer-Verlag, Stuttgart. Chapter 2, pp. 14–15.

Pattee, H. H. (1966). Physical theories, automata and the origin of life. In *Natural Automata and Useful Simulations.* Macmillan, London. pp. 73–108.

Pattee, H. H., Ed. (1973). *Hierarchy Theory: The Challenge of Complex Systems,* Brazillier, New York. 158 pp.

Pattee, H. H., E. A. Eddsock, L. Fein, and A. B. Callahan (1966). *Natural Automata and Useful Simulations,* Spartan Books, Washington, DC. 204 pp.

Patten, B. C. (1961). Competitive exclusion. *Science* **134,** 1599–1601.

Patten, B. C. (1964). Effects of radiation stress on interspecific competition. *Oak Ridge Natl. Lab.* ORNL-**3697,** 104–108.

Patten, B. C. (1965). Community organization and energy relationships in plankton. Oak Ridge National Laboratory. ORNL-3634 UC-48 Biology and Medicine TID-4500.

Patten, B. C. (1966). Biocoenotic processes in an estuarine phytoplankton community. *Oak Ridge Natl. Lab.* ORNL-**3948,** 97 pp.

Patten, B. C. (1968). Mathematical models of plankton production. *Internatl. Revue ges. Hydrobiol.* **53** (3), 357–408.

Patten, B. (1971–1976). *Systems Analysis and Simulation in Ecology,* Vols. 1–4. Academic Press, New York.

Patten, B. C. (1972). Variable-structure aspects of ecosystems. In *Theory and Application of Variable Structure Systems,* R. R. Mohler and A. Ruberti, Eds., Academic Press, New York. pp. 213–232.

Patten, B. C. (1975). Ecosystem linearization: an evolutionary problem. *Am. Natur.* **109,** 529.

Patten, B. C. (1978). Systems approach to the concept of environment. *Ohio J. Sci.* **78** (4), 206–222.

Patten, B. C. and J. T. Finn (1979). Systems approach to continental shelf ecosystems. In *Theoretical Systems Ecology,* E. Halfon, Ed., Academic Press, New York. pp. 184–210.

Patton, D. R. (1975). A diversity index for quantifying habitat edges. *Wildlife Soc. Bull.* **3,** 171–173.

Paulik, G. J. (1967). Digital simulation of natural animal communities. In *Pollution and Marine Ecology,* T. A. Olson and F. J. Burgess, Eds., Interscience, New York. 364 pp.; pp. 67–88.

Paulik, G. J. and J. W. Grennough, Jr. (1966). Management analysis for a salmon resource system. In *Systems Analysis in Ecology,* K. E. F. Watt, Ed., Academic Press, New York. 269 pp., Chapter 9, pp. 215–267.

Pavlidis, T. (1973). *Biological Oscillators, Their Mathematical Analysis,* Academic Press, New York.

Paynter, H. M. (1960). *Analysis and Design of Engineering Systems,* MIT Press, Cambridge, MA.

Pearl, H. W. and L. A. Mackenzie (1977). A comparative study of the diurnal carbon fixation patterns of nannoplankton and net production. *Limnol. Oceanogr.* **22,** 732–738.

Pearl, R. and L. J. Reed (1920). On the rate of growth of the population of the United States since 1790 and its mathematical presentation. *Proc. Natl. Acad. Sci. (USA)* **6,** 275–288.

Petersen, B. J. (1980). Aquatic primary productivity and the ^{14}C-CO_2 method. A history of the productivity problem. Ann. Rev. of Ecology and Systematics. 11, 359–386.

Petersen, C. G. J. (1918). The sea bottom and its production of fishfood. A survey of the work done in connection with the evaluation of the Danish Waters from 1883 to 1917. *Rep. Danish Biol. Stat.* **25,** 1–62.

Peterson, L. and H. Von Foerster (1971). Cybernetics of taxation, the optimization of economic participation. *J. Cybernet.* **1,** 5–22.

Phelps, E. B. (1944). *Stream Sanitation,* Wiley, New York. 276 pp.

Phillips, J. N., Jr., and J. Myers (1953). Measurement of algal growth under controlled steady-state conditions. Growth rate of Chlorella in a flashing light. *Plant Physiol.* **29** (2), 148–152, 152, 161.

Pielou, E. C. (1975). *Ecological Diversity.* Wiley, New York.

Pielou, E. C. (1977). *Mathematical Ecology,* 2nd ed., Wiley, New York. 384 pp.

Pielou, S. C. (1979). *Biogeography,* Wiley, New York. 352 pp.

Pinder, J. E., III, J. G. Wiener, and M. H. Smith (1978). The weibull distribution: a new method of summarizing survivorship. *Ecology* **59,** 175–179.

Pippenger, N. (1978). Complexity theory. *Sci. Am.* (June), 114–124.

Pitts, W. and W. S. McCulloch (1943). A logical calculation of the ideas immersed in nervous activity. *Bull. Math. Biophys.* **5,** 113–133.

Platt, T. (1972). Local phytoplankton abundance and turbulence. *Deep Sea Res.* **19,** 183–187.

Platt, T. and K. L. Denman (1975). Spectral analysis in ecology. *Ann. Rev. Ecol. Syst.* **6,** 189–210.

Platt, T. and C. L. Gallegos (1980). Modelling Primary Production. In *Brookhaven Symposium No. 31,* P. G. Falkowski, Ed., Plenum Press, New York. pp. 339–361.

Poole, R. W. (1974). *An Introduction to Quantitative Ecology,* McGraw-Hill, New York. 332 pp.

Porteous, J. D. (1977). *Environment and Behavior,* Addison-Wesley, Reading, MA.

Poston, T. and I. Stewart (1978). *Catastrophic Theory and Its Application,* Pitman, Boston.

Preston, F. W. (1948). The commonness and rarity of species. *Ecology* **29,** 254–283.

Preston, F. W. (1962). The canonical distribution of commonness and rarity 1. *Ecology* **43,** 185–215; 410–432.

Preston, F. W. (1980). Non-canonical distributions of commonness and rarity. *Ecology* **61,** 88–97.

Prigogine, I. (1947). *Study of Thermodynamics of Irreversible Processes,* 3rd ed., Wiley, New York.

Prigogine, I. (1955). *Introduction to Thermodynamics of Irreversible Processes,* 2nd ed., Interscience, New York. 119 pp.

Prigogine, I. (1978a). Time structure and fluctuations. *Science* **201,** 777–785.

Prigogine, I. (1978b). *From Being to Becoming,* Freeman, San Francisco.

Prigogine, I., and J. M. Wiaume (1946). Biology et thermodynamique des phenomenes irreversibles. *Experientia* **2**, 451–453.

Prigogine, I., P. Allen, and R. Herman (1977). Long term trends and evolution of complexity. pp. 1–63 in *Goals in a Global Community*, Vol. I, E. Laslo and J. Bierman, Eds., Pergamon Press. 333 pp.

Quastler, H. (1953). *Information Theory in Biology*, University of Illinois Press, Urbana.

Quastler, H. (1959). Information theory of biological integration. *Am. Natur.* **93**, 245–254.

Quinlan, A. V. (1978). Thermal sensitivity of Michaelis–Menten kinetics as a function of substrate concentration. *J. Franklin Inst.* **310**, 325–342.

Quinlan, A. V. (1981). The thermal sensitivity of generic Michaelis–Menten processes without catalyst denaturation or inhibition. *J. Therm. Biol.* **6**, 103–114.

Quinlan, A. V. and H. M. Paynter (1976). Some simple non linear dynamic models of interacting element cycles in aquatic ecosystems. *J. Dyn. Syst. Meas. Control* **98**, 1–14.

Rabinowitch, E. T. (1945). *Photosynthesis*, Vol. I, Interscience, New York. pp. 1–603.

Rabinowitch, E. T. (1951). *Photosynthesis*, Vol. II, Interscience, New York. Part 1, pp. 603–1208.

Rabinowitch, E. T. (1956). *Photosynthesis*, Vol. II, Interscience, New York. Part 2, pp. 1208–2087.

Ragotzkie, R. A. (1959). Plankton productivity in estuarine waters of Georgia. *Publ. Inst. Mar. Sci. Univ. Tex.* **6**, 146–158.

Raines, G. E., S. G. Bloom, P. A. McKee, and J. C. Bell (1971). Mathematical simulation of sea otter population dynamics on Amchitka Island, Alaska. *Bioscience* **21**, 686–691.

Rainey, R. H. (1967). Natural displacement of pollution from the Great Lakes. *Science* **155**, 1242–1243.

Ramsey, F. (1976). Preliminary model and metabolism study at the Anclote Power Plant. Course Report ENV 490, Department of Environmental Engineering Science, University of Florida, Gainesville.

Randerson, P. F. (1979). A simulation model of salt marsh development and plant ecology. In *Estuarine and Coastal Land Reclamation and Water Storage*, B. Knights and A. J. Phillips, Eds., 247 pp.; pp. 48–67. Saxon House, Teakfield, Westmead, Farnborough, England.

Ransom, R. (1981). *Computers and Embryos*, Wiley, New York. 212 pp.

Rappaport, R. A. (1971). The flow of energy in an agricultural society. *Sci. Am.* **224**, 116–133.

Rapport, D. J. and J. E. Turner (1977). Economic models in ecology. *Science* **195**, 367–373.

Rashevsky, N. (1938, 1960). *Mathematical Biophysics*, 3rd rev. ed., Dover, New York.

Rashevsky, N. (1961). *Mathematical Principles in Biology and Their Applications*, Thomas, Springfield, IL. 128 pp.

Rashevsky, N. (1968). *Looking at History Through Mathematics*. MIT Press, Cambridge, MA. 199 pp.

Raston, S. (1962). On biological growth. *Bull. Math. Biophys.* **24**, 369–373.

Raunkier, C. (1934). *The Life Form of Plants and Statistical Plant Geography*. Collected papers translated by H. Gilbert-Carter, Clarendon Press, Oxford.

Rawson, D. S. (1953). The bottom fauna of Great Stone Lake. *J. Fish. Res. Board Can.* **10**, 486–518.

Reckshow, K. H. (1979). Empirical lake models for phosphorus: Development, applications, limitations and uncertainty. In *Perspectives on Lake Ecosystem Modelling*, D. Scovia and A. Robertson, Eds., Ann Arbor Science, Ann Arbor, MI. pp. 193–221.

Redfield, A. C. and E. S. Deevey (1952). Temporal Sequences and Biotic Succession, Chap. 4, in *Marine Fouling and Its Prevention*, Woods Hole Oceanographic Institute, U.S. Naval Institute, Annapolis, MD. pp. 42–47.

Regan, E. J. (1977). The natural energy basis for soils and urban growth in Florida. Master's thesis. Department of Environmental Engineering Science, University of Florida, Gainesville. 176 pp.

Reichle, D. E., J. F. Franklin, and D. W. Goodall (1975a). *Productivity of World Ecosystems*, National Academy of Sciences. 166 pp.

Reichle, D. E., D. V. O'Neill, and W. F. Hairs (1975b). Principles of energy and material exchange in ecosystems. In *Unifying Concepts in Ecology*, J. R. H. Love-McConnell and W. H. Van Dobben, Eds., W. Junk, Hague, Netherlands. pp. 27–43.

Reiners, W. A. (1973). A summary of the world carbon cycle and recommendations for critical research. In *Carbon and the Biosphere*, G. M. Woodwell and E. V. Pecan, Eds., Division of Technical Information, U.S. Atomic Energy Commission. pp. 368–382.

Renn, C. E. (1937). Bacteria and the phosphorus cycle in the sea. *Biol. Bull.* **72**, 190–195.

Rescigno, A. (1972). The struggle for life III. Predator prey chain. *Bull. of Math. Biophysics* **34**, 521–522.

Rescigno, A. (1977). The struggle for life. IV. Two predators for a prey. *Bull. Math. Biol.* **39**, 179–185.

Rich, L. G. (1973). *Environmental Systems Engineering*, McGraw-Hill, New York. 448 pp.

Richards, B. N. (1974). *An Introduction to the Soil Ecosystem*, Longmans, London. 266 pp.

Richards, D. W. (1937). The cause of population growth and the size of seeding. *Growth* **1**, 217–227.

Richardson, H. W. (1973). *Regional Growth Theory*, Macmillan, New York.

Richardson, H. W. (1973). *The Economics of Urban Size*, Lexington Press, Lexington, MA.

Richardson, J. R. and H. T. Odum (1981). Power and a pulsing production model, in *Energy and Ecological Modelling*, W. J. Mitsch, R. W. Bosserman, and J. M. Klopatek, Eds., Elsevier, Amsterdam, Neth. pp 641–648.

Richardson, L. F. (1960a). *Arms and Insecurity*, Stevens, London.

Richardson, L. F. (1960b). *Statistics of Deadly Quarrels*, Quadrangle Books, Chicago.

Richey, J. E. (1970). Role of disordering energy in ecosystems. Masters thesis. Dept. Environmental Science and Engineering, University of North Carolina, Chapel Hill.

Richey, J. E., R. C. Wissmar, A. H. Devol, G. E. Likens, J. S. Eaton, R. G. Wetzel, W. E. Odum, N. M. Johnson, O. L. Loucks, R. T. Prentki, and P. H. Rich (1978). Carbon flow in four lake ecosystems: a structural approach. *Science* **202**, 1183–1186.

Richman, S. (1958). The transformation of energy by *Daphnia pulex*. *Ecol. Monogr.* **28**, 273–291.

Ricker, W. E. (1954). Stock and recruitment. *J. Fish Res. Bd. Canada* **11**, 559–623.

Riehl, H. (1979). *Climate and Weather in the Tropics*, Academic Press, New York. 611 pp.

Rifkin, J. (1980). *Entropy: A New World View*, Viking Press, New York. 305 pp.

Riley, G. A. (1946). Factors controlling phytoplankton populations on Georges Bank. *J. Mar. Res.* **6**, 54–73.

Riley, G. A. (1947). A theoretical analysis of the zooplankton population of Georges Bank. *J. Mar. Res.* **6**, 104–113.

Riley, G. A., H. Stommel, and D. W. Bumpus (1949). Quantitative ecology of the plankton of the western north Atlantic. *Bull. Bingham Oceanogr. Coll.* **12**, 160.

Rinaldi, S. (1970). High frequency optimal periodic processes. *IEEE Transact. Automat. Control* **AC-15** (6), 671–673.

Roberts, D. V. (1977). *Enzyme Kinetics*. Cambridge University Press, London. 326 pp.

Roberts, F. S. (1976a). Structural analysis of energy systems. In *Energy, Mathematics and Models*, F. S. Roberts, Ed., SIAM Institute for Mathematics and Society, pp. 84–100.

Roberts, F. S. (1976b). *Discrete Mathematical Numbers with Application to Social, Biological, and Environmental Problems*, Prentice-Hall, Englewood Cliffs, NJ.

Robichaud, L. P. A., M. Boisvert, and J. Robert (1962). *Signal Flow Graphs and Applications,* Prentice-Hall, Englewood Cliffs, NJ. 214 pp.

Roff, D. A. (1974). Analysis of a population model demonstrating the importance of dispersal in a heterogeneous environment. *Oecologia* **15**, 259–277.

Rogers, P. L. (1976). Analog/hybrid computation in biochemical engineering. *Adv. Biochem. Eng.* **4**, 125–153.

Rose, A. H. (1967). *Thermobiology*, Academic Press, New York.

Rosen, R. (1967). *Optimality Principles in Biology*, Plenum Press, New York. 198 pp.

Rosen, R. (1970). *Dynamical System Theory in Biology*, Wiley-Interscience, New York. 302 pp.

Rosenzweig, M. L. (1968). Net primary productivity of terrestrial communities: Prediction from climatological data. *Am. Naturalist* **102**, 57–74.

Rosinsky, N. (1977). Drosophila embryology. The dynamics of evolution. *Fusion Energy Found. Newsl.* **2**, 54–59.

Ross, R. (1911). *The Prevention of Malaria*, 2nd ed., Murray, London.

Royama, T. (1971). A comparative study of models for predation and parasitism. *Rep. Pop. Ecol.*, Supp. No. 1.

Rubey, W. W. (1951). Geologic history of sea water. *Geol. Soc. Am. Bull.* **62**, 1111–1147.

Russell, C. S. (1975). *Ecological Modeling*, Resources for the Future, Johns Hopkins University Press, Baltimore, MD. 394 pp.

Rutledge, R. W., B. L. Basore, and R. J. Mulholland (1976). Ecological stability: An information theory viewpoint. *J. Theor. Biol.* **57**, 355–371.

Rykiel, E. J. (1977). The Okeefenokee Swamp watershed, water balance and nutrient budgets. Ph.D. dissertation. University of Georgia, Athens. 246 pp.

Saaty, T. L. (1968). *Mathematical Models of Arms Control and Disarmament*, Wiley, New York.

Sacher, G. A. and E. F. Staffeldt (1974). Relation of gestation time to brain weight for placental mammals. Implications for the theory of vertebrate growth. *Am. Natur.* **108**, 593–615.

Salisbury, F. B. and C. W. Ross (1978). *Plant Physiology*, 2nd ed., Wadsworth, Belmont, CA.

Samuelsson, G., G. Oquist, and P. Halldal (1978). The variable chlorophyll *a* fluorescence as a measure of photosynthetic capacity in algae. Mitteilungen No. 21. *Internatl. Assoc. Limnol.* **21**, 207–215.

Saunders, H. L. (1968). Marine benthic diversity: A comparative study. *Am. Natur.* **102**, 243–282.

Savageau, M. A. (1976). *Biochemical Systems Analysis*, Addison-Wesley, Reading, MA. 379 pp.

Scavia, D., J. A. Bloomfield, J. S. Fisher, J. Nagy, and R. A. Park. (1974). Documentation of CLEAN X: A generalized model for simulating the open water ecosystems of lakes. *Simulation* **23** (2), 51–56.

Schaefer, G. (1972). *Kybernetik und Biologie*, Metzlersche, Stuttgart. 215 pp.

Schaefer, M. B. (1954). Some aspects of the dynamics of populations important to the management of the commercial marine fisheries. *Inter-Am. Trop. Tuna Commun. Bull.* **1** (2), 27–56.

Schaefer, M. B. (1957). A study of the dynamics of the fishery for yellowfin tuna in the eastern tropical Pacific Ocean. *Bull. Inter.-Am. Trop. Tuna Commun.* **2**, 245–285.

Scheiddeger, A. E. (1970). *Theoretical Geomorphology*, Springer-Verlag, Berlin.

Schlegel, R. (1961). *Time and the Physical World*, Michigan State University. 211 pp.

Schmidt, A. and H. L. List (1962). *Material and Energy Balances*, Prentice-Hall, Englewood Cliffs, NJ.

Schmidt, O. H. (1960). Biophysical and mathematical models of circadian rhythms. *Cold Spring Harbor Symp. Quant. Biol.* **25**, 207–216.

Schnakenberg, J. (1977, 1981). *Thermodynamic Network Analysis of Biological Systems*, 2nd Ed., Springer-Verlag, Berlin.

Scholander, P. F., W. W. Flagg, V. Walters, and L. Irving (1953). Climatic adaptation in arctic and tropical poikilotherms. *Physiol. Zool.* **26**, 67.

Schomer, N. S. (1976). Systems models and simulations of the recovery of Escambia Bay. M.S. thesis. Gamma College, University of West Florida, Pensacola.

Schrodinger, E. (1947). *What Is Life, Mind and Matter*, Cambridge University Press, New York.

Schultz, V., L. L. Eberhardt, J. M. Thomas, and M. I. Cochran (1976). *A Bibliography of Quantitative Ecology*, Dowden, Hutchinson and Ross, Stroudsburg, PA. 362 pp.

Schultze, C. L. (1971). *National Income Analysis*, 3rd ed., Prentice-Hall, Englewood Cliffs, NJ.

Scott, H. (1933). *Introduction to Technocracy*, J. Day, New York.

Searle, S. R. (1966). *Matrix Algebra for the Biological Sciences*, Wiley, New York. 298 pp.

Sebetich, M. J. (1975). Phosphorus kinetics of freshwater microcosms. *Ecology* **56**, 1262–1280.

Sedlik, B. R. (1976). A computer simulation of the spread of the exotic tree species, *Melaleuca quinquenervia* (Cav.) Blake, in South Florida. M.S. thesis. Department of Industrial and Systems Engineering, University of Florida. 124 pp.

Seip, K. L. (1980). A computational model for growth and harvesting of the Maine alga *Ascophyllum nodosum*. *Ecol. Model.* **8**, 189–199.

Sell, M. G. (1977). Modeling the response of mangrove ecosystems to herbicide spraying, hurricanes, nutrient enrichment and economic development. Ph.D. dissertation. Department of Environmental Engineering Science, University of Florida, Gainesville. 389 pp.

Sergin, V. Ya (1979). Numerical modeling of the Glaciers-Ocean-Atmosphere Global System. *J. of Geophysical Research* **84**, 3191–3204.

Sergin, V. Ya. (1980). Origin and mechanism of large scale climatic oscillations. *Science* **209**, 1477–1482.

Servizi, J. A., A. M. Asce, and R. H. Bogan (1963). Free energy as a parameter in biological treatment. *Proc. Am. Soc. Civil Eng.* **3539**, 17–40.

Shabman, L. A. and S. S. Batie (1978). Economic value of natural coastal wetlands. A critique. *Coastal Management J.* **4**, 231–247.

Shannon, C. E. and W. Weaver (1949). *Mathematical Theory of Communication*, University of Illinois Press, Urbana.

Sheldon, R. W., A. Prakash, and W. H. Sutcliffe, Jr. (1972). The size distribution of particles in the ocean. *Limnol. Oceanogr.* **17**, 327–340.

Sheppard, C. W. (1962). *Basic Principles of the Tracer Method*, Wiley, New York.

Shimazu, Y. (1978). *Ecological World View and Global National and Regional Planning* toward the 21st century, preprint Second International Congress of Ecology, Jerusalem. 58 pp.

Shimazu, Y. and T. Urabe (1967). Some numerical experiments on the evolution of the terrestrial atmosphere and hydrosphere. *J. Phys. Earth* **15**, 1–17.

Shimazu, Y., K. Sugiyama, T. Kojima, and E. Tomida (1972). Some problems in ecology oriented environmentology. Terrestrial Environmentology II. *J. Earth Sci. Nagoyo Univ.* **20**, 31–89.

Shinozaki, K. and T. Kira (1956). Intraspecific completion among higher plants. VII. Logistic theory of the C-D effect. *J. Inst. Polytech. Osaka City Univ.* **DY**, 35–72.

Shinozaki, K. and T. Kira (1961). The C-D rule, its theory and practical uses. Intraspecific competition among higher plants. *J. Biol. Osaka City Univ.* **12**, 69–82.

Shoup, C. (1929). The respiration of luminous bacteria and the effect of oxygen tension upon oxygen consumption. *J. Gen. Physiol.* **13**, 27–45.

Shugart, H. H. and J. M. Hett (1975). Succession: Similarities of species turnover rates. *Science* **180**, 1379–1381.

Shugart, H. H., Jr. and D. C. West (1977). Development of an Appalachian deciduous forest succession model and its application to assessment of the impact of the chestnut blight. *J. Envir. Manage.* **5**, 161–179.

Shugart, H. H. and D. C. West (1980). Forest succession models. *Bioscience* **30**, 308–313.

Shugart, H. H., J. M. Klopatek, and W. R. Emanuel (1977). Developments in ecology in systems analysis and land use planning. Environmental Sciences Division Publication No. 1074. Oak Ridge National Laboratories. TM-6009. 43 pp.

Shugart, H. H., R. A. Goldstein, R. V. O'Neill, and J. B. Markin (1974). A terrestrial ecosystem energy model for forests. *Ecol. Plant* **9** (3), 231–264.

Shugart, H. H., D. E. Reichle, N. T. Edwards, and J. R. Kercher (1976). A model of calcium-cycling in an east Tennessee *Liriodendron* forest: Model structure, parameters and frequency response analysis. *Ecology* **57**, 99–109.

Sillén, L. G. (1961). The physical chemistry of sea water. In *Oceanography*, M. Sears, Ed., American Association for the Advancement of Science, Publication No. 67. 654 pp.; pp. 549–581.

Sillén, L. G. (1967). The ocean as a chemical system. *Science* **156**, 1189–1197.

Silvert, W. and T. Platt (1978). Energy flux in the pelagic ecosystem. A time-dependent equation. *Limnol. Oceanogr.* **23**, 813–816.

Simberloff, P. S. and E. O. Wilson (1970). *Ecology* **51**, 934–937.

Simpson, E. H. (1949). Measurement of diversity. *Nature* **163**, 688.

Singh, J. (1966). *Great Ideas in Information Theory, Language, and Cybernetics,* Dover, New York. 338 pp.

Sinha, S. K. and B. K. Kale (1980). *Life Testing and Reliability Estimation.* Halsted (Wiley), New York. 196 pp.

Sipe, N. G. (1978a). A historic and current energy analysis of Florida. M.A. thesis. Department of Urban and Regional Planning, University of Florida, Gainesville.

Sipe, N. G. (1978b). An overview energy analysis of Sarasota County. Urban and Regional Planning Publication Series No. 12. Department of Urban and Regional Planning, University of Florida, Gainesville. 53 pp.

Sitaramiah, P. (1961). Studies on the physiological ecology of a tropical freshwater pond community. Ph.D. dissertation. Sri Venkateswara University, Tirupati, India. 186 pp.

Sjöberg, S. (1977). Are pelagic systems inherently unstable? A model study. *Ecol. Model.* **3**, 17–37.

Sjöberg, S. (1980). Zooplankton feeding and queueing theory. *Ecol. Model.* **10**, 215–225.

Slesser, M. (1978). *Energy in the Economy,* Macmillan, London. 162 pp.

Slobodkin, L. B. (1959). Energetics in daphnia populations. *Ecology* **40**, 232–243.

Slobodkin, L. B. (1960). Ecological energy relationships at the population level. *Am. Natur.* **94** (876), 213, 236.

Slobodkin, L. B. (1961). Preliminary ideas for a predictive theory of ecology. *Am. Natur.* **95** (882), 147–153.

Slobodkin, L. B. (1962a). Predation and efficiency in laboratory populations. In *Exploitation of Natural Animal Populations,* E. D. LeCren and M. W. Holgate, Eds., Wiley, New York, 399 pp.; pp. 223–241.

Slobodkin, L. B. (1962b). *Growth and Regulations of Animal Populations,* Holt, Rinehart, and Winston, New York.

Slobodkin, L. B. (1962c). Energy in animal ecology. In *Advances in Ecological Research,* Vol. 1. Academic Press, New York. pp. 69–101.

Smith, E. L. (1936). Photosynthesis in relation to light and carbon dioxide. *Proc. Natl. Acad. Sci. (USA)* **22**, 504–511.

Smith, F. E. (1952). Experimental models in population dynamics: A critique. *Ecology* **33**, 441–450.

Smith, F. E. (1954). Quantitative aspects of population growth. In *Dynamics of Growth Processes.* E. J. Boell, ed., Princeton University Press, Princeton, NJ. pp. 277–294.

Smith, F. E. (1969). Effects of enrichment in mathematical models. In *Eutrophication,* National Academy of Sciences, Washington, DC. pp. 631–645.

Smith, W. H. B. (1976). Productivity measurements and simulation models of a shallow estuarine ecosystem receiving a thermal plume at Crystal River, Florida. Ph.D. dissertation. Department of Environmental Engineering, University of Florida, Gainesville.

Smuts, J. C. (1926). *Holism and Evolution,* Viking Press, New York. 362 pp.

Socharlan, A. (1967). Preliminary stand table of *Anthocephalus cadamba* MiQ. *Rimbu Indonesia* **12,** 37–46.

Soddy, F. (1912). *Matter and Energy,* Oxford University Press, London.

Soddy, F. (1922). *Cartesian Economics and the Bearing of Physical Science upon State Stewardship,* Hendersons, London.

Soddy, F. (1933). *Wealth, Virtual Wealth, and Debt,* Dutton, New York.

Soemarwoto, O. (1974). Rural ecology and development in Java. In *Unifying Concepts in Ecology,* W. H. Van Dobben and R. H. Lowe-McConnell, Eds., Center for Agricultural Publishing and Documentation, Wageningen, Netherlands.

Soemarwoto, O. (1976). The Japanese home garden as an integrated agroecosystem. In *Science for Better Environment,* HSEC, Science Council of Japan. pp. 193–197.

Solbrig, O. T. and G. H. Orians (1977). The adaptive characteristics of desert plants. *American Scientist* **65,** 412–421.

Sollins, P. (1970). Measurement and simulation of oxygen flows and storages in a laboratory blue–green algal mat ecosystem. M.A. thesis. Department of Zoology, University of North Carolina, Chapel Hill. 185 pp.

Sparrow, A. H. and A. S. Nauman (1976). Evolution of genome size by DNA doubling. *Science* **192,** 524–528.

Spetner, L. M. (1964). Natural selection: an information–transmission mechanism for evolution. *J. Theor. Biol.* **7,** 412–439.

Stafford, H. A., Jr. (1963). The functional bases of small towns. *Econ. Geogr.* **39,** 165–175.

Stanhill, C. (1974). Energy and agriculture. *Agroecosystems,* **1,** 205–217.

Stanhill, G. (1977). An urban agro-ecosystem: The example of 19th century Paris. *Agroecosystems* **3,** 269–284.

Steele, J. H. (1962). Environmental control of photosynthesis in the sea. *Limnol. Oceanogr.* **7,** 137–150.

Steele, J. H. (1965). Notes on some theoretical problems in production ecology. In *On: Primary Production in Aquatic Environments,* C. R. Goldman, Ed., Memoirs Institute Italiano di Idrobiologie, Vol. 18, Supplement; University of California Press, Berkeley. pp. 385–398.

Steele, J. H. (1974). *Structure of Marine Ecosystems,* Harvard University Press, Cambridge, MA.

Steele, J. H. (1976). Application of theoretical models in ecology. *J. Theor. Biol.* **63,** 443–451.

Steele, J. H. (1977). *Spatial Pattern in Plankton Communities,* NATO Conference Series Section IV, Marine Sciences, Plenum Press, New York.

Steele, J. H. (1979). Some problems in the management of marine resources. *Applied Biology* **4,** 103–140.

Steeman-Nielsen, E. (1963). Productivity, definition, and measurement. In *The Sea,* Vol. II, *Fertility of the Ocean,* M. N. Hill, Ed., Interscience, New York. 554 pp.; pp. 129–164.

Steinitz, C., H. J. Brown, and P. Goodale (1977). Managing suburban growth: A modeling approach. Report to RANN, National Science Foundation. Graduate School of Design, Harvard University, Cambridge, MA.

Steller, D. L. (1976). An energy evaluation of residential development alternatives in mangroves. M.S. thesis. Department of Environmental Engineering Sciences, University of Florida. 124 pp.

Stephan, G. E. (1977). Territorial division: The least-time constraint behind the formation of subnational boundaries. *Science* **196,** 523–524.

Sterndl, J. (1965). *Random Processes and the Growth of Firms, a Study of the Pareto Law,* Griffin, London.

Stesen, R. F., J. B. McGuire, and W. A. Hogan (1977). Simulation of nonlinear reaction diffusion equations. *Bull. Math. Biol.* **39,** 391–396.

Stewart, J. Q. (1947). Empirical mathematical rules concerning the distribution and equilibrium of population. *Geogr. Rev.* **37,** 461–485.

Stoekenius, W. (1978). Bioenergetic mechanisms in *Halobacteria.* In *Energetics and Structure of Halophile Microorganisms,* S. R. Caplan and M. Ginsburg, Eds., Elsevier, North Holland. pp. 185–198.

Stoiber, R. E. and A. Jepsen (1973). Sulfur dioxide contributions to the atmosphere by volcanoes. *Science* **182,** 577–578.

Streeter, H. W. and E. B. Phelps (1925). A study of the pollution and natural purification of the Ohio River. *USPHS Bull.* **146,** 1–75.

Strehler, B. L. (1957). Some energy transduction problems in photosynthesis. In *Rhythmic and Synthetic Processes in Growth,* D. Rudnick, Ed., Princeton University Press, Princeton, NJ. pp. 171–199.

Strehler, B. L. (1960). Fluctuating energy demands as determinants of the death process (a parsimonious theory of the Gompertz function). In Strehler et al. (1960), pp. 309–314.

Strehler, B. L. (1962). *Time, Cells, and Aging,* Academic Press, New York.

Strehler, B. L. and A. S. Mildvan (1960). General theory of mortality and aging. *Science* **132,** 14–21.

Strehler, B. L., J. D. Ebert, H. B. Glass, and N. W. Shock (1960). *The Biology of Aging,* American Institute of Biological Sciences, Washington, DC. 364 pp.

Stremler, F. C. (1977). *Introduction to Communication Systems,* Addison-Wesley, Reading, MA.

Sugayima, K. and Y. Shimazu (1972). Some problems in economy oriented environmentology. Terrestrial environmentology I. *J. Earth Sci. Nagoya Univ.* **20** (1), 1–29.

Sugita, M. (1951). Maximum principle in transient phenomena and its application to biophysics. *Bull. Kobayasi Inst.* **1,** 88.

Sugita, M. (1953). Thermodynamical analysis of life. I. Thermodynamics of transient phenomena. II. On the maximum principle of transient phenomena. III. Mathematical analysis of metabolism. *Phys. Soc. Jap.* **8,** 697–714.

Sugita, M. (1961). Functional analysis of chemical systems *in vivo* using a logical circuit equivalent. *J. Theor. Biol.* **1,** 415–430.

Sugita, M. (1963). Functional analysis of chemical systems *in vivo* using a logical circuit equivalent. II. The idea of a molecular automaton. *J. Theor. Biol.* **4,** 179–192.

Sugita, M. and N. Fukuda (1963). Functional analysis of chemical systems *in vivo* using a logical circuit equivalent. III. Analyses using a digital circuit combined with an analogue computer. *J. Theor. Biol.* **5**, 412–425.

Sugiura, Y. (1953). On the diurnal variation of oxygen content in surface layers of the hydrosphere. *Meteorol. Geophys.* **4**, 79.

Sugiura, Y. (1956). Experimental study on the time variation of oxygen content of natural water. *Meteorol. Geophys.* **7**, 42–48.

Summers, J. K., W. M. Kitchens, H. N. McKellar, and R. F. Dame (1980). A simulation model of estuarine subsystem coupling and carbon exchange with the sea. II. North inlet model structure, output, and validation. *Ecol. Model.* **11**, 111–138.

Sussman, M. V. (1981). Availability (Exergy) Analysis. Milliken House, Lexington, MA.

Suttman, C. E. and G. W. Barrett (1979). Effects of sevin on arthropods in an agricultural and an old field plant community. *Ecology* **60**, 628–641.

Sverdrup, H. U. (1953). On conditions for vernal blooming of phytoplankton. *J. Conserv. Perm. Internatl. Explor. Mer.* **18**, 287–295.

Sverdrup, H. U., M. W. Johnson, and R. H. Fleming (1942, 1946). *The Oceans.* Prentice-Hall, Englewood Cliffs, NJ. 1060 pp.

Swallows, C. and H. T. Odum (1976). In Odum (1976a).

Swartz, C. E. (1973). *Used Math.* Prentice-Hall, Englewood Cliffs, NJ.

Swartzman, C. L. and G. M. Van Dyne (1972). An ecologically based simulation optimization approach to natural resource planning. *Am. Rev. Ecol. Syst.* **3**, 347–398.

Swift, J. (1733, 1937). On poetry, A Rhapsody. In *Swift's Poems*, Vol. II. H. Williams, Ed., Oxford University Press, Oxford, U.K. pp. 639–659.

Swift, M. J., O. W. Heal, and J. M. Anderson (1979). *Decomposition in Terrestrial Ecosystems, Studies in Ecology*, Vol. 5. University of California Press, Berkeley. 372 pp.

Taagepera, R. (1968). Growth curves of empires. *Yearbook Gen. Syst.* **8**, 171–175.

Talbert, N. E. and C. B. Osmond (1976). *Photorespiration in Marine Plants*, Univ. Park Press, Baltimore, Maryland.

Talbot, F. H., B. C. Russell, and G. V. Anderson (1976). Coral reef fish communities, unstable high diversity systems. *Ecol. Monogr* **48**, 425–440.

Tamiya, H. (1951). Some theoretical notes on the kinetics of algal growth. *Bot. Mag. Tokyo* **64**, 167–173.

Tanner, J. T. (1975). The stability and intrinsic growth rates of prey and predator populations. *Ecology* **56**, 855–867.

Tansley, A. G. (1935). The use and abuse of vegetational concepts and terms. *Ecology* **16**, 284–307.

Taub, F. (1968). A biological model of a freshwater community: A gnotobiotic Ecosystem. Contribution No. 280, College of Fisheries, University of Washington. 63 pp.

Taub, F. (1969). A biological model of a freshwater community. A gnotobiotic ecosystem. *Limnol. Oceanogr.* **14**, 136–141.

Taub, F. B. and D. H. McKenzie (1973). Continuous cultures of an alga and its grazer. *Bull. Ecol. Res. Commun.* (Stockholm) **17**, 371–377.

Taylor, L. R. (1961). Aggregation, variance and the mean. *Nature* **189**, 732–735.

Taylor, L. W. (1979). A low order systems model of the United States economy. In *Global and Large Scale Systems Models*, A. V. Balakrishnan and M. Thoma, Eds., Lecture Notes on Control and Information Sciences, Springer-Verlag, Berlin. 232 pp.

Teal, J. M. (1957). Community metabolism in a temperate cold spring. *Ecol. Monogr.* **27**, 283–302.

Teal, J. M. (1959). Energy flow in the salt marsh ecosystem. 1958 Salt Marsh Conference, Sapelo Island, Marine Institute of the University of Georgia. pp. 101–103.

Teal, J. T. (1956). Community metabolism in a temperate cold spring. *Ecol. Monogr.* **27**, 282–302.

Teal, M. J. (1962). Energy flow in the salt marsh ecosystem of Georgia. *Ecology* **43**, 614–624.

Tellengen, B. D. H. (1952). A general network theorem with applications. Phelps Research Report No. 7.

Tenore, K. R. (1977). Food chain pathways in detrital feeding benthic communities: A review, with new observations on sediment resuspension and detrital recycling. In *Ecology of Marine Benthos*, Belle W. Baruch Library in Marine Science, No. 6, University of South Carolina Press, Columbia. pp. 37–54.

Terborgh, J. (1974). Preservation of natural diversity, the problem of extinction prone species. *Bioscience* **24**, 715–722.

Terrell, T. J. (1980). *Introduction to Digital Filters*, Wiley, New York. 222 pp.

Thayer, G. W. (1971). Phytoplankton production and the distribution of nutrients in a shallow unstratified estuarine system near Beaufort, NC. *Chesapeake Sci.* **12**, 240–253.

Thieneman, A. (1928). *Der Sauerstoff im Eutrophen und Oligotrophen Seen*. Die Binnengewasser 4, Stuttgart, Schwizerbartsche Verlagsbuchhandlung. 255 pp.

Thirring, H. (1968). *Energy for Man*, Greenwood Press, New York.

Thom, R. (1970). Topological models in biology. In *Towards a Theoretical Biology*, C. H. Waddington, Ed., Edinburg University Press.

Thom, R. (1975). *Structural Stability and Morphogenesis*, Benjamin, Reading, MA.

Thomann, R. V. (1979). Analysis of PCB in Lake Ontario using a size dependent food chain model. In *Lake Ecosystem Modelling*, D. Scavia and A. Robertson, Eds., Ann Arbor Science, Ann Arbor, MI. pp. 293–320.

Thompson, D. A. (1917, 1942). *On Growth and Form*, 2nd ed. Cambridge University Press, London.

Thornes, J. B. and D. Brunsden (1977). *Geomorphology and Time*, Halsted Press (Wiley), New York.

Thornley, J. H. M. (1976). *Mathematical Models in Plant Physiology, Monographs in Experimental Botany*, Vol. 8, Academic Press, New York. 318 pp.

Thron, C. D. (1972). Structure and kinetic behavior of linear multicompartment systems. *Bull. Math. Biophys.* **34**, 277–291.

Tieszen, L. L., Ed. (1978). *Vegetation and Production Ecology of an Alaskan Arctic Tundra*, Springer-Verlag, New York. 686 pp.

Tillman, D. (1977). Resource competition between planktonic algae and experimental and theoretical approach. *Ecology* **58**, 338–348.

Titman, D. (1976). Ecological competition between algae. Experimental confirmation of resource based competition theory. *Science* **192**, 463–465.

Tomovic, R. (1963). *Sensitivity Analysis of Dynamic Systems*, McGraw-Hill, New York.

Tomovic, R. and M. Vukobratovic (1972). *General Sensitivity Theory,* American Elsevier, New York.

Toogepera, R. (1968). Growth curves of empires. *Yearbook Gen. Syst. Theory* **8,** 171–175.

Totsuka, T. (1963). Theoretical analysis of the relationships between water supply and dry matter production of plant communities. *J. Fac. Sci. Univ. Tokyo Sect. III,* **8** (9), 341–375.

Toynbee, A. (1935–1939). *A Study in History,* Vols. 1–6, Oxford University Press, London.

Tribus, M. (1961). *Thermostatics and Thermodynamics,* Van Nostrand, Princeton, NJ. 641 pp.

Tribus, M. (1969a). *Rational Descriptions, Decisions, and Designs,* Pergamon Press, New York. 478 pp.

Tribus, M. (1979b). Thirty years of information theory. In *The Maximum Entropy Formalism,* R. D. Levine and M. Tribus, Eds., MIT Press, Cambridge, MA. pp. 1–14.

Tribus, M. and E. C. McIrvine (1971). Energy and Information. *Sci. Am.* **225,** 179–188.

Trincher, K. S. (1965). *Biology and Information* (Russian translation). N.Y. Consultants Bureau. 93 pp.

Tsuchiya, H. M., J. F. Drake, J. L. Jost, and A. G. Fredrickson (1972). Predator–prey interactions of *dictyostelum* and *escherichia coli* in continuous culture. *J. Bacteriol.* **110,** 1147–1153.

Tucker, V. A. (1975). Energetic cost of moving about. *Am. Sci.* **63,** 413–419.

Turing, A. M. (1936). On computable numbers with an application to the Entscheidungsproblem. *Proc. Lond. Math. Soc.* (Ser. 2), 42.

Turk, A., J. Turk, J. T. Wittes, and R. E. Witts (1978). *Environmental Science,* Saunders, Philadelphia. 597 pp.

Turnbull, C. M. (1963). The lesson of the pygmies. *Sci. Am.* **208** (1), 28–37.

Tygart, F. J. (1960). *Theory and Process of History,* University of California Press, Berkeley.

Tyler, L. W. (1961). Aggregation, variance and the mean. *Nature,* **189,** 732–735.

Ulanowicz, R. E. (1972). Mass and energy flow in closed ecosystems. *J. Theor. Biol.* **34,** 239–253.

Ulanowicz, R. E. (1981). A unified theory of self-organization, in *Energy and Ecological Modelling,* Elsevier, Amsterdam, Netherlands. pp. 649–652.

Ullman, E. L. and M. F. Cacey (1962). The minimum requirements approach to the urban economic base. *Human Geogr.* **24,** 121–143.

Usher, M. B. (1972). Developments in the Leslie Matrix Model. In *Mathematical Models in Ecology,* J. N. R. Jeffers, Ed., Blackwell, Oxford. pp. 29–60.

Usher, M. B. and M. H. Williamson (Eds.) (1974). *Ecological Stability,* Chapman Hall, London.

Valentinizzi, M. and M. E. Valentinizzi (1963). Information content of chemical structures. *Bull. Math. Biophys.* **24,** 11–28.

Van den Ende, P. (1973). Predator–prey interactions in continuous culture. *Science* **181,** 562–564.

Van der Pol, B. (1934). The nonlinear theory of electronic oscillations. *Proc. Inst. Radio Eng.* **22** (9), 1054–1086.

Van der Vaart, H. R. (1978). Conditions for periodic solutions of Volterra differential systems. *Bull. Math. Biol.* **40,** 133–160.

Van Doorn, J. (1975). *Disequilibrium Economics,* Halsted Press (Wiley), New York. 96 pp.

Van Dyne, G. M. (Ed.) (1969). *Ecosystem Concept in National Resource Management,* Academic Press, New York. 383 pp.

Van Loo, J. T. and J. Kepeke (1972). Simulation of model of sacred cow system of India. Unpublished class report.

Van Raag, H. G. T. and A. E. Lugo (1974). *Man and Environment,* Rotterdam Press, The Hague, Netherlands.

Van Valen, L. (1976). Energy and Evolution: *Evolutionary Theory* **1,** 179–229.

Van Voris, P. (1976). Ecological stability: An ecosystem perspective—Classical and current thought. A review of selected literature. Oak Ridge National Laboratory. ORNL/TM-5517. 36 pp.

Van Voris, P., R. V. O'Neill, W. R. Emanuel, and H. H. Shugart (1980). Functional complexity and ecosystem stability, an experimental approach. *Ecology* **61,** 1352–1360.

Vela, G. R. and J. W. Peterson (1969). Azotobacter cysts: Reactivation by white light after inactivation by ultraviolet radiation. *Science* **166,** 1296–1297.

Verhoff, V. and F. J. Smith (1971). Theoretical analysis of a conserved nutrient ecosystem. *J. Theor. Biol.* **33,** 131–147.

Verhulst, P. F. (1845). Recherches mathematiques sur la loi d'accroisment de la population. *Mem. Acad. Roy. Belg.* **18,** 1–38.

Vernadsky, W. I. (1944). Problems of biogeochemistry. II. The fundamental matter–energy difference between the living and the inert natural bodies of the biosphere. *Transact. Conn. Acad. Arts Sci.* **33,** 483–517.

Vestal, A. G. (1949). *Minimum Areas for Different Vegetations,* University of Illinois Press, Urbana. 129 pp. (Reprinted in *Ill. Biol. Monogr.* **20,** 1–129.)

Vinogradov, M. E. and V. V. Menshutkin (1977). The modeling of open-sea ecosystems. In *The Sea,* Vol. 6, *Marine Modeling,* E. D. Goldberg, I. N. McCave, J. J. O'Brien, and J. H. Steele, Eds., Wiley, New York. pp. 891–921.

Voevodsky, J. (1971). Modeling the dynamics of warfare. In *Cybernetics, Simulation and Conflict Resolution,* Spartan Books, New York. pp. 145–170.

Vogel, S. and K. C. Ewel (1972). An electrical analog of a trophic pyramid. In *A Model Menagerie. Laboratory Studies About Living Systems.* Addison-Wesley, Reading, MA. 104 pp.

Vollenweider, R. A. (1966). Calculation models of photosynthesis depth curves and some implications regarding day rate estimates in primary production measurements. In *Primary Production in Aquatic Environments,* C. R. Goldman, Ed., University of California Press, Berkeley. 464 pp.; pp. 425–457.

Vollenweider, R. A. (1969). Moglich gerten und grenzen elemtarer modelle der stoffbilanz Von Seen. *Arch. Hydrobiol.* **66,** 1–36.

Volterra, V. (1926). Variations and fluctuations in the number of individuals in animal species living together. *Atti. Accad. Nazl. Lince. Mem. Cl. Sci. Fish., Mat. Nat.* **6** (2), 31–113. (English translations, R. N. Chapman, 1931, *Animal Ecology,* McGraw-Hill, New York.)

Volterra, V. (1931). Variations and fluctuations in the number of individuals in animal species living together. In *Animal Ecology,* R. N. Chapman, Ed., McGraw-Hill, New York. Appendix.

Von Bertalanffy, L. (1950). The theory of open systems in physics and biology. *Science* **111,** 23–28.

Von Bertalanffy, L. (1962). General systems theory. *Gen. Syst. Yearbook* **7**, 1–20.

Von Bertalanffy, L. (1968). *General Systems Theory*, Brazillier, New York. 289 pp.

Von Brand, R. and N. W. Rakestraw (1941). Decomposition and regeneration of nitrogenous organic matter in sea water. *Biol. Bull. Woods Hole* **81**, 63–72.

Von Foerster, H. and G. W. Zopf, Eds. (1962). *Principles of Self-Organization*, Pergamon Press, New York.

Von Foerster, Heinz (1970). Molecular etiology, an immodest proposal for semantic clarification, in *Molecular Mechanisms in Memory and Learning*, Plenum Press. pp. 213–252.

Von Foerster, H., P. M. Marx, and W. Amiot (1960). Doomsday: Friday, 13 November, A.D. 2026. *Science* **132**, 1291–1295.

Von Thünen, J. H. (1826, 1842). *Der Isolierte Staat in Beziehung auf Landwirtschaft und Nationalökonomie*, 2nd ed, Hamburg, Germany.

Waggoner, P. E. and G. R. Stephens (1970). Transition probabilities for a forest. *Nature*, **255**, 1160–1161.

Waller, W. T., M. L. Dahlberg, R. E. Sparks, and J. Cairns (1971). A computer simulation of the effects of superimposed mortality due to pollutants on populations of fathead minnows (*Pimephales promelas*). *J. Fish Res. Board Can.* **28**, 1107–1112.

Walsh, J. J. (1975). A spatial simulation model of Peru upwelling ecosystem. *Deep Sea Research* **22**, 201–236.

Walsh, J. J. and R. C. Dugdale (1970). A simulation model of the nitrogen flow in the Peruvian upwelling system. *Invest. Pesquera* **35**, 1–21.

Walter, G. (1953). *The Living Brain*, Duckworth, London.

Walter, W. G. (1951). A machine that learns. *Sci. Am.* **185**, 60–63.

Walters, C. (1971). In *Fundamentals of Ecology*, 3rd ed., E. P. Odum, Ed., Saunders, Philadelphia. pp. 276–292.

Walters, R. A. (1980). A time and depth-dependent model for physical, chemical, and biological cycles in temperate lakes. *Ecol. Model.* **8**, 79–96.

Wang, L. K., D. Vielking, and M. H. Wang (1978). Mathematical models of dissolved oxygen concentration in fresh water. *Ecol. Model.* **5**, 115–123.

Wangersky, P. J. and W. J. Cunningham (1956). On time lags in equilibrium of growth. *Proc. Natl. Acad. Sci. (USA)* **43**, 694–702.

Wangersky, P. J. and W. J. Cunningham (1957a). Time lag in population models. *Cold Spring Harbor Symp. Quant. Biol.* **22**, 329–337.

Wangersky, P. J. and W. J. Cunningham (1957b). Time lag in prey–predator models. *Ecology* **38**, 136–139.

Wangersky, P. L. (1978). Lotka–Volterra population models. *Ann. Rev. Ecol. Syst.* **9**, 189–218.

Warren, C. E. and P. Doudoroff (1971). *Biology and Water Pollution Control*, Saunders, Philadelphia.

Warren, C. E., J. H. Wales, G. E. Davis, and P. Doudoroff (1964). Trout production in an experimental stream enriched with sucrose. *J. Wildlife* **28**, 617–660.

Watanabe, M. S. (1951). Reversibility of Quantum electrodynamics. *Physical Review* **84**, 1008–1025.

Watt, K. E. F. (1959). A mathematical model for the effect of densities of attached and attacking species on the number attacked. *Can. Entomol.* **91**, 129–144.

Watt, K. E. F. (1968). *Ecology and Resource Management*, McGraw-Hill, New York.

Watt, K. E. F., J. W. Young, J. L. Mitchiner, and J. W. Brewer (1975). A simulation of the use of energy and land at the national level. *Simulation* (May), 129–143.

Watt, K. E. F., L. F. Molloy, C. K. Varshney, D. Weeks, and S. Wirosardjono (1977). *The Unsteady State*, East–West Center Books, University of Hawaii Press, Honolulu.

Weaver, R. (1965). A mathematical model for circadian rhythms. In *Circadian Clocks*, J. Aschoff, Ed., North-Holland, Amsterdam. pp. 47–63.

Webb, W. L., M. Newton, and D. Starr (1974). Carbon dioxide exchange of *alnus rubra*. A mathematical model. *Oecologia* **17**, 281–291.

Weber, A. (1929). *Alfred Weber's Theory of the Location of Industries*, University of Chicago Press, Chicago. 256 pp.

Webster, J. R. (1979). Hierarchical organization of ecosystems. In *Theoretical Systems Ecology*, E. Halfon, Ed., Academic Press, New York. 516 pp.; p. 119.

Webster, J. R., J. B. Waide, and B. C. Patten (1975). Nutrient recycling and the stability of ecosystems. In *Mineral Recycling in Southeastern Ecosystems*, ERDA Con F-74-513, G. Howell, J. B. Gentry, and M. H. Smith, Eds., National Technical Information Service, U.S. Department of Commerce. pp. 1–27.

Weibull, W. (1951). A statistical distribution function of wide applicability. *J. Appl. Mech.* **18**, 293–296.

Weiss, P. A. (1971). *Hierarchy Organized Systems in Theory and Practice*, Hafner, New York.

Weisz, P. B. (1973). Diffusion and chemical transformation. *Science* **179**, 433–440.

Welch, E. B. C. A. Rock, and J. D. Krull (1975). Long term lake recovery related to available phosphorus. In *Modeling the Eutrophic Process*, E. J. Middlebrooks, D. H. Falkenborg and T. E. Maloney, Eds. Ann Arbor Press, Ann Arbor, MI. pp. 5–14.

Wells, P. V. (1976). A climax index for broadleaf forest. An *n*-dimensional ecomorphological model of succession. In *Proceedings of the Central Hardwood Forest Conference*, J. S. Fralish, G. T. Weaver, and R. C. Schlesinger, Eds. 176 pp.

Wesley, J. P. (1974). *Ecophysics, The Application of Physics to Ecology*, Thomas, Springfield, IL.

Westlake, D. F. (1964). Light extinction, standing crop and photosynthesis within weed beds. *Proc. Int. Assoc. Limnol.* **15**, 415–425.

Wheeler, W. M. (1928). *The Social Insects, Their Origin and Evolution*, Harcourt Brace, New York.

Whipple, G. C. (1927). *The Microscopy of Drinking Water*, 4th ed., revised by G. M. Fair and M. C. Whipple, Wiley, New York. 586 pp.

White, L. A. (1943). Energy and the evolution of culture. *Am. Anthropol.* **14**, 335–356.

White, L. A. (1959). *The Evolution of Culture*, McGraw-Hill, New York. 378 pp.

Whitmore, T. C. (1975). *Tropical Rainforests of the Far East*, Clarendon Press, Oxford.

Whitrow, G. J. (1980). *The Natural Philosophy of Time,* Clarendon Press, Oxford.

Whittaker, R. H. (1961). Experiments with radiophosphorus tracer in aquarium microcosms. *Ecol. Mongr.* **31,** 157–188.

Whittaker, R. H. (1965). Dominance and diversity in land communities. *Science* **143,** 250–260.

Whittaker, R. H. (1972). Evolution and measurement of species diversity. *Taxonomy* **21,** 215–251.

Whyte, L. L., A. G. Wilson, and D. Wilson (1969). *Hierarchical Structures,* American Elsevier, New York. 322 pp.

Widom, B. (1965). Stochastic transitions and chemical reaction rates. *Science* **148,** 1555–1560.

Wiegert, R. G. (1974). Competition: A theory based on realistic general equations of population growth. *Science* **185,** 539–542.

Wiegert, R. G. (1975). Simulation modeling of the algal–fly components of a thermal ecosystem. In *Systems Analysis and Simulation in Ecology,* Vol. III, B. C. Patten, Ed., Academic Press, New York. pp. 157–181.

Wiener, N. (1948). *Cybernetics,* Wiley, New York. 212 pp.

Wiener, N. (1964). *God and Golem,* MIT Press, Cambridge, MA.

Wildi, O. (1978). Simulating the development of peat bogs. *Vegetation* **37,** 1–17.

Williams, C. B. (1964). *Patterns in the Balance of Nature,* Academic Press, London.

Williams, G. R. (1981). Aspects of avian biogeography in New Zealand. *J. Biogeography* **8,** 439–456.

Williams, R. B. (1971). Computer simulation of energy flow in Cedar Bog Lake. In *Systems Analysis and Simulation in Ecology,* Vol. 1, B. C. Patten, Ed., Academic Press, New York. pp. 543–582.

Willis, J. C. (1922). *Age and Area,* Cambridge University Press, Cambridge, U.K. 259 pp.

Wilson, A. G. (1970). *Entropy in Urban and Regional Planning,* Pion, London.

Wilson, A. G. (1974). Urban and Regional Models in Geography and Planning. Wiley, New York.

Wilson, A. G., P. H. Rees, and C. M. Leigh (1977). *Models of Cities and Regions,* Wiley, New York. 535 pp.

Wilson, E. O. (1968). The ergonomics of caste in the social insects. *Am. Natur.* **102,** 41–66.

Wilson, E. O. (1971). *The Insect Societies.* Belknap Press, Harvard University Press, Cambridge, MA. 548 pp.

Wilson, E. O. (1975). *Sociology,* Belknap Press, Harvard University Press, Cambridge, MA. 697 pp.

Wilson, E. O. and W. H. Bossert (1971). *A Primer of Population Biology,* Sinauer Associates, Stamford, CT. 192 pp.

Wilson, R. F. (1963). Studies of organic matter in aquatic ecosystems. Ph.D. dissertation. Department of Zoology, University of Texas, Austin.

Wilson, R. J. (1972). *Introduction to Graph Theory,* Academic Press, New York.

Winarsky, I. (1980). *Art, Information and Energy,* President's Scholars Award Paper for 1980, University of Florida, Gainesville. 103 pp.

Winfree, A. T. (1973). Scroll shaped waves of chemical activity in three dimensions. *Science* **181,** 937–939.

Winfree, A. T. (1980). *The Geometry of Biological Time,* Springer-Verlag, New York. 530 pp.

Wood, A. (1925). *Joule and the Study of Energy,* Bell, London.

Woodruff, L. L. (1912). Observations on the origin and sequence of the protozoan fauna of hay in fusions. *J. Exp. Zool.* **1,** 205–264.

Woodruff, L. L. and M. S. Fine (1910). Biological cycle of hay infusion. *Science,* **31,** 467–468.

Woodwell, G. M. (1967). Radiation and the patterns of nature. *Science,* **156,** 461–470.

Woodwell, G. M., P. P. Craig, and H. A. Johnson (1971). DDT in the biosphere—Where does it go? *Science* **174,** 1101–1107.

Wright, B. (1968). Differentiation in the Cellular Slime Mold

Wroblewski, J. S. (1977). A model of phytoplankton plume formation during variable Oregon upwelling. *Sears Found. J. Mar. Res.* **35,** 357–394.

Wroblewski, J. S. and J. J. O'Brien (1976). A spatial model of phytoplankton patchiness. *Mar. Biol.* **35,** 161–175.

Wroblewski, J. S., J. J. O'Brien, and T. Platt (1975). On the physical and biological scales of phytoplankton patchiness in the ocean. *Mem. Roy. Sci. Liege Series 6* **7,** 43–57.

Wulff, F. V. (1970). Analog computer simulation of seasonal patterns in some brackish water rock pools. Annual report to U.S. Atomic Energy Commission. Contract AT-40-1 3666, University of North Carolina, Chapel Hill.

Wyckoff, R. D. and D. W. Reed (1935). Electrical conduction models for the solution of water seepage problems. *Physics* **6,** 395–401.

Wycoff, D. and R. J. Mulholland (1979). Computer modelling of perturbations of the global nitrogen cycle. *Internatl. J. Syst. Sci.* **10,** 421–436.

Wynn-Edwards, V. C. (1962). *Animal Dispersion in Relation to Social Behavior,* Hafner, New York.

Yakolev, V. I. (1965). In *Encyclopedia of Meteorology,* R. Fairbridge, Ed. (1978), McGraw-Hill, New York.

Yary, C. T. (1970). Potential energy and stream morphology. *Water Resources Res.* **7,** 311–322.

Young, D. L. (1975). Salt marshes and thermal additions at Crystal River. In *Power Plants and Estuaries at Crystal River, Florida,* H. T. Odum, H. N. McKellar, W. Smith, M. Lehman, D. Young, M. Kemp, M. Homer, and T. Gayle. Contract report to Florida Power Corporation, May 1975. pp. 281–371.

Yount, J. L. (1956). Factors that control species numbers in Silver Springs, Florida. *Limnol. Oceanogr.* **1,** 286–295.

Zelaney, M., Ed. (1981). Autopoiesis, a Theory of Living Organization, North Holland, Amsterdam.

Zemanek, H., H. Kretz, and A. J. Agyan (1961). A model for neurophysiological functions. In *Information Theory,* C. Cherry, Ed., Academic Press, New York. pp. 270–284.

Zevallos, A. C. and P. De T. Alvim (1967). Influencia del arbol de sembra *Erythrina glauca* sobre algunos factores edafologicos relacionados con la production del cacaotero. *Dasonomia Interamericana* **17,** 330–336.

Zipf, G. E. (1919). *Human Behavior and the Principle of Least Effort,* Addison-Wesley, Reading, MA.

Zipf, G. K. (1941). *National Unity and Disunity,* Principia Press. Bloomington, IN. 407 pp.

Zipf, G. K. (1949). *Human Behavior and the Principle of Least Effort*, Hafner, New York.

Zucchetto, J. (1975a). Energy basis for Miami, Florida and other urban systems. Ph.D. dissertation. Department of Environmental Engineering Sciences, University of Florida, Gainesville.

Zucchetto, J. (1975b). Energy, economic theory and mathematical models for combining the systems of man and nature. Case study, the urban region of Miami. *Ecol. Model.* **1**, 24–268.

Zucchetto, J. and A. M. Jansson (1979). Total energy analysis of Gotland's agriculture. A northern temperate zone case study. *Agroecosystems* **5**, 329–344.

Periodicals of Systems Ecology, Energy, Environment and Economics

Acta Biotheoretica
Advances in Biophysics
Advances in Control Systems
Advances in Ecological Research
Advances in the Economics of Energy and Resources
Advances in Microbial Ecology
Agroecosystems
American Scientist
Annual Reviews of Ecology and Systematics
Applied Mathematical Modelling
Biomathematics
Biometrics
Bioscience
Biosystems
Biotechnology and Bioengineering
Bulletin of Mathematical Biology
Bulletin of Mathematical Biophysics
Canadian Entomologist
Current Advances in Ecological Sciences
Cybernetics Forum
Ecological Modeling
Ecology
Ecology Abstracts
Energy
Energy Abstracts
Energy Economics
Energy Policy
Energy Research Abstracts
Energy Reviews
Energy Sources
Energy Systems Modeling, Planning and Decisions
Energy Systems and Policy
Environmental Abstracts
Environmental Engineering
Estuarine Coastal and Shelf Science
General Systems
General Systems: Yearbook; Bulletin
Human Ecology
Information Sciences
International Journal of Circuit Theory and Applications
International Journal of Ecological Modelling
International Journal of Ecology and Environmental Science
International Journal of Energy Research
International Journal of Energy Systems
International Journal of General Systems
International Journal of Systems Science
International Journal of Tropical Ecology
Journal of Agricultural Economics
Journal of Animal Ecology
Journal of Chemical Ecology
Journal of Computational Chemistry
Journal of Cybernetics
Journal of Cybernetics and Information Science
Journal of Ecology
Journal of Energy and Development
Journal of Environmental Economics and Management
Journal of Environmental Education
Journal of the Fisheries Research Board of Canada
Journal of Forestry
Journal of Graph Theory
Journal of Interdisciplinary Cycle Research
Journal of Interdisciplinary Modeling and Simulation
Journal of International Economics
Journal International Society of Ecological Modelling
Journal of Mathematical Biology
Journal of Theoretical Biology
Limnology and Oceanography
Mathematical Biosciences
Mathematical Modelling
Microbial Ecology
Natural Resources Journal
Networks
Oikos
Progress in Physical Geography
Progress in Theoretical Biology
Regional Science and Urban Economics
Researches on Population Ecology
Resources and Energy
Science
Simulation
Statistical Ecology
Tellus
Theoretical and Applied Environmental Reviews
Theoretical Population Biology
Urban Ecology
Yearbook of General Systems

Author Index

Abbott, W., 465
Abramson, N., 321
Adams, R. N., 270, 287, 510, 517
Addy, N. D., 457
Adler, J. R., 235
Agee, E. M., 565
Aggarwall, Y. P., 272
Aguilar, R. J., 553
Agyan, A. J., 613
Ahern, J. E., 266, 268
Ahlstrom, T., 603
Ahrens, L. H., 570, 575
Alcock, N. Z., 512
Aldeman, R. A. A., 382
Aldenson, W., 330
Alexander, J. F., 195, 288, 457, 490, 495, 533, 555, 567, 574, 603
Alexandrov, V. Y., 301
All, T. F. H., 511, 513
Allbaugh, C., 603
Allee, W. C., xiii, 88, 150, 350, 433, 508, 593
Allen, E. R., 595
Allen, K. R., 279, 392
Allen, R. C. D., 507
Al-Muqaddasi, 269, 323
Altmon, C., 176
Alvim, P. de T., 519
Amiot, W., 611
Amle, Y., 546
Andersen, N. R., 570
Anderson, G. V., 609
Anderson, J. M., 405, 475, 609
Anderson, J. W., 382
Anderson, R. M., 248, 280, 405
Andrewartha, H. G., 301
Andrews, J. F., 166, 176
Anthony, E. H., 418
Antonini, G. A., 529, 545
Armstrong, J. S., 553
Armstrong, N. E., 187, 190, 369, 423, 603
Arnheim, R., 315
Arnold, G., 582
Arnon, D. I., 358
Arrhenius, S., 278, 289, 291
Ash, R. B., 321
Ashby, M., 350, 475

Ashby, W. R., 13, 72, 94
Assmann, E., 279
Atchison, J., 339, 350
Atkinson, W., 607
Auslander, D., 87
Austin, M. P., 348
Ayres, R. U., 507

Bahm, D., 507
Bailey, N. T. J., 197, 405
Bailey, V. A., 43, 195, 198
Baird, D. C., 268, 476
Baisley, J. R., 597
Bakuzis, E. V., 96, 410, 441
Baltzer, J. O., 146, 441
Banks, F. E., 507
Banks, H. T., 181
Barker, R. R., 415
Barnes, R. S. K., 442
Barnett, G. W., 475
Barney, G. O., 567, 570
Barrett, G. W., 335
Barth, T. F. W., 558, 570
Bartholomay, A. F., 159
Bartlett, M. S., 158, 204, 405
Basore, B. L., 606
Bathersby, A., 553
Batie, S. S., 495
Batty, M., 548, 551
Baule, B., 134
Baumgartner, A., 571
Baumol, W. J., 507, 534
Baustead, I., 265, 268
Bayley, S., 268, 453, 456, 483
Bazin, M. J., 405
Bazzaz, F. A., 475
Begon, M., 159, 405
Behrens, W. W., 600
Bekey, G. A., 45, 582
Bella, D. A., 592
Bell, J. C., 605
Bellman, R., 82, 248
Bender, E. A., 13
Bennett, A. W., 53, 68, 94
Bennett, R. J., 13, 531
Bennett, W. R., 52
Benoit, R. J., 371, 418
Berlinski, D., 79

Berndt, E. R., 268
Bernstein, B. B., 476, 499, 507
Berry, B. J. L., 269, 323, 326, 339, 341, 350, 497, 507
Berry, R. S., 584
Berryman, A. A., 222
Bertine, K. K., 554
Beverton, R. J. H., 386
Beyers, R. J., 291, 293, 465, 467, 526, 527, 588, 603
Bierman, U. J., 373
Biggs, N. L., 87
Bilig, E., 360
Birch, L. C., 301
Birney, E. C., 475
Blackman, R. B., 284, 365
Blackman, V. H., 132
Blalock, H. M., 81, 94
Bledsoe, L. J., 566
Blesser, W. B., 68
Blinchikoff, J. J., 283
Bliss, L. C., 414, 442, 450
Blocker, H. D., 475
Bloom, S. G., 605
Bloomfield, J. A., 606
Bloomfield, P., 287
Blum, H. F., 100, 573
Blum, J. J., 53, 68
Bode, H. W., 40, 182, 183
Bogner, R. E., 23, 45, 283
Bohm, P., 553
Boisvert, M., 606
Bolin, B., 562, 565, 570
Bolt, R., 189
Boltzmann, L., 6, 118, 265, 277, 278, 311, 318, 319, 575
Bondi, H., 576
Bongers, L., 527
Bonham-Carter, E., 13, 84, 557
Bonner, J. T., 287, 323-324, 326
Booth, D. A., 384
Borchers, M., 52
Bormann, F. H., 445, 457, 471, 475
Bosserman, R. W., 23, 302, 507
Bossert, W. H., 33, 45
Botkin, D. B., 388, 469, 472, 475
Boulding, K. E., 531

AUTHOR INDEX

Bowdon, R., 81
Bowen, H. J. M., 563
Box, G. F. P., 282, 287
Boynton, W., 406, 408-409, 495, 536, 538, 603
Boysen-Jensen, P., 364, 374
Bozie, S. M., 582
Bradshaw, A. D., 475, 553
Brady, J., 577
Brainerd, J. G., 587
Brancazio, P., 571
Brebbia, C. A., 531
Breeze, J., xiii
Brelski, B. H. J., 204
Brennan, R., 52
Brewer, J. W., 45, 94, 486, 600, 611
Brian, M. V., 405
Briggs, L. L., 557
Brillouin, L., 306, 311, 321, 343
Brock, T. D., 301
Brocksen, R. W., 423
Brody, S., 113, 120, 475
Broeker, W. S., 557, 560
Brookes, L. G., 285
Browder, J. A., 271, 273, 386, 387, 433, 495, 589, 603
Brower, J. Z., 586
Brown, C. B., 553
Brown, F., 301
Brown, G. S., 94
Brown, H. J. M., 571, 608
Brown, J., 442
Brown, J. A. C., 339, 350
Brown, M. T., 198, 200, 275, 330, 489, 495, 533, 535, 539, 549, 582, 603
Brown, R. Z., 603
Brown, S., 542, 547
Browning, T. O., 405
Brunsden, D., 571, 576
Budd, C. S. J., 215
Bugher, J. C., xiii
Bullard, C. W., 264, 268, 501, 507
Bulloch, T., xiii
Bumpus, D. W., 606
Bungay, H. R., 88
Bunnell, F. L., 442, 600
Burg, J. P., 284
Burk, D. J., 131, 132
Burke, O. W., 43
Burkholder, P. R., 602
Burnett, M. S., 526, 603
Burns, L. A., 134, 395, 397, 412, 430, 468, 561
Burr, D. Y., 495
Burton, A. C., 44
Busch, A. W., 402
Butler, G. C., 553
Butler, T. J., 264, 268, 596, 603
Butterfield, C. T., 447

Cacey, M. F., 539
Cadle, R. D., 595
Cain, S. A., 310, 350
Cairns, J., 475, 611
Cajori, F. A., 587
Cale, W. G., 244
Callahan, A. B., 604
Calow, P., 120, 507
Campbell, G. S., 274, 301
Campbell, I. M., 571
Canale, R. P., 195, 442, 475, 589
Cannon, W. B., 13
Canoy, M., 281
Cantlon, J., 603
Capellos, C., 204
Caplan, S. R., 115
Carlson, A., 68
Carnot, S., 102, 103, 115, 265
Carroll, B. D., 94
Carson, E. R., 52
Cartwright, D., 94, 553
Caspers, H., 419
Cassie, R. M., 338
Casti, J. L., 83, 94, 248, 457
Caswell, H. E., 88-89
Catton, 531, 553, 577, 582
Chadwick, G., 553
Chadwick, M. J., 248, 475
Chamberlin, C. F., 389
Chambers, M. D., 585
Chance, B., 170-171, 175, 176
Chapman, D. G., 222, 405
Charles, J. J., 52
Charlesworth, A. S., 45
Charlesworth, P., 607
Charlsen, R. J., 571
Charnes, A., 531
Charney, J. G., 558
Charpentier, J. P., 489
Chaston, I., 82, 94
Checkland, P., 582
Cherry, C., 341
Chickilly, A. A., 398
Child, G. I., 45
Childers, D. G., 284
Chorley, R. J., 13, 531, 553
Christian, K. R., 553
Christiansen, F. B., 222, 248
Churchmen, C. W., 571
Clapham, W. B., 531
Clark, C., 330
Clark, C. W., 502-503, 507
Clarke, F. E., 597
Clarke, G. L., 284, 285, 349, 438
Cleland, W. W., 175
Clendenning, K. A., 361
Cliffs, A. D., 350
Clymer, A. B., 415
Clymo, R. S., 471
Cobb, C. W., 487
Cobb, J. S., 263, 357, 359-365, 442
Cochran, M. I., 583, 607
Cody, M. L., 350, 475, 499
Cohen, D., 377, 380
Cohen, J. E., 242, 248, 341

Coker, R. E., xiii, 445
Colby, J. A., 438
Cole, L., 405
Cole, S., 571
Collens, T. W., 476
Collier, B. D., 301, 600
Collins, C. T., 586
Conkling, E. C., 350, 507, 585
Conley, C. C., 140, 287
Connell, J. H., 211, 222, 344, 346
Conover, J. T., 366
Constantinides, A. G., 189, 283
Conway, E. J., 557
Cook, E., 265
Cooke, G. D., 449, 465, 467
Coon, C. S., 265
Cooper, D. C., 425, 588
Cooper, R. B., 94, 244, 248
Copeland, B. J., 425, 428, 460, 603
Corbet, A. S., 590
Corey, T., 68
Cormack, R. M., 350
Cornow, R., 571
Costanza, R., 260, 264, 265, 268, 331, 332, 347, 501, 549, 603
Cottrell, F., 265, 507
Cowan, D. J., xiii
Cowles, H. C., 468
Cox, D. R., 405
Cox, G. W., 301
Cox, R., 330
Cragg, J. B., 442
Craig, P. P., 612
Crawford, A. B., 585
Crenson, M. A., 592
Crissey, B. L., 592
Cristaller, W., 323-324
Crossman, J. S., 595
Crowe, A. F., 94
Cuchlaine, A. M. K., 471
Cumming, F. P., 291
Cummins, K. W., 423
Cunningham, G. L., 376
Cunningham, J. P., 151, 153, 154, 487
Curl, H. C., 592
Curran, J. C., 268, 571
Curran, P. F., 120
Curran, S. C., 268, 571
Curtis, J. T., 348-349
Cushing, D. H., 405
Cushing, J. M., 155, 159
Cuzon, R., 603

Dahberg, M. L., 611
Dahlberg, A. O., 507
Dainton, F. S., 174, 175, 181
Dame, R. F., 382, 609
Dansereau, P., 92, 542
D'Arcey, 81
Davidson, J. L., 553
Davidson, R. S., 415
Davis, G. E., 586, 611
Davis, J. C., 327

Davis, K. S., 234
Day, J., 13, 382, 579
Dayton, K., 346
De Amezaga, E., 367
De Angelis, D. L., 321, 402
Deb, A. K., 430
Debach, P., 198
DeBellevue, E., 523, 527
de Chardin, T., 314
Deevey, E. S., 334, 388, 463
Degens, E. T., 570, 585
de Groot, S. R., 114, 265
Deininger, R. A., 442
Delwiche, C. C., 564
Dempster, J. P., 405
Denbigh, K. L., 108, 114, 120
De Nevers, N., 106
Denman, K. L., 283, 287, 382
de Oliviera Castro, G. M., 350
De Palma, L. M., 581
de Pierre, L., 94
De Rosnery, J., 582
Desmaeris, A. P., 279
Desoer, C. A., 11, 169, 235, 248
De Souza, G. R., 531
Devol, A. H., 606
De Wit, C. T., 52, 210, 214, 382, 582
Diamond, J. M., 337-338, 350, 475
Dickerson, R. E., 120
Dickey, J. W., 553
Dickson, K. L., 587
Dillon, J. A., Jr., 507
Di Salvo, L. H., 247, 428, 430, 603
DiToro, D. D., 313, 425, 427
Dodd, J., 571
Domenico, P. A., 94, 531, 542
Donnelly, J. R., 553
Doornkamp, J. C., 471, 571
Doudoroff, P., 308, 423-424, 460, 463, 475, 611
Douglas, I., 553
Douglas, P. H., 487
Doxiadis, C. A., 328, 535, 536
Drake, J. F., 196, 610
Drener, J. J., 571
Dubois, D. M., 117, 199, 306, 349, 376
Duckham, A. N., 270
Dugdale, R. C., 462
Duncan, A., 120, 405
Dwyer, R., 40
Dyckman, J. W., 553
Dyrrsen, D., 571

Eagleman, J. R., 274, 349
Eaton, J. S., 606
Eberhardt, L. L., 405, 583, 607
Ebert, J. D., 609
Eckardt, F. E., 382
Eddsock, E. A., 604
Edmunds, L. N., 301
Edwards, N. T., 607
Ehleringer, J. R., 373

Ehrmantraut, H. C., 361
Eigen, M., 178, 195
Einstein, A., 107, 187
Electronic Associates, 54, 94, 177
Elliott, J. M., 317
Elton, C. S., 194, 224, 350
Emanuel, W. R., 607, 610
Emerson, A. E., 584
Emlen, J. M., 168, 192, 204, 222
Emmanuel, W. R., 469, 472, 591
Engvist, A., 366
Enochson, L., 287
Eppley, R. W., 375
Erdahl, C. A., 592
Esch, G. W., 301
Eschenroeder, A. E., 531
Estabrook, R. W., 587
Etherington, J. R., 475
Evans, C. R., 321
Evans, R., 266
Ewel, J., 81
Ewel, K. C., 34, 421, 433, 441, 529, 603
Eyring, H., 139, 277-278, 293, 301, 595

Fairbanks, D., 594
Fairbridge, R. W., 571, 576
Falk, J. H., 526, 528
Falkenborg, D. H., 382, 442
Falkowski, P. G., 375, 382
Farned, J., 178, 517, 518
Fechner, G. T., 577
Fein, L., 604
Feinstein, A., 321
Fenchel, T., 222, 248, 280, 324
Ferguson, N., 583, 603
Ferrari, T. J., 52
Fey, W. R., 471, 473
Field, B. C., 268
Fields, P. E., 377
Fine, M. S., 465
Finkelstein, L., 52
Finn, J. T., 241
Fisher, A. C., 507
Fisher, J. S., 606
Fisher, R. A., 317, 320, 340, 342, 343
Fitz, R., 529
Flagg, W. W., 607
Fleming, R. H., 357, 609
Fletcher, J. A., 45
Fluck, R. C., 268, 476
Foell, W. K., 553
Fontaine, T. D., 101, 421, 422
Fork, D. C., 374
Forrester, J. W., 52, 80, 85, 87, 93, 94, 185, 393, 492, 547, 550, 568, 569
Fortescue, J. A. G., 571
Fowler, N., 574
Fox, J. F., 457
Frankena, F., 268
Franklin, J. F., 606

Fredrickson, A. G., 222, 589, 610
Freedman, H. I., 13, 405
Freeman, M. C., 567, 571
Freeman, N. J., 14
Freer, M., 553
French, C. S., 374
Frenkel, F. N., 579
Fretwell, S. D., 405
Frey, A., 350
Fukada, N. M., 245
Fuller, B., 313, 572
Fullerton, H. H., 585
Funderlie, R. E., 248
Fusaro, B., xiii

Gaden, E. L., 588
Gallagher, R. G., 305
Gallegos, C. L., 365, 415
Gallo, N., 195
Gallopin, G. C., 87, 145
Gallucci, V., 222, 405
Galomb, S. W., 580
Ganning, B., 464
Garcia-diaz, A., 248
Gardner, G., 589
Gardner, R. H., 565
Garfinkel, D., 195, 200-203
Garrels, R. M., 558, 560, 561, 563
Garrison, R. L., 599
Garrison, W. L., 341
Gates, D. M., 120, 288, 291, 301, 597
Gatlin, L. L., 307, 321
Gaudrain, J., 52
Gause, G. F., 148, 203, 210, 218, 222
Gayle, T. L., 293, 294, 420, 421, 593, 613
Gazis, D. C., 553
Georgescu-Roegen, N., 507, 517, 567
Gerlsbakh, I. B., 287
Getz, W. M., 13
Gibbons, W., 301
Gibbs, W., 3, 265, 266, 311
Gibson, R. D., 45
Giddings, J. M., 451
Giesy, J. P., 442
Gilliland, M. W., 135, 136, 268, 493, 565
Gilmore, R., 582
Gilpin, M. E., 591
Glandsorf, P., 287, 348
Glass, H. B., 609
Glass, L., 154
Gleason, H. A., 337-338
Gloyna, E. F., 423
Glushkev, V. M., 94
Gmitro, J. I., 324
Gödel, K., 3
Goel, N. S., 137, 154, 158, 159, 405, 575
Goguel, J., 301
Goh, B., 204, 405, 582
Gold, H. J., 181, 204, 222
Goldberg, E. D., 240, 382, 554, 571

Goldberg, M. A., 547
Goldbert, M. A., 553
Goldman, C. R., 589
Goldman, S., 317, 321
Goldschmitt, V. M., 555, 557, 559
Goldstein, R. A., 607
Gompertz, B., 156, 390
Gonzalez, J. G., 445
Goodale, P., 608
Goodall, D. W., 523, 579, 580, 603
Goodman, G. T., 531
Goodwin, B. C., 181, 200
Gopal, M., 45
Gordon, G., 13, 52, 68
Gore, A. J. P., 471
Gorham, P., 475
Gorman, M. L., 337, 350
Gotaas, H. B., 604
Goudrian, J., 215
Gowdy, C. M., 565
Gower, J. C., 203
Graham, D., 582
Granger, C. W. J., 492, 507
Grant, J., 582
Grassle, J. F., 350
Gray, D., 586
Gray, I. E., 28
Greenberger, M., 532
Greenstein, C. H., 72, 94
Greenstein, D. S., 587
Gregory, F. G., 373
Greig-Smith, P., 343, 350
Grenney, W. L., 220, 345
Gresham, C. A., 377
Gribbin, J., 577
Grime, J. D., 382, 475
Grinnell, 224
Groden, T. W., 166
Grodins, F. S., 13, 43, 45
Grodzinski, W., 120
Gross, M., 45
Guardabassi, G., 117
Gulland, J. A., 405
Gunther, R., 582
Gurney, W. S. C., 205
Gutierrez, L. T., 471, 473
Gutierrez, R. J., 523, 525-526
Guttenberg, A. Z., 553
Gyllenberg, G., 446

Hadley, E. B., 414, 450
Hagerstrand, T., 85
Haggett, P., 126, 310, 337, 350, 531, 553
Hairs, W. F., 606
Haken, H., 159, 195, 287
Haldane, J. B. S., 166, 176, 311
Haldin, S., 442
Hale, M., 150
Halfon, E., 82, 248
Hall, C. A. S., 13, 23, 382, 400, 461, 507, 565, 579
Hall, W. B., 386

Halldin, S., 301, 531
Halle, F., 382
Hamberg, D., 507, 591
Hamilton, J., 110
Hamilton, P., 233, 442
Hamilton, W. A., III, 405
Hammen, C. S., 287
Hancock, G. F., 265, 268
Hancock, J. C., 321
Hanga, T., 546
Hannauer, G., 68
Hannon, B., 242, 266, 476, 499, 501
Hanshaw, B. B., 597
Harary, F., 87, 94, 553
Harbaugh, J. W., 13, 84, 557
Hardin, G., 210
Hare, V. C., 13
Harris, C. J., 405, 531
Harris, J. P. W., 230
Harris, L. D., 204
Harris, M., 514, 531
Harrison, B., 280
Hart, R. D., 522
Harte, J., 182
Hartney, V. J., 601
Harwell, M. A., 243
Haskell, N. P., 197
Hastings, N. A. J., 94, 287
Hatanaka, M. A., 105
Hatee, N. D., 590
Havlick, S. W., 553
Hawkins, D., 586
Hayes, F. R., 418
Hazen, W. E., 222, 442
Heal, O. W., 405, 442, 475, 609
Heath, M. T., 248
Hedges, H. G., 13, 596
Heilbrun, J., 539
Heinmetz, F., 13, 66, 247
Heinz, E., 120
Heizer, R. F., 515
Helmholtz, H., 265, 294, 318
Henderson, K., 391
Henderson, L. J., 4, 23, 577
Henley, E. J., 94
Herbert, D., 192, 199
Herendeen, R. A., 264, 268, 500, 501, 507
Heron, A. C., 324
Herricks, R. E., 587, 595
Herschman, A., 247
Hesketh, J. D., 382
Hess, S. L., 558
Hesse, R., 350
Hessler, R. R., 346
Hett, J. M., 459-460
Higgins, J., 587
Hilborn, R., 349
Hill, T. C., 120
Hill, T. L., 90, 181
Hillier, F. S., 553
Hinde, R. A., 73, 94
Hiorns, R. W., 158, 204

Hiramatsu, T., 313
Hirshleifer, J., 476, 507
Hobbie, J. E., 131
Hochachka, P. W., 405
Hock, I., 540
Hoffman, E. J., 120, 268
Hohn, F. E., 82
Holderidge, L. R., 410
Holdgate, M. W., 403, 475
Holland, H. D., 563, 571
Holling, C. S., 168, 243, 383, 545, 553, 577, 582, 591, 598
Holsberg, P. J., 68
Holt, S. J., 386
Holtz, K., 442
Homer, M., 297, 595, 613
Hood, D., 465
Hopkins, M., 571
Horn, D. J., 23, 405
Horn, H. S., 451, 475
Hornbach, D. J., 529
Hornberger, G. M., 591
Horrobin, D. F., 45
Horton, R. E., 271, 313
Hoskin, C. M., 339, 363, 450
House, P. W., 507
Howe, C. W., 94, 507, 531
Hubbard, C. J., 349
Hubbert, K., 578
Huber, W. C., 603
Hubin, W., 52
Huettner, D. A., 501
Hughes, B. B., 582
Hughes, J. L., 76, 94
Hulbert, E. M., 343
Hunding, A., 247
Hune, C. W., 507
Hunt, C., 591
Hurlbert, S. H., 446
Hutchings, M. J., 215
Hutchinson, G. E., xiii, 146, 150, 153, 154, 159, 204, 222, 224, 227, 232, 248, 340, 345, 350, 439, 577
Huxley, 117
Huxley, J. S., 136, 139
Hwana, C., 583
Hynes, H. B. N., 424, 460

Iberall, A. S., 287, 582
Ikusima, I., 146, 210
Incarn, J. D., 94
Innes, G. S., 582
Innis, G. S., 442, 471, 582
Ippen, A. T., 187
Irvine, E., 531
Irving, L., 607
Isard, W., 326, 328, 330, 350, 502, 553
Ivlev, V. S., 133, 134, 382, 384

Jacob, N. L., 52
Jacobs, J., 345, 512
Jacobsen, H., 178, 302, 393

Jacquez, J. A., 250
Jaeger, J. C., 40
James, A., 248, 442
James, D. E., 499
James, M. L., 68
Jammer, M., 140
Janak, J. F., 586
Jannasch, J. W., 191, 193
Jansen, H. M. A., 594
Jansson, A. M., 231, 429, 481, 521
Jansson, B. O., 231, 425
Jantsch, E., 195, 321, 582
Jardine, N., 350
Jassby, A. D., 365, 382
Jaynes, E. T., 284, 313
Jeffers, J. N. R., 23, 235, 405, 442
Jeffries, H. P., 243, 345
Jenkins, G. M., 282, 287
Jenness, R. R., 68
Jennings, A., 248
Jenny, H., 43, 430, 468, 513
Jensen, A. L., 148
Jepsen, A., 563
Jobin, W. R., 187
Johnson, A. W., 301
Johnson, C., 405
Johnson, C. L., 68
Johnson, F. H., 289, 293, 301
Johnson, H. A., 612
Johnson, I., 571
Johnson, M. W., 609
Johnson, N. M., 606
Johnson, W. C., 589
Johnston, R. F., 331
Jones, D. D., 598
Jones, F., 386
Jones, G. L., 417, 418
Jones, H. E., 471
Jones, J. G., 589
Jones, J. W., 382
Jorgensen, S. E., 23, 266, 442, 571, 583
Jost, J. L., 589, 610
Joule, P., 265
Juday, C., 265, 269, 438
Judson, D., 252
Juo, W., 582

Kaesler, R. L., 344
Kale, B. K., 286, 287
Kaltun, W. L., 585
Kamenkovich, V. M., 287, 571
Kangas, P. C., 268, 404, 469-471, 603
Kansky, K. J., 310
Kaplan, H. S., 281
Karnopp, D., 87, 94
Karplus, W. J., 45, 68, 235
Karr, J. R., 424
Katchalsky, A., 115, 120
Kauffman, S. A., 348
Kavanaugh, R., 507
Kazemier, B. H., 13
Kearns, P. K., 296, 466
Kedem, O., 115

Keeling, C. D., 562
Keith, A., 531
Kelley, M. G., 591
Kelley, R. A., 291-296, 451
Kelley, R. F., 591
Kellogg, W. W., 557, 563
Kelly, F. P., 321, 350
Kelsey, C. T., 348
Kemeny, J. G., 52, 83, 89
Kemp, W. M., 261, 296, 345, 346, 603
Kempe, S., 570, 585
Kendall, M. G., 287
Kendigh, D. C., 43
Kennish, W. J., 600
Kent, E. W., 94
Kercher, J. R., 607
Kerfoot, W. C., 442
Kerner, E. H., 321, 342, 350
Kerr, S. R., 350
Kershaw, K. A., 348, 350
Kesevan, H. K., 14, 596
Kessler, W. J., 41
Ketner, P., 570, 585
Keyfitz, N., 222, 405
Khodashova, K. S., 405
Kiersted, H., 155
Kilburn, P. D., 337
Kilham, P., 208, 233
King, C. A. M., 571
King, C. E., 393
King, E. L., 176
Kinne, O., 442
Kira, T., 131, 146
Kitaigorodskiy, A. E., 321
Kitchens, W. M., 609
Kitching, R., 333
Kittel, C., 104, 120
Klebuwski, R. Z., 120
Klein, L. R., 507
Kleinrock, L., 248
Klir, G. J., 13, 37, 72, 94
Klir, J., 42
Klopatek, J. M., 23, 521, 607
Klose, P. N., 430
Klotz, I. M., 120
Kluge, M., 374
Kneese, A. V., 507, 553
Knight, R. L., 99, 445, 529
Kobayochi, H., 14
Koch, A. L., 350
Koenig, H. E., 88-89, 93, 169, 587
Koestler, A., 14
Kojima, T., 607
Kolmogorov, A. N., 274
Kondratieff, 492
Kordonskiy, K. B., 287
Kormondy, E. J., 23, 442, 583
Korn, G. A., 68
Korn, J. M., 68
Kornicker, L., 603
Kort, V. G., 287, 571, 600
Kostitzin, V. A., 405, 571, 577

Kostitzin, W., 222
Kowallik, W., 374
Kranz, J., 178, 181, 388
Kraus, E. B., 442, 571
Krebs, C. J., 405
Kremer, J. N., 442
Kremer, P., 281, 401
Krenz, J. H., 264, 268, 501, 507
Kretz, H., 613
Krigman, A., 90
Krol, J. G., 156
Kumar, D., 591
Kumenkovich, V. M., 600
Kurasawa, H., 449
Kurihare, Y., 463
Kurtz, T. E., 52
Kylstra, C., xiii, 603

Labine, P. A., 178
Lack, D., 324
Lam, C. F., 181
Lam, H. Y. F., 45, 287
Lanchester, F. W., 219, 221
Lane, P., 243
Langbein, W. B., 118, 271-272
Langenheim, 5, 23
Langford, T. W., 502
Langmuir, I., 169, 180
Lasker, G. E., 23, 94, 583
Lassiter, R. R., 296, 466, 475
Laszlo, E., 4
Latil, P. de, 94
Launhardt, W., 326
Lavine, M. J., 264, 268, 603
Lawrence, B. A., 381
Lazrus, A. L., 595
Le Chatelier, 118, 311, 313
Le Cren, E. D., 442
Ledley, R. S., 68
Lee, C. L., 551
Lee, R., 94
Lee, R. B., 511, 531
Lehman, M. E., 298, 593, 595, 603, 613
Lehninger, A. L., 120
Leibowitz, S., 566, 603
Leigh, C. M., 507
Leigh, E. G., 350, 577
Lem, P., 603
Lemlich, S., 603
Lemont, B., xiii
Leontief, W. W., 241, 499, 507, 568
Leopold, A. C., 318
Leopold, L. B., 118, 234, 271-272, 495
Lerman, A., 591
Leslie, P. H., 152, 154, 392, 393
Letterman, R. D., 600
Levener, R., 171
Levenspiel, O., 106
Levin, S. A., 351, 442
Levine, O. H., 585
Levine, R. D., 321
Levins, R., 89, 90, 238, 398

Levy, D., 182
Lewis, C. N., 120
Lewis, M. C., 596
Lewontin, R. C., 405, 465
Lieberman, G. J., 553
Liebig, J., 130
Lieth, H. F. F., 382, 410, 555
Likens, G. E., 445, 457, 571, 603
Limburg, K., 230
Lin, S. H., 139
Lin, S. M., 139
Lindell, T., 531
Lindeman, R. L., 265, 269, 439
Lineweaver, H., 131, 132
Linford, S. H., 120
Linnemann, H., 568
Linschitz, H., 311
Liossatos, P., 553
List, H. L., 265
Littlejohn, C., 430, 586, 603
Livingston, D. A., 557, 558
Lloyd, A., 531
Lloyd, E. K., 585
Locatelli, A., 592
Lock, M. A., 442
Locker, A., 322, 405
Loehr, R. C., 420
Lommen, P. W., 376
Longhurst, A. R., 442
Loose, K. D., 245-246, 402
Lorenz, E. N., 557
Lorenz, K. Z., 73
Lorenzen, M. W., 420
Losada, M., 584
Lösch, A., 330
Lotka, A. J., 6, 14, 18, 43, 101, 118, 120, 170, 171, 172, 194, 198, 204, 265, 319, 453, 561, 565, 577
Loucks, O. L., 606
Lovelock, J. E., 577, 583
Lowe-McConnell, R. H., 475
Lucas, R., 513
Lucas, R. F., 507
Ludwig, D., 195, 456, 457, 604
Lugo, A. E., 23, 316, 327, 412, 442, 449, 475, 505, 561
Luikov, A. V., 123
Lumry, R., 167, 356, 358
Lund, J. W. G., 465
Lynch, E. P., 14, 72, 94
Lynch, V., 604
Lynn, W. R., 531

MacArthur, R. H., 33, 44, 222, 276, 307, 309, 340, 343, 346, 350, 410
McCave, I. N., 382, 591
McCormick, J. F., 23, 442, 466, 583
MacCready, P. B., 272
McCulloch, W. S., 78, 79, 94, 327
MacDonald, K. B., 233, 442
McDougall, E., 410
MacFadyen, A., 405
McFarland, D. J., 45

MacFarlane, A. G., 45
McFarlane, R. W., 301
McHarg, I. L., 347
Macias, F. M., 375
McIntire, C. D., 423, 438
McIntosh, R. P., 343, 348, 465, 475
McIrvine, E. C., 266, 319, 322
McKee, M., 585
McKee, P. A., 605
McKellar, H. N., 241, 297, 417, 593, 595, 609, 613
McKendrick, A. G., 145
McKenzie, D. H., 464
MacKenzie, F. T., 558, 591
MacKenzie, L. A., 364
Mackey, M. C., 154
McLaren, I. A., 159, 405
McLeod, J., 68, 90, 507
McMahan, E., 395, 603
McNaught, D. C., 243, 280, 324, 576, 583
McPartlin, M. A., 600
Maguire, B., 459-460, 463
Maguire, L. A., 346
Mahloch, J. L., 402
Maitra, S. C., 159, 405
Makridakis, S., 573
Malahoff, A., 570
Malek, I., 140, 192, 199
Malkevitch, J., 83
Maloney, T. E., 382, 442
Mandelbrot, B. B., 276, 341, 342, 350, 489
Mankin, J. B., 591
Manley, B. F. J., 405
Mann, K., 442
March, L., 553
Margalef, R., 101, 204, 284, 285, 307, 321, 341, 343, 344, 351, 374, 442, 458, 459, 466, 583
Markin, J. B., 607
Marks, P. L., 471
Marler, P., 405
Marline, C. S., 597
Marshall, J. S., 458
Martell, E. A., 595
Martin, C. S., 597
Martin, G. D., 461, 530
Martin, L., 553
Martin, T., 600
Martino, R. L., 245
Marx, K., 265
Marx, P. M., 611
Masaguki, T., 382
Mason, B., 576
Mason, H. L., 5, 23
Mason, R. O., 571
Mason, S. J., 85
Masser, I., 549, 551
Mathews, R., 507
Matis, H., 248
Matsunoya, T., 318
Mattson, W. J., 457

Mau, D., 268
Maxwell, J. C., 100, 101, 265, 277, 278, 311, 312
May, R. M., 154, 155, 158, 159, 197, 204, 243, 280, 326, 341, 475
Mayeda, W., 94
Mayer, R. J., 265
Mazur, P., 114
Meadows, D. H., 80, 368
Meadows, D. L., 600
Medici, J. C., 527
Mees, A. J., 45
Mehnhinick, E. F., 345
Meixner, J., 287
Mejer, H., 266
Mello, J. M. C., 94
Melzak, Z. A., 583
Mendolssohn, K., 515
Menshutkin, V. V., 242, 425
Menten, M. L., 131, 163-169
Mercer, R. L., 553
Merrell, D. J., 405
Mesarovic, M. D., 14, 94, 568
Meyburg, A. H., 264, 268, 596
Meyer, A., 410
Meyer, H. A., 279
Meyer, W., 83
Michaelis, L., 131, 163-169, 175-176, 198-199
Middlebrooks, E. J., 382, 442
Mieghem, J. V., 571
Mihram, G. A., 137
Mikhailov, Y. A., 123
Mikolaj, P. G., 285
Mildvan, A. S., 277, 389
Miles, J., 475
Milhorn, H. T., 43, 45, 185
Miller, A. R., 52
Miller, D. H., 571
Miller, G. T., 287
Miller, J. G., 567, 572
Miller, J. P., 597
Miller, M. A., 545, 603
Miller, P. C., 301, 442, 457, 596
Miller, R. S., 388, 475
Miller, S. L., 178
Millikan, G. A., 587
Mills, J., 350
Milstein, J. S., 549
Milsum, J. H., 14, 42, 43, 45, 68, 116, 152, 183, 185, 205, 211, 287, 364
Minorsky, V., 205
Minsky, M. L., 94
Mishran, E. J., 507
Mitchell, R. 389
Mitchell, R. D., 23, 405, 593
Mitchell, W. C., 549
Mitchiner, J. L., 523-524, 611
Mitsch, W. J., 23, 38, 135, 136, 345, 346, 420, 435, 507, 529, 603
Mitscherlich, L. A., 133, 135, 295, 365
Moder, J. J., 94

Author Index

Molloy, L. F., 611
Monin, A. S., 283, 287, 571
Monod, J., 3, 131, 195, 484
Monroe, J., 400
Monsi, M., 370
Montagg, T., 553
Montague, C. L., 403, 446
Monteith, J. L., 43, 301, 382
Montroll, E. W., 159, 405, 591
Moore, D. T., 205
Moore, J. J., 442
Moran, E. F., 531
Moran, P. A. P., 45
Morgan, J. J., 571
Morishima, M., 507
Morowitz, H. J., 120, 178, 294, 301, 302, 318, 321
Mortimer, M., 159, 405
Morton, J. B., 52
Moses, L. E., 281
Mostow, G. D., 351
Moulton, P. G., 52
Mount, L. E., 301
Mueller, P. T., 78
Mueller-Dumbois, D., 456
Muirhead, V. U., 589
Muitra, S. C., 591
Mulholland, R. J., 565, 606
Muller, P. O., 531
Mullish, R. A., 52
Muraga, S., 94
Murphy, A. T., 94
Murphy, C. E., 377
Murphy, G. I., 371, 373
Murphy, P. E., 466
Murray, B. G., 159, 205, 324
Myers, C., 268
Myers, J., 166, 272
Myers, J. H., 400-401

Nagel, E. J., 3
Nagle, H. T., 94
Nagrath, I. J., 45
Nagy, J., 606
Nahikian, H. N., 94
Naroli, R. S., 282, 337
Nash, L. K., 104, 120, 278
Nauman, A. S., 281
Nauman, E., 408, 418
Naveh, Z., 553
Nayfeh, A. H., 205
Naylor, T. H., 507
Neales, T. F., 376
Neel, R. B., 33, 44, 66, 468
Neill, C., 501
Nessel, J., 268
Neustead, G., 248
Newcombe, C. L., 419
Newman, J. R., 3
Newton, I., 107
Newton, M., 611
Nicholls, D. C., 120
Nicholson, A. J., 43, 195, 198, 204

Nicolis, G., 7, 101, 120, 130, 140, 197, 248, 282, 322
Nihoul, J. C. J., 241
Nijkamp, P., 476, 499, 507
Nikolskei, G. V., 475
Niles, D. M., 595
Nisbet, R. M., 205, 583
Nixon, S. W., 367-368, 374, 442, 462, 464, 589, 603
Norman, R. Z., 94, 553
Novick, A., 192
Noy-Mier, I., 430, 432-433, 450
Nozaki, M., 584

O'Brien, J. J., 283, 349, 375-376, 382, 417, 426, 591, 612
O'Carroll, M. J., 45
O'Connor, D. J., 589
Odum, E. C., 23, 579, 583, 601
Odum, E. P., 23, 345, 449, 458, 466, 475, 495, 586, 588
Odum, H. W., xiii
Odum, W. E., 606
Ogata, K., 14
Olinick, M., 14
Olson, J. S., 33, 44, 66, 258, 468
O'Manique, J., 314
O'Neill, D. V., 606
O'Neill, R. V., 14, 43, 82, 133, 238, 243, 244, 248, 319, 451, 579, 582, 583, 591, 607, 610
Onsager, L., 110, 115, 265
Oparin, A. I., 318
Opshoor, J. B., 594
Ord, J. K., 350, 351
Ordway, J. W., 603
Orians, G. H., 330, 499, 577
Orias, E., 344
Osmond, C. B., 367
Oster, G. F., 12, 87, 101, 169, 235, 395, 396
Osterhelt, D., 367-368
Ostwald, W., 118, 265
O'Sullivan, P., 507
Oswald, W. J., 462
Otnes, R. K., 287
Overman, A. R., 430
Overton, W. S., 438
Oviatt, C., 589
Ozmidov, R. V., 283

Padoch, C., 512, 514
Paine, R. T., 577
Palander, T., 326
Palmer, J. D., 282, 301, 405
Palmisano, J. E., 590
Pamatmat, M. M., 104
Paperna, I., 280
Pareto, W., 285, 336, 489
Park, C. F., 567, 571
Park, O., 584
Park, R. A., 606
Park, T., 584

Parker, R. A., 138, 195, 287, 349, 365, 417
Parrish, W., 265
Parsons, T. R., 280, 382
Parton, W. J., 475
Pasquill, F., 332
Passet, R., 505, 507
Paterson, S. S., 410
Patil, G. P., 350, 351
Pattee, H. H., 23, 178, 287
Patten, A. R., 120
Patten, B. C., 14, 53, 68, 85, 94, 138, 197, 199, 212, 239-243, 248, 307, 356, 382, 611
Patterson, A. A., 601
Patton, D. R., 327
Paulik, G. J., 393
Paulsen, T. L., 585
Pavlidis, T., 176, 181, 282, 577
Paynter, H. M., 87, 169, 171, 173
Pearl, H. W., 364
Pearl, R., 150
Pearman, G. I., 571
Pearson, N. E., 330
Pegeis, E. E., 52
Pencock, J. F., 301
Peng, H. W., 110
Penner, P. S., 268, 507, 586
Perez, K. J., 589
Perkins, D. F., 442
Perrins, C., 351
Pesto, E., 568
Petersen, C. G. J., 269, 366
Peterson, B. J., 382
Peterson, F. M., 507
Peterson, J. W., 315
Peterson, L., 285, 468, 489, 603
Peusner, L., 120
Phelps, E. B., 43, 463
Philips, N. A., 558
Phillips, D. T., 248
Phillips, J. N., Jr., 166
Phillipson, J., 120
Phinney, H. K., 599
Pielou, E. C., 181, 203, 205, 222, 276, 340, 351, 575, 577, 583
Pierce, J. R., 322
Pierrot-Bults, A. C., 405
Pigeon, R. F., 342, 418
Pilati, D. A., 268, 507, 586
Pimentel, D., 531
Pinder, J. E., 388-389
Pinkerton, R. C., 6, 115
Pippenger, N., 311
Pitts, W., 78, 79
Planck, M., 288, 293
Platt, R. B., 466
Platt, T., 279, 283, 284, 287, 327, 365, 382, 612
Plessner, K. W., 360
Polissar, M. J., 293, 301, 595
Pollock, H. N., 557
Pomeroy, L. R., 45

AUTHOR INDEX

Poole, R. W., 14, 140, 159, 205, 222, 248, 351, 393
Porteous, J. D., 532
Porter, J. W., 346
Poston, T., 205
Poule, L., 52
Powers, W. F., 589
Prakash, A., 607
Prentki, R. T., 606
Preston, F. W., 337-338, 340, 343
Prigogine, I., 101, 114, 120, 130, 140, 170, 171, 195, 197, 248, 265, 282, 287, 313, 322, 348, 574
Pugh, R. E., 553
Purdy, W. C., 447
Putzrath, F., 600
Pye, K., 587

Quastler, H., 24, 307, 322
Quinlan, A. V., 171, 173, 294, 297, 301

Rabinowitch, E. T., 133, 356, 361, 365
Ragada, R. K., 507
Ragotzkie, R. A., 371, 372
Raines, G. E., 455, 457
Rainey, R. H., 128
Raisbeck, G., 322
Ramsey, F., 296, 297
Randall, M., 120
Randers, J., 600
Randerson, P. F., 471
Ransom, R., 14, 348, 405
Rappaport, R. A., 512, 514
Rapport, D. J., 476, 487, 507
Rashevsky, N., 33, 43, 79, 94, 120, 134, 140, 181, 517
Rasool, S. I., 571
Rast, W., 597
Raston, S., 154
Raunkier, C., 335
Rawson, D. S., 372, 418, 420
Ray, D. M., 350, 507, 585
Ray, W. H., 140, 287
Rech, J. A., 587
Reckshow, K. H., 420
Redfield, A. C., 463
Reed, L. J., 43, 150
Rees, P. H., 507
Regan, E. J., 216, 348, 468
Reichle, D. E., 555, 607
Reiners, W. A., 475, 561, 565
Renn, C. E., 449
Rescigno, A., 196, 223
Reynolds, J. L., 376
Reza, F. M., 322
Ricci, F. J., 68
Rich, L. G., 442, 461, 475
Rich, P. H., 606
Richards, B. N., 430
Richards, D. W., 146
Richardson, H. H., 94

Richardson, H. W., 549, 553
Richardson, J. C., 195
Richardson, J. L., 24
Richardson, J. R., 117, 574
Richardson, L. F., 218-219, 222, 274, 553
Richey, J. E., 241, 316
Richman, S., 190
Richter-Dyn, N., 137, 158, 405, 575
Ricker, W. E., 386
Ridley, B. K., 571
Riedl, R., 322
Riehl, H., 571
Rieske, J. S., 167, 356
Riley, G. A., 433, 438
Rinaldi, S., 195, 592
Risser, P. G., 404, 475
Rivero, J., 602
Robert, J., 606
Roberts, D. V., 175, 176, 181
Roberts, E. H., 589
Roberts, F. S., 81, 87, 89, 94, 507
Robertson, A., 442, 475
Robertson, A. D. J., 321
Robertson, T., xiii
Robichaud, L. P., 85
Roff, D. A., 399
Rogers, A. E., 68
Rogers, P. H., 14
Rogers, P. L., 164, 189, 195
Rohwer, S. A., 595
Rose, A. H., 295, 301
Rose, J., 140
Rosen, R., 14, 201, 523, 583
Rosenberg, R., 87, 94, 475
Rosenweig, M. L., 346, 531
Rosenzweig, M. L. 410
Rosinsky, W., 348
Ross, J., 590
Ross, R., 172-177, 198, 375, 376
Rothstein, J., 322
Royama, T., 202
Rubey, W. N., 555
Rubinow, S. I., 14, 45, 181, 205, 222
Rublee, P., 131
Rummel, R. J., 553
Russell, B. C., 609
Russell, G. S., 475
Russell, S. R., 350, 442
Rutledge, R. W., 308, 309
Rykiel, E. J., 43, 430

Saaty, T. L., 89
Sacher, G. A., 322, 324
Sage, A. P., 553
Salamon, P., 584
Salisbury, F. B., 375-376
Samuelsson, G., 366
Sanderson, P. C., 52
Saunders, H. L., 337, 345
Savageau, M. A., 90, 181
Scavia, D., 442, 475
Schaefer, G., 94, 520, 521, 579

Schaefer, M. B., 192
Scheidegger, A. F., 287, 332, 334, 351
Schlegel, R., 107, 187, 322
Schlosser, I. J., 424
Schmerl, R. B., 301
Schmidt, A., 265
Schmidt, K. P., 350, 584, 593
Schmidt, O. H., 293
Schnakenberg, J., 195, 247, 268
Schneider, H., 557
Schoffeniels, E., 117
Scholander, P. F., 297
Schomer, N. S., 528
Schreiber, R. K., 589
Schrodinger, E., 306, 318
Schultz, V. L., 579, 583
Schultze, C. L., 490-491
Schuster, P., 178, 195
Schwintzer, C. R., 597
Scott, D., 120
Scott, G., 579, 601
Scott, H., 265
Scriven, L. E., 324
Searl, M. F., 507
Searle, S. R., 82
Seber, C. A. T., 351
Sedlik, B. R., 433
Seip, K. L., 191
Seki, T., 370
Sell, M. G., 436-437, 542-543, 603
Sergin, V. Y., 557, 559
Servizi, J. A., 254, 402
Shabman, L. A., 495
Shannon, C. E., 307, 313, 317, 319, 322, 343
Shapiro, L., 268
Sharitz, B., 301
Sharpe, D. M., 377, 589
Sharpe, W. F., 52
Shearer, J. L., 94
Sheath, P. H. A., 351
Sheldon, R. W., 326
Shen, H. W., 571
Shenton, L. R., 586
Sheppard, C. W., 43, 45
Shimazu, Y., 116, 152, 154, 156, 313, 487, 526, 565, 575
Shinozaki, K., 131, 146
Shock, N. W., 609
Shoup, C., 135
Shugart, H. H., 14, 45, 94, 243, 248, 287, 378, 459-460, 469, 472, 475, 532, 590, 610
Sibseu, R., 350
Siegel, F. R., 571
Siler, W. L., 603
Sillén, L. G., 272, 555
Silvert, W., 279
Simberloff, P. S., 44
Simmons, I. G., 351, 531
Simon, D. E., 52
Simon, H. A., 553
Simon, W., 94

Simpson, E. H., 343
Singer, S. F., 571
Singh, J., 79, 94
Sinha, S. K., 286, 287
Sipe, N. G., 533, 537, 584, 603
Sitaramiah, P., 449
Sjöberg, S., 366, 425
Slater, J. V., 419
Slesser, M., 268, 476
Slobodkin, L. B., 155, 159, 190, 457
Smayda, T. J., 589
Smellie, R. M. S., 301
Smith, E. L., 365
Smith, F. E., 33, 164, 169, 199, 280
Smith, G. M., 68
Smith, H. A., 553
Smith, H. T., 198
Smith, J. M., 442
Smith, K. J., 553
Smith, M. H., 605
Smith, M. J., 14
Smith, N. S., 590
Smith, W. H. B., 150, 295, 297, 400, 593, 595, 613
Smith, W. K., 350
Smithe, R. E., 52
Smuts, J. C., 4, 24, 577
Snedaker, S. C., 442, 475
Snell, J. L., 595
Socharlan, A., 279
Soddy, F., 265, 507
Soemarwoto, O., 515
Sokal, R. R., 351
Solbrig, O. T., 499
Sollins, P., 413, 415
Solomon, M. E., 405
Somero, G. N., 405
Southwood, T. R. E., 351
Spain, J. D., 52
Spanner, D. C., 120
Sparks, R. E., 611
Sparrow, A. H., 281
Spear, I., 592
Spencer, J. L., 588
Spetner, L. M., 211
Spikes, J. D., 356, 358
Staffeldt, E. F., 324
Stafford, H. A., 337
Stairs, G. R., 23, 405, 593
Stanhill, G., 530, 531, 546, 548
Stanton, R. J., 571
Starr, D., 611
Stearns, F. W., 553
Steele, J. H., 24, 243, 280, 287, 323, 326, 365, 366, 382, 386, 405, 576, 591
Steele, J. M., 442
Steeman-Nielsen, E., 373
Steinhart, C. E., 268, 531
Steinhart, J. S., 268, 531
Steinitz, C. H., 547
Stellar, D. L., 541
Stephan, G. E., 330-331

Stephanapoulos, G., 222
Stephens, G. R., 451
Sterndl, J., 341
Stewart, I., 205
Stewart, J. Q., 328
Stewart, W. E., 140, 287
Stiles, E. W., 589
Stoeckinius, W., 367, 374
Stoiber, R. E., 563
Stokes, A. W., 351
Stommel, H., 606
Stonehouse, B., 351
Streeter, H. W., 463
Strehler, B. L., 277, 389-391, 405
Stull, E. A., 589
Stumm, W., 571
Sudo, F. M., 205, 553
Sugayima, K., 116, 607
Sugita, M., 118, 245
Sugiura, Y., 415
Summers, J. K., 430
Sundermann, J., 442
Sussman, M. V., 266
Sutcliffe, W. H., 607
Suttman, C. E., 335
Sverdrup, H. U., 357, 371
Swallows, C., 552, 603
Swamy, M. N. E., 248
Swaney, D., 603
Swartzman, G. L., 475, 524
Swift, J., 274
Swift, M. J., 405
Swinburn, C. D., 507
Sykes, L. R., 272
Szilard, L., 192

Taagepera, R., 517
Tabot, F. H., 460
Tagawa, K., 584
Taillie, C., 350, 351
Takad, Y., 596
Takahara, Y., 14
Takahashi, M., 105
Talbert, N. E., 367
Tallengen, B. D. H., 235
Tamiya, H., 356-357
Tanner, T. T., 203
Tansley, A. G., 453
Tashure, C., 531
Taub, F., 464, 467, 475
Taylor, B. R., 430
Taylor, L. R., 248, 405
Taylor, L. W., 317, 492
Teal, J. M., 225
Tenney, M. W., 585
Tenore, K. R., 411
Terborgh, J., 147, 345
Terrell, T. J., 283
Thayer, G. W., 414
Theineman, A., 418
Thelkeld, J. L., 301
Thibodeaux, R. V., 442
Thiman, K. V., 592

Thirring, H., 265
Thom, R., 195
Thomann, R. V., 442, 529, 589
Thomas, J. M., 583, 607
Thompson, D. A., 279, 289, 351
Thompson, G. L., 94, 595
Thorell, H., 175
Thorndike, E. H., 120
Thornes, J. B., 571, 576
Thornley, J. H. M., 133, 157, 181, 199, 221, 358, 368, 382
Thron, C. D., 234
Thulasiraman, K., 248
Tiezen, L. L., 380, 442
Tillman, D., 233
Tillman, F. A., 583
Tiner, R. K., 301
Ting, I. P., 374
Titman, D., 208, 233
Toates, F. M., 45, 94, 384
Todd, K. K., 94
Todd, W. R., 587
Tokad, Y., 13
Tomida, E., 607
Tomlison, P. B., 382
Tomovic, R., 68, 137
Totsuka, T., 410
Townsend, C. R., 507
Toynbee, A., 328, 517
Tran, K., 531
Tribus, M., 115, 120, 266, 313, 320, 321, 322, 351
Trincher, K. S., 318, 322
Tsuchiya, H. M., 195, 196
Tukey, J. W., 284
Tupper, H. M., 584
Turing, A. M., 79
Turk, A., 12, 610
Turk, J., 610
Turnbull, C. M., 510
Turner, B. W., 248, 405
Turner, J. E., 476, 487, 507
Tygart, F. J., 517, 531

Udvardy, M. D. F., 351, 475
Ulanowicz, R. E., 169, 343, 344
Ulmann, E. L., 539
Urabe, T., 565
Urry, D. W., 277-278, 293
Usher, M. B., 195, 248, 383

Valach, M., 72, 94
Valentinizzi, M. E., 302, 318
Valisalo, P. E., 68
Van den Ende, P., 195
Van den Vaart, H. R., 211
Vandermeer, J., 14
Van der Pol, B., 157, 176-177
Van der Spoel, S., 405
Van Dobben, W. H., 475
Van Dyne, G. M., 380, 475, 524, 531, 581
Van Mieghem, J., 571

Van Raag, H. G. T., 505
Van Valen, L., 465
Van Voris, P., 243, 248, 308, 458, 577
Varshney, C. K., 611
Vasquez, A., 279
Vela, G. R., 315
Velz, C. J., 442, 475
Vemuri, V., 14
Verhoff, F. H., 585
Verhoff, V., 169, 198
Verhulst, P. F., 146
Verkher, R. C., 52
Vernadsky, W. I., 577
Vestal, A. G., 338
Vielking, D., 611
Vinogradov, M. E., 425
Vitousek, P. M., 475
Voevodsky, J., 549
Vogel, S., 34, 433, 437, 441
Vogely, W. A., 507, 531
Volk, B. G., 433
Vollenweider, R. A., 418, 420
Volterra, V., 142, 146, 155, 194
Von Bertalanffy, L., 4, 14, 28, 33, 43, 45, 282, 337
Von Foerster, H., 91, 150, 285, 322, 489, 534
Von Thunen, J. H., 497
Von Uexküll, J., 287
Vukobratovic, 68, 137
Vuysje, D., 13

Waggoner, P. E., 451
Waide, J. B., 611
Waintz, W., 531
Waite, T. D., 14, 45
Wales, J. H., 611
Walford, J. C., 68
Waller, W. T., 388
Wallis, J. R., 586
Wallman, P., 405
Walsh, J. J., 462
Walter, C., 181
Walter, W. G., 78
Walters, C., 235, 380
Walters, R. A., 418
Walters, V., 607
Waltman, P., 205
Wang, F. C., 268, 603
Wang, L. K., 461
Wang, M. H., 611
Wangersky, P. J., 151, 153, 154, 159
Warren, C. E., 308, 423, 424, 460, 463, 475, 586, 599
Wartenberg, D. E., 592
Watanabe, M. S., 110

Waters, T. F., 405
Waters, W. E., 351
Watt, D. E. F., 24
Watt, K. E. F., 24, 133, 134, 140, 145, 159, 181, 192, 205, 384, 567, 571, 583
Watts, J. A., 603
Watts, T. M., 553
Weatherby, A. L., 405
Weaver, R., 176
Weaver, W., 307, 317, 320, 322, 343
Webb, W. L., 294, 295
Webber, M. M., 553
Weber, A., 326
Weeks, D., 611
Weibull, W., 285, 286
Weinberg, D., 583
Weinberg, G. M., 14, 322, 583
Weiss, P. A., 22, 24, 287
Weisz, P. B., 280
Welch, E. B., 420, 531
Wells, P. V., 286
Wesley, J. P., 120, 503, 504, 562, 565
West, D. C., 469, 472, 475, 590
Westlake, D. F., 371
Wetzel, R. G., 606
Wheaton, W. L. C., 553
Wheeler, J. O., 531
Wheeler, W. M., 508
Whipple, 463
White, G. C., 248
White, L. A., 508
White, L. H., 508
Whitmore, T. C., 279
Whitrow, G. J., 28, 576
Whittaker, A. H., 351
Whittaker, R. H., 24, 223, 338, 340, 382, 442, 555
Whyte, L. A., 22
Wiaume, J. M., 7, 118, 313
Wiegert, R. G., 151, 382, 385, 454, 457, 475
Wiener, J. G., 605
Wiener, N., 307, 313, 319, 320, 526, 574
Wiens, J. A., 475
Wildi, O., 430
Wilkenson, T. S., 45
Williams, C. B., 342-343, 351, 590
Williams, D. D., 442
Williams, G. R., 339
Williams, N., 589
Williams, R. A., 94
Williams, R. B., 439
Williams, R. W., 595
Williams, W. D., 442

Williamson, J. R., 587
Williamson, M. H., 195, 248, 351
Willis, J. C., 342
Wilson, A. G., 507, 531, 532, 553
Wilson, C., 582
Wilson, D., 24
Wilson, E. O., 33, 44, 45, 101, 207, 395, 396, 509
Wilson, L. L., 24
Wilson, P. A., 178
Wilson, R. F., 363-364, 366
Wilson, R. J., 87, 140, 585
Winarsky, I., 91
Winbera, G. C., 405
Winfree, A. T., 287, 349, 576
Wirosardjono, S., 611
Wissmar, R. C., 606
Witt, A. A., 218
Wittes, J. T., 610
Wittes, R. E., 610
Witwell, J. C., 301
Woldenberg, M. J., 351
Wolff, P., 531
Wolman, A., 553
Wolman, M. C., 597
Woodman, M. J., 403, 475
Woodruff, L. L., 463, 465
Woodwell, G. M., 43, 462
Wroblewski, J. S., 283, 349, 375, 417, 426, 461
Wulff, F., 425, 464
Wurster, C. B., 553
Wyckoff, R. D., 43
Wycoff, D., 565
Wylie, P. J., 571
Wynne-Edwards, V. C., 453

Yakovlev, V. I., 576
Yerd, C. C., 587
Yocum, C. S., 597
Young, D. L., 297, 435, 586, 593, 595
Young, J. W., 611
Yount, J. L., 346

Zadeh, L. A., 248
Zaret, T. M., 248, 405
Zedler, J., 594
Zeleny, M., 195
Zemanek, H., 78
Zevallas, A. C., 519
Ziegler, J. R., 205, 553
Zipf, G. K., 336, 342-343
Zlotin, R. I., 405
Zopf, G. W., 322
Zucchetto, J., 231, 268, 481, 495, 521, 533, 546, 549, 603
Zverev, A. L., 283

Subject Index

Acceleration, 37
 diagramming, 137-139
Acetylcholine, 118
Action potential of nerves, 117
Activated sludge, 417-418
Adaptation, 457
 light, 364
Additive flow, 9
Additive interaction, 123-124
Additive loop without storage, 172
Adiabatic expansion, 299
Adiabatic process, 102
Adjective, 91
Adsorption: Elovich model, 156
 Langmuir model, 169
Advantage, military, 517
Adverb, 91
Aesthetics and complexity, 324
Affinity, 109
Age and area, 342
Age classes, 279
 graphs, 391-393
Aggregated model: physical, 240
 production function, 321
 United States, 488-491
Aggregation of detail, 19, 26, 355
Agrarian city, 548
 recycle, 530
Agriculture, 18
 industrial, 523
 overstory, 519
 shifting tropical, 513
 subsistence, 523
Agroecosystems, 513-514, 519, 523
Albedo, 577-579
Alders, 294
Algal ecosystem, attached, 429.
 See also Blue green algal mat
Algal yield (brown algae), 191
Algebraic signs for analog computer, 61
Alleles, 397
Allen productivity method, 279
Allometry, 136
Alternation of generations, 395
Alternative meanings for equations, 148
Amchitka Island, Alaska, 455

Amino acid limitation, 191-192
Ampere, 35
Amplifier, 8, 17
 constant gain, 123-124
 effects, 256
 ratio, 256-257
Amplitude ratio, 40-41
Amplitude scaling, 57, 62
Amur, white, 421
Anabolism, 374
Anaerobic digester, 417-418
Analog circuit: additive feedback, 174
 Elovich equation, 156
 materials cycle, 162
 Michaelis-Menten loop, 163
 potential-kinetic oscillator, 188
 stochastic logistic, 157
 two storages, 184
Analog computer, 53, 71, 74
Analog simulation: crop harvest and replanting, 516
 human survivorship, 391
 temperature model, 295-297
Analysis, 4
Anchovies, 461
AND gate, 73-75
Angle, phase lag, 40-41
Angular velocity, 40, 107, 188
Animal managers, 247, 402
Animals: and microbes, 247, 403
 roles in tundra, 446
Anthocephalus chinensis, 216, 319
Ants, 396
Appalachian forest, 472
Appalachicola Bay, Florida, 408-409
Apple, plot command, 49
Approaches to modeling, 579-580
Aquarium, 13, 408
 ecosystem, 19, 413-430, 435-439, 453-455, 460-467
 metabolism, 114
 minimodels, 522
Aquatic producers, 375
Area and age, 342
Arbeit, 109, 293, 294
Arcs, 87
Arctic tundra plants, 377

Arkansas, 527
Arms race, 218-219, 517-518
Arrested succession, 461
Arrhenius temperature effect, 289, 291
Arrow worms, 425
Art and systems, 6, 91-92
Artemia, 466
Artificial stream, 423
Ascendency, 343
Asexual reproduction, 395
Aspirations, 579
Assimilation, 258
Asymptotic growth: recycle limited, 163-165
 source limited, 145
Atmosphere: hierarchy, 555, 558
 pressure distribution, 276-277
ATP, 271
Attractiveness perception, 547
Atwood's Machine, 116
Australian range model, 524
Autecology, 406
Autocatalytic, 6, 7, 101
 conserved material, 170
 chain reaction, 174
Autocatalytic competition, 220
Autocatalytic growth, 141-159
 with backforce, 145
 conserved material loop, 170
 constant gain, 148-149
  quadratic drain, 149
 constant rate, 153
 controlled feeding rates, 151
 controlled flow, 145
 cooperative, 150-151
 cubic drain, 150-151
 cumulative byproduct, 154
 cumulative drain, 154
 density drag on gain, 149
 with diffusion, 155
 discrete, 144
 flow source, 145
 Gompertz equation, 156
 inertial accelerations, 157
 limiting denominator, 154
 linear decay of factor, 156
 quadratic feedback, 150
 quadratic logistic, 147

self dilution, 152-153
steady state storage, 188
threshold, 101, 153
threshold-limited gain, 149
time delay, 152-154
unrenewed storage, 145
Wiegert notation, 153
Autocorrelation, 282-283
Autolysis, 408, 411, 414
Automata, 79
Availability, 105

Back reactions, 127
Backflow, 10
Backforce, 10, 34, 44
autocatalytic growth, 145-146
inertial, 188
through intersections, 125
from storage, 106
Bacteria: green sulfur, 419
luminous, 135
purple sulfur, 419
sulfate-reducing, 419
white sulfur, 419
Bahia Forescente, 445, 466
Balanced aquarium, 408
Balance of forces, 10, 107, 137, 138
inertial backforce, 107, 188
Van der pol oscillator, 177
Balance of payments, 23, 479
Balsa, 216
Baltic Sea, 231, 243, 429
Bank account, 37
Barnacles, 334
Basal area of trees, 340
Basic-non base analysis, 539
BASIC computer language, 48-50
Bathtub, 30
Beef production, 525-526
Beers-Lambert law, 369
Bees, 396
Behavior: and culture, 508-509, 532
as programs, 51
Bibliographies on ecological modeling, 579
Bifurcation, 154, 197
Biochemical network language, 90
Biochemical oxygen demand, 33, 43, 461
Biocoenosis, 96, 224
Biogeochemical cycle, 562
global nitrogen, 564
phosphorus, 563
sulfur, 563
Biological evolution, 464
Biomass: succession, 447, 470-471
tropical forest, 318-319
Biosphere, 17, 18, 20
Biosphere models, 20, 43, 554-571, 572-583
Biostat, 193
Biotic potential, 11
Biotope, 96, 224, 406

Birds: giant extinct, 444
population changes, 85
storage in winter, 43
Bit, 304
Black box, 12
Black hole, 157
Blackman response, 132
Black Sea, 419
Bleedwater lagoon, 414
Blind man and elephant, 406
Block diagram, 30
Bloom-crash, 578
Blue-green algal mat, 367, 369
anaerobic metabolism, 414
maximum power loading, 187
microcosm, 413
microcosm production, 413
BOD, 33, 43, 461
models, 402
Bode plot, 40-41, 182
Bog and peat, 414, 471
Boltzmann's constant, 288, 293, 311
Boltzmann theorem, 311
Bomb calorimeter, 95
Bond graphs, 6, 87-88
Boolean algebra, 77, 78
Boolean matrix, 242
Bottle microcosms, 291-293
Bottom exchange of nutrients, 420
Bottom nutrient regeneration, 420
Boundaries, 11
of diagram, 22
Boundary conditions, 261
Box, 8, 12
concave-sided, 12
modeling, 241
Brachen-blueberry community, 338
Branches (graph theory), 87
Branching chain reaction, 174
Breeding threshold, 73
Brine, 414
Brine bacteria, 367
Brine ecosystem, gnotobiotic microcosm, 466-467
Brookhaven, Long Island, 462
Brownian motion, 281, 576
Brusselator reaction, 197
BTU, 95-96
Bucket brigade, 65
Budworms, 456
Business cycles, 492
Buzzards Bay, Massachusetts, 337
Brazil, 547
Byproduct use, 554

C_3 and C_4 plants, 373, 375
C^{14} uptake in photosynthesis, 366, 367
Cacao, 519, 522
Cactus, 25, 373, 376, 415
Cadam (*Anthocephalus*), 319
Cadmium, 529
in microcosm, 458

Calculation of coefficients, 27, 58, 61
Calibration, 579
Calvin cycle, 375
CAM, 374, 376
Canadian lakes, 420
Canals, 542
Capacitance, 27-28, 35, 43
Capacitor, 34-35, 57
analog integrator, 57
energy storage, 105
Capital: and production, 479, 487
savings, investment, 479
Carbohydrate, 113
Carbon cycle, 560-562
Carbon dioxide, 20
increase, 562
in microcosms, 412
diurnal variation, 412
seasonal and latitudinal variation, 562
Carnot cycle, 102
Carnot efficiency, 102, 115, 266
Carotenoids, 374
Carotenoid and chlorophyll in microcosms, 450
Carotenoid-chlorophyll ratio, 460
Carp pond, 522
Carrier, 208-209
Carrying capacity, 330-331
humans in space, 526
Cash crop, 519
Caste in termites, 397
Casualties equation, 549
Catabolism, 374
Catastrophe, 325
role for humanity, 576
Catastrophe theory, 195, 456
Catastrophic pulsing, 445-446
Cattle, 515, 525
Causal analysis, 244
Causal forces, 21
Causality matrix, 242
Causal word chains, 79-80
Cecropia, 216, 396
Cedar Bog, Minnesota, 439
Cells and time constant, 280
Cellular growth, 43
Central place hierarchy, 323-324
Central places, 539
Centralization during growth, 327-328
Centrifugal force, 138
Certainty, 304
Chain: autocatalytic units with multiplier feedback, 198-200
embodied energy, 253
energy quality, 323, 325
energy transformation, 253, 269-271
with interacting sources, 230
land use transformations, 545
limiting factors, 134-136
linear connected consumers, 197-199
linear storages, 182
mechanical train, 284

Subject Index

Michaelis-Menten loops, 198
   seesaw, 201-202
   two-tank autocatalytic units, 198, 199
Chain reactions, 173-174
   energy and religion, 510
Change of state, 11
Channelization and wood storks, 387
Chaotic growth, 154
Characteristic equation, 183-184
Charge-up, 32
   against backforce, 34
   in photosynthesis, 362-363
Chemical cycle, 17
Chemical equations, energy diagrams, 127
Chemical equilibrium, 108
Chemical hierarchies, 277
Chemical loops, 170-171
Chemical potential, 127
   as thermodynamic force, 114
Chemical potential energy, 106, 109, 113
Chemical reaction: chain, 173-174
   first order, 127
   reversible, 127
   second order, 127
   zero order, 127
Chemical resistance, 289
Chemostat, 191-193
   limiting factor, 191-193
   one source, 193
   prey-predator oscillation, 195-196
   series, 199
   two sources, 193-194
Chestnut, 469
Chicken and egg, 80, 393-394
Chlamydomonas, 190, 467
Chlorella, 166, 371, 374
   in hydra, 411
Chlorophyll: adaptation to light, 165-166
   40,000 foot candles, 418
   light reception, 359
   in microcosms, 450
   upwelling, 462
Chloroplast: efficiency, 117
   photosynthesis, 361
   stacking, 164-166
Choco Indians, Darien, 513
Choices, information, 312
Chromatium, 359
Circadian rhythms, 177
Circular pathways, 110
Circular succession, 445, 453
Cities, 20, 532-553
   high quality consumers, 539
   rank order graph, 336, 342
   regional minimodels, 534
   representative model, 548
   simulation, 546, 549
Citrus agriculture, 523, 526
Civilization, frenzy pulse, 374

Cladocera, 190
Cladophora, 429
Classification, Holderidge, 410
Clay, marl, sand, 560
Climax, 108, 443-449
   noisy, 448
   oscillating, 445
   pond, 449
   self generating matrix, 451
   types, 448
Clock gear train, 284
Closed system, 107, 411
   decomposition in, 449
Clover, 520
Clustering of eggs, 400-401
Coal, energy transformation ratio, 252
Coastal water, 438
Cobb-Douglas function, 263, 357-359, 487
Coefficients: calculations, 27, 58, 61
   consumer preference, 238
   efficiency, 142-143
   energy intensity, 500
   energy turnover, 190
   extinction, 370
   input-output, 241-242, 264, 500
   matrix notations, 237
   relative growth, 137
   storage, 27
Coexistence, 212
   models, 346
Coffee, 519, 522
Cogwheels, 284-285
Cohorts (age), 391
Coin flip, information, 304
Collar symbol, 37
Collier County, Florida, 495
   land development, 545
Colonial stage in Florida, 546
Colonization, 460
   model, 459
Colpidium, 447
Columbia River, 424
Combinations, 302, 304
Communication, information in equipment, 320
Community matrix, 238
Comparator, 72-74
Compartments, 13, 25
   global models, 559-560
   spatial, 233
Compensation point, 370
Competition, 13, 19, 25, 206-219
   autotroph-heterotroph, 426
   biochemical reaction loops, 177
   cooperative, 206
   crops, 214-215
   density, 215
   economic, 492
   energy, 118
   feedback reinforcement, 496
   growth specialists, 216

index of success, 243
   intraspecific, 215
   intra-unit, 206
   nitrogen, phytoplankton, 220
   plankton paradox, 233, 345
   plants and animals for nutrients, 411
   power, 101
   pressure, 211
   recycling resource, 212
   resource dilution, 211
   vapors compared with plants, 213-214
   water, 542
Competitive exclusion, 210-211, 215
Complement, 77
Complex ecosystem models, 433
Complex frequency, 40
Complexity, 302-322
   index, 302, 343
   molecular, 103
Complex models, 581
Complex S plane, 42, 184, 185
Compost, 402
Compressed gas, energy, 105
Compucolor command, 49
Computer commands in BASIC, 49
Computer graphing, 49
Computer program language, 51
Computers and religion, 526
Concentration model, 128
Condenser (capacitor), 35
Conditional entropy, 305
Conditioned reflex, 78
Condition index, 386
Conductivity, 6, 10, 26-28, 35, 43
Conductivity network, 234
Connections, 19
Connectivity, 302
Conquest, 511
Conservation of matter, 132, 162
   autocatalytic loop, 170
   Michaelis-Menten loop, 163
Constant force source, 96-97
Constant gain amplifier, 8, 39, 123, 124, 129
Constant gain module, 148
Constant outflow, 39
Constant production (autocatalytic divisor), 153
Constant rate of change, 152
Consumer preference coefficients, 238
Consumers, 7, 8, 383-405
   host, 195
   households, 491
   intake models, 384
   main pathways, 384
   plants of animals, 411
   producers *vs*., 356
   pulsing, 281
   sludge decomposition, 417
   stabilizer, 244
   strategy, 487
   surge system, 573

Consumption: land, 545
  maximum, 402
Control and culture, 509
Controlled feeding models, 151-152
Continental shelf, 416, 438
Continental species, 339
Continuous functions, 36
Continuum, 348
Control, 15-16, 72
  animals, 403
  arm, 253
  forcing function, 97
  high quality pulsing, 281
  intersection flow, 131
  theory, 30-31, 41-43
Convection, spatial pattern, 324
Convergence, 16-17
Converging web, 223, 226
Converter, 574
Conveyor belt, 100
Coordinates, triangular, 410
Cooperation, 218
Cooperative competition, 206
Cooperative growth, 150-151
Coral reef: animal-microbe relations, 247
  diversity, 346
  microbial detritus, 428, 430
  minimodel, 428
Coral symbiosis, 411
Core conductor, 117
Coriolos force, 240-241
  diagrams, 138
Correlation coefficients, 81
Cost benefit analysis, 493, 498
Coulomb, 35
Counter, 73, 78
Countercurrent, 12
County models, 536-538
Coupled processes, 114
Coupled sources, 98
Coupling: factor, 242
  feedback to drain, 150
  linear, 123-124
Crabs, 297
  and microbes, 402
Cranes, 388
Crassulacean acid metabolism, 374, 376
Creosote bush, 378
Critical depth, 372
Critical materials, 545
Critical size, phytoplankton patch, 417
Crop, cash, 519
Cross catalysis, 171
Crowding, 146, 215
  waste effects, 154
Crystal River estuary, Florida, 296-298, 435
CSMP, 48
Ctenophore, 401
Cube, 28

Cubic function, power dissipation, 309
Cultural eutrophication, 421
Cultural evolution, 510
Cultural flow, 330
Culture, 508-510
Cumulative equations, 30
Cumulative graph of normal distribution, 33, 334
Cumulative species graph, with area, 337
Curvature of gradients, 155
Curve fitting, 32-33
Cusps, 195
Cutting cycle in swamps, 434
Cybernetics, 72
Cycle: biogeochemical, 562
  global carbon, 560-562
  global nitrogen, 564
  global phosphorus, 563
  global sulfur, 563
  hydrologic, 556
  material, 18, 160-162
  sediments, 556, 560
  successional stages, 455
Cycling index, 242
Cycling receptor, 8, 20
  maximum power, 166
Cyclostrophic balance, 138
Cypress ponds, 262
Cypress swamp, 495
  generating basin, 471
  model, 473
  simulation, 434

Daphnia, 178, 190
Darcy's law, 81
Darien, Panama, 513
Darwin-selfish selection, 453
DDT model, 43
Dead-log model, 145
Decline in succession, 443-444
Decomposition: food chain, 402-403
  microcosms, 449
  rate, 32-33, 44
  sludge model, 417-418
Deep sea diversity, 337
  luminescence, 349
Defense, 568
Defense models, 218-219
Deforestation, 529
Degradation, 11
Degree of equation, 30
Demand, 130, 328, 330
  price, 482
  total, 499
De Morgan's principle, 75
Density, 28
  economic, 539
Density-dependent effect, 31
Density-dependent growth, 151
Depreciation, 7, 313
Depth and productivity, 420

Derivative, 32, 43
  diagrammatic representation, 135, 137
  price, 484
  sensor, 37
Deserts, 430, 432
Design features of ecosystems, 407-408
Desulfovibrio, 419
Detritivore, 22, 403
Detritus, 22, 403
  food chain, 440
Development: optimum, 540
  simulation, 543
Diagramming: city in Forrester symbols, 550
  derivative, 135, 137
  environmental impact, 495
  equations of motion, 137-138
  input-output network, 499
  logistic equations, 148
  materials overlay, 160-161
  mortality, 388
  price and demand notations, 482
  system, 21
  temperature, 291, 299
  test questions, 139
Diatoms, 429
Didinium, 203
Dieback, 456
Difference equations, 6, 30, 36, 43, 46
Difference force, 129
Differential equation, 6, 29, 60
Differentiation, 31
Diffusion, 7, 115
  eddy, 234
  equation, 334
  Gaussian model, 331-333
  growth regulation, 155
  limitations, 135
  molecular, 110
  momentum, 138
  Monte Carlo simulation, 85
  oscillation, 348
  oxygen in thin ecosystems, 415
  photosynthate in trees, 156
Digital computation, 36, 46-48
Digital logic, 72-79
Digital nature of limiting factors, 132
Digraphs, 87, 89
Dilution intersection, 127
Diminishing returns, energy model, 483-484
Diodes in analog computer, 64
Directed digraphs, 87
Directedness matrix, 242
Discounting, 498
  energy, 499
Discrete change, 36
Discrete systems, 72-79
  time lag, 154
Discriminant, 184
Disordering energy, stabilizing climax, 446

Subject Index

Disordering process, 314-315
Disorder by ordering, 102
Disorganization and maintenance, 309
Dispersal, 333-334, 398, 399
   energy, 7, 402
   matter, 26
   models, 399
   stabilization, 400
Displacement, equilibrium, 312
Displacement response, 446
Display of analog computer, 59
Dissipation, 26
Distance and feeding, 320
Distribution: age, 392
   exponential, 271, 276
   energy, 26
   gamma, 285
   log normal, 337
   normal (Gaussian), 333
   Pareto, 285, 336
   Poisson, 334
   spatial, 323-351
   spatial pattern of sources, 324
   stable age, 279
   Weibull, 285
Distributor, 574
Diurnal variation, 409, 414
   CO_2 in microcosms, 412
   leaf heat, 290
   leaf water exchange, 290
   oxygen; in blue green algal microcosm, 413
     in thermal microcosm, 292
   oxygen-delayed, 414-415
Diverging energy, 259
Diverging pathways, 128
Diversion flow, 9
Diversity, 323-351
   alternative mechanisms, 345
   biomass in succession, 449
   cumulative species with individuals, 336
   deep sea, 345, 346
   disturbance, 345
   DNA, 281
   dominance, 340
   energy for, 147, 298, 309
   eutrophication, 346, 418
   feedback role, 344-345
   increasing gross productivity, 468
   insect light traps, 342
   kelp system, 426
   logistic model, 147
   log normal summation, 338
   microcosms, 414
   mosaic theory, 455
   oyster reef model, 298
   quadratic minimodel, 344
   sea bottom, 337
   species seeding, 459
   square root of individuals, 345
   succession, 449

   table of indices, 343
Divisor, 124
DNA, 19, 281
   communities, 281
Dollar performance index for pesticide, 524
Domar model, 486
Dominance: diagrams, 88
   diversity, 340
   hierarchical, 515-516
   innovation, 515-517
   relationships, 88
Dominant year class, 279
Dominican Republic, 529, 545
Donor driven, 27
Donor-receiver conventions, 82-83
Dooryard garden system, 513
Do's and don'ts in modeling, 580
Drag, 126
Drain, 25, 29
Driving force, 10, 27, 34
Drought, 432
Dual energy source to reaction, 112
Dual pathway, 126
Duckweed: competition, 210
   logistic growth, 146
Dunaliella, 466
Dynamic equilibrium, 108
Dynamics model, world, 568
Dynamo, 48, 87
   model of grassland succession, 471

Earth: evolution, 560-562
   radioactivity, 555, 560
   terrestrial system, 558-560
Earthquake spectrum, 271-272
Ecological engineering, 18, 529
Ecological microcosms, 414
Ecological system, 4
Ecologic-economic input-output coefficients, 502
Ecology, approaches, 406
Economic competition, models, 493
Economic cycles, 492
Economic overview, 533
Economic sector, 500
Economic systems, 476-507
Economic transfer, 17
Economic vitality without growth, 505
Economic webs and spectra, 480, 487-489
Economy of world, 492
Ecosystems, 13, 17, 406-442
   desert, 432
   features, 406-408
   gnotobiotic, 463
   human, 476-507, 508-531
   lake, 417-421
   littoral, 428-429
   minimodels, 407
   nerves, 78
   oyster reef, 298

   photoelectric, 367
   stream, 421
   terrestrial, 431
   thin, 415
   wetlands, 430, 433-437
Eddy diffusion, 126, 234, 240, 241
   autocatalytic growth, 155
   diagrams, 138
Edges, 87
Efficiency, 15
   food level, 190
   growth of Daphnia, 190
   heat conversion to work, 102
   input force, 117
   loading, 117
   maximum power, 117
   power transformation, 115-117, 186-187
Egg clustering, 400-401
Eggs, 80, 393-394
Egypt, 514, 516
Eigenvalues, 239
   stability criteria, 243
Eigenvector, 239
Ekistics, 535
Elastic prices, 483
Elections as threshold switch, 73
Electrical current, 35
Electrical networks, 34-35, 43
Electricity, energy transformation ratio, 252
Electromagnetic spectra, 283
Electromagnetic waves, 288
Electron-volt, 96
Elemental abundance in universe, 576
Elliptical spatial pattern, 348
Elovich equation, 156-157
Embodied energy, 15, 16, 18, 251-268
   alternative concepts, 501
   comparison of concepts, 264-265
   labor, 490
   matrix concept, 264
   maps, 347
   toxicity, 528
   in trade, 490
   two source web, 254
Embryonic development model, 348
Employment model, 548
Enable control of flip flop, 76
Endothermic reaction, 109
Energy, 95-120, 251-268
   chain reaction, 510
   culture, 510
   diversity, 147, 342, 343
   economic competition, 493
   economic production, 487
   economic vitality, 505
   embodied and actual, 251-257
   evolution, 312
   Florida agriculture, 523-526
   high use, 517
   income, 489
   information, 310-311

landscape, 532-533
loop systems, 161-162
maintenance of pathways, 310
maintenance of units, 310
material flows, 18
money relation, 479, 481
parking trucks, 309
per person, 285
photons, 105
public works, 515
pure, 7
spatial complexity, 324
stirring, 324
types of high quality, 252
used, 141
value, 266
water tanks, 105, 106
Energy of activation, 289
Energy-added factor, 257
Energy-amplifier ratio, 256-257
Energy analysis, 251, 492
 agriculture, 526
 eighteenth century agriculture, 546, 548
 parks, 527
Energy barrier, 9
Energy benefit and cost, 498-499
Energy chain, 15
Energy circuit language, *frontispiece*, 5-8
Energy constraints on mathematics, 4, 37
Energy cost-benefit, 493
Energy degradation, 100
Energy diagrams, 21
 rotifer life-cycle, 394
Energy discounting, 499
Energy dispersal, 7, 101
Energy drain, 25, 29
Energy evolution, 465
Energy excess, 567
Energy flow: global, 556
 laboratory stream microcosm, 423
 structure, 317-319
Energy hierarchy, 269-272
 rank order, 341
Energy history, 265
Energy inhibition of growth, 567
Energy intake by consumers, 383
Energy intensity coefficients, 500
Energy investment ratio, 257, 493
Energy kinetics, 6
Energy language: autocatalytic module, 141-142
 derivative, 135
 diagramming materials, 161, 168
 equations of motion, 136-138
 Michaelis-Menten symbol, 163
Energy laws, 6
Energy matching, 260
 development, 541
Energy of organization, 309-310
Energy potential, 5, 7

Energy quality, 15-16, 251, 252, 259
 altitude, 349
 byproducts, 259
 chains, 102, 323, 325
 chemical reactions, 128
 computers and humans, 526-528
 dispersal, 400
 housing, 542
 information, 320
 life cycles, 393-394
 moss spores, 377
 in series, 202
 spectrum, 270, 325
 toxicity, 528
 transport, 280
Energy receiver, 8
Energy signature, 16, 224, 261, 283
Energy sources, 95
 diminishing returns, 483
 signature, 96
 types, 96
Energy spectra, 269
 two sources, 275
 evolution, 342
Energy storage function, 26
Energy stress, mortality theory, 389-390
Energy transfer through resistances, 186
Energy transformation, 100
 chain, 270, 325
 electricity, 369
 Michaelis-Menten loop, 164-165
 potential to kinetic, 107
Energy transformation ratio, 15-16, 251-253, 478, 480
 local method, 256
 period, 578
Energy transmission: waves, 99
 core conductor, 118
Energy turnover coefficient, 190
Energy units, conversions, 96
Engineering, ecological, 529
English in energy language, 91
English sparrow immigration, 331
Enthalpy, 109
Entrainment of oscillation, 177, 418
Entropy, 103-104
 disorder, 316
 dispersal, 402
 hot source, 312
 information, 306
 landscapes, 118
 microscopic and macroscopic, 312
 minimum principle, 313
 molecular states, 311
 relative, 307
 simulation of drainage basin, 313
Entropy change, 103
 environment, 104, 109-111
 kinds, 312
 sources, 312
Entropy increase, 26, 111

to maintain low entropy, 317-319
 mineral use, 517
 order or disorder, 320
Entropy of state, 103, 104
Entropy tax, 116
Environmental control of culture, 511
Environmental evaluation with embodied energy, 493-494
Environmental limits for development, 541
Environmental loading, 495
Environmental overload, 529
Environmental system, 3, 17
Environmental technology, 529
Enzyme denaturation, 293
Enzyme networks, 200-202, 247
Enzyme recycle, 163
Enzyme-substrate process, 164, 166
Ephemeral plants, 432
Epidemic model, 197
Epiphytes on marsh grass, 435
Epizootic, 456
Equality, 78
Equations: diurnal oxygen, 415
 light and photosynthesis, 365
 networks, 12
 organic balance, 459
Equations of motion, 6, 240, 241
 diagramming, 137-138
Equilibrium, 5, 108
 adsorption, 169
 displacement, 312
 global steady state, 555
 oceanic, 272
 steady state, 313
Equitability, 307
Erosion, 252
Errors: analogs, 65
 intersection, 131
Essergy, 266
Estuarine ecosystem, 408-409
Estuarine microcosm, 425, 428
Estuary, 424-425
 compartments, 233
Euler identity, 184-185
Euler integration, 47
Eutrophication, 346
 cultural, 421
 diversity, 346, 418
 model, 421
Eutrophy, 418, 421, 448
 lake dimensions, 418
 morphometric, 418
Evapotranspiration, 408, 410
Everglades marsh, 453
Everglades National Park, 527
Evolution, 464
 earth, 560-562
 knowledge, 313-314
 loop stages, 178
 order, 573
Ewes, 520
Exchange, 12, 539

Exclusive OR gate, 73-75
Exergy, 266
Exothermic process, 109
Exponentials: age distribution, 279
  asymptotic growth, 33
  attenuation, 125
  charge up, 32-33
  decay, 32-33
  decrease with distance, 330
  derivative, 239
  distribution time lag, 155
  economic growth, 485
  graphical test of growth, 144
  growth, 142-144, 148-149
    trade, 485
  intersection, 124
  light attenuation, 370
  low cost distribution, 341
  model for income, 489
  power spectrum, 271-272
  pumping, 188-189
  semi-log plot of growth, 143
  temperature response, 278, 289
Exports, 18
Externality, 476, 494
  energy evaluation, 493
Extinction coefficient, 370
Extinction rate, 460
Extinction of species, 44
Extremes, ecosystems, 457

Factorials, 304, 306
Failure frequency, 286, 389
Farads, 27, 35
Farm, 7, 9, 20
Fear, 518
Feedback: additive, 174
  autocatalytic, 141
  capacitor, 57
  control, 17
  difference, 152
  inhibition, 200-202
  loops, 160
  negative, 31, 179
  reinforcement, 496, 520
  resistor, 54-55
  reward, 6, 19, 21
  variance, 317
Feeding: autocatalytic rates, 151-152
  models of consumers, 384
Fields, gradients, 136
Filter, 232, 233
  high and low pass, 468
Final demand, 265, 500
Fire, 434
Fire climax, 453, 456
Fire-managed forestry, 522
Fire oscillation, 453
Firmware, 48
First law of thermodynamics, 6, 99
First order chemical reaction, 127
First order equation, 25, 29, 30, 33

  simulation, 47
Fish: herbivores, 421
  migration, 400
  pond, 22
  wood stork model, 387
Fishery yield models, 189-193, 230-231
Fitness of the environment, 577
Fjord, 418
Flax, 511-512
Flexibility, 259
Flip-flop, 48, 76, 452
  harvest, 191
Floodplain river, 423
Florida: agriculture, 523-525
  history, 545-547
  growth and energy, 348
  springs, 424
  state model, 545-546
Flow, 9-10, 25
  without interaction, 98
Flow chart, 47
  insect life cycle, 400
  diversity maintenance, 344
Flow-controlled force, 129
Flowers and fruit, 431
Flow-limited source, 98
Flow proportional to force, 26
Flow sensor, 128-129
Flow per unit threshold, 134
Fluctuating temperature microcosm, 292, 451
Fluid models, 138
Fluid transported source, 98
Fluorescence: mercury tube, 164
  in photosynthesis, 362
Food chains, 271, 273
  cogwheels, 284
Force, 10, 11, 26-27, 35, 105-107
  constant, 142
Force balance, 10, 105
  accelerations, 137-138
  potential-inertial, 177, 188
Force-flux law, 10, 13, 26, 114
Force-opposed, 115
Forcing function, 11, 16, 95, 261
Forest, 23
Forest simulation model, 469
Forest stream, 423
Forest wood production, 522
Form, chemical reactions, 349
Forrester diagram: blue green mat, 413
  city, 550
  desert ecosystem, 432
Forrester symbols, 6, 85-87
Fort Myers, Florida, 549
Fortran, 48
Fouling community, 334
Fouling surface, 463
Fourier transform, 282
Fourth law of thermodynamics, 101
Fractional distillation, 214-215

Franklin County, Florida, 408-409, 538
Free energy, 109, 111, 127, 311
  chemical reaction, 109-112
  decomposition rate, 254
  irreversible process, 108
  metabolism, 114
Free energy of activation, 278
Frenzy system, 444, 445
Frequency: failure, 389
  filtered sources, 233
  maximum power, 117
  response, 39-41
  signature, 283
  sources, 327
  spectra, 282
  variation, 137
Frequency analysis, 282-283
  microcosms, 458
  stream photosynthesis, 365, 415
Frequency domain, 31, 39-42, 327
  P—R model, 365, 415
Friction, 10
Fucus, 429
Function, 19

Gaia hypothesis, 577
Gain, 7
  analog computer, 55, 58
  constant, 148
Game fish pond, 522
Game theory, 85
  countries, 549
Gamma distribution, 285
Gamma irradiation: forest at Brookhaven, 462
  microcosm bacteria, 467
  microcosms, 316, 412
  rainforest, 465
Gaseous exchange, 20
Gas law, 106
Gates, logic, 72, 74-75
Gaussian distribution, 97, 158, 275, 317, 331-333
  energy, 574
  information, 317
  mortality, 389
  noise in autocatalytic model, 154-155
  pathway properties, 137-138
  time lag, 155
Gear chain, 284
General circulation, 558
General systems plan, 574
General systems theory, 4
Generation time and size, 280
Genes, 19
  segregation and recombination, 398
  transport, 330
  turnover period, 465
Genetic disruption, 467
Genetic systems, 395-396
Geography, economic, 326
Geologic uplift, 18

Geometric series and species, 340-341
Geomorphology, energy, 326, 327
Georges Bank, 438
Geostrophic balance, 138
Geothermal net energy, 493
Gibbs free energy, 112
 decomposition rate, 252
 sewage, 402
Gills of fish, 28
Global cycles, 554, 562-565
Global solar calories, 252
Glucose, 114
Gnotobiotic ecosystem, 447, 463
GNP growth per capita, 567
Godel's theorem, 3
Golden Gate development, Florida, 545
Goldschmidt reaction, 555-556
Gompertz growth equation, 33, 156
Gompertz mortality equation, 390
Gotland, Sweden, 481
Government, 490-491
Gradients, 155
 of forest properties, 348
Gradient wind diagram, 138
Grammar equivalents in energy language, 77-78, 90-91
Granite outcrops, 450
Graphs: commonness and rarity, 334
 functions, 32
 microcomputers, 49
 rank order, 335-336, 339-340
 time, 6, 11, 32
Graph theory, 6, 83
Grass, lawn system, 526
Grass grubs, 520
Grassland: fire model, 76
 model, 471
Gravity: equation in star model, 157
 model, 328-329
Grazing models, 523
Great Lakes shores, 468
Greece, 517-518
Greed, 518
Greenhouse effect of CO_2, 565
Green sulfur bacteria, 419
Green Swamp, Florida, 542
Grindstones, 187
Gross National Product (GNP), 489, 499
Gross production, 257
 square wave response, 413
Group selection, 453
Group symbol, maintenance, 142
Growth: accelerator, 477
 constant force, 142
 diffusion oscillation, 348
 embodied energy, 347-348
 Malthusian, 142
 optimum percent, 150
 priorities, 327-328
 regional stages, 542

relative, 136
sigmoid, 147
source-limited economic, 485-486
stored resources, 532-533
superaccelerated, 149, 170
temperature, 296
United States, 150
Growth efficiency, 190
Growth rate, specific, 143
Grubs, 520
Guano: birds, 18
 deposits, 461
 effect on tide pools, 464
Guilds, 493
Gulf of Mexico shelf, 415-417
Gulf stream, 241
Gyrator, 87-88

Half-life, 32-33
Halobacterium, 367, 374
Hamiltonian principle, 118, 186
Hand simulation, 46
Hardware: computer, 528
 electrical-passive networks, 34-35
Hardy-Weinberg law, 398
Hartley, 304
Harvest: crops, 516
 exponential growth unit, 189
 fish, 189-193, 230-231, 503, 521-522
 floodplain trees, 434
Hawaiian rain forest, 456
Hay infusion succession, 463
Heart rates, 282
Heat, 95, 110
 budget, 288-290
 conversion to mechanical work, 102
 definition, 6
 energy source, 300
 flow, 115
 leaf, 290
Heat engine, efficiency and power, 504
Heat sink, 8, 26, 35, 100-102, 104, 289
 intersections, 125
 tax, 104
 transaction, 477
Helmholtz free energy, 294, 318
Hendry County, Florida, 523
Herbivore chain, 403-404
Herring, 230-231
Heterotrophic spring, 424
Hexagonal spatial pattern, 323-324
Hierarchy, 15-17
 computers and humans, 526-527
 cosmic rays, 575
 chain, 200
 global pulsing, 555
 Horton stream, 271-272
 income, 489
 individuals, 341

landscape systems, 534-535
 living systems web, 572-574
 model structure, 438
 period between pulses, 578
 poems, 274
 political units, 510
 pulses, 573
 spatial, 323-325
 taxonomic, 342
 urban, 323
High energy times, 517
High intensity agroecosystem, 519
High quality energy: flexibility, 259
 types, 252
Highways, 516
Hill reaction, 381
History: ecosystem models, 433, 438-440
 energy and value, 265
 Florida, 545-547
 hierarchy, 269
 information and entropy, 320
 production models, 356
 simulation of, 517
 single storage, 43
Hodgkin-Huxley model, 117-118
Holism, 577
Holling limiting factor model, 168
Homeostasis: earth, 577
 temperature, 291, 295, 297
Horse and buggy, 546
Horses, 530
Horton stream hierarchy, 271-272
Host-parasite model, 172, 197, 203
Hot springs, 454
 global cycle, 555
Housefly oscillation, 198
Households, 490-491
Housing, density, 543
H_2S, *see* Hydrogen sulfide
Hubbard Brook, New Hampshire, 469, 471
Humans, 3
 action on biosphere, 554
 demand, 505
 effect on ecosystems, 495, 529
 energy transformation ratio, 252
 global economic pulse, 574
 mortality, 391
 settlements, 536-538
 use of ecosystems, 493, 509, 521-529
 within ecosystems, 508-521
Hunger, 74, 384
Hunting and gathering, 512, 523
Hurricanes, 240
 effect on mangroves, 436
 frequency, 327
Husan, Korea, 438
Hybrid system, 73-74
Hydra with Zoochlorellae, 411
Hydraulic conductivity, 81
Hydrilla, 421

Hydrogen: reaction with bromine, 174
 reaction with hydrogen, 174
Hydrogen sulfide, 419
 blue green algal mat, 414
 equilibrium, 563
 stratified lakes, 419
Hydrogenomonas, 526-527
Hydrologic cycle, 273, 556, 559
Hydrology and energy language, 81
Hydroperiod: desert adaptation, 433
Hydrostatic balance, 138
Hyperbola, limiting factor, 129-131
Hypercycle, 178
Hypervolume, niche, 224-227
Hypoteneuse of a right triangle, 39-40
Hypotrichs, 465

Index: competition success, 243
 complexity, 302, 343
 crowding, 334
 environmental performance, 530
 models, 262
i, square root of minus one, 40
IBP, see International Biological Program
Ice, 299
 ages, 557-559
 value, 261
If and only if, 78
IFIAS conventions on energy, 264
IF statements, 46, 49
Image, 505
Imaginary number, 40, 184
Immigrants, 325, 330
Impact: diagram, 495, 541
 mangroves, 541
Implies, 78
Incidence matrix, 83
Income, 479
 density, 539
 personal, 491
Indian agriculture, 514
Indian stage, Florida, 545-547
Industrial revolution, 511, 532-553
Inelastic prices, 483
Inertial force, 105
Infection-immunity loops, 171-172
Infection model, 145
Infinity, 40
Inflation, 476-477
 energy-induced, 21
Information, 18, 19, 302-322, 343
 chain reaction, 510
 communication equipment, 320
 connections, 306, 343
 decisions, 303
 diversity, 343-344
 DNA, 281
 entropy, 306
 flow, 18, 302-322
 Gaussian distribution, 317

high quality energy, 99
 organic chemical structure, 318
 per individual, 306, 343
 required energy, 310-311
 species, 307, 343
 to specify unit, 303, 343
 symmetry, 307
 units and connectors, 306, 343
 variance, 317
Information theory, 303-308
Ingester, 574
Inhibition, 176
 biochemical models, 177
 feedback, 200-202
 high energy growth, 567
 Michaelis-Menten unit, 166
Initial conditions, 32
 analog computer, 57
 competition, 210
 growth, 146
Innovation, 328, 515
 earth evolution, 561
Input loading, 116
Input-output: analysis, 499
 coefficients, 500
 ecologic-economic mixture, 502
 error, 501
 matrix, 241, 359
Insects: castes, 395-396
 light trap diversity, 342
 plant symbiosis, 400
Instant ecosystem, 431, 448, 450
Institutions and culture, 509-510
Integral: equation, 6, 30-31
 log normal species index, 343
Integrated equation, 31-33, 41
Integration, 30-41
 analog computer, 56-58
Integrator in analog computer, 54, 56-59
Integrodifferential equations, 154-155
Intelligent systems, 402
Interaction, 8-10, 123-140. See also Intersection
Interface: principle, 529
 technology, 524
 waste, 529
Intermediate technology, 524-527
International Biological Program, 379
 grassland models, 471
Inter-organized sectors, 534
Intersections, 77, 123-140
 accelerations, 136-138
 controlled by switches, 132
 drag, 126
 energy quality convention, 128
 parallel, 208
 renewable and nonrenewable sources, 229
 resistance model, 133-134
 sources, 326
 switching, 132
 symbol, 123-124

two or three variables, 128, 133-136
Intrinsic rate of natural increase, 143-144, 148-149, 280
Inventory cycle, 492
Inverse square decrease, 9-10, 328
Inverter: analog, 54-55
 logic, 73-74
Investment, 479
Investment ratio, 493, 494
 definition, 257, 260
 environmental loading, 529
 Everglades National Park, Florida, 527
 Florida agriculture, 526
 regional support, 540
Irreversible: entropy change, 109, 111
 heat, 109
 process, 104, 108
 thermodynamics, 114-117
Island populations, 44
Islands species, 338
Isostasy, 252, 557
Iteration, 36-37, 46
Ivlev-Watt model, 134

j, square root of minus one, 40
Jade, 512
Japan, model, 487
Jitter, 96-97
Joules, suitability as an energy unit, 102

Kadam, 216
Kalman filter, 581
Kelp, 426
Kelvin temperature, 108, 288
Kilocalorie, 6
Kilowatt-hour, 96
Kinetic energy, 19, 107
Kinetic fields, 535-536
King-Altman procedure, 176
King Hubbert's blip, 575, 578
Kissimmee pastures, Florida, 523
K limited growth, 148
Koenig diagrams, 88-89
Kolmogorov, turbulence, 274
Kondratieff cycle, 492
Kootenay Lake, British Columbia, 138
Krill, 425
Kumara potato, 512

L function, 293-294, 318
Labeled digraphs, 80-81
Labile organic pool in plants, 87
Labor: city model, 540
 embodied energy, 490
 production, 487
Lab, time, 9, 39, 152-155
Lag distribution functions, 155
Lag-lead, oxygen metabolism, 415
Lagrangian, 186
Lake: Byelovod, 419
 Mendota, Wisconsin, 438
 Okeechobee, Florida, 420-421

SUBJECT INDEX

Lakes, 417-421
  dimensions, 418
  marl, 135
  stratified, 418-419
  water hyacinths, 420
Lambs, 520
Landform, plant succession, 471
Landscape ecosystems, 532
Landscape hierarchy, 436
Land tax, environmental impact, 497
Land uplift energy transformation ratio, 252
Land use, 535
Land use transformation, 545
Land value, 252, 497
Langley, 96
Langmuir adsorption model, 169
Language precision and power, 91
Laplace transform, 40-41
  two storage system, 184
Larval set, 298
Lawns, 526
Leaf area index, 81
Leaf fall, 26, 44
Leaf temperature, 290
Learning: loop, 226
  network, 402
  web, 245-246
Le Chatelier's principle, 118, 312, 313
Lee County, Florida, 539
Length of life, 281
Leslie matrix, 392
Less than the best, 580
Levee, 73
Level of size, 4
Levin's digraphs, 89-90
  prey-predator models, 195
Lichens, 414
Liebig law of minimum, 130
Life cycles, 178, 377, 393, 394
Life span, size, 324, 326
Life support in space, 527
Light adaptation, 364
  attenuation, 369-371
  bright, 40,000 foot candles, 418
  photons as oscillators, 188
  photosynthesis, 363, 366
  plant transformation, 259-364
  pressure, 105, 117
  transformation in photosynthesis, 359-361
Limestone, 560
Limit cycle, 171, 173, 182
Limiter: analog, 64
  simulation, 144-145
Limiting factors, 20
  chemostat, 191-193
  digital, 132
  external source, 129-131
  sharply responding intersections, 132
Linear: cascade, 183, 185-186

circle, 162
coupling, 123-124
employment model, 551
laws, 9
parallel pathways, 209
pathway, 27
programming, 207-208, 232, 485
steady state, 238
system, 41
system of energy flow, 235
systems stability, 183
web, 234
work pathways, 129
Linearizing models, 239
Line source, 325-326
Lineweaver-Burk plot, 129-132, 176
Linguistics, 91
Litter, 44
Litter fall model, 66
Littoral ecosystem, 428-429
Living systems model, 572, 574
Load, 115-116
  adaptation, 374
  command, 49
  Michaelis-Menten loop, 164-165
  nutrients, 420
  resistive, 186-187
Loans, 498
Log, see Logarithm
Log normal: conversion, cumulative species graph, 338
  distribution, 285
  species distribution, 337-338
Logarithm: base e, 304, 306
  coordinates for classification, 410
  conversion of series to rank order, 340
  information, 303-308
  series, 342
  slope, 32
Logic, 48, 72-94
  circuits, 6, 36-37, 72-94
  continuous process, fast time constants, 245
  inversion, 73-75
  on-off, 73
  seed as control, 514
  subsystem symbol, 8, 36-37, 73
  symbols, 72-73, 75-77
  systems, 72-79
  timers, 69-70, 73, 77
  turtle, 78
  web, 245-246
Logistic, 33
  backforce type, 146
  conserved matter, 170
  fish yield, 520
  history of, 146
  loop, conserved storage, 149
  maximum profit, 503
  model of history, 517
  pairs, 197, 210
  quadratic drain type, 146

  stochastic, 157-158
  time delay, 154
  Volterra form, 146
Longitudinal succession, 460
Loops, 160-181
  additive without storage, 172-174
  autocatalytic conservation, 170
  bi-bi, 175
  bi-uni, 175
  chain reaction, 173-174
  choice order type, 175
  closed system, 169
  digraphs, 89-90
  embodied energy, 254
  enzyme types, 175
  equal value, 254
  feedback, 160, 170
  flow chart, 179, 344
  generalized pattern, 161
  logic, 76, 79, 179
  Lotka, 162, 182
  matrix, 161-162, 170
  Michaelis-Menten, 162-169
  other languages, 179
  ping-pong, 175
  pulse effect on, 161
  quadratic, 179
  reproduction, 178
  signal flow, 85-86, 179
  stability, 179
  three storage, 162, 169, 177
  two energy sources, 167-168
  two tank growth, 164-165
  without storage, 172-173
  uni-uni, 175
Lotka loop, 162, 182
Lotka-Volterra models, 171, 195, 211, 212, 218, 349
  competition, 211-212
  prey-predator oscillation, 194-195
Low energy times, 511
Luminescence pulses in deep sea, 349
Lyapunov function, 182, 195

Machine language, 48
Machine unit, analog computer, 55, 57-58
Macroeconomics model, 86, 477-478
Macroscope, 581
Macroscopic entropy, 312
Macroscopic and microscopic structure, 317
Macroscopic minimodel, 579
Magans, 517
Maintenance, 142, 418
Malaria, 198, 398
  model, 171-172
Malate, 375
Malthusian growth, 142
Mangrove swamps, 402
  hurricanes, 327
  simulation, 436
Manning equation, 38
Maoris, New Zealand, 444

culture and diagram, 511-512
Maps, energy distribution, 347
Marcali scale, 272
Marco Island, Florida, 542
Marginal utility, 484
  spatial model, 332
Marine, ecosystems, 424-426
Marine resources: development, 541-543
  larvae, 425
Markov chains, 6, 84
  stability, 309
Marsh: crabs, 446
  grass decomposition, 414
  simulation, 435
  succession, 471
  temperature, 298
Mass of cities, 546
Mass expulsion, star model, 157
Master-slave flip flop, 452, 455
Matching energy, 260
  centers and surroundings, 330
Materials as energy, 18
  modeling disadvantages, 161
Matrix: addition, 82-83
  adjacency, 302
  age classes, 392
  algebra, 6, 82-83, 235
  Boolean, 242
  causality, 242
  community, 238, 243
  economic, ecologic input-output, 502
  embodied energy partition 284
  environmental impact, 495
  game theory, 85
  identity, 236
  input-output, 241-242, 499-500
  money circulation, 359
  multiplication, 235
  notation, 83, 235-239
  notation for production, 501
  plus minus effects in Sahel, 529
  relations of nations, 549
  seed dispersal, 401
  substitute, 239
  successional transitions, 451
  system web, 235
  urban spatial development, 551
Matter dispersal, 26
Maturity concept, 459
Mauna Loa, Hawaii, 562
Maximum biomass, 101, 458
Maximum efficiency, 575
Maximum entropy, 284, 313
  spatial, 272
Maximum metabolism, 418
Maximum power: autocatalytic feedback, 141
  competition, 118
  efficiency, 116, 503-504
  loading in photosynthesis, 360
  money circulation, 476

optimum speed arrangements, 308
  photosynthetic pigments, 374
  principle, 6-7, 101, 102, 118, 262, 575
  priorities, 327
  profit, 503-504, 524
  pulsing, 282, 574
  pulsing frequency, 195
  resistive load transfer, 186-187
  sludge decomposition, 418
  storage in succession, 451
  tropical forest, 319
  water use, 542
  world trade and hierarchy, 568
Maximum production, 207
Maximum reproduction rate, 101
Maximum yield, 503
  logistic, 192-193
Maxwell-Boltzmann distribution, 277, 289
Maxwell demon, 311
  as natural selection, 576
Meander, 138
Mechanical work, 100, 266
Megaohm, 36, 57
Memory, 48
Metabolism, 11
  biomass, 28
  city, 539, 546
  concentrated, 402
  maximum, 418
  microcosm, 291
  overall equation, 114
  specific rate, 28
  temperature, 297
  tropical forest, 318
Miami, Florida, 546
Mho (conductivity), 13
Michaelis-Menten: bond graph, 88
  closed microcosm, 411
  diminishing returns, 484
  equations, 360
  examples, 164
  flow-limited source, 166-167
  inhibition, 166
  interaction of two units, 135
  King-Altman procedure, 176
  loading, 164-165
  loop, 88, 162-169
  model, 163, 357
  pairs, 199-201
  parallel, 221
  within P—R model, 171-173
  within producers, 356
  simulations, 165, 167
  stacking, 164-166
  temperature effect, 292, 294
  two sources, 168
Microampere, 35
Microbes: animals, 403
  chemostats, 191-193
  coral reef detritus, 428, 430
  logic, 245

  manager animals, 247
  networks, 245
Microcomputer, 46-52
Microcosms: bamboo, 463
  blue-green algal mat, 413
  brine, 462
  CO_2 homeostasis, 561
  decomposition, 449
  diversity, 414
  ecological, 414
  epilimnion-hypolimnion, 465
  fishing, 190
  gamma irradiation, 316
  Goldschmidt reaction, 465
  laboratory stream, 423
  marine, 414, 425
  order-disorder, 317
  phosphorus cycle, 223
  plankton in vertical tube, 465
  P/R ratio, 450
  sealed monoculture, 411
  stream, 450
  succession in bottles, 451
  temperature effects, 291-292, 465
  terrestrial, 411
  terrestrial rainforest, 412
Microecosystem, see Microcosm
Microfarad, 35, 36, 57
Microphone, 123
Microscopic reversibility, 110-111
Migration, 147-148
Military components, 517-518
  hierarchy, 15
Milliampere, 35
Mimicry, 226, 396
Mineral cycle, 162
Mineral processing system, 517-519, 557, 564-566
Miniac analog computer, 54, 56, 68-71, 74
Minimodels: agrarian economy, 511
  biosphere, 555-557
  carp pond, 522
  classification of ecosystems, 408
  consumer population, 385-386
  consumers, 383
  desert ecosystem, 432
  diversity and stress, 344
  ecosystems, 407
  fire-managed forestry, 522
  fishing, 520-521
  Florida history, 547
  forestry yield, 521
  forest wood production, 522
  game-fish pond, 522
  growth and exchange, 486
  hunting-gathering, 511
  kelp, urchins, otters, sea bass, 426
  logic switching of components, 469
  marine ecosystems, 426
  oyster culture, 522
  pesticide agriculture, 521
  plankton and exchange, 426

production, 355
regional systems, 553-554
salt balance, 426
soil, 431
solar based urban system, 511
stream, 423
succession, 444
terrestrial ecosystems, 430-431
toxic action, 529
United States, 489
world economy, 492, 564, 566-567
world prices, 492
world pulsing, 575
Minimum entropy generation, 101, 118
Minimum entropy principle, 7
Minimum power principle, 7, 101
Mining system, 566
Mining town, 519
Mitscherlich equation, 135, 295
Mixing depth, 371
Mixing stability, 417
Moas, 512
Model: age classes, 392
 aggregated United States, 487-490
 algae and flies, 454
 BOD, 402
 bogs, 471
 broken stick, 340
 Cobb-Douglas production, 487
 colonization, 459
 combining measurements, 402
 complex, 581
 county, 536-538
 cypress swamp, 435, 473
 definition, 3
 demand, 481
 dispersal, 399
 disturbance and diversity, 345
 diversity maintenance, 344
 earth crust, 560
 economic competition, 493
 employment in city, 548
 energy and information, 314-315
 exponential economic growth, 485
 fire in Everglades, 453
 fishing effort, 520
 Florida phosphate mining, 566
 generalized landscape, 537
 genetic process, 398
 grass, rabbits, foxes, 202
 grassland succession, 471-472
 gravity, 328-329
 grazing yield, 523
 historical, 438
 humanity and nature, 477
 hunting and gathering, 511-512
 innovation, 515-517
 input-output, 242
 Japan, 487
 leaf temperature, 290
 learning, 245-246

life cycles, 395
macroscopic, 579
mass and number, 383
maximum profit, 502
money circulation, 476
mortality, 388, 389
nutrient loading, 420
optimum development, 540
oxygen sag, 463
oyster reef, 298
pigment control, 376
population support area, 330
producers, 356-364
production and diversity, 468
recruitment, 388
reproduction, 385
Sacramento-San Joaquin delta, 427
salt marsh, 435
sea otters, 455
self pulsing, 445
Silver Springs, 440
simplified producer, 356
sludge decomposition, 417-418
spatial competition, 346
spatial distribution, 548-551
spatial energy exchange, 331
spatial oscillation, 348
spruce budworm epidemics, 456
state of Florida, 545
stratified lake, 419
succession and diversity, 544
supply and demand, 481-483
synthesizing empirical relations, 379
terrestrial plant, 377-378
tide pools, 464
tree populations, 469
tropical forest energetics, 319
unstratified lake, 417
upper thousand meters of the sea, 425
wetlands, 430, 433-437
Modeling: approaches, 355
 discussion, 578
 spatial compartments, 233
Models, purpose of, 579
Modules, autocatalytic, 141-159
Molecular complexity, 311
Molecular cycles, 160
Molecular hierarchy, 274-277, 288
Moment of inertia, 107
Money, 12
 circulation, 21
 drive, 97
 shortages, 484
Monoculture: diversity, 335
 energetics, 319
Monod limiting factor model, 131-133, 195-196, 484
Monostable flip flop, 69, 70, 73, 78
Monsoon climate, 514-515
Monte Carlo process, 85
Montreal, Canada, 542
Moran plot, 144

Mormoniella, 203
Morphometric entropy, 418
Mortality: discrete model, 144
 human, 391
 models, 389
 pathways, 178
 theory, 390
  humans, 277
 webs, 388
 Weibull, 387-388
Mosaic theory of succession, 455
Mosquitoes in microcosms, 463
Moss model, 377, 380
Mossdale, California, 427
Motivation models, 73, 245-246
Motomura geometric series, 340
Motor efficiency and power, 116
Multiplication, vector-matrix, 82
Multiplicative input, 97
Multiplier, 125
 analog computer, 56
 compared to switch, 133
 energy sources, 125-126
 feedback, 141-159
 intersection, 125, 142
 loop of, 172-173
 negative, 125-126
Muscle efficiency, 116
Mutation, 39
Mutualism, 218

NAND gates, 75, 452
Nannoplankton, 43
Nannoplankton photosynthesis, 364
National income, 491
 quarrels, 549
 systems, 476-507
Natural selection, 118
 developing order, 175
 information, 311-312
 Maxwell demon, 576
 radiation in space, 576
Nature and culture, 508
Negated OR, 76
Negative feedback, 174
 stabilization, 31
Negative multiplier, 125-126
Negentropy, 306-307
Neive impulse, model, 117-118
Net energy, 494
 coal, 495
 concentration, 105
 definition, 257
 economic activity, 492-493
 geothermal steam, 493
 solar energy use, 523
 uses, 258-259
 yield ratio, 257
Net national product, 491
Net production, 257, 258
 aquatic environment, 371
 growth, 366
 rank order, 340

Subject Index

square wave response, 413
succession, 459
Networks, 12
chemical reaction, 6
coherence, 343
electricity, 34-35, 43
Goodwin, 202
heterogeneous, 247
homogeneous, 234
Neual nets, 6, 78-79
New England coastal water, 438
New England growth, 547
New York City, 539
New Zealand: sheep pastoral system, 520-521
simulation, 568
Niche, 224
breadth, 243
dimensions, 243
vectors, 224
Nicholson-Bailey model, 195-198
Nit, 304
Nitrogen: diurnal variation, 409
global cycle, 564
limiting factor, 136
phytoplankton competition, 220
succession, 468
Nodes, 85, 87
Noise, 325
autocatalytic model, 157
non-random pulse hierarchy, 574
red, 283
variation, 317
white, 282
Non-renewable source, 229
Noosphere, 577
NOR, 75
Normal distribution, 97, 275, 276, 333
Normalization, 57
Notations, economic, 479
Not, logic inversion, 75
Nuclear reaction, star model, 157
Nutrients: algal ecosystem, 429
capture by animals, 411
effects in swamps, 434
grazing yield model, 525
oceanic cycle, 425
succession, 450
water exchange in mangroves, 437
Nyquist plot, 42-43

Object, 91
Obsidian, 512
Occupation-niche, 224
Ocean: generalized model, 425
gyral, 241
system, 557-558
Ohio River, 463
Ohms, 28, 35
Ohms law, 10, 11, 29
Okeefenokee Swamp, 43
Oklahoma, 521

Old field: rank order, 335
succession, 470
Oligotrophic systems, 418, 448
Onsager's principle, 115, 124
Open and closed systems, 4, 272
Open questions, 582
Open sea, 425
Operational amplifier, 53, 55
Optical density, 81, 370, 371
Optimization, 523, 581
Optimum: density for maximum power, 541
development, 540
efficiency for maximum profit, 503
speed for arranging, 308
temperature, 297
Order, 302-322
conflicting concepts, 315-316
disorder, 314-316
oscillating reactions, 282
Ordination, 348
Oregon streams, 424
Organic matter, 20
mangrove swamp, 436
sea, 105
Organization: society, 515
water flows, 535
OR gate, 73-75
Origin of life, 178
Oscillation, 348
adaptive control, 171
carbon dioxide, 565
climax, 445
defense preparation and war, 512
diffusion models, 155, 348
digital model, 452
dominants, 455
economic, 492
energy controlled, 153
enzyme web, 247
flip flop, 452-453
harmonic, 184, 188
hot springs ecosystem, 454
ice ages, 559
inductance-capacitance, 179
inertial, 179, 187-188
inhibition, 247
order, 574
prey-predator, 194-196
reactions, 282
stability characteristics, 183-185
steady state, 117
systems, 182
time delay, 153
vacuum tube, 177
Van der pol, 176-177
see also Pulses
Oscilloscope display, 53, 60, 68
Osmotroph, 403
Outcrops (rock), 450
Output of force difference, 129
Overhang, 195

Overlay: diagrams, 160
maps, 347
money, 356
Overshoot, 449
Overstory agroecosystem, 519
Overview, economic, 533
Oxidation-reduction, global process, 557
Oxygen, 19, 20
consumption, 297
cycle in microcosm, 413
diurnal variation, 409
limits, 135
metabolism, 114
sag, 460-461, 463
swamps, 430
Oyster: culture, 522
energy evaluation, 495
industry, 538
Oyster reef, 298

Paddock, New Zealand sheep, 520-521
Paper industry production, 487
Parabola: power-efficiency, 115-117, 186-187
stock and yield, 192-193
temperature effects, 294-295, 297
yield and fishing pressure, 192-193
Paradox: plankton, 232-233, 345
Parallel: consumption, 10, 26
elements, 206-222
intersections, 208
negative interaction pairs, 211
Paramecium, 203
Parasite control, 226
Parasite-host: discrete model, 195, 198
relations, 395
spectrum, 280
Pareto distribution, 285, 336, 489
Paris, France, 530, 546, 548
Parks, 527
Partition: function, 278
information, 306-307
Passive analogs, 6, 34-35, 43, 90, 433, 440-441
autocatalytic growth with backforce, 146
ecosystem web, 441
nerve, 118
photorespiration-photosynthesis, 368
P–R, 167
pulsing simulator, 440
Passive energy storage, 105
Pastoral system, 520
Patch boards, 65
Patching analog computer, 65
Path coefficients, 6, 85
Pathways: diverging, 128
energy to maintain, 310
importance index, 303
information, 305, 309
languages, 79

linear, 129
linear circle, 162
probabilities, 83-84
stochastic, 137-138
types, 9, 112
Patterns, common, 572
Payback, 497
PCB's, 529
Peace and war, 512, 514, 518-519
Peat: bog, 471
    wetland storage, 433
Peck order, 88-89, 508, 509
pE equilibrium, 563
Pelagic ecosystem, 425
Pendelum, 105
Penicillin model, 189
Per capita growth, 567
Perception, system behavior, 228
Percolation, soil, 430
Period: pulse and size, 577
    sine wave, 188
Periodic harvest model, 190-191
Periphyton, 450
Permutation, 302
Peruvian upwelling, 461
Pesticide models, 521, 524
Pharoah, 516
Phase lag, 40-41
Phase plane, 171-173, 177, 182
    competition equations, 212
    Lotka-Volterra, 194
pH—CO_2, equilibrium, 562
Ph.d production model, 189
pH feedback by ecosystem, 526
Phosphate deposits, 18
Phosphorus, 18, 20
    cycle, Lake Okeechobee, Florida, 421
    ecosystems, 88
    global cycle, 563
    limiting factor, 135-136
    sediments, 420
Photocell, 164, 360
Photoconductor, 368, 441
Photoelectric ecosystem, 367, 369
Photon, 288-289
Photooxidation, 368
Photoperiod control, 73
Photorespiration, 367
Photosynthesis, 113
    adaptations, 372
    chain of transformations, 362
    effect of temperature, 293-298
    half saturation to light, 363
    inflow-outflow, 362
    overall reaction, 114, 361, 362
    production, 19-20, 357-367
    response to light, 363, 366
    terrestrial and marine, 562
    types, 359
Photovoltaic cells, 360
    chloroplast, 117
    maximum power loading, 186-187

Physical equilibrium, 107
Phytoplankton, 15
    bloom, 417
    early models, 357
    models, 438, 451
    net production, 367
    spectra, 284
    zooplankton, 427
Piezoelectric crystal, 164
Pigs for the ancestors, 512
Pigs and chickens, 513-514
Pipe flow, 26
Pitcher plants, 414
Pitfalls in modelling, 580
Planar source, 325-326
Planck distribution, 288
Planck's constant, 293
Plankton, 43
    oscillation, 415-416
    patch model, 155-156, 348-349
    size distribution, 326
    succession, 466
Plantation, forest, 319
Plant nutrient storage, 421
Plant organic growth, 87
Planting, 516
Plants, carnivorous, 411
Playas, 450
Plotting, computer, 49
Plus-minus diagram, hunting-gathering, 512
Plus and minus pathways, 80
Poems, hierarchy, 274
Point source, 325-326
Poison: mimic, 226
    shared, 206
Poisson distribution, 331
Polar coordinates, 40, 42
Poles, control theory, 184
Pollution model, 128
Polymerization, 178
Ponds, 22
    carp, 522
    game fish, 522
    longitudinal series, 460
    sewage, 462
Pophyridium, 374
Population: disorder energy, 458
    distance, 330, 333
    force, 13, 27
    minimodels, 385-386
    pool, 147-148
    potential, 328-329, 441
Possibilities, 303
Pot analog unit, 55
Potential, 104
    elevated water, 81
Potential energy, 95
    concentration, 104, 106
    storage, 106, 115
    transformation to kinetic energy, 186-188
Potential generating work, 100, 104

Potential production, multiple sources, 263
Power, 5, 6, 100
    analog exponents, 69
    energy storage, 119
    equation, 115
    oscillation, 117
Power and efficiency, 115, 117
    photosynthesis, 360-361
Power, maximum, *see* Maximum power
Power plants, 535
Power spectra, 269, 282-283
    cosmic rays, 576
    economic cycles, 492
Power supply for control networks, 245
P—R model, 19-20, 167, 363, 364, 408, 411-414
    frequency domain, 365, 415
    Michaelis-Menten, 171-173
    oxygen sag, 463
    symbiosis, 411
P/R ratio: lichens, 450
    longitudinal succession, 460
    maturity, 460
    microcosm, 450
    succession, 459
Precautions, analog computers, 65
Predation efficiency, 190
Predation and stock, 190
Pressure, 10, 81
Pressure gradient force, 138, 240-241
Prey-predator: alternate food mechanisms, 202
    chain, 198
    compared to economic demand, 481
Price, 8-9, 12
    control of fishing, 520
    coupling, 495, 496
    inverse to supply, 481
    marginal utility, 484
    models, 478
    notations, 477, 479
    stimulating supply, 484
    transaction, 124
    world minimodels, 492
Primitive photosynthesis, 359
Primitive systems and trade, 511
Principles of energy design, 573
Probability: density, 285
    diagrams, 6
    matrix, 83-84
    pathways, 83-84
    possibilities, 305
    successive sampling, 343
Process analysis, 265
Producer, 8, 355-382
    aquatic, 375
    with money overlay, 356
    simplified models, 356
    terrestrial, 376
    typical, 356
Production, 11, 20, 91

autocatalysis, 141
beef, 525-526
Cobb-Douglas, 357-359, 487
constant, 152-153
depth, 371-373, 420
diurnal pattern, 364
economic, 356, 479, 487
economic, consumption, 486
functions, 230-231
global model, 562
GNP, 491
index, 410
measurement methods, 362, 366
potential of sources, 263
shortages, 232
vertical water column, 370
world, 554-555
Productivity, *see* Production
Profit, 495
efficiency, maximum power, 503, 524
maximization, 502
Programming, 46-52, 72
Projection: vector, 227
wheel, 39, 40, 188
Proton pump, 368
Protozoa, 465
Public works, 514-516
Pulses, 36, 73, 79, 444
adaptive source use, 450
age class model, 279
combination, 573
consumers, 197
frequency, hierarchy, 578
global, 555
hierarchy, 325
high quality control, 281
ice age, 559
maximum power, 574
recycle, 76, 195-197
retrogression, 444
war in New Guinea, 512, 514
yield, 522
Pulsing, *see* Pulses
Pumping: autocatalytic yield, 189-190
downhill, 168
intersections, 8, 21
proportional to drain, 150
from sources, 130
Purple sulfur bacteria, 419
Push-pull, temperature action, 291, 296-297
Pygmies, 510
Pyramids, 514
Pythagorean theorem, 42

Q_{10}, 291
Quadratic: cost, 533-534
decay, 37-38
drain, 37-38
feedback, 150
logistic drain, 146
loops, 179
migration, 147-148
minimodel of diversity, 344
pathway, 126
rank order model, 342
self interaction, 124
Quality of energy, *see* Energy quality
Quality and frequency, 282-284
Quantum hierarchy, 278
Quasar, 576
Quarter-square analog multiplier, 56
Quenching, 174
Queues, 6, 244
Quinone, 361
Quotient interaction, 63

Radians, 41, 188
Radiation, 288, 290
gamma rays on microcosms, 412
star model, 157
Radiation cycle in universe, 576
Radiation sensitivity, 281
Radioactive decay, 11, 33
Radioactive tracers, 208-209
Radioshack microcomputer, 48, 49
Radius of movement, 436
Rainfall distribution, 272-274
Rainfall use in classification, 410
Rainforest, Hawaii, 456
Rainforests, 455
Rain forest species diversity graphs, 342
r and K specialist, 217
r limited growth, 144, 149
Ramp, 34, 36, 96
discrete, 144
Random partition model, 340
Random spatial distribution, 334
Random walk, 333
Range of analog computer, 64
Rangelands, 526
Rank of cities, 336
Rank-frequency graph, 339
Rank order: and energy cost, 341
graph, 335-336, 339-340
and hierarchy, 341
net production, 340
quadratic energy cost, 342
Raoults law, 213
Rapid response networks, 245
Rate of change on analog computer, 60
Rate equation, 29
Rate of generation of entropy, 115
Rate principles, 118
Ratio: benefit to cost, 498
feedback from storage to renewable energy, 495
mean to variance, 401
money flow to energy flow, 478
Ratios: useful, 256

Raunkier distribution, 335
RC circuits, 28, 34-35, 58, 182, 234, 440, 441
Reactions chain, 173-174
Reading energy diagrams, 21
Reaeration of oxygen, 418
Real component, 42
Recapitulation on modeling, 579
Reciprocal function of time, 37-38
Recolonization, 44
Recorder patching for analog computer, 68
Recovery: waste effect in streams, 461
Recruitment, 386
Recycle: animal wastes from city, 530
critical materials, 545
of wastes, 18
Recycling, 19-20
Recycling consumer time, 168
Recycling receptor, 165
Red algae, 374
Redox potential, 419
Red tide, 156
Red tide fish kill, 456
Redundancy, 307
Reefs, 428
Refinery waste ponds, 460
Reflex, 78
Refrigeration system, 299
Refuge in autocatalytic unit, 147-148
Regeneration, 20
Regional models and water, 542
Regional system generalized model, 533
Regional system model, 537
Regional water model, 544
Regions, 532-553
Regression, 443-444, 455
Reinforcement of extreme, 457
Reinforcement feedback, 520
Relationship of countries, 549
Relaxation, 33
Reliability, 286
Rents, 497
Repop, repeat operation, 53, 60, 68
Repressor, 247
Reproduction loop, 178
Research approaches to systems ecology, 582
Reset, 73, 76
Resiliency, 243
Resistance, 10, 26-28, 35, 43
chemical, 289
model of intersection, 133-134
in series, 186
Resistor, 35
network, 56
in series, 186-187
Resonance, 42
Resource limited logistics, 148, 170
Resource limited models, 145
Respiration, 19-20, 113
as measure of structure, 318

tropical forests, 319
Restaurants and guilds, 404
Retrogression, 443-444, 449
Reversible heat change, 109
Reversible process, 108, 110
Reversible state change, 103, 108
Reversing multipliers, 125
Reward loop, 520
Reynolds stresses, 317
Rhodopsin, 368
Rice paddy, 449
Richter scale, 272
Ring waves, 349
Roentgens, microcosm irradiation, 411
Rome, 517
Rootcrops, 512
Rooted plants in lakes, 421
Root locus, 243
Roots of characteristic equation, 183
Root Spring, 225, 424
Ross malaria model, 171-172, 198
Rotation of land, 510, 511, 513
Rotating wheel projection, 40
Rotifer life history model, 393-394
Ruminant stomach, 402
Run button on analog computer (logic), 76
Ryegrass, 520

Sacred cow, 514-515
Sacramento-San Joaquin delta, 427
Sahel, Africa, 529
Salinity: distribution, 313
  energy, 106
  mangrove swamp model, 437
Salt: balance, 426
  concentration, 128
  effect on mangroves, 437
  global cycle, 556-557
  microcosm, 429
Salmon, 230-231
Salt marsh model, 225
Sand dunes, 468
Sandstone, 560
San Francisco Bay, 425, 427
Sarasota County, Florida, 537
Saturation deficit, classification, 410
Saving programs, disc, 49
Savings, 479
Sawtooth wave for analog computer, 69, 96
Scaled fraction, 57
Scales of succession, 447
Scaling, computer, 57-61
Scaling short cuts, 62, 74
Scarcity and strategy, 487
Scheduling tasks, 245
Schögl chemical reaction, 150
Schrodinger ratio, 318
Sculpin, 423-424
Sea, upper thousand meters, 425

Sea compartments, 560
Sea models, see Marine
Sea otters, 426, 455
Sea shell system, 67
Seasonal carbon dioxide, 562, 565
Seasonal function, analog, 69
Seasonal patterns, 415
Seasonal work, 516
Second derivative: analog circuit, 188
Second law analysis, 266
Second law of thermodynamics, 6, 11, 100
  causing order, 573
  self organization, 262
  stabilizing influence, 243
Second order chemical kinetics, 37-38, 127
Second order equation, 183
Sector, 500
Sedimentary cycle, 556, 560
Seeding, 143
Seeds, 431
  agroecosystems, 513-514, 516
  germination, 377, 380
  predation, 452, 455
  spectrum, 286
Segregation, genetic, 398
Selection, see Natural selection
Selenium cell, 360
Self dilution, 152-153
Self interaction, 124
Selfish selection, 453
Self limiting, see Michaelis-Menten
Self optimization, 523, 581
Self organization, 258, 407, 443
  algorithm, 101
  maximum power, 262
  parallel components, 450
Self-regulation, 33
  cycles, 577
Self-reproducing machines, 79
Self sufficiency, agroecosystems, 526-527
Senescence and yield, 190
Sensitivity: analysis, 137-138, 242
  linear and non linear models, 565-566
Sensor, 112
  flow, 128-129
  rate of change, 37
Sentence structure, 91
Sequences, see Succession, 445
Seres, 444
Series, 182-205
  autocatalytic pairs, 194-203
  autocatalytic unit and tank, 189
  chemostats, 199
  exponential growth and flip flop, 191
  exponential growth unit and harvest, 189-190
  information content, 304
  logistic units, 195, 197

Michaelis-Menten and autocatalytic units, 199
Michaelis-Menten units, 199-201
  logistic unit and tank, 189
  pairs with positive feedback, 195
  tank and Michaelis-Menten units, 202
Set pot analog mode, 57
Sets, 73, 76-78
Sewage: effect in stream, 463
  ponds, 462
  waste plants, 418
Sexual life cycle, 395
Shade adaptation, 165-166, 374
Shade trees, 519
Shale, 560
Shannon-Weaver-Wiener expression, 307-308, 343
Sheep, 520
Shelf, 438
Shelf ecosystem, 417
Shifting agriculture, 513
Shipping and distance, 328-329
Shrimp, 297
  food web, 244
Sickle cell anemia, 398
Sigmoid growth, 147
Signal flow diagrams, 6, 85
Signature, 16
  energy sources, 96, 261
Signed digraphs, 87
Silicon solar cells, 360
Silver Springs, Florida, 424, 445
  diagram, 99
  input-output analysis, 501
Simscript, 48
Simulation, 46-52, 53-71
  city models, 546, 549
  complex world models, 568
  development in mangroves, 542
  Dynamics, city, 547, 550
  economic growth in wetlands, 543
  energy spectral chain, 275
  Everglades fire, 453
  export, analog computer, 69
  Franklin County, Florida, 538
  formation of phosphate in Florida, 565
  harvest and replanting, 516
  history, 517
  hot springs ecosystem, 454
  hurricanes and mangroves, 436
  linear global models, 561, 565
  Miami, Florida, 548
  military roles, 518
  Pigs for the ancestors, 514
  P—R in microcosm, 412
  procedural steps, 50-66
  pulsing, 440
  RC circuit, 90
  salt and nutrients, mangroves, 437
  seasonal plankton, 415-416
  successional model with seeding, 470

termites, 397
upwelling, 462
war, 549, 552
world minimodel, 492
world pulsing, 575
Sine: angle, 39
  wave, 39-40, 96
  wave, analog computer, 69
  wave generator, 188
  wave seasons, 69
Singular point, 171
Sink, 11
Sink water, 207
Sinusoidal oscillators: in series, 200
  single, 188
Size: depreciation, 313
  energy hierarchy, 325
  length of life, 281, 324-326
  metabolism, 280
  temperature, 298
  territory, 325
  time, 28
  turnover time, 324, 326, 576
  variation, 317
Skeletal deposition, carbon, 565
Sky scrapers, 516
Slope, 484
Sludge, 417-418
Snail resisting current, 187
Social insects, 395
Social power, influence, 89
Sociobiology, 508-509
Sodon Lake, Michigan, 419
Software, 48
Soils, 18, 23, 43, 430
  genesis, 430, 468
  non renewable source, 444
  restoration, 513
Solar: energy use, 523
  energy, wild ecosystems, 523
  equivalents, 15-16
  technology, 523
  tower, 523
  voltaic cells, 523
  wind, 555
Solute, energy language, 128
Solution: energy, 105
Source, 8, 11, 16, 26
  analog computer, 60, 63
  constant force with limit, 132
  controlled, 130, 145
  coupled, 98
  divisor feedback, 152-153
  double webs, 228
  energy distribution, 97
  flow-controlled, 98
  limited and Michaelis-Menten unit, 167
  line, 325-326
  linear pathways, 129
  planar type, 325-326
  point, 325-326
  spatial distribution, 324-326

temperature effect, 292
threshold limited, 132
types, 96
waves, 99
South Florida, embodied energy, 347
Space: carrying capacity, humans, 526
  dust, star fuel, 157
  time equivalence, 253
Spartina salt marsh, 225
  temperature, 298
Spatial: cluster, 331-332
  convergence, 15, 271
  distribution, 323-351
  growth, 332, 346
  hierarchy, 15-16, 323-325
  matrix, 551
  oscillation fields, 417
  pulsing sources, 327
  scan and frequency, 327
Species: islands and continents, 338-339
  per 1000 individuals, 81
  replacement rate, 459
  seeding, diversity, 459
Species substitution, 408
  microcosm, 451
  plankton, 466
Specific activity, 209
Specific growth rate, 143
Specific heat, 103
Specific metabolism, 28
Spectra, 269-287
  economic webs, 487
  electromagnetic, 283
  energy, 269
  energy per person, 285
  energy quality, 325
  energy sources, 283
  frequency, 282, 284
  income, 489
  land use, 535
  reproduction, 286
Spending rate, 476
Spending, storage, 478
Sphaerotilus, 424
Sponges, green fresh water, 414
Spontaneous processes, 314
Spring, 23
Spring bloom, 415
Spring, root, 225
Spring run, 423, see Silver Springs
Spruce budworms, 456
  model, 195
Squared functions of energy, 275-276
Square root, diversity index, 345
Square root of minus one, 40
Square wave, 39-40
Stability, 243
  concepts, 577
  disordering, 446
  linear systems, 183-185

loops, 179
matrix criteria, 234
mean-variance, 401
non-linear systems, 182
Nyquist plot, 42
pathway information, 309
stress microcosm, 458
survival, 577
Stable age distribution, 279, 392
Stable point: cooperative autocatalysis, 150-151
  logistic growth, 151
Stages, succession, 445, 542
Stairsteps, 36, 47
Standard analog computer, 53
Standard deviation, 317, 331
Standard error of mean, 331
Star model, 157
State: change, 144
  diagram, 172-173, 177
  from rates, 82-83, 236
  space plot, 171-173, 176
  surface, 172, 195
  transition, 36-37, 82-83
  variable, 4, 11, 25, 29, 36, 82
Static balance, 105
Statistical: mechanisms, species, 342
  models, 157
  path coefficients, 81
  pathways, 26
  structure, 574
Steady state, 5, 234
  climax, 445, 447
  equation, 176
  global, 555, 558, 560
  oscillating, 445
  phase plane graph, 171
  plankton, 445
  Silver Springs, 445
  similarity to equilibria, 320
  thermodynamics notation, 11
  war, 552
Step: down, 96
  function, 39, 42
  surges, 561
  up, 96
  up, loops, 161
Steps in modeling, 579
Stirling's formula, 306, 308
Stochastic models, 157
  seed dispersal, 380
Stockpile, 545
Stomata, 373, 377
  daytime closure, 415
Stone flies, 423
Storage, 8, 19, 25-45
  adaptive role, 25
  analogs, 57-63, 90
  capacitance, 27, 81
  coefficient, 27
  energy transformation ratio, 252
  energy units and flow, 119
  following autocatalytic unit, 189

integral equation, 31
inverse flow, 162
low energy state, 26
nutrients in aquatic plants, 421
perched, 164
steady state, 234
succession, 444
water tank, 26, 81
Storms, 558
Stratification, 431
Streams, 23, 421-424
   capture, 118
   forest, 423
   generalized model, 422
   gradients, 118
   hierarchy, 272, 313
   longitudinal succession, 460
   microcosm, 450
   minimodels, 423
   mountain, 423
   overflow, 73
   sucrose, 423
   turbid, 423
Stress: aging, 390
   diversity, 344
   oscillation, 458
   removal time, 457
Strip mining, succession, 469
Structure, 19
   brine microcosm, 462
   economic, 500
   measured by energy flow, 317-319
   removal rate, 457
Struggle: for energy, 118
   for existence, 6, 575
Subject, 91
Subset, 78
Subsidy, agricultural, 520
Substitution, succession, 445
Substrate, biochemical, 175
Subsystem, 3, 13, 17
Suburbia, 526
   lawn system, 528
Success and failure, 245-246
Succession, 108, 443-475
   aquatic microcosm, 463
   arrested, 461
   bottle microcosms, 451
   circular, 445, 453
   component tree model, 469
   cutting, 471
   displacement, 446
   diversity, 449
   energy sources, 447
   extinction rate, 460
   global system, 561
   hay infusion, 465
   irradiation delay, microbes, 465
   lake sand dunes, 468
   learning, 245
   longitudinal, 460
   marsh, 471

   mineral use, 545
   minimodels, 468
   model, 216
   nutrient supply, 450
   old field, 470
   oligotrophy-eutrophy, 448
   outcrops, 450
   phosphate strip mining, 469-470
   plankton, 466
   regional, 542
   species substitution, microcosm, 451
   using storage, 444
   yield systems, 522
Succulents, 373, 415
Sucrose in streams, 309, 424
Sugar use in plants, 87
Sulfate-reducing bacteria, 419
Sulfur cycle, global, 563
Summary: steps in modeling, 579
   steps in simulation, 50, 66
   see ends of chapters
Summer, analog computer, 54-55
Summing, time period, 30
Summing junction, 55, 56, 57, 70
Sunspot cycles, 565
Superaccelerated growth, 149
Supply and demand, coupling, 486
Support area, city, 539
Supporter, 574
Surface relation to volume, 27-28, 35
Survival: information transfer, 465
   maximum power, 252
Survivorship curves, 388
Swamp: floodplain, 433
   mangrove and housing, 541
   waste receiving, 529-530
Sweden: income distribution, 489
   sheep agroecosystem, 521
   tide pools, 464
Switch, 8, 72, 76-79
   compared with multiplier, 133
   protecting analog, 65
   threshold limit, 133
Switching, 8, 36
   control path, 124
   intersections, 132
   systems, 75, 79
Symbiosis, 206-207, 394-395
   animal-microbe, 247
   city and farms, 530
   consumer models, 396
   interactions, 173
   lichens, 414
   order and disordered matter, 164
   pairs, 217-218
   P and R, 411
Symbolic structure, 512, 515
Symbols: energy language, frontispiece, 3-14
   hexagon, 141
   producer or consumer, 355

Synchronous, 79
Syntax, 91
Systems: closed, 4, 271
   definition, 4
   earth terrestrial, 558-560
   education, 579
   elementary education, 579
   languages, 5-6, 72-94
   matrix language, 82-83
   oceanic, 557-558
   open, 4, 271
   second order, 182

Tabanuco forest, 319
Tampa, Florida, 542
Tank, 7, 8, 25-45
   comparison with autocatalytic, 147
   in hexagon, 147
Tax, 491
   land, 497
Taxonomic hierarchy, 342
   information, 302
Taylor farm, Arkansas, 527
Technocracy, 265
Technoecosystem, 526-527
Technology: intermediate, 524
   solar, 523
Teleological mechanisms, 577
Temperature, 95, 288-301
   alder photosynthesis, 294
   depreciation, 295
   entropy, 306
   enzymes, 293
   leaf, 290
   microcosm succession, 451, 465
   models, 290-298
   optimum, 297
   regulation, 295-298
   size, 298
   use in classification, 410
Ten gain, analogs, 55, 58
Tension and war, 517-518
Terminating processes, 173
Terminology, production, 257-258
Termites, 395, 397
Terrestrial ecosystems, 428-433
Territory, 15
   centers, 324
   city, 330
   size, 279
Tetrahymena, 467
Texas bays, 428
Thermal diffusion, 115
Thermochemistry, 112-113
Thermodynamic force, 105
Thermodynamics, irreversible, 6
Third law of thermodynamics, 103
Three dimensional spatial form, 349
Threshold, 73
   energy use, 132
   flow, 132-133
   hunger in autocatalytic growth, 151-152

Subject Index

switch, 133
Thunderstorm structure, 349
Tide, energy transformation ratio, 252
Tide pools, succession, 464
Tiki, 512
Time: constant, 26, 28, 33, 35, 41-43
   constant and size, 280
   delay, 73
   discrete systems, 154
   domain, 6, 31, 40
   as energy, 168
   feeding limitation, 168
   graphs of data, 180
   hierarchy, 576
   lag, 10, 39
   scaling, analogs, 58-59
   size, 28
   space equivalence, 252
   steps, 37
Time series analysis, 282-283, 327
   tree population simulation, 469
Timers, 282
Times arrow, 100
Timing oscillators, 176
Top consumer, 386
Tourist facilities in strips, 326
Toxic action, 528-529
Tracers, 28, 43, 208, 209
Track store: analogs, 64-65
   harvest and replanting, 516
Trade, 511-512
   balance, 490
   embodied energy effect, 568
Transaction, 477
   symbol, 8, 12
Transfer coefficients, 26, 27-28, 36, 82-83
Transfer function: frequency domain, 41-42
   time domain, 30-31
Transfer payments, 491
Transformation, 15-16, 37
Transformation efficiency, 115
Transformer, 87-88
Transition probability matrix, 84
Transition state, 5, 37
   stability, 83
Transpiration, 290
Transport: city hierarchies, 535
   energy quality, 280
   resistance, 328-329
Travel, radius, 328
Trees: crown and trunk, 279
   graph theory, 87
   models, 379
Trial and error, 17
Triangular coordinates, 410
Triangular webs, 229
Trigger, ephemeral ecosystem, 432
Trim pot for sine wave, analogs, 69
Trophic, chain, 223

Trophic levels, 269, 439
   equations, 439
   nomenclature, 408
Tropical agriculture: shifting agriculture, 513
   successional position of crops, 522
Tropical forest: comparison of plantation and climax, 319
   forest entropy evaluation, 318
   respiration, 318
Trucks, parking energy, 309
Truth tables, 74-75
Tube animals, 430
Tundra: arctic plants, 377
   ecosystems, 457
Turbulence: eddies, 274
   phytoplankton, 284, 426
   variance, 317
Turing machine, 79
Turnover: species, 415, 459
   time and size, 28, 576
Two limiting factors, 134-136
Two storage system, 183
Types of climax, 448

Ultrasonic disordering, 316
Ultraviolet inactivation, 317
Unbalanced production arms, 260
Uncertainty, 304
Uniforce, 97
Unit circle, 42
United States, 534
   aggregated model, 489
   net yield ratios, 493
Unity of systems, 572
Universe, 575
Unrenewed storage, autocatalytic growth, 145
Unstable balance, exponential growth, 143
Urban dynamics, 547, 551
Urban ecosystem, Paris in 1882, 530
Urban recycle, 530
Urban system, 18, 548
Urchins, 426
Used energy, 141
Use of fuels and minerals, 517
Useful power, 101
Use of models, 579
Use of parks, 527

Validation, 579
Value: added, 493
   distance from center, 497
   embodied energy, 251, 265, 266
   scarcity, 261
Valve, 10, 97
Van der Pol oscillator, 176-177
Vaporization, 108
Vapor-liquid partition, 213
Variability: eutrophy, 418
Variable function generator, 71
Variable iteration interval, 50

Variance, 137-138, 282, 317, 331-333
   energy, 26, 317
   information, 317
   power flow, 317
   state variable, 317
   stock, 317
Vector, 42, 82, 83
   addition, 82, 227
   components, 227
   niche, 224-225
Velocity, 10
   biochemical, 175
Venezuela, 547
Venn diagrams, 77
Venus flytraps, 414
Verbal language, 77-78, 90-91
Verbs, 77-78, 90-91
Verification, 579
Vertices, 87
Vietnam, simulation, 549, 552
Volcanic materials in stratosphere, 21
Volcanoes, 555-556, 560
Voltage, analog simulation, 53
Volume relation to surface, 27-28, 35
Von Bertalanffy growth, 33
Vortices, fluid convection, 324

Wages, city model, 540
Walrus model, 486
Walter's turtle, 78
War, 511, 549, 552, 568
   energy control, 511
   low energy times, 570
   models, 219, 221
   order disorder, 568
Washout, chemostats, 193
Waste accumulation, autocatalytic growth, 154
Waste interface, 529
Water: channel flow, 82
   cycle, 556, 559
   desert ecosystem, 432
   energy content, 252
   energy storage, 105-106
   fish pool system, 67
   flows, 72, 190
   grazing model, 525
   mangroves, 437
   regional models, 542
Water hyacinth, 22, 420
Water-vapor equilibrium, 108, 110, 273
Watt-Ivlev function, 384
Waves, tide pools, 164
Weathering, 430-431
Weber's theory, industrial plant location, 326
Webs, 223-248, 269-287, 302-322
   artificial stream, 423
   embodied energy, 255
   general ecosystem form, 407
   heterogeneous, 247
   inhibition, 247
   money and energy, 480

motion, 240
two source, 228-229
Weibull distribution, 285
survivorship, 388
Weighted digraphs, 81
Wetlands, 430, 433-437
coastal development, 543
Whales, 425
White amur, 421
White box, 12
White sulfur bacteria, 419
Wiens law, 289
Window of interest, 578
Winds, energy transformation ratio, 252
Winter crops, Florida, 523
Wood burning, 160
Wood production, 522
Wood storks, 273
simulation, 386-387
Wood sugar in streams, 424
Word chains, 6
Word diagram, atmospheric system, 557
Work, 5, 6, 100
energy transformation, 7, 124
friction, 100, 105
gate, 125, 126
mechanical, 266
seasonal, 516
World: computer models, 567-568
Dynamics model, 568
fuel use, 574
mass flow, 560
patterns, 554-571

Yeast growth, 146
Yellowstone Park, Wyoming, 454
Yield: minimodels, 520-522
models, 187-193
net energy ratio, 257
successional position, 522
use, 521

Zener diodes, 433, 441
Zero order chemical reaction kinetics, 38-39, 127
Zooplankton, 15
Zooxanthellae, 411

DATE DUE

GAYLORD PRINTED IN U.S.A.